AF559962

Encyclopaedia

of

BIOINFORMATICS SCIENCE, TECHNOLOGY AND ENGINEERING

Encyclopaedia of Bioinformatics Science, Technology and Engineering

Volume 1

Prof. Nirmal Chandra Pradhan

ANMOL PUBLICATIONS PVT. LTD.
NEW DELHI-110 002 (INDIA)

ANMOL PUBLICATIONS PVT. LTD.
Regd. Office: 4360/4, Ansari Road, Daryaganj,
New Delhi-110 002 (India)
Ph.: 23278000, 23261597
Branch Office: No. 1015, Ist Main Road, BSK IIIrd Stage
IIIrd Phase, IIIrd Block,
Bangalore-560 085 (India)
Tel.: 080-41723429
Visit us at: www.anmolpublications.com

Encyclopaedia of Bioinformatics Science, Technology and Engineering

© Reserved

First Edition, 2009

ISBN 978-81-261-4120-3 (Set)

PRINTED IN INDIA

Printed at Mehra Offset Press, Delhi.

Contents

Volume 3

Preface

Bioinformatics is the application of computer technology to the management of biological information. Computers are used to gather, store, analyze and integrate biological and genetic information which can then be applied to gene-based drug discovery and development. The need for Bioinformatics capabilities has been precipitated by the explosion of publicly available genomic information resulting from the Human Genome Project. The goal of this project - determination of the sequence of the entire human genome.

The science of Bioinformatics, which is the melding of molecular biology with computer science, is essential to the use of genomic information in understanding human diseases and in the identification of new molecular targets for drug discovery. In recognition of this, many universities, government institutions and pharmaceutical firms have formed bioinformatics groups, consisting of computational biologists and bioinformatics computer scientists. Such groups will be key to unraveling the mass of information generated by large scale sequencing efforts underway in laboratories around the world. Recent years have seen an explosive growth in biological data. Large sequencing projects are producing increasing quantities of nucleotide sequences. The latest release of GenBank exceeded one billion base pairs. Not only the size of sequence data is rapidly increasing, but also the number of characterized genes from many organisms and protein structures doubles about every two years. To cope with this great quantity of data, a new scientific discipline has emerged: *bioinformatics, biocomputing* or *computational biology*.

Author

Chapter 1

Introduction to Bioinformatics

Bioinformatics is the application of information technology to the field of molecular biology. Bioinformatics entails the creation and advancement of databases, algorithms, computational and statistical techniques, and theory to solve formal and practical problems arising from the management and analysis of biological data.

Over the past few decades rapid developments in genomic and other molecular research technologies combined developments in information technologies have combined to produce a tremendous amount of information related to molecular biology.

It is the name given to these mathematical and computing approaches used to glean understanding of biological processes.

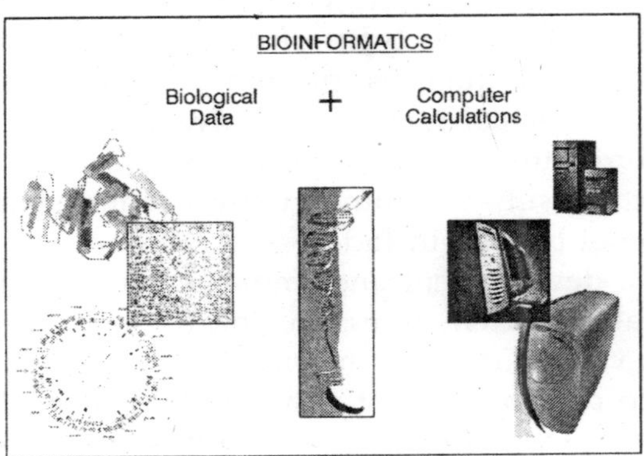

Fig. Bioinformatics

Common activities in Bioinformatics include mapping and

analyzing DNA and protein sequences, aligning different DNA and protein sequences to compare them and creating and viewing 3-D models of protein structures.

The primary goal of bioinformatics is to increase our understanding of biological processes. What sets it apart from other approaches, however, is its focus on developing and applying computationally intensive techniques (e.g., data mining, and machine learning algorithms) to achieve this goal. Major research efforts in the field include sequence alignment, gene finding, genome assembly, protein structure alignment, protein structure prediction, prediction of gene expression and protein-protein interactions, and the modeling of evolution.

Bioinformatics was applied in the creation and maintenance of a database to store biological information at the beginning of the "genomic revolution", such as nucleotide and amino acid sequences.

Development of this type of database involved not only design issues but the development of complex interfaces whereby researchers could both access existing data as well as submit new or revised data.

In order to study how normal cellular activities are altered in different disease states, the biological data must be combined to form a comprehensive picture of these activities. Therefore, the field of bioinformatics has evolved such that the most pressing task now involves the analysis and interpretation of various types of data, including nucleotide and amino acid sequences, protein domains, and protein structures. The actual process of analyzing and interpreting data is referred to as computational biology.

Important sub-disciplines within bioinformatics and computational biology include:

- The development and implementation of tools that enable efficient access to, and use and management of, various types of information.
- The development of new algorithms (mathematical formulas) and statistics with which to assess relationships among members of large data sets, such as methods to locate a gene within a sequence,

predict protein structure and/or function, and cluster protein sequences into families of related sequences.

MAJOR RESEARCH AREAS

Since the Phage Φ-X174 was sequenced in 1977, the DNA sequences of hundreds of organisms have been decoded and stored in databases. The information is analyzed to determine genes that encode polypeptides, as well as regulatory sequences. A comparison of genes within a species or between different species can show similarities between protein functions, or relations between species (the use of molecular systematics to construct phylogenetic trees).

With the growing amount of data, it long ago became impractical to analyse DNA sequences manually. Today, computer programs are used to search the genome of thousands of organisms, containing billions of nucleotides.

These programs would compensate for mutations (exchanged, deleted or inserted bases) in the DNA sequence, in order to identify sequences that are related, but not identical. A variant of this sequence alignment is used in the sequencing process itself.

The so-called shotgun sequencing technique (which was used, for example, by The Institute for Genomic Research to sequence the first bacterial genome, *Haemophilus influenzae*) does not give a sequential list of nucleotides, but instead the sequences of thousands of small DNA fragments (each about 600-800 nucleotides long). The ends of these fragments overlap and, when aligned in the right way, make up the complete genome. Shotgun sequencing yields sequence data quickly, but the task of assembling the fragments can be quite complicated for larger genomes.

In the case of the Human Genome Project, it took several days of CPU time (on one hundred Pentium III desktop machines clustered specifically for the purpose) to assemble the fragments. Shotgun sequencing is the method of choice for virtually all genomes sequenced today, and genome assembly algorithms are a critical area of bioinformatics research. Another aspect of bioinformatics in sequence analysis is the automatic search for genes and regulatory sequences within a genome.

Not all of the nucleotides within a genome are genes. Within the genome of higher organisms, large parts of the DNA do not serve any obvious purpose. This so-called junk DNA may, however, contain unrecognized functional elements. Bioinformatics helps to bridge the gap between genome and proteome projects—for example, in the use of DNA sequences for protein identification.

GENOME ANNOTATION

In the context of genomics, annotation is the process of marking the genes and other biological features in a DNA sequence. The first genome annotation software system was designed in 1995 by Dr. Owen White, who was part of the team that sequenced and analyzed the first genome of a free-living organism to be decoded, the bacterium *Haemophilus influenzae*.

Dr. White built a software system to find the genes (places in the DNA sequence that encode a protein), the transfer RNA, and other features, and to make initial assignments of function to those genes. Most current genome annotation systems work similarly, but the programs available for analysis of genomic DNA are constantly changing and improving.

COMPUTATIONAL EVOLUTIONARY BIOLOGY

Evolutionary biology is the study of the origin and descent of species, as well as their change over time. Informatics has assisted evolutionary biologists in several key ways; it has enabled researchers to:

- Trace the evolution of a large number of organisms by measuring changes in their DNA, rather than through physical taxonomy or physiological observations alone.
- More recently, compare entire genomes, which permits the study of more complex evolutionary events, such as gene duplication, horizontal gene transfer, and the prediction of factors important in bacterial speciation.

- Build complex computational models of populations to predict the outcome of the system over time.
- Track and share information on an increasingly large number of species and organisms.

Future work endeavours to reconstruct the now more complex tree of life. The area of research within computer science that uses genetic algorithms is sometimes confused with computational evolutionary biology, but the two areas are unrelated.

MEASURING BIODIVERSITY

Biodiversity of an ecosystem might be defined as the total genomic complement of a particular environment, from all of the species present, whether it is a biofilm in an abandoned mine, a drop of sea water, a scoop of soil, or the entire biosphere of the planet Earth. Databases are used to collect the species names, descriptions, distributions, genetic information, status and size of populations, habitat needs, and how each organism interacts with other species.

Specialized software programs are used to find, visualize, and analyse the information, and most importantly, communicate it to other people. Computer simulations model such things as population dynamics, or calculate the cumulative genetic health of a breeding pool (in agriculture) or endangered population (in conservation). One very exciting potential of this field is that entire DNA sequences, or genomes of endangered species can be preserved, allowing the results of Nature's genetic experiment to be remembered *in silico*, and possibly reused in the future, even if that species is eventually lost.

ANALYSIS OF GENE EXPRESSION

The expression of many genes can be determined by measuring mRNA levels with multiple techniques including microarrays, expressed cDNA sequence tag (EST) sequencing, serial analysis of gene expression (SAGE) tag sequencing, massively parallel signature sequencing (MPSS), or various applications of multiplexed in-situ hybridization.

All of these techniques are extremely noise-prone and/or subject to bias in the biological measurement, and a major research area in computational biology involves developing statistical tools to separate signal from noise in high-throughput gene expression studies. Such studies are often used to determine the genes implicated in a disorder: one might compare microarray data from cancerous epithelial cells to data from non-cancerous cells to determine the transcripts that are up-regulated and down-regulated in a particular population of cancer cells.

ANALYSIS OF REGULATION

Regulation is the complex orchestration of events starting with an extracellular signal such as a hormone and leading to an increase or decrease in the activity of one or more proteins. Bioinformatics techniques have been applied to explore various steps in this process. For example, promoter analysis involves the identification and study of sequence motifs in the DNA surrounding the coding region of a gene. These motifs influence the extent to which that region is transcribed into mRNA.

Expression data can be used to infer gene regulation: one might compare microarray data from a wide variety of states of an organism to form hypotheses about the genes involved in each state. In a single-cell organism, one might compare stages of the cell cycle, along with various stress conditions (heat shock, starvation, etc.).

One can then apply clustering algorithms to that expression data to determine which genes are co-expressed. For example, the upstream regions (promoters) of co-expressed genes can be searched for over-represented regulatory elements.

ANALYSIS OF PROTEIN EXPRESSION

Protein microarrays and high throughput (HT) mass spectrometry (MS) can provide a snapshot of the proteins present in a biological sample.

Bioinformatics is very much involved in making sense of

protein microarray and HT MS data; the former approach faces similar problems as with microarrays targeted at mRNA, the latter involves the problem of matching large amounts of mass data against predicted masses from protein sequence databases, and the complicated statistical analysis of samples where multiple, but incomplete peptides from each protein are detected.

ANALYSIS OF MUTATIONS IN CANCER

In cancer, the genomes of affected cells are rearranged in complex or even unpredictable ways. Massive sequencing efforts are used to identify previously unknown point mutations in a variety of genes in cancer. Bioinformaticians continue to produce specialized automated systems to manage the sheer volume of sequence data produced, and they create new algorithms and software to compare the sequencing results to the growing collection of human genome sequences and germline polymorphisms.

New physical detection technology are employed, such as oligonucleotide microarrays to identify chromosomal gains and losses (called comparative genomic hybridization), and single nucleotide polymorphism arrays to detect known *point mutations*. These detection methods simultaneously measure several hundred thousand sites throughout the genome, and when used in high-throughput to measure thousands of samples, generate terabytes of data per experiment.

Again the massive amounts and new types of data generate new opportunities for bioinformaticians. The data is often found to contain considerable variability, or noise, and thus Hidden Markov model and change-point analysis methods are being developed to infer real copy number changes.

Another type of data that requires novel informatics development is the analysis of lesions found to be recurrent among many tumors.

PREDICTION OF PROTEIN STRUCTURE

Protein structure prediction is another important

application of bioinformatics. The amino acid sequence of a protein, the so-called primary structure, can be easily determined from the sequence on the gene that codes for it.

In the vast majority of cases, this primary structure uniquely determines a structure in its native environment. (Of course, there are exceptions, such as the bovine spongiform encephalopathy-aka Mad Cow Disease-prion.) Knowledge of this structure is vital in understanding the function of the protein. For lack of better terms, structural information is usually classified as one of *secondary, tertiary* and *quaternary* structure. A viable general solution to such predictions remains an open problem. As of now, most efforts have been directed towards heuristics that work most of the time.

One of the key ideas in bioinformatics is the notion of homology. In the genomic branch of bioinformatics, homology is used to predict the function of a gene: if the sequence of gene *A*, whose function is known, is homologous to the sequence of gene *B*, whose function is unknown, one could infer that B may share A's function.

In the structural branch of bioinformatics, homology is used to determine which parts of a protein are important in structure formation and interaction with other proteins. In a technique called homology modeling, this information is used to predict the structure of a protein once the structure of a homologous protein is known. This currently remains the only way to predict protein structures reliably.

One example of this is the similar protein homology between hemoglobin in humans and the hemoglobin in legumes (leghemoglobin). Both serve the same purpose of transporting oxygen in the organism. Though both of these proteins have completely different amino acid sequences, their protein structures are virtually identical, which reflects their near identical purposes. Other techniques for predicting protein structure include protein threading and *de novo* (from scratch) physics-based modeling.

Chapter 2

Genetic Molecules

The discovery that DNA is the prime genetic molecule, carrying all the hereditary information within chromosomes, immediately focused attention on its structure. It was hoped that knowledge of the structure would reveal how DNA carries the genetic messages that are replicated when chromosomes divide to produce two identical copies of themselves. During the late 1940s and early 1950s, several research groups in the United States and in Europe engaged in serious efforts—both cooperative and rival—to understand how the atoms of DNA are linked together by covalent bonds and how the resulting molecules are arranged in three-dimensional space.

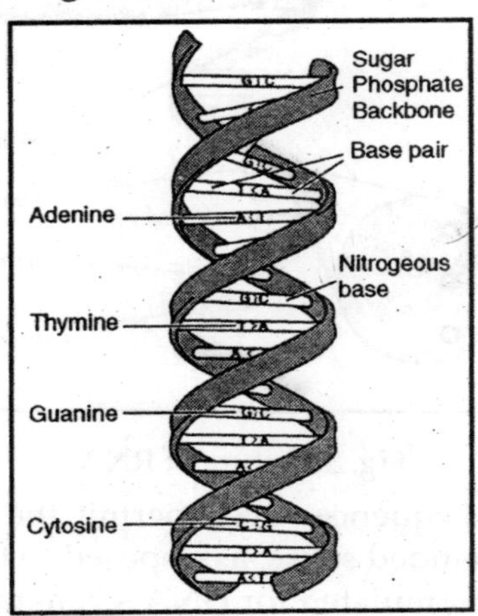

Fig. Structure of DNA

Not surprisingly, there initially were fears that DNA

might have very complicated and perhaps bizarre structures that differed radically from one gene to another. Great relief, if not general elation, was thus expressed when the fundamental DNA structure was found to be the double helix. It told us that all genes have roughly the same three-dimensional form and that the differences between two genes reside in the order and number of their four nucleotide building blocks along the complementary strands.

Now, some 50 years after the discovery of the double helix, this simple description of the genetic material remains true and has not had to be appreciably altered to accommodate new findings. Nevertheless, we have come to realise that the structure of DNA is not quite as uniform as was first thought. For example, the chromosome of some small viruses have single-stranded, not double-stranded, molecules. Moreover, the precise orientation of the base pairs varies slightly from base pair to base pair in a manner that is influenced by the local DNA sequence.

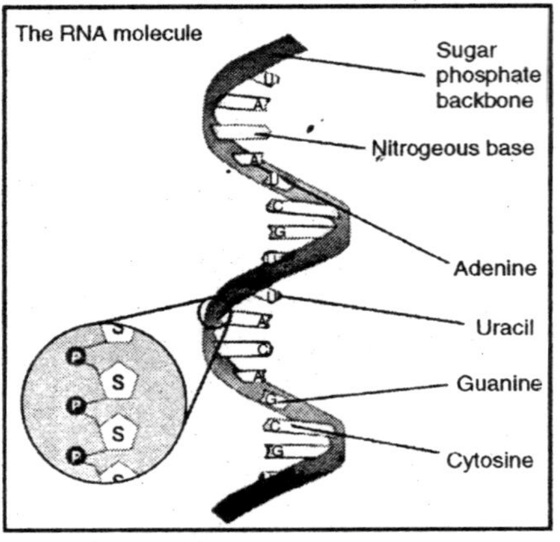

Fig. Structure of RNA

Some DNA sequences even permit the double helix to twist in the left-handed sense, as opposed to the right-handed sense originally formulated for DNA's general structure. And while some DNA molecules are linear, others are circular. Still additional complexity comes from the supercoiling (further twisting) of the double helix, often around cores of DNA-binding proteins. Likewise, we now realise that RNA, which at first glance appears to be very similar to DNA, has its own distinctive structural features. It is principally found as a single-stranded molecule.

Yet by means of intra-strand base pairing, RNA exhibits extensive double-helical character and is capable of folding into a wealth of diverse tertiary structures. These structures are full of surprises, such as non-classical base pairs, base-backbone interactions, and knot-like configurations. Most remarkable of all, and of profound evolutionary significance, some RNA molecules are enzymes that carry out reactions that are at the core of information transfer from nucleic acid to protein. Clearly, the structures of DNA and RNA are richer and more intricate than was at first appreciated. Indeed, there is no one generic structure for DNA and RNA.

DNA IS COMPOSED OF POLYNUCLEOTIDE CHAINS

The most important feature of DNA is that it is usually composed of two polynucleotide chains twisted around each other in the form of a double helix. The upper part of the figure presents the structure of the double helix shown in a schematic form. Note that if inverted 180°, the double helix looks superficially the same, due to the complementary nature of the two DNA strands.

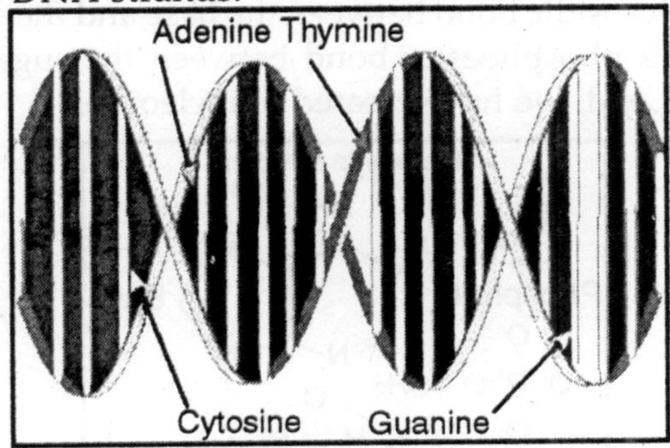

Fig. Polynucleotide Chains

The space-filling model of the double helix, in the lower

part of the figure, shows the components of the DNA molecule and their relative positions in the helical structure. The backbone of each strand of the helix is composed of alternating sugar and phosphate residues; the bases project inward but are accessible through the major and minor grooves.

Let us begin by considering the nature of the nucleotide, the fundamental building block of DNA. The nucleotide consists of a phosphate joined to a sugar, known as 2′-deoxyribose, to which a base is attached. The sugar is called 2′-deoxyribose because there is no hydroxyl at position 2′ (just two hydrogens).

Note that the positions on the ribose are designated with primes to distinguish them from positions on the bases. We can think of how the base is joined to 2′-deoxyribose by imagining the removal of a molecule of water between the hydroxyl on the 1′ carbon of the sugar and the base to form a glycosidic bond.

The sugar and base alone are called a nucleoside. Likewise, we can imagine linking the phosphate to 2′-deoxyribose by removing a water molecule from between the phosphate and the hydroxyl on the 5′ carbon to make a 5′ phosphomonoester. Adding a phosphate (or more than one phosphate) to a nucleoside creates a nucleotide. Thus, by making a glycosidic bond between the base and the sugar, and by making a phosphoester bond between the sugar and the phosphoric acid, we have created a nucleotide.

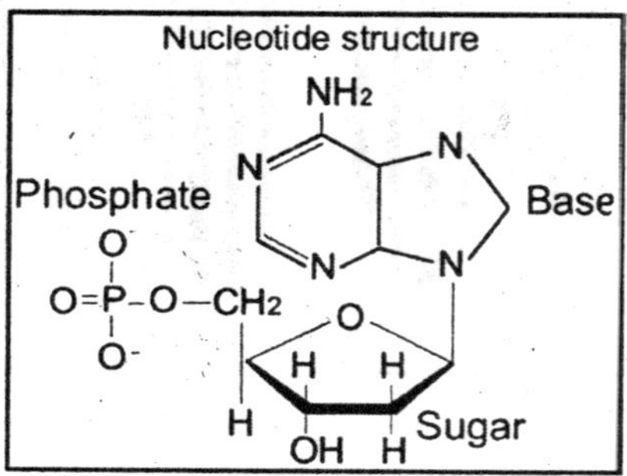

Fig. Nucleotide

Nucleotides are, in turn, joined to each other in polynucleotide chains through the 3′ hydroxyl of 2′-deoxyribose of one nucleotide and the phosphate attached to the 5′ hydroxyl of another nucleotide.

This is a phosphodiester linkage in which the phosphoryl group between the two nucleotides has one sugar esterified to it through a 3′ hydroxyl and a second sugar esterified to it through a 5′ hydroxyl. Phosphodiester linkages create the repeating, sugar-phosphate backbone of the polynucleotide chain, which is a regular feature of DNA. In contrast, the order of the bases along the polynucleotide chain is irregular. This irregularity as well as the long length is, as we shall see, the basis for the enormous information content of DNA. The phosphodiester linkages impart an inherent polarity to the DNA chain.

This polarity is defined by the asymmetry of the nucleotides and the way they are joined. DNA chains have a free 5′ phosphate or 5′ hydroxyl at one end and a free 3′ phosphate or 3′ hydroxyl at the other end. The convention is to write DNA sequences from the 5′ end (on the left) to the 3′ end, generally with a 5′ phosphate and a 3′ hydroxyl.

BASE HAS ITS PREFERRED TAUTOMERIC FORM

The bases in DNA fall into two classes, purines and pyrimidines. The purines are adenine and guanine, and the pyrimidines are cytosine and thymine. The purines are derived from the double-ringed structure. Adenine and guanine share this essential structure but with different groups attached. The bases are attached to the deoxyribose by glycosidic linkages at N1 of the pyrimidines or at N9 of the purines.

Each of the bases exists in two alternative tautomeric states, which are in equilibrium with each other. The equilibrium lies far to the side of the conventional structures, which are the predominant states and the ones important for base pairing.

The nitrogen atoms attached to the purine and pyrimidine rings are in the amino form in the predominant state and only rarely assume the imino configuration.

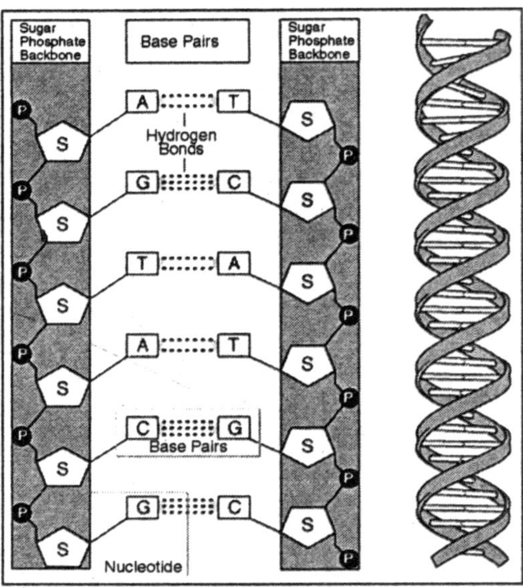

Fig. Polynucleotide

Likewise, the oxygen atoms attached to the guanine and thymine normally have the keto form and only rarely take on the enol configuration. As we shall see, the capacity to form an alternative tautomer is a frequent source of errors during DNA synthesis.

STRANDS OF THE DOUBLE HELIX

The double helix is composed of two polynucleotide chains that are held together by weak, non-covalent bonds between pairs of bases. Adenine on one chain is always paired with thymine on the other chain and, likewise, guanine is always paired with cytosine. The two strands have the same helical geometry but base pairing holds them together with the opposite polarity.

That is, the base at the 5′ end of one strand is paired with the base at the 3′ end of the other strand. The strands are said to have an anti-parallel orientation. This anti-parallel orientation is a stereochemical consequence of the way that adenine and thymine and guanine and cytosine pair with each together.

CHAINS OF THE DOUBLE HELIX

The pairing between adenine and thymine and between guanine and cytosine results in a complementary relationship between the sequence of bases on the two intertwined chains and gives DNA its self-encoding character. For example, if we have the sequence 5′-ATGTC-3′ on one chain, the opposite chain must have the complementary sequence 3′-TACAG-5′. The strictness of the rules for this "Watson-Crick" pairing derives from the complementarity both of shape and of hydrogen bonding properties between adenine and thymine and between guanine and cytosine. Adenine and thymine match up so that a hydrogen bond can form between the exocyclic amino group at C6 on adenine and the carbonyl at C4 in thymine; and likewise, a hydrogen bond can form between N1 of adenine and N3 of thymine.

A corresponding arrangement can be drawn between a guanine and a cytosine, so that there is both hydrogen bonding and shape complementarity in this base pair as well. A G:C base pair has three hydrogen bonds, because the exocyclic NH_2 at C_2 on guanine lies opposite to, and can hydrogen bond with, a carbonyl at C_2 on cytosine. Likewise, a hydrogen bond can form between N_1 of guanine and N_3 of cytosine and between the carbonyl at C_6 of guanine and the exocyclic NH_2 at C_4 of cytosine.

Watson-Crick base pairing requires that the bases are in their preferred tautomeric states. An important feature of the double helix is that the two base pairs have exactly the same geometry; having an A:T base pair or a G:C base pair between the two sugars does not perturb the arrangement of the sugars. Neither does T:A or C:G. In other words, there is an approximately twofold axis of symmetry that relates the two sugars and all four base pairs can be accommodated within the same arrangement without any distortion of the overall structure of the DNA.

HYDROGEN BONDING

The hydrogen bonds between complementary bases are a fundamental feature of the double helix, contributing to the thermodynamic stability of the helix and providing the information content and specificity of base pairing. Hydrogen bonding might not at first glance appear to contribute importantly to the stability of DNA for the following reason. An organic molecule in aqueous solution has all of its hydrogen bonding properties satisfied by water molecules that come on and off very rapidly.

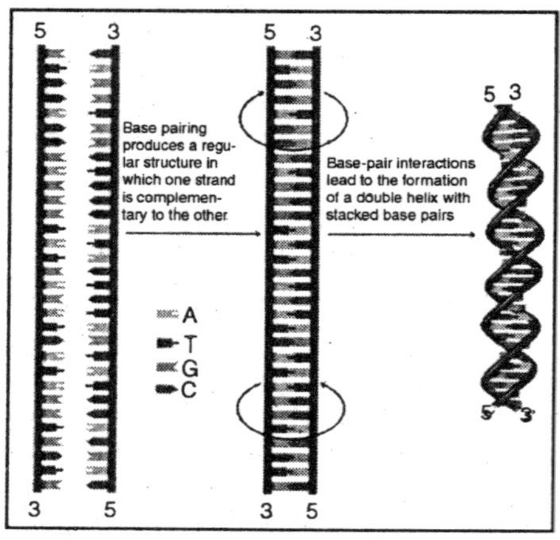

Fig. Hydrogen Bonds Between Complementary Cases

As a result, for every hydrogen bond that is made when a base pair forms, a hydrogen bond with water is broken that was there before the base pair formed. Thus, the net energetic contribution of hydrogen bonds to the stability of the double helix would appear to be modest.

However, when polynucleotide strands are separate, water molecules are lined up on the bases. When strands come together in the double helix, the water molecules are displaced from the bases. This creates disorder and increases entropy, thereby stabilizing the double helix. Hydrogen bonds are not the only force that stabilizes the double helix.

A second important contribution comes from stacking interactions between the bases. The bases are flat, relatively waterinsoluble molecules, and they tend to stack above each other roughly perpendicular to the direction of the helical axis. Electron cloud interactions (ð–ð) between bases in the helical stacks contribute significantly to the stability of the double helix. Hydrogen bonding is also important for the specificity of base pairing. Suppose we tried to pair an adenine with a cytosine.

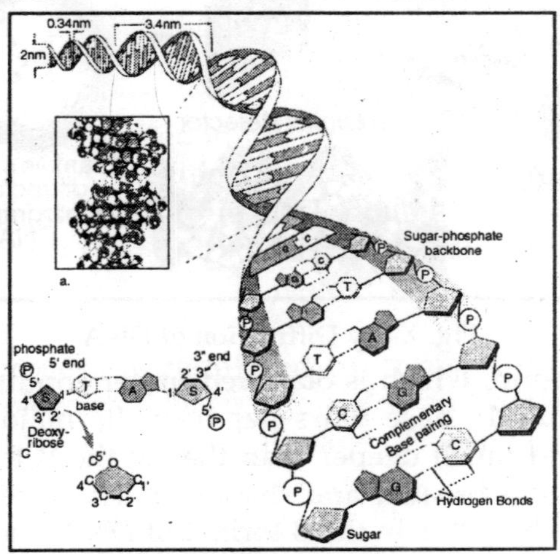

Figa. hydrogen bonds between Double Helix

Then we would have a hydrogen bond acceptor (N_1 of adenine) lying opposite a hydrogen bond acceptor (N_3 of cytosine) with no room to put a water molecule in between to satisfy the two acceptors. Likewise, two hydrogen bond donors, the NH_2 groups at C_6 of adenine and C_4 of cytosine, would lie opposite each other. Thus, an A:C base pair would be unstable because water would have to be stripped off the donor and acceptor groups without restoring the hydrogen bond formed within the base pair.

THE DOUBLE HELIX EXISTS IN MULTIPLE CONFORMATIONS

Early X-ray diffraction studies of DNA, which were carried out using concentrated solutions of DNA that had been drawn out into thin fibers, revealed two kinds of structures, the B and the A forms of DNA. The B form, which is observed at high humidity, most closely corresponds to the average structure of DNA under physiological conditions. It has 10 base pairs per turn, and a wide major groove and a narrow minor groove.

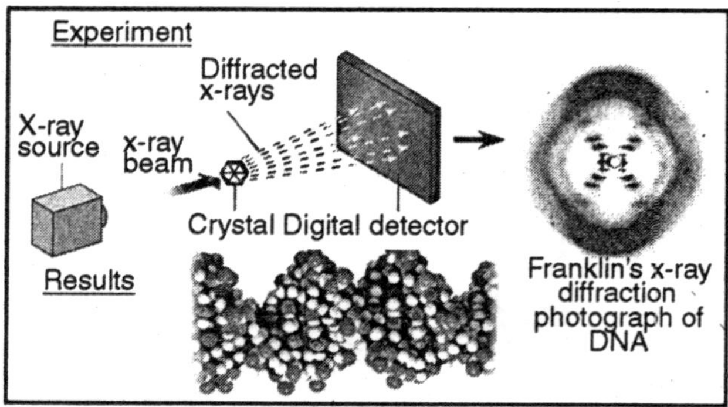

Fig. X-ray Diffraction of DNA

The A form, which is observed under conditions of low humidity, has 11 base pairs per turn. Its major groove is narrower and much deeper than that of the B form, and its minor groove is broader and shallower. The vast majority of the DNA in the cell is in the B form, but DNA does adopt the A structure in certain DNA-protein complexes. Also, as we shall see, the A form is similar to the structure that RNA adopts when double helical.

The B form of DNA represents an ideal structure that deviates in two respects from the DNA in cells. First, DNA in solution, as we have seen, is somewhat more twisted on average than the B form, having on average 10.5 base pairs per turn of the helix. Second, the B form is an average structure whereas real DNA is not perfectly regular. Rather, it exhibits variations in its precise structure from base pair to base pair. This was revealed by comparison of the crystal structures of individual DNAs of different sequences. For example, the two members of each base pair do not always lie exactly in the same plane.

Rather, they can display a "propeller twist" arrangement in which the two flat bases counter rotate relative to each other along the long axis of the base pair, giving the base pair a propeller-like character. Moreover, the precise rotation per base pair is not a constant. As a result, the width of the major and minor grooves varies locally. Thus, DNA molecules are never perfectly regular double helices. Instead, their exact conformation depends on which base pair (A:T, T:A, G:C, or

C:G) is present at each position along the double helix and on the identity of neighboring base pairs. Still, the B form is for many purposes a good first approximation of the structure of DNA in cells.

LEFT-HANDED HELIX

DNA containing alternative purine and pyrimidine residues can fold into left-handed as well as right-handed helices. To understand how DNA can form a left-handed helix, we need to consider the glycosidic bond that connects the base to the 1′ position of 2′-deoxyribose. This bond can be in one of two conformations called *syn* and *anti*. In right-handed DNA, the glycosidic bond is always in the *anti* conformation.

In the left-handed helix, the fundamental repeating unit usually is a purine-pyrimidine dinucleotide, with the glycosidic bond in the *anti* conformation at pyrimidine residues and in the *syn* conformation at purine residues. It is this *syn* conformation at the purine nucleotides that is responsible for the left-handedness of the helix.

The change to the *syn* position in the purine residues to alternating *anti–syn* conformations gives the backbone of left-handed DNA a zigzag look, which distinguishes it from right-handed forms.

The rotation that effects the change from *anti* to *syn* also causes the ribose group to undergo a change in its pucker. In solution alternating purine–pyrimidine residues assume the left-handed conformation only in the presence of high concentrations of positively charged ions (e.g., Na^+) that shield the negatively charged phosphate groups. At lower salt concentrations, they form typical right-handed conformations. The physiological significance of Z DNA is uncertain and left-handed helices probably account at most for only a small of proportion of a cell's DNA.

DNA STRANDS CAN DENATURE AND REASSOCIATE

Because the two strands of the double helix are held together by relatively weak (non-covalent) forces, we might expect that the two strands could come apart easily. Indeed, the original structure for the double helix suggested that DNA replication would occur in just this manner. The complementary strands of double helix can also be made to come apart when a solution of DNA is heated above physiological temperatures (to near 100°C) or under conditions of high pH, a process known as denaturation. However, this complete separation of DNA strands by denaturation is reversible.

When heated solutions of denatured DNA are slowly cooled, single strands often meet their complementary strands and reform regular double helices. The capacity to renature denatured DNA molecules permits artificial hybrid DNA molecules to be formed by slowly cooling mixtures of denatured DNA from two different sources. Likewise, hybrids can be formed between complementary strands of DNA and RNA.

The ability to form hybrids between two single-stranded nucleic acids (hybridization) is the basis for several indispensable techniques in molecular biology, such as Southern blot hybridization and DNA microarrays. Important insights into the properties of the double helix were obtained from classic experiments carried out in the 1950s in which the denaturation of DNA was studied under a variety of conditions. In these experiments DNA denaturation was monitored by measuring the absorbance of ultraviolet light passed through a solution of DNA. DNA maximally absorbs ultraviolet light at a wavelength of about 260 nm.

It is the bases that are principally responsible for this absorption. When the temperature of a solution of DNA is raised to near the boiling point of water, the optical density (absorbance) at 260 nm markedly increases.

The explanation for this increase is that duplex DNA is hypochromic; it absorbs less ultraviolet light by about 40 per cent than do individual DNA chains. The hypochromicity is due to base stacking, which diminishes the capacity of the bases in duplex DNA to absorb ultraviolet light. If we plot the optical density of DNA as a function of temperature, we observe that the increase in absorption occurs abruptly over a relatively narrow temperature range.

The midpoint of this transition is the melting point or Tm. Like ice, DNA melts: it undergoes a transition from a highly ordered double-helical structure to a much less ordered structure of individual strands. The sharpness of the increase in absorbance at the melting temperature tells us that the denaturation and renaturation of complementary DNA strands is a highly cooperative, zippering-like process. Renaturation, for example, probably occurs by means of a slow nucleation process in which a relatively small stretch of bases on one strand find and pair with their complement on the complementary strand. The remainder of the two strands then rapidly zipper-up from the nucleation site to reform an extended double helix.

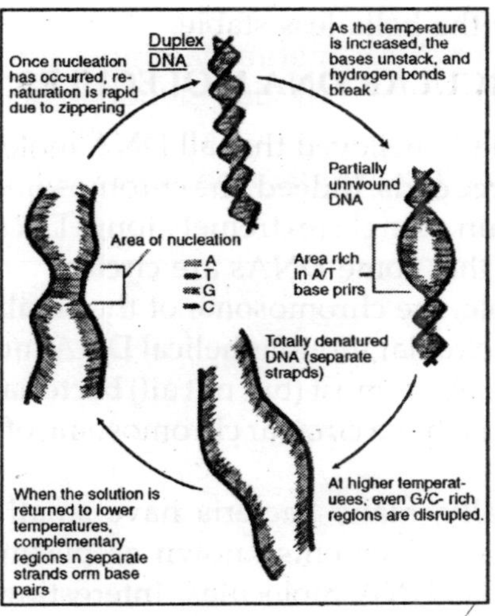

Fig. DNA Denaturation

The melting temperature of DNA is a characteristic of each DNA that is largely determined by the G:C content of the DNA and the ionic strength of the solution. The higher the per cent of G:C base pairs in the DNA (and hence the lower the content of A:T base pairs), the higher the melting point. Likewise, the higher the salt concentration of the solution the greater the temperature at which the DNA denatures. How do we explain this behaviour? G:C base pairs contribute more to the stability of DNA than do A:T base pairs because of the greater number of hydrogen bonds for the former (three in a G:C base pair versus two for A:T) but also importantly because the stacking interactions of G:C base pairs with adjacent base pairs are more favorable than the corresponding interactions of A:T base pairs with their neighboring base pairs. The effect of ionic strength reflects another fundamental feature of the double helix.

The backbones of the two DNA strands contain phosphoryl groups, which carry a negative charge. These negative charges are close enough across the two strands that if not shielded they tend to cause the strands to repel each other, facilitating their separation. At high ionic strength, the negative charges are shielded by cations, thereby stabilizing the helix. Conversely, at low ionic strength the unshielded negative charges render the helix less stable.

CIRCLES CIRCULAR DNA MOLECULES

It was initially believed that all DNA molecules are linear and have two free ends. Indeed, the chromosomes of eukaryotic cells each contain a single (extremely long) DNA molecule. But now we know that some DNAs are circles.

For example, the chromosome of the small monkey DNA virus SV40 is a circular, double-helical DNA molecule of about 5,000 base pairs. Also, most (but not all) bacterial chromosomes are circular; *E. coli* has a circular chromosome of about 5 million base pairs.

Additionally, many bacteria have small autonomously replicating genetic elements known as plasmids, which are generally circular DNA molecules. Interestingly, some DNA molecules are sometimes linear and sometimes circular. The most well-known example is that of the bacteriophage ', a DNA virus of *E. coli*.

The phage 'genome is a linear double-stranded molecule in the virion particle. However, when the 'genome is injected into an *E. coli* cell during infection, the DNA circularizes. This occurs by base-pairing between single-stranded regions that protrude from the ends of the DNA and that have complementary sequences ("sticky ends").

DNA TOPOLOGY

As DNA is a flexible structure, its exact molecular parameters are a function of both the surrounding ionic environment and the nature of the DNA-binding proteins with which it is complexed. Because their ends are free, linear DNA molecules can freely rotate to accommodate changes in the number of times the two chains of the double helix twist about each other.

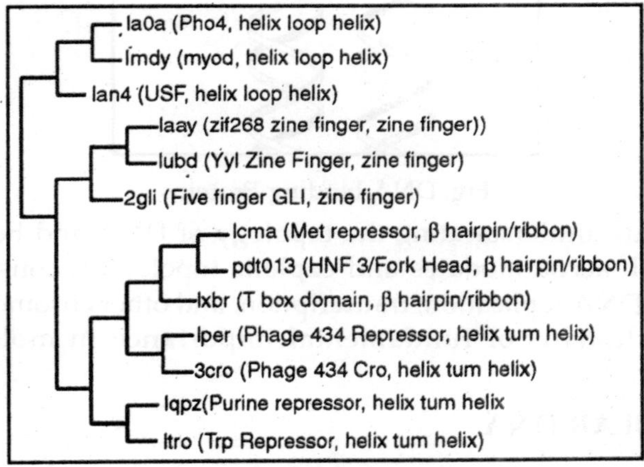

Fig. DNA-Proteins

But if the two ends are covalently linked to form a circular DNA molecule and if there are no interruptions in the sugar phosphate backbones of the two strands, then the absolute number of times the chains can twist about each other cannot change. Such a covalently closed, circular DNA is said to be topologically constrained. Even the linear DNA molecules of eukaryotic chromosomes are subject to topological constraints due to their entrainment in chromatin and interaction with other cellular components.

Despite these constraints, DNA participates in numerous dynamic processes in the cell. For example, the two strands of the double helix, which are twisted around each other, must rapidly separate in order for DNA to be duplicated and to be transcribed into RNA.

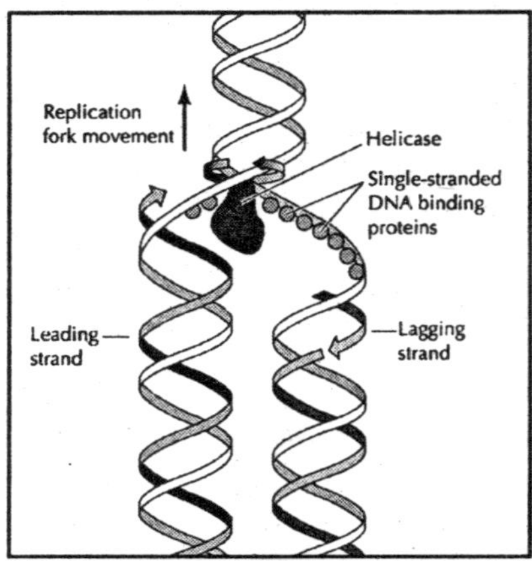

Fig. DNA-binding Proteins

Thus, understanding the topology of DNA and how the cell both accommodates and exploits topological constraints during DNA replication, transcription, and other chromosomal transactions is of fundamental importance in molecular biology.

CIRCULAR DNA

Let us consider the topological properties of covalently closed, circular DNA, which is referred to as cccDNA. Because there are no interruptions in either polynucleotide chain, the two strands of cccDNA cannot be separated from each other without the breaking of a covalent bond.

If we wished to separate the two circular strands without permanently breaking any bonds in the sugar phosphate backbones,we would have to pass one strand through the other strand repeatedly. The number of times one strand would have to be passed through the other strand in order for the two strands to be entirely separated from each other is called the linking number. The linking number, which is always an integer, is an invariant topological property of cccDNA, no matter how much the DNA molecule is distorted.

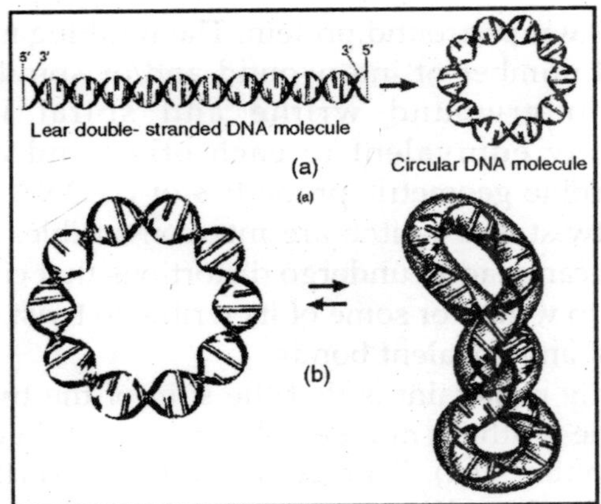

Fig. Closed Circular DNA

Linking Number Is Composed of Twist and Writhe

The linking number is the sum of two geometric components called the twist and the writhe. Let us consider twist first. Twist is simply the number of helical turns of one strand about the other, that is, the number of times one strand completely wraps around the other strand. Consider a cccDNA that is lying flat on a plane.

In this flat conformation, the linking number is fully composed of twist. Indeed, the twist can be easily determined by counting the number of times the two strands cross each other. The helical crossovers (twist) in a right-handed helix are defined as positive such that the linking number of DNA will have a positive value. But cccDNA is generally not lying flat on a plane.

Rather, it is usually torsionally stressed such that the long axis of the double helix crosses over itself, often repeatedly, in three-dimensional space. This is called *writhe*. To visualize the distortions caused by torsional stress, think of the coiling of a telephone cord that has been overtwisted. Writhe can take two forms. One form is the interwound or plectonemic writhe, in which the long axis is twisted around itself.

The other form of writhe is a toroid or spiral in which the long axis is wound in a cylindrical manner, as often occurs when DNA wraps around protein. The writhing number (*Wr*) is the total number of interwound and/or spiral writhes in cccDNA. Interwound writhe and spiral writhe are topologically equivalent to each other and are readily interconvertible geometric properties of cccDNA.

Also, twist and writhe are interconvertible. A molecule of cccDNA can readily undergo distortions that convert some of its twist to writhe or some of its writhe to twist without the breakage of any covalent bonds.

The only constraint is that the sum of the twist number (*Tw*) and the writhing number (*Wr*) must remain equal to the linking number (*Lk*). This constraint is described by the equation: $Lk = Tw + Wr$.

Lk^O Is the Linking Number of Fully Relaxed cccDNA

Consider cccDNA that is free of supercoiling (that is, it is said to be relaxed) and whose twist corresponds to that of the B form of DNA in solution under physiological conditions (about 10.5 base pairs per turn of the helix). The linking number (*Lk*) of such cccDNA under physiological conditions is assigned the symbol Lk^o. Lk^o for such a molecule is the number of base pairs divided by 10.5.

For a cccDNA of 10,500 base pairs, $Lk = +1{,}000$. (The sign is positive because the twists of DNA are right-handed.) One way to see this is to imagine pulling one strand of the 10,500 base pair cccDNA out into a flat circle. If we did this, then the other strand would cross the flat circular strand 1,000 times.

How can we remove supercoils from cccDNA if it is not already relaxed? One procedure is to treat the DNA mildly with the enzyme DNase I, so as to break on average one phosphodiester bond (or a small number of bonds) in each DNA molecule. Once the DNA has been "nicked" in this manner, it is no longer topologically constrained and the strands can rotate freely, allowing writhe to dissipate.

If the nick is then repaired, the resulting cccDNA molecules will be relaxed and will have on average an *Lk* that is equal to Lk^o. (Due to rotational fluctuation at the time the nick is repaired, some of the resulting cccDNAs will have an

Lk that is somewhat greater than Lk^o and others will have an Lk that is somewhat lower. Thus, the relaxation procedure will generate a narrow spectrum of topoisomers whose average Lk is equal to Lk^o).

Nucleosomes Introduce Negative Supercoiling in Eukaryotes

DNA in the nucleus of eukaryotic cells is packaged in small particles known as nucleosomes in which the double helix is wrapped almost two times around the outside circumference of a protein core. We will be able to recognize this wrapping as the toroid or spiral form of writhe. Importantly, it occurs in a lefthanded manner. It turns out that writhe in the form of left-handed spirals is equivalent to negative supercoils. Thus, the packaging of DNA into nucleosomes introduces negative superhelical density.

Topoisomerases Can Relax Supercoiled DNA

As we have seen, the linking number is an invariant property of DNA that is topologically constrained. It can only be changed by introducing interruptions into the sugar-phosphate backbone. A remarkable class of enzymes known as topoisomerases are able to do just that by introducing transient nicks or breaks into the DNA. Topoisomerases are of two broad types.

Type II topoisomerases change the linking number in steps of two. They make transient double-stranded breaks in the DNA, through which they pass a region of uncut duplex DNA before resealing the break.

Type II topoisomerases require energy from ATP hydrolysis for their action. Type I topoisomerases, in contrast, change the linking number of DNA in steps of one. They make transient singlestranded breaks in the DNA, allowing one strand to pass through the break in the other before resealing the nick. Type I topoisomerases relax DNA by removing supercoils.

They can be compared to the protocol of introducing nicks into cccDNA with DNase and then repairing the nicks, which as we saw can be used to relax cccDNA, except that type I topoisomerases relax DNA in a controlled and concerted manner. In contrast to type II topoisomerases, type I topoisomerases do not require ATP. Both type I and type II topoisomerases work through an intermediate in which the enzyme is covalently attached to one end of the broken DNA.

Supercoils

Both prokaryotes and eukaryotes have type I and type II topoisomerases, which are capable of removing supercoils from DNA. In addition, however, prokaryotes have a special type II topoisomerase known as DNA gyrase that is able to introduce negative supercoils, rather than remove them. DNA gyrase is responsible for the negative supercoiling of chromosomes in prokaryotes, which facilitates unwinding of the DNA duplex during transcription and DNA replication.

Electrophoresis

Covalently closed, circular DNA molecules of the same length but of different linking numbers are called DNA topoisomers. Even though topoisomers have the same molecular weight, they can be separated from each other by electrophoresis through a gel of agarose. The basis for this separation is that the greater the writhe the more compact the shape of a cccDNA. Once again, think of how supercoiling a telephone cord causes it to become more compact. The more compact the DNA, the more easily (up to a point) it is able to migrate through the gel matrix.

Thus, a fully relaxed cccDNA migrates more slowly than a highly supercoiled topoisomer of the same circular DNA. Molecules in adjacent rungs of the ladder differ from each other by a linking number difference of just one. Obviously, electrophoretic mobility is highly sensitive to the topological state of DNA.

DNA Has a Helical Periodicity

The observation that DNA topoisomers can be separated from each other electrophoretically is the basis for a simple experiment that proves that DNA has a helical periodicity of about 10.5 base pairs per turn in solution. Consider three cccDNAs of sizes 3990, 3995, and 4011 base pairs that were relaxed to completion by treatment with topoisomerase I. When subjected to electrophoresis through agarose, the 3990- and 4011-base-pair DNAs exhibit essentially identical mobilities.

Due to thermal fluctuation, topoisomerase treatment actually generates a narrow spectrum of topoisomers, but for simplicity let us consider the mobility of only the most abundant topoisomer (that corresponding to the cccDNA in its most relaxed state). The mobilities of the most abundant topoisomers for the 3990- and 4011-base-pair DNAs are indistinguishable because the 21-base-pair difference between them is negligible compared to the sizes of the rings. The most abundant topoisomer for the 3995-base-pair ring, however, is found to migrate slightly more rapidly than the other two rings even though it is only 5 base pairs larger than the 3990-base-pair ring.

How are we to explain this anomaly? The 3990- and 4011-base-pair rings in their most relaxed states are expected to have linking numbers equal to Lk^o, that is, 380 in the case of the 3990-base-pair ring (dividing the size by 10.5 base pairs) and 382 in the case of the 4011-base-pair ring. Because Lk is equal to Lk^o, the linking difference ('$Lk = Lk - Lk^O$) in both cases is zero and there is no writhe. But because the linking number must be an integer, the most relaxed state for the 3995-base-pair ring would be either of two topoisomers having linking numbers of 380 or 381. However, Lk^O for the 3995-base-pair ring is 380.5.

Thus, even in its most relaxed state, a covalently closed circle of 3995 base pairs would necessarily have about half a unit of writhe (its linking difference would be 0.5), and hence it would migrate more rapidly than the 3990- and 4011-base-pair circles. In other words, to explain how rings that differ in length by 21 base pairs (two turns of the helix) have the same mobility whereas a ring that differs in length by only 5 base pairs (about half a helical turn) exhibits a different mobility, we must conclude that DNA in solution has a helical periodicity of about 10.5 base pairs per turn.

Ethidium Ions Cause DNA to Unwind

Ethidium is a large, flat, multi-ringed cation. Its planar shape enables ethidium to slip (intercalate) between the stacked base pairs of DNA. Because it fluoresces when exposed to ultraviolet light, and because its fluorescence increases dramatically after intercalation, ethidium is used as a stain to visualize DNA. When an ethidium ion intercalates between two base pairs, it causes the DNA to unwind by 26°, reducing the normal rotation per base pair from 36° to 10°. In other words, ethidium decreases the twist of DNA.

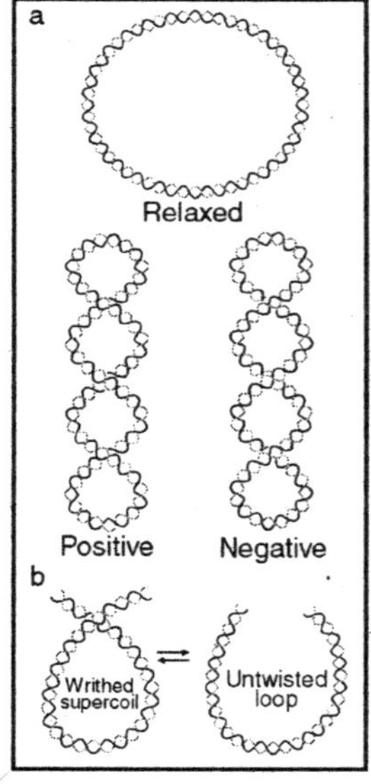

Fig. Supercoiled DNA

Imagine the extreme case of a DNA molecule that has an ethidium ion between every base pair. Instead of 10 base pairs per turn it would have 36! When ethidium binds to linear DNA or to a nicked circle, it simply causes the helical pitch to increase. But consider what happens when ethidium binds to covalently closed, circular DNA. The linking number of the cccDNA does not change (no covalent bonds are broken and resealed), but the twist decreases by 26° for each molecule of ethidium that has bound to the DNA.

Because $Lk = Tw + Wr$, this decrease in Tw must be compensated for by a corresponding increase in Wr. If the circular DNA is initially negatively supercoiled (as is normally the case for circular DNAs isolated from cells), then the addition of ethidium will increase Wr.

In other words, the addition of ethidium will relax the DNA. If enough ethidium is added, the negative supercoiling will be brought to zero, and if even more ethidium is added, Wr will increase above zero and the DNA will become positively supercoiled.

Because the binding of ethidium increases Wr, its presence greatly affects the migration of cccDNA during gel electrophoresis. In the presence of non-saturating amounts of ethidium, negatively supercoiled circular DNAs are more relaxed and migrate more slowly, whereas relaxed cccDNAs become positively supercoiled and migrate more rapidly.

Chapter 3

DNA Structure

DNA stands for deoxyribonucleic acid. DNA is pretty unusual in that it is about the only common molecule capable of directing its own synthesis. The processes of mitosis and meiosis were discovered in the 1870s and 1890s. It was observed that, as cells divided, chromosomes moved around in a cell, and people began to wonder what their function was.

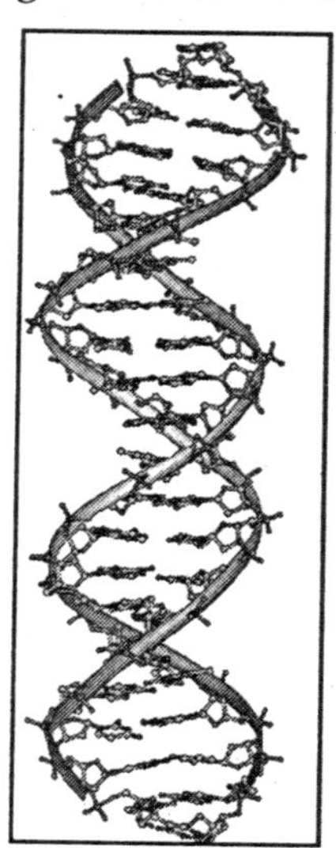

Fig. Deoxyribonucleic Acid

It was determined that chromosomes were made of protein and DNA, about which people knew almost nothing. People began to suspect that chromosomes had something to do with genetics, but couldn't explain what/how. When enough evidence was accumulated to confirm that chromosomes did, indeed, have something to do with genetics, most people thought that in some way the protein in the chromosomes served as the genetic material.

People knew that DNA was also in the chromosomes, but because its structure was unknown and people didn't know much about it, few people thought it was the genetic material. Frederick Griffith performed an experiment using pneumonia bacteria and mice. This was one of the first experiments that hinted that DNA was the genetic code material. He used two strains of *Streptococcus pneumoniae*: a "smooth" strain which has a polysaccharide coating around it that makes it look smooth when viewed with a microscope, and a "rough" strain which doesn't have the coating, thus looks rough under the microscope.

When he injected live S strain into mice, the mice contracted pneumonia and died. When he injected live R strain, a strain which typically does not cause illness, into mice, as predicted they did not get sick, but lived. Thinking that perhaps the polysaccharide coating on the bacteria somehow caused the illness and knowing that polysaccharides are not affected by heat, Griffith then used heat to kill some of the S strain bacteria and injected those dead bacteria into mice.

This failed to infect/kill the mice, indicating that the polysaccharide coating was not what caused the disease, but rather, something within the living cell. Since Griffith had used heat to kill the bacteria and heat denatures protein, he next hypothesized that perhaps some protein within the living cells, that was denatured by the heat, caused the disease.

He then injected another group of mice with a mixture of heat-killed S and live R, and the mice died! When he did a necropsy on the dead mice, he isolated live S strain bacteria from the corpses.

Griffith concluded that the live R strain bacteria must have absorbed genetic material from the dead S strain bacteria, and since heat denatures protein, the protein in the bacterial chromosomes was not the genetic material. This evidence pointed to DNA as being the genetic material. Transformation is the process whereby one strain of a bacterium absorbs genetic material from another strain of bacteria and "turns into" the type of bacterium whose genetic material it absorbed. Because DNA was so poorly understood, scientists remained skeptical up through the 1940s.

Alfred Hershey and Martha Chase did an experiment which is so significant, it has been nicknamed the "Hershey-Chase Experiment". At that time, people knew that viruses were composed of DNA (or RNA) inside a protein coat/shell called a capsid. It was also known that viruses replicate by taking over the host cell's metabolic functions to make more virus. We are used to thinking and talking about viruses which invade our bodies and make us sick, but there are other, different kinds of viruses that infect other kinds of animals, still other viruses which infect plants, and even some viruses that infect bacteria.

A virus which infects a bacterium is called a bacteriophage because the host bacterium cell is killed as the new virus particles leave the bacterial cell. In order to do all this, the virus must inject whatever is the viral genetic code into the host cell.

Thus, people realized that the viral genetic code material had to be either its DNA or its protein capsid. Hershey and Chase sought an answer to the question, "Is it the viral DNA or viral protein coat (capsid) that is the viral genetic code material which gets injected into a host bacterium cell? To try to answer this question, Hershey and Chase performed an experiment using a bacterium named *Escherichia coli*, or *E. coli* for short (named after a scientist whose last name was Escher) and a virus called T2 that is a bacteriophage that infects *E. coli*. Isolated T2, like other viruses, is just a crystal of DNA and protein, so it must live inside *E. coli* in order to make more virus like itself.

When the new T2 viruses are ready to leave the host *E. coli* cell (and go infect others), they burst the *E. coli* cell open, killing it (hence the name "bacteriophage"). The results that Hershey and Chase obtained indicated that the viral DNA, not the protein, is its genetic code material.

Hershey and Chase used radioactive chemicals to distinguish between ("label") the protein capsid and the DNA in T2 virus so they could tell which of those molecules entered the *E. coli* cells. Since some amino acids contain sulfur in their side chains, if T2 is grown in *E. coli* with a source of radioactive sulfur, the sulfur will be incorporated into the T2 protein coat making it radioactive. Since DNA has lots of phosphorus in its phosphate (–PO_4) groups, if T2 is grown in *E. coli* with a source of radioactive phosphorus, the phosphorus will be incorporated into the viral DNA, making that radioactive.

Hershey and Chase grew two batches of T2 and *E. coli*: one with radioactive sulfur and one with radioactive phosphorus to get batches of T2 "labeled" with either radioactive S or radioactive P. Then, these radioactive T2 were placed in separate, new batches of *E. coli*, but were left there only 10 minutes. This was to give the T2 time to inject their genetic material into the bacteria, but not reproduce. In the next step, still in separate batches, the mixtures were agitated in a kitchen blender to knock loose any viral parts not inside the *E. coli* but perhaps stuck on the outer surface.

Hopefully, this would differentiate between the protein and DNA portions of the virus. Then, each mixture was spun in a centrifuge to separate the heavy bacteria (with any viral parts that had gone into them) from the liquid solution they were in (including any viral parts that had not entered the bacteria). The centrifuge causes the heavier bacteria to be pulled to the bottom of the tube where they form a pellet, while the light-weight viral "left-overs" stay suspended in the liquid portion called the supernatant.

In the subsequent step, the pellet and supernatant from each tube were separated and tested for the presence of radioactivity. Radioactive sufur was found in the supernatant, indicating that the viral protein did not go into the bacteria. Radioactive phosphorus was found in the bacterial pellet, indicating that viral DNA did go into the bacteria.

Based on these results, Hershey and Chase concluded that DNA must be the genetic code material, not protein as many poeple believed. When their experiment was published and people finally acknowledged that DNA was the genetic material, there was a lot of competition to be the first to discover its chemical structure. What was known is that DNA contains a nitrogenous base.

DNA IS A DOUBLE HELIX

The outer edges are formed of alternating ribose sugar molecules and phosphate groups. The two strands go in opposite directions (1 "up" and 1 "down"). The nitrogenous bases are "inside" like rungs on a ladder. Adenine on one side pairs with thymine (uracil in RNA) on the other by hydrogen bonding, and cytosine pairs with guanine. Note that the C-G pair has three hydrogen bonds while the A-T pair has only two, which keeps them from pairing wrong.

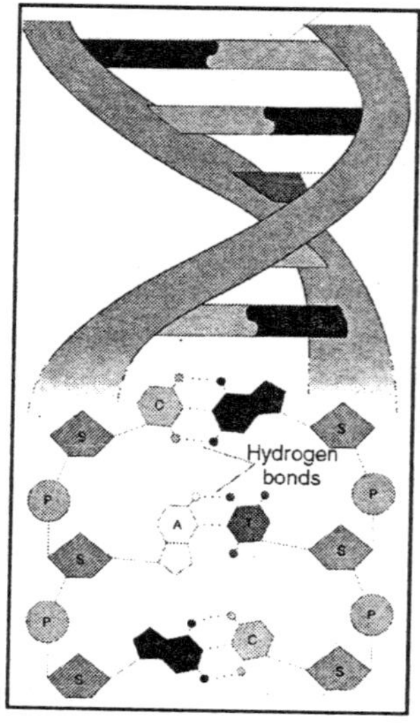

Fig. Double Helix DNA

This dictates side-to-side pairing, but says nothing about the order *along* the molecule. Watson and Crick said this variability along the molecule can account for the variety in the genetic code. Their model also accounts for how DNA can replicate itself. They said the molecule "unzips" and new matching bases are added in to create two new molecules. They called this semiconservative replication because each new molecule has one "old" and one "new" strand of DNA.

DNA codes for protein synthesis by first coding for RNA. First, the DNA code is transcribed to RNA code, which is still in the "language" of nitrogenous bases, except that adenine on the DNA pairs with uracil (in place of thymine) on the RNA. The RNA code is then translated to protein code, which is a different "language."

This process involves ribosomes and two kinds of RNA: mRNA and tRNA. The mRNA codes for the gene in question and is copied off the DNA, while tRNA matches a specific group of nucleotides with a specific amino acid. A "unit" of three nucleotides on the tRNA codes for one amino acid. Each of these "units" is called an anticodon. These match up with corresponding three-nucleotide sequences on the mRNA called codons, and in this manner the amino acids are organized into the correct sequence to build a protein. The ribosome works with the mRNA and tRNA to hook the amino acids together to form a protein.

Here is a list of the mRNA codons and the corresponding amino acids for which they code. Mutations can be caused by a change in the sequence of the nucleotides. Some mutations have more effect than others, depending on where in the code they are and how important that area is to the code. While mutations in some areas of some genes have little effect, sickle cell anemia is caused by a mutation in only one nucleotide. This changes the codon at that location to code for a different amino acid, and that, in turn, significantly changes the shape of the hemoglobin molecules in that person's blood.

When some viruses (especially *Herpes* viruses, including Chicken Pox and Cold Sores) infect us, they insert their DNA into our cells' DNA, and stay resident in our cells for the rest of our lives. These can potentially become active again either making a person sick again or just being shed from a person's body (to infect others) without obvious symptoms of illness (like Mononucleosis). Some kinds of cancer may be caused this way. For example, there is some pretty strong evidence linking genital warts (human papillomavirus, HPV) and cervical cancer.

The AIDS virus does things "backwards." This virus contains RNA rather than DNA, yet when it gets into someone's cells, it can do reverse transcription and code from its RNA to make DNA which, then, can code to make more virus.

GENETIC ENGINEERING

We now have the knowledge and ability to transfer genes from one organism to another, which seems to have some benefits associated with it, but may also have many yet-to-be-discovered problems associated with it. Because this is all so new, not enough time has elapsed to allow scientists to study/look for any possible long-term effects of genetically-modified organisms (GMOs).

Many medicines are now made by GMOs. For example, insulin was formerly extracted from the pancreas of animals after they were slaughtered for meat. However, now most insulin is produced by bacteria with the insulin gene spliced onto their chromosome. Using this method of production, drug companies can make more insulin, faster.

In theory, insulin produced in this way should be more "pure" — someone who could not use pig insulin due to a pork allergy may be able to tolerate insulin made in this way (but someone could, potentially, be allergic to some component of the bacteria present in the refined insulin). In the relatively short time that this form of insulin has been available, there is no doubt that it has saved many people's lives — let's hope that some unforseen, long-term, deleterious effects are not discovered later on.

Experimentation is being done to investigate the possible use of genetically-engineered viruses to treat genetic diseases such as cystic fibrosis. In this "treatment," a kind of virus that infects our lungs is used. The genes that enable it to infect our cells are kept while the genes which make us sick are (hopefully) all removed, and the missing human (normal) gene that relieves cystic fibrosis is then inserted into the virus' genome. These viruses are then sprayed into the lungs of a person with CF and allowed to "infect" the cells in that person's lungs.

When the normal gene is inserted into the genetic make-up of the cells lining that person's lungs, those cells function normally and the CF symptoms are alleviated. However, this treatment doesn't last because only that layer of cells is "infected," and when those cells die and are replaced by new cells, the new cells do not contain the genetic code to overcome the CF gene, and the person must inhale more genetically-engineered virus. While this seems to be a promising, life-giving, technique, no data are yet available on long-term effects and safety.

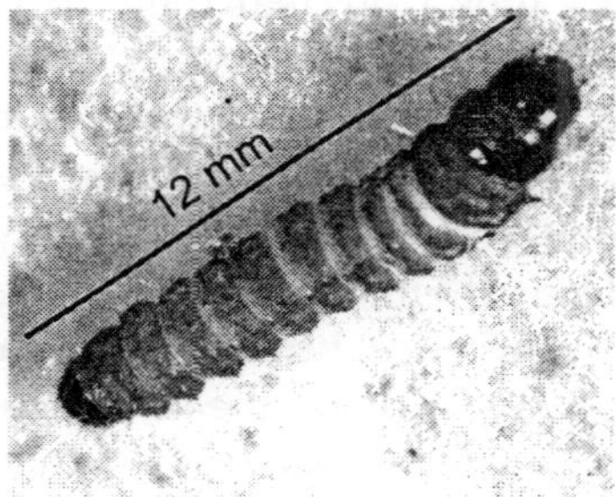

Fig. Bacillus Thuringiensis

Bacillus thuringiensis (BT) is a species of bacterium that infects and kills a number of species of caterpillars (Remember that caterpillars turn into butterflies or moths when they grow up.). There is a species of moth whose caterpillar is a "pest" in corn plants, and insects of any kind are more successful when we humans plant huge monocultures of their favourite foods. For a number of years, now, people have realized that in certain situations, by judiciously applying BT, "pest" species of caterpillars can be infected and killed without the use of man-made, chemical insecticides.

More recently, a major US chemical company came up with the idea of creating genetically-engineered corn containing BT genes, which was good news to agri-business firms who plant huge areas of land with monocultures of corn. Because corn is wind-pollinated and therefore makes lots of pollen which, in this case, contains BT genes, many scientists are concerned about the effects of this corn on local butterfly populations. Some research has indicated that when this pollen settles on nearby caterpillar host plants and is, therefore, consumed by caterpillars, this might cause an increase in mortality (therefore fewer "good" butterflies such as Monarchs). It is assumed that these BT genes in corn should, they think, have no adverse effects on people or cattle who consume this corn, but no long-term testing has been done. Also, this company has convinced the government that this corn should be marketed without any labeling indicating that it is genetically-engineered — they're scared that if we are given a choice, we won't buy/eat their corn if we know it is genetically engineered.

Additionally, to increase their profits, this company has also put "suicide" genes into this corn so that farmers cannot save seed from one year to plant the next year, and have to buy more seed from them, instead. For big agri-businesses, this is of small consequence, but for small, family farmers this is total disaster. To save money, the latter often save seed from one year to plant the next, and even if they don't plant this genetically-engineered corn, if their corn is pollinated by genetically-engineered corn from a neighboring field, it will not produce viable seed.

This same chemical company came up with the idea of "transplanting" what they think is the cold-tolerance gene from a species of cold-water-inhabiting fish into tomatoes, thereby hoping to "invent" cold-tolerant tomatoes, and again, has convinced the government that these tomatoes should not be labeled in any way to indicate that they have fish genes in them, and that we should not have a choice about what we eat. At the very least, this would be a problem for someone who is a vegetarian and chooses to not eat fish.

A more serious consequence, however, would be that a person who is severely allergic to fish could also have an allergic reaction to these tomatoes and end up in the hospital. Again, no tests were done before the government approved these tomatoes and no data are available on the long-term effects on humans (or anything else).

This same chemical company manufactures a widely-used herbicide, and came up with the idea to genetically engineer cotton, soybeans, and other crop plants so they would be immune to the effects of that herbicide. That way, farmers can buy their seed from that company, then spray their fields with herbicide also purchased from that company to (hopefully) kill all local plants except the immune crop, simultaneously putting all their money in the corporation's pockets.

Preliminary research has shown that these resistant genes have already begun to "jump" into local weeds, thereby making them resistant to that herbicide, and again, no data are available on long-term effects on humans, other animals, or the environment in general.

DISCOVERY OF THE STRUCTURE OF DNA

Most biological experiments are done with samples of living matter- cells, tissues, whole organisms or extracts of these materials. Only a few experiments are done with mathematical or physical models of biological components. Ocassionally, however, a model experiment gives an insight that would be difficult to obtain in any other way. This was true of the discovery of the structure of DNA. A crucial experiment was done by James Watson in 1952 using nothing more complicated than chemical models cut out of cardboard.

DNA AND MOLECULAR GENETICS

CARRIER OF INHERITANCE

While the period from the early 1900s to World War II has been considered the "golden age" of genetics, scientists still had not determined that DNA, and not protein, was the hereditary material. However, during this time a great many genetic discoveries were made and the link between genetics and evolution was made. Friedrich Meischer in 1869 isolated DNA from fish sperm and the pus of open wounds. Since it came from nuclei, Meischer named this new chemical, nuclein. Subsequently the name was changed to nucleic acid and lastly to deoxyribonucleic acid (DNA).

Robert Feulgen, in 1914, discovered that fuchsin dye stained DNA. DNA was then found in the nucleus of all eukaryotic cells. During the 1920s, biochemist P.A. Levene analyzed the components of the DNA molecule. He found it contained four nitrogenous bases: cytosine, thymine, adenine, and guanine; deoxyribose sugar; and a phosphate group. He concluded that the basic unit (nucleotide) was composed of a base attached to a sugar and that the phosphate also attached to the sugar. He (unfortunately) also erroneously concluded that the proportions of bases were equal and that there was a tetranucleotide that was the repeating structure of the molecule.

The nucleotide, however, remains as the fundemantal unit (monomer) of the nucleic acid polymer. There are four nucleotides: those with cytosine (C), those with guanine (G), those with adenine (A), and those with thymine (T).

During the early 1900s, the study of genetics began in earnest: the link between Mendel's work and that of cell biologists resulted in the chromosomal theory of inheritance; Garrod proposed the link between genes and "inborn errors of metabolism"; and the question was formed: what is a gene?

The answer came from the study of a deadly infectious disease: *pneumonia.*

During the 1920s Frederick Griffith studied the difference between a disease-causing strain of the pneumonia causing bacteria (*Streptococcus peumoniae*) and a strain that did not cause pneumonia.

The pneumonia-causing strain (the S strain) was surrounded by a capsule. The other strain (the R strain) did not have a capsule and also did not cause pneumonia. Frederick Griffith was able to induce a nonpathogenic strain of the bacterium *Streptococcus pneumoniae* to become pathogenic. Griffith referred to a transforming factor that caused the non-pathogenic bacteria to become pathogenic.

Griffith injected the different strains of bacteria into mice. The S strain killed the mice; the R strain did not. He further noted that if heat killed S strain was injected into a mouse, it did not cause pneumonia.

When he combined heat-killed S with Live R and injected the mixture into a mouse (remember neither alone will kill the mouse) that the mouse developed pneumonia and died.

Bacteria recovered from the mouse had a capsule and killed other mice when injected into them!

Hypotheses:

- The dead S strain had been reanimated/resurrected.
- The Live R had been transformed into Live S by some "transforming factor".

Further experiments led Griffith to conclude that number 2 was correct.

In 1944, Oswald Avery, Colin MacLeod, and Maclyn McCarty revisited Griffith's experiment and concluded the transforming factor was DNA. Their evidence was strong but not totally conclusive. The then-current favourite for the hereditary material was protein; DNA was not considered by many scientists to be a strong candidate.

The breakthrough in the quest to determine the hereditary material came from the work of Max Delbruck and Salvador Luria in the 1940s. Bacteriophage are a type of virus that attacks bacteria, the viruses that Delbruck and Luria worked with were those attacking *Escherichia coli*, a bacterium found in human intestines. Bacteriophages consist of protein coats covering DNA. Bacteriophages infect a cell by injecting DNA into the host cell. This viral DNA then "disappears" while taking over the bacterial machinery and beginning to make new virus instead of new bacteria. After 25 minutes the host cell bursts, releasing hundreds of new bacteriophage.

Phages have DNA and protein, making them ideal to resolve the nature of the hereditary material. In 1952, Alfred D. Hershey and Martha Chase conducted a series of experiments to determine whether protein or DNA was the hereditary material. By labeling the DNA and protein with different (and mutually exclusive) radioisotopes, they would be able to determine which chemical (DNA or protein) was getting into the bacteria.

Such material must be the hereditary material (Griffith's transforming agent). Since DNA contains Phosphorous (P) but no Sulfur (S), they tagged the DNA with radioactive Phosphorous-32. Conversely, protein lacks P but does have S, thus it could be tagged with radioactive Sulfur-35. Hershey and Chase found that the radioactive S remained outside the cell while the radioactive P was found inside the cell, indicating that DNA was the physical carrier of heredity.

THE STRUCTURE OF DNA

Erwin Chargaff analyzed the nitrogenous bases in many different forms of life, concluding that the amount of purines does not always equal the amount of pyrimidines (as proposed by Levene). DNA had been proven as the genetic material by the Hershey-Chase experiments, but how DNA served as genes was not yet certain.

DNA must carry information from parent cell to daughter cell. It must contain information for replicating itself. It must be chemically stable, relatively unchanging. However, it must be capable of mutational change. Without mutations there would be no process of evolution.

Many scientists were interested in deciphering the structure of DNA, among them were Francis Crick, James Watson, Rosalind Franklin, and Maurice Wilkens. Watson and Crick gathered all available data in an attempt to develop a model of DNA structure.

Franklin took X-ray diffraction photomicrographs of crystalline DNA extract, the key to the puzzle. The data known at the time was that DNA was a long molecule, proteins were helically coiled (as determined by the work of Linus Pauling), Chargaff's base data, and the x-ray diffraction data of Franklin and Wilkens.

DNA is a double helix, with bases to the centre (like rungs on a ladder) and sugar-phosphate units along the sides of the helix (like the sides of a twisted ladder). The strands are complementary (deduced by Watson and Crick from Chargaff's data, A pairs with T and C pairs with G, the pairs held together by hydrogen bonds).

Notice that a double-ringed purine is always bonded to a single ring pyrimidine. Purines are Adenine (A) and Guanine (G). We have encountered Adenosine triphosphate (ATP) before, although in that case the sugar was ribose, whereas in DNA it is deoxyribose.

Pyrimidines are Cytosine (C) and Thymine (T). The bases are complementary, with A on one side of the molecule we only get T on the other side, similarly with G and C. If we know the base sequence of one strand we know its complement.

DNA REPLICATION

DNA was proven as the hereditary material and Watson et al. had deciphered its structure. What remained was to determine how DNA copied its information and how that was expressed in the phenotype. Matthew Meselson and Franklin W. Stahl designed an experiment to determine the method of DNA replication. Three models of replication were considered likely.

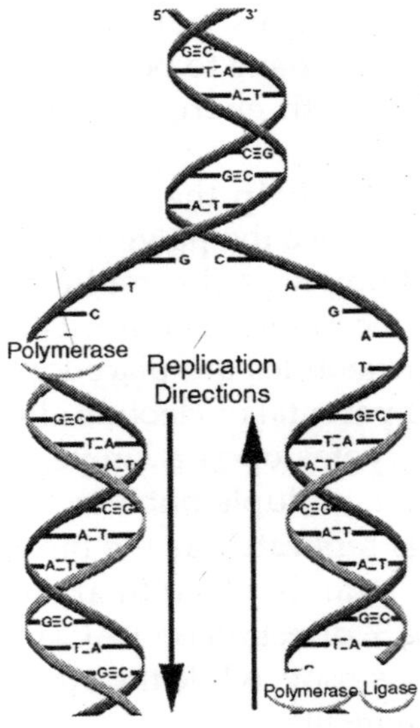

Fig. DNA Replication

Dispersive Replication

Involved the breaking of the parental strands during replication, and somehow, a reassembly of molecules that were a mix of old and new fragments on each strand of DNA. The Meselson-Stahl experiment involved the growth of *E. coli* bacteria on a growth medium containing heavy nitrogen (Nitrogen-15 as opposed to the more common, but lighter molecular weight isotope, Nitrogen-14). The first generation of bacteria was grown on a medium where the sole source of N was Nitrogen-15. The bacteria were then transferred to a medium with light (Nitrogen-14) medium. Watson and Crick had predicted that DNA replication was semi-conservative. If it was, then the DNA produced by bacteria grown on light medium would be intermediate between heavy and light.

It was. DNA replication involves a great many building blocks, enzymes and a great deal of ATP energy (remember that after the S phase of the cell cycle cells have a G phase to regenerate energy for cell division). Only occurring in a cell once per (cell) generation, DNA replication in humans occurs at a rate of 50 nucleotides per second, 500/second in prokaryotes. Nucleotides have to be assembled and available in the nucleus, along with energy to make bonds between nucleotides.

DNA polymerases unzip the helix by breaking the H-bonds between bases. Once the polymerases have opened the molecule, an area known as the replication bubble forms (always initiated at a certain set of nucleotides, the origin of replication). New nucleotides are placed in the fork and link to the corresponding parental nucleotide already there (A with T, C with G). Prokaryotes open a single replication bubble, while eukaryotes have multiple bubbles. The entire length of the DNA molecule is replicated as the bubbles meet.

Since the DNA strands are antiparallel, and replication proceeds in thje 5' to 3' direction on EACH strand, one strand will form a continuous copy, while the other will form a series of short Okazaki fragments.

ENZYME STRUCTURE PROVIDES CLUES TO DNA

Before a cell can begin to divide or differentiate, the genetic information within the cell's DNA must be copied, or "transcribed," onto complementary strands of RNA. RNA polymerase II (pol II) is an enzyme that, by itself, can unwind the DNA double helix, synthesize RNA, and proofread the result. When combined with other molecules that regulate and control the transcription process, pol II is the key to successful interpretion of an organism's genetic code.

However, the size, complexity, scarcity, and fragility of pol II complexes have made analysis of these macromolecules by x-ray crystallography a formidable challenge. A team of structural biologists has met this challenge using data obtained from both the Stanford Synchrotron Radiation Laboratory and the Macromolecular Crystallography Facility at the ALS. The resultant high-resolution model of a 10-subunit pol II complex suggests roles for each of the subunits and will allow researchers to begin unraveling the intricacies of DNA transcription and its role in gene expression.

In this work, the researchers studied the pol II enzyme from the yeast *Saccharomyces cerevisiae,* which is likely to be an excellent model for the human enzyme in light of its highly similar gene sequences.

It is also the best-characterized form of the pol II enzyme, having been the subject of many biochemical and low-resolution structural studies in the past. To obtain a high-resolution structure, the research team drew on its considerable expertise in the preparation of protein crystals: two-dimensional crystals of pol II (minus two small subunits found to impede crystal growth) were used as seeds for growing three-dimensional crystals.

These crystals, when produced in an inert atmosphere to prevent oxidation, enabled the collection of data to 3.5-angstrom resolution. The addition of a final soaking procedure to produce uniform crystals, combined with high-brightness x-ray sources, resulted in a resolution of 3.0 angstroms.

The current results bring into focus the somewhat fuzzy features previously observed in or inferred from earlier experiments.

More importantly, the structural details suggest possible explanations for some of the unusual characteristics of this enzyme, which include a high processivity (the ability to synthesize very long strands of RNA) and the tendency to work in periodic spurts separated by pauses.

While it is known that additional proteins (transcription factors) play a role in controlling the activity of pol II, scientists have yet to understand how such proteins interact with pol II binding sites to perform their various functions. The pol II model reported here establishes the positions of the various subunits and provides detailed information about the DNA/RNA binding domains.

The data reveal two main subunits separated by a deep cleft where DNA can enter the complex. At the end of the cleft is the active site, where the DNA can be unwound for a short distance (the "transcription bubble") and a DNA/RNA hybrid can be produced.

Two prominent grooves lead away from the active site, either of which could accommodate the exiting RNA transcript. An opening below the active site may allow the entry of nucleotides (for manufacturing RNA) and transcription factors (for regulating the process).

The same opening may provide room for the leading end of the RNA strand during "backtracking" maneuvers, which are important for proofreading and for traversing obstacles such as DNA damage.

Other notable features that might help account for the great stability of this transcribing complex include a pair of "jaws" that appear to grip the DNA strands as they enter the complex and, closer to the active site, a clamp on the DNA that could possibly be locked in the closed position by the presence of RNA.

The high-resolution pol II structure reported here is a landmark achievement, pulling together threads from numerous diverse research efforts into a cohesive whole. Further study should yield many new insights into the detailed mechanisms of pol II and its transcription factors. Construction of an atomic model is already well underway.

UNRAVELING DNA

Encoded into the double-helical strands of DNA, the human genome is the complete set of instructions required to make a human being. While the mapping of the human genome (roughly three billion components) is certainly a Herculean accomplishment, it is only the first step toward realizing the full potential of genomic medicine in the diagnosis, monitoring, and treatment of disease. Beyond knowing what the genetic blueprint says, scientists must understand how that blueprint gets interpreted, or "expressed" as an individual with unique traits. The pol II enzyme is the catalyst for a major step in this process

As a pol II molecule slides along a DNA molecule, it "unzips" the strands of the DNA double helix, synthesizes a complementary strand of RNA (which will carry the genetic information to where it is needed), and verifies that no mistakes have occurred.

This process is regulated by transcription factors—separate molecules that bind to pol II and determine which genes are expressed, at what stage of development, and in which tissue.

Done correctly, this process results in healthy cell growth and differentiation; otherwise, aberrations such as cancer can be the result.

Thus, details of the structure of pol II, including information about its binding sites and how they interact with transcription factors, will provide valuable insight into the detailed mechanisms underlying the flow of genetic information from DNA to RNA to protein, which is necessary for life and health.

HYDROGEN BONDING IN DNA

Without hydrogen bonds there could be no life because they hold the double helix of DNA together, and this they do by charge attractions.

A hydrogen bonded to oxygen, or nitrogen, becomes slightly positively charged which enables it to attract a centre of negative charge on another molecule, such as another oxygen or nitrogen atom. The hydrogen bond is then written, e.g., O–H...N, with the dotted line signifying the hydrogen bond. There are also O–H...O, N–H...O and N–H...N bonds, the last being among the weakest.

The secondary effects they have on structures, molecular vibrations, etc., can be used to infer hydrogen bonding, but there is no primary way of observing them because of their inherent weakness.

NMR appears to be the least useful technique because neither of the common isotopes, oxygen-16 or nitrogen-14, has a magnetic nucleus. However, nitrogen-15 has a magnetic moment, and by replacing ^{14}N by N, Grzesiek has opened up a new area of investigating these enigmatic bonds.

Paper reports for the first time the direct observation by NMR of an N–H...N hydrogen bond between nucleic acids enriched with ^{15}N, by measuring the coupling of the nitrogen atoms. Grzesiek, working with Andrew Dingley of the Heinrich-Heine University in Düsseldorf, has been able to do this and show that the coupling is surprisingly large. Normally atomic nuclei only couple with each other if they are linked by normal chemical bonds, and in theory hydrogen bonds have neither the strength nor stability for this to occur.

The German researchers studied an ^{15}N enriched sample of the T1 domain of the potato spindle tuber viroid and were able to prove that N–H...N hydrogen bonding was present between the base pairs, uridine...adenosine and guanosine...cytidine, with couplings of approximately 7 Hz. How could they be certain that the signals they were observing are due to N–H...N hydrogen bonds?

The answer was to use triple resonance techniques to examine base pairs that hydrogen bond only via O–H...N hydrogen bonds, and show that the signal they had previously observed was absent.

Grzesiek's second paper on hydrogen bonds, coauthored by Florence Cordier, Heinrich-Heine University, extends the work in an even more remarkable way by measuring the NMR coupling between nitrogens and carbons in the backbone hydrogen bonds of the human protein ubiquitin. The carbons are part of a carbonyl (C=O) group, so are one removed from the hydrogen bond, i.e., N–H...O=C.

This time they used material enriched with ^{15}N and ^{13}C (normal ^{12}C has no nuclear magnet), and there too was the evidence for these hydrogen bonds, albeit with an interaction an order of magnitude weaker (at –0.25 to –0.9 Hz) than the N–H…N coupling.

Nevertheless, the couplings correlate with the strength of the hydrogen bond, being stronger in the stronger bonds. These findings were confirmed by paper #9, which is from the researchers of Ad Bax's group based at NIH, Bethesda, Maryland.

Grzesiek's third paper, was done in conjunction with Dingley and researchers at UCLA. Together they studied not only the hydrogen bonding of Watson-Crick base pairs but also of Hoogsten base pairs within a DNA triplex consisting of one purine and two pyrimidine strands.

Four different base pairs were identified, their various couplings distinguished–including those of the weaker interactions at the "frayed ends" of the DNA chains–and relationships with other hydrogen bonding parameters, such as bond length, were established.

In addition they were able to show that density functional computer simulations by computer could reproduce these findings exactly.

Chapter 4

RNA Structure

We now turn our attention to RNA, which differs from DNA in three respects. First, the backbone of RNA contains ribose rather than 2'-deoxyribose. That is, ribose has a hydroxyl group at the 2' position. Second, RNA contains uracil in place of thymine. Uracil has the same single-ringed structure as thymine, except that it lacks the 5' methyl group. Thymine is in effect 5'methyl-uracil. Third, RNA is usually found as a single polynucleotide chain.

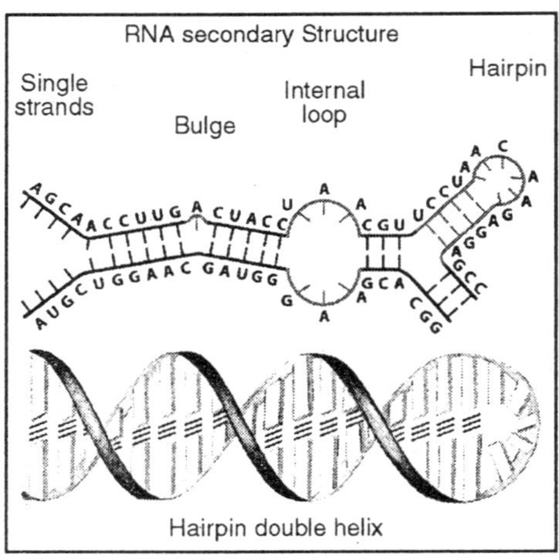

Fig. RNA Structure

Except for the case of certain viruses, RNA is not the genetic material and does not need to be capable of serving as a template for its own replication. Rather, RNA functions as the intermediate, the mRNA, between the gene and the protein-synthesizing machinery. Another function of RNA is as an adaptor, the tRNA, between the codons in the mRNA and amino acids. RNA can also play a structural role as in the case of the RNA components of the ribosome.

Yet another role for RNA is as a regulatory molecule, which through sequence complementarity binds to, and interferes with the translation of, certain mRNAs. Finally, some RNAs (including one of the structural RNAs of the ribosome) are enzymes that catalyze essential reactions in the cell. In all of these cases, the RNA is copied as a single strand off only one of the two strands of the DNA template, and its complementary strand does not exist. RNA is capable of forming long double helices, but these are unusual in nature.

RNA CHAINS

Despite being single-stranded, RNA molecules often exhibit a great deal of double-helical character. This is because RNA chains frequently fold back on themselves to form base-paired segments between short stretches of complementary sequences.

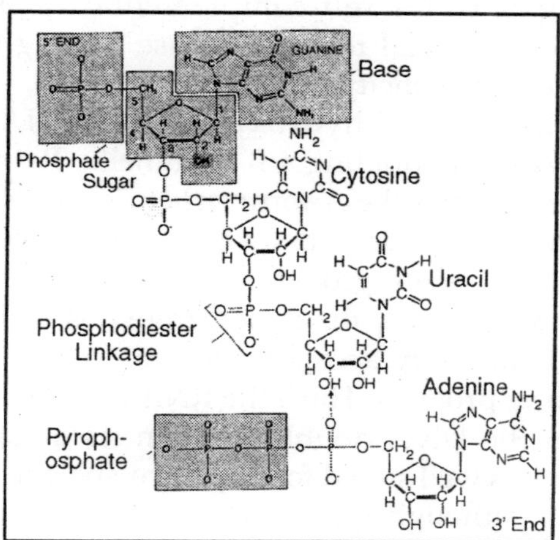

Fig. RNA Chains

If the two stretches of complementary sequence are near each other, the RNA may adopt one of various stem-loop structures in which the intervening RNA is looped out from the end of the double-helical segment as in a hairpin, a bulge, or a simple loop.

The stability of such stem-loop structures is in some instances enhanced by the special properties of the loop. For example, a stem-loop with the "tetraloop" sequence UUCG is unexpectedly stable due to special base-stacking interactions in the loop. Base pairing can also take place between sequences that are not contiguous to form complex structures aptly named pseudoknots. The regions of base pairing in RNA can be a regular double helix or they can contain discontinuities, such as noncomplementary nucleotides that bulge out from the helix.

A feature of RNA that adds to its propensity to form double-helical structures is an additional, non-Watson-Crick base pair. This is the G:U base pair, which has hydrogen bonds between N3 of uracil and the carbonyl on C6 of guanine and between the carbonyl on C2 of uracil and N1 of guanine. Because G:U base pairs can occur as well as the four conventional, Watson-Crick base pairs, RNA chains have an enhanced capacity for self-complementarity. Thus, RNA frequently exhibits local regions of base pairing but not the long-range, regular helicity of DNA.

The presence of 2'-hydroxyls in the RNA backbone prevents RNA from adopting a B-form helix. Rather, double-helical RNA resembles the A-form structure of DNA. As such, the minor groove is wide and shallow, and hence accessible, but recall that the minor groove offers little sequence-specific information. Meanwhile, the major groove is so narrow and deep that it is not very accessible to amino acid side chains from interacting proteins. Thus, the RNA double helix is quite distinct from the DNA double helix in its detailed atomic structure and less well suited for sequence-specific interactions with proteins (although some proteins do bind to RNA in a sequence-specific manner).

COMPLEX TERTIARY STRUCTURES

Freed of the constraint of forming long-range regular

helices, RNA can adopt a wealth of tertiary structures. This is because RNA has enormous rotational freedom in the backbone of its non-base-paired regions. Thus, RNA can fold up into complex tertiary structures frequently involving unconventional base pairing, such as the base triples and base-backbone interactions seen in tRNAs. Proteins can assist the formation of tertiary structures by large RNA molecules, such as those found in the ribosome.

Proteins shield the negative charges of backbone phosphates, whose electrostatic repulsive forces would otherwise destabilize the structure.

Researchers have taken advantage of the potential structural complexity of RNA to generate novel RNA species (not found in nature) that have specific desirable properties. By synthesizing RNA molecules with randomized sequences, it is possible to generate mixtures of oligonucleotides representing enormous sequence diversity.

For example, a mixture of oligoribonucleotides of length 20 and having four possible nucleotides at each position would have a potential complexity of 420 sequences or 1012 sequences! From mixtures of diverse oligoribonucleotides, RNA molecules can be selected biochemically that have particular properties, such as an affinity for a specific small molecule.

Some RNAs Are Enzymes

It was widely believed for many years that only proteins could be enzymes. An enzyme must be able to bind a substrate, carry out a chemical reaction, release the product and repeat this sequence of events many times. Proteins are well suited to this task because they are composed of many different kinds of amino acids and they can fold into complex tertiary structures with binding pockets for the substrate and small molecule cofactors and an active site for catalysis.

Now we know that RNAs, which as we have seen can similarly adopt complex tertiary structures, can also be biological catalysts. Such RNA enzymes are known as ribozymes, and they exhibit many of the features of a classical enzyme, such as an active site, a binding site for a substrate and a binding site for a cofactor, such as a metal ion. One of the first ribozymes to be discovered was RNase P, a ribonuclease that is involved in generating tRNA molecules from larger, precursor RNAs. RNase P is composed of both RNA and protein; however, the RNA moiety alone is the catalyst.

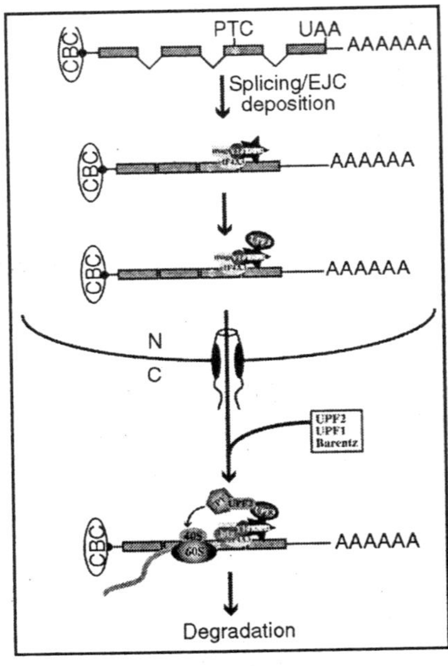

Fig. RNA Splicing

The protein moiety of RNase P facilitates the reaction by shielding the negative charges on the RNA so that it can bind effectively to its negatively charged substrate. The RNA moiety is able to catalyze cleavage of the tRNA precursor in the absence of the protein if a small, positively charged counter ion, such as the peptide spermidine, is used to shield the repulsive, negative charges. Other ribozymes carry out trans-esterification reactions involved in the removal of intervening sequences known as introns from precursors to certain mRNAs, tRNAs, and ribosomal RNAs in a process known as RNA splicing.

HAMMERHEAD RIBOZYME

Before concluding our discussion of RNA, let us look in more detail at the structure and function of one particular ribozyme, the hammerhead. The hammerhead is a sequence-specific ribonuclease that is found in certain infectious RNA agents of plants known as *viroids,* which depend on self-cleavage to propagate. When the viroid replicates, it produces multiple copies of itself in one continuous RNA chain. Single viroids arise by cleavage, and this cleavage reaction is carried out by the RNA sequence around the junction.

One such self-cleaving sequence is called the *hammerhead* because of the shape of its secondary structure, which consists of three base-paired stems surrounding a core of non-complementary nucleotides required for catalysis. The tertiary structure of the ribozyme, however, looks more like a wishbone. To understand how the hammerhead works, let us first look at how RNA undergoes hydrolysis under alkaline conditions.

At high pH, the 2′ hydroxyl of the ribose in the RNA backbone can become deprotonated, and the resulting negatively charged oxygen can attack the scissile phosphate at the 3′ position of the same ribose. This reaction breaks the RNA chain, producing a 2′, 3′ cyclic phosphate and a free 5′ hydroxyl. Each ribose in an RNA chain can undergo this reaction, completely cleaving the parent molecule into nucleotides. Many protein ribonucleases also cleave their RNA substrates via the formation of a 2′, 3′ cyclic phosphate. Working at normal cellular pH, these protein enzymes use a metal ion, bound at their active site, to activate the 2′ hydroxyl of the RNA. The hammerhead is a sequencespecific ribonuclease, but it too cleaves RNA via the formation of a 2′, 3′ cyclic phosphate.

Hammerhead-mediated cleavage involves a ribozyme-bound Mg+ ion that deprotonates the 2′ hydroxyl at neutral pH, resulting in nucleophilic attack on the scissile phosphate. Because the normal reaction of the hammerhead is self-cleavage, it is not really a catalyst; each molecule normally promotes a reaction one time only, thus having a turnover number of one. But the hammerhead can be engineered to function as a true ribozyme by dividing the molecule into two portions—one, the ribozyme, that contains the catalytic core and the other, the substrate, that contains the cleavage site. The substrate binds to the ribozyme at stems I and III. After cleavage, the substrate is released and replaced by a fresh uncut substrate, thereby allowing repeated rounds of cleavage.

LIFE EVOLVE FROM AN RNA

The discovery of ribozymes has profoundly altered our view of how life might have evolved. We can now imagine that there was a primitive form of life based entirely on RNA. In this world, RNA would have functioned as the genetic material and as the enzymatic machines. This RNA world would have preceded life as we know it today, in which information transfer is based on DNA, RNA, and protein.

A hint that the protein world might have arisen from an RNA world is the discovery that the component in the ribosome that is responsible for the formation of the peptide bond, the peptidyl transferase, is an RNA molecule. Unlike RNase P, the hammerhead, and other previously known ribozymes which act on phosphorous centers, the peptidyl transferase acts on a carbon centre to create the peptide bond.

It thus links RNA chemistry to the most fundamental reaction in the protein world, peptide bond formation. Perhaps then the ribosome ribozyme is a relic of an earlier form of life in which all enzymes were RNAs. DNA is usually in the form of a right-handed double helix. The helix consists of two polydeoxynucleotide chains.

Each chain is an alternating polymer of deoxyribose sugars and phosphates that are joined together via phosphodiester linkages. One of four bases protrudes from each sugar: adenine and guanine, which are purines, and thymine and cytosine, which are pyrimidines. While the sugar phosphate backbone is regular, the order of bases is irregular and this is responsible for the information content of DNA.

Each chain has a 5′ to 3′ polarity, and the two chains of the double helix are oriented in an antiparallel manner—that is, they run in opposite directions. Pairing between the bases holds the chains together. Pairing is mediated by hydrogen bonds and is specific: Adenine on one chain is always paired with thymine on the other chain, whereas guanine is always paired with cytosine. This strict base-pairing reflects the fixed locations of hydrogen atoms in the purine and pyrimidine bases in the forms of those bases found in DNA. Adenine and cytosine almost always exist in the amino as opposed to the imino tautomeric forms, whereas guanine and thymine almost always exist in the keto as opposed to enol forms.

The complementarity between the bases on the two strands gives DNA its self-coding character. The two strands of the double helix fall apart (denature) upon exposure to high temperature, extremes of pH, or any agent that causes the breakage of hydrogen bonds. Upon slow return to normal cellular conditions, the denatured single strands can specifically reassociate to biologically active double helices (renature or anneal).

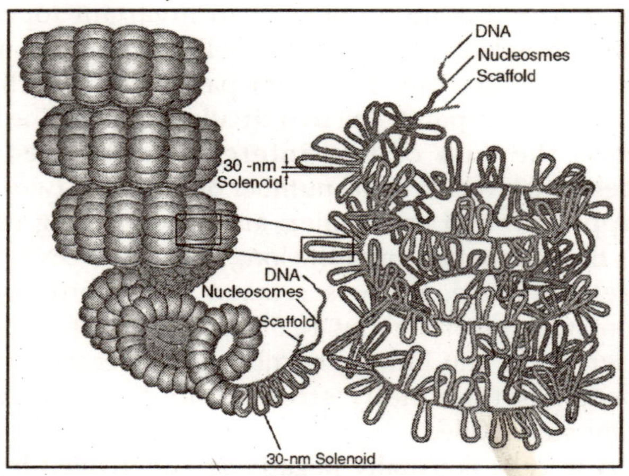

Fig. Nucleosomes

DNA in solution has a helical periodicity of about 10.5 base pairs per turn of the helix. The stacking of base pairs upon each other creates a helix with two grooves. Because the sugars protrude from the bases at an angle of about 120°, the grooves are unequal in size. The edges of each base pair are exposed in the grooves, creating a pattern of hydrogen bond donors and acceptors and of van der Waals surfaces that identifies the base pair.

The wider—or *major*—groove is richer in chemical information than the narrow (*minor*) groove and is more important for recognition by nucleotide sequence-specific binding proteins. Almost all cellular DNAs are extremely long molecules, with only one DNA molecule within a given chromosome. Eukaryotic cells accommodate this extreme length in part by wrapping the DNA around protein particles known as nucleosomes. Most DNA molecules are linear but some DNAs are circles, as is often the case for the chromosomes of prokaryotes and for certain viruses. DNA is flexible. Unless the molecule is topologically constrained, it can freely rotate to accommodate changes in the number of times the two strands twist about each other.

DNA is topologically constrained when it is in the form of a covalently closed circle, or when it is entrained in chromatin. The linking number is an invariant topological property of covalently closed circular DNA. It is the number of times one strand would have to be passed through the other strand in order to separate the two circular strands. The linking number is the sum of two interconvertible geometric properties: twist, which is the number of times the two strands are wrapped around each other; and the writhing number, which is the number of times the long axis of the DNA crosses over itself in space. DNA is relaxed under physiological conditions when it has about 10.5 base pairs per turn and is free of writhe. If the linking number is decreased, then the DNA becomes torsionally stressed, and it is said to be negatively supercoiled.

DNA in cells is usually negatively supercoiled by about 6%. The left-handed wrapping of DNA around nucleosomes introduces negative supercoiling in eukaryotes. In prokaryotes, which lack histones, the enzyme DNA gyrase is responsible for generating negative supercoils. DNA gyrase is a member of the type II family of topoisomerases. These enzymes change the linking number of DNA in steps of two by making a transient break in the double helix and passing a region of duplex DNA through the break. Some type II topoisomerases relax supercoiled DNA, whereas DNA gyrase generates negative supercoils. Type I topoisomerases also relax supercoiled DNAs but do so in steps of one in which one DNA strand is passed through a transient nick in the other strand.

RNA differs from DNA in the following ways: its backbone contains ribose rather than 2′-deoxyribose; it contains the pyrimidine uracil in place of thymine; and it usually exists as a single polynucleotide chain, without a complementary chain.

As a consequence of being a single strand, RNA can fold back on itself to form short stretches of double helix between regions that are complementary to each other. RNA allows a greater range of base pairing than does DNA. Thus, as well as A:U and C:G pairing, U can also pair with G. This capacity to form a non-Watson-Crick base pair adds to the propensity of RNA to form doublehelical segments. Freed of the constraint of forming longrange regular helices, RNA can form complex tertiary structures, which are often based on unconventional interactions between bases and between bases and the sugarphosphate backbone.

Some RNAs act as enzymes—they catalyze chemical reactions in the cell and in vitro. These RNA enzymes are known as ribozymes. Most ribozymes act on phosphorous centers, as in the case of the ribonuclease RNase P. RNase P is composed of protein and RNA, but it is the RNA moiety that is the catalyst. The hammerhead is a self-cleaving RNA, which cuts the RNA backbone via the formation of a 2′, 3′ cyclic phosphate in a reaction that involves an RNA-bound Mg″ ion. Peptidyl transferase is an example of a ribozyme that acts on a carbon centre. This ribozyme, which is responsible for the formation of the peptide bond, is one of the RNA components of the ribosome. The discovery of RNA enzymes that can act on phosphorous or carbon centers suggests that life might have evolved from a primitive form in which RNA functioned both as the genetic material and as the enzymatic machinery.

Chapter 5

DNA Replication

The process of copying a double-stranded DNA molecule to form two double-stranded molecules DNA replication. The process of DNA replication is a fundamental process used by all living organisms as it is the basis for biological inheritance. As each DNA strand holds the same genetic information, both strands can serve as templates for the reproduction of the opposite strand. The template strand is preserved in its entirety and the new strand is assembled from nucleotides. This process is called "semiconservative replication".

The resulting double-stranded DNA molecules are identical; proofreading and error-checking mechanisms exist to ensure near perfect fidelity. In a cell, DNA replication must happen before cell division can occur. DNA synthesis begins at specific locations in the genome, called "origins", where the two strands of DNA are separated.

RNA primers attach to single stranded DNA and the enzyme DNA polymerase extends the primers to form new strands of DNA, adding nucleotides matched to the template strand. The unwinding of DNA and synthesis of new strands forms a replication fork.

In addition to DNA polymerase, a number of other proteins are associated with the fork and assist in the initiation and continuation of DNA synthesis. DNA replication can also be performed artificially, using the same enzymes used within the cell. DNA polymerases and artificial DNA primers are used to initiate DNA synthesis at known sequences in a template molecule. The polymerase chain reaction (PCR), a common laboratory technique, employs artificial synthesis in a cyclic manner to rapidly and specifically amplify a target DNA fragment from a pool of DNA.

ORIGINS OF REPLICATION

The first step in DNA replication is the separation of the two DNA strands that make up the helix that is to be copied. DNA Helicase untwists the helix at locations called replication origins.

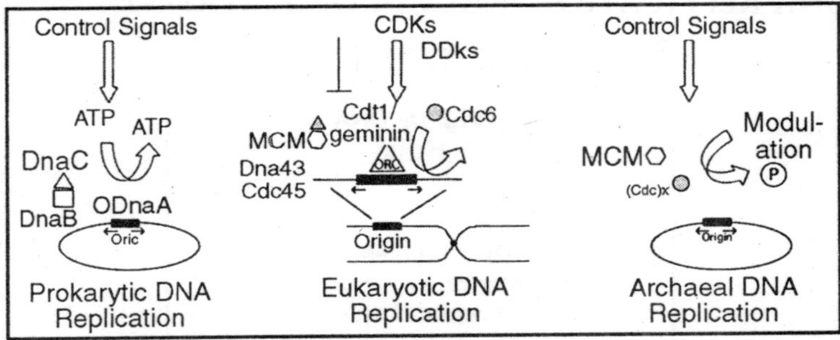

Fig. DNA replication

The replication origin forms a Y shape, and is called a replication fork. The replication fork moves down the DNA strand, usually from an internal location to the strand's end. The result is that every replication fork has a twin replication fork, moving in the opposite direction from that same internal location to the strand's opposite end. Single-stranded binding proteins (SSB) work with helicase to keep the parental DNA helix unwound.

It works by coating the unwound strands with rigid subunits of SSB that keep the strands from snapping back together in a helix. The SSB subunits coat the single-strands of DNA in a way as not to cover the bases, allowing the DNA to remain available for base-pairing with the newly synthesized daughter strands.

The two parent strands of DNA are separated to begin replication, one strand is oriented in the 5' to 3' direction while the other strand is oriented in the 3' to 5' direction. DNA replication, however, is inflexible: the enzyme that carries out the replication, DNA polymerase, only functions in the 5' to 3' direction. This characteristic of DNA polymerase means that the daughter strands synthesize through different methods, one adding nucleotides one by one in the direction of the replication fork, the other able to add nucleotides only in chunks. The first strand, which replicates nucleotides one by one is called the leading strand; the other strand, which replicates in chunks, is called the lagging strand.

DNA REPLICATION IS SEMI-CONSERVATIVE

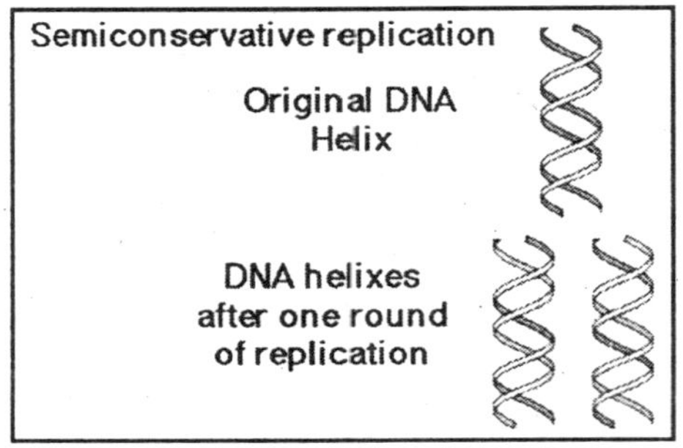

Fig. Semi-Conservative Nature of DNA Replication

DNA replication of one helix of DNA results in two identical helices. If the original DNA helix is called the "parental" DNA, the two resulting helices can be called "daughter" helices. Each of these two daughter helices is a nearly exact copy of the parental helix (it is not 100% the same due to mutations). DNA creates "daughters" by using the parental strands of DNA as a template or guide.

Each newly synthesized strand of DNA (daughter strand) is made by the addition of a nucleotide that is complementary to the parent strand of DNA. In this way, DNA replication is semi-conservative, meaning that one parent strand is always passed on to the daughter helix of DNA.

THE LAGGING STRAND

Whereas the DNA polymerase on the leading strand can simply follow the replication fork, because DNA polymerase must move in the 5' to 3' direction, on the lagging strand the enzyme must move away from the fork. But if the enzyme moves away from the fork, and the fork is uncovering new DNA that needs to be replicated, then how can the lagging strand be replicated at all? The problem posed by this question is answered through an ingenious method.

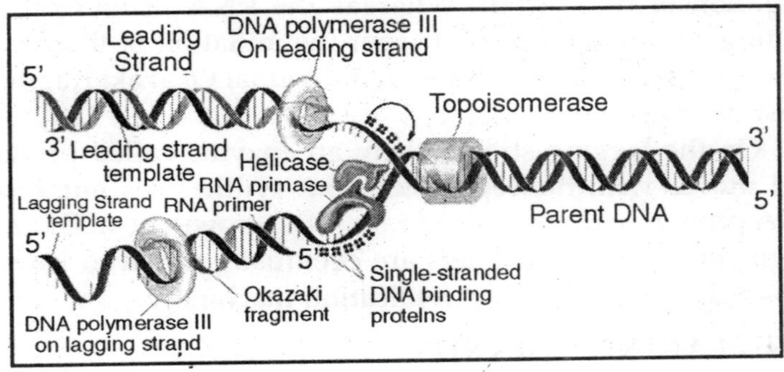

Fig. Leading and Lagging Strands

The lagging strand replicates in small segments, called Okazaki fragments. These fragments are stretches of 100 to 200 nucleotides in humans (1000 to 2000 in bacteria) that are synthesized in the 5' to 3' direction away from the replication fork. Yet while each individual segment is replicated away from the replication fork, each subsequent Okazaki fragment is replicated more closely to the receding replication fork than the fragment before. These fragments are then stitched together by DNA ligase, creating a continuous strand. This type of replication is called discontinuous

The first synthesized Okazaki fragment on the lagging strand is the furthest away from the replication fork, which is itself receding to the right. Each subsequent Okazaki fragment starts at the replication fork and continues until it meets the previous fragment. The two fragments are then stitched together by DNA ligase.

Because synthesis on the lagging strand takes place in a "backstitching" mechanism, its replication is slightly delayed in relation to synthesis on the leading strand. The lagging strand must wait for a patch of the parent helix to open up a short distance in front of the newly synthesized strand before it can begin its synthesis back to the end of the daughter strand.

This "Lag" time does not occur in the leading strand because it synthesizes the new strand by following right behind as the helix unwinds at the replication fork. Another complication to replication on the lagging strand is the initiation of replication. Whereas the RNA primer on the leading strand only has to trigger the initiation of the strand once, on the lagging strand each individual Okazaki fragment must be triggered.

On the lagging strand, then, an enzyme called primase that moves with the replication fork synthesizes numerous RNA primers, each of which triggers the growth of an Okazaki fragment. The RNA primers are eventually removed leaving gaps that are filled by the replication machinery.

THE LEADING STRAND

Since DNA replication moves along the parent strand in the 5' to 3' direction, replication can occur very easily on the leading strand. The nucleotides are added in the 5' to 3' direction. Triggered by RNA primase, which adds the first nucleotide to the nascent chain, the DNA polymerase simply sits near the replication fork, moving as the fork does, adding nucleotides one after the other, preserving the proper anti-parallel orientation. This sort of replication, since it involves one nucleotide being placed right after another in a series, is called continuous.

DNA STRUCTURE

DNA usually exists in a double-stranded structure, with both strands coiled together to form the characteristic double-helix. Each single strand of DNA is a chain of four types of nucleotide: adenine, cytosine, guanine, and thymine. A nucleotide consists of a phosphate and a deoxyribose sugar forming the backbone of the DNA double helix plus a base that points inwards. Nucleotides are matched between strands through hydrogen bonds to form base pairs.

Adenine pairs with thymine and cytosine pairs with guanine. The physical pairing of bases in DNA means that the information contained within each strand is redundant. The nucleotides on a single strand can be used to reconstruct nucleotides on a newly synthesized partner strand. DNA strands have a directionality, and the different ends of a single strand are called the "3' end" and the "5' end" (these refer to the carbon atom in ribose that the next phosphate in the chain attaches to).

In addition to being complementary, the two strands of DNA are antiparallel: they are orientated in opposite directions. This directionality has consequences in DNA synthesis, because DNA polymerase can only synthesize DNA in one direction by adding nucleotides to the 3' end of a DNA strand.

DNA POLYMERASE

DNA polymerases are a family of enzymes critical for all forms of DNA replication. A DNA polymerase synthesizes a new strand of DNA by extending the 3' end of an existing nucleotide chain, adding new nucleotides matched to the template strand one at a time.

Some DNA polymerases may also have some proofreading ability, removing nucleotides from the end of a strand in order to remove any mismatched bases. DNA polymerases are generally extremely accurate, making less than one error for every million nucleotides added. The energy for the process of DNA polymerization comes from the two additional phosphates attached to each of the unincorporated nucleotides.

These free nucleotides, also known as nucleoside triphosphates, contain a total of three phosphates. When a nucleotide is being added to a growing DNA strand, two of the phosphates are removed and the energy produced is used to attach the remaining phosphate to the growing chain. The energetics of this process may also explain the directionality of synthesis - if DNA were synthesized in the 3' to 5' direction, the energy for the process would come from the 5' end of the growing strand rather than from free nucleotides.

During proofreading, if the 5' nucleotide needed to be removed this triphosphate end would be lost, losing the energy source required to add a new nucleotide to the end. DNA polymerase can only extend an existing DNA strand paired with a template strand, it cannot begin the synthesis of a new strand. To do this a short fragment of DNA or RNA, called a primer, must be created and paired with the template strand before DNA polymerase can synthesize new DNA.

DNA REPLICATION WITHIN THE CELL

Origins of Replication

For a cell to divide, it must first replicate its DNA. This process is initiated at particular points within the DNA, known as "origins", which are targeted by proteins that separate the two strands and initiate DNA synthesis. Origins contain DNA sequences recognized by replication initiator proteins (eg. dnaA in *E coli'* and the Origin Recognition Complex in yeast).

These initiator proteins recruit other proteins to separate the two strands and initiate replication forks. Initiator proteins recruit other proteins to separate the DNA strands at the origin, forming a bubble. Origins tend to be "AT-rich" (rich in adenine and thymine bases) to assist this process because A-T base pairs have two hydrogen bonds (rather than the three formed in a C-G pair)—strands rich in these nucleotides are generally easier to separate. Once strands are separated, RNA primers are created on the template strands and DNA polymerase extends these to create newly synthesized DNA.

As DNA synthesis continues, the original DNA strands continue to unwind on each side of the bubble, forming replication forks. In bacteria, which have a single origin of replication on their circular chromosome, this process eventually creates a "theta structure". In contrast, eukaryotes have longer linear chromosomes and initiate replication at multiple origins within these.

THE REPLICATION FORK

The replication fork is a structure which forms when DNA is being replicated. It is created through the action of helicase, which breaks the hydrogen bonds holding the two DNA strands together. The resulting structure has two branching "prongs", each one made up of a single strand of DNA.

Leading Strand Synthesis

In DNA replication, the leading strand is defined as the new DNA strand at the replication fork that is synthesized in the 5'→3' direction in a continuous manner. When the enzyme helicase unwinds DNA, two single stranded regions of DNA (the "replication fork") form. On the leading strand DNA polymerase III is able to synthesize DNA using the free 3' OH group donated by a single RNA primer and continuous synthesis occurs in the direction in which the replication fork is moving.

Lagging Strand Synthesis

The lagging strand is the DNA strand at the opposite side of the replication fork from the leading strand, running in the 3' to 5' direction. Because DNA polymerase cannot synthesize in the 3'→5' direction, the lagging strand is synthesized in short segments known as Okazaki fragments. Along the lagging strand's template, primase builds RNA primers in short bursts.

DNA polymerases are then able to use the free 3' OH groups on the RNA primers to synthesize DNA in direction. The RNA fragments are then removed (different mechanisms are used in eukaryotes and prokaryotes) and new deoxyribonucleotides are added to fill the gaps where the RNA was present. DNA ligase then joins the deoxyribonucleotides together, completing the synthesis of the lagging strand.

DYNAMICS AT THE REPLICATION FORK

As helicase unwinds DNA at the replication fork, the DNA ahead is forced to rotate. This process results in a build-up of twists in the DNA ahead. This build-up would form a resistance that would eventually halt the progress of the replication fork. DNA topoisomerases are enzymes that solve these physical problems in the coiling of DNA. Topoisomerase I cuts a single backbone on the DNA, enabling the strands to swivel around each other to remove the build-up of twists.

Topoisomerase II cuts both backbones, enabling one double-stranded DNA to pass through another, thereby removing knots and entanglements that can form within and between DNA molecules.

Bare single-stranded DNA has a tendency to fold back upon itself and form secondary structures; these structures can interfere with the movement of DNA polymerase. To prevent this, single-strand binding proteins bind to the DNA until a second strand is synthesized, preventing secondary structure formation.

Clamp proteins form a sliding clamp around DNA, helping the DNA polymerase maintain contact with its template and thereby assisting with processivity. The inner face of the clamp enables DNA to be threaded through it. Once the polymerase reaches the end of the template or detects double stranded DNA, the sliding clamp undergoes a conformational change which releases the DNA polymerase. Clamp-loading proteins are used to initially load the clamp, recognizing the junction between template and RNA primers.

Eukaryotes

Within eukaryotes, DNA replication is controlled within the context of the cell cycle. As the cell grows and divides, it progresses through stages in the cell cycle; DNA replication occurs during the S phase (Synthesis phase). The progress of the eukaryotic cell through the cycle is controlled by cell cycle checkpoints.

Progression through checkpoints is controlled through complex interactions between various proteins, including cyclins and cyclin-dependent kinases.

The G1/S checkpoint (or restriction checkpoint) regulates whether eukaryotic cells enter the process of DNA replication and subsequent division. Cells which do not proceed through this checkpoint are quiescent in the "G0" stage and do not replicate their DNA.

Replication of chloroplast and mitochondrial genomes occurs independent of the cell cycle, through the process of D-loop replication.

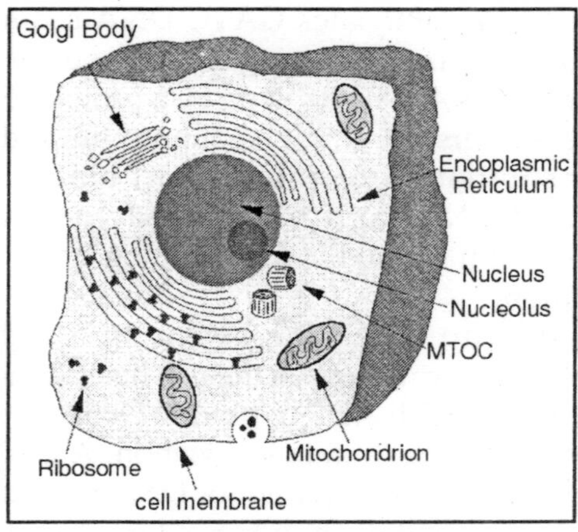

Fig. Eukaryotic Cell

Bacteria

Most bacteria do not go through a well-defined cell cycle and instead continuously copy their DNA; during rapid growth this can result in multiple rounds of replication occurring concurrently. Within *E coli*, the most well-characterized bacteria, regulation of DNA replication can be achieved through several mechanisms, including: the hemimethylation and sequestering of the origin sequence, the ratio of ATP to ADP, and the levels of protein DnaA. These all control the process of initiator proteins binding to the origin sequences.

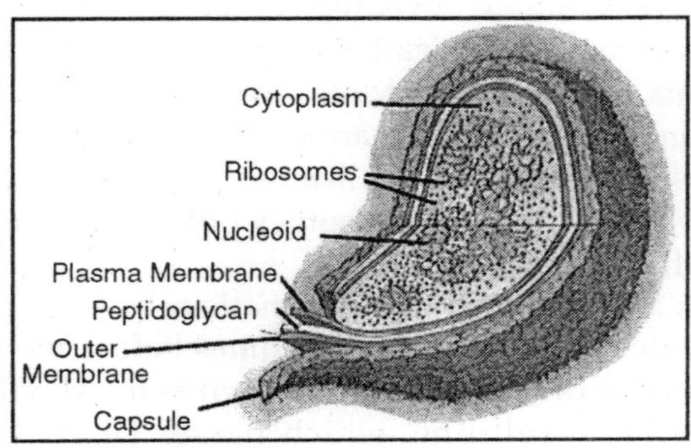

Fig. Bacteria

Because *E coli* methylates GATC DNA sequences, DNA synthesis results in hemimethylated sequences. This hemimethylated DNA is recognized by a protein (SeqA) which binds and sequesters the origin sequence; in addition, dnaA (required for initiation of replication) binds less well to hemimethylated DNA. As a result, newly replicated origins are prevented from immediately initiating another round of DNA replication.

ATP builds up when the cell is in a rich medium, triggering DNA replication once the cell has reached a specific size. ATP competes with ADP to bind to DnaA, and the DNA-ATP complex is able to initiate replication. A certain number of DnaA proteins are also required for DNA replication — each time the origin is copied the number of binding sites for DnaA doubles, requiring the synthesis of more DnaA to enable another initiation of replication.

TERMINATION OF REPLICATION

Because bacteria have circular chromosomes, termination of replication occurs when the two replication forks meet each other on the opposite end of the parental chromosome. *E coli* regulate this process through the use of termination sequences which, when bound by the Tus protein, enable only one direction of replication fork to pass through. As a result, the replication forks are constrained to always meet within the termination region of the chromosome.

Eukaryotes initiate DNA replication at multiple points in the chromosome, so replication forks meet and terminate at many points in the chromosome; these are not known to be regulated in any particular manner. Because eukaryotes have linear chromosomes, DNA replication often fails to synthesize to the very end of the chromosomes (telomeres), resulting in telomere shortening.

This is a normal process in somatic cells — cells are only able to divide a certain number of times before the DNA loss prevents further division. (This is known as the Hayflick limit.) Within the germ cell line, which passes DNA to the next generation, the enzyme telomerase extends the repetitive sequences of the telomere region to prevent degradation. Telomerase can become mistakenly active in somatic cells, sometimes leading to cancer formation.

ROLLING CIRCLE REPLICATION

Another method of copying DNA, sometimes used *in vivo* by bacteria and viruses, is the process of rolling circle replication. In this form of replication, a single replication fork progresses around a circular molecule to form multiple linear copies of the DNA sequence. In cells, this process can be used to rapidly synthesize multiple copies of plasmids or viral genomes.

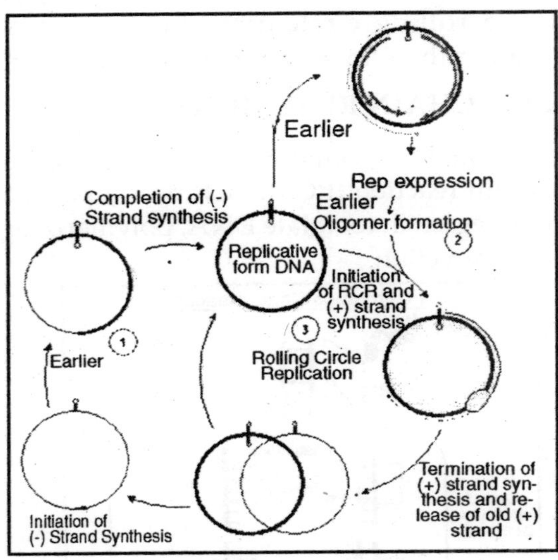

Fig. Rolling Circle Replication

In the cell, rolling circle replication is initiated by an initiator protein encoded by the plasmid or virus DNA. This protein is able to nick one strand of the double-stranded, circular DNA molecule at a site called the double-strand origin (DSO) and remains bound to the 5' phosphate end of the nicked strand. The free 3' hydroxyl end is released and can serve as a primer for DNA synthesis. Using the unnicked strand as a template, replication proceeds around the circular DNA molecule, displacing the nicked strand as single-stranded DNA. Continued DNA synthesis produces multiple single-stranded linear copies of the original DNA in a continuous head-to-tail series. *In vivo* these linear copies are subsequently converted to double-stranded circular molecules.

Rolling circle replication can also be performed *in vitro* and has found wide uses in academic research and biotechnology, often used for amplification of DNA from very small amounts of starting material. Replication can be initiated by nicking a double-stranded circular DNA molecule or by hybridizing a primer to a single-stranded circle of DNA. The use of a reverse primer (or random primers) produces hyperbranched rolling circle amplification, resulting in exponential rather than linear growth of the DNA molecule.

POLYMERASE CHAIN REACTION

In vitro, researchers commonly replicate DNA using the polymerase chain reaction (PCR). PCR uses a pair of primers to span a target region in template DNA, polymerizing partner strands in each direction.

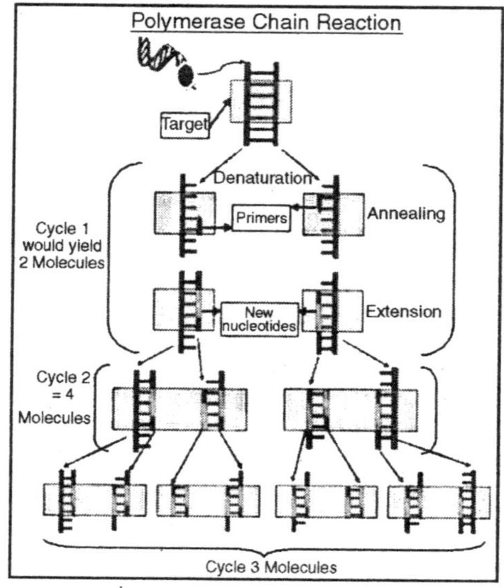

Fig. Polymerase Chain Reaction

This process can be repeated through multiple cycles through the use of a thermostable polymerase. At the start of each cycle, the mixture of template and primers is heated, separating the newly synthesized molecule and template. Then, as the mixture cools, both of these become templates for new primers to anneal to, and the polymerase extends from these. As a result the number of copies of the target region doubles each round, growing exponentially.

PROKARYOTIC DNA REPLICATION

Duplication of DNA is one of the fundamental properties of "molecule of life". Duplication or replication takes place once in a cell cycle. The duration and initiation point differs from one system to another. When conditions are favorable cell cytoplasmic mass increases and when cytoplasmic mass reaches a ratio to that of the cell size to 2L, bacterial cell division is triggered.

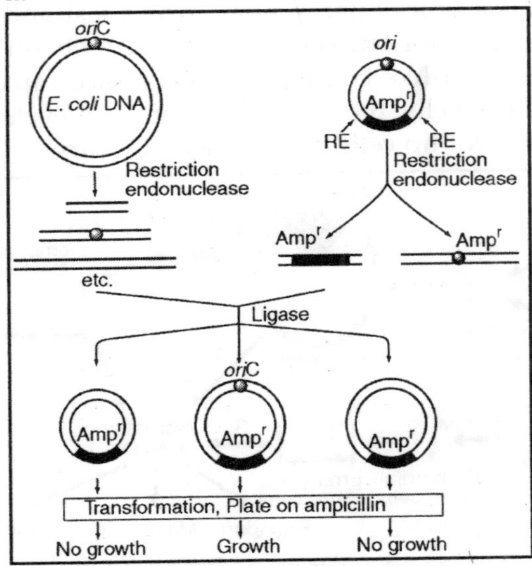

Fig. Prokaryotic DNA Replication

To complete its cell cycle it requires hardly 30 -40 minutes. Several genes involved in different steps of cell division have been characterized. Par excellent system to understand molecular process of prokaryotic cell division is E.coli. The size of the E.coli chromosome is 4.6×10^6 base pairs, but it is compacted by the binding of histone like proteins. First the nucleoid i.e the chromosome frees from the membrane and unwinds. Such DNA initiates replication and completes in a matter of 15-20 minutes.

Then the cytoplasm starts dividing almost in the centre of the cell by central septal ring formation. It is at this time the daughter DNA molecules are partitioned and segregated to the two compartments. The central periseptal ring that started 1-2 minutes after DNA replication now furrows inwards and divides the cytoplasm in the middle region. Then mid-wall splits in the middle and daughter cell are set free, hence the Schizomycetes to bacteria (E.coli)

Bacterial cell division requires the participation of the following genes and gene products. Eukaryotic cell division is more complex and highly regulated. The paradigm for it is yeast, for it is a unicellular system, grows fast and requires hardly 60 minutes for one cell cycle. One can obtain different kinds of mutation like conditional and null or auxotrophic kinds. In culture cells, the duration of one cell cycle is about 12 hrs. Onion root tip cells divide once in 12 hrs.

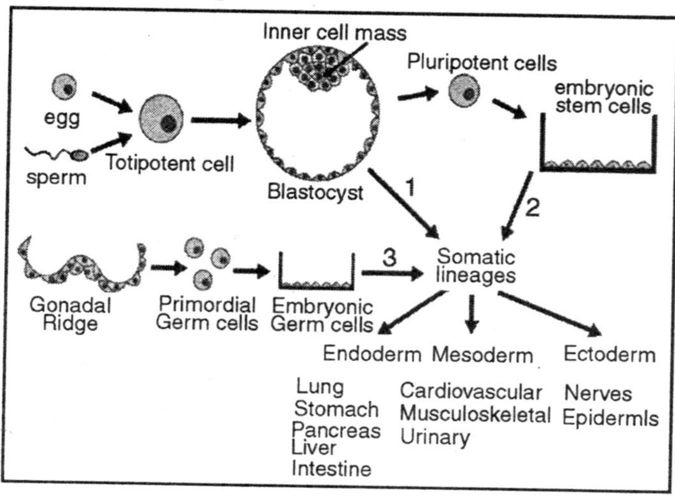

Fig. Embryonic Cells

Whereas embryonic cells in its early stages require just 30 minutes for one cell cycle and they go through 6 to 8 such cell divisions very fast. Replication of DNA in cell division is an important phase. The replication is very accurate and it is nucleotide by nucleotide and there is no provision for making any mistakes, even if mistakes are made they are immediately corrected or repaired, otherwise consequences are serious and deleterious. In general DNA replication is precise, exact and regulated and involves initiation, elongation and termination steps, but mechanisms and rate of replication vary from one system to the other.

REPLICATION IS SEMI CONSERVATIVE

Meselson and Stahl showed that DNA replicates in semi conservative mode. There were some doubts about the mode whether it is conservative, dispersive or semi conservative. But Meselson and Stahl used nonradioactive heavy Nitrogen isotope (^{15}N), as the source of Nitrogen. When cells were grown on such source, ^{15}N incorporated nitrogen bases get incorporated into DNA, thus it becomes heavier than the DNA that is grown in normal (^{14}N) nitrogen media.

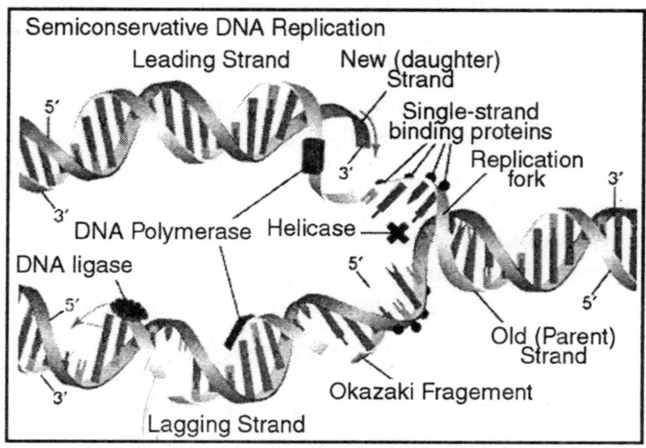

Fig. Replication is Semi Conservative

When two such DNAs, one grown in the presence of ^{15}N source and the other in ^{14}N (normal), are extracted separately and stained with ethidium Bromide, then subjected to equilibrium density centrifugation at ultra speed of 42000rpm for about 24 –36 hrs, the DNA, separates as two distinct bands, the one that is heavier (^{15}N) is the lower band than the normal (^{14}N) upper DNA.

If cells grown in heavy isotopes are shifted to normal Nitrogen source and grown for one or two cell generation and if DNA is isolated from it and if one combine this with the first two DNA sources and subject them to equilibrium density ultra centrifugation (using cesium chloride or cesium sulphate), one can observe three bands, the top DNA band is from cells grown in ^{14}N without isotope, the second band is the hybrid band with one strand ^{15}N and the other with normal 14N-Nitrogen and the third band is with heavy nitrogen where both the strands are labeled with ^{15}N.

This experiment virtually and unambiguously proved that DNA replicates in semi conservative mode. Taylor using root tips, labeled with ^{32}P radioisotopes, of Vicia faba demonstrated semi conservative mode of replication at chromosomal level.

Semi conservative means that the two daughter molecules produced at the end of replication, each of the molecules retains one parental strand and the other one is the newly made one. So it is semi conservative.

The mode of replication is absolutely semi conservative irrespective the kind of DNA, whether it is ds DNA or ssDNA, whether it is linear or circular, and the basic mechanism is more or less similar; each of them uses their own specialized mechanisms.

Experimentally semiconservative mode of replication has been shown by density gradient centrifugation of non-radioactive isotopes such as ^{15}N and ^{14}N sources.

DNA SYNTHESIS

DNA Replication (also known as DNA synthesis) is a process where the double stranded Deoxyribonucleic Acid (DNA) is copied.

Replication is an important first step in cell division, as cells must duplicate their entire genetic constitution before they can divide into two daughter cells. DNA replication is also a critical requirement for DNA repair.

REQUIREMENTS OF DNA REPLICATION

Replication requires three important components in order for this process to work.

Template Strand

The DNA serves as a template to guide the incoming nucleotides. The template strand attaches to the RNA primer strand in order to replicate the strand of DNA. The template just guides the nucleotides to the place that they need to go in. This is another component that is needed in order for DNA replication to even take place. If the DNA only had the template strand and none of the rest of the things then the strand could not replicate. The template starnd is very thin. Through an enzyme called primase the DNA template strand is synthesized one nucleotide at a time.

DNA Polymerase

There are many DNA polymerases within a cell, but only one of which is used in the replication of DNA. In order for DNA to replicate, it must have a primer, which is a small strand of RNA. The DNA template strand synthesizes one nucleotide at a time by the use of primase. The adding of nucleotides to the 3′ end, which is continued until that section of DNA is replicated. The other DNA polymerases that are within the cell are used in other ways besides replication. DNA polymerase III is one certain polymerase, which aids in replication of what is known as the "lagging strand". DNA ligase is the final enzyme, which combines the lagging strand

Free 3' Hydroxyl

The free 3' hydroxyl is the starter strand called a primer which is required in order for DNA to be replicated. The primer strand is not very big, it is shorter then a single strand of RNA. The primer is the complementary strand to the template strand of DNA. This is another one of the things that are needed in order to make the process of DNA replication even possible. The DNA polymerase is added to the 3' end of the primer, which keeps growing until the section is finally complete. After a while the RNA primer is degraded and disappears.

Types of Replication

During the discovery of DNA replication they also discovered that there are three modes of replication that a DNA strand can take.

The three types of DNA replication are semiconservative, conservative, and dispersive. These three types of replication were tested on in order to see the possible patterns that would result in the complementary base pairing.

Semiconservative Replication

Through this type of replication the parent strand serves as a template for the new strand. The two offsprings would have one of the parent strand and one new strand.

Conservative Replication

Through this type of replication the double helix serves as a template. This although does not contribute to the new double helix.

Dispersive Replication

In this type of replication the fragments from the parent DNA molecule. The DNA serves as a template for the assembly of the new molecule. The new double helix contains old and new parts of the DNA strands.

DNA Repairing

After the DNA is replicated, there are three DNA repair mechanisms.

They are to be used when they feel necessary. This will lessen the chance of human mutation. The rate of error in which the DNA polymerases makes a mistake is very high. This is why the DNA has mechanisms to cause fewer mutations. The three mechanisms are:

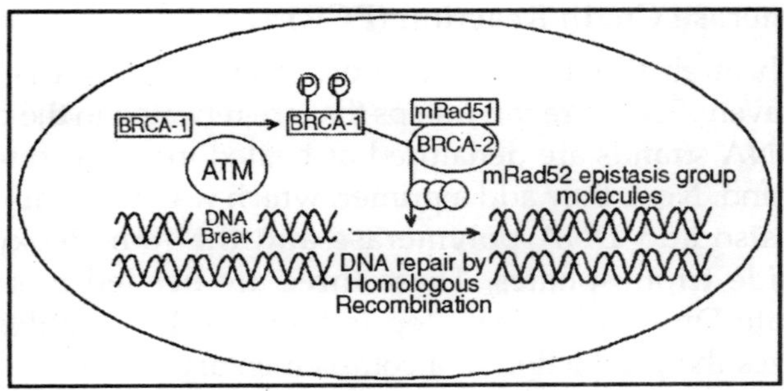

Fig. DNA Repairing

Proofreading

Every time a nucleotide is introduced to an already growing chain, the DNA polymerase checks the connection. If the pair is mismatched, it will be removed and redone. This lowers the overall rate of mutations.

Mismatch Pair

After the replication takes place, another set of proteins check to make sure that there are no more mismatched base pairs and also which strand is the wrong one.

Excision Pair

There can still be damage while the cell is living. Because of this, there are enzymes constantly checking the cell. If they detect a problem, the enzyme cuts the strand and also cuts away the opposite base. DNA polymerase and ligase then fix this base sequence.

LABORATORY TECHNIQUE

Scientists use two techniques in a laboratory that involve DNA replication. They use these techniques to observe genes and genomes.

One of the techniques can make multiple copies of DNA from only a short piece. Another technique allows scientists to determine the base sequence of DNA. The two techniques are:

Polymerase Chain Reaction (PCR)

Through this process, a short strand of DNA is copied repetitively. There are three steps that are repeated in the process. The DNA strands are denatured or heated in order to separate the strand. Next, they add a primer, which was artificially made. They also add DNA polymerase and the four deoxyribonucleotide triphosphates. These three are needed in order to replicate DNA. The final step is then the DNA polymerase catalyzes the production of a complementary strand.

DNA Sequencing

This allows scientists to determines the base sequence of a DNA. The process is the DNA is first denatured and a single strand is placed in a test tube. They then add DNA polymerase, primer, the four dNTP's, and small amounts of ddNTP's. The DNA replicates within the tube and after the DNA fragments are denatured. The fragments then go through electrophoresis, which sorts the lengths of DNA fragments. The fragments then pass through a laser beam which excites the fluorescent tags. This information is fed into a computer at which tells the sequence.

Mendelson-Stahl Experiment

Through this experiment Meselson- Stahl convinced the scientists that the correct model for DNA replication was the semiconservative replication. They used density labeling in order to distinguish the old and new strands of DNA. In their experiment they used "heavy" isotopes of nitrogen. This nitrogen is a nonradioactive isotope that made the molecule more dense. The chemically identical molecules containing an isotope known as ^{14}N. They needed a way to distinguish the different DNA densities.

Because of this they needed to come up with something to measure the densities of solutions. Meselson, Stahl, and Jerome Vanguard came up with centrifuge, which is a procedure that involves the spinning of solutions at high speed, which causes the particles to separate and form gradient according to the different densities of the solution. The first part of their experiment they grew cultures of Escherichia coli. One of the cultures was grown in ^{15}N, which made the DNA "heavy".

The other culture was placed in a medium using ^{14}N rather than ^{15}N, which made all the DNA "light". In order to see if the use of centrifuge worked they combined the two cultures and when centrifuged it formed two DNA bands. This showed them the different densities with in the two cultures. The next part of their experiment they grew E. Coli in the ^{15}N medium and then transferred the bacteria from that medium to the ^{14}N.

Through this they came up with the conclusion that through E. Coli DNA replicates every 20 minutes. They then took DNA samples from each generation through, which they found that the density gradient was different each generation.

Their observations could only be explained with the semiconservative model of DNA replication. Their results showed that when the DNNA first underwent replication, it was in 15N, which caused the DNA to be heavy.

Because in the semiconservative model of DNA replication, the one strand acts as a template for a second strand which the DNA was in a ^{14}N DNA strand and a ^{15}N and they were of intermediate density.

Chapter 6

Gene Expression

GENE REGULATION

Gene Regulation refers to the cellular control of the amount and timing of changes to the appearance of the functional product of a gene. Although a functional gene product may be an RNA or a protein, the majority of known mechanisms regulate protein coding genes. Any step of the gene's expression may be modulated, from DNA-RNA transcription to the post-translational modification of a protein.

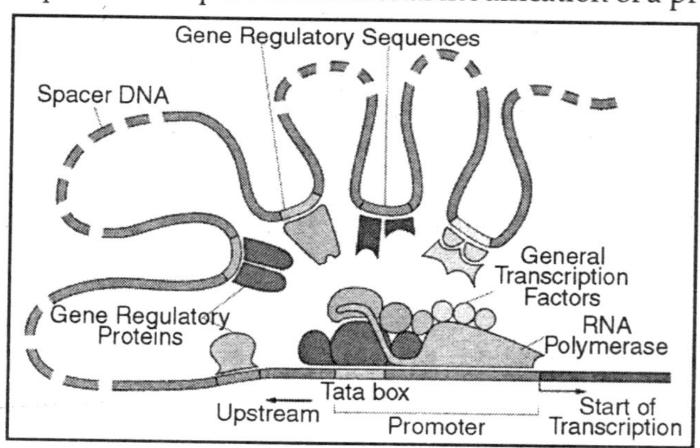

Fig. Gene Regulation

Gene regulation is essential for viruses, prokaryotes and eukaryotes as it increases the versatility and adaptability of an organism by allowing the cell to express protein when needed. The first example of gene regulation system was the lac operon, discovered by Jacques Monod, in which protein involved in lactose metabolism are expressed by E.coli only in the presence of lactose and absence of glucose.

Furthermore, gene regulation allows the presence in a multicellular organism of different cells types arranged in a complex pattern, hence different transcriptomes despite them having all the same genome and the generation of patterns by cellular differentiation and morphogenesis.

REGULATED STAGES OF GENE EXPRESSION

Any step of gene expression may be modulated, from the DNA-RNA transcription step to post-translational modification of a protein. The following is a list of stages where gene expression is regulated:

- Chromatin domains
- Transcription
- Post-transcriptional modification
- RNA transport
- Translation
- mRNA degradation
- Post-translational modifications

MODIFICATION OF DNA

In eukaryotes, the accessibility of large regions of DNA can depend on its chromatin structure which can be altered as a result of histone modifications which are directed by DNA methylation, ncRNA or DNA binding protein.

Chemical

Methylation of DNA is a common method of gene silencing. DNA is typically methylated by methyltransferase enzymes on cytosine nucleotides in a CpG dinucleotide sequence (also called "CpG islands" when densely clustered). Analysis of the pattern of methylation in a given region of DNA (which can be a promoter) can be achieved through a method called bisulfite mapping.

Methylated cytosine residues are unchanged by the treatment, whereas unmethylated ones are changed to uracil. The differences are analyzed by DNA sequencing or by methods developed to quantify SNPs, such as Pyrosequencing (Biotage) or MassArray (Sequenom), measuring the relative amounts of C/T at the CG dinucleotide. Abnormal methylation patterns are thought to be involved in carcinogenesis.

Structural

Transcription of DNA is dictated by its structure. In general, the density of its packing is indicative of the frequency of transcription. Octameric protein complexes called histones are responsible for the amount of supercoiling of DNA, and these complexes can be temporarily modified by processes such as phosphorylation or more permanently modified by processes such as methylation.

Such modifications are considered to be responsible for more or less permanent changes in gene expression levels. Histone acetylation is also an important process in transcription. Histone acetyltra-nsferase enzymes (HATs) such as CREB-binding protein also dissociate the DNA from the histone complex, allowing transcription to proceed.

Often, DNA methylation and histone deacetylation work together in gene silencing. The combination of the two seems to be a signal for DNA to be packed more densely, lowering gene expression.

REGULATION OF TRANSCRIPTION

Regulation of transcription controls when transcription occurs and how much RNA is created. Transcription of a gene by RNA polymerase can be regulated by at least five mechanisms:

- *Specificity factors* alter the specificity of RNA polymerase for a given promoter or set of promoters, making it more or less likely to bind to them (i.e. sigma factors used in prokaryotic transcription).
- *Repressors* bind to non-coding sequences on the DNA strand that are close to or overlapping the promoter region, impeding RNA polymerase's progress along the strand, thus impeding the expression of the gene.
- *General transcription factors* These transcription factors

position RNA polymerase at the start of a protein-coding sequence and then release the polymerase to transcribe the mRNA.

- *Activators* enhance the interaction between RNA polymerase and a particular promoter, encouraging the expression of the gene. Activators do this by increasing the attraction of RNA polymerase for the promoter, through interactions with subunits of the RNA polymerase or indirectly by changing the structure of the DNA.
- *Enhancers* are sites on the DNA helix that are bound to by activators in order to loop the DNA bringing a specific promoter to the initiation complex.

POSTTRANSCRIPTIONAL REGULATION

After the DNA is transcribed and mRNA is formed there must be some sort of regulation on how much the mRNA is translated into Proteins.

Cells do this by modulating the Capping, Splicing, addition of a Poly(A) Tail, the sequence-specific nuclear export rates and in several contexts sequestration of the RNA transcript. These processes occur in eukaryotes but not in prokaryotes. This modulation is a result of a protein or transcript which in turn is regulated and may have an affinity for certain sequences.

- *Capping* changes the five prime end of the mRNA to a three prime end by 5'-5' linkage, which protects the mRNA from 5' exonuclease, which degrades foreign RNA. The cap also helps in ribosomal binding.
- *Splicing* removes the introns, noncoding regions that are transcribed into RNA, in order to make the mRNA able to create proteins. Cells do this by spliceosome's binding on either side of an intron, looping the intron into a circle and then cleaving it off. The two ends of the exons are then joined together.
- *Addition of poly(A) tail* otherwise known as poly-adenylation. Junk RNA is added to the 3' end, and

acts as a buffer to the 3' exonuclease in order to increase the half life of mRNA.

In both prokaryotes and eukaryotes a large number of RNA binding protein exist, with often are directed to their target sequence by the secondary structure of the transcript, which may change depending on certain conditions, such as temperature or presence of a ligand (aptamer), some transcripts act as ribozymes and self-regulate their expression.

EXAMPLES OF GENE REGULATION

- Enzyme induction is a process in which a molecule (e.g. a drug) induces (i.e. initiates or enhances) the expression of an enzyme.
- The induction of heat shock proteins in the fruit fly *Drosophila melanogaster*.
- The Lac operon is an interesting example of how gene expression can be regulated.
- Viruses despite having only a few genes, possess mechanisms to regulate their gene expression, typically into a early and late phase, using collinear systems regulated by anti-terminators (lambda phage) or splicing modulators (HIV)

CIRCUITRY

UP-REGULATION AND DOWN-REGULATION

Up-regulation is a process which occurs within a cell triggered by a signal (originating internal or external to the cell) which results in increased expression of one or more genes and as a result the protein(s) encoded by those genes. Conversely down-regulation is a process resulting in decreased gene and corresponding protein expression.

Up: Regulation occurs for example when a cell is deficient in some kind of receptor. In this case, more receptor protein is synthesized and transported to the membrane of the cell and thus the sensitivity of the cell is brought back to normal reestablishing homeostasis.

Down: Regulation occurs for example when a cell is

overly stimulated by a neurotransmitter, hormone, or drug for a prolonged period of time and the expression of the receptor protein is decreased in order to protect the cell.

INDUCIBLE VS. REPRESSIBLE SYSTEMS

Gene Regulation can be summarized as how they respond:

Inducible systems: An inducible system is off unless there is the presence of some molecule (called an inducer) that allows for gene expression. The molecule is said to "induce expression". The manner in which this happens is dependent on the control mechanisms as well as differences between prokaryotic and eukaryotic cells.

Repressible Systems: A repressible system is on except in the presence of some molecule (called a corepressor) that suppresses gene expression. The molecule is said to "repress expression". The manner in which this happens is dependent on the control mechanisms as well as differences between prokaryotic and eukaryotic cells.

DEVELOPMENTAL BIOLOGY

A large number of studied regulatory systems come from developmental biology. Examples include:

The collinearity of the Hox gene cluster with their nested antero-posterior patterning.

It has been speculated that pattern generation of the hand (digits - interdigits) The gradient of Sonic hedgehog (secreted inducing factor) from the zone of polarizing activity in the limb which creates a gradient of active Gli3 which activates Gremlin which inhibits BMPs also secreted in the limb resulting in the formation of an alternating pattern of activity as a result of this reaction-diffusion system.

Somatogenesis is the creation of segmatation (somites) form a uniform tissue (PSM) sequentially from anterior to posterior, this is a achieved in amniotes possibly by means of two opposing gradients, Retinoic acid in the anterior (wavefront) and an oscillating gradient in the posterior (clock) composed of FGF + Notch and Wnt in antiphase. Sex determination in the soma of a Drosophila requires the sensing of the ratio of autosomal genes to sex chromosome encoded genes, which results in the production of sexless splicing factor in females resulting in the female isoform of doublesex.

THEORETICAL CIRCUITS

- Repressor/Inducer: a activation of a sensor results in the change of expression of a gene
- Negative feedback: the gene product downregulates its own production directly or indirectly, which can result in
  - Keeping transcript levels constant/proportional to a factor
  - Inhibition of run-away reactions when coupled with a positive feedback loop
  - Creating an oscillator by taking advantage in the time delay of transcription and translation, given that the mRNA and protein half-life is shorter
- Positive feedback: the gene product upregulates its own production directly or indirectly, which can result in
  - Signal amplification
  - Bistable switches when two genes inhibit each other and have both positive feedback
  - Pattern generation

Methods

Generally, most experiments investigating differential expression used whole cell extracts of RNA, called steady-state levels, to determined which genes changed and by how much they did, these are however not informative of where the regulation has occurred and may actually mask conflicting regulatory processess, it is the most commonly analysed (QPCR and DNA microarray).

When studying gene expression there are several methods to look at the various stages. In eukaryotes these include:

The chromatin conformation of the region can be determined by ChIP-chip analysis by pulling down RNA Polymerase II, Histone 3 modifications, Trithorax-group protein,Polycomb-group protein or any other DNA binding element to which a good antibody is available.

Due to post-transcriptional regulation, transcription rates and total RNA levels differ significantly, to measure the transcription rates nuclear run-on assays can be done and newer high-throughput methods are being developed, using thiol labelling instead of radioactivity.

Only 5 per cent of the RNA polymerised in the nucleus actually exists and not only introns, abortive products and non-sense transcripts are degradated therefore the differences in nuclear and cytoplasmic levels can be see by separating the two fractions by gentle lysis.

Alternative splicing can be analysed with a splicing array or with a tiling array. All in vivo RNA is complexed as RNPs. The quantity of transcripts bound to specific protein can be also analysed by RIP-Chip, for example DCP2 will give an indication of sequestered protein, ribosome bound gives and indication of transcripts active in transcription (although it should be noted that a more dated method, called polysome fractionation, is still popular in some labs)

Protein levels can be analysed by Mass spectrometry, which can only be compare to QPCR data as microarray data is relative and not absolute. RNA and protein degradation rates are measured by means of transcription inhibitors or translation inhibitors (Cycloheximide) respectively.

GENE EXPRESSION

Gene expression is the process by which inheritable information from a gene, such as the DNA sequence, is made into a functional gene product, such as protein or RNA. Several steps in the gene expression process may be modulated, including the transcription step and the post-translational modification of a protein. Gene regulation gives the cell control over structure and function, and is the basis for cellular differentiation, morphogenesis and the versatility and adaptability of any organism.

Gene regulation may also serve as a substrate for evolutionary change, since control of the timing, location, and amount of gene expression can have a profound effect on the functions (actions) of the gene in the organism. Non-protein coding genes (e.g. rRNA genes, tRNA genes) are transcribed, but not translated into protein. Genes are expressed by being transcribed into RNA, and this transcript may then be translated into protein.

MEASUREMENT

The expression of many genes is regulated after transcription (i.e., by microRNAs or ubiquitin ligases), so an increase in mRNA concentration need not always increase expression.

Nevertheless, mRNA levels can be quantitatively measured by Northern blotting, a process in which a sample of RNA is separated on an agarose gel and hybridized to a radio-labeled RNA probe that is complementary to the target sequence. Northern blotting requires the use of radioactive reagents and can have lower data quality than more modern methods (due to the fact that quantification is done by measuring band strength in an image of a gel), but it is still often used. It does, for example, offer the benefit of allowing the discrimination of alternately spliced transcripts.

A more modern low-throughput approach for measuring mRNA abundance is real-time polymerase chain reaction (The term RT-PCR is used to refer to both reverse transcription PCR as well as real-time PCR, which is also known as quantitative RT-PCR or quantitative PCR (qPCR). With a carefully constructed standard curve qPCR can produce an absolute measurement such as number of copies of mRNA per nanolitre of homogenized tissue. The lower level of noise in data obtained via qPCR often makes this the method of choice, but the price of the required equipment and reagents can be prohibitive.

In addition to low-throughput methods, transcript levels for many genes at once (expression profiling) can be measured with DNA microarray technology or "tag based" technologies like Serial analysis of gene expression (SAGE) or the more advanced version Super SAGE, which can provide a relative measure of the cellular concentration of different messenger RNAs. Recent advances in microarray technology allow for the quantification, on a single array, of transcript levels for every known gene in the human genome.

The great advantage of tag-based methods is the "open architecture", allowing for the exact measurement of any transcript, known or unknown. Especially SuperSAGE recommends itself therefore also for studying organisms with unknown genomes.

Protein levels themselves can be estimated by a number of means. The most commonly used method is to perform a Western blot against the protein of interest, whereby cellular lysate is separated on a polyacrylamide gel and then probed with an antibody to the protein of interest. The antibody can either be conjugated to a fluorophore or to horseradish peroxidase for imaging or quantification. Another commonly used method for assaying the amount of a particular protein in a cell is to fuse a copy of the protein to a reporter gene such as Green fluorescent protein, which can be directly imaged using a fluorescent microscope.

Because it is very difficult to clone a GFP-fused protein into its native location in the genome, however, this method often cannot be used to measure endogenous regulatory mechanisms (GFP-fusions are therefore most often expressed on extra-genomic DNA such as an expression vector). Fusing a target protein to a reporter can also change the protein's behaviour, including its cellular localization and expression level. The pattern of detection of a gene or gene product may be described using terms such as facultative, constitutive, circadian, cyclic, housekeeping, or inducible.

REGULATION OF GENE EXPRESSION

Regulation of gene expression is the cellular control of the amount and timing of appearance of the functional product of a gene. Any step of gene expression may be modulated, from the DNA-RNA transcription step to post-translational modification of a protein. Gene regulation gives the cell control over structure and function, and is the basis for cellular differentiation, morphogenesis and the versatility and adaptability of any organism.

Expression System

An expression system consists, minimally, of a source of DNA and the molecular machinery required to transcribe the DNA into mRNA and translate the mRNA into protein using the nutrients and fuel provided. In the broadest sense, this includes every living cell capable of producing protein from DNA. However, an expression system more specifically refers to a laboratory tool, often artificial in some manner, used for assembling the product of a specific gene or genes. It is defined as the "combination of an expression vector, its cloned DNA, and the host for the vector that provide a context to allow foreign gene function in a host cell, that is, produce proteins at a high level".

In addition to these biological tools, certain naturally observed configurations of DNA (genes, promoters, enhancers, repressors) and the associated machinery itself are referred to as an expression system, as in the simple repressor 'switch' expression system in Lambda phage. It is these natural expression systems that inspire artificial expression systems, (such as the Tet-on and Tet-off expression systems).

Each expression system has distinct advantages and liabilities, and may be named after the host, the DNA source or the delivery mechanism for the genetic material. For example, common expression systems include bacteria (such as E.coli), yeast (such as S.cerevisiae), plasmid, artificial chromosomes, phage (such as lambda), cell lines, or virus (such as baculovirus, retrovirus, adenovirus).

Overexpression

In the laboratory, the protein encoded by a gene is sometimes expressed in increased quantity. This can come about by increasing the number of copies of the gene or increasing the binding strength of the promoter region.

Often, the DNA sequence for a protein of interest will be cloned or subcloned into a plasmid containing the *lac* promoter, which is then transformed into the bacterium *Escherichia coli*. Addition of IPTG (a lactose analog) causes the bacteria to express the protein of interest. However, this strategy does not always yield functional protein, in which case, other organisms or tissue cultures may be more effective.

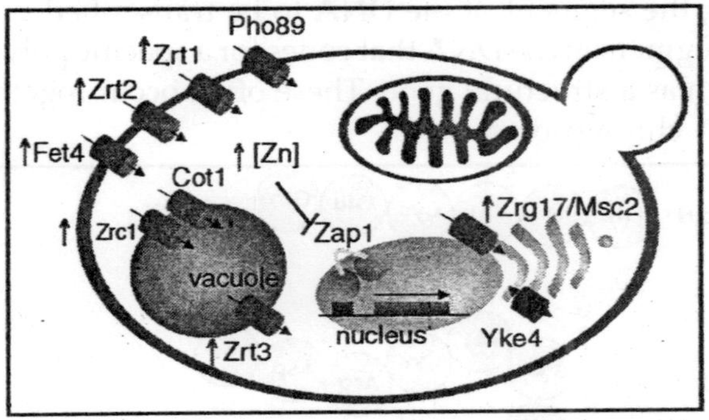

Fig. The Yeast Saccharomyces cerevisiae

For example, the yeast *Saccharomyces cerevisiae* is often preferred to bacteria for proteins that undergo extensive posttranslational modification. Nonetheless, bacterial expression has the advantage of easily producing large amounts of protein, which is required for X-ray crystallography or nuclear magnetic resonance experiments for structure determination.

GENE NETWORKS AND EXPRESSION

Genes have sometimes been regarded as nodes in a network, with inputs being proteins such as transcription factors, and outputs being the level of gene expression. The node itself performs a function, and the operation of these functions have been interpreted as performing a kind of information processing within cell and determine cellular behaviour.

CONTROL OF GENE EXPRESSION

The Chromosome of *E. Coli*

The single chromosome of the common intestinal

bacterium *E. coli* is circular and contains some 4.7 million base pairs. It is nearly 1 mm long, but only 2nm wide. The chromosome replicates in a bidirectional method, producing a figure resembling the Greek letter theta. The promoter is the part of the DNA to which the RNA polymerase binds before opening the segment of the DNA to be transcribed.

A segment of the DNA that codes for a specific polypeptide is known as a structural gene. These often occur together on a bacterial chromosome.

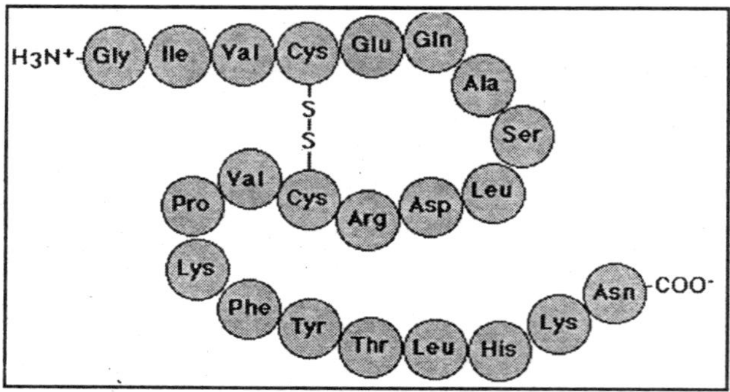

Fig. Polypeptides

The location of the polypeptides, which may be enzymes involved in a biochemical pathway, for example, allows for quick, efficient transcription of the mRNAs. Often leader and trailer sequences, which are not translated, occur at the beginning and end of the region. *E. coli* can synthesize 1700 enzymes. Therefore, this small bacterium has the genes for 1700 different mRNAs.

Fig. Enzyme B-galactosidase

Lactose, milk sugar, is split by the enzyme b-galactosidase. This enzyme is inducible, since it occurs in large quantities only when lactose, the substrate on which it operates, is present. Conversely, the enzymes for the amino acid tryptophan are produced continuously in growing cells unless tryptophan is present.

If tryptophan is present the production of tryptophan-synthesizing enzymes is repressed.

THE OPERON MODEL

The operon model of prokaryotic gene regulation was proposed by Fancois Jacob and Jacques Monod. Groups of genes coding for related proteins are arranged in units known as operons.

An operon consists of an operator, promoter, regulator, and structural genes. The regulator gene codes for a repressor protein that binds to the operator, obstructing the promoter (thus, transcription) of the structural genes. The regulator does not have to be adjacent to other genes in the operon. If the repressor protein is removed, transcription may occur.

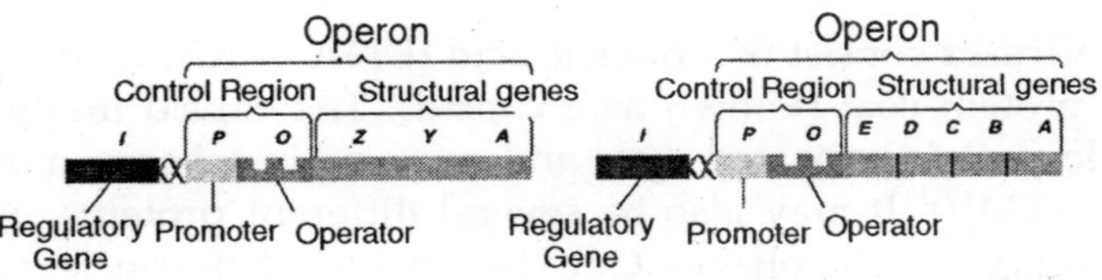

Fig. Operon Model of Prokaryotic Gene Regulation

Operons are either inducible or repressible according to the control mechanism. Seventy-five different operons controlling 250 structural genes have been identified for *E. coli*. Both repression and induction are examples of negative control since the repressor proteins turn off transcription.

Plasmids, small DNA fragments, are known from almost all bacterial cells. Plasmids carry between 2 and 30 genes. Some seem to have the ability to move in and out of the bacterial chromosome.

An episome is a plasmid incorporated in the bacterial chromosome. Plasmids are self-replicating in a manner like the bacterial chromosome. Many plasmids have been recognized for *E. coli*, including the F plasmids ("sex factors") and R plasmids (drug/antibiotic resistance). The F plasmid contains 25 genes, some of which control the production of F pili (proteins which extend from the surface of F^+, or male, cells to the surface of F^-, or female, cells).

The R plasmid conveys drug resistance on cells having it. As many as 10 resistance genes can be contained on a single R plasmid. The R plasmids can be transferred to other bacteria of the same species, to viruses, and even to bacteria of different species. Drug (antibiotic) resistance has been found among pathogens causing the diseases typhoid fever, gastroenteritus, plague, undulant fever, meningitis, and gonorrhea.

In addition to the more common modes of transfer, R plasmids may be passed through the cell membrane. The resistance genes appear to operate by either breaking down the antibiotics or by circumventing the block the antibiotic places on a key bacterial metabolic pathway.

VIRUSES

Viruses consist of a nucleic acid (DNA or RNA) enclosed in a protein coat (known as a capsid). The capsid may be a single protein repeated over and over, as in tobacco mosaic virus (TMV). It may also be several different proteins, as in the T-even bacteriophages. Once inside the cell, the nucleic acid follows one of two paths: lytic or lysogenic.

Retroviruses, such as Human Immunodifficiency Virus (HIV), also include the enzyme reverse transcriptase with the viral RNA. Reverse transcriptase makes a single-stranded viral DNA copy of the single-stranded viral RNA. The single stranded viral DNA is subsequently turned into a double-stranded DNA.

The lytic cycle occurs when the viral DNA immediately takes over the host cell (remember that viruses are obligate intracellular parasites) and begins making new viruses. Eventually the new viruses cause the rupture (or lysis) of the cell, releasing those new viruses to continue the infection cycle. The lysogenic cycle occurs when the viral DNA is incorporated into the host DNA as a prophage. When the cell replicates the prophage is passed along as if it were host DNA. Sometimes the prophage can emerge from the host chromosome and enter the lytic cycle spontaneously once every 10,000 cell divisions. Ultraviolet light and x-rays may also trigger emergence of the prophage.

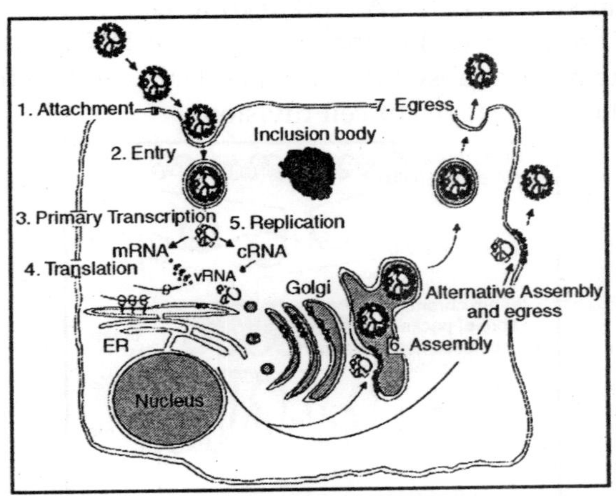

Fig. Viral Replication Cycle

Transduction is the transfer of host DNA from one cell to another by a virus. Some bacteriophages are temperate since they tend to go lysogenic rather than lytic. These types of viruses are able to transduce fragments of the host DNA.

Transposons are DNA fragments incorporated into the chromosomal DNA.

Unlike episomes and prophages, transposons contain a gene producing an enzyme that catalyzes insertion of the transposon at a new site. They also have repeated sequences 20-40 nucleotides in length at each end. Insertion sequences are short (600-1500 base pairs long) simple transposons that do not carry genes beyond those essential for insertion of the transposon into *E. coli*. Complex transposons are much larger and carry additional genes.

Genes incorporated in a complex transposon are known as jumping genes since they can move about on the chromosome (even from chromosome to chromosome). Often the complex transposons are flanked by simple transposons.

THE EUKARYOTIC CHROMOSOME

The eukaryotic chromosome consists of DNA and proteins that appear to play a major role in regulation of eukaryote genes. The DNA of each chromosome is a long single molecule of double stranded DNA. Eukaryotic DNA comes in two forms. Chromatin is the uncoiled form of DNA and is over 50% protein. Chromosomes are coiled DNA/protein that form during the early stages of cell division.

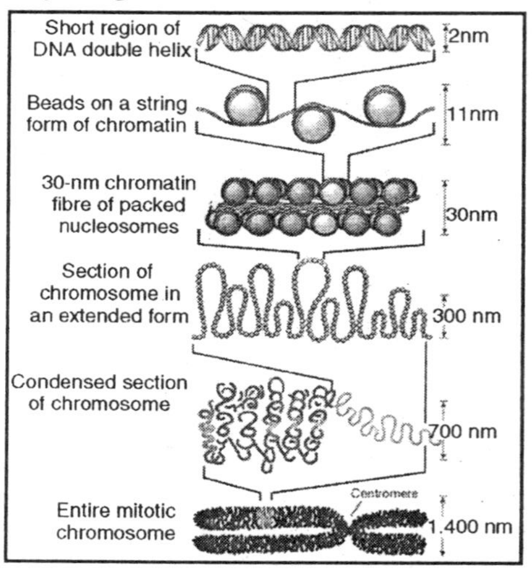

Fig. Eukaryotic Chromosome

The proteins associated with DNA are collectively known as histones. They are relatively short polypeptides which are positively charged (basic) and thus are attracted to the negatively charged (acidic) DNA. Histones are synthesized in quantity during the S-phase of the cell cycle.

One function of theses proteins seems to be the folding and packaging of DNA into chromosome form: the 2 m of DNA in a human cell are packaged into 46 chromosomes with a combined length of 200nm (a nm remember is 10-6m). Some 90 million molecules of histones occur in a single cell, with the majority (30 million) being H1 histones. Five types of histone are known (H1, H2A, H2B, H3, and H4); with the exception of H1 most eukaryote histones are very similar.

A nucleosome is the fundamental packing unit of eukaryotic DNA. The core consists of two molecules each of H2A, H2B, H3, and H4; around which the DNA is wound twice. The H1 histone is outside the core. Between 150-200 nucleotide pairs are associated with the core and linker DNA. This level of packing is known as "beads on a string".

The next level are known as the 30nm strand, whose details of organization are not yet well known. 30 nm strands are further condensed into 300nm wide looped domains. Looped domains are part of condensed sections of chromosomes (the chromosome being 1400nm wide at Metaphase I).

Replication of the Eukaryotic Chromosome

Nucleotide triphosphates (in the forms ATP, GTP, CTP and TTP) are assembled according to the semi-conservative model. Other details of DNA replication are consistent with what we know for prokaryotes.

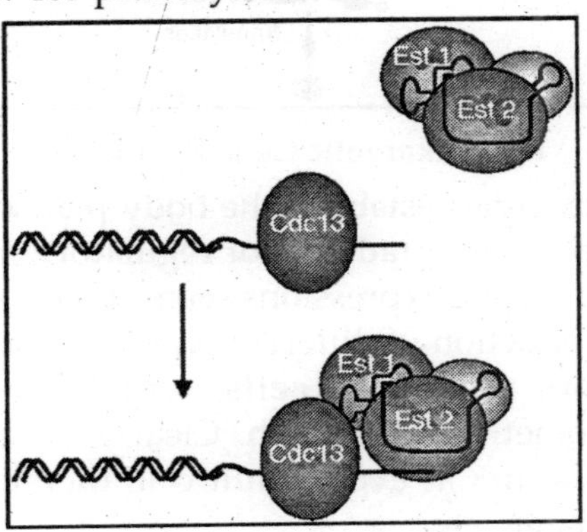

Fig. Replication of the Eukaryotic Chromosome

Eukaryotic DNA has many replication forks and also bidirectional synthesis, contrasting to the unidirectional rolling theta prokaryotic method. Eukaryotic DNA is synthesized much slower than prokaryotic DNA, in humans 50 nucleotides per second per replication fork. After replication the new DNA is immediately associated with histones.

Regulation of Eukaryotic Gene Expression

Eukaryotic gene regulation, especially in multicellular organisms, is complicated by the process of development unique to multicellular organisms. Each multicellular organism begins as a single-celled zygote which divides by mitosis. Cells differentiate into functional types by using some genes but ignoring others.

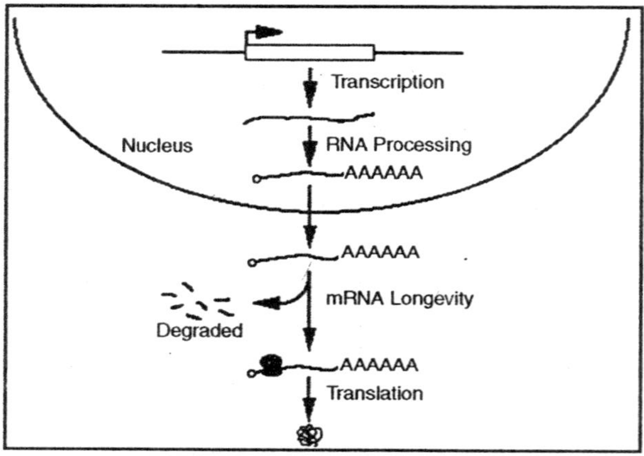

Fig. Eukaryotic Gene Regulation

Homeobox genes establish the body plan and position of organs in response to gradients of regulatory molecules. The timing of certain gene expressions seems to follow a sequence, such as the production of different types of fetal hemoglobins by mammalian red blood cells, which switch to adult hemoglobin sometime after birth. Clearly the inactivation of certain genes occurs in every adult cell; therein lies the cure for cancer, old age, etc.

TYPES OF CHROMATIN

Heterochromatin stains more strongly and is a more condensed chromatin. Euchromatin stains weakly and is more open (less condensed). Euchromatin remains dispersed (uncondensed) during Interphase, when RNA transcription occurs. Some regions of heterochromatin appear to be structural (as in the heterochromatin near the centromere region). Barr bodies, irreversibly inactivated X-chromosomes, are also condensed heterochromatin. Other heterochromatin regions vary from cell to cell. As the cell differentiates, the proportion of heterochromatin to euchromatin increases, reflecting increased specialization of the cell as it matures.

Loops (or puffs) in insect chromosomes are areas of active RNA synthesis, suggesting again the functional genes are located in open areas of the chromatin (or, the euchromatin). Eukaryotes also have specific binding proteins working in a similar fashion to prokaryotic mechanisms, however the eukaryotes, as one would expect, have a much more complicated process.

EUKARYOTIC GENOME

We use the term genome to refer to all of the alleles possessed by an organism (or by a population, species, or larger taxonomic group). While the amount of DNA for a diploid cell is constant within a species, the differences can be great between species.

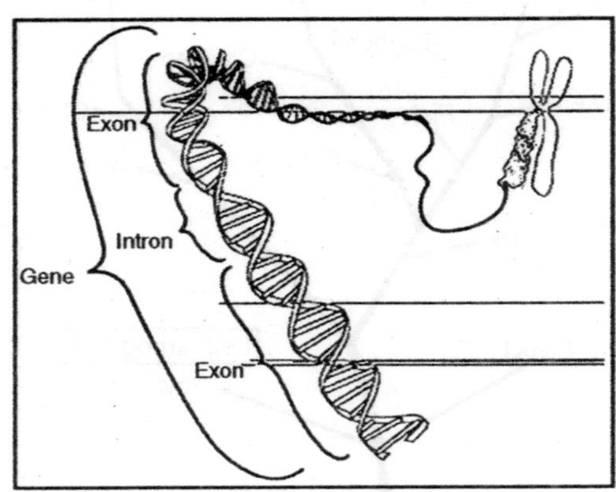

Fig. Introns and Exons

Humans have 3.5×10^9 base pairs, *Drosophila* has 1.5×10^8, toads have 3.32×10^9, and salamanders have 8×10^{10} base pairs per haploid genome. Much of the DNA in each cell either has no function or has a function not yet known. Eukaryotes have only 10 per cent of their DNA coding for proteins. Humans may have a little as 1 per cent coding for proteins. Viruses and prokaryotes use a great deal more of their DNA.

Almost half the DNA in eukaryotic cells is repeated nucleotide sequences. Protein-coding sequences are interrupted by non-coding regions. Non-coding interruptions are known as intervening sequences or introns. Coding sequences that are expressed are exons.

Most, but not all structural eukaryote genes contain introns. Although transcribed, these introns are excised (cut out) before translation (a seemingly energy inefficient process). The number of introns varies with the particular gene, even occurring in tRNAs, rRNAs and viral genes! Generally the more complex and recently evolved the organism, the more numerous and larger the introns.

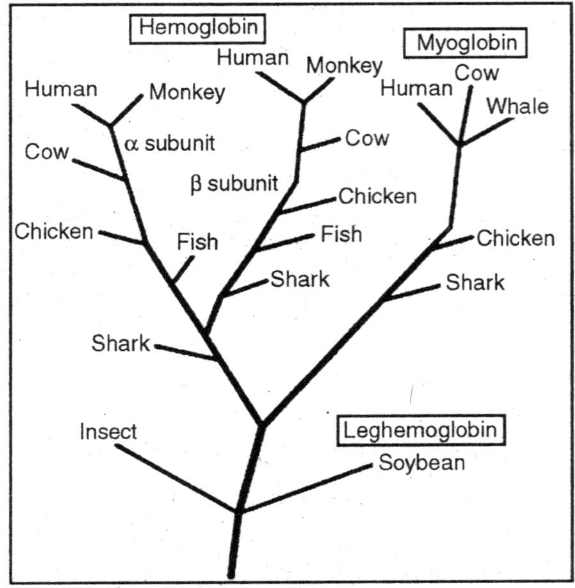

Fig. Globin Family

Which came first: continuous genes lacking introns or interrupted genes containing introns? Introns have been hypothesized to promote genetic recombination (via crossing-over), thus speeding up the evolution of new proteins. Exons are also thought to code for different functional regions of proteins. Gene families are made up of similar, but not identical, genes. The globin family is the best studied gene family.

Hemoglobin consists, in humans, of 2 a-chains and 2 b-chains clustered about a common heme. Human beta-globin genes are scattered at five loci on human chromosome 11. These genes are expressed sequentially during development, and are similar with same-length introns in similar positions in each gene. Some of the genes are inactivated copies, others are functional only during certain phases of development. The actin family of genes also exhibits a similar pattern.

TRANSCRIPTION AND PROCESSING OF mRNA

The process of transcription in eukaryotes is similar to that in prokaryotes, although there are some differences. Eukaryote genes are not grouped in operons as are prokaryote genes. Each eukaryote gene is transcribed separately, with separate transcriptional controls on each gene.

Whereas prokaryotes have one type of RNA polymerase for all types of RNA, eukaryotes have a separate RNA polymerase for each type of RNA. One enzyme for mRNA-coding genes such as structural proteins. One enzyme for large rRNAs. A third enzyme for smaller rRNAs and tRNAs.

Prokaryote translation begins even before transcription has finished, while eukaryotes have the two processes separated in time and location (remember the nuclear envelope). After eukaryotes transcribe an RNA, the RNA transcript is extensively modified before export to the cytoplasm. A cap of 7-methylguanine (a series of an unusual base) is added to the 5' end of the mRNA; this cap is essential for binding the mRNA to the ribosome. A string of adenines (as many as 200 nucleotides known as poly-A) is added to the 3' end of the mRNA after transcription.

The function of a poly-A tail is not known, but it can be used to capture mRNAs for study. Introns are cut out of the message and the exons are spliced together before the mRNA leaves the nucleus. There are several examples of identical messages being processed by different methods, often turning introns into exons and vice-versa. Protein molecules are attached to mRNAs that are exported, forming ribonucleo-protein particles (mRNPs) which may help in transport through the nuclear pores and also in attaching to ribosomes.

ANTIBODY-CODING GENES

Antibodies are globular complex proteins made by multicellular individuals in response to a specific antigen. The cells making antibodies are lymphocytes, better known as white blood cells. Antibodies immobilize and destroy their specific antigens. A mouse can make 10,000,000 antibodies, more than the total genes of the mouse would suggest. Antibody proteins are composed of two long and two short chains.

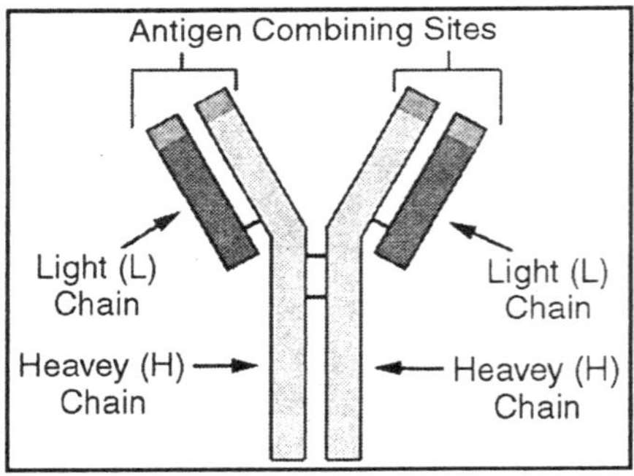

Fig. Antibody proteins

Each species has a constant region characteristic for the species and type of antibody. The other region is the variable in which antigen-specific amino-acids are located. Susumu Tonegawa tested (and proved) an old hypothesis that the constant and variable regions of antibodies were coded for by different genes. Tonegawa's work further demonstrated that gene fragments in embryos are rearranged to form the variable (functional) genes.

VIRUSES AND EUKARYOTES

The viruses of eukaryotes are similar to prokaryote-infecting viruses. Proviruses are viral DNA integrated into the host cell. Some of the DNA viruses can either initiate an infection (lytic in prokaryotes) cycle or can form proviruses. Simian Virus 40 (SV40) causes cancer in hamsters but not in its normal hosts. SV40 can introduce new functional genes into the host DNA, as can a number of other viruses.

Retroviruses can also insert their nucleic acid into host DNA via the reverse transcriptase mode mentioned previously. Reverse transcriptase is carried with the RNA into the host cell. In the process of making the cDNA strand for the viral RNA, the enzyme also makes long terminal repeats (LTRs) sequences at the terminal ends of the cDNA.

These LTRs may also make insertion of the viral DNA into the host DNA easier. The inserted viral DNA makes RNA transcripts which are packaged with viral protein coats and reverse transcriptase.

The viral DNA may, depending on its insertion point, cause mutations of the host DNA. Most viral DNA insertions do not damage the host, but rather become part of the host genome and can be passed on if they have managed to infect a germ-line cell. From 0.5 to 1.0 per cent of mouse DNA may be of viral origin.

EUKARYOTIC TRANSPOSONS

Eukaryotic transposons resemble prokaryotic transposons in many features. Many eukaryotic transposons are first copied to RNA and then back to DNA before being inserted in a new location.

GENES, VIRUSES AND CANCER

Cancer is a disease in which cells escape the restraints on normal cell growth. Cancer is an inheritable disease (at least from cell to daughter cells). Once a cell has become cancerous, all of its descendant cells are cancerous. Gross chromosomal abnormalities are often visible in cancerous cells. Most carcinogens (cancer-generating factors) are also mutagens (mutation-generating factors). Oncogenes are genes resembling normal genes but in which something has gone wrong, resulting in a cancer. Fifty oncogenes have thus far been discovered. Viruses seem able to cause cancer in three ways. Presence of the viral DNA may disrupt normal host DNA functions.

Viral proteins needed for virus replication may also affect normal host gene regulation. Since most cancer-causing viruses are retroviruses, the virus may serve as a vector for oncogene insertion. Transfers of genes between eukaryotic cells will allow doctors, who have historically been limited to phenotypic cures, to attack disease at the genotypic level. SV40 virus has been used to inject the rabbit beta-globin gene into monkeys. Viruses can thus serve as a possible vector to place healthy (non-mutated) alleles into eggs.

GENE REGULATION IN EUKARYOTES

Like prokaryotes, eukaryotic organisms do not want to express all of their genes all of the time. Given the complexity of multicellular eukaryotes, gene regulation in these organisms needs to be very complex.

This module provides a brief overview of the various levels of regulation of gene expression in eukaryotes, and takes a look at the basics of transcriptional regulation; a detailed look at eukaryotic transcriptional regulation is beyond the scope of this course.

THE NEED FOR GENE REGULATION IN EUKARYOTES

Eukaryotes need to regulate their genes for different reasons than prokaryotes. In prokaryotes, gene regulation allowed them to respond to their environment efficiently and economically. While eukaryotes can respond to their environment, the main reason higher eukaryotes need to regulate their genes is cell specialization. Whereas prokaryotes are (relatively speaking) simple, unicellular organisms, multicellular eukaryotes consist of hundreds of different cell types, each differentiated to serve a different specialized function.

Each cell type differentiates by activating a different subset of genes. Because of the multitude of cell types, the regulation of gene expression required to bring about such differentiation is necessarily complex. One way this complexity is demonstrated is in multiple levels of regulation of gene expression.

LEVELS OF REGULATION

Before we discuss the specifics of regulation, it is necessary to understand that "gene expression" covers the entire process from transcription through protein synthesis. The final measure of whether or not a gene is "expressed" is if the protein is produced, because it is protein that will ultimately carry out the function specified by the gene.

We've seen numerous examples of how eukaryotic cells are more complex than prokaryotic cells. One obvious example of this is the presence of a nucleus in eukaryotic cells, which separates transcription from translation in a way not seen in prokaryotes. Furthermore, eukaryotic transcripts must be processed before they can be translated. Here is a diagram outlining the steps involved in the production of a protein in eukaryotic cells:

Regulation can occur at any point in this pathway; specifically, it occurs at the levels of transcription, RNA processing, mRNA lifetime (longevity), and translation. Each of these types of regulation will be considered in turn.

REGULATION OF RNA PROCESSING

After transcription, the RNA must be processed before it can be translated. As described elsewhere, RNA processing involves addition of a 5' cap, addition of a 3' poly (A) tail, and removal of introns. This processing represents another level of regulation of gene expression, particularly in regard to splicing out of introns. Regulation can be of two types:

- Whether an RNA gets processed;
- Which exons are retained in the mRNA.

The first type of regulation can determine whether or not an mRNA gets translated. If an RNA is not processed, it will not be transported out of the nucleus, and will not be translated.

The second type of regulation can affect the function of the protein produced. Some genes have exons that can be exchanged in a process known as exon shuffling. For example, a gene with four exons might be spliced differently in two different cell types. In each of these cases, the polypeptide produced could have a different function. In mammals, for example, the calcitonin gene produces a hormone in one cell type, and a neurotransmitter in another cell type, due to exon shuffling.

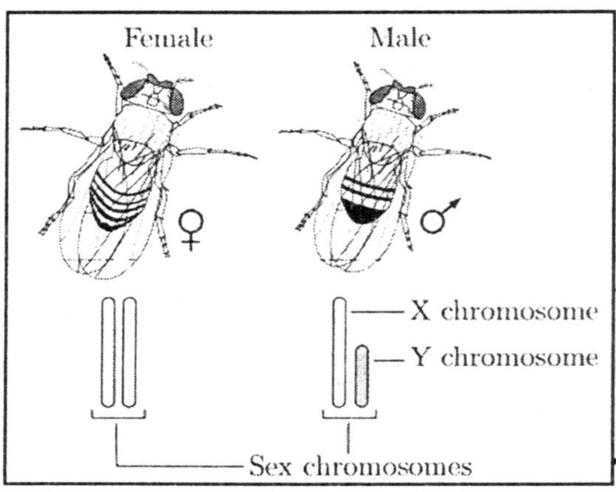

Fig. Drosophila

In *Drosophila,* alternate splicing of the *sex-lethal* RNA can produce an mRNA encoding a functional polypeptide, or one with a premature stop codon that encodes a short, nonfunctional polypeptide.

REGULATION OF RNA LONGEVITY

Imagine two mRNA molecules: one lasts for five minutes in the cytoplasm before being degraded, while the other one manages to linger for an hour before being degraded. If both are translated continually while they exist, it is obvious that more of the second polypeptide will be produced than the first. This is the principle behind regulation of RNA longevity.

mRNAs from different genes have their approximate lifespan encoded in them; this serves to help regulate how much of each polypeptide is produced. The information for lifespan is found in the 3' UTR. The sequence AUUUA, when found in the 3' UTR, is a signal for early degradation (and therefore short lifetime). The more times the sequence is present, the shorter the lifespan of the mRNA. Because it is encoded in the nucleotide sequence, this is a set property of each different mRNA; the longevity of an mRNA can't be varied.

REGULATION OF TRANSLATION

Whether or not an mRNA molecule is translated can be regulated as well. The various mechanisms of translational regulation are incompletely understood, but there are many documented examples (particularly in embryonic development) of mRNA molecules that are present routinely, but are only translated under certain circumstances. For example, many animals sequester large amounts of mRNA in their eggs, and those mRNA molecules are not translated unless the egg is fertilized.

REGULATION OF TRANSCRIPTION

Whether or not a gene is transcribed is the major way that gene expression is regulated in eukaryotes, as it was in prokaryotes. There are some major differences between transcriptional regulation in prokaryotes and eukaryotes. For one thing, because of the complexity of eukaryotic patterns of gene expression, each eukaryotic gene needs its own promoter. In other words, eukaryotic genes are not organized into operons.

Another difference is that prokaryotic genes are regulated primarily by repressors. Although repressors occasionally play a role in eukaryotes, eukaryotic genes are primarily regulated by transcriptional activators. These activators are transcription factors.

REGULATORY ELEMENTS OF EUKARYOTIC GENES

As discussed in the module on transcription, eukaryotic genes have promoters that are recognized by basal transcription factors. In addition to the promoter, eukaryotic genes have one or more enhancers.

These are DNA sequences associated with the gene being regulated, and whereas the promoter is responsible for initiating low levels of transcription and determining the transcription start site, enhancers are responsible for increasing ("enhancing") transcription levels, and they are responsible for regulating cell- or tissue-specific transcription (i.e. the transcription responsible for differentiation). There are some other basic differences between enhancers and promoters, which are outlined in the module on transcription.

Enhancers function by being recognized and bound to by transcription factors. These are not the basal-type transcription factors (such as TFIID) discussed elsewhere. These are specialized transcription factors, of which there are very many types (all are proteins). A very large number of enhancer elements has been identified and characterized, and each different enhancer has its own transcription factor that it binds to.

TISSUE-SPECIFIC GENE EXPRESSION

How is a gene turned on in one cell type, and not in another? It depends primarily on whether the transcription factor for the gene's enhancer is active or not in a cell. To illustrate this, let's consider two different genes - one is regulated by enhancer A, which is recognized by transcription factor A, and the other gene is regulated by enhancer B, which is recognized by transcription factor B.

In one cell type (let's say muscle, for the sake of argument), transcription factor A might be active whereas transcription factor B might not.

So how is it determined which transcription factors are active in which cell types? The answer to this question is very complicated, and not completely understood. However, there are a number of mechanisms by which transcription factors can be regulated: Often, the presence or absence of a transcription factor in a cell is the determining step. If the factor is present, the gene is transcribed. If not, then the gene is not transcribed. The presence of a transcription factor, of course, depends upon the activity of the gene encoding that transcription factor, which forces us to ask how the gene encoding the transcription factor is regulated, which pretty much brings us back to where we began.

Transcription factors can be activated by environmental signals. For example, virtually all organisms have a set of genes called heat shock genes that encode proteins that help the organism survive heat stress. These genes are activated under conditions of heat stress, under the control of a transcription factor called heat shock transcription factor. This factor is always present, but is only activated when greatly increased temperatures are detected.

Transcription factors can be activated by signals from other cells in the same organism. Such signals include hormones and growth factors. Hormones must bind to a specific receptor on the target cell, and the receptor mediates the cellular effects of the signal. There are two basic mechanisms used, one for steroid hormones, and one for peptide hormones:

- Steroid hormones
- Peptide hormones

Steroid Hormones

Steroid hormones are lipid (actually cholesterol) derivatives, such as testosterone and progesterone. These hormones can cross the cytoplasmic membrane into a cell, where they bind to their specific receptor. Steroid receptors are transcription factors, and when they bind to their ligand, they become activated and initiate transcription of a specific set of genes.

Peptide Hormones

Peptide hormones cannot cross the cytoplasmic membrane, and so must bind to a receptor on the cell surface. When bound to its ligand, these receptors initiate a series of biochemical reactions inside the cell, with the ultimate result being the activation of a transcription factor (often by phosphorylation of the transcription factor), which initiates transcription of a specific set of genes.

DEVELOPMENTAL GENETICS

The primary function of the majority of genes in eukaryotic organisms is to coordinate the development of embryos. This module takes a look at some of the basic principles underlying the role of genes in embryonic development.

In most of the other modules of this course, the effects of genes are examined with regard to their effect on the phenotype of the juvenile or adult organism. In reality, however, the vast majority of these phenotypes are established during embryonic development, because most genes function to generate the pattern (the 'shape') of the developing embryo. Loss of function of a particular gene results in an abnormality in development, which manifests itself as a phenotype in the adult. In this sense, virtually all of eukaryotic genetics is developmental genetics, even though we haven't considered it as such.

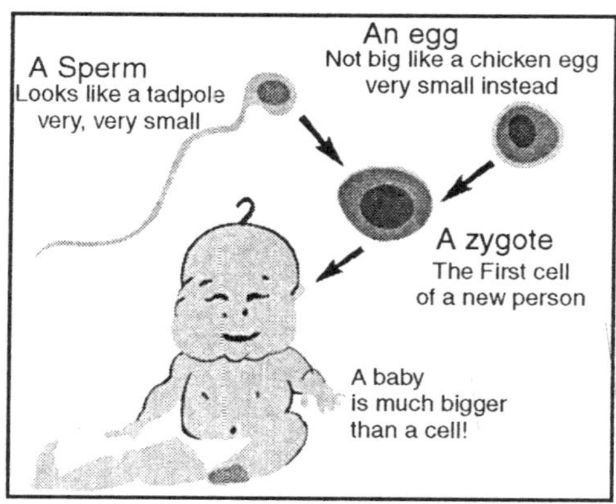

Fig. Zygote

The field of developmental biology is concerned with the function of genes in embryogenesis. How does a single cell, the fertilized egg (or zygote), manage to produce an extremely complex adult organism, which is composed not only of trillions of cells, but of thousands of different types of cells (such as nerve cells, muscle cells, etc.)? There is a simple answer to this question, and that is that cells differentiate. In other words, different cells become specialized to carry out different functions by following different developmental pathways. But how does this occur? How do cells know what pathway to follow?

THE MOSAIC THEORY OF DEVELOPMENT

One early theory that attempted to explain the process of differentiation was the mosaic theory, as outlined by Wilhelm Roux and August Weismann. This theory proposed that there were determinants that specified the various differentiation pathways. The theory stated that these determinants would all be present in the zygote, but as cell division occurred after fertilization, the determinants would be unequally inherited by the offspring cells.

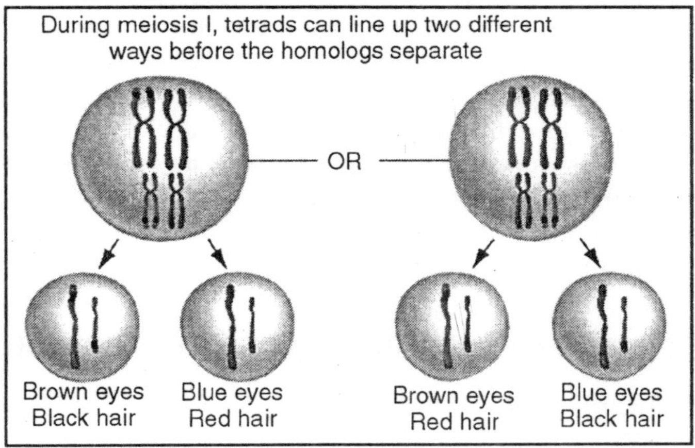

Fig. Qualitative Cell Division

This is known as qualitative cell division. The set of determinants would thus be divided up until each cell contained only one type of determinant, and that determinant would determine the fate of that cell. When Mendel's work was rediscovered in the early 1900's and the concept of the 'gene' was developed, it was believed that the genes were the determinants that were divided up during cell division.

Evidence was accumulated over sixty years that disproved the notion that genes are divided up in development. The final nail in the coffin of this theory came from nuclear transplantation studies done using amphibians in the early 1960's. In these studies, nuclei were isolated from tadpole intestinal cells (differentiated cells) and injected into eggs that had their own haploid nuclei removed. A small percentage of the injected eggs developed into completely normal adult frogs! These frogs had developed using only the genetic information found in a tadpole intestinal nucleus.

Therefore, a tadpole intestinal nucleus must contain all of the genetic information necessary to allow the differentiation of every cell type in an adult frog. If the mosaic theory were correct, this would not be the case; a tadpole intestinal nucleus would have only the genetic information necessary to cause the differentiation of an intestinal cell.

Note: The frogs produced by this procedure would be genetically identical to the frog from which the intestinal cell nucleus was obtained. In other words, they would be clones. In fact, these experiments comprised the earliest successful cloning of vertebrate organisms. The recent cloning of Dolly the sheep was done for the same reasons as outlined above: demonstrating that differentiated mammalian nuclei (from mammary gland in this case) have all of the genetic information necessary to drive the development of a normal adult organism.

THE THEORY OF DIFFERENTIAL GENE EXPRESSION

If all nuclei in an organism contain the same genetic information, then how does differentiation occur? The explanation for this is found in the Theory of Differential Gene Expression. This theory states that differentiation occurs as a result of expression in a particular cell of only a subset of the total genes present. For example, if a cell expresses only the set of genes that causes muscle differentiation, then that cell will differentiate into a muscle cell.

Alternatively, if the cell expresses only the set of genes that causes spleen cell differentiation, then the cell will differentiate into a spleen cell. This is a fairly simple concept, but it hasn't really answered the question. The question has now become: how do cells activate only a certain set of genes (inactivating all other genes), and how do they know which set of genes to activate? The answer to this question is fairly straightforward as well: the set of genes activated in a cell is dependent on the set of transcription factors found in the cell.

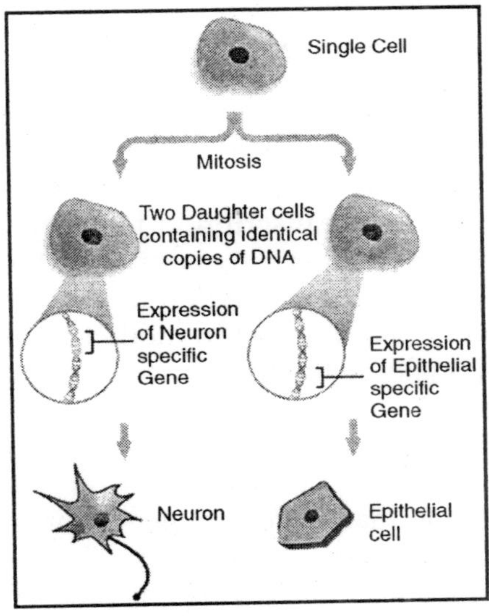

Fig. Differential Gene Expression

Using the muscle cell example from above, the cell activates the muscle-specific genes because it contains transcription factors that specifically activate the muscle-specific genes. If we think about this for a minute, you'll realise that we still haven't answered the question; we've only changed it again.

Now the question becomes: how did the cell come to contain only that specific set of transcription factors? Well, those transcription factors are encoded by genes, and those genes are regulated by other specific transcription factors. Those transcription factors are in turn encoded by other genes, which are regulated by still other transcription factors, etc. As we can see, there is a hierarchy of genes within each cell. Genes are expressed that encode transcription factors, which activate other genes that encode transcription factors, which activate other genes that bring about differentiation along a specific pathway. How many levels are there to the hierarchy? From what we've seen so far, the hierarchy seems to go on forever. Each set of transcription factors is encoded by genes that are activated by yet another set of transcription factors. It must end somewhere, but where?

MASTER CONTROL GENES

In one sense, it ends with master control genes. A master control gene is the first gene activated in a hierarchy that leads to differentiation along a particular pathway. Master control genes encode the first transcription factor in a hierarchy; a master control gene product activates the next set of genes that encodes the next set of transcription factors, and the cascade of gene expression has been set in motion.

Let's look again at the muscle cell example considered earlier. The master control 'gene' for muscle development is actually a family of genes called the MyoD gene family. These genes encode HLH-type transcription factors. How do we know these are the master control genes for muscle? Master control genes have a particular property. Because they initiate muscle differentiation, if the MyoD genes are activated in other cell types, they cause those cells to transdifferentiate into muscle. For example, hepatocytes (liver cells) or adipocytes (fat cells) that are caused to express the MyoD genes (using tricks of molecular biology) will change from their normal phenotype into muscle cells.

So how are the master control genes regulated? There are two basic ways that this can happen. One way is through the process of induction. Induction occurs when once cell sends a signal to another cell, telling it to differentiate a certain way. The signal, usually a diffusable protein, causes the recipient cell to activate the appropriate master control gene product. This is how muscle differentiation occurs. Signals from other cells cause the MyoD genes to become active in the cells receiving the signal, and muscle differentiation is initiated.

Chapter 7

Biophysics

Biophysics is the application of physics to biological systems. This interdisciplinary field is quickly growing in the modern scientific community. The Department of Biophysics, is a centre of drug discovery and clinical proteomics. It has combined the fields of structural biology, bioinformatics and proteomics seamlessly. The goal of modern research in Drug Discovery is to develop drugs that will act in a specific way with minimal side effects and also are demonstrably better than the existing therapies.

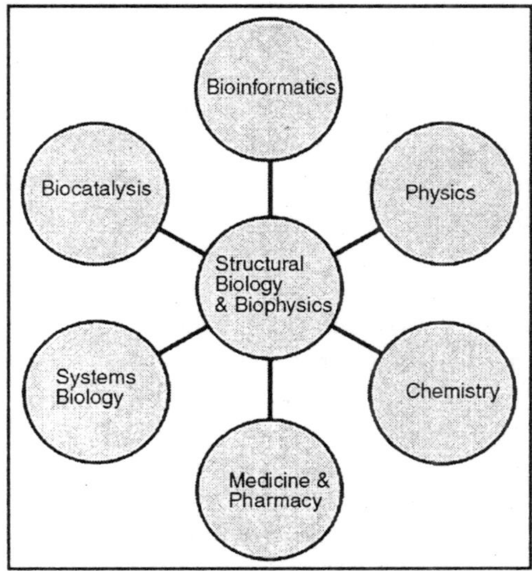

Fig. Biophysics

The conventional approaches of Drug discovery render

it a long and an expensive process. Biophysics is an interdisciplinary field which applies techniques from the physical sciences to understanding biological structure and function at the molecular level.

In simpler terms we use a range of scientific techniques to try to understand what macromolecules like DNA, proteins, fats and sugars look like, and how they interact to form the biological systems around us. Plants, animals and even seemingly simple organisms like bacteria and archae consist of complex biological networks of chemical signals and molecular interactions, which enable them to do such things as move, respire or reproduce. Biophysics is a varied field which draws on biology, physics, chemistry, mathematics, engineering, genetics, physiology and medicine, all with the aim to understand these systems using experiments or theoretical and computational modeling. It is also a young science and is still rapidly developing–bionanotechnology and biosensors are just some of the latest fields to emerge in the last few years. It is also known as biological physics that applies the theories and methods of the physical sciences to questions of biology.

Biophysics apply to theories of the human species and methods of the way we question problems. Biophysics research today is comprised of several specific biological studies which neither share a unique identifying factor nor subject themselves to clear and concise definitions. The studies included under the umbrella of biophysics range from sequence analysis to neural networks. Biophysics is also concerned with creating mechanical limbs and nanomachines to regulate biological functions, although currently these are more commonly referred to as belonging to the fields of bioengineering and nanotechnology respectively. Biophysics typically addresses biological questions that are similar to those in biochemistry, but the questions are asked at a molecular level.

Traditional studies in biochemistry and molecular biology are conducted using statistical ensemble experiments, typically using pico- to micro-molar concentrations of macromolecules. Because the molecules that comprise living cells are so small, techniques such as PCR amplification, gel blotting, fluorescence labeling and in vivo staining are used so that experimental results are observable with an unaided eye or, at most, optical magnification.

Using these techniques, researchers in these subjects attempt to elucidate the complex systems of interactions that give rise to the processes that make life possible. By drawing knowledge and experimental techniques from a wide variety of disciplines, biophysicists are able to indirectly observe or model the structures and interactions of *individual* molecules or complexes of molecules. They require screening of hundreds of thousands of samples before reaching some potential compounds with desired properties. By some estimates, it takes dozens of years and millions of dollars.

However, with the advances in protein structure determination, structure based drug design has emerged as a powerful tool for developing new drugs with specific properties and minimal side effects. This technique is usually faster than the conventional methods. In structure based drug design, the three-dimensional structure of a drug target interacting with small molecules is used for drug discovery. Structure-based drug design represents the idea that one can see exactly how the ligand molecule interacts with its target protein. The designed compounds that have affinities in the acceptable pharmacological range can be further processed for other biological assays and clinical trials.

BIOPHYSICS: NEW CHALLENGES FOR PHYSICS

Recent progress in life sciences has demonstrated that the first decades of the new century are likely to be dominated by developments in this field. Since physics forms the basis of the life science evolution, the visionary guidance and assistance of physicists who have enjoyed a versatile training will be needed. Perhaps the most easily recognizable example of this need for an interdisciplinary approach is the astounding revelation of the human genetic code, the self-organized plan according to which self-reproducing entities, such as cells, form complex organisms.

The underlying interactions are governed by physical principles. An equally important challenge to physics is to clarify the way in which ensembles of cells, cells as individual entities, and their molecular constituents, function in their respective surroundings. Biophysics has thus become a central theme of the physics department, offering young students unique opportunities for study and ample openings for top-level research.

Excellent job opportunities and the dynamics of Munich as a centre of biotechnology with its many startup companies underline the attractiveness of the biophysics programme of the TU Munich (TUM). The role of physical methods in life sciences is manifested by modern techniques such as ultrasound, positron emission, X-ray and nuclear magnetic resonance tomography, light and electron microscopy, laser spectroscopy, X-ray structure analysis, and electrophysiological techniques such as patch clamp.

These techniques have benefited from more than half a century of research in physics and are the result of combining classical instrumentation with computational physics. The discoveries of these new physical methods have triggered off dramatic progress in life sciences. At the same time there are also numerous examples originating from biology that have inspired new developments in physics.

The most prominent one is the discovery of the general energy conservation law by Robert Meyer and Hermann von Helmholtz and the theory of Brownian motion by Albert Einstein. Einstein's ingenious interpretation of the observation of the botanist Robert Brown that seeds perform random walks in water influenced the development of modern physics at the beginning of the last century nearly as much as Planck's equation describing black body radiation. Nowadays, physicists do not content themselves with being the designers of new instrumentation but strive for a more active role in the search for universal physical principles governing the assembly and function of biomaterials.

In order to be successful, it is absolutely necessary that physicists accept the complexity of biomaterials and become familiar with the principal questions of biology. The physicist's capacity to unravel universal laws governing complex processes or to develop new measuring techniques to test laws predicted by theory is urgently needed in life sciences.

BIOPHYSICS AT THE TUM

As early as 1980, to meet the challenges of modern life sciences, the physics department of the TUM decided to strengthen its resources in the field of biophysics. There are now 5 full professors at the TUM, complemented by 3 adjunct professors, all working in the field of biophysics. Recently, a junior professorship and a junior group funded by the Emmy Noether programme were established so as to broaden the scope of the research activities.

The TUM was, in fact, the first university in Germany to introduce a full biophysics programme. The combination of experimental and theoretical studies on single molecules, living cells, and cellular networks is a unique strength of the biophysical research efforts of the physics department. The goal is not only to obtain a detailed understanding of biophysical principles but also to develop new tools with which to study and quantify biological processes. It is this combination of different approaches and foci that makes the study of biophysics at the TUM an outstanding experience. The biophysics research groups of the physics department engage in a very broad spectrum of activities and are complemented by several special areas of research.

BIOCHEMICAL TECHNIQUE

FUSION PROTEIN

Fusion proteins, also known as chimeric proteins, are proteins created through the joining of two or more genes which originally coded for separate proteins. Translation of this *fusion gene* results in a single polypeptide with function properties derived from each of the original proteins. *Recombinant fusion proteins* are created artificially by recombinant DNA technology for use in biological research or therapeutics. Chimeric mutant proteins occur naturally when a large-scale mutation, typically a chromosomal translocation, creates a novel coding sequence containing parts of the coding sequences from two different genes. Naturally occurring fusion proteins are important in cancer, where they may function as oncoproteins. The bcr-abl fusion protein is a well-known example of an oncogenic fusion protein, and is considered to be the primary oncogenic driver of chronic myelogenous leukemia.

Properties of Fusion Proteins

The functionality of fusion proteins is made possible by the fact that many protein functional domains are *modular*. In other words, the linear portion of a polypeptide which corresponds to a given domain, such as a tyrosine kinase domain, may be removed from the rest of the protein without destroying its intrinsic enzymatic capability.

Recombinant Fusion Proteins

A recombinant fusion protein is a protein created through genetic engineering of a fusion gene. This typically involves removing the stop codon from a cDNA sequence coding for the first protein, then appending the cDNA sequence of the second protein in frame through ligation or overlap extension PCR. That DNA sequence will then be expressed by a cell as a single protein.

The protein can be engineered to include the full sequence of both original proteins, or only a portion of either. If the two entities are proteins, often linker (or "spacer") peptides are also added which make it more likely that the proteins fold independently and behave as expected.

Especially in the case where the linkers enable protein purification, linkers in protein or peptide fusions are sometimes engineered with cleavage sites for proteases or chemical agents which enable the liberation of the two separate proteins.

This technique is often used for identification and purification of proteins, by fusing a GST protein, FLAG peptide, or a hexa-his peptide (aka: a 6 × his-tag) which can be isolated using nickel or cobalt resins (affinity chromatography). Chimeric proteins can also be manufactured with toxins or anti-bodies attached to them in order to study disease development.

Chimeric Protein Drugs

Several drugs made from chimeric proteins are currently available for medical use. Several chimeric protein drugs are TNFá blockers, such as Etanercept, Infliximab, and Adalimumab.

NATURALLY OCCURRING FUSION PROTEINS

Naturally occurring fusion genes are most commonly created when a chromosomal translocation replaces the terminal exons of one gene with intact exons from a second gene. This creates a single gene which can be transcribed, spliced, and translated to produce a functional fusion protein. Many important cancer-promoting oncogenes are fusion genes produced in this way.

Examples include:

- Gag-onc fusion protein
- Bcr-abl fusion protein
- Tpr-met fusion protein

The polymerase chain reaction (PCR) is a technique widely used in molecular biology. It derives its name from one of its key components, a DNA polymerase used to amplify a piece of DNA by *in vitro* enzymatic replication. As PCR progresses, the DNA thus generated is itself used as a template for replication. This sets in motion a chain reaction in which the DNA template is exponentially amplified.

With PCR it is possible to amplify a single or few copies of a piece of DNA across several orders of magnitude, generating millions or more copies of the DNA piece. PCR can be extensively modified to perform a wide array of genetic manipulations. Almost all PCR applications employ a heat-stable DNA polymerase, such as Taq polymerase, an enzyme originally isolated from the bacterium *Thermus aquaticus*. This

DNA polymerase enzymatically assembles a new DNA strand from DNA building blocks, the nucleotides, using single-stranded DNA as template and DNA oligonucleotides (also called DNA primers) required for initiation of DNA synthesis.

The vast majority of PCR methods use thermal cycling, i.e., alternately heating and cooling the PCR sample to a defined series of temperature steps. These thermal cycling steps are necessary to physically separate the strands (at high temperatures) in a DNA double helix (DNA melting) used as template during DNA synthesis (at lower temperatures) by the DNA polymerase to selectively amplify the target DNA.

The selectivity of PCR results from the use of primers that are complementary to the DNA region targeted for amplification under specific thermal cycling conditions. Developed in 1983 by Kary Mullis, PCR is now a common and often indispensable technique used in medical and biological research labs for a variety of applications.

These include DNA cloning for sequencing, DNA-based phylogeny, or functional analysis of genes; the diagnosis of hereditary diseases; the identification of genetic fingerprints (used in forensic sciences and paternity testing); and the detection and diagnosis of infectious diseases. In 1993 Mullis won the Nobel Prize in Chemistry for his work on PCR.

PCR PRINCIPLES AND PROCEDURE

PCR is used to amplify specific regions of a DNA strand (the DNA target). This can be a single gene, a part of a gene, or a non-coding sequence. Most PCR methods typically amplify DNA fragments of up to 10 kilo base pairs (kb), although some techniques allow for amplification of fragments up to 40 kb in size. A basic PCR set up requires several components and reagents. These components include:

- *DNA template* that contains the DNA region (target) to be amplified.
- Two *primers,* which are complementary to the DNA regions at the 5' (five prime) or 3' (three prime) ends of the DNA region.
- A DNA polymerase such as *Taq polymerase* or another

DNA polymerase with a temperature optimum at around 70°C.

- *Deoxynucleoside triphosphates* (dNTPs; also very commonly and erroneously called deoxynucleotide triphosphates), the building blocks from which the DNA polymerases synthesizes a new DNA strand.
- *Buffer solution,* providing a suitable chemical environment for optimum activity and stability of the DNA polymerase.
- *Divalent cations,* magnesium or manganese ions; generally Mg^{2+} is used, but Mn^{2+} can be utilized for PCR-mediated DNA mutagenesis, as higher Mn^{2+} concentration increases the error rate during DNA synthesis
- *Monovalent cation* potassium ions.

The PCR is commonly carried out in a reaction volume of 10-200 ìl in small reaction tubes (0.2-0.5 ml volumes) in a thermal cycler.

The thermal cycler heats and cools the reaction tubes to achieve the temperatures required at each step of the reaction. Many modern thermal cyclers make use of the Peltier effect which permits both heating and cooling of the block holding the PCR tubes simply by reversing the electric current.

Thin-walled reaction tubes permit favorable thermal conductivity to allow for rapid thermal equilibration. Most thermal cyclers have heated lids to prevent condensation at the top of the reaction tube. Older thermocyclers lacking a heated lid require a layer of oil on top of the reaction mixture or a ball of wax inside the tube.

Procedure

Denaturing at 94-96°C. Annealing at ~65°C Elongation at 72°C. The blue lines represent the DNA template to which primers anneal that are extended by the DNA polymerase to give shorter DNA products, which themselves are used as templates as PCR progresses.

The PCR usually consists of a series of 20 to 40 repeated temperature changes called cycles; each cycle typically consists of 2-3 discrete temperature steps. Most commonly PCR is carried out with cycles that have three temperature steps. The cycling is often preceded by a single temperature step (called *hold*) at a high temperature (> 90°C), and followed by one hold at the end for final product extension or brief storage. The temperatures used and the length of time they are applied in each cycle depend on a variety of parameters.

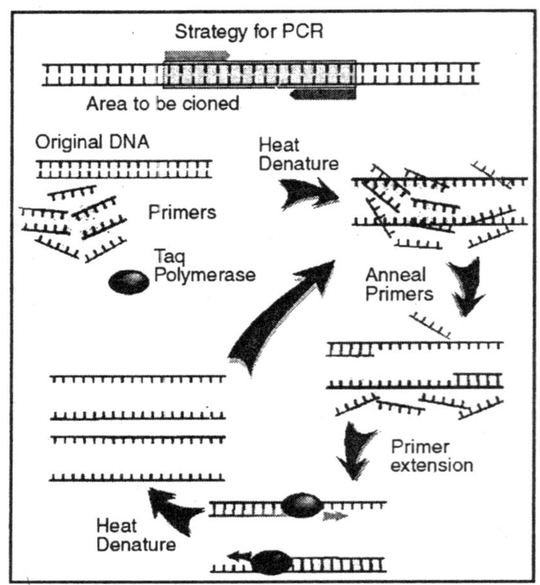

Fig. The PCR Cycle

These include the enzyme used for DNA synthesis, the concentration of divalent ions and dNTPs in the reaction, and the melting temperature (Tm) of the primers.

Initialization Step

This step consists of heating the reaction to a temperature of 94-96°C (or 98°C if extremely thermostable polymerases are used), which is held for 1-9 minutes. It is only required for DNA polymerases that require heat activation by hot-start PCR.

Denaturation Step

This step is the first regular cycling event and consists of heating the reaction to 94-98°C for 20-30 seconds. It causes melting of DNA template and primers by disrupting the hydrogen bonds between complementary bases of the DNA strands, yielding single strands of DNA.

Annealing Step

The reaction temperature is lowered to 50-65°C for 20-40 seconds allowing annealing of the primers to the single-stranded DNA template. Typically the annealing temperature is about 3-5 degrees Celsius below the Tm of the primers used. Stable DNA-DNA hydrogen bonds are only formed when the primer sequence very closely matches the template sequence. The polymerase binds to the primer-template hybrid and begins DNA synthesis.

Extension/elongation Step

The temperature at this step depends on the DNA polymerase used; Taq polymerase has its optimum activity temperature at 75-80°C, and commonly a temperature of 72°C is used with this enzyme. At this step the DNA polymerase synthesizes a new DNA strand complementary to the DNA template strand by adding dNTPs that are complementary to the template in 5' to 3' direction, condensing the 5'-phosphate group of the dNTPs with the 3'-hydroxyl group at the end of the nascent (extending) DNA strand.

The extension time depends both on the DNA polymerase used and on the length of the DNA fragment to be amplified. As a rule-of-thumb, at its optimum temperature, the DNA polymerase will polymerize a thousand bases per minute. Under optimum conditions, i.e., if there are no limitations due to limiting substrates or reagents, at each extension step, the amount of DNA target is doubled, leading to exponential (geometric) amplification of the specific DNA fragment.

Final Elongation

This single step is occasionally performed at a temperature of 70-74°C for 5-15 minutes after the last PCR cycle to ensure that any remaining single-stranded DNA is fully extended.

Final Hold

This step at 4-15°C for an indefinite time may be employed for short-term storage of the reaction. To check whether the PCR generated the anticipated DNA fragment (also sometimes referred to as the amplimer or amplicon), agarose gel electrophoresis is employed for size separation of the PCR products. The size(s) of PCR products is determined by comparison with a *DNA ladder* (a molecular weight marker), which contains DNA fragments of known size, run on the gel alongside the PCR products.

PCR STAGES

The PCR process can be divided into three stages:

Exponential amplification: At every cycle, the amount of product is doubled (assuming 100% reaction efficiency). The reaction is very specific and precise.

Levelling off stage: The reaction slows as the DNA polymerase loses activity and as consumption of reagents such as dNTPs and primers causes them to become limiting.

Plateau: No more product accumulates due to exhaustion of reagents and enzyme.

PCR Optimization

In practice, PCR can fail for various reasons, in part due to its sensitivity to contamination causing amplification of spurious DNA products. Because of this, a number of techniques and procedures have been developed for optimizing PCR conditions. Contamination with extraneous DNA is addressed with lab protocols and procedures that separate pre-PCR mixtures from potential DNA contaminants.

This usually involves spatial separation of PCR-setup areas from areas for analysis or purification of PCR products, and thoroughly cleaning the work surface between reaction setups. Primer-design techniques are important in improving PCR product yield and in avoiding the formation of spurious products, and the usage of alternate buffer components or polymerase enzymes can help with amplification of long or otherwise problematic regions of DNA.

APPLICATION OF PCR

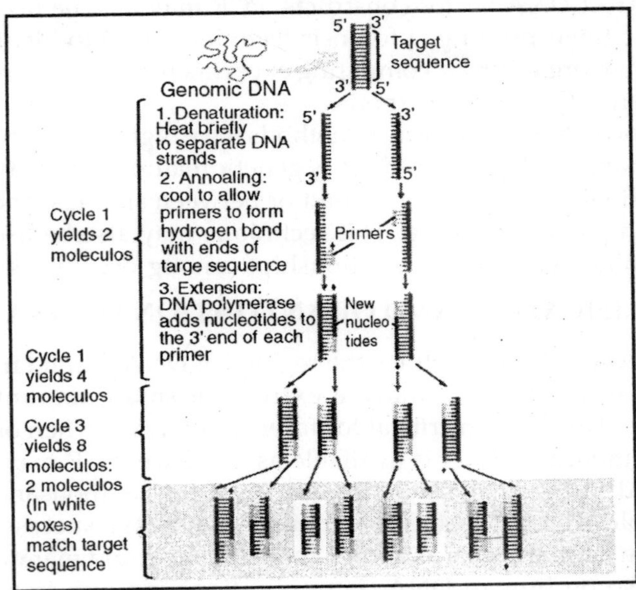

Fig. Application of PCR

ISOLATION OF GENOMIC DNA

PCR allows isolation of DNA fragments from genomic DNA by selective amplification of a specific region of DNA. This use of PCR augments many methods, such as generating hybridization probes for Southern or northern hybridization and DNA cloning, which require larger amounts of DNA, representing a specific DNA region.

PCR supplies these techniques with high amounts of pure DNA, enabling analysis of DNA samples even from very small amounts of starting material. Other applications of PCR include DNA sequencing to determine unknown PCR-amplified sequences in which one of the amplification primers may be used in Sanger sequencing, isolation of a DNA sequence to expedite recombinant DNA technologies involving the insertion of a DNA sequence into a plasmid or the genetic material of another organism.

Bacterial colonies (E.coli) can be rapidly screened by PCR for correct DNA vector constructs. PCR may also be used for genetic fingerprinting; a forensic technique used to identify a person or organism by comparing experimental DNAs through different PCR-based methods.

Some PCR 'fingerprints' methods have high discriminative power and can be used to identify genetic relationships between individuals, such as parent-child or between siblings, and are used in paternity testing. This technique may also be used to determine evolutionary relationships among organisms.

AMPLIFICATION AND QUANTITATION OF DNA

Because PCR amplifies the regions of DNA that it targets, PCR can be used to analyse extremely small amounts of sample. This is often critical for forensic analysis, when only a trace amount of DNA is available as evidence. PCR may also be used in the analysis of ancient DNA that is thousands of years old. These PCR-based techniques have been successfully used on animals, such as a forty-thousand-year-old mammoth, and also on human DNA, in applications ranging from the analysis of Egyptian mummies to the identification of a Russian Tsar.

Quantitative PCR methods allow the estimation of the amount of a given sequence present in a sample – a technique often applied to quantitatively determine levels of gene expression. Real-time PCR is an established tool for DNA quantification that measures the accumulation of DNA product after each round of PCR amplification.

PCR IN DIAGNOSIS OF DISEASES

PCR allows early diagnosis of malignant diseases such as leukemia and lymphomas, which is currently the highest developed in cancer research and is already being used routinely. PCR assays can be performed directly on genomic DNA samples to detect translocation-specific malignant cells at a sensitivity which is at least 10,000 fold higher than other methods. PCR also permits identification of non-cultivatable or slow-growing microorganisms such as mycobacteria, anaerobic bacteria, or viruses from tissue culture assays and animal models. The basis for PCR diagnostic applications in microbiology is the detection of infectious agents and the discrimination of non-pathogenic from pathogenic strains by virtue of specific genes.

Viral DNA can likewise be detected by PCR. The primers used need to be specific to the targeted sequences in the DNA of a virus, and the PCR can be used for diagnostic analyses or DNA sequencing of the viral genome. The high sensitivity of PCR permits virus detection soon after infection and even before the onset of disease. Such early detection may give physicians a significant lead in treatment. The amount of virus ("viral load") in a patient can also be quantified by PCR-based DNA quantitation techniques.

GEL ELECTROPHORESIS

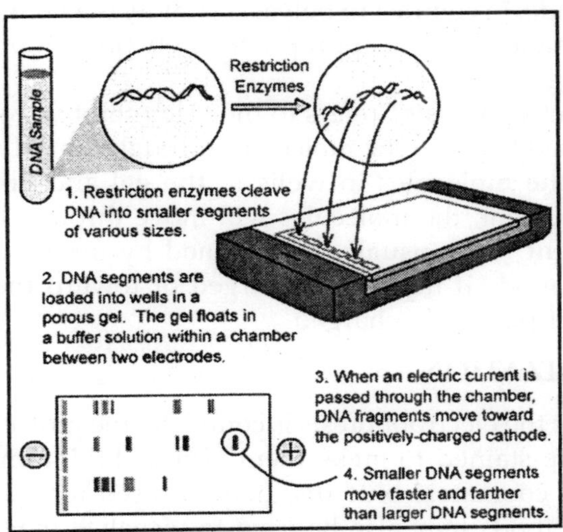

Fig. Gel Electrophoresis

Gel electrophoresis is a technique used for the separation of deoxyribonucleic acid, ribonucleic acid, or protein molecules using an electric current applied to a gel matrix. It is usually performed for analytical purposes, but may be used as a preparative technique prior to use of other methods such as mass spectrometry, RFLP, PCR, cloning, DNA sequencing, or Southern blotting for further characterization.

SEPARATION

The term "gel" in this instance refers to the matrix used to contain, then separate the target molecules. In most cases the gel is a crosslinked polymer whose composition and porosity is chosen based on the specific weight and composition of the target to be analyzed.

When separating proteins or small nucleic acids (DNA, RNA, or oligonucleotides) the gel is usually composed of different concentrations of acrylamide and a cross-linker, producing different sized mesh networks of polyacrylamide.

When separating larger nucleic acids (greater than a few hundred bases), the preferred matrix is purified agarose. In both cases, the gel forms a solid, yet porous matrix. Acrylamide, in contrast to polyacrylamide, is a neurotoxin and must be handled using appropriate safety precautions to avoid poisoning.

"Electrophoresis" refers to the electromotive force (EMF) that is used to move the molecules through the gel matrix. By placing the molecules in wells in the gel and applying an electric current, the molecules will move through the matrix at different rates, usually determined by mass, toward the positive anode if negatively charged or toward the negative cathode if positively charged.

VISUALIZATION

After the electrophoresis is complete, the molecules in the gel can be stained to make them visible. Ethidium bromide, silver, or coomassie blue dye may be used for this process. Other methods may also be used to visualize the separation of the mixture's components on the gel. If the analyte molecules fluoresce under ultraviolet light, a photograph can be taken of the gel under ultraviolet lighting conditions. If the molecules to be separated contain radioactivity added for visibility, an autoradiogram can be recorded of the gel. If several mixtures have initially been injected next to each other, they will run parallel in individual lanes. Depending on the number of different molecules, each lane shows separation of the components from the original mixture as one or more distinct bands, one band per component. Incomplete separation of the components can lead to overlapping bands, or to indistinguishable smears representing multiple unresolved components.

Bands in different lanes that end up at the same distance from the top contain molecules that passed through the gel with the same speed, which usually means they are approximately the same size. There are molecular weight size markers available that contain a mixture of molecules of known sizes. If such a marker was run on one lane in the gel parallel to the unknown samples, the bands observed can be compared to those of the unknown in order to determine their size. The distance a band travels is approximately inversely proportional to the logarithm of the size of the molecule.

APPLICATIONS

Gel electrophoresis is used in forensics, molecular biology, genetics, microbiology and biochemistry. The results can be analyzed quantitatively by visualizing the gel with UV light and a gel imaging device. The image is recorded with a computer operated camera, and the intensity of the band or spot of interest is measured and compared against standard or markers loaded on the same gel. The measurement and analysis are mostly done with specialized software.

Depending on the type of analysis being performed, other techniques are often implemented in conjunction with the results of gel electrophoresis, providing a wide range of field-specific applications.

Nucleic Acids

In the case of nucleic acids, the direction of migration, from negative to positive electrodes, is due to the naturally-occurring negative charge carried by their sugar-phosphate backbone. Double-stranded DNA fragments naturally behave as long rods, so their migration through the gel is relative to their radius of gyration, or, for non-cyclic fragments, size. Single-stranded DNA or RNA tend to fold up into molecules with complex shapes and migrate through the gel in a complicated manner based on their tertiary structure. Therefore, agents that disrupt the hydrogen bonds, such as sodium hydroxide or formamide, are used to denature the nucleic acids and cause them to behave as long rods again.

Gel electrophoresis of large DNA or RNA is usually done by agarose gel electrophoresis.

Proteins

Proteins, unlike nucleic acids, can have varying charges and complex shapes, therefore they may not migrate into the gel at similar rates, or at all, when placing a negative to positive EMF on the sample. Proteins therefore, are usually denatured in the presence of a detergent such as sodium dodecyl sulfate/ sodium dodecyl phosphate (SDS/SDP) that coats the proteins with a negative charge. Generally, the amount of SDS bound is relative to the size of the protein (usually 1.4g SDS per gram of protein), so that the resulting denatured proteins have an overall negative charge, and all the proteins have a similar charge to mass ratio.

Since denatured proteins act like long rods instead of having a complex tertiary shape, the rate at which the resulting SDS coated proteins migrate in the gel is relative only to its size and not its charge or shape. Proteins are usually analyzed by sodium dodecyl sulfate polyacrylamide gel electrophoresis (SDS-PAGE), by native gel electrophoresis, by quantitative preparative native continuous polyacrylamide gel electrophoresis (QPNC-PAGE), or by 2-D electrophoresis.

DNA SEQUENCING

The term DNA sequencing encompasses biochemical methods for determining the order of the nucleotide bases, adenine, guanine, cytosine, and thymine, in a DNA oligonucleotide. The sequence of DNA constitutes the heritable genetic information in nuclei, plasmids, mitochondria, and chloroplasts that forms the basis for the developmental programs of all living organisms. Determining the DNA sequence is therefore useful in basic research studying fundamental biological processes, as well as in applied fields such as diagnostic or forensic research. The advent of DNA sequencing has significantly accelerated biological research and discovery. The rapid speed of sequencing attained with modern DNA sequencing technology has been instrumental in the large-scale sequencing of the human genome, in the Human Genome Project.

Related projects, often by scientific collaboration across continents, have generated the complete DNA sequences of many animal, plant, and microbial genomes.

Maxam-Gilbert Sequencing

In 1976-1977, Allan Maxam and Walter Gilbert developed a DNA sequencing method based on chemical modification of DNA and subsequent cleavage at specific bases. Although Maxam and Gilbert published their chemical sequencing method two years after the ground-breaking paper of Sanger and Coulson on plus-minus sequencing, Maxam-Gilbert sequencing rapidly became more popular, since purified DNA could be used directly, while the initial Sanger method required that each read start be cloned for production of single-stranded DNA.

However, with the development and improvement of the chain-termination method, Maxam-Gilbert sequencing has fallen out of favour due to its technical complexity, extensive use of hazardous chemicals, and difficulties with scale-up. In addition, unlike the chain-termination method, chemicals used in the Maxam-Gilbert method cannot easily be customized for use in a standard molecular biology kit.

In brief, the method requires radioactive labelling at one end and purification of the DNA fragment to be sequenced. Chemical treatment generates breaks at a small proportion of one or two of the four nucleotide bases in each of four reactions (*G, A + G, C, C + T*). Thus a series of labelled fragments is generated, from the radiolabelled end to the first 'cut' site in each molecule.

The fragments are then size-separated by gel electrophoresis, with the four reactions arranged side by side. To visualize the fragments generated in each reaction, the gel is exposed to X-ray film for autoradiography, yielding an image of a series of dark 'bands' corresponding to the radiolabelled DNA fragments, from which the sequence may be inferred.

Also sometimes known as 'chemical sequencing', this method originated in the study of DNA-protein interactions (footprinting), nucleic acid structure and epigenetic modifications to DNA, and within these it still has important applications.

Chain-termination Methods

While the chemical sequencing method of Maxam and Gilbert, and the plus-minus method of Sanger and Coulson were orders of magnitude faster than previous methods, the chain-terminator method developed by Sanger was even more efficient, and rapidly became the method of choice. The Maxam-Gilbert technique requires the use of highly toxic chemicals, and large amounts of radiolabeled DNA, whereas the chain-terminator method uses fewer toxic chemicals and lower amounts of radioactivity.

The key principle of the Sanger method was the use of dideoxynucleotides triphosphates (ddNTPs) as DNA chain terminators. The classical chain-termination or Sanger method requires a single-stranded DNA template, a DNA primer, a DNA polymerase, radioactively or fluorescently labeled nucleotides, and modified nucleotides that terminate DNA strand elongation. The DNA sample is divided into four separate sequencing reactions, containing all four of the standard deoxynucleotides (dATP, dGTP, dCTP and dTTP) and the DNA polymerase. To each reaction is added only one of the four dideoxynucleotides (ddATP, ddGTP, ddCTP, or ddTTP).

These dideoxynucleotides are the chain-terminating nucleotides, lacking a 3'-OH group required for the formation of a phosphodiester bond between two nucleotides during DNA strand elongation. Incorporation of a dideoxynucleotide into the nascent (elongating) DNA strand therefore terminates DNA strand extension, resulting in various DNA fragments of varying length. The dideoxynucleotides are added at lower concentration than the standard deoxynucleotides to allow strand elongation sufficient for sequence analysis.

The newly synthesized and labeled DNA fragments are heat denatured, and separated by size (with a resolution of just one nucleotide) by gel electrophoresis on a denaturing polyacrylamide-urea gel. Each of the four DNA synthesis reactions is run in one of four individual lanes (lanes *A, T, G, C*); the DNA bands are then visualized by autoradiography or UV light, and the DNA sequence can be directly read off the X-ray film or gel image.

In the image on the right, X-ray film was exposed to the gel, and the dark bands correspond to DNA fragments of different lengths.

A dark band in a lane indicates a DNA fragment that is the result of chain termination after incorporation of a dideoxynucleotide (ddATP, ddGTP, ddCTP, or ddTTP). The terminal nucleotide base can be identified according to which dideoxynucleotide was added in the reaction giving that band. The relative positions of the different bands among the four lanes are then used to read (from bottom to top) the DNA sequence as indicated.

There are some technical variations of chain-termination sequencing. In one method, the DNA fragments are tagged with nucleotides containing radioactive phosphorus for radiolabelling. Alternatively, a primer labeled at the 5′ end with a fluorescent dye is used for the tagging. Four separate reactions are still required, but DNA fragments with dye labels can be read using an optical system, facilitating faster and more economical analysis and automation.

This approach is known as 'dye-primer sequencing'. The later development by L Hood and coworkers of fluorescently labeled ddNTPs and primers set the stage for automated, high-throughput DNA sequencing.

The different chain-termination methods have greatly simplified the amount of work and planning needed for DNA sequencing. For example, the chain-termination-based "Sequenase" kit from USB Biochemicals contains most of the reagents needed for sequencing, prealiquoted and ready to use.

Some sequencing problems can occur with the Sanger method, such as non-specific binding of the primer to the DNA, affecting accurate read-out of the DNA sequence. In addition, secondary structures within the DNA template, or contaminating RNA randomly priming at the DNA template can also affect the fidelity of the obtained sequence. Other contaminants affecting the reaction may consist of extraneous DNA or inhibitors of the DNA polymerase.

Dye-terminator Sequencing

An alternative to primer labelling is labelling of the chain terminators, a method commonly called 'dye-terminator sequencing'. The major advantage of this method is that the sequencing can be performed in a single reaction, rather than four reactions as in the labelled-primer method. In dye-terminator sequencing, each of the four dideoxynucleotide chain terminators is labelled with a different fluorescent dye, each fluorescing at a different wavelength.

This method is attractive because of its greater expediency and speed and is now the mainstay in automated sequencing with computer-controlled sequence analyzers. Its potential limitations include dye effects due to differences in the incorporation of the dye-labelled chain terminators into the DNA fragment, resulting in unequal peak heights and shapes in the electronic DNA sequence trace chromatogram after capillary electrophoresis.

This problem has largely been overcome with the introduction of new DNA polymerase enzyme systems and dyes that minimize incorporation variability, as well as methods for eliminating "dye blobs", caused by certain chemical characteristics of the dyes that can result in artifacts in DNA sequence traces.

The dye-terminator sequencing method, along with automated high-throughput DNA sequence analyzers, is now being used for the vast majority of sequencing projects, as it is both easier to perform and lower in cost than most previous sequencing methods.

Challenges

Modern sequencing typically produces a sequence that has poor quality in the first 15-40 bases, a high quality region of 700-900 bases, and then quickly deteriorating quality. Base calling software typically outputs an estimate of quality along with the sequence to aid in quality trimming. Before the DNA can be sequenced, linker sequences are attached to its ends, and it is inserted into a cloning vector. The resulting sequence can therefore often contain parts of the vector or the linker sequences, which must be filtered out prior to analysis.

In contrast, emerging sequencing technologies based on pyrosequencing often avoid using cloning vectors. During PCR amplification, unrelated sequences can hybridize, and the resulting clone can be a chimaeric sequence, containing fragments from both sequences. Another problem is polymerase stuttering, where the polymerase repeatedly outputs the same fragments, giving an artificially long low-complexity part of the sequence.

Automation and Sample Preparation

Modern automated DNA sequencing instruments (DNA sequencers) can sequence up to 384 fluorescently labelled samples in a single batch (run) and perform as many as 24 runs a day. However, automated DNA sequencers carry out only DNA-size-based separation (by capillary electrophoresis), detection and recording of dye fluorescence, and data output as fluorescent peak trace chromatograms. Sequencing reactions by thermocycling, cleanup and re-suspension in a buffer solution before loading onto the sequencer are performed separately.

In the past, an operator had to trim the low quality ends of every sequence manually in order to remove the sequencing errors. However, today, software like Chromatogram Explorer or DNA Baser can automatically trim the ends at batch. Such programs score the peaks of each base for quality and remove low-quality base peaks(generally located at the ends of the sequence). If too many bases have a quality score value below a certain threshold, those bases will be automatically deleted. The accuracy of such algorithms is currently below visual examination by a human operator, but is high enough for processing of sequence data sets that are too large for manual examination.

LARGE-SCALE SEQUENCING STRATEGIES

Current methods can directly sequence only relatively short (300-1000 nucleotides long) DNA fragments in a single reaction. The main obstacle to sequencing DNA fragments above this size limit is insufficient power of separation for resolving large DNA fragments that differ in length by only one nucleotide.

Large-scale sequencing aims at sequencing very long DNA fragments. Even relatively small bacterial genomes contain millions of nucleotides, and the human chromosome 1 alone contains about 246 million bases. Therefore, some approaches consist of cutting (with restriction enzymes) or shearing (with mechanical forces) large DNA fragments into shorter DNA fragments. The fragmented DNA is cloned into a DNA vector, usually a bacterial artificial chromosome (BAC), and amplified in *Escherichia coli*.

The amplified DNA can then be purified from the bacterial cells (a disadvantage of bacterial clones for sequencing is that some DNA sequences may be inherently *un-clonable* in some or all available bacterial strains, due to deleterious effect of the cloned sequence on the host bacterium or other effects). These short DNA fragments purified from individual bacterial colonies are then individually and completely sequenced and assembled electronically into one long, contiguous sequence by identifying 100 per cent-identical overlapping sequences between them (shotgun sequencing).

This method does not require any pre-existing information about the sequence of the DNA and is often referred to as *de novo* sequencing. Gaps in the assembled sequence may be filled by Primer walking, often with sub-cloning steps (or transposon-based sequencing depending on the size of the remaining region to be sequenced). These strategies all involve taking many small *reads* of the DNA by one of the above methods and subsequently assembling them into a contiguous sequence.

The different strategies have different tradeoffs in speed and accuracy; the shotgun method is the most practical for sequencing large genomes, but its assembly process is complex and potentially error-prone–particularly in the presence of sequence repeats. Because of this, the assembly of the human genome is not literally complete—the repetitive sequences of the centromeres, telomeres, and some other parts of chromosomes result in gaps in the genome assembly.

Despite having only 93 per cent of the full genome assembled, the Human Genome Project was declared complete because their definition of human genome sequencing was limited to euchromatic sequence (99 per cent complete at the time), excluding these intractable repetitive regions. The human genome is about 3 billion (3,000,000,000) bp long; if the average fragment length is 500 bases, it would take a minimum of six million (3 billion/500) to sequence the human genome (not allowing for overlap = 1-fold coverage).

Keeping track of such a high number of sequences presents significant challenges, only held down by developing and coordinating several procedural and computational algorithms, such as efficient database development and management. *Resequencing* or *targeted sequencing* is utilized for determining a change in DNA sequence from a "reference" sequence. It is often performed using PCR to amplify the region of interest (pre-existing DNA sequence is required to design the PCR primers). Resequencing uses three steps, extraction of DNA or RNA from biological tissue; amplification of the RNA or DNA (often by PCR); followed by sequencing. The resultant sequence is compared to a reference or a normal sample to detect mutations.

Chapter 8

Bioprocess

Bioprocess technology is currently employed for the production of several economically important commodity and fine chemicals, enzymes and therapeutically active recombinant proteins. To give an indication of the market volume of the production of recombinant proteins, the top selling biopharmaceutical product of the year 2000.

Due to economic needs and because of the complex nature of microbial growth and product formation in batch and fed-batch cultivations, the monitoring and control of bioprocesses represents an ever-increasing engineering challenge.

In order to achieve optimal exploitation of the particular production organism, the advancement of bioprocesses to improve monitoring and control capabilities is pivotal to achieve reduction of production costs and increase of yield while at the same time maintaining the quality of the individual metabolic product. However, a great number of the current processes are still far from being optimal mainly due to limited process monitoring capabilities.

For further optimisation of bioprocesses as well as for ensuring high, consistent product quality, the lack of accurate real-time monitoring of different physical, chemical, and biological parameters will be the bottleneck.

Therefore, in the years to come, increasing focus has to be given to online/inline techniques for process monitoring, driven by the industry's never-ending need for process optimisation. Further requirements will arise by regulatory affairs like the process analytical technology (PAT) initiative issued by the FDA.

The major goal of PAT is to improve the understanding and control of the manufacturing process: quality cannot be tested in products, it should be built-in or should be achieved by design. Process Analytical Technology is seen as a system for designing, analyzing, and controlling manufacturing through timely measurements (i.e. during processing) of critical quality and performance attributes of raw and in-process materials and processes with the goal of ensuring final product quality.

The observation of variables related to the biological system, representing the production entity, is one of the key requirements to enable controlled gene expression and optimal operation of the host cell. On the contrary, key variables of cultivation processes are still beyond direct measurement despite the progress in monitoring and control of bioprocesses in the last years. A direct reading of biological key process variables, such as biomass, has not yet been achieved, although online sensor systems that provide different types of signals are available.

However, the inability to directly measure these key process variables does not imply one cannot extract valuable information from the bioprocess. In order to develop novel monitoring concepts, it is necessary to scrutinise the role and properties of the different cellular compartments to the synthesis process of economically important biotechnological products at first.

Bioprocesses are ubiquitous-they range from the production of food in a kitchen to the synthesis of sophisticated, extremely high-value, cutting-edge therapeutics. Bioprocesses involve the use of cells (or in certain cases, cellular molecules) as micro-factories to manufacture the product of interest. Natural cellular products can be directly utilized or, alternatively, the cells can be engineered to produce products of interest. In either case, production levels can be improved by employing several strategies for formation or processing of the product.

Although bioprocesses have a wide range of applications, their governing principles are beautifully simple. These principles have been obtained with inputs from chemical engineering, molecular biology, molecular genetics, physics, chemistry, mathematics, computer science and other engineering disciplines. This course is designed to introduce the participant to these principles; both, theoretical aspects and laboratory work at hands-on level, will be covered.

BIOPROCESS TECHNOLOGY

When our early ancestors made alcoholic beverages, they used a bioprocess. The combination of yeast cells and nutrients (cereal grains) formed a fermentation system in which the organisms consumed the nutrients for their own growth and produced by-products (alcohol and carbon dioxide gas) that helped to make the beverage.

Bioprocess technology is an extension of ancient techniques for developing useful products by taking advantage of natural biological activities such as production of enzymes (used, for example, in food processing and waste management) and antibiotics.

Use of living material offers several advantages over conventional chemical methods of production. They usually require lower temperature, pressure, and pH. They can use renewable resources as raw materials and greater quantities can be produced with less energy consumption. With the development of automated and computerized equipment, it is becoming much easier to accurately monitor reaction conditions and thus increase production efficiency.

As advances in bioprocess technology, particularly separation and purification techniques, are made, commercial firms will be able to economically produce these substances in large amounts, and thus make them available for use in medical research, food processing, agriculture, pharmaceutical development, waste management, and numerous other fields of science and industry. Knowledge of the metabolism of microbial cells is indispensable for:

- The optimization of bioreactor operation.
- The selection of the best microorganism for a given biochemical conversion.

- Targeted or evolutionary improvement of an existing microorganism.

Design (metabolic engineering) of microorganisms with new reactivities and/or morphology. This helps improve the efficiency of bioprocesses by focusing on microbial performance. Growth and production characteristics of the microorganisms or cell culture must be measured as a function of environment.

Shake flasks of 250 ml to 1 liter capacity are used for these experiments. Varying conditions for optimal growth of microorganisms and their productivity are medium composition, pH, temperature and other environmental conditions. How is the working of the bench top fermenter which is also a bioreactor works?

Bench Top Fermenter (bioreactor) This is the first step of the scale up process for large-scale operation of the bioprocess for which optimum conditions for the growth are established in the shake flask experiments above. The bench top fermenter usually is of 1 to 2 liter capacity. It is equipped with instruments for measuring and adjusting

- Temperature
- pH
- Dissolve O_2
- Concentration of substrate
- Concentration of product
- Agitator speed
- Other process variables

Process can be more closely controlled than in Shake flask experiments. Information is collected on

- O_2 requirements of the cells
- Sensitivity of the cells to shear
- Foaming characteristics

Many proteins and polymers produced will cause extensive foaming. The main concern of this study is whether optimal conditions can be provided or not for the cell growth in a reactor and their determination. Information gathered from this study is used for the further scale-up of the bioreactor. The information obtained includes:

- Mass transfer coefficients (concentration gradients)
- Mixing time (a parameter used for determining mixing efficiency)
- Gas hold up

$$\varepsilon = \frac{v_g}{v_g + v_2}$$

V_g is the volume of gas bubbles in the reactor
V_2 is the volume of liquid

$$\text{Power number} \frac{P}{rN^3D^3}$$

P is the power required,
r is the viscosity,
D_1^5 is the impeller diameter

Impeller Shear Rate

It is also decided whether the bioreactor must be operated for batch process or continuous process. At this stage, the commercial viability of the process is also evaluated.

CONCEPT OF FERMENTATION

Fermentation is the process of deriving energy from the oxidation of organic compounds, such as carbohydrates, using an endogenous electron acceptor, which is usually an organic compound. Sugars are the common substrate of fermentation, and typical examples of fermentation products are ethanol, lactic acid, and hydrogen. However, more exotic compounds can be produced by fermentation, such as butyric acid and acetone. Yeast carries out fermentation in the production of ethanol in beers, wines and other alcoholic drinks, along with the production of large quantities of carbon dioxide. Fermentation occurs in mammalian muscle during periods of intense exercise where oxygen supply becomes limited.

Decomposition of foodstuffs generally accompanied by the evolution of gas. The best-known example is alcoholic fermentation, in which sugar is converted into alcohol and carbon dioxide. During fermentation organic matter is decomposed in the absence of air (oxygen); hence, there is always an accumulation of reduction products, or incomplete oxidation products. Some of these products (for example, alcohol and lactic acid) are of importance to humans, and fermentation has therefore been used for their manufacture on an industrial scale.

There are also many microbiological processes that go on in the presence of air while yielding incomplete oxidation products. Good examples are the formation of acetic acid (vinegar) from alcohol by vinegar bacteria, and of citric acid from sugar by certain molds (for example, *Aspergillus niger*). These microbial processes, too, have gained industrial importance, and are often referred to as fermentations, even though they do not conform to L. Pasteur's concept of fermentation as a decomposition in the absence of air.

Anaerobic metabolism. Used generally of alcohol fermentation of sugars, also production of lactic acid, citric acid, etc., by micro-organisms, which may be yeasts, bacteria or fungi. The breakdown of organic substances by organisms to release energy in the absence of oxygen. It is especially applied to the anaerobic breakdown of carbohydrates by yeasts to produce alcohol and carbon dioxide, and the bacterial breakdown of milk sugar to give lactic acid (as in the production of cheese and yoghurt).

Process by which the living cell is able to obtain energy through the breakdown of glucose and other simple sugar molecules without requiring oxygen. Fermentation is achieved by somewhat different chemical sequences in different species of organisms.

Two closely related paths of fermentation predominate for glucose. When muscle tissue receives sufficient oxygen supply, it fully metabolizes its fuel glucose to water and carbon dioxide.

However, at times of strenuous activity, muscle tissue uses oxygen faster than the blood can supply it. During this anaerobic condition, the six-carbon glucose molecule is only partly broken down to two molecules of the three-carbon sugar called lactic acid.

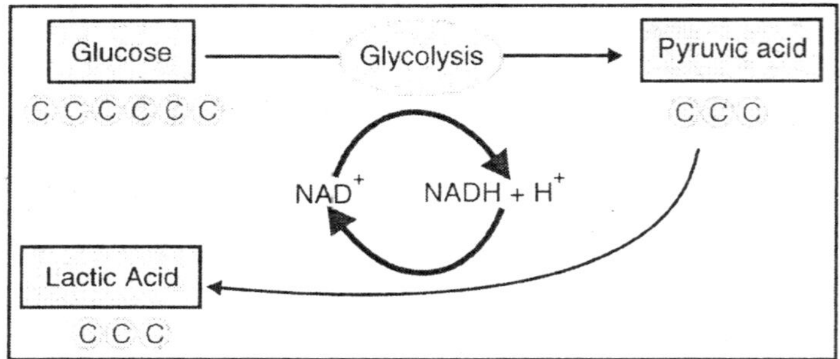

Fig. Lactic Acid Fermentation

This process, called lactic acid fermentation, also occurs in many microorganisms and in the cells of most higher animals. In alcoholic fermentation, such as occurs in brewer's yeast and some bacteria, the production of lactic acid is bypassed, and the glucose molecule is degraded to two molecules of the two-carbon alcohol, ethanol, and to two molecules of carbon dioxide.

Many of the enzymes of lactic acid and alcoholic fermentation are identical to the enzymes that bring about the metabolic conversion known as glycolysis. Alcoholic fermentation is a process that was known to antiquity. Before 2000 B.C. the Egyptians apparently knew that crushed fruits stored in a warm place would produce a substance with a pleasant intoxicating power.

By 1500 B.C. the production of beer from germinating cereals (malt) and the preparation of wines from crushed grapes were established arts in most of the Middle East. Aristotle believed that grape juice was an infantile form of wine and that fermentation was, therefore, the maturation of the grape extract.

Interest in the process of fermentation has continued through the ages, and much of modern biochemistry, especially enzyme studies, has emerged directly from early studies on the fermentation process. One of the earliest laboratories established for the study of biological chemistry was that founded in Copenhagen in 1875 and financed by the brewing family of Jacob Christian Jacobsen.

Fermentation is one of the oldest known food preservation techniques. Along with drying and salting, fermentation was a key method of extending the life of foods, allowing them to be available, and eaten safely, in times of scarcity or seasonal nonavailability. These methods helped allow the transition from hunting and gathering to organized food cultivation and storage, which took place some ten to fifteen thousand years ago in the Middle East.

Fermentation involves the action of desirable microorganisms, or their enzymes, on food ingredients to make biochemical changes, which cause significant modification to the food. Often lactic-acid bacteria convert the carbohydrate energy source of food, such as lactose in milk, to lactic acid; examples are yogurt and cheeses from milk, and pickles from fruits and vegetables. Alternatively, yeasts, often of the *Saccharomyces* species, may convert the glucose to ethanol and carbon dioxide in leavened breads, or the sugars in grain or fruit beverages to beers and wines.

Molds also can be active in certain fermentations, such as Stilton cheese and soy sauce. It is estimated that about one-third of all the food we consume is fermented. World estimates for beer consumption are about 22 million gallons, and a total of 15 million tons of some one thousand varieties of cheese are eaten annually.

FERMENTED BEVERAGES AND FOODS

Fermentation is often the key to the safe, enjoyable consumption of perishable food materials, as it changes their composition, flavour, and texture. For example, milk is a nutritious but highly perishable beverage. Originally, in the Middle East, milk carried in animal-skin containers, often on horseback, would sour naturally, to produce acidic fermented milk. The combined action of the two lactic-acid bacteria, *Streptococcus lactis*, producing lactic acid, and *Lactobacillus bulgaricus*, producing lactic acid and acetaldehyde, a major contributor to flavour, are involved in yogurt production.

The Tartars of Central Asia used the milk of horses, donkeys, or camels to produce a fizzy, gray acidic and alcoholic drink, *kumiss,* in which yeasts were active. In acid conditions, the milk protein, casein, denatures and is precipitated to form a curd, producing cottage and soft cheese. By stirring and pressing, whey is removed and a more solid curd is produced, which by ripening or maturation produces semi-hard or hard cheeses.

Surface-active bacteria of *Brevibacterium linens* are active in producing the aroma of Limburger type cheeses, while the blue molds of the genus *Penicillium* give Stilton and Gorgonzola cheeses their character.

The use of *Saccharomyces* yeasts has allowed the production of a range of fermented beverages, enabling safe consumption of liquid when fresh water supplies are not available. Lagers, the light golden, gassy beverage made by "bottom" yeast fermentation of cereal extracts, were first made in the regions of Germany and Czechoslovakia, but are now produced and consumed throughout the world.

In Africa, a thick, sour alcoholic beverage is made from sorghum or millet, or sometimes maize or banana. These sorghum beers are important sources of nutrients, particularly B vitamins, to people on marginal diets in these regions. The Romans planted extensive vineyards in North Africa to harvest and ferment their grapes into wine, thereby producing a fermented beverage that could be readily stored, transported, and consumed when and where required.

Distillation of these alcoholic beverages, such as whiskey from beers, brandies from grape wines, or arrack from palm or rice wine, further extend our range of drinks and play important cultural roles in festivities.

FERMENTATION VESSELS

Art meets science in the production of fermented foods. Traditional practices are passed down through generations of producers, often small in scale, and consumption patterns often have great cultural importance. In Scandinavia, traditionally the brides and mothers jealously guard their own supplies of sourdough starters, so that they can always make the desired bread for their partners and families.

In West Africa, a homeowner keeps a supply of *dawadawa*, a dried fermented African locust bean paste (*Parkia* species); it is used to give everyday soups and stews the desired "meaty" flavour, while also providing important nutrients, such as riboflavin, the B vitamin that protects against blindness, which is endemic to the region due to nutritional deficiency.

In Korea, few meals are complete without *kimchi*, a pickled fermented cabbage, which may also contain fish and other components. The practice of every home having their own *kimchi* jars, often on their verandahs, originated as a way of preserving vegetables through the cold winter season, providing year-round vitamin C. *Kimchi* together with *kochujang*, the fermented red pepper paste, give Korean preparations a unique and characteristic attractive colour and flavour. Where food fermentation occurred naturally as conditions favored particular organisms, an important art arose to encourage the desired fermentation organisms, while preventing undesirable microorganisms from developing, for successful fermented food production.

Food storage often took place in earthenware vessels, whose semipermeable inner walls were difficult to clean completely.

This allowed a biofilm of desirable microorganisms to remain, to initiate a successful fermentation of the next batch of food. Because of their significance, the vessels themselves were artistically designed and treasured. Interesting examples can be seen in museum collections, such as the Nezu Museum in Tokyo, Japan, and a museum dedicated to *kimchi* in Seoul, South Korea.

In Europe, the fermented meat producers, while using ceramic or metallic vats with smoother, more easily cleaned surfaces, developed the technique of "backslopping" to introduce a small quantity of the fermenting liquor from the previous batch of meat to initiate successful fermentation.

In many cases, dried grains or balls of the derived fermenting microorganisms on cereal or other substrates would be used to start fermentation. Baker's yeast may be used in this work. Kefir grains are used in North Africa, the Middle East, and Russia for production of kefir, *laban*, or *leben* fermented milks. *Ragi* is used in Indonesia and throughout East and Southeast Asia as *inoculum* for *lao-chao* and other fermented foods.

CULTURAL DIVERSITY

The production, consumption, and enjoyment of different fermented foods reflects the diversity of cultures and cuisines that make up our varied world. In Chinese and Japanese cuisines, *shoyu*, or soy sauce, is added almost universally to dishes, while the Indian vegetarian diet depends on fermented cereals and legumes, often in combinations, as in *dosas* and *vadas*. The art and science of fermenting meat to a wide range of salamis are vital to the enjoyment of Eastern and Central Europeans, while Italian food market stall holders proudly display their mold-covered fermented sausages and traditional cheeses. As people migrate, they normally carry their traditional fermented food practices with them.

The range of fermented cheeses and meats in Latin America reflects the European origins of these populations, and the wineries of Chile were originally established by French families. Consumers of imported wine, chocolate, coffee, or tea are all beneficiaries of the internationalism and significance of fermented foods.

MICROBIAL GROWTH KINETICS

Microbial growth can be defined as an orderly increase in cellular components, resulting in cell enlargement and eventually leading to cell division. This definition is not strictly accurate as it implies that a consequence of growth is always an increase in cell numbers.

However, under certain conditions growth can occur without cell division, for example, when cells are synthesizing storage compounds, e.g. glycogen or poly-b-hydroxybutyrate. In this situation the cell numbers remain constant, but the concentration of biomass continues to increase. This is also true for coenocytic organisms, such as some fungi, that are not divided into separate cells. Their growth results only in increased Size.

BACTERIAL GROWTH KINETICS

Growth kinetics of homogeneous unicellular suspension. cultures can be modeled using differential equations in a continuum model. However, filamentous growth and growth in heterogeneous cell aggregates and assemblage, particularly biofilms, colonies, flocs, mats and pellicles, is much more complex. In fact, heterogeneous systems require a very different approach using cellular automaton and Swarm system models, e.g. BacSim. The growth kinetics of filamentous organisms and heterogeneous systems will not be discussed here.

MODELING THE MICROBIAL GROWTH KINETICS

The growth model that will be examined is bacterial binary fission in homogeneous suspension cultures, where cell division produces identical daughter cells. Each time a cell divides is called a generation and the time taken for the cell to divide is referred to as the generation time. Therefore, the generation time or doubling time *(td)* is the time required for a microbial population to double. Theoretically, after one generation, both the microbial cell population and biomass concentration have doubled.

However, as previously stated, under certain conditions growth can be associated with an increase in biomass and not cell numbers. Also, the generation time recorded during microbial growth is in reality an average value, as the cells will not be dividing at exactly the same rate. At anyone time there are cells at different stages of their cell cycle. This is termed asynchronous growth. However, under certain conditions synchronous growth can be induced so that all cells divide simultaneously, which is a useful research tool in the study of microbial physiology.

Microbial fermentations in liquid media can be carried out under different operating conditions, i.e. batch growth, fed-batch growth or continuous growth. Batch growth involves a closed system where all nutrients are present at the start of the fermentation within a fixed volume. The only further additions may be acids or bases for pH control, or gases (e.g. aeration, if required).

In fed-batch systems fresh medium or medium components are fed continuously, intermittently or are added as a single supplement and the volume of the batch increases with time. Continuous fermentations are open systems where fresh medium is continuously fed into the fermentation vessel, but the volume remains constant as spent medium and cells are removed at the same rate.

BATCH GROWTH

During batch fermentations the population of microorganisms goes through several distinct growth phases: lag, acceleration, exponential growth, deceleration, stationary and death. In the lag phase virtually no growth occurs and the microbial population remains relatively constant. Nevertheless, it is a peripd of intense metabolic activity as the microbial inoculum adapts to the new environment. When cells are inoculated into fresh medium they may be deficient in essential enzymes, vitamins or cofactors, etc., that must be synthesized in order to utilize available nutrients, prior to cell division taking place.

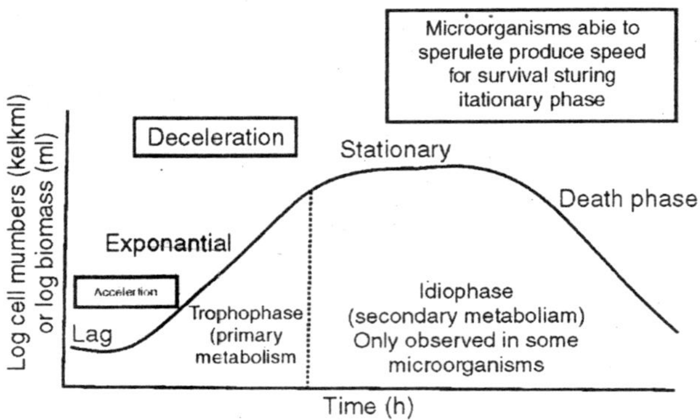

The chemical composition of the fermentation media influences the length of the lag phase. It is usually longer if the inoculum was grown up using a carbon source different from that in the fresh medium, because the cells must synthesize enzymes required to catabolize the new substrate. Physiological stress may also have an effect, especially as cells are often transferred from an inoculum medium of low osmotic pressure (low solute concentration) to fresh medium of higher osmotic pressure (high solute concentration). Other factors influencing the length of the lag phase are the age, concentration, viability and morphology of the inoculum.

Generally, inocula prepared from cells harvested in the exponential growth phase (period of most rapid growth) exhibit shorter lag phases than those harvested from subsequent stages. Once the cells have adapted to their new environment they enter the acceleration phase. Cell division occurs with increasing frequency until the maximum growth rate (mmax) for the specific conditions of the batch fermentation is reached.

At this point exponential growth begins and cell numbers/ biomass increase at a constant rate. Mathematically, this exponential growth can be described by two methods; one is related to biomass *(x)* and the other to cell numbers *(N)*. For cell biomass, growth can be considered as an autocatalytic reaction. Therefore, the rate of growth is dependent on the biomass concentration, i.e. catalyst, that is present at any given time. This can be described as follows: rate of change of biomass is $dx/dt = \mu x$ where x = concentration of biomass (g/ L), μ = specific growth rate (per hour) and t = time (h).

When a graph is plotted of cell biomass against time, the product is a curve with a constantly increasing slope.

Equation can also be rearranged to estimate the specific growth rate (μ): $m = 1/x * dx/dt$ During any period of true exponential growth, equation can be integrated to provide the following equation: $xt = xoemt$ where xt = biomass concentration after time' t, xo = biomass concentration at the start exponential growth, and e = base of the natural logarithm.

Taking natural logarithms, loge (In), gives

$\text{In } Xt = \text{In } X_0 + \mu t$

This equation is of the form $y = c$ (intercept on y axis) + mx where m = gradient, which is the general equation for a straight-line graph. For cells in exponential phase, a plot of natural log of biomass concentration against time, a semilog plot, should yield a straight line with the slope (gradient) equal to μ, or m = (In xt – In x_0)/t

(***Note:*** when plotting log10 values instead of the natural log, the gradient of the semilog plot is equal to m/2.303;).

A second approach is to examine growth in relation to cell number, where the number of cells at the start of exponential growth is No.

If we consider a hypothetical case where the microbial cell population at this point of exponential growth is one (No = 1), we can describe binary fission as follows:

No. of divisions 0, 1, 2, 3...n

No. of cells 1, 2, 4, 8... 2n

Mathematically No N 2 No 2 * 2 No 2 * 2 * 2–

No21 No22 No23 ... No2n

Consequently, after a period of exponential growth, time *(t)*, the number of cells *(Nt)* is given by

Nt = No 2n

where n=the number of divisions, No = initial cell number.

Taking natural logarithms gives

In Nt = lnNo + n_1n_2

Therefore, the number of divisions *(n)* that have taken place is given by In Nt-In No n = In_2

The number of divisions per unit time during this period of exponential growth is determined by dividing by the time period *(t)*:

$$n = \text{lnNt} - \text{lnNo}$$

$$t = tln_2$$

where n/t = division rate constant (average number of generations per hour).

Often, we are not really interested in the number of divisions per hour, unit time, but rather in the mean generation time or doubling time *(td)*, that is, the time required to undergo a single generation that doubles the population. Thus,

$$td = t = t\ \ln_2 n \text{ In } Nt - \text{In No}$$

During exponential growth, when all nutrients are supplied in excess and are therefore non-limiting, there is a direct relationship between cell numbers and biomass concentration, assuming that mean cell size is constant. This is balanced growth, and a direct relationship between specific growth rate and doubling time can also be established. However, under conditions where an essential nutrient becomes limiting, unbalanced growth arises, and variations in cell numbers *(N)* and biomass *(x)* concentration occur, as during the synthesis of cell storage compounds.

If we consider a situation where at time zero, the cell biomass is x_0, then after a fixed period of time *(t)* of exponential growth, equivalent to one doubling time *(td)*, the microbial biomass will double to $2x_0$, i.e. $xt = 2x_0$, when $t = td$, Substituting these parameters into equation gives $2x_0 = x_0e\ mt\ d$

Taking natural logarithms produces

In $2x_0$ = In $x_0 + mtd$

or

$$\mu td = \text{In } 2.$$

Therefore, in this case

$$\mu td = 0.693$$

Equation, rate of change of biomass *(dx/dt=*mx*)*, predicts that growth will occur indefinitely. However, during batch growth the microorganisms are continuously metabolizing the finite supply of nutrients available in the fermentation broth. After a certain time the growth rate decreases and eventually stops.

This cessation of growth can be due to depletion of essential nutrients (carbon source, essential amino acids, etc.) or the build-up of toxic metabolites, such as ethanol and lactic acid, or a combination of nutrient depletion and toxin accumulation.

Monod showed that growth rate is an approximate hyperbolic function of the concentration of the growth-limiting nutrient(s).

This impact of essential nutrient depletion on-growth can be described mathematically by the Monod equation, in a form similar to that used in biochemistry, where Michaelis-Menten kinetics define the rate of an enzymecatalysed reaction in relation to its substrate concentration.

$$\mu = \mu_{max} \frac{S}{Ks + S}$$

where μ_{max} = maximum specific growth (per hour) of the cells, i.e. when substrate concentration is not limiting; S = concentration of limiting substrate (*g/L*); *Ks* = saturation constant, concentration (*g/L*) of limiting nutrient enabling growth at half the maximum specific growth rate, i.e. *m* = 1/2 mmax and is a measure of the affinity of the cells for this nutrient.

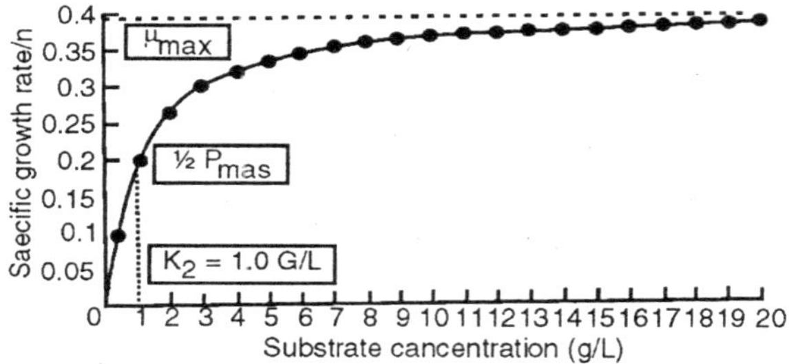

Fig. The Relationship Between Substrate Concentration and Specific Growth Rate.

When a microorganism is provided with the limiting substrate at a concentration much greater than the *Ks,* and with all other nutrients in excess, the microorganism will grow exponentially at its maximum rate, i.e. when S» *Ks,* then $\mu = \mu_{max}$. However, as the level of this substrate decreases, it eventually becomes limiting and can no longer sustain μ_{max}.

This is the beginning of the deceleration phase. As the residual concentration of the limiting substrate approaches *Ks* and then falls below this concentration, there is an accompanying gradual decrease in growth rate (μ). The growth rate of a microorganism with a very high affinity for a rate-limiting substrate (i.e. a low *Ks)* will not be affected until the substrate concentration becomes very low.

However, where there is a low affinity for the limiting substrate (i.e. a high *Ks),* the growth rate will begin to fall even at relatively high substrate concentrations and the organism exhibits a longer deceleration phase. The specific growth rate of the microorganism continues decelerating until all of the available limiting substrate is metabolized. Growth is no longer sustainable and the cells enter the stationary phase. At this point, the overall growth rate has declined to zero and there is no net change in cell numbers/biomass (rate of cell division equals rate of cell death).

However, the microorganisms are still metabolically active, involved in metabolizing intracellular storage compounds, utilizing nutrients released from lysed cells, and in some cases producing secondary metabolites. The duration of the stationary phase varies depending on the microorganisms involved and the environmental conditions. For cells unable to survive by forming spores, this is followed by an exponential death phase when the cells die at a constant rate and often undergo lysis.

APPLICATION OF BATCH FERMENTATIONS

During batch fermentations certain environmental conditions continually change, particularly nutrient and product concentrations, as does the specific growth rate, because the cells must pass through the sequence of growth phases described above. Consequently, the system never achieves steady-state conditions. A further disadvantage is that several distinct practical stages are associated with the operation of a batch fermentation:

- Charging of the fermenter with fresh medium;
- Sterilization of the fermenter and medium;
- Inoculation of the fermenter;
- Production of microbial products;
- Harvesting of biomass and spent fermentation broth; and finally
- Cleaning of the vessel.

This has major economic implications for industrial processes. For a considerable period of time, the fermentation vessel is not producing microbial products, but is being cleaned, filled, sterilized, etc. The non-productive period is referred to as the down-time of the fermenter.

OPTIMIZATION OF BATCH FERMENTATIONS

During the development of a batch process, key growth

parameters can be determined that enable the production of a given microbial product to be optimized, whether it is the biomass itself or a specific metabolite. An important parameter is the yield coefficient (Y), which is determined on the basis of the quantity of rare limiting nutrient, normally the carbohydrate source, converted into the microbial product. In respect of biomass, it is defined as

$$x = Y\ x/s(S - Sr)$$

where x = biomass concentration (g/L), $Y\ x/s$ = *yield* coefficient (g biomass/g substrate utilized), S = initial substrate concentration (g/L), and Sr = residual substrate concentration (g/L).

In the case of biomass production, the yield coefficient relates to the quantity of biomass produced per gram of substrate utilized. Therefore, the higher the yield coefficient, the greater the percentage of the original substrate converted into microbial biomass. For microbial metabolic products *(p)* the yield coefficient is related to the quantity of metabolite produced in relation to the quantity of substrate used $Y\ p/s$.

Determination of yield coefficients is vitally important because the cost of the fermentation medium, particularly the carbon source, can be a significant proportion of the overall production cost.

By performing a range of experiments under different operating conditions, varying medium constituents and component concentrations, pH, temperature, etc., optimum growth/ production conditions can be established.

It is also important to determine the maximum specific growth rate (m max) of the production organism. This is particularly true for primary metabolites, where product formation is related to growth. To optimize the overall productivity of the system, the microorganism must usually be grown at its mmax. As previously stated, the operating substrate concentration has a major effect on the growth rate of a microorganism.

By performing a series of batch fermentations, each with a different initial concentration of the limiting substrate, the specific growth rate (m) for each experiment can be determined. These data can then be used to estimate both mmax and the saturation constant *(Ks)* by simply taking the reciprocal values in the Monod equation and rearranging equation to give

$$1 = Ks + S$$
$$\mu = \mu_{max} S$$

or

$$1 = Ks + S$$
$$\mu = \mu_{max} S \ \mu_{max} S$$

and

$$1 = Ks * 1 + S$$
$$\mu = \mu_{max} S \ \mu_{max} S$$

A plot of 1/m against 1/*S* should produce a straight line with the intercept on the y-axis at $1/\mu_{max}$ and a gradient equal to K/μ_{max}. Therefore, an the key kinetic parameters can be readily determined and when the values of *K*, mmax and Y are known, a complete quantitative description can be given of the growth events occurring during a batch culture.

CONTINUOUS GROWTH KINETICS

Continuous culture fermentations have many applications for both laboratory research and industrial-scale processes. Studies can be performed on all aspects of cell growth, physiology and biochemistry. They are useful for ecological studies and as a genetics tool for the examination of mutation rates, mutagenic effects, etc. Application in industrial fermentations overcomes many limitations of batch processes. Initially, continuous fermentations start as batch cultures, but exponential growth can then be extended indefinitely, in theory, through the continuous addition of fresh fermentation medium.

The reactor is continuously stirred and a constant volume is maintained by incorporating an overflow weir or other levelling device. Fresh medium is continuously added and displaces an equal volume of spent fermentation broth and cells at the same rate as fresh medium is introduced. Steadystate conditions prevail, where the rate of microbial cell growth equals the rate at which the cells are displaced from the vessel. As with batch fermentations, the specific rate at, which the microorganism grows in continuous culture is controlled by the availability of the rate-limiting nutrient. Therefore, the rate of addition of fresh medium controls the rate at which the microorganisms grow.

However, the actual rate of growth depends not only on the volumetric flow rate of the medium into the reactor, but also on the dilution rate *(D)*.

This equals the number of reactor volumes passing through the reactor per unit time and is expressed in units of reciprocal time, per hour.

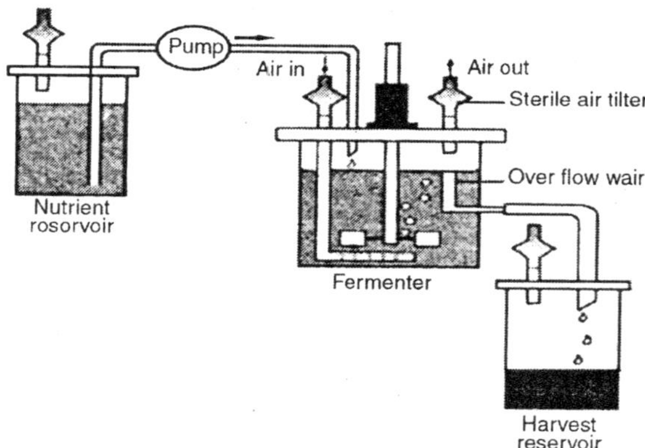

Fig. Continuous Culture Apparatus.

$$D = F \times V$$

where D = dilution rate (per hour), F = flow (L/h) and V = reactor volume (L).

The term D is the reciprocal of the mean residence time or hydraulic retention time, as used in waste-water treatment.Addition of fresh medium into the reactor can be controlled at a fixed value, therefore the rate of addition of the rate-limiting nutrient is constant.

Within certain limits, the growth rate and the rate of loss of cells from the fermenter will be determined by the rate of medium input.

Therefore, under steady state conditions the net biomass balance can be described as

dx = rate of growth-rate of loss from reactor
dt in reactor (wash-out)
or

$$dx/dt = \mu x - Dx$$

Under steady-state conditions the rate of growth = rate of loss, hence $dx/dt = 0$ and therefore

$$\mu x = Dx$$

and

$$m = D$$

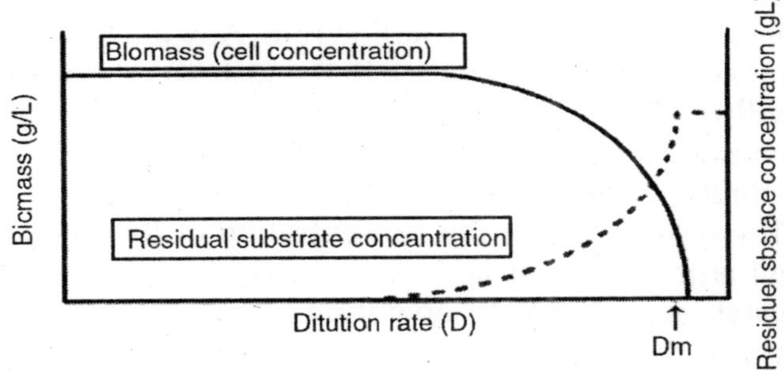

Fig. Growth of a Microorganism in Continuous Chemostat culture.

Consequently, at fixed flow rates and dilution rates under constant physical and chemical operating conditions, i.e. under steady-state conditions, the specific growth rate of the microorganism is dependent on the operating dilution rate, up to a maximum value equal to μ_{max}.

If the dilution rate is increased above μ_{max} complete wash-out of the cells occurs, as the cells have insufficient time to 'double' before being washed out of the reactor via the overflow.

The point at which this is just avoided is referred to as the critical dilution rate (Dcri). For any given dilution rate, under steady-state conditions, the residual substrate concentration in the reactor can be predicted by substituting D for /l in the Monod equation:

$$D = \mu_{max}\ Sr$$
$$Ks + Sr$$

where *Sr* is the steady-state residual substrate concent -ration in the reactor at a fixed dilution rate. Rearrangment gives

$D(Ks + Sr) = \mu_{max}\ Sr$ or dividing by *Sr* then gives

$$D\ Ks + D\ Sr = \mu_{max}\ Sr$$

$$D\ Ks + D = \mu_{max}\ Sr$$

Hence,

$$Sr = DKs$$

$$\mu_{max} - D$$

Consequently, the residual substrate concentration in the reactor is controlled by the dilution rate. Any alteration to this dilution rate results in a change in the growth rate of the cells that will be dependent on substrate availability at the new dilution rate. Thus, growth is controlled by the availability of a rate-limiting nutrient.

This system, where the concentration of the ratelimiting nutrient entering the system is fixed, is often described as a chemostat, as opposed to operation as a turbidostat, where nutrients in the medium are not limiting. In this case turbidity or absorbance of the culture is monitored and maintained at a constant value by regulating the dilution rate, i.e. cell concentration is held constant.

At low dilution rates with fixed substrate concentrations, the residual substrate concentration will be low. However, as *D* approaches μ_{max} the residual substrate concentration increases along with the growth rate of the microorganism. Beyond Dcrit, input substrate concentration will equal output concentration, as all the cells have been lost from the system. Thus, this continuous reactor can be described as a selfregulating nutrient-limited chemostat.

The concentration of biomass or microbial metabolite in a continuous fermenter under steady-state conditions can be related to the yield coefficient, as described in the batch fermentation section.

Inserting the equation for residual substrate into the biomass or metabolic product yield coefficient equation gives, in this case for steadystate biomass *(x)*,

$$-x = Yx/s\ SR - DKs$$

$$\mu_{max} - D$$

where *SR* is the substrate concentration of inflowing medium

or

$$-x = Yx/S(SR\text{-}Sr)$$

Therefore, the biomass concentration under steady state conditions is controlled by the substrate feed concentration and the operating dilution rate. Under non-inhibitory conditions, where there is no substrate or product inhibition, the higher the feed concentration, the greater the biomass concentration and residual substrate concentration remains constant.

However, the higher the dilution rate, the faster the cells grow, which results in a simultaneous increase in the residual substrate concentration and a consequent reduction in the steady-state biomass concentration.

As D approaches μ_{max}, the biomass concentration becomes even lower, yet the cells grow faster and there is a concurrent increase in the residual substrate concentration.

Chapter 9

Bioprocess Based Products

ANTIBIOTICS

Antibiotics are chemical substances that can inhibit the growth of, and even destroy, harmful microorganisms. They are derived from special microorganisms or other living systems, and are produced on an industrial scale using a fermentation process.

Although the principles of antibiotic action were not discovered until the twentieth century, the first known use of antibiotics was by the Chinese over 2,500 years ago. Today, over 10,000 antibiotic substances have been reported. Currently, antibiotics represent a multibillion dollar industry that continues to grow each year.

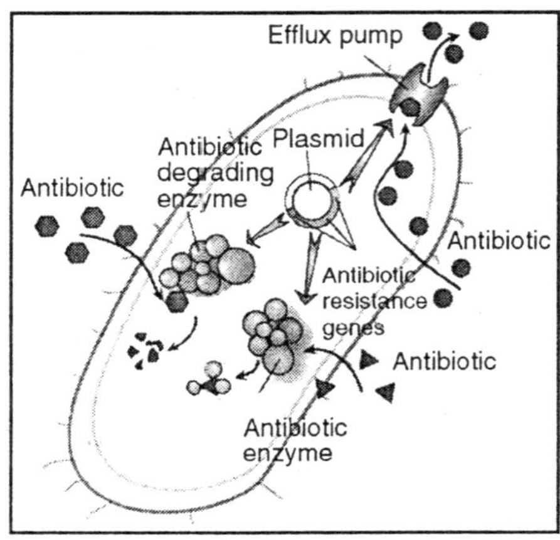

Fig. Antibiotics

Antibiotics are used in many forms—each of which imposes somewhat different manufacturing requirements. For bacterial infections on the skin surface, eye, or ear, an antibiotic may be applied as an ointment or cream. If the infection is internal, the antibiotic can be swallowed or injected directly into the body. In these cases, the antibiotic is delivered throughout the body by absorption into the bloodstream.

Antibiotics differ chemically so it is under-standable that they also differ in the types of infections they cure and the ways in which they cure them. Certain antibiotics destroy bacteria by affecting the structure of their cells. This can occur in one of two ways. First, the antibiotic can weaken the cell walls of the infectious bacteria, which causes them to burst. Second, antibiotics can cause the contents of the bacterial cells to leak out by damaging the cell membranes.

Another way in which antibiotics function is by interfering with the bacteria's metabolism. Some antibiotics such as tetracycline and erythromycin interfere with protein synthesis. Antibiotics like rifampin inhibit nucleic acid biosynthesis. Still other antibiotics, such as sulfonamide or trimethoprim have a general blocking effect on cell metabolism.

The commercial development of an antibiotic is a long and costly proposal. It begins with basic research designed to identify organisms, which produce antibiotic compounds. During this phase, thousands of species are screened for any sign of antibacterial action. When one is found, the species is tested against a variety of known infectious bacteria. If the results are promising, the organism is grown on a large scale so the compound responsible for the antibiotic effect can be isolated.

This is a complex procedure because thousands of antibiotic materials have already been discovered. Often, scientists find that their new antibiotics are not unique. If the material passes this phase, further testing can be done. This typically involves clinical testing to prove that the antibiotic works in animals and humans and is not harmful. If these tests are passed, the Food and Drug Administration (FDA) must then approve the antibiotic as a new drug. This whole process can take many years. The large-scale production of an antibiotic depends on a fermentation process. During fermentation, large amounts of the antibiotic-producing organism are grown. During fermentation, the organisms produce the antibiotic material, which can then be isolated for use as a drug. For a new antibiotic to be economically feasible, manufacturers must be able to get a high yield of drug from the fermentation process, and be able to easily isolate it. Extensive research is usually required before a new antibiotic can be commercially scaled up.

While our scientific knowledge of antibiotics has only recently been developed, the practical application of antibiotics has existed for centuries. The first known use was by the Chinese about 2,500 years ago. During this time, they discovered that applying the moldy curd of soybeans to infections had certain therapeutic benefits. It was so effective that it became a standard treatment.

Evidence suggests that other cultures used antibiotic-type substances as therapeutic agents. The Sudanese-Nubian civilization used a type of tetracycline antibiotic as early as 350 A.D. In Europe during the Middle Ages, crude plant extracts and cheese curds were also used to fight infection. Although these cultures used antibiotics, the general principles of antibiotic action were not understood until the twentieth century.

The development of modern antibiotics depended on a few key individuals who demonstrated to the world that materials derived from microorganisms could be used to cure infectious diseases. One of the first pioneers in this field was Louis Pasteur. In 1877, he and an associate discovered that the growth of disease-causing anthrax bacteria could be inhibited by a saprophytic bacteria.

They showed that large amounts of anthrax bacilli could be given to animals with no adverse affects as long as the saprophytic bacilli were also given. Over the next few years, other observations supported the fact that some bacterially derived materials could prevent the growth of disease-causing bacteria.

In 1928, Alexander Fleming made one of the most important contributions to the field of antibiotics. In an experiment, he found that a strain of green *Penicillium* mold inhibited the growth of bacteria on an agar plate. This led to the development of the first modern era antibiotic, penicillin.

A few years later in 1932, a paper was published which suggested a method for treating infected wounds using a penicillin preparation. Although these early samples of penicillin were functional, they were not reliable and further refinements were needed.

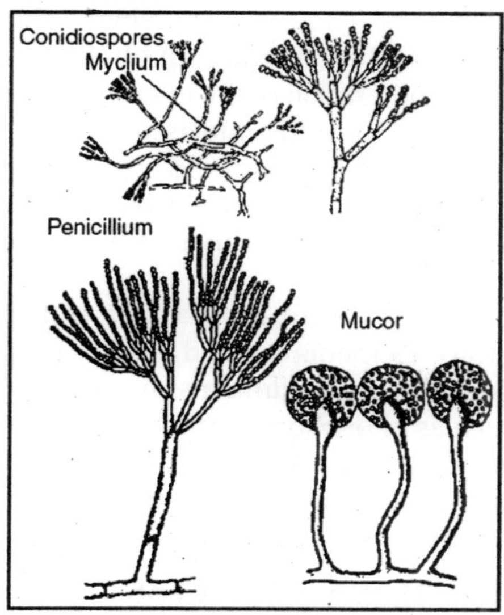

Fig. Penicillium

These improvements came in the early 1940s when Howard Florey and associates discovered a new strain of *Penicillium,* which produced high yields of penicillin. This allowed large-scale production of penicillin, which helped launch the modern antibiotics industry. After the discovery of penicillin, other antibiotics were sought. In 1939, work began on the isolation of potential antibiotic products from the soil bacteria streptomyces. It was around this time that the term antibiotic was introduced. Selman Waxman and associates discovered streptomycin in 1944.

Subsequent studies resulted in the discovery of a host of new, different antibiotics including actinomycin, streptothricin, and neomycin all produced by *Streptomyces*. Other antibiotics that have been discovered since include bacitracin, polymyxin, viomycin, chloramphenicol and tetracyclines. Since the 1970s, most new antibiotics have been synthetic modifications of naturally occurring antibiotics.

RAW MATERIALS

The compounds that make the fermentation broth are the primary raw materials required for antibiotic production. This broth is an aqueous solution made up of all of the ingredients necessary for the proliferation of the microorganisms. Typically, it contains a carbon source like molasses, or soy meal, both of which are made up of lactose and glucose sugars. These materials are needed as a food source for the organisms. Nitrogen is another necessary compound in the metabolic cycles of the organisms.

For this reason, an ammonia salt is typically used. Additionally, trace elements needed for the proper growth of the antibiotic-producing organisms are included. These are components such as phosphorus, sulfur, magnesium, zinc, iron, and copper introduced through water soluble salts. To prevent foaming during fermentation, anti-foaming agents such as lard oil, octadecanol, and silicones are used.

THE MANUFACTURING PROCESS

Although most antibiotics occur in nature, they are not normally available in the quantities necessary for large-scale production. For this reason, a fermentation process was developed. It involves isolating a desired microorganism, fueling growth of the culture and refining and isolating the final antibiotic product. It is important that sterile conditions be maintained throughout the manufacturing process, because contamination by foreign microbes will ruin the fermentation.

Starting the Culture

Before fermentation can begin, the desired antibiotic-producing organism must be isolated and its numbers must be increased by many times.

To do this, a starter culture from a sample of previously isolated, cold-stored organisms is created in the lab. In order to grow the initial culture, a sample of the organism is transferred to an agar-containing plate. The initial culture is then put into shake flasks along with food and other nutrients necessary for growth. This creates a suspension, which can be transferred to seed tanks for further growth.

The seed tanks are steel tanks designed to provide an ideal environment for growing microorganisms.

They are filled with the all the things the specific microorganism would need to survive and thrive, including warm water and carbohydrate foods like lactose or glucose sugars. Additionally, they contain other necessary carbon sources, such as acetic acid, alcohols, or hydrocarbons, and nitrogen sources like ammonia salts.

Growth factors like vitamins, amino acids, and minor nutrients round out the composition of the seed tank contents. The seed tanks are equipped with mixers, which keep the growth medium moving, and a pump to deliver sterilized, filtered air. After about 24-28 hours, the material in the seed tanks is transferred to the primary fermentation tanks.

Fermentation

The fermentation tank is essentially a larger version of the steel, seed tank, which is able to hold about 30,000 gallons. It is filled with the same growth media found in the seed tank and also provides an environment inducive to growth. Here the microorganisms are allowed to grow and multiply. During this process, they excrete large quantities of the desired antibiotic. The tanks are cooled to keep the temperature between 73-81° F (23-27.2°C).

It is constantly agitated, and a continuous stream of sterilized air is pumped into it. For this reason, anti-foaming agents are periodically added. Since pH control is vital for optimal growth, acids or bases are added to the tank as necessary.

Isolation and Purification

After three to five days, the maximum amount of antibiotic will have been produced and the isolation process can begin. Depending on the specific antibiotic produced, the fermentation broth is processed by various purification methods. For example, for antibiotic compounds that are water soluble, an ion-exchange method may be used for purification. In this method, the compound is first separated from the waste organic materials in the broth and then sent through equipment, which separates the other water-soluble compounds from the desired one.

To isolate an oil-soluble antibiotic such as penicillin, a solvent extraction method is used. In this method, the broth is treated with organic solvents such as butyl acetate or methyl isobutyl ketone, which can specifically dissolve the antibiotic. The dissolved antibiotic is then recovered using various organic chemical means. At the end of this step, the manufacturer is typically left with a purified powdered form of the antibiotic, which can be further refined into different product types.

Refining

Antibiotic products can take on many different forms. They can be sold in solutions for intravenous bags or syringes, in pill or gel capsule form, or they may be sold as powders, which are incorporated into topical ointments. Depending on the final form of the antibiotic, various refining steps may be taken after the initial isolation.

For intravenous bags, the crystalline antibiotic can be dissolved in a solution, put in the bag, which is then hermetically sealed. For gel capsules, the powdered antibiotic is physically filled into the bottom half of a capsule then the top half is mechanically put in place. When used in topical ointments, the antibiotic is mixed into the ointment.

From this point, the antibiotic product is transported to the final packaging stations. Here, the products are stacked and put in boxes. They are loaded up on trucks and transported to various distributors, hospitals, and pharmacies.

The entire process of fermentation, recovery, and processing can take anywhere from five to eight days.

QUALITY CONTROL

Quality control is of utmost importance in the production of antibiotics. Since it involves a fermentation process, steps must be taken to ensure that absolutely no contamination is introduced at any point during production. To this end, the medium and all of the processing equipment are thoroughly steam sterilized.

During manufacturing, the quality of all the compounds is checked on a regular basis. Of particular importance are frequent checks of the condition of the microorganism culture during fermentation. These are accomplished using various chromatography techniques. Also, various physical and chemical properties of the finished product are checked such as pH, melting point, and moisture content.

In the United States, antibiotic production is highly regulated by the Food and Drug Administration (FDA). Depending on the application and type of antibiotic, more or less testing must be completed. For example, the FDA requires that for certain antibiotics each batch must be checked by them for effectiveness and purity. Only after they have certified the batch can it be sold for general consumption.

Since the development of a new drug is a costly proposition, pharmaceutical companies have done very little research in the last decade. However, an alarming development has spurred a revived interest in the development of new antibiotics. It turns out that some of the disease-causing bacteria have mutated and developed a resistance to many of the standard antibiotics.

This could have grave consequences on the world's public health unless new antibiotics are discovered or improvements are made on the ones that are available. This challenging problem will be the focus of research for many years to come.

PRODUCTION OF ANTIBIOTICS

The mass production of antibiotics began during World

War II with streptomycin and penicillin. Now most antibiotics are produced by staged fermentations in which strains of microorganisms producing high yields are grown under optimum conditions in nutrient media in fermentation tanks holding several thousand gallons. The mold is strained out of the fermentation broth, and then the antibiotic is removed from the broth by filtration, precipitation, and other separation methods.

In some cases new antibiotics are laboratory synthesized, while many antibiotics are produced by chemically modifying natural substances; many such derivatives are more effective than the natural substances against infecting organisms or are better absorbed by the body, e.g., some semisynthetic penicillins are effective against bacteria resistant to the parent substance. Antibiotic, any of a variety of substances, usually obtained from microorganisms, that inhibit the growth of or destroy certain other microorganisms.

TYPES OF ANTIBIOTICS

The great number of diverse antibiotics currently available can be classified in different ways, e.g., by their chemical structure, their microbial origin, or their mode of action. They are also frequently designated by their effective range.

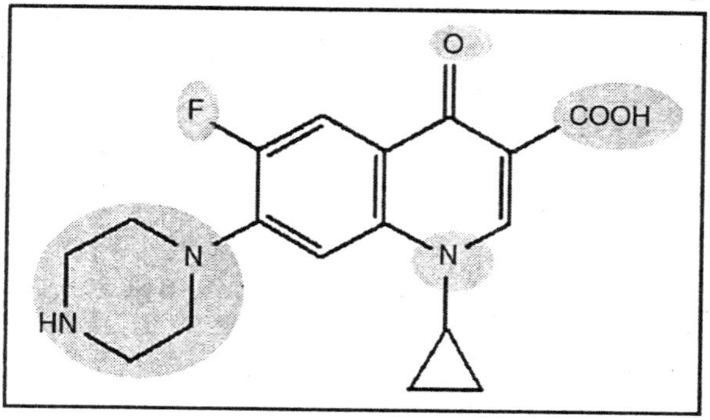

O
F
COOH
N
N
HN

Fig. Ciprofloxacin

Tetracyclines, the most widely used broad-spectrum antibiotics, are effective against both Gram-positive and Gram-negative bacteria, as well as against rickettsias and psittacosis-causing organisms. Ciprofloxacin (Cipro) is another broad-spectrum antibiotic, effective in the treatment of mild infections of the urinary tract and sinuses. The medium-spectrum antibiotics bacitracin, the erythromycins, penicillin, and the cephalosporins are effective primarily against Gram-positive bacteria, although the streptomycin group is effective against some Gram-negative and Gram-positive bacteria. Polymixins are narrow-spectrum antibiotics effective against only a few species of bacteria.

Antibiotics are either injected, given orally, or applied to the skin in ointment form. Many, while potent anti-infective agents, also cause toxic side effects. Some, like penicillin, are highly allergenic and can cause skin rashes, shock, and other manifestations of allergic sensitivity. Others, such as the tetracyclines, cause major changes in the intestinal bacterial population and can result in superinfection by fungi and other microorganisms. Chloramphenicol, which is now restricted in use, produces severe blood diseases, and use of streptomycin can result in ear and kidney damage.

Many antibiotics are less effective than formerly because antibiotic-resistant strains of microorganisms have emerged. Antibiotics have found wide nonmedical use. Some are used in animal husbandry, along with vitamin B_{12}, to enhance the weight gain of livestock.

However, some authorities believe the addition of antibiotics to animal feeds is dangerous because continuous low exposure to the antibiotic can sensitize humans to the drug and make them unable to take the substance later for the treatment of infection. In addition low levels of antibiotics in animal feed encourage the emergence of antibiotic-resistant strains of microorganisms.

Drug resistance has been shown to be carried by a genetic particle transmissible from one strain of microorganism to another, and the presence of low levels of antibiotics can actually cause an increase in the number of such particles in the bacterial population and increase the probability that such particles will be transferred to pathogenic, or disease-causing, strains. Antibiotics have also been used to treat plant diseases such as bacteria-caused infections in tomatoes, potatoes, and fruit trees. The substances are also used in experimental research. Although for centuries preparations derived from living matter were applied to wounds to destroy infection, the fact that a microorganism is capable of destroying one of another species was not established until the latter half of the 19th cent. when Pasteur noted the antagonistic effect of other bacteria on the anthrax organism and pointed out that this action might be put to therapeutic use.

Meanwhile the German chemist Paul Ehrlich developed the idea of selective toxicity: that certain chemicals that would be toxic to some organisms, e.g., infectious bacteria, would be harmless to other organisms, e.g., humans. In 1928, Sir Alexander Fleming, a Scottish biologist, observed that *Penicillium notatum*, a common mold, had destroyed staphylococcus bacteria in culture, and in 1939 the American microbiologist René Dubos demonstrated that a soil bacterium was capable of decomposing the starchlike capsule of the pneumococcus bacterium, without which the pneumococcus is harmless and does not cause pneumonia.

Dubos then found in the soil a microbe, *Bacillus brevis*, from which he obtained a product, tyrothricin, that was highly toxic to a wide range of bacteria. Tyrothricin, a mixture of the two peptides gramicidin and tyrocidine, was also found to be toxic to red blood and reproductive cells in humans but could be used to good effect when applied as an ointment on body surfaces. Penicillin was finally isolated in 1939, and in 1944 Selman Waksman and Albert Schatz, American microbiologists, isolated streptomycin and a number of other antibiotics from *Streptomyces griseus*.

ETHANOL

```
    H    H
    |    |
H— C — C —OH
    |    |
    H    H
```

Fig. Ethanol

Ethanol fermentation is the biological process by which sugars such as glucose, fructose, and sucrose are converted into cellular energy and thereby producing ethanol and carbon dioxide as metabolic waste products. Yeasts carry out ethanol fermentation on sugars in the absence of oxygen. Because the process does not require oxygen, ethanol fermentation is classified as anaerobic. Ethanol fermentation is responsible for the rising of bread dough, the production of ethanol in alcoholic beverages, and for much of the production of ethanol for use as fuel.

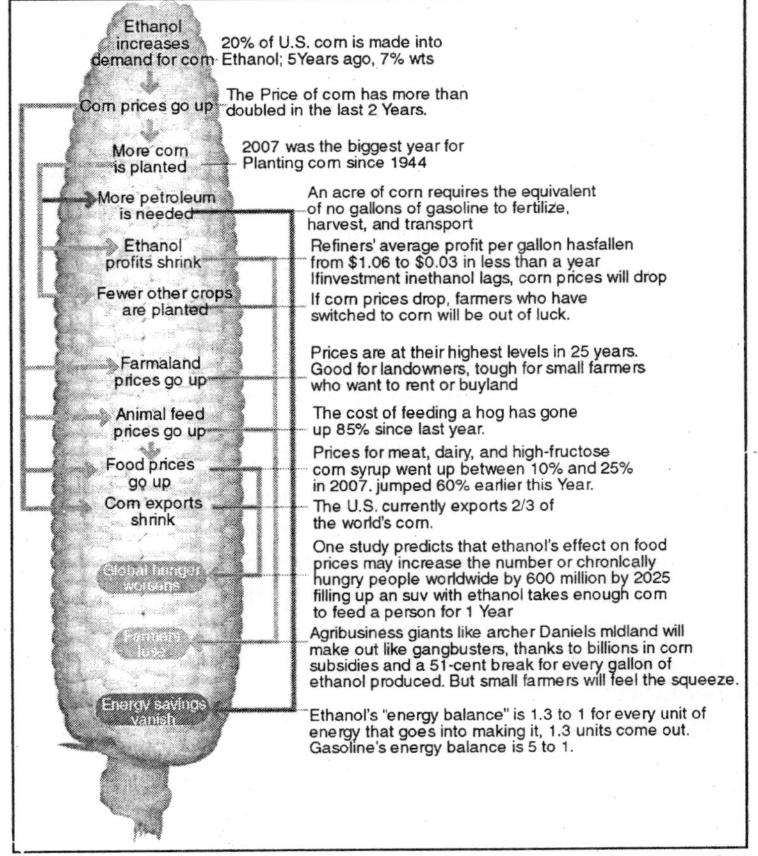

Fig. Ethanol or Grain Alcohol

Ethanol (ethyl alcohol, grain alcohol) is a clear, colorless liquid with a characteristic, agreeable odor. In dilute aqueous solution, it has a somewhat sweet flavour, but in more concentrated solutions it has a burning taste. Ethanol, CH_3CH_2OH, is an alcohol, a group of chemical compounds whose molecules contain a hydroxyl group, –OH, bonded to a carbon atom. The word *alcohol* derives from Arabic *al-kuhul*, which denotes a fine powder of antimony used as an eye makeup. *Alcohol* originally referred to any fine powder, but medieval alchemists later applied the term to the refined products of distillation, and this led to the current usage.

Ethanol melts at –114.1°C, boils at 78.5°C, and has a density of 0.789 g/mL at 20°C. Its low freezing point has made it useful as the fluid in thermometers for temperatures below –40°C, the freezing point of mercury, and for other low-temperature purposes, such as for antifreeze in automobile radiators.

Ethanol has been made since ancient times by the fermentation of sugars. All beverage ethanol and more than half of industrial ethanol is still made by this process. Simple sugars are the raw material. Zymase, an enzyme from yeast, changes the simple sugars into ethanol and carbon dioxide. The fermentation reaction, represented by the simple equation

$$C_6H_{12}O_6 \rightarrow 2CH_3CH_2OH + 2CO_2$$

Is actually very complex, and impure cultures of yeast produce varying amounts of other substances, including glycerine and various organic acids. In the production of beverages, such as whiskey and brandy, the impurities supply the flavour. Starches from potatoes, corn, wheat, and other plants can also be used in the production of ethanol by fermentation.

However, the starches must first be broken down into simple sugars. An enzyme released by germinating barley, diastase, converts starches into sugars. Thus, the germination of barley, called malting, is the first step in brewing beer from starchy plants, such as corn and wheat. The ethanol produced by fermentation ranges in concentration from a few per cent up to about 14 per cent. Above about 14 per cent, ethanol destroys the zymase enzyme and fermentation stops. Ethanol is normally concentrated by distillation of aqueous solutions, but the composition of the vapour from aqueous ethanol is 96 per cent ethanol and 4 per cent water. Therefore, pure ethanol cannot be obtained by distillation. Commercial ethanol contains 95 per cent by volume of ethanol and 5 per cent of water. Dehydrating agents can be used to remove the remaining water and produce absolute ethanol.

Much ethanol not intended for drinking is now made synthetically, either from acetaldehyde made from acetylene, or from ethylene made from petroleum. Ethanol can be oxidized to form first acetaldehyde and then acetic acid. It can be dehydrated to form ether. Butadiene, used in making synthetic rubber, may be made from ethanol, as can chloroform and many other organic chemicals. Ethanol is used as an automotive fuel by itself and can be mixed with gasoline to form gasohol.

Ethanol is miscible (mixable) in all proportions with water and with most organic solvents. It is useful as a solvent for many substances and in making perfumes, paints, lacquer, and explosives. Alcoholic solutions of nonvolatile substances are called tinctures; if the solute is volatile, the solution is called a spirit. Most industrial ethanol is denatured to prevent its use as a beverage. Denatured ethanol contains small amounts, 1 or 2 per cent each, of several different unpleasant or poisonous substances. The removal of all these substances would involve a series of treatments more expensive than the federal excise tax on alcoholic beverages.

These denaturants render ethanol unfit for some industrial uses. In such industries undenatured ethanol is used under close federal supervision. When an alcoholic beverage is swallowed, it passes through the stomach into the small intestine where the ethanol is rapidly absorbed and distributed throughout the body.

The ethanol enters body tissues in proportion to their water content. Therefore, more ethanol is found in the blood and the brain than in muscle or fat tissue. The ethanol is greatly diluted by body fluids. For example, a 1-ounce shot of 100-proof whiskey, which contains 0.5 fluid ounces of ethanol (about 15 mL), is diluted 5000-fold in a 150-pound human, producing a 0.02 per cent blood alcohol concentration.

The Chemical Process of Fermentation

The chemical equation below summarizes ethanol fermentation, in which one hexose molecule is converted into two ethanol molecules and two carbon dioxide molecules:

$$C_6H_{12}O_6 \rightarrow 2C_2H_5OH + 2CO_2$$

The process begins with a molecule of glucose being broken down by the process of glycolysis into pyruvate:

$$C_6H_{12}O_6 \rightarrow 2CH_3COCOO^- + 2H^+$$

This reaction is accompanied by the reduction of two molecules of NAD^+ to NADH and a net of two ADP molecules converted to two ATP plus the two water molecules.

Pyruvate is then converted to acetaldehyde and carbon dioxide. The acetaldehyde is subsequently reduced to ethanol by the NADH from the previous glycolysis, which is returned to NAD^+:

$$CH_3COCOO^- + H^+ \rightarrow CH_3CHO + CO_2$$
$$CH_3CHO + NADH \rightarrow C_2H_5OH + NAD^+$$

Yeast will perform the above two reactions only if oxygen is excluded from the environment. Otherwise yeast will oxidize pyruvate completely to carbon dioxide and water.

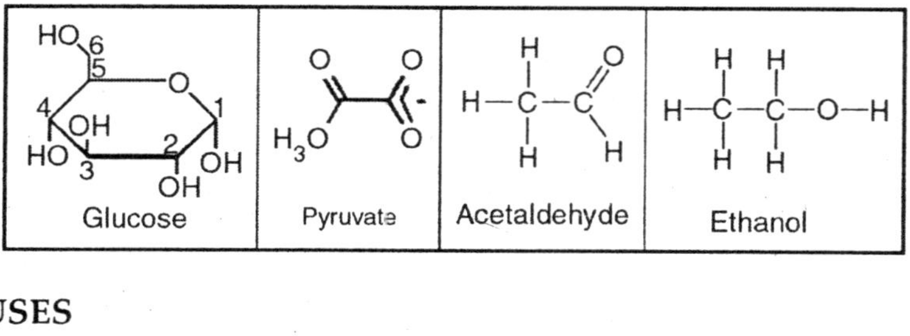

Glucose	Pyruvate	Acetaldehyde	Ethanol

USES

Ethanol fermentation is responsible for the rising of bread dough. Yeast organisms consume sugars in the dough and produce ethanol and carbon dioxide as waste products. The carbon dioxide forms bubbles in the dough, expanding it into something of a foam. Nearly all the ethanol evaporates from the dough when the bread is baked.

Fig. Alcoholic Beverages

The production of all alcoholic beverages employs ethanol fermentation by yeast. Wines and brandies are produced by fermentation of the natural sugars present in fruits, especially grapes. Beers, ales, and whiskeys employ fermentation of grain starches that have been converted to sugar by the application of the enzyme, amylase, which is present in grain kernels that have been germinated.

Amylase-treated grain or amylase-treated potatoes are fermented for the production of vodka. Fermentation of cane sugar is the first step in producing rum. In all cases, the fermentation must take place in a vessel that is arranged to allow carbon dioxide to escape, but that prevents outside air from coming in, as exposure to oxygen would prevent the formation of ethanol.

Similar yeast fermentation of various carbohydrate products is used to produce much of the ethanol used for fuel.

SINGLE CELL PROTEIN

Single-cell proteins (SCP) refer to the dried cells of microorganisms. SCP's are used as protein sources in human foods or animal feeds. Many raw materials have been considered as carbon and energy sources for SCP production. In many cases, these raw materials have been hydrolyzed by physical, chemical and enzymatic methods before use. Ram horns are significant waste products of the meat industry in Turkey.

For example, the slaughterhouse in Erzurum directly

discharges about 25 tons a year. Fibrous proteins such as horn, feather, nail and hair are also abundant waste products. These waste products can be converted to biomass, protein concentrate or amino acids using proteases derived from certain microorganisms.

Horns consist of a-keratin, which has a very high content of cysteine and they also contain most of the common amino acids. Horns have the components of bone and blood tissues and are rich in some growth factors required by microorganisms. The main aim of this study was to investigate the suitability of horn hydrolysate as a substrate in SCP production.

SCP is the name given to a variety of microbial products, that are produced by fermentation. When properly produced, this materials make satisfactory proteinaceous ingredients for animal feed or human food. The production of protein from hydrocarbon wastes of the petroleum industry is the most recent microbiological industry.

Yeast, fungi, bacteria, and algae are grown on hydrocarbon wastes, and cells are harvested as sources of protein. It has been calculated that 100 lbs of yeast will produce 250 tons of proteins in 24 hours, whereas a 1000 lbs steer will synthesize only 1 lb of protein 24 hours and this after consuming 12 to 20 lbs of plant proteins. Similar, algae grown in ponds can produce 20 tons (dry weight) of protein, per acre, per year.

This yield is 10 to 15 times higher than soybean and 25 to 50 times higher than corn. There are both advantages and disadvantages in using microorganisms for animal or human consumption. Bacteria are usually high in protein (50 to 80 per cent) and have a rapid growth rate. The principal disadvantages are as follows:

- Bacterial cells have small size and low density, which makes harvesting from the fermented medium difficult and costly.
- Bacterial cells have high nucleic acid content relative to yeast and fungi This can be detrimental to human beings, tending to increase the uric acid level in

blood. This may cause uric acid poising or gout. To decrease the nucleic acid level additional processing step has to be introduced, and this increases the cost.

- The general public thinking is that all bacteria are harmful and produce disease. An extensive education programme is1required to remove this misconception and to make the public accept bacterial protein.

Yeasts have as advantages their larger size (easier to harvest), lower nucleic add content, high lysine content and ability to grow at acid pH. However the most important advantage is familiarity and acceptability because of the long history of its use in traditional fermentations. Disadvantages include lower growth rates, lower protein content (45 to 65 per cent), and lower methionine content than in bacteria.

Filamentous fungi have advantages.in ease of harvesting, but have their limitations in lower growth rates, lower protein content, and acceptability. Algae have disadvantages of having cellulosic cell walls which are not digested by human beings. Secondly, they also concentrate heavy metals.

Single cell protein basically comprises proteins, fats carbohydrates, ash ingredients, water, and other elements such as phosphorus and Potassium. The composition depends upon the organism and the substrate which it grows. some typical compositions which are compared with soymeal and fish meal. If SCP is to be used successfully, there are five main criteria to be satisfied;

- The SCP must be safe to eat.
- The nutritional value dependent on the amino acid composition must be high.
- It must be acceptable to the general public.
- It must have the functionality, i.e. characteristics, which are found in common staple foods.
- The economic viability of the SCP process is extremely complex and is yet to be demonstrated

Chapter 10

Biomass and Single Cell Protein

Single Cell Protein must be competitive with commercial animal and plant proteins, in terms of proce and nutritional value and must conform with human and animal food safety requirements. Productivity, yield and selling price are the major factors affecting the economics of SCP production. Microbial inoculants, which are used as a process aid, generally have a higher value.

In this case, the objective of the production process is to optimiseyield of viable cells of defined biological activity with good shelf life characteristics.

Saccharomyces cerevisiae, is categorised primarily as a microbial inoculant. Inactive dried brewer's or baker's yeast is also used as a dietary source of vitamins and trace minerals in specific medical conditions. Considerable amounts of yeast extract are produced from bakers yeast as a source of flavour and vitamins.

Microbial cells are produced for two main applications:

- As a source of protein for animal or human food,
- For use as a commercial inoculum in food fermentations and for agriculture and waste treatment.

SINGLE CELL PROTEIN PRODUCTION

Product Safety and Quality

SCP has applications in animal feed, human food and as functional protein concentrates. Some bacterial SCP have amino acid profiles similar to animal/plant protein. Yeast, fungal and soya bean proteins tend to be deficient in methionine. Ingestion of RNA from non-conventional sources should be limited to 50g per day.

Ingestion of purine compounds arising from RNA breakdown, leads to increased plasma levels of uric acid, which can cause gout and kidney stones.

High content of nucleic acids causes no problems to animals since uric acid is converted to allantoin which is readily excreted in urine. Nucleic acid removal is not necessary from animal feeds but is from human foods. A temperature hold at 64C inactivates fungal proteases and allows RNA-ases to hydrolyse RNA with release of nucleotides from cell to culture broth. A 30 min stand at 64C reduces intracellular RNA levels in Fusarium graminearum from 80 mg/g to 2 mg/g.

Production

Large scale fermenters are required High biomass productivity requires high oxygen transfer rates which promotes high respiration rates which in turn increase metabolic heat production and the need for an efficient cooling system. In order to maximise fermentation productivity it is essential to operate continuous fermentation processes. Different processes have adopted different fermenter designs with respect to process requirements.

BP Process – Candida on *n*-paraffin

Mechanically agitated fully baffled fermenters with turbine mixers. Reactor feed consists of paraffin, gaseous ammonia and other salts. Oxygen requirements per unit biomass produced by aerobic micro-organisms grown on n-hexadecane is 2.5 times higher rhan that required for growth on glucose and amounts to 2.2 kg O_2 per kg biomass. Heat produced was 25000 kJ/kg biomass. Fermenters require substantial agitation. Because of the insoluble nature of alkanes, they exist in agiteated fermenter as suspensions of alkane drops 1-100 mm in diameter. Cell recovery is by centrifugation, producing 15 per cent dry solids, evaporation to 25 per cent dry solids and spray drying.

ICI Process – Methylophilus Methylotrophus from Methanol

Pressure recycle fermentor. A combination of an airlift and loop reactor consisting of an airlift column, a down-flow tube with heat removal and a gas-release space. The nitrogen source is ammonia gas and pH is controlled between 6 and 7. Cell specific growth rate is approximately 0.5h-1 and cell yield of 0.5 g/g. Methanol is oxidised via dehydrogenation to formaldehyde which can either be assimilated for conversion to cell mass or further oxidised to CO_2 with concomitant energy production. Cells are recovered by agglomeration followed by centrifugation, flash dried and ground.

Bel Fromageries Process: Kluyveromyces Marxianus from whey

Whey which contains about 5 per cent lactose, 0.8 per cent protein and 0.2-0.6 per cent lactic acid, is used as a substrate. Biomass production requires an aerobic fermentation whereas aeration is minimal for ethanol production. For feed grade biomass, the entire fermentation contents, containing yeast, residual whey proteins, minerals and lactic acid may be recovered. For preparation of food grade material, cells are harvested by centrifugation, washed and dried. Cell yield is 0.45-0.55 g/g based on lactose consumed.

RHM Mycoprotein Process: Fusarium Graminearum from Glucose

Medium components include food grade glucose syrup, gaseous ammonia, salts, and biotin. Fermentation pH is controlled at 6 by gaseous ammonia addition, fed into the air inlet stream. Cell concentrations are 15-20g/L and a specific growth rate of up to 0.2h-1 is achieved. Following cyclone separation and an RNA reduction step, cells are recovered by rotary vacuum filtration and formulated into a range of products.

POTENTIAL SUBSTRATES FOR SCP PRODUCTION SULPHITE WASTE LIQUOR

Candida utilis has been produced as a protein supplement

by fermentation of sulphite waste liquor in Germany during both world wars. More recently a Finnish company developed a fungal SCP production process, the Pekilo Process, to grow *Paecilomyces varioti* using sulphite waste liquor.

CELLULOSE

Cellulose from natural sources and waste wood is an attractive starting material for SCP production because of its abundance. The association of cellulose with lignin in wood makes it somewhat intractable to microbial degradation. Thermal or chemical pretreatment, used in combination with enzymatic hydrolysis, is usually required. Systems using cellulolytic organisms appear to have promise, but economic viability has yet to be achieved.

WHEY

Whole milk whey or deproteinised whey is a carbohydrate source, which creates disposal problems. (High BOD) Problems associated with whey for SCP production are usually insifficient substrate, seasonal supply variations and its high water content (> 90 per cent) which makes transport prohibitively expensive. While most organisms do not grow on lactose as a carbon source, strains of the yeast Kluyveromyces marxianus readily grow on lactose.

STARCH

The symba process was developed in Sweden to produce SCP from potato starch using two yeast strains. *Saccharomyces fibuligera* produces the enzyme necessary for starch degradation enabling co-growth of *Candida utilis*. This process is known as simultaneous saccharification and fermentation.

GLUCOSE

Food grade glucose was the substrate chosen by RHM for production of fungal SCP using *Fusarium graminearum*. The strategy adopted was to take advantage of mycelial fibre content to produce a range of high added value products including meat analogues for human consumption.

HIGHER ALKANES

The original alkane SCP fermentation process, developed by BP in France used 10-20 per cent wax contained in gas oil. Substrate costs were very low, however due to their crude nature, exhaustive processing was required to recover the yeast free of a gas-oil flavour taint. Other alkane based SCP processes were developed in Italy, Japan and Romania but many of them suffered from the problems of potential carcinogenic residues and most of the plants have never run on full capacity or have been closed.

METHANE/METHANOL

Methane was initially considered as a SCP raw material because, as a gas product, purification problems after fermentation would be minimal. Disadvantages associated with methane-based processes are related to:

- The greater oxygen requirements necessary to fully oxidise methane compared with paraffins,
- The low solubility of methane in water
- The requirement that the fermentation plant be flame proof as methane-oxygen mixtures are highly explosive.

Methane is however easily converted to methanol which requires less oxygen, less fermenter cooling, is highly water soluble and has minimal explosion risks.

ICI, which manufactures bulk methanol, chose this substrate for backterial SCP production for animal feed using the trade name Pruteen. The company designed a non-mechanical "pressure cycle fermenter" which uses air for both agitation and aeration in the worlds largest single aerobic fermenter of 3000m^3 capacity.

The process which produces 50-60,000 tonnes SCP per year, using the organism Methylophilus methylotrophus, was comissioned in 1979-1980, but has suffered from dramatic increases in methanol prices.

The economic difficulties encountered by ICI with animal feed processes lead to a joint venutre with RHM to produce *Fusarium* SCP in the ICI plant.

CHOICE OF MICROORGANISM

The key criteria used in selecting suitable strains for SCP production should consider the following: The substrates to be used as carbon energy and nitrogen source and the need for nutrient supplementation. High specific growth rates, productivity and yields on a given substrate. pH and temperature tolerance.

Aeration requirements and foaming characteristics. Growth morphology in the reactor. Safety and acceptability – non pathogenic, absence of toxins. Ease of recovery. Protein, RNA and nutritional composition of the product. Structural properties of the final product. In general, fungi have the capacity to degrade a wider range of complex plant materials, particularly plant polysaccharides. They can tolerate low pH which contributes to reducing fermenter infections.

Growth of fungi as short, highly branched filaments rather than in pellets is essential in order to optimise growth rate. Bacteria, in general, have faster growth rates than fungi and grow at higher temperatures, thereby reducing fermenter cooling requirements. Bacterial and yeast fermentations are easier to aerate. In contrast to fungi, which are easily recovered by filtration, bacteria and yeast require the use of sedimentation techniques and centrifugation. Bacteria, in general produce a more favourable protein composition than yeast or fungi. Protein content in bacterial can range from 60-65 per cent whereas fungi selected for biomass production and yeast have protein contents in the range of 33-45 per cent. However, associated with the higher bacterial protein levels is a much higher level of nutritionally undesirable RNA content of 15-25 per cent Microorganisms involved in SCP production must be safe and acceptable for use in food. Organisms should be stable genetically so that the strain with optimal biochemical and physiological characteristics may be maintained in the process through many hundreds of generations.

ECONOMICS OF SCP PRODUCTION

The initial reasoning by companies such as BP and ICI to

enter SCP production was to produce, at low cost, high value SCP from petroleum, for addition to animal feed. The intention was to replace imported protein additives such as soyabean meal. Factors which contributed to the failure of hydrocarbon SCP to make a major commercial impact uncluded the 1973 dramatic oil price increases, which raised feedstock and energy costs and the lower price increases achieved by agricultural products such as soyabean.

When one considers that crude oil prices increased by a factor of 6 in 1973, and that the cost of substrate for SCP processes represents 40-60 per cent of total manufacturing costs, the negative impact on hydrocarbon based SCP processes may be understood.

Agricultural crops, the major competitor to SCP for animal feed, manifest a remarkable ability to respond to market forces and maintain price stability. By producing human protein supplements, the end value of the product can be appreciated, however the production process has to utilise more conventional and costly substrates for production (glucose). These products take advantage of the high fibre content in certain fungi to produce a product that is a good meat analogue. This product is also low in sodium and fat. The product Quorn is a fungal protein.

Single Cell Protein in Solid State Fermentation

Solid state fermentation (SSF) is growth of microorganisms on predominantly insoluble substrate where there is no free liquid. Generally, under combined conditions of low water activity and presence of intractable solid substrate, fungi show luxuriant growth. Mycelial tips of fungi have immense turgor pressure, which assist in their penetration of hard substrate.

Hence, proper growth of fungi in SSF gives much higher concentration of the biomass and higher yield when compared to submerged fermentation. The advantage in SSF process is the unique possibility of efficient utilization of waste as the substrate to produce commercially viable products.

The process does not need elaborate prearrangements for

media preparation. The process of SSF initially concentrated on enzyme production. But presently, there is worldwide interest for single cell protein (SCP) production due to the dwindling conventional food resources.

Materials and Methods

Microorganism

A local isolate of *Aspergillus niger,* from decaying wood in the soil, was used for the experimental studies. The culture was maintained on PDA slants at 28+2°C. The subculturing was done once in a fortnight.

Substrate Preparation

Oil free rice bran waste was procured from a local rice mill. It was pretreated to remove lignin and to expose the inner cellulose fibers to cellulolytic attack by the organism. The pretreatment process was carried out by adding one liter of 1% sodium hydroxide to 100 g of rice bran waste and autoclaving at 121°C and at 15 psi for 30 min. The pretreated rice bran was allowed to cool and subsequently filtered and washed to neutral pH. Then it was then dried at 60°C in an oven for 12 h. The dried bran was milled and sieved to obtain 50-mesh size and kept ready for SSF process.

Inoculum

Spores were harvested from a week old *A. niger* in five ml of sterile distilled water. Two ml of spore suspension was added to the pretreated rice bran based medium.

FERMENTATION MEDIUM

Different buffers namely; citrate (pH 5.0), phosphate (pH 7.0) and carbonate-bicarbonate (pH 10.0) were used for the extraction of total proteins from *A. niger*. The organism was cultured by submerged fermentation in Modified Czapeck Dox Medium (MCD medium) in order to obtain the mycelium. The medium contained the following ingredients (g/100 ml): K_2HPO_4, 0.12; $MgSO_4$, 0.06; $FeSO_4$, 0.05; KCl, 0.02; $NaNO_3$, 0.3

and sucrose, 3.0. Four sets of 100 ml each of MCD media was inoculated by *A. niger*. These were kept on a shaker at 120-150 rpm and 28°C for eight days before harvesting.

The fungal mycelia were harvested by centrifugation at 2500 rpm. It was then homogenized at 6000 rpm using citrate buffer and centrifuged subsequently at 6000 rpm for 45 min. The temperature during the course of extraction was maintained at 4°C. The supernatant obtained was used for estimating protein content.

Same procedure was adopted for carbonate-bicarbonate buffer and also for phosphate buffer. Sodium hydroxide solution was also used for extraction of proteins in addition to the buffers. 0.1N sodium hydroxide was added to the weighed mycelial mass followed by boiling at 80°C for 5 min and then was centrifuged at 6000 rpm and 4°C to obtain the supernatant.

GROWTH OF ASPERGILLUS NIGER

The mineral solution and different nitrogen sources were added in various combinations to 500 ml Erlenmeyer's flasks containing 10 g of pretreated rice bran. Mineral solution was prepared by mixing $ZnSO_4.7H_2O$, 0.008 g; $FeSO_4. 7H_2O$, 0.008 g and $CuSO_4. 7H_2O$, 0.008 g in 100 ml of 0.2 N HCl. The nitrogen stock solutions were prepared by adding 0.05 g of the compound i.e., sodium nitrate, ammonium nitrate and ammonium sulphate and 10 ml of distilled water respectively.

The flasks were made in triplicates and autoclaved at 121°C at 15 psi for 20 min. These were inoculated with 2 ml of spore suspension of *A. niger*. The contents of the flask were tapped gently to mix the spore suspension in the fermentation medium. All these flasks were incubated at 28°C for eight days.

The contents of each flask were harvested, weighed separately and homogenized in a homogeniser. Carbonate-bicarbonate buffer was used during the homogenization. The homogenized contents of sample from each flask were centrifuged at 6000 rpm for 45 min. The supernatant volume was measured and used for further analysis.

For estimation of total sugars, one g of the substrate was

suspended in 60 ml of distilled water. This was kept at an ambient temperature for 12 h for the extraction of sugars as soaking makes the bran softer. The filtrate was then analyzed for sugars by using anthrone reagent. The biomass was expressed in terms of total protein content. Protein estimation was done by Folin method of Lowry *et al.*

Results

Rice bran is one of the most abundant and locally available agricultural wastes. Rice bran showed higher carbon content after pretreatment with alkali (carbon content = 3.458 mg/g of rice bran). Among the extraction buffers used carbonatebicarbonate buffer at pH 10 was found to be most efficient in extracting out total proteins of *A. niger*. The protein yield with various buffers was as follows (mg/g of sample): citrate buffer, 3.403; phosphate buffer, 6.351; carbonate-bicarbonate, 6.884 and sodium hydroxide (0.1N), 3.154.

Rice bran when supplemented with mineral solution and nitrogen sources individually or together improved the biomass yield (expressed in terms of total protein yield). Among all these combinations, the supplementation of rice bran with sodium nitrate gave the highest protein yield. Supplementation of rice bran with mineral solution individually also improved the biomass yield.

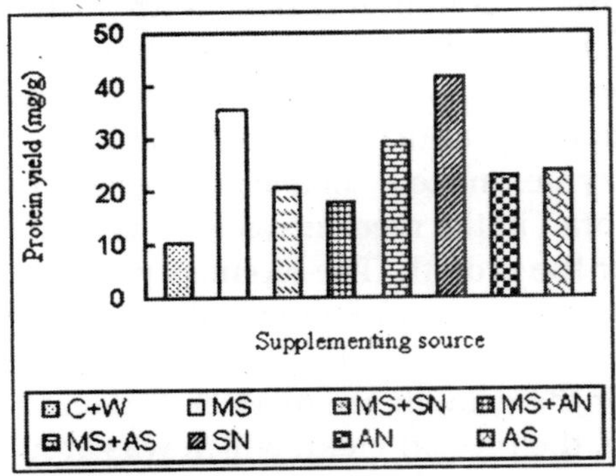

Fig. Effect of Supplementation on the Final Protein Yield

But the biomass yield was not as high as that with the

sodium nitrate supplementation. Individual nitrogen sources in combination with mineral solution supported the biomass yield to a lesser extent as compared to sodium nitrate or mineral solution.

Where, C+W (Control + Water), MS (Mineral solution), MS+SN (Mineral solution + Sodium nitrate), MS+AN (Mineral solution + Ammonium nitrate), MS+AS (Mineral solution + Ammonium sulfate), SN (Sodium nitrate), AN (Ammonium nitrate) and AS (Ammonium sulfate) are different media compositions provided for the cultivation of the biomass

The change in C/N ratio was observed due to supplementation. The C/N for pretreated rice bran was 1.729. The highest biomass yield was obtained when the C/N was 1.387. The high biomass yield was also observed for C/N 1.729 when mineral solution supplemented the medium.

But the results indicated that C/N ratio was not the only factor controlling the final biomass yield. This was because when same C/N ratios were maintained by various sources the protein yield was not similar. Even under the conditions of same C/N, sodium nitrate still favours the organism to grow better giving a higher biomass yield when compared to other nitrogen sources.

From the above observations it is clear that the availability of the nitrogen is the major controlling factor in the final biomass yield. Supplementation with mineral solution individually, gives higher yield because it provides continuous acidic medium to support the growth. As the organism grows in this condition it produces more of cellulases.

These enzymes make available more of glucose monomers from the rice bran in the medium as well as help in the release of nitrogen for the growth. The biomass yield was the highest when sodium nitrate supplements the medium alone. But when it was used along with the mineral solution its free availability for the growth seems to be hindered.

Therefore, the protein yield, which was around 41.68 mg/g of rice bran, drops to almost half to 22.88 mg/g of rice bran. The contributing factor for the drastic reduction in the biomass yield could be traced to the reactivity of sodium ion. Sodium is placed higher than all the metals in mineral solution viz., zinc, ferrous, cuprous and hydrogen ions in the activity series of metals. Hence, it displaces all these metals from their salts.

The nitrate ion released from sodium nitrate in the medium (due to the above principle of displacement) reacted with the free cations of the mineral solution and formed nitrates. As a result, free nitrogen and also mineral solution controlling the pH was not available as before to support the growth of *A. niger*. Ammonium nitrate or ammonium sulphate when supplemented the medium along with mineral solution gave higher protein yield when compared to medium supplemented by sodium nitrate along with mineral solution.

This was because the activity of mineral solution was not hindered, as ammonium ion formed was weak and hence unable to replace the metal ions in mineral solution. So the mineral solution controlled the pH and helped in the release of glucose as explained above and ammonium ion supplemented the nitrogen simultaneously. A higher yield of SCP production from *A. niger* was possible by SSF of rice bran.

Though, utilization of SSF for SCP production is emerging field, encouraging results are obtained and some success is achieved in improving the overall protein yield by supplementation of rice bran based SSF medium. The C/N ratio was not very effective in controlling biomass yield but the supplementation of rice bran with various nitrogen sources or mineral solution separately or in combination improved the *A. niger* growth. The biomass yield was best with rice bran based medium with sodium nitrate as the nitrogen source for single cell protein production.

DOWNSTREAM PROCESS

Traditional chemical analysis in downstream biotechnology manufacturing operations has been performed in off-line laboratories which are physically and organizationally remote, from the manufacturing site. Results of these assays are often not provided in a timeframe that is useful in controlling and monitoring the downstream process with maximum efficiency. Reports of delays (ranging from hours to days) in obtaining results of analytical monitoring of downstream processes often relegates these functions and the corresponding data to a compilation of historical databases rather than to its intended use as a tool to manage and control the real-time process under observation.

Modern manufacturing processes in other industries, such as semiconductor and petrochemical processing, have long recognized and employed the advantages of performing analytical measurements at the production line and have widely used these techniques to enjoy advantages in production efficiency and product quality.

The comparison and contrast to the current state of biotechnology manufacturing process efficiency and quality is particularly evident. FDA has long considered application of on-line analytical techniques as credible opportunities to improve product quality and efficiency in pharmaceutical production, including biotechnology and biological production. As a result, FDA began investigating technologies and prospects for application of automation in pharmaceutical, biotechnology, and biological manufacturing processes in ten years ago.

The purpose of the PAT initiative is to move analytical laboratory functions close to the manufacturing process and to improve manufacturing efficiencies and product quality. This would be accomplished by providing real time support and control of manufacturing processes through analysis of the process stream coupled with statistical process control and tight feedback control loops.

THE PAT INITIATIVE

The FDA PAT initiative was created formally by the Centre for Drug Evaluation and Research (CDER) branch of the FDA in 2002 and sought to provide a framework to employ these techniques for small molecule pharmaceutical production under guidance from FDA. PAT processes are becoming an established tool to promote and improve quality and production efficiency in the pharmaceutical manufacturing process. The successes of early implementation in small molecule manufacturing processes led early adopters in the biotech community to explore PAT techniques and processes for potential use in biotechnology applications.

It has been demonstrated that the same PAT technology used in traditional pharmaceutical manufacturing can be employed in biotechnology manufacturing with the same positive effect. To that end, FDA is now focusing its PAT initiative on biotechnology and biologicals manufacturing through its Centre for Biologics Evaluation and Research (CBER) and PAT offices.

Successful PAT programs are defined by a careful combined application of statistical process control (chemometrics) and use of on-line, at-line, or near-line sensors and instrumentation capable of measuring key parameters of the manufacturing system in real time. These capabilities are arguably most critical and useful in the final stages of biotechnology manufacturing, where the value of the product is at its highest.

Downstream processing applications are a specific target for implementation of PAT processes because this production stage is where the desired product of the fermentation or cell culture process is separated from the complex production matrix, then concentrated, identified, quantitated, and collected for final processing and formulation. These operations are particularly suited for use of on-line, at-line, and near-line sensors and analytical instrumentation associated with corresponding chemometric and process control systems.

Statistical process control and chemometric systems are employed across a broad spectrum of industries and are not further explored here. References to general chemometrics and chemometrics specific to biotechnology and biologicals are widely available. The focus of this article is on-line sensors and instrumentation specific to downstream processing.

PRIMARY ANALYSIS REQUIREMENT

The primary analysis requirement in biotechnology downstream processing is molecular identification and quantitation; used to identify and quantitate the product to determine its purity after separation from its production matrix by process liquid chromatography. Therefore, the most powerful and useful analytical techniques to consider for downstream processing in the PAT context are spectroscopy and analytical scale separation methods that can characterize analytes on a molecular basis.

A typical downstream process scale chromatographic technique not only purifies but concentrates the analyte and is operated at high load factors and high flow rates, therefore, the associated analytical technique must have a wide dynamic range and a fast cycle time.

Typical cycle time requirements for on-line downstream assays are under 15 min, with many under 10 min. This cycle time requirement plus the need to separate, characterize, and identify the desired product in an often complex matrix composed of chemical species similar to the desired product, requires precise, high speed, purpose-designed instrumentation characterized by high accuracy, reproducibility, and resolving power. This combination of requirements limits the discussion of applicable downstream biotech PAT instrumentation to near infrared spectroscopy (NIR) and other specialty optical spectroscopy techniques, such as mass spectroscopy, liquid chromatography, and electrophoretic techniques.

NIR has enjoyed some success in pharmaceutical manufacturing operations, primarily in blending and formulations of dry bulk materials. Its use in on-line downstream processing applications for identification and quantitation of product is less established, but its possibilities should not be overlooked.

Practical on-line NIR analyses for biotechnology applications have been demonstrated. Factors that will control use of NIR for downstream process applications include probe design and placement, calibration methodology, and heavy reliance of NIR on statistical analysis to calibrate and report concentration and identity of analytes. In many applications, a major advantage of NIR is non-contact. In these applications, the spectrometer is not in physical contact with the sample and, therefore enjoys long and reliable operation. In the case of a downstream process application however, the spectrometer samples the analyte by inserting an optical probe into a product transfer line at an appropriate sample point. This has the advantage that the sterility of the downstream system is maintained and that there is no physical withdrawal of sample from the process line. However, with the probe in direct contact with the fluid sample matrix, proper orientation and maintenance of this probe is critical to the technique's success. Fouling resulting from deposition of biological material on the probe quickly renders the measurement unusable.

Furthermore, the low frequency characteristics of the spectra produced in NIR requires heavy reliance on calibration by multivariate solution of simultaneous equations. The sample and its matrix must be fully characterized and non-variate for the calibration to remain valid.

This is not a simple task in a complex sample and matrix characteristic of biotechnology production. The speed and simplicity of instrumentation do, however, make it appropriate for consideration for downstream process applications and early applications of this technology appear promising.

Mass spectroscopy (MS) is a powerful and well appreciated tool, finding broad acceptance in a number of non-traditional applications including biotechnology. For example, mass spectrometers have been installed and used in upstream applications for measurement of headspace gas analysis in cell culture and fermentation reactors, to monitor O_2 and CO_2.

Mass spectrometry can be used to determine absolute analyte identity and for quantitation—both desirable attributes. It is rapid and has high resolving capability. In downstream processing, a mass spectrometer could however, acquire a sample stream from an automated sample port or sampling system at the appropriate point in the process stream.

Mass spectrometer applications in a downstream process are effectively prohibited by typical high salt content found in many biotech process streams. All known commercial fluid interfaces for mass spectrometers are characteristically intolerant of salt content in the fluid stream. Development of a successful salt resistant or salt remediating interface would open downstream processing to the power of MS.

This would appear to be the primary and only limiting factor in successful, robust application of MS in biotech process applications, notwithstanding the high acquisition and operating cost of a mass spectrometer.

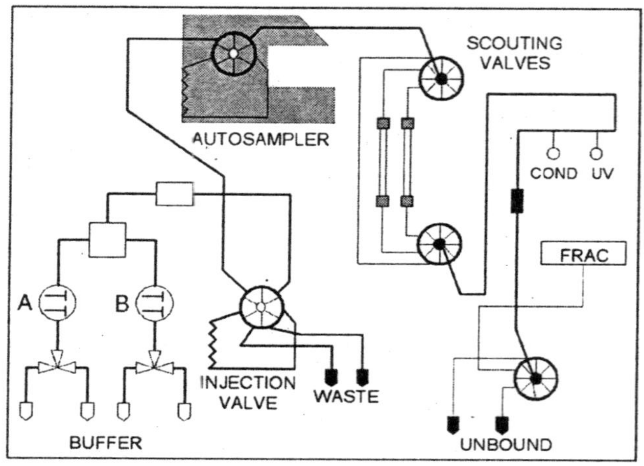

Fig. Liquid Chromatography

Liquid chromatography has been successfully employed in pharmaceutical PAT applications for decades. Of particular note are the pioneering examples of on-line high performance liquid chromatography (HPLC) applications by Eli Lilly in biotechnology applications.

On-line downstream applications of HPLC for biotechnology are now widely accepted and deployed. Commercial systems for biotech process monitoring are available from Dionex and other suppliers.

This technique is mature, well characterized, and well accepted in all fields of chemistry separation including biotechnology.

The system also requires capture of a physical sample from the process stream while maintaining sterility of the stream. Recent advances in on-line sampling for biotechnology, including introduction of commercial products that maintain process sterility, will further increase utilization of the HPLC technique in downstream applications. The chief drawback of this technique for process characterization appears to be the complexity of fully automated HPLC systems and their attendant control processes.

These include column switching, fast column equilibration, and high flow systems, all of which are strategies employed to reduce cycle times to remain consistent with downstream needs.

ELECTROPHORETIC TECHNIQUES

Electrophoretic techniques, most notably slab gel electrophoresis for protein and genetics analysis, have been a mainstay of biotechnology development analytical laboratories since its commercial inception in the 1970s. Slab gel electrophoresis is simple in concept, inexpensive to acquire, does have high operating costs, and has cycle times inconsistent with the needs of downstream processing.

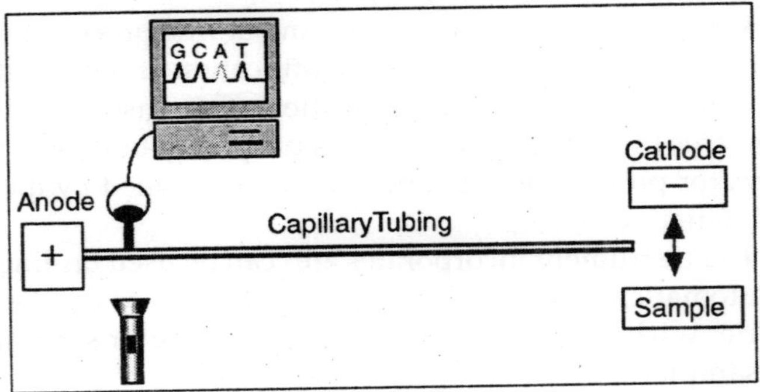

Fig. Capillary Electrophoresis

Capillary electrophoresis (HPCE), was introduced in the late 1980s and automates the electrophoresis process in a format similar to HPLC.

Many commercial instruments are available today. Data are quantitative in contrast to slab gel techniques. It is rapid. It has recently found widespread application for protein analysis in off line laboratories operated in support of downstream applications, specifically for protein product quality, characterization, and quantitation. The technique is inherently fast, versatile in the biotechnology environment, and amenable to the demands of protein assay in the downstream process environment.

The technique is employed daily in this capacity, but still has failed to gain wider appreciation and acceptance for downstream processing assay requirements because commercial instruments have been designed exclusively for standard benchtop laboratory use.

Manual sample acquisition, preparation, sample introduction, and data reduction are the rule as is the fact that most HPCE systems must be operated by HPCE specialists. These instruments have remained exclusively in the province of remote laboratories operating in support of downstream operations and these instruments do not promote the advantages of PAT technology in downstream operations.

NEW DOWNSTREAM TECHNOLOGY

We are introducing an on-line and at-line process analysis technology specific for protein identification and quantitation in a downstream process application. This instrument fully automates the entire process and is purposely built for on-line assays for protein identity and titer as envisioned by the FDA PAT initiative.

The instrument incorporates and can be used on-line with an automatic, sanitary sample port interface, or can be used at-line with samples introduced by the instrument's autosampler.

In either case, the entire sample introduction, preparation, separation, detection, and data reduction are performed automatically by the instrument without operator input. Once the sample is introduced to the instrument it is transferred directly to the sample preparation module, which delivers an appropriately prepared sample to a HPCE module that automatically performs the separation. Detection is performed by an ultraviolet absorbance detector. Data reduction and reporting are performed by a data system based on standard

HPLC data and user interfaces. Molecular weight range for the instrument is 10 to 200 kD. Calibration of the instrument is through a timed, automatic introduction of calibration standards in a manner typical of calibration for similar HPLC systems. Cycle times are consistent with the needs for most downstream processing.

Biotechnology manufacturing, particularly downstream processes, has many operations and opportunities suited to the premise of the PAT initiative—namely, improving production efficiency, yield, and product quality. Several technologies have been implemented in downstream processing for the purpose of on-line assays that have proven the concept.

New, versatile instrumentation technologies for downstream process applications will continue to emerge as the PAT initiative becomes more fully implemented in biotechnology and biological manufacturing.

Chapter 11

Enzymes

Enzymes are catalysts. Most are proteins. A few ribonucleoprotein enzymes have been discovered and, for some of these, the catalytic activity is in the RNA part rather than the protein part. Link to discussion of these ribozymes.

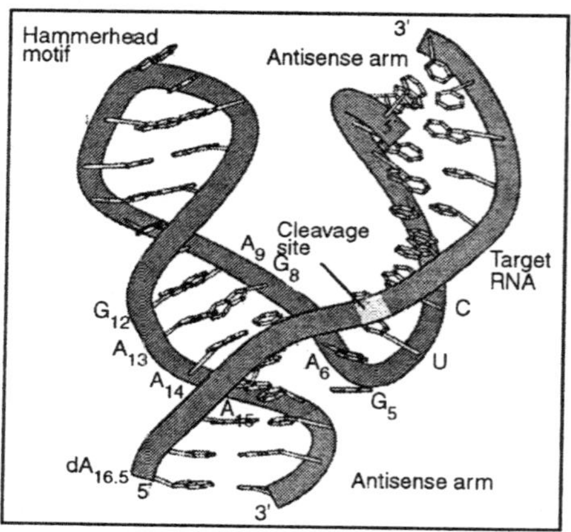

Fig. Ribozymes

Enzymes bind temporarily to one or more of the reactants of the reaction they catalyze. In doing so, they lower the amount of activation energy needed and thus speed up the reaction. Enzymes are biomolecules that catalyze chemical reactions. Almost all enzymes are proteins. In enzymatic reactions, the molecules at the beginning of the process are called substrates, and the enzyme converts them into different molecules, the products.

Almost all processes in a biological cell need enzymes in

order to occur at significant rates. Since enzymes are extremely selective for their substrates and speed up only a few reactions from among many possibilities, the set of enzymes made in a cell determines which metabolic pathways occur in that cell. Like all catalysts, enzymes work by lowering the activation energy (E_a or ÄG) for a reaction, thus dramatically increasing the rate of the reaction. Most enzyme reaction rates are millions of times faster than those of comparable uncatalyzed reactions.

As with all catalysts, enzymes are not consumed by the reactions they catalyze, nor do they alter the equilibrium of these reactions. However, enzymes do differ from most other catalysts by being much more specific. Enzymes are known to catalyze about 4,000 biochemical reactions.

A few RNA molecules called ribozymes catalyze reactions, with an important example being some parts of the ribosome. Synthetic molecules called artificial enzymes also display enzyme-like catalysis. Enzyme activity can be affected by other molecules. Inhibitors are molecules that decrease enzyme activity; activators are molecules that increase activity. Many drugs and poisons are enzyme inhibitors.

Activity is also affected by temperature, chemical environment (e.g. pH), and the concentration of substrate. Some enzymes are used commercially, for example, in the synthesis of antibiotics.

In addition, some household products use enzymes to speed up biochemical reactions (*e.g.*, enzymes in biological washing-powders break down protein or fat stains on clothes; enzymes in meat tenderizers break down proteins, making the meat easier to chew). The use of enzymes in the diagnosis of disease is one of the important benefits derived from the intensive research in biochemistry since the 1940's.

Enzymes have provided the basis for the field of clinical chemistry. It is, however, only within the recent past few decades that interest in diagnostic enzymology has multiplied. Many methods currently on record in the literature are not in wide use, and there are still large areas of medical research in which the diagnostic potential of enzyme reactions has not been explored at all. This section has been prepared by

Worthington Biochemical Corporation as a practical introduction to enzymology.

Because of its close involvement over the years in the theoretical as well as the practical aspects of enzymology, Worthington's knowledge covers a broad spectrum of the subject. Some of this information has been assembled here for the benefit of laboratory personnel. This section summarizes in simple terms the basic theories of enzymology.

ENZYMOLOGY

As early as the late 1700s and early 1800s, the digestion of meat by stomach secretions and the conversion of starch to sugars by plant extracts and saliva were known. However, the mechanism by which this occurred had not been identified. In the 19th century, when studying the fermentation of sugar to alcohol by yeast, Louis Pasteur came to the conclusion that this fermentation was catalyzed by a vital force contained within the yeast cells called "ferments", which were thought to function only within living organisms. He wrote that "alcoholic fermentation is an act correlated with the life and organization of the yeast cells, not with the death or putrefaction of the cells."

In 1878 German physiologist Wilhelm Kühne first used the term *enzyme,* which comes from Greek *åíæõìïí* "in leaven", to describe this process. The word *enzyme* was used later to refer to nonliving substances such as pepsin, and the word *ferment* used to refer to chemical activity produced by living organisms.

In 1897 Eduard Buchner began to study the ability of yeast extracts that lacked any living yeast cells to ferment sugar. In a series of experiments at the University of Berlin, he found that the sugar was fermented even when there were no living yeast cells in the mixture.

He named the enzyme that brought about the fermentation of sucrose "zymase". In 1907 he received the Nobel Prize in Chemistry "for his biochemical research and his discovery of cell-free fermentation". Following Buchner's example; enzymes are usually named according to the reaction they carry out. Typically the suffix-*ase* is added to the name of the substrate (*e.g.*, lactase is the enzyme that cleaves lactose) or the type of reaction.

Having shown that enzymes could function outside a living cell, the next step was to determine their biochemical nature. Many early workers noted that enzymatic activity was associated with proteins, but several scientists argued that proteins were merely carriers for the true enzymes and that proteins *per se* were incapable of catalysis. However, in 1926, James B. Sumner showed that the enzyme urease was a pure protein and crystallized it; Sumner did likewise for the enzyme catalase in 1937.

The conclusion that pure proteins can be enzymes was definitively proved by Northrop and Stanley, who worked on the digestive enzymes pepsin, trypsin and chymotrypsin. These three scientists were awarded the 1946 Nobel Prize in Chemistry. This discovery that enzymes could be crystallized eventually allowed their structures to be solved by x-ray crystallography.

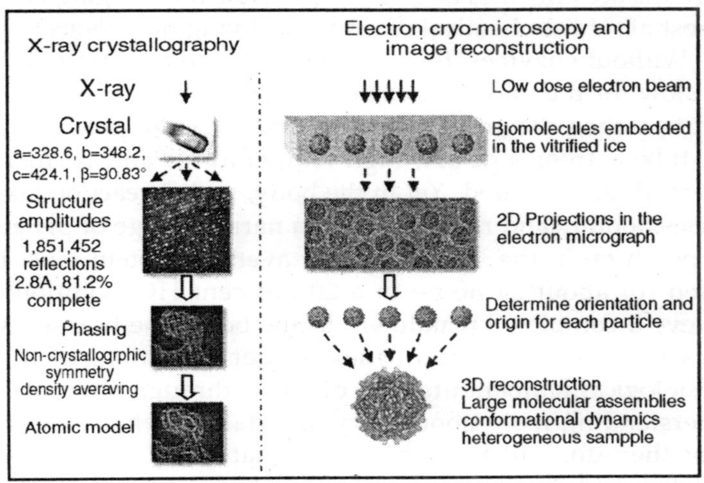

Fig. X-ray Crystallography

This was first done for lysozyme, an enzyme found in tears, saliva and egg whites that digests the coating of some bacteria. This high-resolution structure of lysozyme marked the beginning of the field of structural biology and the effort to understand how enzymes work at an atomic level of detail.

ENZYMES AND LIFE PROCESSES

The living cell is the site of tremendous biochemical activity called metabolism. This is the process of chemical and physical change which goes on continually in the living organism. Build-up of new tissue, replacement of old tissue, conversion of food to energy, disposal of waste materials, reproduction-all the activities that we characterize as "life." This building up and tearing down takes place in the face of an apparent paradox. The greatest majority of these biochemical reactions do not take place spontaneously. The phenomenon of catalysis makes possible biochemical reactions necessary for all life processes.

Catalysis is defined as the acceleration of a chemical reaction by some substance which itself undergoes no permanent chemical change. The catalysts of biochemical reactions are enzymes and are responsible for bringing about almost all of the chemical reactions in living organisms.

Without enzymes, these reactions take place at a rate far too slow for the pace of metabolism. The oxidation of a fatty acid to carbon dioxide and water is not a gentle process in a test tube-extremes of pH, high temperatures and corrosive chemicals are required. Yet in the body, such a reaction takes place smoothly and rapidly within a narrow range of pH and temperature. In the laboratory, the average protein must be boiled for about 24 hours in a 20 per cent HCl solution to achieve a complete breakdown. In the body, the breakdown takes place in four hours or less under conditions of mild physiological temperature and pH. It is through attempts at understanding more about enzyme catalysts-what they are, what they do, and how they do it-that many advances in medicine and the life sciences have been brought about.

SPECIFICITY OF ENZYMES

One of the properties of enzymes that makes them so important as diagnostic and research tools is the specificity they exhibit relative to the reactions they catalyze. A few enzymes exhibit absolute specificity; that is, they will catalyze only one particular reaction.

Other enzymes will be specific for a particular type of chemical bond or functional group. In general, there are four distinct types of specificity:

- *Absolute specificity*-the enzyme will catalyze only one reaction.
- *Group specificity*-the enzyme will act only on molecules that have specific functional groups, such as amino, phosphate and methyl groups.
- *Linkage specificity*-the enzyme will act on a particular type of chemical bond regardless of the rest of the molecular structure.
- *Stereochemical specificity*-the enzyme will act on a particular steric or optical isomer.

Though enzymes exhibit great degrees of specificity, cofactors may serve many apoenzymes.

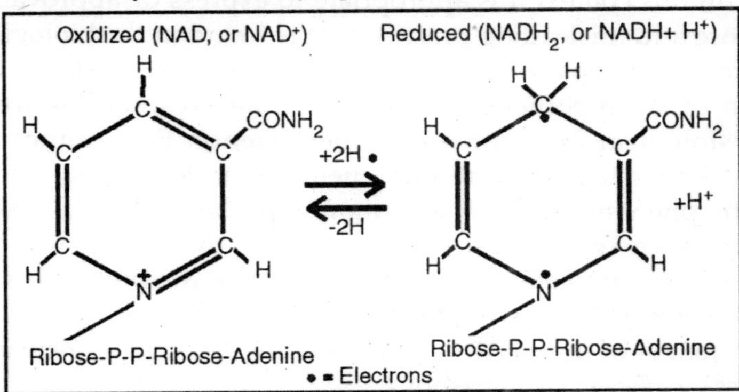

Fig. Nicotinamide Adenine Dinucleotide

For example, nicotinamide adenine dinucleotide (NAD) is a coenzyme for a great number of dehydrogenase reactions in which it acts as a hydrogen acceptor. Among them are the alcohol dehydrogenase, malate dehydrogenase and lactate dehydrogenase reactions.

ENZYMES

Because of their close interdependence, it is convenient

to deal with the classification and nomenclature together. The *first general principle* of these 'Recommendations' is that names purporting to be names of enzymes, especially those ending in-*ase*, should be used only for single enzymes, *i.e.* single catalytic entities.

They should not be applied to systems containing more than one enzyme. When it is desired to name such a system on the basis of the overall reaction catalysed by it, the word *system* should be included in the name.

For example, the system catalysing the oxidation of succinate by molecular oxygen, consisting of succinate dehydrogenase, cytochrome oxidase, and several intermediate carriers, should not be named *succinate oxidase*, but it may be called the *succinate oxidase system*. Other examples of systems consisting of several structurally and functionally linked enzymes (and cofactors) are the *pyruvate dehydrogenase system*, the similar *2-oxoglutarate dehydrogenase system*, and the *fatty acid synthase system*.

In this context it is appropriate to express disapproval of a loose and misleading practice that is found in the biological literature. It consists in designation of a natural substance (or even of an hypothetical active principle), responsible for a physiological or biophysical phenomenon that cannot be described in terms of a definite chemical reaction, by the name of the phenomenon in conjugation with the suffix-*ase*, which implies an individual enzyme. Some examples of such *phenomenase* nomenclature, which should be discouraged even if there are reasons to suppose that the particular agent may have enzymic properties, are: *permease, translocase, reparase, joinase, replicase, codase, etc.*.

The *second general principle* is that enzymes are principally classified and named according to the reaction they catalyse. The chemical reaction catalysed is the specific property that distinguishes one enzyme from another, and it is logical to use it as the basis for the classification and naming of enzymes.

Several alternative bases for classification and naming had been considered, *e.g.* chemical nature of the enzymes (whether it is a flavoprotein, a hemoprotein, a pyridoxal-phosphate protein, a copper protein, and so on), or chemical nature of the substrate (nucleotides, carbohydrates, proteins, *etc.*). The first cannot serve as a general basis, for only a minority of enzymes have such identifiable prosthetic groups.

The chemical nature of the enzyme has, however, been used exceptionally in certain cases where classification based on specificity is difficult, for example, with the peptidases. The second basis for classification is hardly practicable, owing to the great variety of substances acted upon and because it is not sufficiently informative unless the type of reaction is also given. It is the overall reaction, as expressed by the formal equation, that should be taken as the basis.

Thus, the intimate mechanism of the reaction, and the formation of intermediate complexes of the reactants with the enzyme is not taken into account, but only the observed chemical change produced by the complete enzyme reaction. For example, in those cases in which the enzyme contains a prosthetic group that serves to catalyse transfer from a donor to an acceptor (*e.g.* flavin, biotin, or pyridoxal-phosphate enzymes) the name of the prosthetic group is not normally included in the name of the enzyme.

Nevertheless, where alternative names are possible, the mechanism may be taken into account in choosing between them. A consequence of the adoption of the chemical reaction as the basis for naming enzymes is that a systematic name cannot be given to an enzyme until it is known what chemical reaction it catalyses.

This applies, for example, to a few enzymes that have so far not been shown to catalyse any chemical reaction, but only isotopic exchanges; the isotopic exchange gives some idea of one step in the overall chemical reaction, but the reaction as a whole remains unknown. A second consequence of this concept is that a certain name designates not a single enzyme protein but a group of proteins with the same catalytic property.

Enzymes from different sources (various bacterial, plant or animal species) are classified as one entry. The same applies to isoenzymes. However, there are exceptions to this general rule. Some are justified because the mechanism of the reaction or the substrate specificity is so different as to warrant different entries in the enzyme list. This applies, for example, to the two cholinesterases, the two citrate hydro-lyases, and the two amine oxidases. Others are mainly historical, *e.g.* acid and alkaline phosphatases.

A *third general principle* adopted is that the enzymes are divided into groups on the basis of the type of reaction catalysed, and this, together with the name(s) of the substrate(s) provides a basis for naming individual enzymes. It is also the basis for classification and code numbers.

Special problems attend the classification and naming of enzymes catalysing complicated transformations that can be resolved into several sequential or coupled intermediary reactions of different types, all catalysed by a single enzyme (not an enzyme system). Some of the steps may be spontaneous non-catalytic reactions, while one or more intermediate steps depend on catalysis by the enzyme.

Wherever the nature and sequence of intermediary reactions is known or can be presumed with confidence, classification and naming of the enzyme should be based on the first enzyme-catalysed step that is essential to the subsequent transformations, which can be indicated by a supplementary term in parentheses, e.g. acetyl-CoA:glyoxylate C-acetyltransferase (thioester-hydrolysing, carboxymethyl-forming).

To classify an enzyme according to the type of reaction catalysed, it is occasionally necessary to choose between alternative ways of regarding a given reaction. Some considerations of this type are outlined in section of this chapter. In general, that alternative should be selected which fits in best with the general system of classification and reduces the number of exceptions.

One important extension of this principle is the question of the direction in which the reaction is written for the purposes of classification. To simplify the classification, the direction chosen should be the same for all enzymes in a given class, even if this direction has not been demonstrated for all. Thus the *systematic* names, on which the classification and code numbers are based, may be derived from a written reaction, even though only the reverse of this has been actually demonstrated experimentally.

In the list in this volume, the reaction is written to illustrate the classification, *i.e.* in the direction described by the systematic name. However, the *common* name may be based on either direction of reaction, and is often based on the presumed physiological direction.

Many examples of this usage are found in section of the list. The reaction is written as an oxidation of xylitol by NAD^+, in parallel with all other oxidoreductases in subgroup, and the systematic name is accordingly, *xylitol:NAD^+ 2-oxidoreductase* (D-*xylulose-forming)*. However, the common name, based on the reverse direction of reaction, is D-*xylulose reductase.*

SYSTEMATIC NAMES OF ENZYMES

The first Enzyme Commission gave much thought to the question of a systematic and logical nomenclature for enzymes, and finally recommended that there should be two nomenclatures for enzymes, one systematic, and one working or trivial.

The systematic name of an enzyme, formed in accordance with definite rules, showed the action of an enzyme as exactly as possible, thus identifying the enzyme precisely. The trivial name was sufficiently short for general use, but not necessarily very systematic; in a great many cases it was a name already in current use.

The introduction of (often cumbersome) systematic names was strongly criticised. In many cases the reaction catalysed is not much longer than the systematic name and can serve just as well for identification, especially in conjunction with the code number.

The Commission for Revision of Enzyme Nomenclature discussed this problem at length, and a change in emphasis was made. It was decided to give the trivial names more prominence in the Enzyme List; they now follow immediately after the code number, and are described as Common Name.

Also, in the index the common names are indicated by an asterisk. Nevertheless, it was decided to retain the systematic names as the basis for classification for the following reasons:

- The code number alone is only useful for identification of an enzyme when a copy of the Enzyme List is at hand, whereas the systematic name is self-explanatory;
- The systematic name stresses the type of reaction, the reaction equation does not;
- Systematic names can be formed for new enzymes by the discoverer, by application of the rules, but code numbers should not be assigned by individuals;
- Common names for new enzymes are frequently formed as a condensed version of the systematic name; therefore, the systematic names are helpful in finding common names that are in accordance with the general pattern.

It is recommended that for enzymes that are not the main subject of a paper or abstract, the common names should be used, but they should be identified at their first mention by their code numbers and source. Where an enzyme is the main subject of a paper or abstract, its code number, systematic name, or, alternatively, the reaction equation and source should be given at its first mention; thereafter the common name should be used.

In the light of the fact that enzyme names and code numbers refer to reactions catalysed rather than to discrete proteins, it is of special importance to give also the source of the enzyme for full identification; in cases where multiple forms are known to exist, knowledge of this should be included where available. When a paper deals with an enzyme that is not yet in the Enzyme List, the author may introduce a new name and, if desired, a new systematic name, both formed according to the recommended rules. A number should be assigned only by the Nomenclature Committee of IUBMB.

The Enzyme List contains one or more references for each enzyme. It should be stressed that no attempt has been made to provide a complete bibliography, or to refer to the first description of an enzyme. The references are intended to provide sufficient evidence for the existence of an enzyme catalysing the reaction as set out. Where there is a major paper describing the purification and specificity of an enzyme, or a major review article, this has been quoted to the exclusion of earlier and later papers. In some cases separate references are given for animal, plant and bacterial enzymes.

CLASSIFICATION OF ENZYMES

The first Enzyme Commission, in its report in 1961, devised a system for classification of enzymes that also serves as a basis for assigning code numbers to them. These code numbers, prefixed by EC, which are now widely in use, contain four elements separated by points, with the following meaning:

- The first number shows to which of the six main divisions (classes) the enzyme belongs,
- The second figure indicates the subclass,
- The third figure gives the sub-subclass,
- The fourth figure is the serial number of the enzyme in its sub-subclass.

The subclasses and sub-subclasses are formed according to principles indicated below.

The main divisions and subclasses are:

OXIDOREDUCTASES

To this class belong all enzymes catalysing oxidoreduction reactions. The substrate that is oxidized is regarded as hydrogen donor. The systematic name is based on donor:acceptor oxidoreductase. The common name will be dehydrogenase, wherever this is possible; as an alternative, reductase can be used. Oxi*dase* is only used in cases where O_2 is the acceptor.

The second figure in the code number of the oxidoreductases, indicates the group in the hydrogen (or electron) donor that undergoes oxidation: 1 denotes a-CHOH-group, 2 a-CHO or-CO-COOH group or carbon monoxide, and so on, as listed in the key. The third figure, except in subclasses, indicates the type of acceptor involved: 1 denotes $NAD(P)^+$, 2 a cytochrome, 3 molecular oxygen, 4 a disulfide, 5 a quinone or similar compound, 6 a nitrogenous group, 7 an iron-sulfur protein and 8 a flavin.

In subclasses a different classification scheme is used and sub-subclasses are numbered from 11 onwards. It should be noted that in reactions with a nicotinamide coenzyme this is always regarded as acceptor, even if this direction of the reaction is not readily demonstrated. The only exception is the subclass, in which NAD(P)H is the donor; some other redox catalyst is the acceptor. Although not used as a criterion for classification, the two hydrogen atoms at carbon-4 of the dihydropyridine ring of nicotinamide nucleotides are not equivalent in that the hydrogen is transferred stereospecifically.

TRANSFERASES

Transferases are enzymes transferring a group, *e.g.* a methyl group or a glycosyl group, from one compound (generally regarded as donor) to another compound (generally regarded as acceptor). The systematic names are formed according to the scheme donor:acceptor grouptransferase.

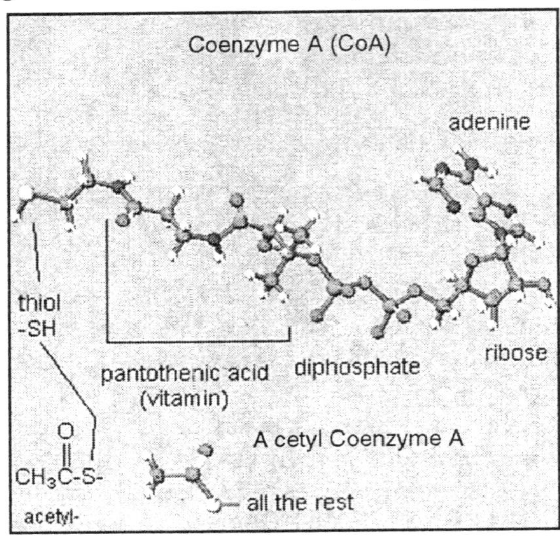

Fig. Coenzyme

The common names are normally formed according to

acceptor grouptransferase or donor grouptransferase. In many cases, the donor is a cofactor (coenzyme) charged with the group to be transferred.

A special case is that of the transaminases. Some transferase reactions can be viewed in different ways. For example, the enzyme-catalysed reaction X-Y + Z = X + Z-Y may be regarded either as a transfer of the group Y from X to Z, or as a breaking of the X-Y bond by the introduction of Z. Where Z represents phosphate or arsenate, the process is often spoken of as 'phosphorolysis' or 'arsenolysis', respectively, and a number of enzyme names based on the pattern of *phosphorylase* have come into use.

These names are not suitable for a systematic nomenclature, because there is no reason to single out these particular enzymes from the other transferases, and it is better to regard them simply as *Y-transferases*. In the above reaction, the group transferred is usually exchanged, at least formally, for hydrogen, so that the equation could more strictly be written as:

$$X - Y + Z - H = X - H + Z - Y.$$

Another problem is posed in enzyme-catalysed transaminations, where the-NH_2 group and-H are transferred to a compound containing a carbonyl group in exchange for the = O of that group, according to the general equation:

$$R^1\text{-CH(-}NH_2\text{)-}R^2 + R^3\text{-CO-}R^4 \rightarrow R^1\text{-CO-}R^2 + R^3\text{-CH(-}NH_2\text{)-}R^4.$$

The reaction can be considered formally as oxidative deamination of the donor (*e.g.* amino acid) linked with reductive amination of the acceptor (*e.g.* oxo acid), and the transaminating enzymes (pyridoxal-phosphate proteins) might be classified as oxidoreductases.

However, the unique distinctive feature of the reaction is the transfer of the amino group (by a well-established mechanism involving covalent substrate-coenzyme intermediates), which justified allocation of these enzymes among the transferases as a special subclass (*transaminases*). The second figure in the code number of transferases indicates the group transferred; a one-carbon group, an aldehydic or ketonic group, an acyl group and so on. The third figure gives further information on the group transferred; *e.g.* subclass is subdivided into methyltransferases, hydroxymethyl-and formyltransferases and so on; only in subclass does the third figure indicate the nature of the acceptor group.

HYDROLASES

These enzymes catalyse the hydrolytic cleavage of C–O, C–N, C–C and some other bonds, including phosphoric anhydride bonds. Although the systematic name always includes *hydrolase,* the common name is, in many cases, formed by the name of the substrate with the suffix-*ase*. It is understood that the name of the substrate with this suffix means a hydrolytic enzyme.

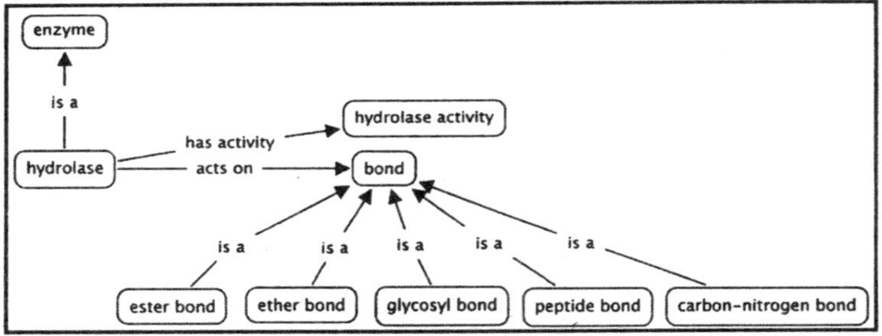

Fig. Hydrolase

A number of hydrolases acting on ester, glycosyl, peptide, amide or other bonds are known to catalyse not only hydrolytic removal of a particular group from their substrates, but likewise the transfer of this group to suitable acceptor molecules. In principle, all hydrolytic enzymes might be classified as transferases, since hydrolysis itself can be regarded as transfer of a specific group to water as the acceptor. Yet, in most cases, the reaction with water as the acceptor was discovered earlier and is considered as the main physiological function of the enzyme.

This is why such enzymes are classified as hydrolases rather than as transferases. Some hydrolases (especially some of the esterases and glycosidases) pose problems because they have a very wide specificity and it is not easy to decide if two preparations described by different authors (perhaps from different sources) have the same catalytic properties, or if they should be listed under separate entries. An example is *vitamin A esterase*. To some extent the choice must be arbitrary; however, separate entries should be given only when the specificities are sufficiently different.

Another problem is that proteinases have 'esterolytic' action; they usually hydrolyse ester bonds in appropriate substrates even more rapidly than natural peptide bonds. In this case, classification among the peptide hydrolases is based on historical priority and presumed physiological function.

The second figure in the code number of the hydrolases indicates the nature of the bond hydrolysed; the *esterases*; the *glycosylases*, and so on.

The third figure normally specifies the nature of the substrate, e.g. in the esterases the carboxylic ester hydrolases, thiolester hydrolases, phosphoric monoester hydrolases; in the glycosylases the O-glycosidases, N-glycosylases, etc. Exceptionally, in the case of the peptidyl-peptide hydrolases the third figure is based on the catalytic mechanism as shown by active centre studies or the effect of pH.

LYASES

Lyases are enzymes cleaving C–C, C–O, C–N, and other bonds by elimination, leaving double bonds or rings, or conversely adding groups to double bonds. The systematic name is formed according to the pattern *substrate group-lyase.* The hyphen is an important part of the name, and to avoid confusion should not be omitted, *e.g. hydro-lyase* not 'hydrolyase'. In the common names, expressions like *decarboxylase, aldolase, dehydratase* (in case of elimination of CO_2, aldehyde, or water) are used.

In cases where the reverse reaction is much more important, or the only one demonstrated, *synthase* (not synthetase) may be used in the name. Various subclasses of the lyases include pyridoxal-phosphate enzymes that catalyse the elimination of a b-or g-substituent from an a-amino acid followed by a replacement of this substituent by some other group. In the overall replacement reaction, no unsaturated end-product is formed; therefore, these enzymes might formally be classified as *alkyl-transferases*.

However, there is ample evidence that the replacement is a two-step reaction involving the transient formation of enzyme-bound a,b(or b,g)-unsaturated amino acids. According to the rule that the first reaction is indicative for classification, these enzymes are correctly classified as *lyases*. Examples are *tryptophan synthase* and *cystathionine b-synthase*.

The second figure in the code number indicates the bond broken: carbon-carbon lyases, carbon-oxygen lyases and so on. The third figure gives further information on the group eliminated (*e.g.* CO_2, H_2O).

ISOMERASES

These enzymes catalyse geometric or structural changes within one molecule. According to the type of isomerism, they may be called *racemases, epimerases, cis-trans-isomerases, isomerases, tautomerases, mutases* or *cycloisomerases*.

In some cases, the interconversion in the substrate is brought about by an intramolecular oxidoreduction; since hydrogen donor and acceptor are the same molecule, and no oxidized product appears, they are not classified as oxidoreductases, even though they may contain firmly bound $NAD(P)^+$. The subclasses are formed according to the type of isomerism, the sub-subclasses to the type of substrates.

LIGASES

Ligases are enzymes catalysing the joining together of two molecules coupled with the hydrolysis of a diphosphate bond in ATP or a similar triphosphate. The systematic names are formed on the system *X:Y ligase (ADP-forming)*. In earlier editions of the list the term *synthetase* has been used for the common names. Many authors have been confused by the use of the terms *synthetase* and *synthase* (used throughout the list when it is desired to emphasis the synthetic nature of the reaction). Consequently NC-IUB decided in 1983 to abandon the use of synthetase for common names, and to replace them with names of the type *X–Y ligase*. In a few cases in Group 6, where the reaction is more complex or there is a common name for the product, a synthase name is used.

It is recommended that if the term *synthetase* is used by authors, it should continue to be restricted to the ligase group.

The second figure in the code number indicates the bond formed: EC 6.1 for C–O bonds (enzymes acylating tRNA), EC 6.2 for C–S bonds (acyl-CoA derivatives), *etc*. Sub-subclasses are only in use in the C–N ligases. In a few cases it is necessary to use the word *other* in the description of subclasses and sub-subclasses. They have been provisionally given the figure in order to leave space for new subdivisions.

From time to time, some enzymes have been deleted from the List, while some others have been renumbered. However, the old numbers have *not* been allotted to new enzymes; rather the place has been left vacant and cross-reference is made according to the following scheme: Entries for reclassified enzymes transferred from one position in the List to another are followed, for reference, by a comment indicating the former number. It is regarded as important that the same policy be followed in future revisions and extensions of the Enzyme List, which may become necessary from time to time.

RULES FOR NOMENCLATURE

General Rules for Systematic Names and Guidelines for Common Names

Rule 1.

Generally accepted trivial names of substrates may be used in enzyme names. The prefix D-should be omitted for all D-sugars and L-for individual amino acids, unless ambiguity would be caused. In general, it is not necessary to indicate positions of substituents in common names, unless it is necessary to prevent two different enzymes having the same name. The prefix *keto* is no longer used for derivatives of sugars in which-CHOH-has been replaced by-CO-; they are named throughout as dehydro-sugars.

To produce usable systematic names, accepted trivial names of substrates forming part of the enzyme names should be used. Where no accepted and convenient trivial names exist, the official IUPAC rules of nomenclature should be applied to the substrate name. The 1,2,3 system of locating substituents should be used instead of the a,b,g system, although group names such as b-aspartyl-, g-glutamyl-, and also b-alanine and g-lactone are permissible; a,b should normally be used for indicating configuration, as in a-D-glucose. For nucleotide groups, *adenylyl* (not adenyl), *etc.* should be the form used. The name oxo acids (not keto acids) may be used as a class name, and for individual compounds in which-CH_2-has been replaced by-CO-, oxo should be used.

Rule 2.

Where the substrate is normally in the form of an anion, its name should end in-*ate* rather than-*ic; e.g. lactate dehydrogenase,* not 'lactic dehydrogenase' or 'lactic acid dehydrogenase'.

Rule 3.

Commonly used abbreviations for substrates, *e.g.* ATP, may be used in names of enzymes, but the use of new abbreviations (not listed in recommendations of the IUPAC-IUB Commission on Biochemical Nomenclature) should be discouraged. Chemical formulae should not normally be used instead of names of substrates. Abbreviations for names of enzymes, *e.g.* GDH, should not be used.

Rule 4.

Names of substrates composed of two nouns, such as glucose phosphate, which are normally written with a space, should be hyphenated when they form part of the enzyme names, and thus become adjectives, *e.g. glucose-6-phosphate 1-dehydrogenas.*

Rule 5.

The use as enzyme names of descriptions such as

condensing enzyme, acetate-activating enzyme, pH 5 enzyme should be discontinued as soon as the catalysed reaction is known. The word *activating* should not be used in the sense of converting the substrate into a substance that reacts further; all enzymes act by activating their substrates, and the use of the word in this sense may lead to confusion.

Rule 6.

If it can be avoided, a common name should not be based on a substance that is not a true substrate, *e.g.* enzyme should not be called 'crotonase', since it does not act on crotonate.

Rule 7.

Where a name in common use gives some indication of the reaction and is not incorrect or ambiguous, its continued use is recommended. In other cases a common name is based on the same general principles as the systematic name but with a minimum of detail, to produce a name short enough for convenient use. A few names of proteolytic enzymes ending in-*in* are retained; all other enzyme names should end in-*ase*.

Systematic names consist of two parts. The first contains the name of the substrate or, in the case of a bimolecular reaction, of the two substrates separated by a colon. The second part, ending in-*ase*, indicates the nature of the reaction.

Rule 8.

A number of generic words indicating a type of reaction may be used in either common or systematic names: *oxidoreductase, oxygenase, transferase* (with a prefix indicating the nature of the group transferred), *hydrolase, lyase, racemase, epimerase, isomerase, mutase, ligase.*

Rule 9.

A number of additional generic words indicating reaction types are used in common names, but not in the systematic nomenclature, *e.g. dehydrogenase, reductase, oxidase, peroxidase, kinase, tautomerase, deaminase, dehydratase, etc.*.

Rule 10.

Where additional information is needed to make the

reaction clear, a phrase indicating the reaction or a product should be added in parentheses after the second part of the name *e.g. (ADP-forming), (dimerizing), (CoA-acylating).*

Rule 11.

The direct attachment of-*ase* to the name of the substrate will indicate that the enzyme brings about hydrolysis.

The suffix-*ase* should never be attached directly to the name of the substrate.

Rule 12.

The name 'dehydrase' which was at one time used for both dehydrogenating and dehydrating enzymes, should not be used. *Dehydrogenase* will be used for the former and *dehydratase* for the latter.

Rule 13.

Where possible, common names should normally be based on a reaction direction that has been demonstrated, *e.g. dehydrogenase* or *reductase, decarboxylase* or *carboxylase.*

In the case of reversible reactions, the direction chosen for naming should be the same for all the enzymes in a given class, even if this direction has not been demonstrated for all. Thus, systematic names may be based on a written reaction, even though only the reverse of this has been actually demonstrated experimentally.

Rule 14.

When the overall reaction includes two different changes, *e.g.* an oxidative demethylation, the classification and systematic name should be based, whenever possible, on the one (or the first one) catalysed by the enzyme; the other function(s) should be indicated by adding a suitable participle in parentheses, as in the case of *sarcosine:oxygen oxidoreductase (demethylating)*; D-*aspartate:oxygen oxidoreductase (deaminating)*; L-*serine hydro-lyase (adding indoleglycerol-phosphate).*

Other examples of such additions are *(decarboxylating), (cyclizing), (acceptor-acylating), (isomerizing).*

Rule 15.

When an enzyme catalyses more than one type of reaction,

the name should normally refer to one reaction only. Each case must be considered on its merits, and the choice must be, to some extent, arbitrary. Other important activities of the enzyme may be indicated in the List under 'Reaction' or 'Comments'.

Similarly, when any enzyme acts on more than one substrate (or pair of substrates), the name should normally refer only to one substrate (or pair of substrates), although in certain cases it may be possible to use a term that covers a whole group of substrates, or an alternative substrate may be given in parentheses.

Rule 16.

A group of enzymes with closely similar specificities should normally be described by a single entry. However, when the specificity of two enzymes catalysing the same reactions is sufficiently different (the degree of difference being a matter of arbitrary choice) two separate entries may be made. Separate entries are also appropriate for enzymes having similar catalytic functions, but known to differ basically with regard to reaction mechanism or to the nature of the catalytic groups, e.g. amine oxidase (flavin-containing) and amine oxidase (copper-containing).

(b) Rules and Guidelines for Particular Classes of Enzymes

Rule 17.

The terms dehydrogenase or reductase will be used much as hitherto. The latter term is appropriate when hydrogen transfer from the substance mentioned as donor in the systematic name is not readily demonstrated.

Transhydrogenase may be retained for a few well-established cases. Oxidase is used only for cases there O_2 acts as an acceptor, and oxygenase only for those cases where the O_2 molecule (or part of it) is directly incorporated into the substrate. Peroxidase is used for enzymes using H_2O_2 as acceptor. Catalase must be regarded as exceptional. Where no ambiguity is caused, the second reactant is not usually named; but where required to prevent ambiguity, it may be given in parentheses, e.g alcohol dehydrogenase and alcohol dehydrogenase ($NADP^+$). All enzymes catalysing oxidoreductions should be named oxidoreductases in the systematic nomenclature, and the names formed on the pattern donor:acceptor oxidoreductase.

Rule 18.

For oxidoreductases using NAD^+ or $NADP^+$, the coenzyme should always be named as the acceptor except for the special case of Section (enzymes whose normal physiological function is regarded as reoxidation of the reduced coenzyme). Where the enzyme can use either coenzyme, this should be indicated by writing $NAD(P)^+$.

Rule 19.

Where the true acceptor is unknown and the oxidoreductase has only been shown to react with artificial acceptors, the word *acceptor* should be written in parentheses, *succinate:(acceptor) oxidoreductase.*

Rule 20.

Oxidoreductases that bring about the incorporation of molecular oxygen into one donor or into either or both of a pair of donors are named *oxygenase.* If only one atom of oxygen is incorporated the term *monooxygenase* is used; if both atoms of O_2 are incorporated, the term *dioxygenase* is used.

Oxidoreductases bringing about the incorporation of oxygen into one of paired donors should be named on the pattern *donor,donor:oxygen oxidoreductase (hydroxylating).*

Rule 21.

Only one specific substrate or reaction product is generally indicated in the common names, together with the group donated or accepted. The forms *transaminase, etc.,* may be replaced if desired by the corresponding forms *aminotr ansferase, etc..* A number of special words are used to indicate reaction types, *e.g. kinase* to indicate a phosphate transfer from ATP to the named substrate (not 'phosphokinase'), *diphosphokinase* for a similar transfer of diphosphate.

Enzymes catalysing group-transfer reactions should be named *transferase* and the names formed on the pattern *donor:acceptor group-transferred-transferase, e.g. ATP:acetate phosphotransferas*. A figure may be prefixed to show the position to which the group is transferred, *e.g. ATP:*D-*fructose 1-phosphotransferase*. The spelling 'transphorase' should not be used. In the case of the phosphotransferases, ATP should always be named as the donor. In the case of the transaminases involving 2-oxoglutarate, the latter should always be named as the acceptor.

Rule 22.

The prefix denoting the group transferred should, as far as possible, be non-committal with respect to the mechanism of the transfer, *e.g. phospho-*, rather than *phosphate-*.

Rule 23.

The direct addition of-*ase* to the name of the substrate generally denotes a hydrolase. Where this is difficult, *e.g.* for EC 3.1.2.1, the word *hydrolase* may be used. Enzymes should not normally be given separate names merely on the basis of optimal conditions for activity. The acid and alkaline phosphatases should be regarded as special cases and not as examples to be followed. The common name *lysozyme* is also exceptional.

Hydrolysing enzymes should be systematically named on the pattern *substrate hydrolase*. Where the enzyme is specific for the removal of a particular group, the group may be named as a prefix, *e.g. adenosine aminohydrolase* (EC 3.5.4.4). In a number of cases this group can also be transferred by the enzyme to other molecules, and the hydrolysis itself might be regarded as a transfer of the group to water.

Rule 24.

The old names *decarboxylase, aldolase, etc.*, are retained; and *dehydratase* (not 'dehydrase') is used for the hydro-lyases. 'Synthetase' should not be used for any enzymes in this class. The term *synthase* may be used instead for any enzyme in this class (or any other class) when it is desired to emphasize the synthetic aspect of the reaction.

Enzymes removing groups from substrates non-hydrolytically, leaving double bonds (or adding groups to double bonds) should be called *lyases* in the systematic nomenclature. Prefixes such as *hydro-*, *ammonia*-should be used to denote the type of reaction, *e.g.* *(S)-malate hydro-lyase*. Decarboxylases should be regarded as *carboxy-lyases*. A hyphen should always be written before *lyase* to avoid confusion with hydrolases, carboxylases, *etc.*

Rule 25.

Where the equilibrium warrants it, or where the enzyme has long been named after a particular substrate, the reverse reaction may be taken as the basis of the name, using *hydratase, carboxylase, etc., e.g. fumarate hydratase* (in preference to 'fumarase', which suggests an enzyme hydrolysing fumarate).

The complete molecule, not either of the parts into which it is separated, should be named as the substrate.

The part indicated as a prefix to-*lyase* is the more characteristic and usually, but not always, the smaller of the two reaction products. This may either be the removed (saturated) fragment of the substrate molecule, as in *ammonia-, hydro-, thiol-lyases, etc.* or the remaining unsaturated fragment, *e.g.* in the case of *carboxy-, aldehyde*-or *oxo-acid-lyases*.

Rule 26.

Various subclasses of the lyases include a number of strictly specific or group-specific pyridoxal-5-phosphate enzymes that catalyse *elimination* reactions of b-or g-substituted a-amino acids. Some closely related pyridoxal-5-phosphate-containing enzymes, *e.g. tryptophan synthase* and *cystathionine* b-*synthase* catalyse *replacement* reactions in which a b-or g-substituent is replaced by a second reactant without creating a double bond.

Formally, these enzymes appear to be transferases rather than lyases. However, there is evidence that in these cases the elimination of the b-or g-substituent and the formation of an unsaturated intermediate is the first step in the reaction. Thus, applying rule 14, these enzymes are correctly classified as lyases.

Rule 27.

In this class, the common names are, in general, similar to the systematic names which indicate the basis of classification.

Rule 28.

Isomerase will be used as a general name for enzymes in this class. The types of isomerization will be indicated in systematic names by prefixes, *e.g. maleate cis-trans-isomerase, phenylpyruvate keto-enol-isomerase, 3-oxosteroid* D^5-D^4*-isomerase*. Enzymes catalysing an aldose-ketose interconversion will be known as *aldose-ketose-isomerases, e.g.* L-*arabinose aldose-ketose-isomerase*. When the isomerization consists of an intramolecular transfer of a group, the enzyme is named a *mutase, e.g.* the *phosphomutases*; when it consists of an intramolecular lyase-type reaction, *e.g.* it is systematically named a *lyase (decyclizing)*.

Rule 29.

Isomerases catalysing inversions at asymmetric centres should be termed *racemases* or *epimerases,* according to whether the substrate contains one, or more than one, centre of asymmetry: compare,. A numerical prefix to the word *epimerase* should be used to show the position of the inversion.

Rule 30.

Common names for enzymes of this class were previously of the type *XY synthetase*. However, as this use has not always been understood and synthetase has been confused with synthase, it is now recommended that as far as possible the common names should be similar in form to the systematic names.

The class of enzymes catalysing the linking together of two molecules, coupled with the breaking of a diphosphate link in ATP, *etc.* should be known as *ligases*. These enzymes were often previously known as 'synthetases'; however, this terminology differs from all other systematic enzyme names in that it is based on the product and not on the substrate. For these reasons, a new systematic class name was necessary.

Rule 31

The common names should be formed on the pattern *X – Y ligase,* where X – Y is the substance formed by linking X and Y. In certain cases, where a trivial name is commonly used for XY, a name of the type *XY synthase* may be recommended.

The systematic names should be formed on the pattern *X:Y ligase (ADP-forming),* where X and Y are the two molecules to be joined together. The phrase shown in parentheses indicates both that ATP is the triphosphate involved, and also that the terminal diphosphate link in broken. Thus, the reaction is X + Y + ATP = X – Y + ADP + P_i.

Rule 32.

In the special case where glutamine acts as an ammonia-donor, this is indicated by adding in parentheses (*glutamine-hydrolysing*) to a ligase name. In this case, the name *amido-ligase* should be used in the systematic nomenclature.

Enzymes Activity

Enzymes are catalysts that optimize cell activity while minimizing the amount of energy needed to achieve a specific reaction. Enzymes are also energized protein molecules found in every living cell, and are necessary for life. There are over 2000 known enzymes, each of which is involved with specific chemical reaction.

They are any of various proteins, originating from living cells and capable of producing certain chemical changes in organic substances by catalytic action, such as digestion. These proteins, and their function (s), are determined by their shape 1, 2. In cells and organisms, most reactions are catalyzed by enzymes, which are regenerated during the course of a reaction.

Biological catalysts are physiologically important because

they speed up rates of reactions that would otherwise be too slow to support life. Our bodies naturally produce digestive and metabolic enzymes as they are needed.

Specifically, the pancreas produces enzymes that break down foods into nutrients the body can use for energy and other bodily functions.The names of enzymes often include the substrate or substance on which they act, joined with an-ase ending. For example, lactase acts upon lactose and maltase acts on maltose to produce glucose.

Sometimes they are named for their reaction product, for example, sucrase is often called invertase, because invertase is the result of the reaction of sucrose. Enzymes can also bear a name the describes the reaction that is catalyzed. An example of this includes the use of the name oxidase, because oxidase is involved in an oxidation reaction.

TWO TYPES OF ENZYMES:

- Simple enzymes are those composed completely of proteins
- Complex enzymes are those composed of protein and a small organic molecule(s) (also known as holoenzymes)

Enzymes are Categorized by their Function (s):

- Metabolic enzymes catalyze and regulate every biochemical reaction that occurs in our body and are essential to cellular function and health.
- Digestive enzymes are secreted along the digestive tract, break down food into nutrients and waste, and allow the nutrients found in foods to be absorbed into the bloodstream. The pancreas produces most digestive enzymes, but the liver, gallbladder, small intestines, stomach and colon also play critical roles in the production of these enzymes. Examples of digestive enzymes: lipase, protease, amylase, ptyalin, pepsin and trypsin.
- Food enzymes enter the body through the consumption of raw foods or enzyme supplements.

Cooking and processing destroys most enzymes in food.

- Plant Based enzymes are synthetically grown, help break down fat, protein and carbohydrates and function within a broad pH range. Plant enzymes, commonly bromelin (pineapple) and papain (papaya) play an important role in more complete digestion of all foods. They are activated at a temperature higher than normal body temperature, also making them a good anti-inflammatory.
- Proteoplytic enzymes are enzymes that digest proteins, including Trypsin, Chymotrypsin, Pancreatin, Bromelin and Papain. Some are produced by the pancreas and others are supplemental from animals or plants. The primary use of proteolytic enzymes in supplements is, are as digestive aids. They are also used in drugs as anti-inflammatory agents and pain relievers.

Enzymes in Nature

Enzymes are one of the most interesting and important substances found in nature. Firts, it's important to realise that enzymes are not living things. They are inanimate-like minerals. But unlike minerals, they are made by living cells. If we were to look inside a cell we would see many different activities going on. There would be some molecules joining together and others breaking apart. These activities keep the cell alive and enzymes make these activities possible. That's why every cell of every living creature on Earth produces enzymes.

In a broad sense, there are two types of enzyme. Some that help join specific molecules together to form new molecules. Others that help break specific molecules apart into separate molecules. Enzymes play many important roles ouside the cell as well.

One of the best examples of this is the digestive system. For instance, it is enzymes in our digestive system that break food down in our digestive system break food down into small molecules that can be absorbed by the body. Some enzymes in our digestive system break down starch, some proteins and others break down fats.

Four Things to Remember about Enzymes

- *Enzymes are specific*: An enzyme that is able to break fat down would not be able to dissolve protein or starch. Enzymes perform only one specific job. That means an enzyme can do its job with very few side effects. It also explains why there are so many different types of enzyme. To date, 3,000 different types have been identified and there are many more waiting to be discovered.
- *Enzymes are catalysts*: While it is true that an enzyme can only perform one speific job, it is important to know that one enzyme can perform that same job over and over again, millions of times, without being consumed in the process. And enzymes do their job best in the mild ph and temperature conditions found in nature.
- *Enzymes are efficient*: Not only do enzymes work hard, they also work with blinding speed. For instance, there is an enzyme in the liver that helps hydrogern peroxide break down into water and oxygen. What's amazing is that one enzyme can process 5 million hydrogen peroxide molecules in one minute.
- *Enzymes are natural*: Enzymes are proteins. Like all other proteins, enzymes are organic. Once they have done their job, enzymes break down swiftly and can be absorbed back into nature.

ENZYMES IN INDUSTRY

Many people believe enzyme technology is fairly new. However, that's not the case. Enzymes have been used by man from the dawn of civilization. As long as people have been eating bread and cheese, and drinking wine and beer-they have been using enzymes. That's because enzymes help to make these products.

Cheese

Take cheese for example. Most likely, cheese was discovered by ancient hunters who stored milk in the stomachs of slaughtered calves. When exposed to heat, the milk inside the container would turn into a solid-cheese. The reason for this is simple. Calves have an enzyme called chymosin in their stomachs. At ambient temperature chymosin will break down milk protein, causing milk to separate into curd and whey. If we strain away the watery whey we are left with a soft tangy curd or cheese.

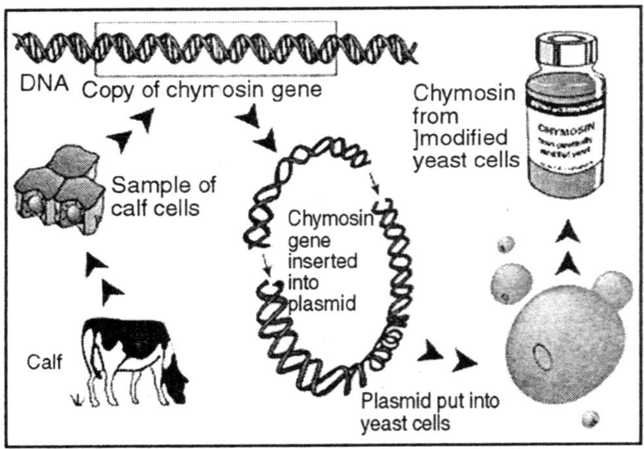

Fig. Chymosin

Chymosin derived from calf stomachs has been used in cheese making ever since.

Detergent

In more recent times, other commercial uses for enzymes have been found. Laundry detergents are a good example. As mentioned, there are certain enzymes which dissolve proteins and others which dissolve fat. Proteins and fat make up two of the major causes of stains on cloting. Grass, blood and egg are all protein stains. Lipstick, frying oil, butter, sauces, and tough stains on cuffs and collars are fat stains. to remove these stains without enzymes is difficult and requires a lot of washing at high temperatures with a lot of detergent.

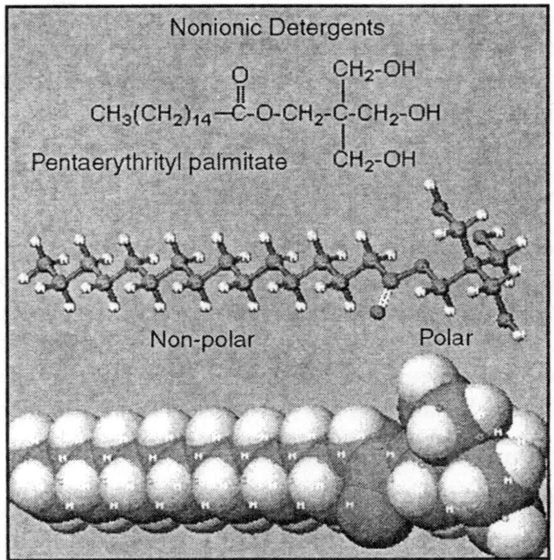

Fig. Detergent

Enzymes can remove these stains better, faster and in a way that is a lot kinder to the clothing and the environment. The enzymes in detergent dissolve stains in the same natural way they help digest food in our stomach-specifically and efficintly. Just a tiny amount of enzymes can dissolve stains that would require much larger amounts of detergent alone.

Because of this they can reduce the amount of detergent actually found in laundry detergent. And since enzymes work best in mild conditions, washing machines using enzymatic detergants can be set at lower tempreatures, reducing electricity consumption by as much as a third. In the future, enzymes could replace more and more of the harsh chemicals found in detergents.

Many other Industries

These two examples represent just the tip of the iceberg. Enzymes are contributing to industry in many other areas. They are starting to replace petroleum-based solvents used to make vegetable seed oil; replacing harsh acids used in the production of glucose products such as corn syrup; taking over part of the role played by clorine in the paper industry; and replacing sulphides used in the tanning industry. Enzymes are also being used instead of pumice stones in the stonewashing of jeans.

Enzymes help Industry and Nature

Today enzymes are helping industry make the products used by society in a way that is less harmful to the environment. In the cases mentioned above, enzymatic processes are, in general, safer and more environment-friendly than the traditional processes they replace. That is why enzymes in particular and biotechnology in general hold such promise.

How Enzymes are Produced

Looking back to the cheese example above, chymosin is an enzyme used to turn milk into cheese. And chymosin is only found in the stomachs of young calves (as well as young sheep, goats, and a few other farm animals). Up until the 60s all the world's cheese was made from chymosin extracted from the stomachs of slaughtered calves. Then two things happened. The demand for cheese increased and the demand for alf meat decreased.

Soon there were not enough slaughtered calf stomachs to produce enough high-quality chymosin required by the expanding cheese industry. To solve the chymosin shortage, it would have been very wasteful and very expensive to raise and slaughter calves just to extract a relatively small amount of chymosin from their stomachs. The cheese industry and the enzyme researchers began to look for another type of organism that produced chymosin.

Microorganisms are Natural Enzyme Factories

The researchers wanted to find an organism that was easier and less expensive to raise than a calf. An organism that reproduced quickly and wouldn't require a lot of space and food. So they began looking among the smallest and simplest organisms they knew: micoorganisms. Microorganisms are very small living organisms. Some examples are bacteria, fungi and yeast. They live in soil or water in every corner of the earth. Because they are so small, they are obviously a lot less complex than a calf.

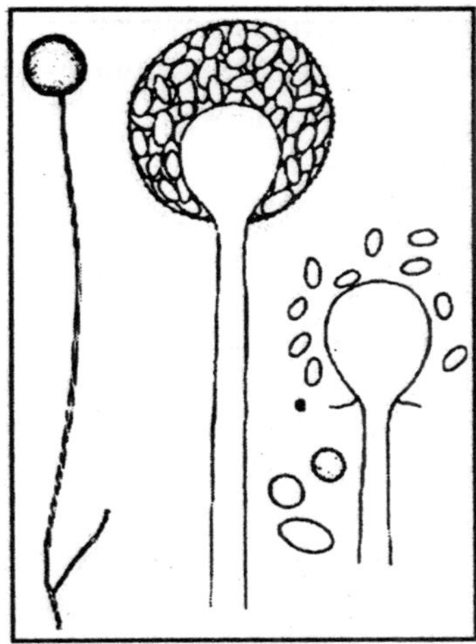

Fig. Mucor

Even so, a simple microorganism can produce many different types of enzyme. In fact, it doesn't take a complex organism to create a complex enzyme-as was proved by a simple fungus named Mucor. This fungus produced a chymosin-like enzyme that was almost the same as that produced by calves. But finding the right enzyme is only half the battle.

Microorganisms Often aren't Suited for *"Life on the Farm"*

Once identified a microorganism needs to be "farmed" in a controlled setting. Most microorganisms can grow and multiply quite well in liquid. So large tanks are used called fermentation tanks to grow microorganisms.

Ideally, a microorganism will grow fast and produce a lot of the desired enzyme at mild temperatures consuming inexpensive nutrients. But like most things in life, the ideal microorganism is hard to come by. As it is, most microorganisms found in the wild are not so well suited for domestication in fermentation tanks.

Some only produce tiny quantities of the desired enzyme, or they may produce undesirable by-products, or take a long time to grow, or require a growth environment that is very difficult and expensive to maintain. The result is that although we often find ourselves with a microorganism that can produce the enzyme needed the cost for cultivating it is prohibitive.

Function and Structure

Enzymes are very efficient catalysts for biochemical reactions. They speed up reactions by providing an alternative reaction pathway of lower activation energy. Like all catalysts, enzymes take part in the reaction-that is how they provide an alternative reaction pathway. But they do not undergo permanent changes and so remain unchanged at the end of the reaction. They can only alter the rate of reaction, not the position of the equilibrium.

Most chemical catalysts catalyse a wide range of reactions. They are not usually very selective. In contrast enzymes are usually highly selective, catalysing specific reactions only. This specificity is due to the shapes of the enzyme molecules. Many enzymes consist of a protein and a non-protein (called the cofactor). The proteins in enzymes are usually globular. The intra-and intermolecular bonds that hold proteins in their secondary and tertiary structures are disrupted by changes in temperature and pH. This affects shapes and so the catalytic activity of an enzyme is pH and temperature sensitive.

Cofactors may be:

- Organic groups that are permanently bound to the enzyme (prosthetic groups)
- Cations-positively charged metal ions (activators), which temporarily bind to the active site of the enzyme, giving an intense positive charge to the enzyme's protein
- Organic molecules, usually vitamins or made from vitamins (coenzymes), which are not permanently bound to the enzyme molecule, but combine with the enzyme-substrate complex temporarily.

Chapter 12

How Enzymes Work

For two molecules to react they must collide with one another. They must collide in the right direction (orientation) and with sufficient energy. Sufficient energy means that between them they have enough energy to overcome the energy barrier to reaction. This is called the activation energy. Enzymes have an active site.

This is part of the molecule that has just the right shape and functional groups to bind to one of the reacting molecules. The reacting molecule that binds to the enzyme is called the substrate. An enzyme-catalysed reaction takes a different 'route'. The enzyme and substrate form a reaction intermediate. Its formation has a lower activation energy than the reaction between reactants without a catalyst.

A simplified picture

Route A reactant 1 + reactant 2 → product

Route B reactant 1 + enzyme → intermediate

intermediate + reactant 2 → product + enzyme

So the enzyme is used to form a reaction intermediate, but when this reacts with another reactant the enzyme reforms.

LOCK AND KEY HYPOTHESIS

This is the simplest model to represent how an enzyme works. The substrate simply fits into the active site to form a reaction intermediate.

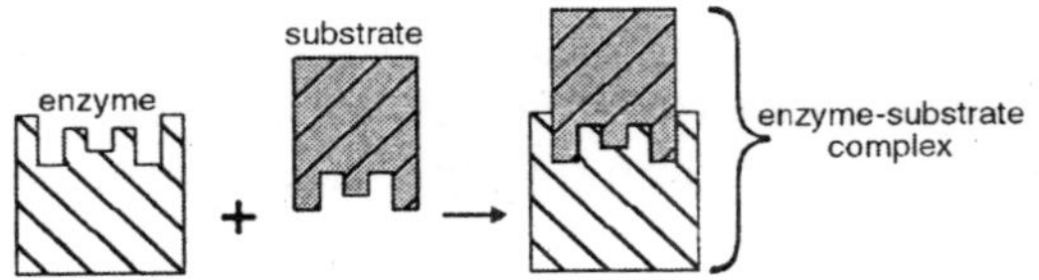

Induced Fit Hypothesis

In this model the enzyme molecule changes shape as the substrate molecules gets close. The change in shape is 'induced' by the approaching substrate molecule. This more sophisticated model relies on the fact that molecules are flexible because single covalent bonds are free to rotate.

Factors Affecting Catalytic Activity of Enzymes

Temperature

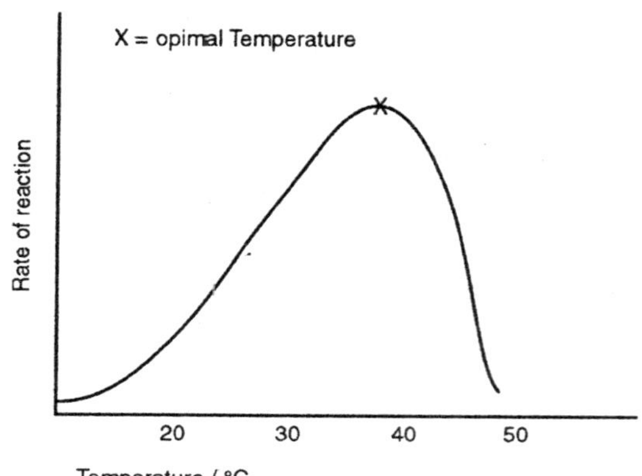

As the temperature rises, reacting molecules have more and more kinetic energy. This increases the chances of a successful collision and so the rate increases. There is a certain temperature at which an enzyme's catalytic activity is at its greatest. This optimal temperature is usually around human body temperature (37.5°C) for the enzymes in human cells.

pH

Above this temperature the enzyme structure begins to break down (denature) since at higher temperatures intra-and intermolecular bonds are broken as the enzyme molecules gain even more kinetic energy.

Each enzyme works within quite a small pH range. There is a pH at which its activity is greatest (the optimal pH). This is because changes in pH can make and break intra-and intermolecular bonds, changing the shape of the enzyme and, therefore, its effectiveness.

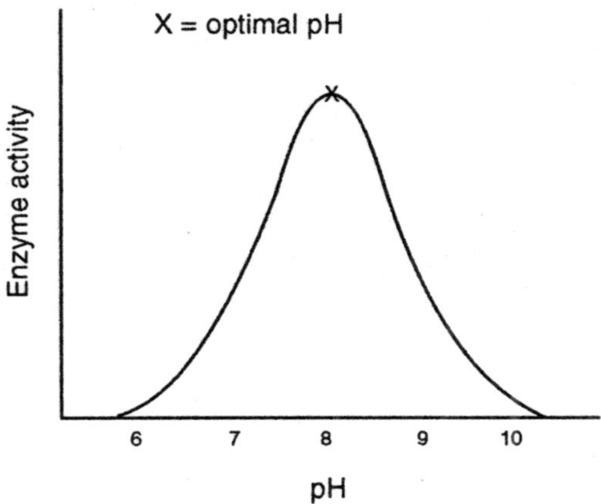

Concentration of Enzyme and Substrate

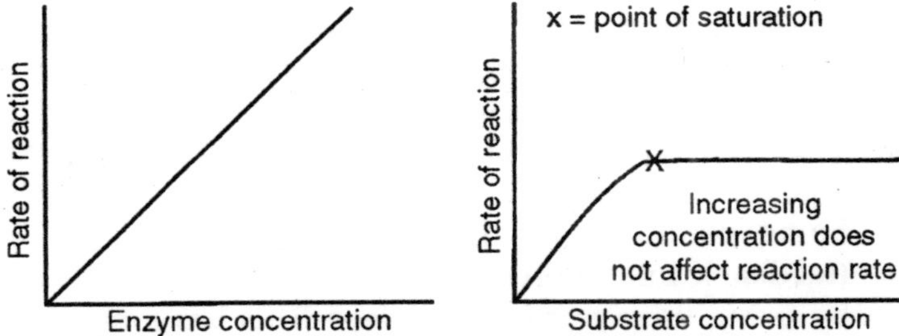

The rate of an enzyme-catalysed reaction depends on the concentrations of enzyme and substrate. As the concentration of either is increased the rate of reaction increases.

For a given enzyme concentration, the rate of reaction increases with increasing substrate concentration up to a point, above which any further increase in substrate concentration produces no significant change in reaction rate. This is because the active sites of the enzyme molecules at any given moment are virtually saturated with substrate. The enzyme/substrate complex has to dissociate before the active sites are free to accommodate more substrate

Provided that the substrate concentration is high and that temperature and pH are kept constant, the rate of reaction is proportional to the enzyme concentration.

Inhibition of Enzyme Activity

Some substances reduce or even stop the catalytic activity of enzymes in biochemical reactions. They block or distort the active site. These chemicals are called inhibitors, because they inhibit reaction. Inhibitors that occupy the active site and prevent a substrate molecule from binding to the enzyme are said to be active site-directed (or competitive, as they 'compete' with the substrate for the active site). Inhibitors that attach to other parts of the enzyme molecule, perhaps distorting its shape, are said to be non-active site-directed (or non competitive).

Immobilized Enzymes

Enzymes are widely used commercially, for example in the detergent, food and brewing industries. Protease enzymes are used in 'biological' washing powders to speed up the breakdown of proteins in stains like blood and egg. Pectinase is used to produce and clarify fruit juices. Problems using enzymes commercially include:

- They are water soluble which makes them hard to recover
- Some products can inhibit the enzyme activity (feedback inhibition) Enzymes can be immobilized by fixing them to a solid surface. This has a number of commercial advantages:
- The enzyme is easily removed
- The enzyme can be packed into columns and used over a long period
- Speedy separation of products reduces feedback inhibition
- Thermal stability is increased allowing higher temperatures to be used
- Higher operating temperatures increase rate of reaction

There are four principal methods of immobilization currently in use:

- Covalent bonding to a solid support
- Adsorption onto an insoluble substance

- Entrapment within a gel.
- Encapsulation behind a selectively permeable membrane.

WORKING PROCESS OF ENZYMES

CATALYTIC MECHANISMS

Open any textbook of biochemistry and we will be presented with an overwhelming number of figures depicting the crystal structures of a multitude of enzymes. Of course, the three-dimensional structures of enzymes are crucial to their functions.

Many enzymes are just regular protein molecules, composed of nothing else than the 20 standard amino acids. If we mix of all these amino acids in free form at, say, 10 mM each, this mixture will not have any significant catalytic activity. It therefore is the precise arrangement of the amino acids in the enzyme molecule that brings about the function.

As the α-amino and α-carboxyl groups of the amino acids are hooked up to each other in the polypeptide chain, they usually do not directly contribute to the catalytic effect of the enzyme. Instead, it is the side chains that are directly engaged win the reaction.

A very good example of this is chymotrypsin. Chymotrypsin is one of the major proteases in the human digestive tract, where its job is to knock down large protein molecules into small peptides that are then further processed by peptidases.

How does chymotrypsin do that? The enzyme-catalyzed reaction is similar to alkaline hydrolysis. In alkaline hydrolysis, a hydroxide ion, which is a strong nucleophile, attacks the carbon in the peptide bond that carries a partial positive charge. In the enzyme-catalyzed reaction, a deprotonated serine residue of the enzyme plays a role similar to that of the hydroxide ion. Now, we know that serine normally is a neutral amino acid – its OH–group does not spontaneously dissociate, no more than the –OH group of alcohol does. The question therefore is, how does the enzyme deprotonate its own serine side chain?

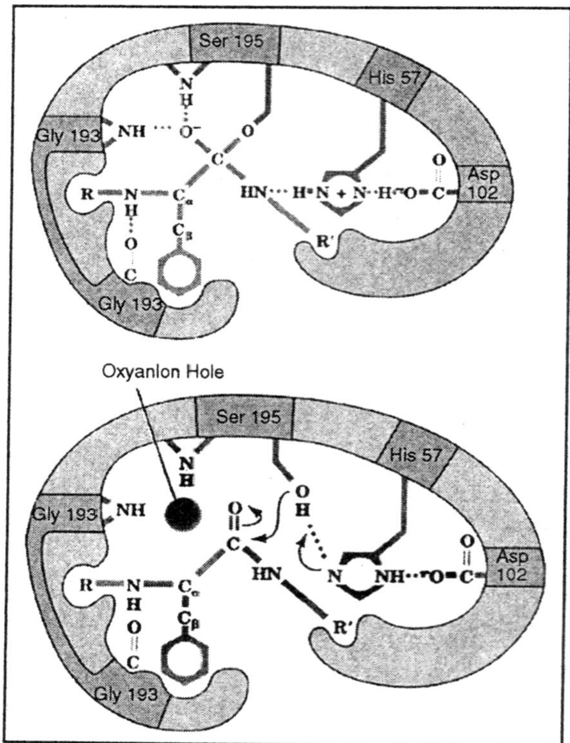

Fig. Chymotrypsin

This is brought about by placing the serine next to a histidine and an aspartate inside the active centre. Aspartate deprotonates histidine, which in turn deprotonates the serine residue. This motive – asp, his, ser – is very widespread among proteases and esterases, so much so that it is commonly called 'the catalytic triad'. E.g., the protease trypsin and several lipases that occur in human metabolism have this motif and share the same mechanism of catalysis.

While with many enzymes the protein molecule and its amino acid side chains are sufficient for catalysis, many others require co-enzymes for their catalytic activity. Very often, both active-site amino acid side chains and coenzymes are required. For example, amino acid transaminases have a molecule of the coenzyme pyridoxalphosphate bound to the active site, which co-operates with an active site lysine in the enzyme's catalytic activity. Most enzymes have just one active site, or if they are multimeric one active site per subunit.

OH

R — C — CH_2 — N NH

O

SER- 195

HIS-57

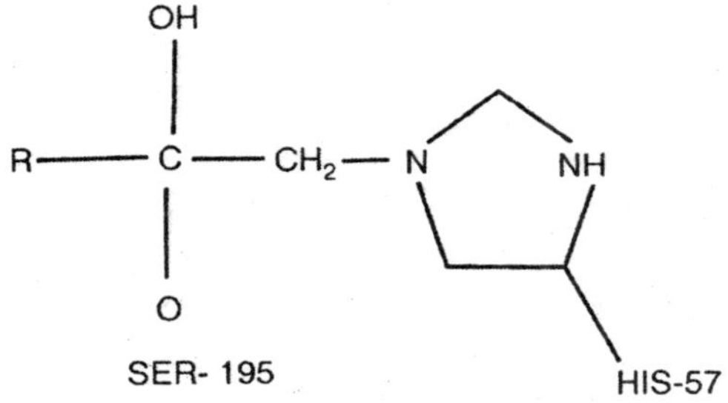

Fig. Structure of Chymotrypsin

However, there are exceptions: Fatty acid synthase has as many as eight different active sites on each subunit. Multi-enzyme complexes such as pyruvate dehydrogenase have one active site per subunit but combine different types of subunits and enzyme activities in one functional assembly.

TYPES OF ENZYMATIC REACTIONS

When looking at enzyme names such as 'transketolase' or 'phosphorylase', we will note that these names don't tell we exactly what reactions the enzymes may catalyze. A complete description should mention the coenzymes required, the substrates and the particular bonds in the substrates that are being severed or created. A nomenclature that meets these criteria has been developed by the Enzyme Commission of the IUBMB (Interna-tional Union of Biochemistry and Molecular Biology). In the IUBMB nomenclature, the enzyme transketolase bears the formidable name.

Such names, of course, are rather lengthy, and their use is not very widespread. To make the tasks of tracking and bookkeeping more manageable, these names are supplemented with numeric codes. In the IUBMB scheme, enzymes are put into one out of six classes according to the reactions they catalyze. These classes are:

- *Oxidoreductases*. These catalyze redox reactions, frequently involving one of the coenzymes NAD^+, $NADP^+$, or FAD.

- *Transferases*. These bring about the transfer of functional groups – e.g., phosphate groups from ATP to another metabolite, which activates the latter and sets it up for subsequent reaction steps.
- *Hydrolases*. These catalyze hydrolysis reactions – e.g., such as those involved in digestion of foodstuffs.
- *Lyases* – these effect elimination reactions that result in the formation of double bonds.
- *Isomerases*. These facilitate the interconversion of isomers. We will meet two examples as soon as we get into glycolysis.
- Ligases, which form new covalent bonds at the expense of ATP hydrolysis.

Of course, within each of these main classes, there are subclasses and sub-sub classes that correspond to details of substrates and mechanisms of the enzyme reactions. Each individual enzyme activity is assigned an individual number within a sub-sub class, so that we wind up with a four-figure designation, which is preceded by the letters 'EC' (Enzyme Commission). One good thing about this classification is that it is rarely used – the 'recommended names', which most of the time happen to be the traditional ones, are used instead. The other good thing is that it has a sound appreciation of priorities.

The single most important enzyme in student lifestyle – namely, alcohol dehydrogenase (or, as IUBMB puts it, alcohol:NAD oxidoreductase). This beneficial enzyme, residing in the liver, degrades ethanol, and without it we would be drunk all the time!

Enzyme Kinetics: Basic Enzyme Reactions

Enzymes are catalysts and increase the speed of a chemical reaction without themselves undergoing any permanent chemical change. They are neither used up in the reaction nor do they appear as reaction products. The basic enzymatic reaction can be represented as follows where E represents the enzyme catalyzing the reaction, S the substrate, the substance being changed, and P the product of the reaction.

Enzyme Kinetics: Energy Levels

Chemists have known for almost a century that for most chemical reactions to proceed, some form of energy is needed. They have termed this quantity of energy, "the energy of activation." It is the magnitude of the activation energy which determines just how fast the reaction will proceed. It is believed that enzymes lower the activation energy for the reaction they are catalyzing. The enzyme is thought to reduce the "path" of the reaction. This shortened path would require less energy for each molecule of substrate converted to product. Given a total amount of available energy, more molecules of substrate would be converted when the enzyme is present (the shortened "path") than when it is absent. Hence, the reaction is said to go faster in a given period of time.

Enzyme Kinetics

A theory to explain the catalytic action of enzymes was proposed by the Swedish chemist Savante Arrhenius in 1888. He proposed that the substrate and enzyme formed some intermediate substance which is known as the enzyme substrate complex. The reaction can be represented as:

$$\underset{\text{substrate}}{S} + \underset{\text{enzyme}}{E} \longrightarrow \underset{\text{enzyme substrate complex}}{ES}$$

If this reaction is combined with the original reaction equation, the following results:

$$\underset{\text{substrate}}{S} + \underset{\text{enzyme}}{E} \longrightarrow \underset{\text{enzyme substrate complex}}{ES} \longrightarrow \underset{\text{Product}}{P} + \underset{\text{enzyme}}{E}$$

The existence of an intermediate enzyme-substrate complex has been demonstrated in the laboratory, for example, using catalase and a hydrogen peroxide derivative. This experimental evidence indicates that the enzyme first unites in some way with the substrate and then returns to its original form after the reaction is concluded.

Chemical Equilibrium

The study of a large number of chemical reactions reveals

that most do not go to true completion. This is likewise true of enzymatically-catalyzed reactions. This is due to the reversibility of most reactions. In general:

$$A + B \xrightarrow{K+1} C + D \quad \text{forward reaction}$$

$$C + D \xrightarrow{K-1} A + B \quad \text{reverse reaction}$$

where K^{+1} is the forward reaction rate constant and K^{-1} is the rate constant for the reverse reaction.

Combining the two reactions gives:

$$A + B \underset{K-1}{\overset{K+1}{\rightleftarrows}} C + D$$

Applying this general relationship to enzymatic reactions allows the equation:.

$$E + S \underset{K-1}{\overset{K+1}{\rightleftarrows}} ES \underset{K-2}{\overset{K+2}{\rightleftarrows}} p + E$$

Equilbrium, a steady state condition, is reached when the forward reaction rates equal the backward rates. This is the basic equation upon which most enzyme activity studies are based.

Factors Affecting Enzyme Activity

Knowledge of basic enzyme kinetic theory is important in enzyme analysis in order both to understand the basic enzymatic mechanism and to select a method for enzyme analysis.

The conditions selected to measure the activity of an enzyme would not be the same as those selected to measure the concentration of its substrate. Several factors affect the rate at which enzymatic reactions proceed-temperature, pH, enzyme concentration, substrate concentration, and the presence of any inhibitors or activators.

Enzyme Concentration

In order to study the effect of increasing the enzyme concentration upon the reaction rate, the substrate must be present in an excess amount; i.e., the reaction must be independent of the substrate concentration. Any change in the amount of product formed over a specified period of time will be dependent upon the level of enzyme present. Graphically this can be represented as:

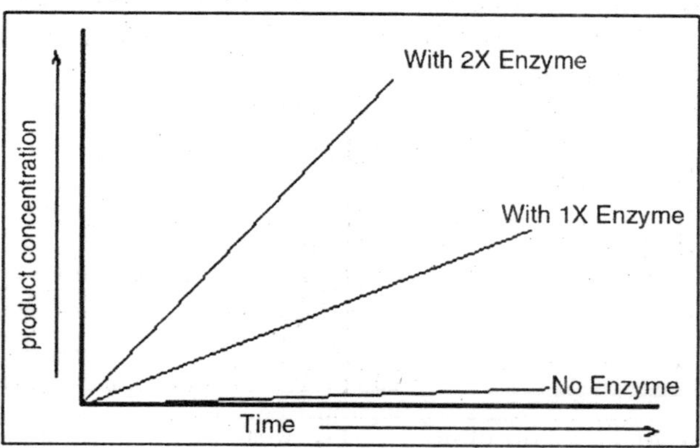

Fig. "Zer order" Reaction Rate is Independent of Substrate Concentration.

These reactions are said to be "zero order" because the rates are independent of substrate concentration, and are equal to some constant k. The formation of product proceeds at a rate which is linear with time. The addition of more substrate does not serve to increase the rate. In zero order kinetics, allowing the assay to run for double time results in double the amount of product.

Table. Reaction Orders with Respect to Substrate Concentration

Order	Rate Equation	Comments
Zero	Rate = k	Rate is independent of substrate concentration
First	Rate = $k[S]$	Rate is proportional to the first power of substrate concentration
Second	Rate = $k[S][S]=k[S]^2$	Rate is proportional to the square of the substrate concentration
Second	Rate = $k[S_1][S_2]$	Rate is proportional to the first power of each of two reactants

The amount of enzyme present in a reaction is measured by the activity it catalyzes. The relationship between activity and concentration is affected by many factors such as temperature, pH, etc. An enzyme assay must be designed so that the observed activity is proportional to the amount of enzyme present in order that the enzyme concentration is the only limiting factor.

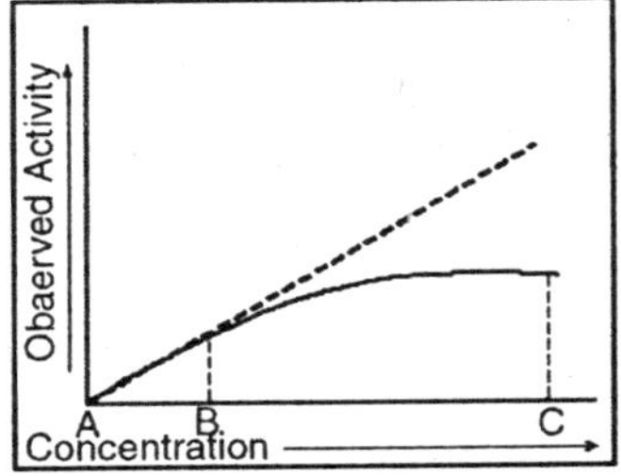

Fig. Activity Concentration.

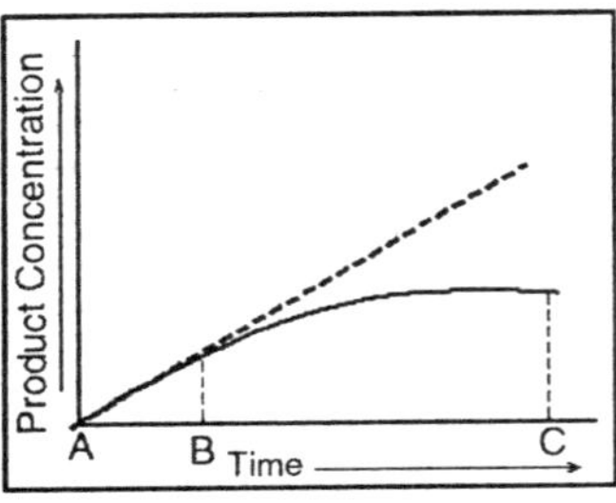

Fig. Reaction Rate Limited by Substrate Concentration

It is satisfied only when the reaction is zero order. In Figure, activity is directly proportional to concentration in the area AB, but not in BC. Enzyme activity is generally greatest when substrate concentration is unlimiting.

When the concentration of the product of an enzymatic Reaction is plotted against time, a similar curve results. Between A and B, the curve represents a zero order reaction; that is, one in which the rate is constant with time. As substrate is used up, the enzyme's active sites are no longer saturated, substrate concentration becomes rate limiting, and the reaction becomes first order between B and C.

To measure enzyme activity ideally, the measurements must be made in that portion of the curve where the reaction is zero order. A reaction is most likely to be zero order initially since substrate concentration is then highest. To be certain that a reaction is zero order, multiple measurements of product (or substrate) concentration must be made. Three types of reactions which might be encountered in enzyme assays and shows the problems which might be enountered if only single measurements are made.

B is a straight line representing a zero order reaction which permits accurate determination of enzyme activity for part or all of the reaction time. This reaction is zero order initially and then slows, presumably due to substrate exhaustion or product inhibition. This type of reaction is sometimes referred to as a "leading" reaction. True "potential" activity is represented by the dotted line.

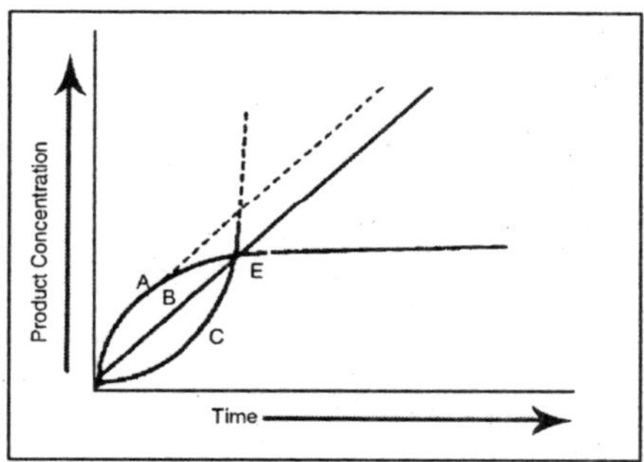

Fig. Leading, Lagging and Liner Reactions

Curve C represents a reaction with an initial "lag" phase. Again the dotted line represents the potentially measurable activity. Multiple determinations of product concentration enable each curve to be plotted and true activity determined. A single end point determination at E would lead to the false conclusion that all three samples had identical enzyme concentration.

Substrate Concentration

It has been shown experimentally that if the amount of the enzyme is kept constant and the substrate concentration is then gradually increased, the reaction velocity will increase until it reaches a maximum. After this point, increases in substrate concentration will not increase the velocity (delta A/ delta T).

It is theorized that when this maximum velocity had been reached, all of the available enzyme has been converted to ES, the enzyme substrate complex. This point on the graph is designated Vmax. Using this maximum velocity and equation, Michaelis developed a set of mathematical expressions to calculate enzyme activity in terms of reaction speed from measurable laboratory data.

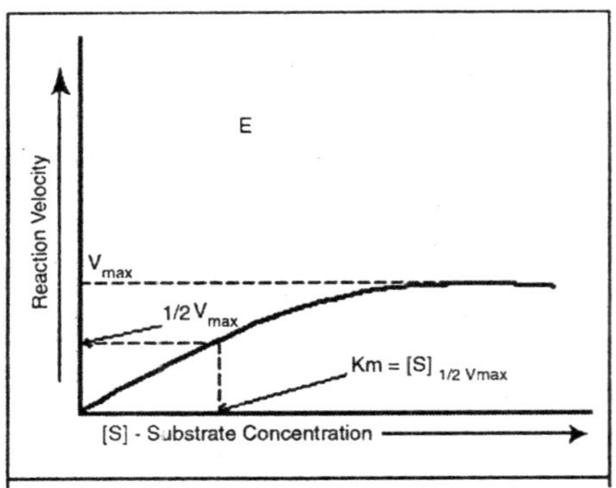

Fig. Effect of SConcentration

$$E + S \underset{K-1}{\overset{K+1}{\rightleftharpoons}} ES \underset{K-2}{\overset{K+2}{\rightleftharpoons}} p + E$$

The Michaelis constant Km is defined as the substrate concentration at 1/2 the maximum velocity. Using this constant and the fact that Km can also be defined as:

$$K_m = K_{-1} + K_2/K_{+1}$$

K^{+1}, K^{-1} and K^{+2} being the rate constants from equation. Michaelis developed the following

$$V_1 \frac{V_{mx}[S]}{K_m + [S]}$$

V_1 = The velocity at any time

[S] = The substrate concentration at this time

V_{mx} = The highest under this set of experimental condition (pH, temperature, etc.)

K_M = The Michaselis constant for the particular enzyme being investigated Michaelis constants have been determined for many of the commonly used enzymes. The size of Km tells us several things about a particular enzyme.

- A small Km indicates that the enzyme requires only a small amount of substrate to become saturated. Hence, the maximum velocity is reached at relatively low substrate concentrations.
- A large Km indicates the need for high substrate concentrations to achieve maximum reaction velocity.

- The substrate with the lowest Km upon which the enzyme acts as a catalyst is frequently assumed to be enzyme's natural substrate, though this is not true for all enzymes.

Effects of Inhibitors on Enzyme Activity

Enzyme inhibitors are substances which alter the catalytic action of the enzyme and consequently slow down, or in some cases, stop catalysis. There are three common types of enzyme inhibition-competitive, non-competitive and substrate inhibition.

Most theories concerning inhibition mechanisms are based on the existence of the enzyme-substrate complex ES. As mentioned earlier, the existence of temporary ES structures has been verified in the laboratory. Competitive inhibition occurs when the substrate and a substance resembling the substrate are both added to the enzyme. A theory called the "lock-key theory" of enzyme catalysts can be used to explain why inhibition occurs.

The lock and key theory utilizes the concept of an "active site." The concept holds that one particular portion of the enzyme surface has a strong affinity for the substrate. The substrate is held in such a way that its conversion to the reaction products is more favorable. If we consider the enzyme as the lock and the substrate the key-the key is inserted in the lock, is turned, and the door is opened and the reaction proceeds.

However, when an inhibitor which resembles the substrate is present, it will compete with the substrate for the position in the enzyme lock. When the inhibitor wins, it gains the lock position but is unable to open the lock. Hence, the observed reaction is slowed down because some of the available enzyme sites are occupied by the inhibitor.

If a dissimilar substance which does not fit the site is present, the enzyme rejects it, accepts the substrate, and the reaction proceeds normally. Non-competitive inhibitors are considered to be substances which when added to the enzyme alter the enzyme in a way that it cannot accept the substrate.

Substrate inhibition will sometimes occur when excessive amounts of substrate are present. Figure shows the reaction velocity decreasing after the maximum velocity has been reached. Additional amounts of substrate added to the reaction mixture after this point actually decrease the reaction rate.

This is thought to be due to the fact that there are so many substrate molecules competing for the active sites on the enzyme surfaces that they block the sites and prevent any other substrate molecules from occupying them. This causes the reaction rate to drop since all of the enzyme present is not being used.

Temperature Effects

Like most chemical reactions, the rate of an enzyme-catalyzed reaction increases as the temperature is raised. A ten degree Centigrade rise in temperature will increase the activity of most enzymes by 50 to 100 per cent. Variations in reaction temperature as small as 1 or 2 degrees may introduce changes of 10 to 20 per cent in the results. In the case of enzymatic reactions, this is complicated by the fact that many enzymes are adversely affected by high temperatures.

The reaction rate increases with temperature to a maximum level, then abruptly declines with further increase of temperature. Because most animal enzymes rapidly become denatured at temperatures above 40°C, most enzyme determinations are carried out somewhat below that temperature. Over a period of time, enzymes will be deactivated at even moderate temperatures. Storage of enzymes at 5°C or below is generally the most suitable. Some enzymes lose their activity when frozen.

Effects of *pH*

Enzymes are affected by changes in *pH*. The most favorable pH value-the point where the enzyme is most active-is known as the optimum *pH*.

Extremely high or low *pH* values generally result in complete loss of activity for most enzymes. *pH* is also a factor in the stability of enzymes. As with activity, for each enzyme there is also a region of *pH* optimal stability. The optimum pH value will vary greatly from one enzyme to another:

Table. *pH* for Optimum Activity

Enzym	*pH* Optimum
Lipase (pancreas)	8.0
Lipase (stomach)	4.0-5.0
Lipase (castor oil)	4.7
Pepsin	1.5-1.6
Trypsin	7.8-8.7
Urease	7.0
Invertase	4.5
Maltase	6.1-6.8
Amylase (pancreas)	6.7-7.0
Amylase (malt)	4.6-5.2
Catalase	7.0

In addition to temperature and *pH* there are other factors, such as ionic strength, which can affect the enzymatic reaction. Each of these physical and chemical parameters must be considered and optimized in order for an enzymatic reaction to be accurate and reproducible.

The term enzyme kinetics implies a study of the speed, rate or velocity of an enzyme catalysed reaction, and of the various factors which may affect this. At the heart of any study of enzyme kinetics is a knowledge of the way in which reaction velocity is altered by changes in the concentration of the enzyme's substrate and of the simple mathematics underlying this. To ease ourselves gently into this we will assume that the enzyme that we are discussing has no special features, such as allosteric properties, and catalyses the conversion of just one substrate to one product.

This may seem unrelated to real enzymes, very few have just one substrate after all, but it will provide a basis which we can expand on later when we study more complex systems. Enzyme kinetics is the study of the chemical reactions that are catalysed by enzymes, with a focus on their reaction rates. The study of an enzyme's kinetics reveals the catalytic mechanism of this enzyme, its role in metabolism, how its activity is controlled, and how a drug or a poison might inhibit the enzyme. Enzymes are usually protein molecules that manipulate other molecules—the enzymes' substrates. These target molecules bind to an enzyme's active site and are transformed into products through a series of steps known as the enzymatic mechanism.

These mechanisms can be divided into single-substrate and multiple-substrate mechanisms. Kinetic studies on enzymes that only bind one substrate, such as triosephosphate isomerase, aim to measure the affinity with which the enzyme binds this substrate and the turnover rate. When enzymes bind multiple substrates, such as dihydrofolate reductase (shown right), enzyme kinetics can also show the sequence in which these substrates bind and the sequence in which products are released. An example of enzymes that bind a single substrate and release multiple products are proteases, which cleave one protein substrate into two polypeptide products. Others join two substrates together, such as DNA polymerase linking a nucleotide to DNA. Although these mechanisms are often a complex series of steps, there is typically one rate-determining step that determines the overall kinetics.

This rate-determining step may be a chemical reaction or a conformational change of the enzyme or substrates, such as those involved in the release of product(s) from the enzyme. Knowledge of the enzyme's structure is helpful in interpreting the kinetic data. For example, the structure can suggest how substrates and products bind during catalysis; what changes occur during the reaction; and even the role of particular amino acid residues in the mechanism.

Some enzymes change shape significantly during the mechanism; in such cases, it is helpful to determine the enzyme structure with and without bound substrate analogs that do not undergo the enzymatic reaction. Not all biological catalysts are protein enzymes; RNA-based catalysts such as ribozymes and ribosomes are essential to many cellular functions, such as RNA splicing and translation.

The main difference between ribozymes and enzymes is that the RNA catalysts perform a more limited set of reactions, although their reaction mechanisms and kinetics can be analysed and classified by the same methods. The reaction catalysed by an enzyme uses exactly the same reactants and produces exactly the same products as the uncatalysed reaction.

Like other catalysts, enzymes do not alter the position of equilibrium between substrates and products. However, unlike normal chemical reactions, enzymes are saturable. This means as more substrate is added, the reaction rate will increase, because more active sites become occupied. This can continue until all the enzyme becomes saturated with substrate and the rate reaches a maximum. The two most important kinetic properties of an enzyme are how quickly the enzyme becomes saturated with a particular substrate, and the maximum rate it can achieve.

Knowing these properties suggests what an enzyme might do in the cell and can show how the enzyme will respond to changes in these conditions.

Enzyme assays are laboratory procedures that measure the rate of enzyme reactions. Because enzymes are not consumed by the reactions they catalyse, enzyme assays usually follow changes in the concentration of either substrates or products to measure the rate of reaction. There are many methods of measurement. Spectrophotometric assays observe change in the absorbance of light between products and reactants; radiometric assays involve the incorporation or release of radioactivity to measure the amount of product made over time.

Spectrophotometric assays are most convenient since they allow the rate of the reaction to be measured continuously. Although radiometric assays require the removal and counting of samples (i. e., they are discontinuous assays) they are usually extremely sensitive and can measure very low levels of enzyme activity. An analogous approach is to use mass spectrometry to monitor the incorporation or release of stable isotopes as substrate is converted into product.

The most sensitive enzyme assays use lasers focused through a microscope to observe changes in single enzyme molecules as they catalyse their reactions. These measurements either use changes in the fluorescence of cofactors during an enzyme's reaction mechanism, or of fluorescent dyes added onto specific sites of the protein to report movements that occur during catalysis.

These studies are providing a new view of the kinetics and dynamics of single enzymes, as opposed to traditional enzyme kinetics, which observes the average behaviour of populations of millions of enzyme molecules. On the left is shown a typical progress curve for an enzyme assay. The enzyme produces product at a linear initial rate at the start of the reaction. Later in this progress curve, the rate slows down as substrate is used up or products accumulate.

The length of the initial rate period depends on the assay conditions and can range from milliseconds to hours. Enzyme assays are usually set up to produce an initial rate lasting over a minute, to make measurements easier. However, equipment for rapidly mixing liquids allows fast kinetic measurements on initial rates of less than one second. These very rapid assays are essential for measuring pre-steady-state kinetics.

Most enzyme kinetics studies concentrate on this initial, linear part of enzyme reactions. However, it is also possible to measure the complete reaction curve and fit this data to a non-linear rate equation. This way of measuring enzyme reactions is called progress-curve analysis. This approach is useful as an alternative to rapid kinetics when the initial rate is too fast to measure accurately.

MICHAELIS-MENTEN EQUATION

Study of the impact made on the rate of an enzyme-catalysed reaction by changes in experimental conditions is known as enzyme kinetics. Knowledge of kinetics can be a very useful tool in understanding the mechanism by which an enzyme carries out its catalytic activity. The effect of substrate concentration on the initial rate of an enzyme-catalysed reaction is a central concept in enzyme kinetics.

When data are generated from experiments of this type and the results plotted as a graph of initial rate (v, y-axis) against substrate concentration ([*S*], *x*-axis), many enzymes exhibit a rectangular hyperbolic curve like the one shown in the diagram below.

Note: the use of square brackets, as for [*S*] above is shorthand notation for "concentration of *S*", a convention that will be used extensively in the derivation below.

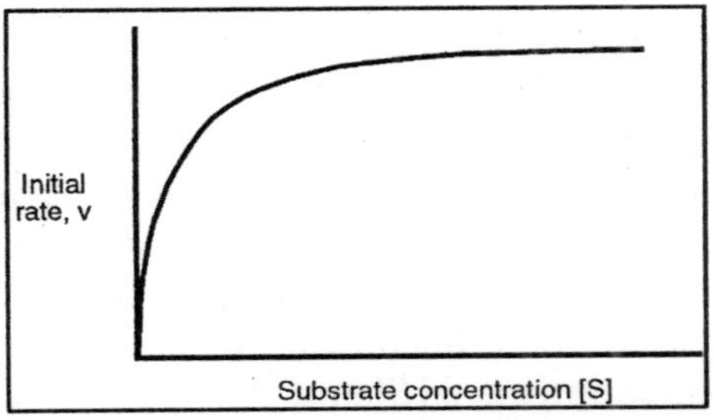

Observations of this type set Leonor Michaelis and Maud Menten thinking about the underlying reasons why a curve should follow this shape and led them to derive an algebraic equation that now bears their names. There are several modern ways to explain the way in which the Michaelis-Menten equation is derived, and one is spelt out below.

DERIVING THE MICHAELIS-MENTEN EQUATION

Start with the generalised scheme for enzyme-catalysed production of a product(*P*) from substrate(*S*). The enzyme(*E*) does not magically convert *S* into *P*, it must first come into physical contact with it, i. e. E binds S to form an enzyme-substrate complex (*ES*). Michaelis and Menten therefore set out the following scheme:

$$E + S \underset{k_1}{\overset{k_1}{\rightleftharpoons}} ES \xrightarrow{k_2} E + P$$

The terms k_1, k_{-1} and k_2 are rate constants for, respectively, the association of substrate and enzyme, the dissociation of unaltered substrate from the enzyme and the dissociation of product (= altered substrate) from the enzyme. Note that there is the theoretical possibility of a reverse reaction, with *ES* complex forming from *E* and *P*, but this can be ignored because we are considering *initial* rates of reaction, i.e., when the enzyme is first provided with substrate, so there should not be any product available to combine with enzyme.

The overall rate of the reaction is limited by the step *ES* to $E + P$, and this will depend on two factors-the rate of that step (i.e., k_2) and the concentration of enzyme that has substrate bound, i. e. $[ES]$. This can be written as:

$$v = k_2 [ES]$$

At this point it is important to draw our attention to two assumptions that are made in this scheme. The first is the availability of a vast excess of substrate, so that $[S]>>[E]$. Secondly, it is assumed that the system is in steady-state, i. e. that the *ES* complex is being formed and broken down at the same rate, so that overall $[ES]$ is constant. The formation of *ES* will depend on the rate constant K_1 and the availability of enzyme and substrate, i.e., $[E]$ and $[S]$. The breakdown of $[ES]$ can occur in two ways, either the conversion of substrate to product or the non-reactive dissociation of substrate from the complex.

In both instances the $[ES]$ will be significant. Thus, at steady state we can write:

$$K_1[E][S] = k_{-1}[ES] + K_2[ES]$$

The next couple of steps are rearrangements of this equation. First of all we can collect together the rate constants on the right-hand side because they are both multiplied by $[ES]$, this gives us:

$$K_1[E][S] = k_{-1} + k_2 [ES]$$

Then dividing both sides by $(k_{-1} + k_2)$, this becomes:

$$\frac{k_1 [E][S]}{(k_{-1} + k_2)} = [ES]$$

Notice that the three rate constants are now on the same side of the equation. As the name implies, these terms are constants, so we can actually combine them into one term. This new constant is termed the Michaelis constant and is written K_M.

$$\frac{(k_{-1} + k_2)}{k_1} = k_M$$

Notice that the three rate constants in the definition of K_M are actually inverted (the other way up) compared with our previous equation.

This is a 'trick' that makes for easier calculation at a later stage. Substituting this definition of K_M into our previous equation now gives us:

$$\frac{[E][S]}{K_M} = [ES]$$

The total amount of enzyme in the system must be the same throughout the experiment, but it can either be free (unbound) *E* or in complex with substrate, *ES*. If we term the total enzyme E_0, this relationship can be written out:

$$[E_0] = [E] + [ES]$$

This can be rearranged (by subtracting [*ES*] from each side) to give:

$$[E] = [E_0] - [ES]$$

So, the [*E*] free in solution is equal to the total amount of enzyme minus the amount that has substrate bound. Substituting this definition of [*E*] back into equation gives us:

$$\frac{([E_0] - [ES])[S]}{K_M} = [ES]$$

This can now be rearranged in several steps. First of all, open the bracket so that the terms [E_0] and [*ES*] are separately multiplied by [*S*]

$$\frac{[E_0][S] - [ES][S]}{K_M} = [ES]$$

Next, multiply each side by K_M, this gives us:

$$[E_0][S] - [ES][S] = k_M[ES]$$

Then collect the two [*ES*] terms together on the same side (we can either think of this as adding [*ES*][*S*] to both sides or as 'carry over and change the sign'-our preference will probably be an indication of how long ago we went to school). This gives:

$$[E_0][S] = K_M[ES] + [ES][S]$$

Then because both terms on the right-hand side are multiplied by [*ES*] we can collect them together into a bracket:

$$[E_0][S] = (K_M + [S])[ES]$$

Dividing both sides by (K_M + [*S*]) now gives us:

$$\frac{[E_0][S]}{[S] + K_M} = [ES]$$

Substituting this left-hand side into Equation 1 in place of [*ES*] results in:

$$v = K_2 \frac{[E_0][S]}{[S] + K_M}$$

The maximum rate, which we can call V_{max}, would be achieved when all of the enzyme molecules have substrate bound.

Under conditions when [*S*] is much greater than [*E*], it is fair to assume that all E will be in the form *ES*.

Therefore [E_0] = [*ES*]. Thinking again about Equation, we could substitute the term V_{max} for v and [E_0] for [*ES*]. This would give us:

$$V_{max} = k_2 [E_0]$$

Notice that $k_2[E_0]$ was present in our previous equation, so we can replace this with V_{max}, giving a final equation:

$$v = \frac{v_{max}[S]}{k_M + [S]}$$

This final equation is actually called the Michaelis-Menten equation. Perhaps this derivation still leaves we puzzled about the importance of the Michaelis-Menten equation. The significance becomes clearer when we consider the case when the rate of reaction is exactly half of the maximal reaction rate (V_{max}). Under those circumstances, the Michaelis-Menten equation could be written:

$$\frac{v_{max}}{2} = \frac{v_{max}[S]}{k_M + [S]}$$

On dividing both sides by V_{max} this becomes:

$$\frac{1}{2} = \frac{[S]}{K_M + [S]}$$

Multiplying both sides by (*KM* + [*S*]) gives:

$$\frac{k_M + [S]}{2} = [S]$$

And then multiplying both sides by 2 further resolves the equation to:

$$k_M + [S] = 2[S]$$

2[*S*] on the right-hand side is the same as [*S*] + [*S*], so we can take away one [*S*] from each side. Thus when the rate of the reaction is half of the maximum rate:

$$k_M = [S]$$

The K_M of an enzyme is therefore the substrate concentration at which the reaction occurs at half of the maximum rate. If we now reconsider the graph that came at the start of this tutorial it could be written:

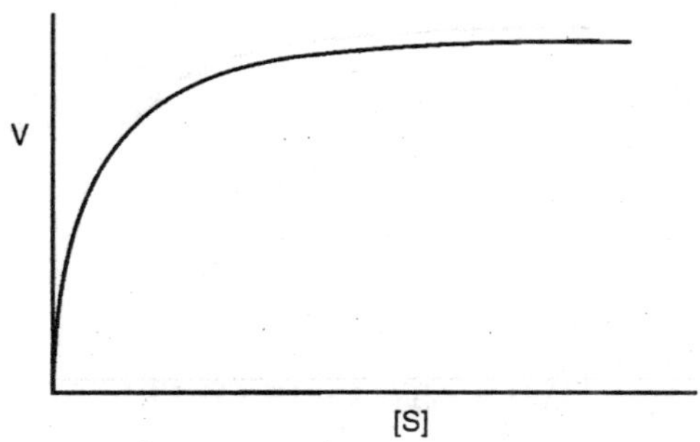

What does this all mean in physical terms? K_M is an indicator of the affinity that an enzyme has for a given substrate, and hence the stability of the enzyme-substrate complex.

Look at the shape of the graph. At low [*S*], it is the availability of substrate that is the limiting factor. Therefore as more substrate is added there is a rapid increase in the initial rate of the reaction-any substrate is rapidly mopped up and converted to product. At the K_M, 50 per cent of active sites have substrate bound. At higher [*S*] a point is reached (at least theoretically) where all of the enzyme has substrate bound and is working flat out. Adding more substrate will not increase the rate of the reaction, hence the levelling out observed in the graph. There are limitations in the quantitative (i.e., numerical) interpretation of this type of graph, known as a Michaelis plot.

The V_{max} is never really reached and therefore V_{max} and hence K_M values calculated from this graph are somewhat approximate. A more accurate way to determine V_{max} and K_M (though still not perfect) is to convert the data into a linear Lineweaver-Burk plot.

"Experimental calculation of V_{max} and K_M"

If, incidentally, we consider that the rate constant k_2 for the conversion of *ES* to *E* + *P* in the initial scheme is the step determining the overall rate of production of *P* (as we have in deriving the Michaelis-Menten equation) then k_2 is actually the same term as k_{cat}, the turnover number.

$$E + S \underset{k_{-1}}{\overset{k_1}{\rightleftharpoons}} ES \xrightarrow{k_2} E + P$$

INTRACTEION WITH K_M AND V_{MAX}

Living systems depend on chemical reactions which, on their own, would occur at extremely slow rates. Enzymes are catalysts which reduce the needed activation energy so these reactions proceed at rates that are useful to the cell.

PRODUCT ACCUMULATION IS OFTEN LINEAR WITH TIME

In most cases, an enzyme converts one chemical (the *substrate*), into another (the *product*). A graph of product concentration vs. time follows three phases as shown in the following graph.

At very early time points, the rate of product accumulation increases over time. Special techniques are needed to study the early kinetics of enzyme action, since this transient phase usually lasts less than a second. For an extended period of time, the product concentration increases linearly with time.

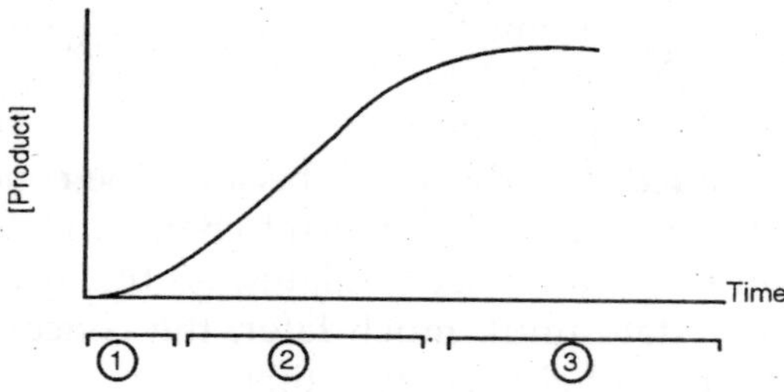

At later times, the substrate is depleted, so the curve starts to level off. Eventually the concentration of product reaches a plateau and doesn't change with time. It is difficult to fit a curve to a graph of product as a function of time, even if we use a simplified model that ignores the transient phase and assumes that the reaction is irreversible. The model simply cannot be solved to an equation that expresses product concentration as a function of time.

To fit these kind of data (called an *enzyme progress curve*) we need to use a programme that can fit data to a model defined by differential equations or by an implicit equation. Prism cannot do this. Rather than fit the enzyme progress curve, most analyses of enzyme kinetics fit the initial velocity of the enzyme reaction as a function of substrate concentration. The velocity of the enzyme reaction is the slope of the linear phase, expressed as amount of product formed per time.

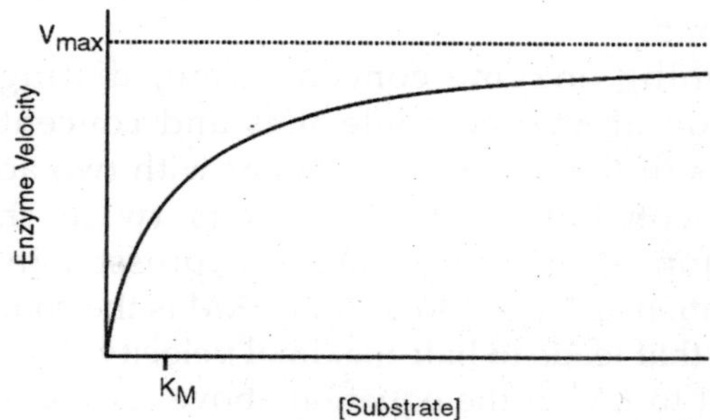

If the initial transient phase is very short, we can simply measure product formed at a single time, and define the velocity to be the concentration divided by the time interval. This chapter considers data collected only in the second phase

The terminology describing these phases can be confusing. The second phase is often called the "initial rate", ignoring the short transient phase that precedes it.

It is also called "steady state", because the concentration of enzyme-substrate complex doesn't change. However, the concentration of product accumulates, so the system is not truly at steady state until, much later, the concentration of product truly doesn't change over time.

Enzyme Velocity as a Function of Substrate Concentration

If we measure enzyme velocity at many different concentrations of substrate, the graph generally looks like this:

Enzyme velocity as a function of substrate concentration often follows the Michaelis-Menten equation:

$$\text{Velocity} = v = \frac{V_{max}[S]}{[S] + K_M}$$

Vmax is the limiting velocity as substrate concentrations get very large. Vmax (and V) are expressed in units of product formed per time. If we know the molar concentration of enzyme, we can divide the observed velocity by the concentration of enzyme sites in the assay, and express Vmax as units of moles of product formed per second per mole of enzyme sites. This is the *turnover number*, the number of molecules of substrate converted to product by one enzyme site per second.

In defining enzyme concentration, distinguish the concentration of enzyme molecules and concentration of enzyme sites (if the enzyme is a dimer with two active sites, the molar concentration of sites is twice the molar concentration of enzyme). *KM* is expressed in units of concentration, usually in Molar units. *KM* is the concentration of substrate that leads to half-maximal velocity. To prove this, set [*S*] equal to *KM* in the equation above. Cancel terms and we' ll see that $V = V_{max}/2$.

THE MEANING OF K_M

To understand the meaning of *Km*, we need to have a model of enzyme action. The simplest model is the classic model of Michaelis and Menten, which has proven useful with many kinds of enzymes.

$$E + S \underset{k-1}{\overset{k_1}{\rightleftarrows}} ES \xrightarrow{K_2} E = P$$

The substrate (*S*) binds reversibly to the enzyme (*E*) in the first reaction. In most cases, we can't measure this step. What we measure is production of product (*P*), created by the second reaction.

From the model, we want to derive an equation that describes the rate of enzyme activity (amount of product formed per time interval) as a function of substrate concentration.

The rate of product formation equals the rate at which ES turns into *E* + *P*, which equals k_2 times [*ES*]. This equation isn't helpful, because we don't know ES. We need to solve for *ES* in terms of the other quantities. This calculation can be greatly simplified by making two reasonable assumptions.

First, we assume that the concentration of *ES* is steady during the time intervals used for enzyme kinetic work. That means that the rate of *ES* formation, equals the rate of *ES* dissociation (either back to *E* + *S* or forward to *E* + *P*). Second, we assume that the reverse reaction (formation of ES from *E* + *P*) is negligible, because we are working at early time points where the concentration of product is very low.

Rate of ES formation = Rate of ES dissolution

$$k_1.[S].[E_{free}] = k_{-1}[ES] + k_2.[ES]$$

We also know that the total concentration of enzyme, Etotal, equals *ES* plus *E*. So the equation can be rewritten.

Solving for *ES*:

$$[ES] = \frac{K_1.[E_{total}][S]}{k_1.[S] + k_2 + k_{-1}} = \frac{[E_{total}][S]}{[S] + \frac{k_2 + k_{-1}}{k_1}}$$

The velocity of the enzyme reaction therefore is:

$$\text{Velocity} = k_2[ES] = \frac{k_2[E_{total}][S]}{[S] + \frac{k_2 + k_{-1}}{k_1}}$$

Finally, define V_{max} (the velocity at maximal concentrations of substrate) as k_2 times Etotal, and KM, the Michaelis-Menten constant, as $(k_2 + k - 1)/k_1$. Substituting:

$$\text{Velocity} = V = \frac{V_{max}[S]}{[S] + K_M}$$

Note that Km is not a binding constant that measures the strength of binding between the enzyme and substrate. Its value includes the affinity of substrate for enzyme, but also the rate at which the substrate bound to the enzyme is converted to product. Only if k_2 is much smaller than $k - 1$ will *KM* equal a binding affinity. The Michaelis-Menten model is too simple for many purposes. The Briggs-Haldane model has proven more useful:

$$E + S \rightleftarrows ES \rightleftarrows EP \rightarrow E + P$$

Under the Briggs-Haldane model, the graph of enzyme velocity vs. substrate looks the same as under the Michaelis-Menten model, but *KM* is defined as a combination of all five of the rate constants in the model.

ASSUMPTIONS OF ENZYME KINETIC ANALYSES

Standard analyses of enzyme kinetics (the only kind discussed here) assume:

- The production of product is linear with time during the time interval used.
- The concentration of substrate vastly exceeds the concentration of enzyme. This means that the free concentration of substrate is very close to the concentration we added, and that substrate concentration is constant throughout the assay.
- A single enzyme forms the product.
- There is negligible spontaneous creation of product without enzyme
- No cooperativity. Binding of substrate to one enzyme binding site doesn't influence the affinity or activity of an adjacent site.
- Neither substrate nor product acts as an allosteric modulator to alter the enzyme velocity.

TO DETERMINE V_{MAX} AND K_M WITH PRISM:

- Enter substrate concentrations into the X column and velocity into the Y column (entering replicates if we have them).
- Pick nonlinear regression from the list of curves and regressions.
- Choose more equations. Enter this equation as a new equation, or choose from the enzyme kinetics equation library.

$$Y = (V_{max} * X)/(Km + X)$$

Variable	Comment
X	Substrate concentration. Usually expressed in µM or mM.
Y	Enzyme velocity in units of concentration of product per time. It is sometimes normalized to enzyme concentration, so concentration of product per time per concentration of enzyme.
Vmax	The maximum enzyme velocity. A reasonable rule for choosing an initial value might be that Vmax equals 1.0 times YMAX. Vmax is expressed in the same units as the Y values.
Km	The Michaelis-Menten constant. A reasonable rule for choosing an initial value might be 0.2*XMAX

COMPARISON OF ENZYME KINETICS WITH RADIOLIGAND BINDING

The Michaelis-Menten equation for enzyme activity has a form similar to the equation describing equilibrium binding.

$$\text{EnzymeVelocity} = V = \frac{V_{max}[S]}{[S] + K_M}$$

$$\text{SpecificBinding} = B = \frac{B_{max[L]}}{[L] + K_D}$$

Note these differences between binding experiments and enzyme kinetics.

- It usually takes many minutes or hours for a receptor incubation to equilibrate. It is common (and informative) to measure the kinetics prior to equilibrium. Enzyme assays reach steady state (defined as constant rate of product accumulation) typically in a few seconds. It is uncommon to measure the kinetics of the transient phase before that, although we can learn a lot by studying those transient kinetics.

- The equation used to analyse binding data is valid at equilibrium-when the rate of receptor-ligand complex formation equals the rate of dissociation. The equation used to analyse enzyme kinetic data is valid when the rate of product formation is constant, so product accumulates at a constant rate. But the overall system is not at equilibrium in enzyme reactions, as the concentration of product is continually increasing.
- *KD* is a dissociation constant that measures the strength of binding between receptor and ligand. KM is not a binding constant. Its value includes the affinity of substrate for enzyme, but also the kinetics by which the substrate bound to the enzyme is converted to product
- Bmax is measured as the number of binding sites normalized to amount of tissue, often fmol per milligram, or sites/cell. Vmax is measured as moles of product produced per minute.

Displaying Enzyme Kinetic Data on a Lineweaver-Burk plot

The best way to analyse enzyme kinetic data is to fit the data directly to the Michaelis-Menten equation using nonlinear regression. Before nonlinear regression was available, investigators had to transform curved data into straight lines, so they could analyse with linear regression.

One way to do this is with a Lineweaver-Burk plot. Take the inverse of the Michaelis-Menten equation and simplify:

$$\frac{1}{v} = \frac{[S]+K_M}{V_{max}[S]} = \frac{[S]}{V_{max}[S]} + \frac{k_M}{V_{max}[S]} = \frac{1}{V_{max}} + \frac{1}{V_{max}} \cdot \frac{1}{[S]}$$

Ignoring experimental error, a plot of $1/V$ *vs.* $1/S$ will be linear, with a Y-intercept of $1/V_{max}$ and a slope equal to Km/V_{max}. The X-intercept equals $1/Km$. Use the Lineweaver-Burk plot only to display our data. Don't use the slope and intercept of a linear regression line to determine values for V_{max} and *KM*. If we do this, we won't get the most accurate values for V_{max} and *KM*.

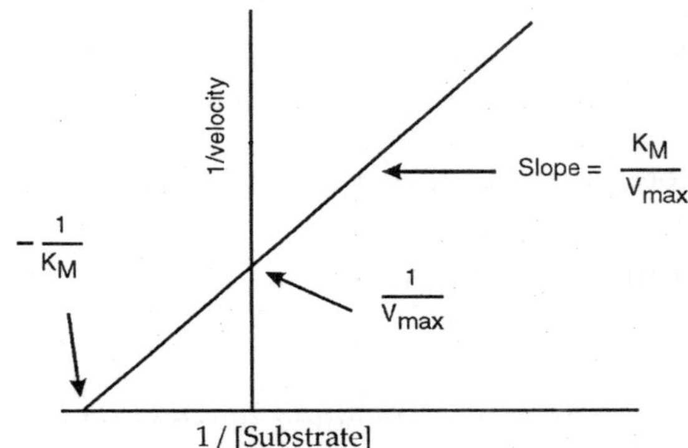

The problem is that the transformations (reciprocals) distort the experimental error, so the double-reciprocal plot does not obey the assumptions of linear regression. Use nonlinear regression to obtain the most accurate values of *KM* and V_{max}. Tip. we should analyse enzyme kinetic data with nonlinear regression, not with Lineweaver-Burk plots. Use Lineweaver-Burk plots to display data, not to analyse data. To create a Lineweaver-Burk plot with Prism, start from a table where *X* is substrate concentration and *Y* is velocity. Then choose Transformations from the list of data manipulations. Check the option boxes to transform both *X* to be $1/X$, and *Y* to be $1/Y$. Be sure to check the option to create a new graph of the results.

From that graph, Analyse and choose linear regression to superimpose the regression line. This linear regression line should not be used to obtain values for Vmax and *Km*. The X-intercept of the regression line will be near-$1/KM$, and the negative inverse of the slope will be near the V_{max}. However, the V_{max} and *KM* values determined directly with nonlinear regression will be more accurate. It is better to draw the line that corresponds to the nonlinear regression fit.

To create a Lineweaver-Burk line corresponding to the nonlinear regression fit, follow these steps:

- Create a new data table, with numerical *X* values and single *Y* values.
- Into row 1 enter $X = -1/KM$ (previously determined by nonlinear regression), Y = 0.
- Into row 2 enter $X = 1/S_{max}$ (S_{max} is the largest value of [substrate] we want to include on the graph) and $Y = (1/V_{max})(1.0 + KM/S_{max})$.
- Note the name of this data table. Perhaps rename it to something appropriate.

ALLOSTERIC ENZYMES

One of the assumptions of Michaelis-Menten kinetics is that there is no cooperativity. If the enzyme is multimeric, then binding of a substrate to one binding site should have no effect on the activity of neighboring sites.

This assumption is often not true. If binding of substrate to one binding site increases the activity of neighboring sites, the term *positive cooperativity* is used. Activity is related to substrate concentration by this equation:

$$\text{Velocity} = v = \frac{V_{max}[S]^h}{[S]^h + K_{0.5}^h}$$

When the variable h equals 1.0, this equation is the same as the Michaelis-Menten equation. With positive cooperativity, h will have a value greater than 1.0.

If there are two interacting binding sites, *h* will equal two (or less, depending on the strength of the cooperativity). If there are three interacting binding sites, *h* will equal 3 (or less). Note that the denominator has the new variable *K*0.5 instead of *KM*.

To fit data to the equation for enzyme activity with positive cooperativity, use the equation below.

For initial values, try these rules: $V_{max} = Y_{max}$, $K = .5 \times \text{mid}$, and $h = 1.0$

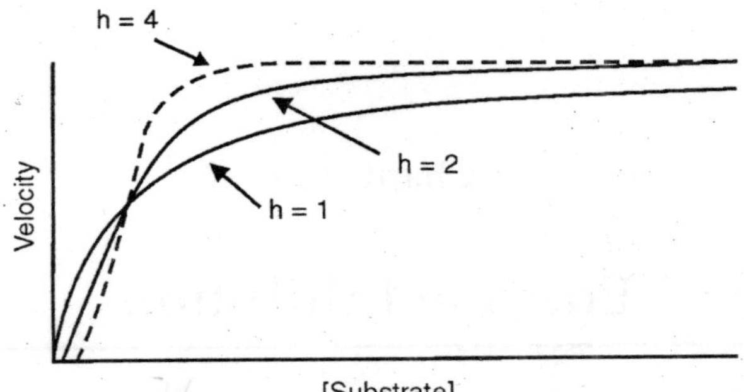

$$Y = V_{max} \times h/(K^h + X^h)$$

The variable h does not always equal the number of interacting binding sites (although h can not exceed the number of interacting sites). Think of h as an empirical measure of the steepness of the curve and the presence of cooperativity.

Chapter 13

Enzyme Inhibition

Enzyme inhibitors are molecules that reduce or abolish enzyme activity. These are either *reversible* (i.e., removal of the inhibitor restores enzyme activity) or *irreversible* (i.e., the inhibitor permanently inactivates the enzyme).

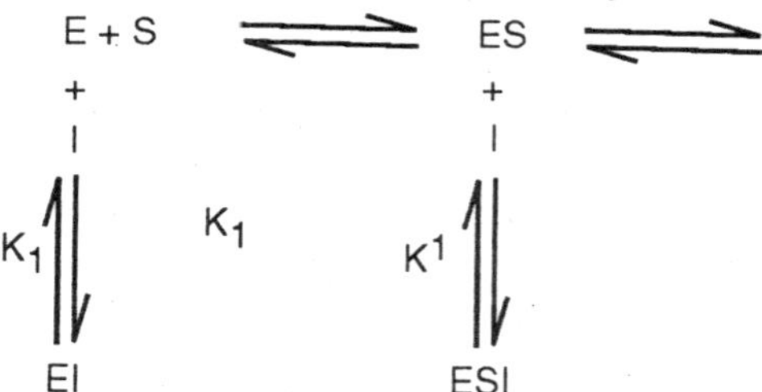

Fig. Kinetic Scheme for Reversible Enzyme Inhibitors.

REVERSIBLE INHIBITORS

Reversible enzyme inhibitors can be classified as competitive, uncompetitive, non-competitive or mixed, according to their effects on K_m and V_{max}. These different effects result from the inhibitor binding to the enzyme *E*, to the enzyme–substrate complex *ES*. The particular type of an inhibitor can be discerned by studying the enzyme kinetics as a function of the inhibitor concentration. The four types of inhibition produce Lineweaver–Burke and Eadie–Hofstee plots that vary in distinctive ways with inhibitor concentration. For brevity, two symbols are used:

$$\alpha = 1 + \frac{[I]}{K_i} \quad \text{and} \quad \alpha' = 1 + \frac{[I]}{k_i'}$$

where K_i and K'_i are the dissociation constants for binding to the enzyme and to the enzyme–substrate complex, respectively. In the presence of the reversible inhibitor, the enzyme's apparent K_m and V_{max} become $(\alpha/\alpha')K_m$ and $(1/\alpha')V_{max}$, respectively, as shown below for common cases.

		Type of Inhibition	**K_m Apparent**	**V_{max} Apparent**
K_i only	$(\alpha' = 1)$	Competitive	$K_m\alpha$	V_{max}
K_i' only	$(\alpha = 1)$	Uncompetitive	$\frac{K_m}{\alpha'}$	$\frac{V_{max}}{\alpha'}$
$K_i = K_i'$	$(\alpha = \alpha')$	Non-competitive	K_m	$\frac{V_{max}}{\alpha'}$
$K_i \neq K_i'$	$(\alpha \neq \alpha')$	Mixed	$\frac{K_m\alpha}{\alpha'}$	$\frac{V_{max}}{\alpha'}$

Non-linear regression fits of the enzyme kinetics data to the rate equations above can yield accurate estimates of the dissociation constants K_i and K'_i.

IRREVERSIBLE INHIBITORS

Enzyme inhibitors can also irreversibly inactivate enzymes, usually by covalently modifying active site residues. These reactions, which may be called suicide substrates, follow exponential decay functions and are usually saturable. Below saturation, they follow first order kinetics with respect to inhibitor.

ENZYME KINETICS IN THE PRESENCE OF AN INHIBITOR

Competitive Inhibitors

If an inhibitor binds reversibly to the same site as the substrate, the inhibition will be competitive. Competitive inhibitors are common in nature.

One way to measure the effect of an inhibitor is to measure enzyme velocity at a variety of substrate concentrations in the presence and absence of an inhibitor. As the graph below shows, the inhibitor substantially reduces enzyme velocity at low concentrations of substrate, but doesn't alter velocity very much at very high concentrations of substrate.

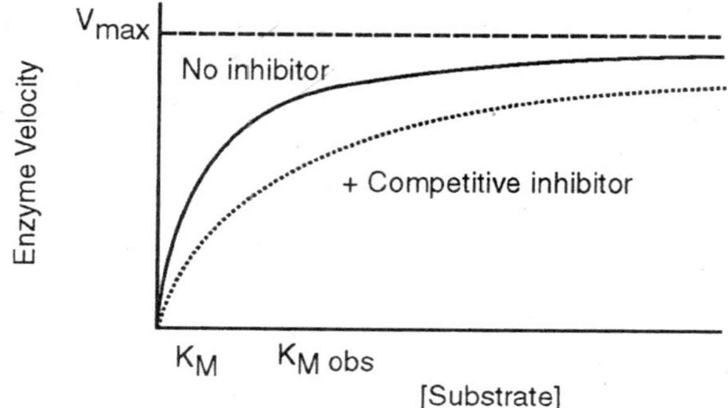

As the graph above shows, the inhibitor does not alter Vmax, but increases the observed KM (concentration of substrate that produces half-maximal velocity, in the presence of a competitive inhibitor). The observed KM is defined by the following equation, where Ki is the dissociation constant for inhibitor binding (in the same concentration units as [Inhibitor]):

$$k_{M,obs} = k_M \left[1 + \frac{\text{Inhibitor}}{K_i}\right]$$

If we have determined the Km plus and minus a single concentration of inhibitor, we can rearrange that equation to determine the Ki.

$$k_i = \frac{[\text{Inhibitor}]}{\dfrac{K_{M,obs}}{K_M} - 1.0}$$

We'll get a more reliable determination of *Ki* if we determine the observed *Km* at a variety of concentrations of inhibitor.

Fit each curve to determine the observed *Km*. Enter the results onto a new table, where *X* is the concentration of inhibitor, and *Y* is the observed *Km*.

If the inhibitor is competitive, the graph will be linear.

Use linear regression to determine the *X*-and *Y*-intercepts. The *Y*-axis intercept equals the *KM* and the *X*-axis intercept equals the negative *Ki*.

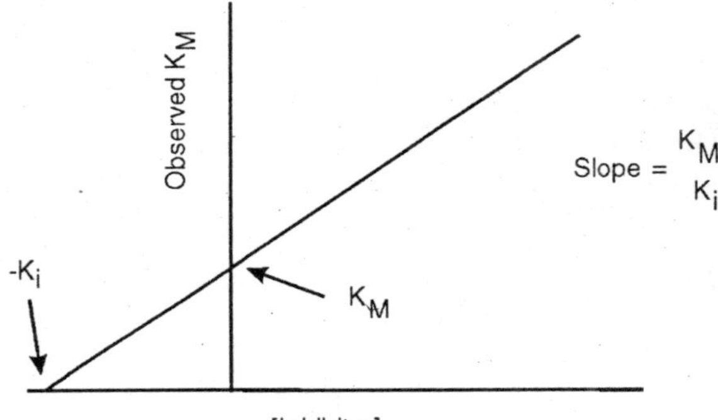

Another experimental design is to measure enzyme velocity at a single concentration of substrate with varying concentrations of a competitive inhibitor. The results will look like this.

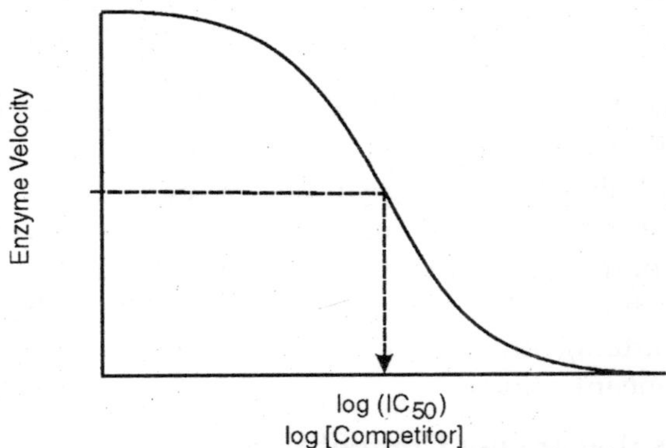

The concentration of competitor that reduces enzyme velocity by half is called the EC50 or IC50. Its value is determined by three factors:

- The dissociation constant for binding of inhibitor to enzyme, the *Ki*. If the *Ki* is low (the affinity is high), the EC50 will be low.

  The subscript i is used to indicate that the competitor inhibited enzyme activity. It is the concentration of the competitor that will bind to half the enzyme sites at equilibrium in the absence of substrate or other competitors.

- The concentration of the substrate. If we choose to use a higher concentration of substrate, it will take a larger concentration of inhibitor to compete for 50 per cent of the activity.
- The *KM*. It takes more inhibitor to compete for a substrate with a low *KM* than for a substrate with a high *KM*. Prism calculates the *Ki*, using the equation of Cheng and Prusoff.

$$k_i = \frac{IC_{50}}{1 + \frac{[\text{Substrate}]}{K_M}}$$

TO DETERMINE THE Ki WITH PRISM

- Enter data with *X* equal to logarithm of inhibitor concentration and *Y* equal to velocity.
- Press Analyse, and choose built-in analyses. From the curves section, choose nonlinear regression.
- Choose the one-site competitive binding equation.
- On the nonlinear regression dialog, choose the option "*Ki* from IC50". This option is usually used for radioligand binding studies, so enter the *KM* where it asks for the *Kd*, and enter substrate concentration where it asks for radioligand concentration. Enter both in *mM* (or any concentration units; only the ratio matters). Enter concentrations, not the logarithm of concentrations.

Inhibitors that are not Competitive

If an inhibitor binds to a site on the enzyme distinct from the site that binds substrate, the inhibition cannot be overcome by increasing the concentration of substrate. The inhibition is not competitive, and the inhibitor decreases the observed V_{max}.

Competitive and non-competitive inhibitors bind reversibly. An inhibitor that binds covalently to irreversibly inactivate the enzyme is called an *irreversible inhibitor* or *inactivator*.

ENZYMES AND INDUSTRIAL APPLICATIONS

Maps produces industrial enzymes originating from microorganisms in the soil. Microorganisms are usually bacteria, fungi or yeast. One microorganism contains over 1, 000 different enzymes. A long period of trial and error in the laboratory is needed to isolate the best microorganism for producing a particular type of enzyme. When the right microorganism has been found, it has to be modified so that it is capable of producing the desired enzyme at high yields.

What Do Enzymes Do for us?

Sector	Application Area	Benefits
Detergents	Household washing and cleaning agents	Wash our clothes in cold water; make our teeth cleaner
Textiles	Denim washing, silk polishing, leather goods softening	Stonewash our jeans; make cotton look and feel like silk; make our leather soft
Food	Baking, brewing, fruit juice processing	Clarify our juice and beer; make bread better; turn corn starch into sugar syrup
Pulp and	Stach conversion, pitch control, bleach-boosting,	Reduce production costs and improve quality deinking, stickies control, slime control

Then the microorganism is 'grown' in trays or huge fermentation tanks where it produces the desired enzyme. With the latest technological advancements of fermenting microorganisms, it possible to produce enzymes economically and in virtually unlimited quantities.

The end product of fermentation is a broth from which the enzymes are extracted.

After this, the remaining fermentation broth is centrifuged or filtered to remove all solid particles. The resulting biomass, or sludge in everyday language, contains the residues of microorganisms and raw materials, which can be a very good natural fertiliser.

The enzymes are then, used for various industrial applications.

ENZYMES IN TEXTILES

Biotechnology is the application of living organisms and their components to industrial products and processes. In 1981 the European federation of biotechnology defined Biotechnology as. Integrated use of Biochemistry, Microbiology and Chemical Engineering in order to achieve the technological application of the capacities of microbes and cultured tissue cells. The rapid developments in the field of genetic engineering have given a new impetus to bio-technology.

This introduces the possibility of. tailoring. organisms in order to optimize the production of established or novel metabolites of commercial importance and of transferring genetic material (genes) from one organism to another. Biotechnology also offers the potential for new industrial processes that require less energy and are based on renewable raw materials. It is important to note that biotechnology is not just concerned with biology, but it is a truly interdisciplinary subject involving the integration of natural and engineering sciences.

Defining the scope of biotechnology is not easy because it overlaps with so many industries, such as, the chemical industry or food industry being the majors, but biotechnology has found many applications in textile industry also, especially in genetic engineering, textile processing and effluent management.

FARMING WITH BUGS

Simple cellular organisms, such as, yeast have been used for millennia, knowingly or otherwise, to make bread, beer and wine. Most industrial applications of biotechnology are still based upon fermentation processes using bacteria and enzymes to digest, transform and synthesize natural materials from one form into another. It is not surprising perhaps that biotechnology is often described as farming with bugs.

For example, the active agent in many transformation processes is an enzyme rather than the cellular living organism itself. Enzymes are not alive themselves but are complex chemical catalysts, which can, in principle, be produced by a number of different methods, including non-biological synthetic routes.

SYSTEMATIC APPLICATION OF BIOLOGICAL SCIENCE

The contribution of science has been to understand to a much greater extent what exactly are the active components and mechanisms of the. Bugs. and their derivatives and therefore to begin to control, manipulate and reproduce their capabilities in a more systematic and intelligent fashion. Modern biotechnology has also brought forward a number of techniques, which do not rely on microbes and enzymes at all, but which directly modify the power of DNA molecule first and foremost among these is genetic engineering.

GENETIC ENGINEERING

With an improved understanding of how different genes are responsible for the various characteristics and properties of a living organism, techniques have been developed for isolating these active components (in particular, the DNA which carries the genetic code) and manipulating them outside of the cell. The next step has been to introduce fragments of DNA obtained from one organism into another, thereby transferring some of the properties and capabilities of the first to the second. For example, scientists working for the leading enzyme producer, Novo of Denmark, discovered that an enzyme produced in minute quantities by one particular fungus had very desirable properties for dissolving fats.

The relevant genes where. spliced. into another micro-organism which was capable of producing the desired enzyme at much higher yield. Genetic engineering methods are being investigated for their potential to produce new kinds of textile fibres. The systems fall into two main groups.

There are those systems that can produce monomeric protein molecules in solution from appropriately engineered genes and include expression in bacteria, cell cultures or in the milk of transgenic animals such as goats or sheep.

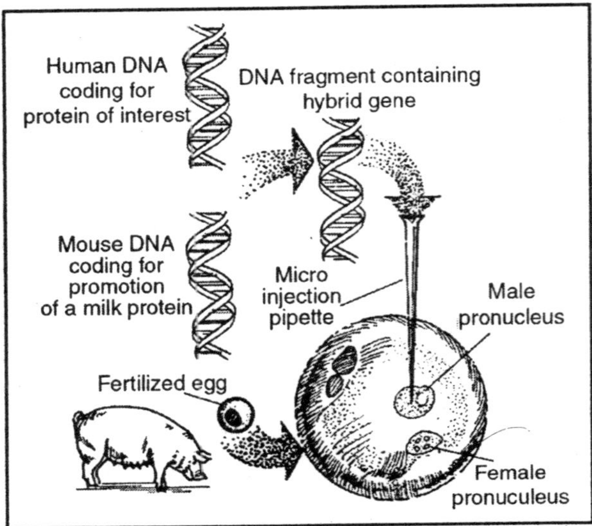

Fig. Genetic Engineering

The protein monomers are then isolated from the chosen system and spun and drawn into fibres. The other approach is to modify keratin fibres such as wool by expressing other proteins in the internal components by trangenesis. Biopesticides based on a strain of soil bacteria known as Bt are already being used for control of caterpillar and beetle pests in a wide variety of fruits, vegetables and crops. More stable, longer lasting and more active Bts are now being developed for the suppression of loopers, bollworms and budworms in cotton. The next stage will be to introduce greater insect and herbicide resistance by direct genetic engineering of the cotton plant itself. Practical results achieved so far include development of a cotton fibre with 50 per cent greater strength than its. parent.

Coloured cottons are also being developed, not only by conventional genetic selection but also by direct DNA engineering to produce, for example, deep blue cotton for denim production. The prospect is even being held out of encouraging natural polyesters such as polyhydroxy butyrate (PHB) to grow within the central hollow channel of the cotton fibre, thereby creating. natural. polyester-cotton.

MONOCLONAL ANTIBODIES

Monoclonal antibodies are protein molecules with an amazing ability to. recognize. specific substances, even at extremely low concentrations. They were first developed for use in medicine to detect and target cancer cells so called. magic bullet approach.; they have also been used for pregnancy testing. Recently, Biocode Company has developed monoclonal antibodies as very sensitive marking tool for the prevention of counterfeiting.

The markers themselves are cheap and safe substances, which can be applied to foodstuffs, drinks and textiles in concentrations of a few parts per million or less. The. codes. embodied in these markers are completely secure but can readily be detected by customs or trading standards inspectors using simple equipment in the field.

The technology has already been evaluated for the marking of branded denims. Methods have been perfected for use in nylon and acrylic resins and markers can also be incorporated into dyestuffs or applied to surfaces using ink jet printers.

DNA PROBES

DNA probes are another technology, which has grown out of genetic engineering. Short pieces of DNA can be designed to stick very specifically to other pieces of DNA and thereby, to help identify target species. The technique can be applied, for example, to distinguish Cashmere from Wool and other goat fibres.

The initial impetus for application of DNA probes in the textile industry has come from importers and processors of specialty animal hairs who have seen a surge in trading and labelling fraud, especially in the wake of recent high fibre prices. Now, similar probes are being identified to distinguish between cotton, ramie, kapok, coir, flax, jute and hemp.

BIOSENSORS

Another way in which biological systems can be used as extremely sensitive analytical and control tools is biosensors.

These employ some change produced by very small quantities of biologically active agents to measure and therefore, in principle, to control chemical and physical reactions. BTTG has been working on the use of certain fungi, which are capable of absorbing and concentrating heavy metal ions such as lead, copper and cadmium.

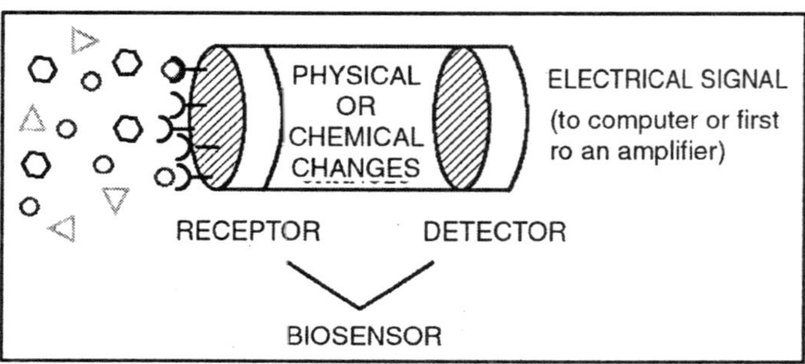

Fig. Biosensor

Resultant changes in the conductivity and dielectric properties of the fungi can be used to measure these species in a process or effluent stream relatively cheaply and easily. Application can be envisaged which incorporate biosensitive materials into textiles, for example, to produce. intelligent. filter media or protective clothing which detects as well as protects against chemicals, gases and biological agents.

APPLICATIONS TO PROCESSING

The use of enzymes in textile processing and after care is already the best established example of the application of biotechnology to textiles and is likely to continue to provide some of the most immediate and possibly dramatic illustrations of its potential in the near-to medium-term future.

Fibre Preparation

Linen is a cellulosic fibre obtained from the flax plant. These fibres are formed in the cortex between the lignified core and the outer layers of the stem, they are separated from the stems by retting, in which matrix components, mainly pectin and lignin are removed and the fibres are separated. Recently, considerable efforts have been put to use enzymes in the retting process to control the process to produce linen fibres of consistent quality. Pre-treatment of the flax with sulphur dioxide gas brings about sufficient breakdown of the woody straw material to speed up enzyme retting whilst preventing excessive bacterial or fungal deterioration of the fibre.

The carbonization process in which vegetable matter in wool is degraded by treatment with strong acid and then subjected to mechanical crushing can, in principle, be replaced by selective enzyme degradation of the impurities.

Fabric Preparation

Desizing using amylase enzymes has been well established for many years. However, there is still considerable scope for improving the speed, economics and consistency of the process, including the development of more temperature stable enzymes as well as a better understanding of how to characterize their activity and performance with respect to different fabrics, sizes, and processing conditions, eg, for pad batch as opposed to jigger desizing. The current application in the textile industry involves mainly hydrolases and now to some extent is Oxidoreductase.

Another desirable development would be enzymes capable of destroying honeydew sugars, insect secretions that cause stickiness and severe processing problems for cotton spinners. An already established application is the use of catalase enzymes to breakdown residual hydrogen peroxide after, for example, pre-bleach of cotton that is to be dyed a pale or medium shade. Reactive dyes are especially sensitive to peroxide and currently require extended rinsing and/or use of chemical scavengers. The enzyme catalase is added after oxidative bleaching and allowed to react for 15 minutes at 30° C-40°C. It degrades the residual peroxide in water and oxygen.

The results obtained were compared with the conventional process and it was found that the outcome of the enzymatic process was excellent. The best suitable conditions are the temperature range of 20°C–60°C, *pH* 5–10 and the application time is 10 min to 15 min.

FINISHING

Biostoning and the closely related process of bio-polishing are perhaps attracting most current attention in the area of enzyme processing.

They are also an excellent illustration of how different industry structural and market considerations can affect the uptake of enzyme technology. Conventional stone washing uses abrasive pumice stones in a tumbling machine to abrade and remove particles of indigo dyestuff from the surfaces of denim yarns and fabric. Cellulase enzymes can also cut through cotton fibres and achieve much the same effect without the damaging abrasion of the stones on both garment and machine.

Disadvantages can include degradation of the fabric and loss of strength as well as. back staining.. A slight reddening of the original indigo shade can also occur. Now processors are learning to play more sophisticated tunes such as achieving a peach skin finish by use of a combination of stones and natural cellulase. Bio-polishing employs basically the same cellulose action to remove fine surface fuzz and fibrils from cotton and viscose fabrics.

The polishing action thus achieved helps to eliminate pilling and provides better print definition, colour brightness, surface texture, drapeability, and softness without any loss of absorbency. Bio-polishing can be used to clean up the fabric surface after the primary fibrillation of a peach skin treatment and prior to a secondary fibrillation process which imparts interesting fabric aesthetics.

A weight loss in the base fabric of some 3%-5% is typical but reduction in fabric strength can be controlled to within 2%-7% by terminating the treatment after about 30 min-40 min using a high temperature or low pH. enzyme stop.. One area that still poses problems is that of tubular cotton finishing. Here, the fibre residues tend to be trapped inside the fabric rather than washed away.

WOOL PROCESSING APPLICATIONS

The international wool secretariat (IWS) together with,

Novo, been developing the use of protease enzymes for a range of wool finishing treatments aimed at increased comfort (reduced prickle, greater softness) as well as improved surface appearance and pilling performance. The basic mechanisms closely parallel those of bio-polishing.

The improved enzyme treatments will allow more selective removal of parts of the wool cuticle, there by modifying the luster, handle and felting characteristics without degradation or weakening of the wool fibre as a whole and without the need for environmentally damaging pre-chlorination treatment.

OTHER PROTEASE APPLICATIONS

Protease enzymes similar to those being developed for wool processing are already being used for the degumming of silk and for producing sand washed effects on silk garments. Treatment of Silk-Cellulosic blend is claimed to produce some unique effects. Proteases are also being used to wash down printing screens after use in order to remove the proteinaceous gums, which are used for thickening of printing pastes.

TEXTILE AFTER-CARE

Enzymes have been widely used in domestic laundering detergents since the 1960s.

Early problems of allergic reactions to some of these enzymes have now largely been overcome by the use of advanced granulation technology. Modern enzyme systems have reduced the use of sodium perborate in detergents by 25% along with the release of harmful salts into the environment. However, enzymes still have to make a corresponding impact upon the commercial laundering market. One of the problems here has been the level of investment in. continuous-batch. or tunnel washers.

These typically afford a residence time of 6 min-12 min which is not long enough for present enzyme systems to perform adequately. More efficient methods of. enzyme kill. are also required because of the extent of water recycling in modern washers.

Table: Types of enzymes and their Effectiveness Against various stains

Enzyme	Effective for
Proteases	Grass, Blood, Egg, Sweat stains
Lipases	Lipstick, Butter, Salad oil, Sauces
Amylases	Spaghetti, Custard, chocloate
Celluloses	Colour brighterning, Softening, Soil removal

Further developments in the field of textile after-care may include treatments to reverse wool shrinkage as well as alternatives to dry cleaning.

CARING FOR THE ENVIRONMENT

Natural and enhanced microbial process have been used to treat waste materials and effluent streams from the textile industry. Conventional activated sludge and other systems are generally well able to meet BOD and related discharge limits on most cases.

The industry faces some specific problems like colour removal from dyestuff effluent and handling of toxic wastes including PCPs and heavy metals.

The synthetic dyes are designed in such a way that they become resistant to microbial degradation under the aerobic conditions. Also, the water solubility and the high molecular weight inhibit the permeation through biological cell membranes.

Anaerobic processes convert the organic contaminants principally into methane and carbon dioxide, usually occupy less space, treat wastes containing up to 30 000 mg/l of COD, have lower running costs and produce less sludge. A novel approach to promoting aerobic degradation in contaminated lagoons and preventing the development of malodorous and unpleasant anaerobic processes.

The development based on a 3-D. biomat. of knitted polyester monofilament is used as a support for the micro-organisms. The mat is stable and resistant to compression; its open supporting structure counteracts the build-up of anaerobic sludges on the bottom of the lagoon.

NEW FIBRE SOURCES

Several possibilities exist for producing entirely new fibre materials, so called biopolymers, using biotechnological process routes, Naturally occurring polyester, PHB is produced by bacterial fermentation of a sugar feed stock and commercially available as. Biopol.

The polymer is stable under normal conditions but biodegrades completely in any microbially active environment. Other biopolymers with textile potential include polylactates and polycaprolactones, which are investigated for medical applications.

Bacterial Cellulose The speciality papers and nonwovens are produced based on bacterially grown cellulose fibres these are extremely fine and resilient and are used as specialized filters, odour absorbers and reinforcing blends with aramids. Genetically Modified Micro-Organisms Attempts have been made to transfer certain advantageous textile properties into micro-organisms where they can be more readily reproduced by bulk· fermentation processes. The spider DNA is transferred into bacteria with the air of manufacturing proteins with the strength and resilience of spider silk for use in bulletproof vests.

DYESTUFFS AND INTER MEDIATES

Attempts have been made to synthesize bacterial forms of indigo as well as fungal pigments for use in the textile industry. Certain micro fungi are capable of yielding up to 30% of their biomass as pigment. Potential non-textile applications include food industry colorants This note of caution needs to be echoed across the whole spectrum of biotechnology developments.

Although biological systems after many attractive possibilities and new approaches to all sorts of problems and needs, considerable advances are still being made in. conventional. technologies, such as, catalysis, chemical synthesis and physical fibre modification which need to be kept in perspective. There is also still great concern in society about the unbridled advance of biotechnology, especially with regard to the modification of natural species with possible unknown long-term consequences.

NATURE'S ENZYMES AND BIOTECHNOLOGIC INDUSTRIAL ENZYMES

Campo Research-JTC Enzymes Laboratories of Japan produces the same important types of enzyme as those found in malt and apples, for instance, as well as a wide range of novel enzymes specialities from various array of temperature and tropical fruits, for use in the cosmetic industry. Campo's enzymes are just as natural, but they come from a different source. Almost all of Campo's cosmetic industrial enzymes originate from microorganisms' genetic extrapolated into fruit cells. Microorganisms can be bacteria, funds or yeasts and different species thrive in different conditions.

The search for a new enzyme can start by examining water and soil samples natural products samples and botanical samples from the far corners of the world. Just one microorganism can contain over 1,000 different enzymes. A long period of trial and error in the laboratory is needed to isolate the best microorganism for producing a particular type of enzyme. When the right microorganism has been found, the work is still not over. The microorganism's genes has to be modified and transcripted into the biotechnologic fruit cells, so that it (micro-organisms genetic transcription combined in the biotechnologic fruits' genes) is capable of producing the desired enzyme at high yields in the biotechnologic fruit cells. Then the fruit cells is 'grown' in huge mass-tissue culture vats or tanks where it produces the desired enzyme. The technique of mass-tissue culture fruit cells has made it possible to produce enzymes economically and in virtually unlimited quantities.

The end-product of mass-tissue-culture is primed matured fruit cells in a broth from which the enzymes are extracted, After this, the remaining media broth is centrifuged or filtered to remove all solid particles. The resulting biomass, or sludge in everyday language, contains the residues of fruit cells and raw materials. For past 25 years, tobacco in Japan has been produced in this way and now, this technology is used in converting this techniques to produce cosmetic enzymes such as fruit enzymes with detergency properties, and skin-care enzymes, sunscreen and UV-filter enzymes.

The by-products (sludge)of this technology is disposed in the best way to minimize impact on the environment. A recycling project was started as early as 1963, converting a large part of the waste sludge into pulverized anhydrous powder for agriculture industry to be used as fertilizer. This is an example of industry giving back to nature, rather than depleting natural resources.

Nature, it seems, has provided all the enzymes the world is likely to need; literally millions, each with a specific role in nature. By isolating them and producing them on a large scale, they can be used in the service of mankind for a vast range of applications. This can be illustrated briefly by going back to the apple and malt mentioned earlier above, as an comparison:

Campo produces a range of sunscreen/solar UV-filtering enzymes called UV-zymes and UV-extremo-zymes,; Pseudo-Phyto-p53 enhancer that is of the same type as that found in all the various plants/herbs to protect themselves from solar UV-radiation. By adding this enzyme, a sunscreen topical formulation the SPF and UV filtering and absorbing properties can dramatically increase and the less or none of any other chemical or organic sunscreen are needed. In this way, industrial enzymes can be used to give nature a helping hand.

Many of the enzymes contained in fruit can also be produced bio-technologically via biotechnological techniques.. In fact, in some cases the industrial enzymes are an improvement on the fermentation-obtained enzymes because they are better adapted to the conditions found in industrial applications. An other example is fruit enzymes with detergency properties, these used to replace the harsh irritant potential surfactants, and consequently these fruit enzyme can be replaced as detergents and its content, can vary considerably 0.1-50 per cent in a concentrate. These are a few of the possibilities opened up by biotechnologic industrial enzymes. There are many more and the main applications of enzymes in industry are described under Industrial Applications.

Chapter 14

Microbial Growth

Contamination of the environment with hazardous and toxic chemicals is one of the major problems facing the industrialized nations today. The petroleum industry is responsible for the generation of high amounts of petroleum hydrocarbons and their derivatives as well as for the pollution of air, soils, rivers, seas and underground water. These compounds undergo modifications by either physico-chemical or biological processes.

Diverse metabolic capabilities of microorganisms have been exploited by man in diverse ways in the biodegradation of waste materials. Microbial activities allowed the mineralization of some petroleum components into carbon dioxide and water, and microbial transformation is considered a major route for complete degradation of petroleum components.

The potentiality of microbes as agents of degradation of several compounds thus indicates biological treatment as the major promising alternative to attenuate environmental impact caused by pollutants. Many scientific approaches have been used in the *in situ* and *ex situ* biodegradation of organic pollutants.

However, the extent of biodegradation is critically dependent on salinity, temperature, *pH*, heavy metals surfactants, nutrients and presence of readily assimilable carbon sources. Many methods such as oxidation, precipitation, ion exchange, solvent extraction, enzyme treatment and adsorption have been used for removing both organic and inorganic materials from aqueous and non-aqueous solution. A variety of microbial growth and biodegradation kinetic models have been developed, proposed and used by many researchers. Such models allow prediction of chemicals that remain at a certain time, calculation of the time required to reduce chemical to certain concentration, estimation of how long it will take before a certain chemical concentration will be attained at a certain point (e.g., a case of aquifer, soil or surface water) and design of bioremediation schemes *in situ* or *ex situ* to remove chemical contaminant to a designed concentration.

On the other hand, it can be used to predict the amount of biomass production achievable at a given time. This review gives an overview of the kinetic models as applied in the prediction of microbial growth and degradation of organic substances. Substrate inhibition and interactions during biodegradation of pollutant mixtures are also discussed.

MICROBIAL GROWTH KINETICS

The relation between the specific growth rate (m) of a population of microorganisms and the substrate concentration (*S*) is a valuable tool in biotechnology. This relationship is represented by a set of empirically derived rate laws referred to as theoretical models. These models are nothing but mathematical expressions generated to describe the behaviour of a given system.

The classical models, which have been applied to microbial population growth, include the Verhulst and Gompertz function. The Gompertz function was originally formulated for actuarial science for fitting human mortality data but it has also been applied deterministically to organ growth. The Gompertz function is based on an exponential relationship between specific growth rate and population density. Equation represents one of its parameterization.

$$N_{(t)} = C\exp\{\exp[-B(t-M)]\}$$

where t = time, $N(t)$ = population density at time t, C = upper asymptotic value, that is; the maximum population density, M = time at which the absolute growth rate is maximal, and B = relative growth rate at M. time. Gibson et al. modified the

Gompertz function to a function which could be applied to the description of cell density versus time in bacterial growth curves in terms of exponential growth rates and lag phase duration

$$\text{Log}\, N_{(t)} = A + D\exp\{-\exp[-B(t - M)]\}$$

Where N(t) = population density at time t, A = value of the lower asymptote (Log *N*(–∞)), *D* = difference in value of the upper and lower asymptote [Log *N*(∞) – log *N*(–∞)], *M* = time at which the exponential growth rate is maximal. The idea of microbial growth kinetics has been dominated by an empirical model originally proposed by Monod. The Monod model introduced the concept of a growth limiting substrate.

$$\mu = \mu_{\max} \frac{S}{K_s + S}$$

Where *m* = specific growth rate, mmax = maximum specific growth rate, *S* = substrate concentration, *Ks* = substrate saturation constant (i.e. substrate concentration at half m_{max}). In Monod's model, the growth rate is related to the concentration of a single growth-limiting substrate through the parameters m_{max} and *Ks*. In addition to this, Monod also related the yield coefficient (*Yx*/*s*) to the specific rate of biomass growth(*m*) and the specific rate of substrate utilization (*q*).

$$Y_{x/s} = \frac{dx}{ds}$$

$$\mu = \frac{Y_{x/s}}{X} \cdot \frac{ds}{dt} \cong Y_{x/s} q$$

DERIVATIVES OF THE MONOD KINETIC MODEL

Penfold and Norris proposed the first kinetic principle for microbial growth. They stated that the relationship between *m* and *S* is best described by a "saturation" type of curve where at high concentration of substrate, the organism grows at a maximum rate (m_{max}) independent of the substrate concentration. Monod's model satisfies this requirement, but it has been criticized particularly because of derivations of m at low substrate concentration.

Owing to the limitations of the Monod's model, a number of structured and unstructured kinetic expressions were put forward to describe the hyperbolic curve characteristic of microbial growth. However, the development of structured models had suffered serious setback due to the complexity of cell growth. Thus, most proposed growth models are unstructured. Three approaches were used to develop the equations for growth kinetics of cells in suspension:

- Describing the influence of physicochemical factors on Monod growth parameters.
- Inclusion of additional constants into the original Monod model to correct for substrate or product inhibition, substrate diffusion, maintenance or effects of cell density on mmax.
- Proposing different kinetic theories, which result in both empirical and mechanistic models.

Like the Monod kinetics, Gompertz function has also been modified to generate models that describe the effect of intrinsic factors such as temperature and oxygen availability on microbial growth parameters. Studies have been carried out on the combined effects of several controlling factors on bacterial growth for example papers by Sutherland et al, McMeekin et al, Wijtzes et al. and Adams et al.

These models are mostly used to predict the change in quality of a food over time and can therefore be applied to estimate the shelf-life of foods.

KINETICS OF BIODEGRADATION

The basic hypothesis of biodegradation kinetics is that substrates are consumed via catalyzed reactions carried out only by the organisms with the requisite enzymes. Therefore, rates of substrate degradation are generally proportional to the catalyst concentration (concentration of organisms able to degrade the substrate) and dependent on substrate concentration characteristic of saturation kinetics (e.g., Michaelis-Menten and Monod kinetics).

Saturation kinetics suggests that at low substrate concentrations (relative to the half-saturation constant), rates are approximately proportional to substrate concentration (first order in substrate concentration), while at high substrate concentrations, rates are independent of substrate concentration (zeroorder in substrate concentration).

In the case of substrates that contribute to the growth of the organisms, rates of substrate degradation are linked to rates of growth (i.e. the concentration of the biomass increases with substrate depletion). The mathematical analysis of such growth-linked systems is more complex than those situations where growth can be ignored. There are a number of situations where it may not be possible to quantify the concentration of substrate-degrading organisms in a heterogeneous microbial community.

However, the rate of substrate depletion can be measured. There are also situations in which the organism concentration remains essentially constant even as the substrate is degraded (i.e. no growth situation). Given these various features of biodegradation kinetics, different models including first-order, zero-order, logistic, Monod (with and without growth) and logarithmic models can be used to describe biodegradation.

Biodegradation kinetics is used to predict concentrations of chemical substances remaining at a given time during *ex situ* and *in situ* bioremediation processes. In most cases, information is based on loss of parent molecule targeted in the process.

The key interest is frequently the decrease in toxicity concentration. Nevertheless, toxicity measurements require bioassays, which are always very difficult and tedious.

Therefore, efficacy of biodegradation is based on chemical measurements, e.g. disappearance of parent molecule, appearance of mineralization products or disappearance of other compounds used stoichiometrically during biodegradation of a compound, for instance, electron acceptors. There are several scenarios by which a compound can be transformed biologically. This includes when the compounds serve as:

- Carbon and energy source

- Electron acceptor
- Source of other cell components.

Other scenarios are the transformation of a compound by non-growing cells (the compound does not support growth) and the transformation of a compound by cometabolism, that is; transformation of a compound by cells growing on other substrate.

The simplest case is where the compound serves as source of carbon and energy for the growth of a single bacterial species. The compound is assumed to be water-soluble, non-toxic and other substrates or growth factors are limiting. In the case of single-substrate limited process, the Monod equation is often used to describe microbial growth and biodegradation processes.

$$\mu = \frac{\mu_{max} S}{K_s + S}$$

$$q = \frac{q_{max} S}{K_s + S}$$

where μ = specific growth rate ($1/X.dX/dt$), q = specific substrate utilization/removal rate (1/X.dS/dt), and $\mu = Yq$, with Y = true growth yield [mass of biomass (X) synthesized per unit of substrate(S) utilized or removed], S = aqueous phase concentration of the compound, Ks = affinity constant or half saturation constant for the compound (meaning the concentration of compound when m or q is maximum).

The hyperbolic equation proposed by Monod was modified by Lawrence and McCarty to describe the effects of substrate concentration (S) on the rate at which a given microbial concentration (X) removes the target substrate ($-dS/dt$). Alternatively, Monod equation can be written in terms of microbial growth by incorporating the net yield coefficient (Y).

$$\frac{dS}{dt} = -\frac{q_{max} SX}{k_s + S}$$

$$\frac{dX}{dt} = -Y\frac{dS}{dt} = \frac{Yq_{max} SX}{K_s + S}$$

The Monod equation has frequently been simplified to an equation, which is either zero or first order in substrate concentration and the kinetics, has been widely used to describe biodegradation of organic contaminations in aquifer systems. The versatility of Monod's equation is attributed to its ability to describe biodegradation rates that follow zero-to first-order kinetics with respect to the concentration of the target substrate.

Moreso, Monod's model describes the dependence of biodegradation rate on the concentration of biomass.

SUBSTRATE INHIBITION OF BIODEGRADATION

When a substrate inhibits its own biodegradation, the original Monod model becomes unsatisfactory. In this case, Monod derivatives that provided corrections for substrate inhibition (by incorporating the inhibition constant K_i) can be used to describe the growth-linked biodegradation kinetics. Among the substrate inhibition models, the Andrew's equation is most widely used.

It is also a good representation of experimental data sets examined in the study of Goudar et al.

$$\mu = \mu_{max} \frac{S}{K_s + S + \frac{S^2}{K_i}}.$$

$$q = q_{max} \frac{S}{K_s + S + \frac{S^2}{K_i}}$$

A generalized Monod type model originally proposed by Han and Levenspiel has been used to account for substrate stimulation at low concentration and substrate inhibition at high concentration.

$$q = \frac{q_{max}\left(1 - \frac{S}{S_m}\right)^n}{S + K_s - \left(1 - \frac{S}{S_m}\right)^m}$$

Where μ = specific substrate consumption rate of cells, q_{max} = maximum consumption rate constant, S = substrate concentration, K_s = the Monod constant, S_m = critical inhibitor concentration above which reaction stops, n and m are constants. Information available on the substrate inhibition of biodegradation are mostly those that described microbial degradation of phenol.

Thus in this review, substrate inhibition of biodegradation is discussed with particular reference to phenol. The inhibitory nature of phenol at high concentrations is well known, and the kinetics of pure and mixed culture microbial growth on phenol have been described by a variety of substrate inhibition models. Most of these models are empirical. However, they are able to provide satisfactory description of phenol biodegradation data, thus providing a convenient means of modelling phenol biodegradation.

Rozich et al. examined 113 microbial curves and reported that among 5 different models, Andrew's provided the best description of observed data.

However, the superiority of the Andrew's equation is not a consistent feature in literatures. Pawlowsky and Howell observed statistically insignificant difference between 5 inhibition models. Yang and Humphrey made similar observation with Andrew's equation and 2 other models in describing phenol degradation by *Pseudomonas putida* and *Trichosporon cutaneum*.

A two-parameter derivative of Andrew's equation was reported to be a better representation of experimental data obtained from mixed culture biodegradation of phenol. When different substrate inhibition models are used to describe experimental data, it becomes difficult to compare kinetic parameters across different studies.

This complicates the application of laboratory kinetic information in the design of biological treatment systems for inhibitory waste.

In the study of Goudar et al, a theoretical basis for selection of an appropriate substrate inhibition model (to solve this problem) was discussed. In this regard, the generalized substrate inhibition model (GSIM) of Tan et al, which describes substrate inhibition of microbial growth using a statistical thermodynamics, was used.

ESTIMATION OF MODEL KINETIC PARAMETERS

Kinetic equations, which describe the activity of an enzyme or a microorganism on a particular substrate, are crucial in understanding many phenomena in biotechnological processes. Quantitative experimental data is required for the design and optimization of biological transformation processes. A variety of mathematical models have been proposed to describe the dynamics of metabolism of compounds exposed to pure cultures of microorgainsms or microbial populations of natural environment.

The Monod equation has been widely used to describe growth-linked substrate utilization. Characterization of the enzyme or microbe-substrate interactions involves estimation of several parameters in the kinetic models from experimental data. In order to describe the true behaviour of the system, it is important to obtain accurate estimates of the kinetic parameters in these models.

Both derivative and integrated forms of equations derived for enzyme catalyzed reactions have been used to estimate kinetic parameters of microbiological processes. Estimates of kinetic parameters Vmax and Km have been calculated by fitting data to either integrated or derivative forms of Michaelis-Menten and Monod equation. Different approaches have been proposed for estimating the kinetic parameters, but progress curve analysis is the most popular because substrate depletion or product formation data from a single experiment are enough for parameter estimation.

In this approach, substrate depletion or product formation-time course is used in the integrated form of the kinetic model for parameter estimation. Some of these differential and integral equations can be found in the papers of Gouder and Delvin, Schmidt et al. and Simkins and Alexander. It is important to note that most kinetic models and their integrated forms are nonlinear. This makes parameter estimation relatively difficult. However, some of these models can be linearized. Various linearized forms of the integrated expressions have been used for parameter estimation. However, the use of linearized expression is limited because it transforms the error associated with the dependent variable making it not to be normally distributed, thus inaccurate parameter estimates.

Therefore, nonlinear least-squares regression is often used to estimate kinetic parameters from nonlinear expressions. However, the application of nonlinear least-squares regression to the integrated forms of the kinetic expressions is complicated. This problem and solutions were discussed by Goudar and Delvin. The parameter estimates obtained from the linearized kinetic expressions can be used as initial estimates in the iterative nonlinear least-squares regression using the Levenberg-Marquardt method.

The kinetic parameters of the Andrew's equation (m_{max}, q_{max}, K_s or K_i) can be estimated with the application of reduced form of the generalized substrate inhibition model (GLIM), reduced to the form of Andrews equation. The linearized expression of this model was used to obtain initial parameter estimates for use in nonlinear regression.

SUBSTRATE INTERACTIONS

Wastewaters from industrial and municipal sources are characterized by presence of mixtures of chemicals. Pollutant mixtures may contain only organic chemicals or may also include inorganic substances such as heavy metals. Co-contamination of natural environments with mixtures of pollutants is an important problem. In biodegradation or bioremediation investigations and projects, it is important to understand and be able to model the fate of specific chemicals.

Development of treatment strategies for soil or water contamination requires consideration of interactions among substrates to control the concentration of individual pollutants to meet regulatory standards. Single substrate kinetic parameters alone cannot describe the phenomena observed with degradation of mixtures. It is important therefore to predict the biodegradation kinetics of pollutant mixtures in a given system. The removal of one component may be inhibited by other components in the mixture and different conditions may be required to degrade different compounds within the mixture. Biodegradation patterns of a compound as component of pollutant mixture and as a single component have been shown to be different. Strong interactions among components of a pollutant mixture have been reported.

In the case of homologous mixture (mixture of substrates serving the same purpose) of carbon and energy substrates, the effect of other compounds in a mixture can be positive or negative due to competitive, and the formation of toxic intermediates by non-specific enzymes. The utilization pattern can change with different mixture compositions, depending on the chemical nature and concentration of the substrate, oxygen concentration and microbial growth rates. Arvin et al. observed both substrate inhibition and stimulation interactions during the aerobic degradation of mixtures of benzene, toluene and *o*-xylene. When toluene or *o*-xylene was degraded in the presence of benzene, the degradative ability of toluene and *o*-xylene by the microorganisms was stimulated. When *p*-xylene and toluene were both present, an inhibition effect on benzene degradation was observed.

Similar inhibition and stimulation of biodegradation have been observed with mixtures of benzene, toluene and *p*xylene. In addition to biodegradation stimulation due to increased growth at low substrate concentrations, stimulation of one compound by another in a mixture can be by induction of catabolic enzymes required for degradation of the second pollutant. This mechanism produces simultaneous degradation of pollutants in mixtures and has been reported for pentachlorophenol and chlorinated aromatics, toluene and *p*-xylene, and toluene. Moreso, one component of a mixture can be degraded in the presence of another by co-metabolism.

In the literature, most studies on the kinetics of biodegradation were on single substrate utilization. However, models of mixed homologous substrate utilization and microbial growth have been proposed. Most of these models have been tested with only two substrates. However, in recent times, models have been proposed and tested for larger mixtures.

Typical examples include the growth of *Escherichia coli* on six sugars, the growth of a mixed culture on five BTEX compounds, and the biodegradation of three polycyclic aromatic hydrocarbons. Like with the homologous substrates, some efforts have also been made to develop kinetic models that described multiple-nutrient-controlled growth with heterologous substrate (substrate that serves different purposes, e.g. carbon and nitrogen mixtures). The heterologous substrate concept assumes that the growth rate can be affected simultaneously by more than one substrate. A "Double Monod" model originally proposed by McGee et al. was used to describe this phenomenon.

$$\mu = \mu_{max} \frac{S_1}{K_1 S_1} \frac{S_2}{K_s + S_2}$$

Where 1 and 2 represent the substrates. However, this multiplicative model has narrow range of utility. Mankad and Bungay had expressed growth rates under dual substrate limitation in terms of weighted average of rates under individual nutrient limitations.

$$\frac{\mu}{\mu_{max}} = (W1)\frac{S_1}{K_1 + S_1} + (W2)\frac{S_2}{k_2 + S_2}$$

Where W (i) is the weight assigned to nutrient i. substituting the functional dependence for the weight functions $W(i)$ into equation yields the Mankad and Bungay's expression for growth rate.

$$W1 = \frac{\frac{K_1}{S_1}}{\frac{K_1}{S_1} + \frac{K_2}{S_2}};\ W2 = \frac{\frac{K_2}{S_2}}{\frac{K_1}{S_1} + \frac{K_2}{S_2}}$$

$$\frac{\mu}{\mu_{max}} = \frac{\frac{K_1}{S_1}}{\frac{K_1}{S_1} + \frac{K_2}{S_2}}\left(\frac{S_1}{K_1 + S_1}\right) + \frac{\frac{K_2}{S_2}}{\frac{K_1}{S_1} + \frac{K_2}{S_2}}\left(\frac{S_2}{K_2 + S_2}\right)$$

For the homologous substrate, sum kinetic model incorporating purely competitive substrate kinetics was proposed by Yoon et al.

$$\mu(S_1, S_2) = \mu_{max} \frac{\mu_{max,1} S_1}{K_{S,1} + S_1 + \left(\frac{K_{S,1}}{K_{S,2}}\right) S_2} + \frac{\mu_{ma,2x} S_2}{K_{s2} + S_2 + \left(\frac{K_{S,2}}{K_{s,1}}\right) S_1}$$

Equation indicates that each substrate exhibits a competitive inhibition effect on the utilization of the other substrate. The competitive substrate kinetics can be used to describe simultaneous and sequential substrate consumption for mixtures of substrate. Another form of dual-substrate interaction with an enzyme is noncompetitive inhibition, characterized by the formation of a non-reactive complex when both substrates are simultaneously bound to the enzyme. The cell growth model based on this type of interaction is expressed mathematically.

$$\mu = \frac{\mu_{max1} S_1}{(K_{S1} + S_1)\left(1 + \frac{S_2}{ks2}\right)} + \frac{\mu_{max2} S_2}{(K_s 2 + S_2)\left(1 + \frac{S_1}{K_{S1}}\right)}$$

Uncompetitive enzyme inhibition model has also been used to describe dual substrate interaction. It differs from non-competitive inhibition in that one of the compounds (the inhibitor) can bind only to the enzyme substrate complex and not the free enzyme. A cell growth model based on uncompetitive substrate interaction is

$$\mu = \frac{\mu_{max1} S_1}{K_{s1} + S_1\left(1 + \frac{S_2}{K_{s2}}\right)} + \frac{\mu_{max2} S_2}{K_{s2} + S_2\left(1 + \frac{S_1}{K_{S1}}\right)}$$

In the sum kinetic models, kinetic parameters determined in the single substrate experiments are used for curve fitting. These models were evaluated by Reardon et al. for biodegradation of benzene, toluene and phenol mixtures, and found that the interactions between these substrates could not be described by sum kinetics models using only parameters determined in a single substrate experiment. An alternative model was formulated by adding an unspecified type of interaction into the sum kinetics model to produce the sum kinetics with interaction parameter (SKIP) model first proposed by Yoon et al.

$$\mu = \frac{\mu_{max1} S_1}{K_{S1} + S_1 + 1_{2,1} S_1} + \frac{\mu_{max2} S_2}{k_{s2} + S_2 1_{1,2} S_1}$$

The interaction parameter Ii, j indicates the degree to which substrate i affects the biodegradation of substrate j. The larger the value, the stronger the inhibition. The SKIP model form for a three-compound mixture is

$$\mu = \left[\frac{\mu_{max1} S_1}{K_{s3} + S_1 + 1_{2,1} S_2 + I_{3,1} S_3}\right] + \left[\frac{\mu_{max2} S_2}{K_{s2} + S_2 \, I_{1,2} S_1 + I_{3,2} S_3}\right] + \left[\frac{\mu_{max3} S_3}{K_{S3} + S_3 \, I_{1,3} S_1 + I_{2,3} S_2}\right]$$

where the subscripts 1, 2 and 3 denote parameters for three different substrates. The extended SKIP model for

N substrates is expressed as

$$\mu = (S_1 S_2 \ldots, S_N) = \sum_{i=1}^{N} \frac{\mu_{max,i} S_i}{K_{si} + S_i + \sum_{j=1,\, j=1}^{N} S_j I_{j.i}}$$

The effect of one substrate on the degradation of another is given by the SjIj,i terms. The values of the interaction coefficients, Ij,i, represent the degree of inhibition exerted by substrate j on substrate i. In a dualsubstrate system, sequential substrate utilization is represented by a large value of I1, 2 and a small value of I 2, 1. The SKIP model satisfactorily described simultaneous and sequential degradation patterns in two different biological systems.

PESTICIDE BIODEGRADATION KINETICS

Increased agricultural practice and pesticide application had resulted in the contamination of natural environments with different kinds of pesticides. Wolt et al. described the design and interpretation of biodegradation studies conducted globally for the purpose of regulatory decision making with respect to pesticide use. Emphasis was placed on the various approaches utilized for addressing degradation studies in soil and the variability in pesticide soil fate parameters.

Understanding pesticide risks requires characterizing pesticide exposure within the environment in a manner that can be broadly generalized across widely varied and soil degradation are especially important for understanding the potential environmental exposure of pesticides. The data obtained from degradation studies are inherently variable and when limited in extent, lend uncertainty to exposure characterization and risk assessment. Pesticide decline in soils reflects dynamically coupled processes of sorption and degradation that add complexity to the treatment of soil biodegradation data from a kinetic perspective.

Additional complexity arises from study design limitations that may not fully account for the decline in microbial activity of test systems or that may be inadequate for considerations of all potential dissipation routes for a given pesticide. Accordingly, kinetic treatment of data must accommodate a variety of differing approaches starting with very simple assumptions as to reaction dynamics and extending to more involved treatments if warranted by the available experimental data.

Selection of the appropriate kinetic model to describe pesticide degradation should rely on statistical evaluation of the data fit to ensure that the models used are not over parameterized. Recognizing the effects of experimental conditions and methods for kinetic treatment of degradation data is critical for making appropriate comparisions among pesticide biodegradation data sets. Statistical evaluation of measures of central tendency for multisoil kinetic studies shows that geometric means better represent the distribution in soil half-lives than do the arithmetic or harmonic means.

METAL INHIBITION OF BIODEGRADATION

In sites co-contaminated with metals and organic compounds, metal toxicity inhibits the activity of organicdegrading microorganisms, impacting both their physiology and ecology, thus reducing the rate of biodegradation of the organic compounds.

Metal inhibition of a broad range of microbial processes including methane metabolism, growth, nitrogen and sulphur conversions, dehalogenation and reductive processes in general is well documented. Thorough reviews of the impacts of metals on many of these processes are available.

The toxicity of metals to microorganisms is dependent on its bioavailabity. Quantification of bioavailable metal concentration is an important step in the process of standardizing experiments to determine the impact of metals on organic pollutant biodegradation. Concentrations of bioavailable metals (metal speciation) can be estimated from solution phase using ion-selective electrodes and atomic absorption spectroscopy.

Biological systems involving immunoassay or bioreporters have been used for mercury. However, the use of immunoassay and bioreporters is limited because of variation in measurements depending on the metal resistance of the bioreporter system used. The application and limitations of immunoassay and bioreporters for metal detection have been reviewed by Neilson and Maier.

As the alternative, bioavailable metal concentrations as a

function of pH and ionic strength can be predicted using geochemical modeling software's. A number of computational models have been developed to predict the impact of metal on organic biodegradation. These models accounted for metal inhibition by incorporating metal inhibition constant (*Ki*) to conventional growth or degradation model.

For example, Amor et al. used a form of the Andrew's equation (originally used to describe substrate inhibition of microbial growth or substrate degradation) to model the effect of cadmium, zinc and nickel on rates of alkyl benzene biodegradation.

BIOREACTOR

A bioreactor may refer to any device or system that supports a biologically active environment. In one case, a bioreactor is a vessel in which is carried out a chemical process which involves organisms or biochemically active substances derived from such organisms.

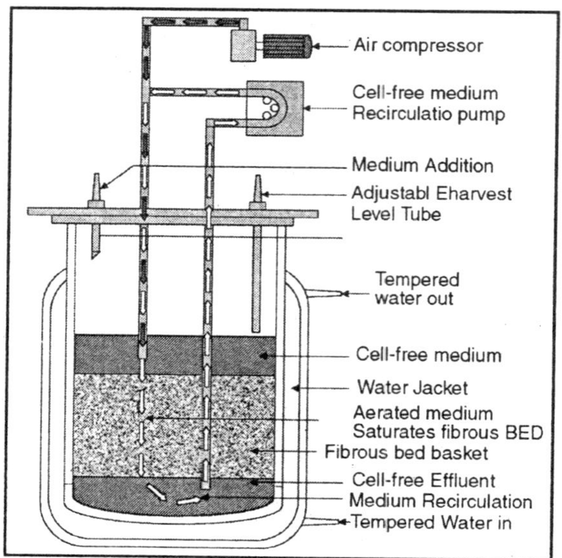

Fig. Bioreactor

This process can either be aerobic or anaerobic. These

bioreactors are commonly cylindrical, ranging in size from liters to cubic meters, and are often made of stainless steel.

BIOREACTORS IN BIOTECHNOLOGY

Bioreactors are the places where proteins for biotechnology are made during upstream processing. Today, bioreactors can be vessels made of plastic, glass or steel in which cells are cultured or, alternatively, whole plants or animals can be genetically engineered to produce a particular protein (transgenics). In this discussion we will focus on the types and characteristics of the glass, steel or plastic bioreactors used to house suspension or fixed cells for the purpose of making proteins during upstream processing.

Bioreactor vessels (also referred to as fermenters) can be almost any vessel capable of holding cells. In this course we will be using 50ml and/or 500ml glass Bellco Spinner Bottles with a plastic cap for holding a shaft with an impeller for mixing and a glass 5000ml (5liter) Bioflo 3000 bioreactor with a stainless steel headplate for holding a shaft with an impeller for mixing, various sensors, a sparging tube and a sampling port for culturing suspension cells for upstream processing of proteins. In addition, during the scale-up of bacterial and yeast cell cultures for upstream processing, we will also be using shake flasks and/or test tubes which also could be termed "bioreactors". Other simple bioreactors include t-flasks and roller bottles.

Once assembled, the bioreactor is sterilized either before or after the addition of media. The sterilization is either in place (SIP) or the bioreactor is placed in an autoclave and sterilized there. Following sterilization of the first bioreactor used in the scale-up process, one milliliter of cells to be cultured is added to the media in this bioreactor. Typically, the one milliliter of cells comes from a master cell bank laid down during process development and stored in liquid nitrogen or in a-86 degree Centigrade freezer.

This one milliliter of cells from the master cell bank ordinarily contains about 1 million live cells. After these cells have grown and multiplied a number of generations so that the concentration of cells reaches 1 million cells/ml again, this batch of cells is added to a larger bioreactor vessel containing a larger volume of media. Scale-up continues until the cells reach the final bioreactor where the final stage of upstream processing occurs. The media or the cells are then harvested for the downstream processing of proteins produced during culture in the final bioreactor.

The final bioreactor used in industry for upstream processing of proteins is far larger than the final bioreactors used in this class. Typically, for the production of human therapeutic proteins, a stainless steel bioreactor constructed of the best quality virgin steel with a working volume of 2000 to 5000 liters is used.

However, stainless steel bioreactors with working volumes of up to 100,000 liters are also in use today. The most common shape for these stainless steel bioreactors is a vessel that is about as wide as it is tall containing a Rushton impeller which looks a bit like a boat's propeller. This is a stirred tank bioreactor. An alternative to this shape is the airlift bioreactor which is quite a lot longer than it is wide (a kind of bullet shape) that uses air sparged at its bottom as a mixing device.

There are many more alternatives to these typical final bioreactors. At one company, the final bioreactor for the manufacture of the protein, tissue plasminogen activator (tPA), is carried out in multiple 10 liter glass Bellco Spinner Bottles, with previous scale-up steps carried out in 1000 ml and 100ml Bellco Spinner Bottles. (Bellco Spinner Bottles are available from 25 ml to 36,000 ml.) For industrial upstream processing of proteins various types of plastic bioreactors are also in use including fixed cells in a fluidized-bed bioreactor. In this bioreactor the cells are fixed in a compartment and media is constantly pumped to the cells in the fixed compartment and constantly withdrawn and pooled for downstream processing.

Another plastic bioreactor is the hollow fibre bioreactor in which cells grow inside hollow fibers or tubes and media is pumped either through the hollow fibers or over the surface of the hollow fibers and constantly withdrawn and pooled for downstream processing. Another sort of bioreactor is the perfusion bioreactor. In industry, where quality control in the form of cGMP is a necessity, many parameters of bioreactors must be controlled and recorded. It is typical to use biosensors connected to computers to control and record these parameters. This is referred to as process control.

This allows for the monitoring of various parameters and provides negative feedback to insure that the bioreactor will operate within pre-defined limits. This is not un-like our bodies which have sensors and negative feedback mechanisms to keep us operating in homeostasis (or within pre-determined limits).

The physical parameters that can be measured on-line during fermentation or cell culture include pH, temperature, dissolved oxygen (DO), dissolved CO_2, impeller speed (or agiation rate), torque, power, foaming, volume of liquid, CO_2 and O_2 in exit gasses, substrate concentration, product concentration, and biomass concentration (or turbidity).

Some of these parameters are still measured by fermentation or upstream processing operators or technicians, by "hand". For instance, it is common for the operator or technician to remove samples at periodic intervals and check the biomass concentration or turbidity by taking an OD reading on the spectrophotometer and also by staining the cells in the sample with vital dyes to determine the live cell number/ ml of medium.

pH Control Loop

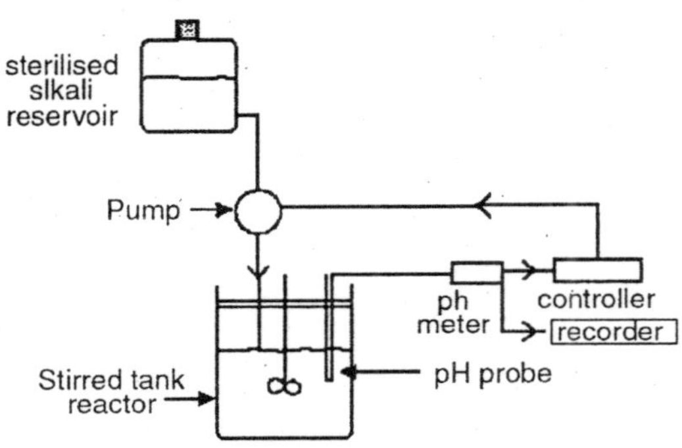

Fig. Dissolved Oxygen Control Loop

Quality Control technicians often measure the concentration of the protein of interest in such samples. Many sensors are still under development inlcuding those that would measure the concentration of various enzymes, substrates, ions, ATP, DNA and RNA and others.

In our laboatory, the process controlled New Brunswick Bioflo 3000 bioreactor is connected to a computer loaded with New Brunswick AFS Software. In our yeast culture protocol for the making of human serum albumin (HSA) control of *pH*, DO, and the addition of media will take place automatically. A schematic representation of a *pH* control loop and a dissolved oxygen control loop taken from one of our texts is shown above.

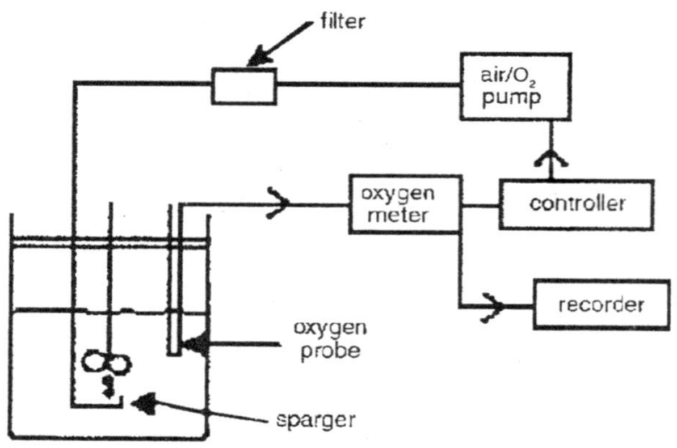

Encyclopaedia

of

BIOINFORMATICS SCIENCE, TECHNOLOGY AND ENGINEERING

Encyclopaedia
of
Bioinformatics Science, Technology and Engineering

Volume 2

Prof. Nirmal Chandra Pradhan

ANMOL PUBLICATIONS PVT. LTD.
NEW DELHI-110 002 (INDIA)

ANMOL PUBLICATIONS PVT. LTD.
Regd. Office: 4360/4, Ansari Road, Daryaganj,
New Delhi-110 002 (India)
Ph.: 23278000, 23261597
Branch Office: No. 1015, Ist Main Road, BSK IIIrd Stage
IIIrd Phase, IIIrd Block,
Bangalore-560 085 (India)
Tel.: 080-41723429
Visit us at: www.anmolpublications.com

Encyclopaedia of Bioinformatics Science, Technology and Engineering

© Reserved

First Edition, 2009

ISBN 978-81-261-4120-3 (Set)

PRINTED IN INDIA

Printed at Mehra Offset Press, Delhi.

Contents

Volume 3

Preface

Bioinformatics is the application of computer technology to the management of biological information. Computers are used to gather, store, analyze and integrate biological and genetic information which can then be applied to gene-based drug discovery and development. The need for Bioinformatics capabilities has been precipitated by the explosion of publicly available genomic information resulting from the Human Genome Project. The goal of this project - determination of the sequence of the entire human genome.

The science of Bioinformatics, which is the melding of molecular biology with computer science, is essential to the use of genomic information in understanding human diseases and in the identification of new molecular targets for drug discovery. In recognition of this, many universities, government institutions and pharmaceutical firms have formed bioinformatics groups, consisting of computational biologists and bioinformatics computer scientists. Such groups will be key to unraveling the mass of information generated by large scale sequencing efforts underway in laboratories around the world. Recent years have seen an explosive growth in biological data. Large sequencing projects are producing increasing quantities of nucleotide sequences. The latest release of GenBank exceeded one billion base pairs. Not only the size of sequence data is rapidly increasing, but also the number of characterized genes from many organisms and protein structures doubles about every two years. To cope with this great quantity of data, a new scientific discipline has emerged: *bioinformatics, biocomputing* or *computational biology*.

Author

Chapter 15

Immune System

The immune system is a network of cells, tissues, and organs that work together to defend the body against attacks by "foreign" invaders. These are primarily microbes—tiny organisms such as bacteria, parasites, and fungi that can cause infections. Viruses also cause infections, but are too primitive to be classified as living organisms.

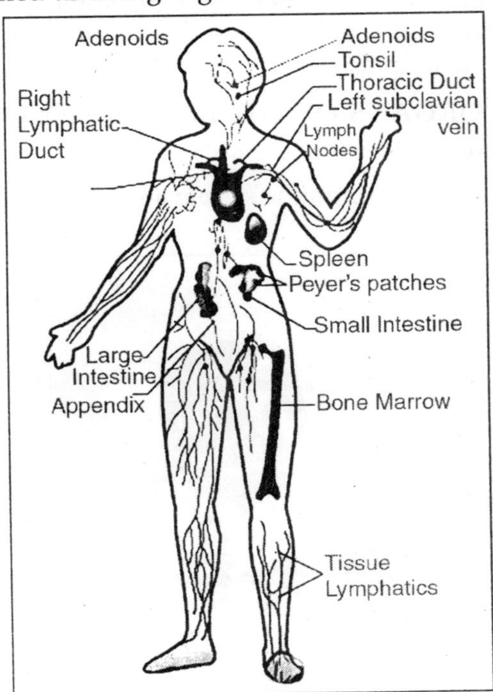

Fig. Immune System

The human body provides an ideal environment for many microbes. It is the immune system's job to keep them out or, failing that, to seek out and destroy them. When the immune system hits the wrong target, however, it can unleash a torrent of disorders, including allergic diseases, arthritis, and a form of diabetes. If the immune system is crippled, other kinds of diseases result. The immune system is amazingly complex. It can recognize and remember millions of different enemies, and it can produce secretions (release of fluids) and cells to match up with and wipe out nearly all of them.

The secret to its success is an elaborate and dynamic communications network. Millions and millions of cells, organized into sets and subsets, gather like clouds of bees swarming around a hive and pass information back and forth in response to an infection. Once immune cells receive the alarm, they become activated and begin to produce powerful chemicals. These substances allow the cells to regulate their own growth and behaviour, enlist other immune cells, and direct the new recruits to trouble spots.

THE ORGANS OF THE IMMUNE SYSTEM

Bone Marrow

All the cells of the immune system are initially derived from the bone marrow.

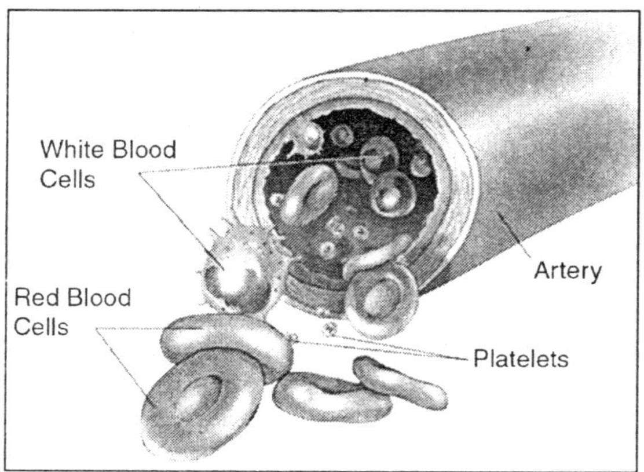

Fig. Bone Marrow

They form through a process called hematopoiesis. During hematopoiesis, bone marrow-derived stem cells differentiate into either mature cells of the immune system or into precursors of cells that migrate out of the bone marrow to continue their maturation elsewhere.

The bone marrow produces B cells, natural killer cells, granulocytes and immature thymocytes, in addition to red blood cells and platelets.

Thymus

The function of the thymus is to produce mature T cells. Immature thymocytes, also known as prothymocytes, leave the bone marrow and migrate into the thymus.

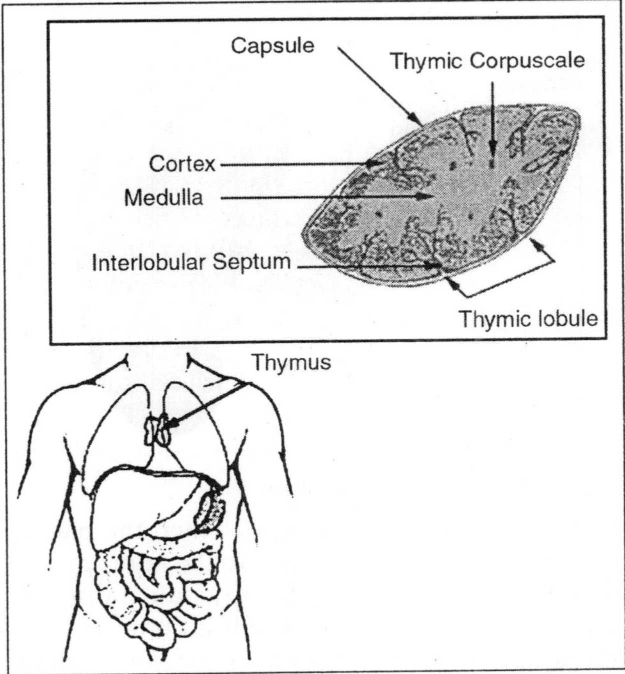

Fig. Thymus

Through a remarkable maturation process sometimes referred to as thymic education, T cells that are beneficial to the immune system are spared, while those T cells that might evoke a detrimental autoimmune response are eliminated. The mature *T* cells are then released into the bloodstream.

Spleen

The spleen is an immunologic filter of the blood. It is made up of B cells, T cells, macrophages, dendritic cells, natural killer cells and red blood cells.

In addition to capturing foreign materials (antigens) from the blood that passes through the spleen, migratory macrophages and dendritic cells bring antigens to the spleen via the bloodstream.

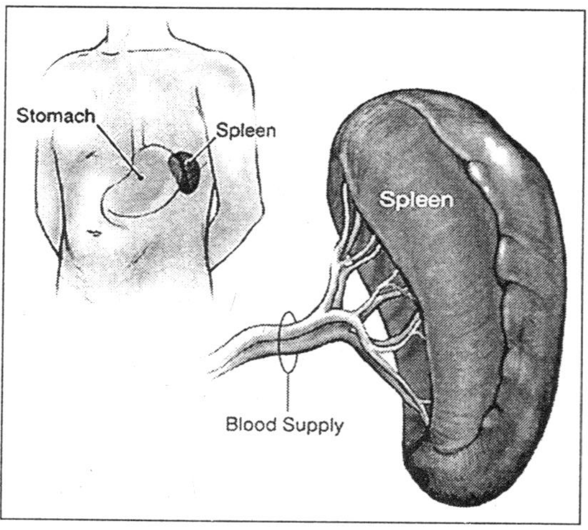

Fig. Spleen

An immune response is initiated when the macrophage or dendritic cells present the antigen to the appropriate *B* or *T* cells. This organ can be thought of as an immunological conference centre. In the spleen, *B* cells become activated and produce large amounts of antibody. Also, old red blood cells are destroyed in the spleen.

Lymph Nodes

The lymph nodes function as an immunologic filter for the

bodily fluid known as lymph. Lymph nodes are found throughout the body. Composed mostly of *T* cells, *B* cells, dendritic cells and macrophages, the nodes drain fluid from most of our tissues.

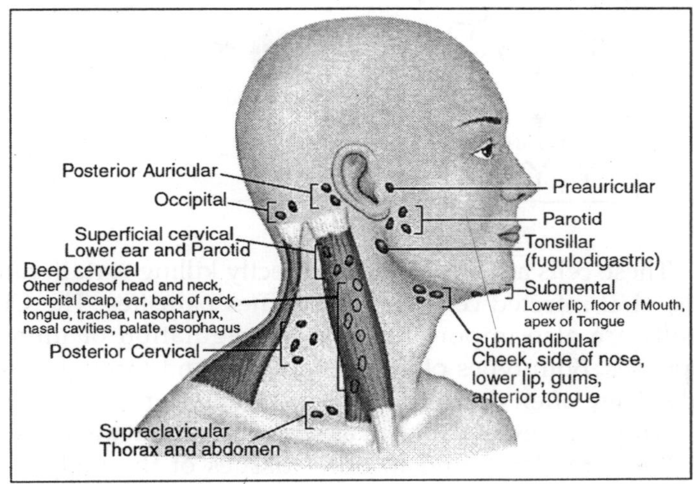

Fig. Lymph Nodes

Antigens are filtered out of the lymph in the lymph node before returning the lymph to the circulation. In a similar fashion as the spleen, the macrophages and dendritic cells that capture antigens present these foreign materials to *T* and *B* cells, consequently initiating an immune response.

THE CELLS OF THE IMMUNE SYSTEM

T-Cells

Tlymphocytes are usually divided into two major subsets that are functionally and phenotypically (identifiably) different. The *T* helper subset, also called the CD^{4+} *T* cell, is a pertinent coordinator of immune regulation.

The main function of the *T* helper cell is to augment or potentiate immune responses by the secretion of specialized factors that activate other white blood cells to fight off infection. Another important type of T cell is called the *T* killer/ suppressor subset or CD^{8+} *T* cell.

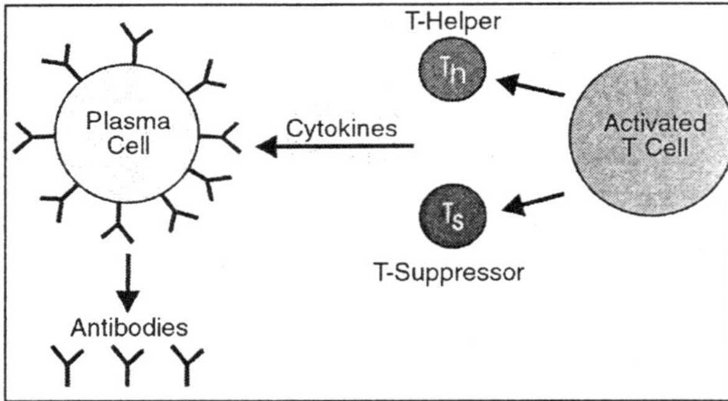

Fig. *T*-Cells

These cells are important in directly killing certain tumor cells, viral-infected cells and sometimes parasites. The CD8+ *T* cells are also important in down-regulation of immune responses. Both types of *T* cells can be found throughout the body. They often depend on the secondary lymphoid organs (the lymph nodes and spleen) as sites where activation occurs, but they are also found in other tissues of the body, most conspicuously the liver, lung, blood, and intestinal and reproductive tracts.

Natural Killer Cells

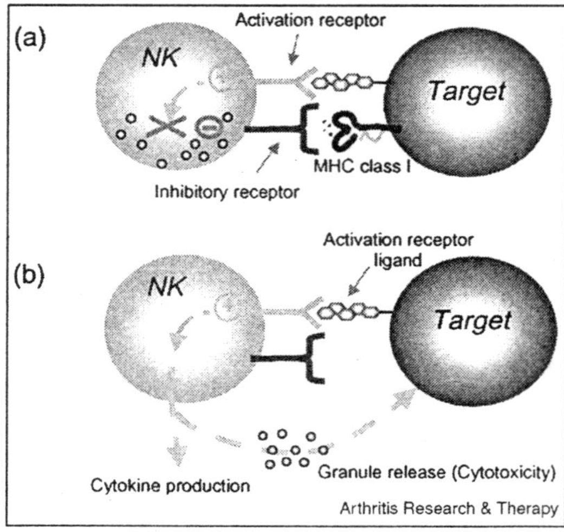

Fig. Natural Killer Cells

Natural killer cells, often referred to as *NK* cells, are similar to the killer *T* cell subset (CD^{8+} *T* cells). They function as effector cells that directly kill certain tumors such as melanomas, lymphomas and viral-infected cells, most notably herpes and cytomegalovirus-infected cells.

NK cells, unlike the CD^{8+} (killer) *T* cells, kill their targets without a prior "conference" in the lymphoid organs. However, *NK* cells that have been activated by secretions from CD^{4+} *T* cells will kill their tumor or viral-infected targets more effectively.

B Cells

The major function of B lymphocytes is the production of antibodies in response to foreign proteins of bacteria, viruses, and tumor cells. Antibodies are specialized proteins that specifically recognize and bind to one particular protein that specifically recognize and bind to one particular protein.

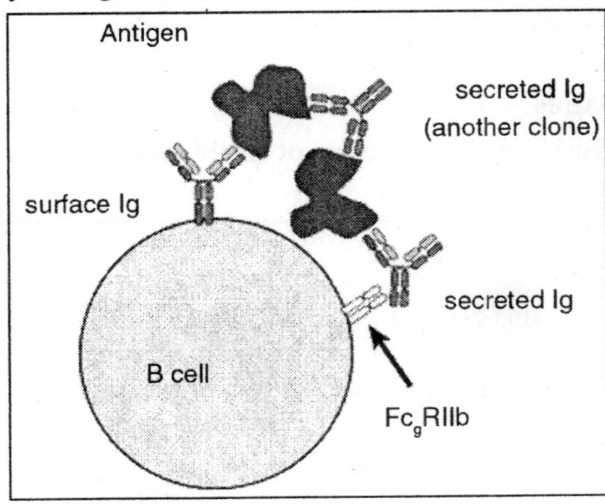

Fig. *B* Cells

Antibody production and binding to a foreign substance or antigen, often is critical as a means of signaling other cells to engulf, kill or remove that substance from the body.

Granulocytes

Another group of white blood cells is collectively referred to as granulocytes or polymorphonuclear leukocytes (PMNs).

Granulocytes are composed of three cell types identified as neutrophils, eosinophils and basophils, based on their staining characteristics with certain dyes.

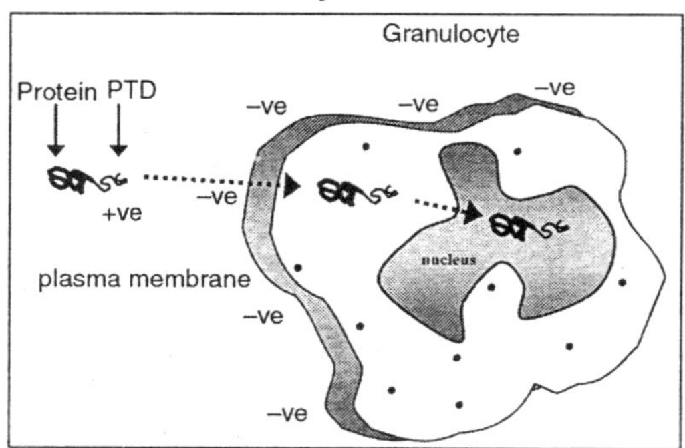

Fig. Granulocytes

These cells are predominantly important in the removal of bacteria and parasites from the body. They engulf these foreign bodies and degrade them using their powerful enzymes.

Macrophages

Macrophages are important in the regulation of immune responses.

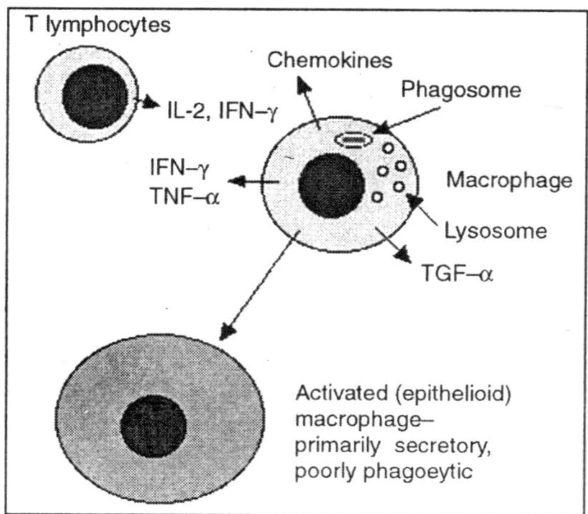

Fig. Macrophages

They are often referred to as scavengers or antigen-presenting cells (APC) because they pick up and ingest foreign materials and present these antigens to other cells of the immune system such as *T* cells and *B* cells.

This is one of the important first steps in the initiation of an immune response. Stimulated macrophages exhibit increased levels of phagocytosis and are also secretory.

Dendritic Cells

Another cell type, addressed only recently, is the dendritic cell. Dendritic cells, which also originate in the bone marrow, function as antigen presenting cells (APC). In fact, the dendritic cells are more efficient apcs than macrophages. These cells are usually found in the structural compartment of the lymphoid organs such as the thymus, lymph nodes and spleen. However, they are also found in the bloodstream and other tissues of the body.

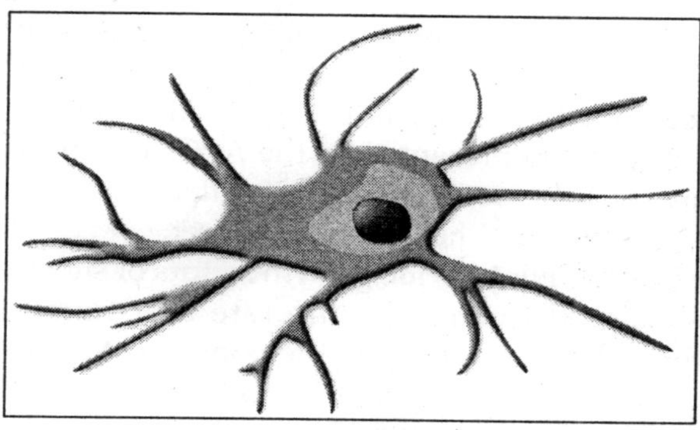

Fig. Dendritic Cells

It is believed that they capture antigen or bring it to the lymphoid organs where an immune response is initiated. Unfortunately, one reason we know so little about dendritic cells is that they are extremely hard to isolate, which is often a prerequisite for the study of the functional qualities of specific cell types. Of particular issue here is the recent finding that dendritic cells bind high amount of HIV, and may be a reservoir of virus that is transmitted to CD^{4+} *T* cells during an activation event.

THE IMMUNE RESPONSE

An immune response to foreign antigen requires the presence of an antigen-presenting cell (APC), (usually either a macrophage or dendritic cell) in combination with a *B* cell or *T* cell. When an APC presents an antigen on its cell surface to a *B* cell, the *B* cell is signalled to proliferate and produce antibodies that specifically bind to that antigen. If the antibodies bind to antigens on bacteria or parasites it acts as a signal for pmns or macrophages to engulf (phagocytose) and kill them.

Another important function of antibodies is to initiate the "complement destruction cascade." When antibodies bind to cells or bacteria, serum proteins called complement bind to the immobilized antibodies and destroy the bacteria by creating holes in them. Antibodies can also signal natural killer cells and macrophages to kill viral or bacterial-infected cells.

If the APC presents the antigen to *T* cells, the *T* cells become activated. Activated *T* cells proliferate and become secretory in the case of CD^{4+} *T* cells, or, if they are CD^{8+} *T* cells, they become activated to kill target cells that specifically express the antigen presented by the APC. The production of antibodies and the activity of CD^{8+} killer *T* cells are highly regulated by the CD^{4+} helper *T* cell subset.

The CD^{4+} *T* cells provide growth factors or signals to these cells that signal them to proliferate and function more efficiently. This multitude of interleukins or cytokines that are produced and secreted by CD^{4+} *T* cells are often crucial to ensure the activation of natural killer cells, macrophages, CD^{8+} *T* cells, and PMNs is listed in the chart below.

One possible line of therapy is to reintroduce some of these cytokines to people who have severe immune deficiencies. This approach can be tricky because large amounts of any particular cytokine can have serious side effects. Furthermore, their half-life in the body is usually relatively short. Another short-coming of "replacement" therapy is that many cytokines will activate the CD^{4+} *T* cells or macrophages harboring HIV, and this could lead to faster rates of HIV production by those cells.

Theoretically, this could lead to progression of HIV rather than prophylaxis against opportunistic infections. However, recent progress in this area warrants attention and further study. Lack of interleukin-2 (IL-2) is believed to be one of the major causes of immune deficiency in AIDS. In recent studies where low dose IL-2 was administered to people with HIV, CD^{4+} *T* cell counts rose, as did anti-viral specific immunity, and natural killer cell cytotoxic activity. Administration of IL-3 to people with HIV is currently under investigation as a treatment for HIV associated cytopenia (low production of cells from the bone marrow). IL-4, a cytokine that activates *B* cells and also has inhibitory effects on the production of TNF, is currently under investigation for the treatment of Kaposi's sarcoma.

GM-CSF (granulocyte macrophage-colony stimulating factor), another growth factor produced by CD^{4+} *T* cells, is under investigation for the treatment of decreased white blood cell production for people on ganciclovir therapy. Additionally, IFN-gamma (gamma interferon) is under investigation as a treatment for people with PCP. Hopefully, these studies will lead to cures for certain opportunistic infections, or for use in inhibiting HIV production and ultimately saving lives.

SELF AND NONSELF

The key to a healthy immune system is its remarkable ability to distinguish between the body's own cells, recognized as "self," and foreign cells, or "nonself."

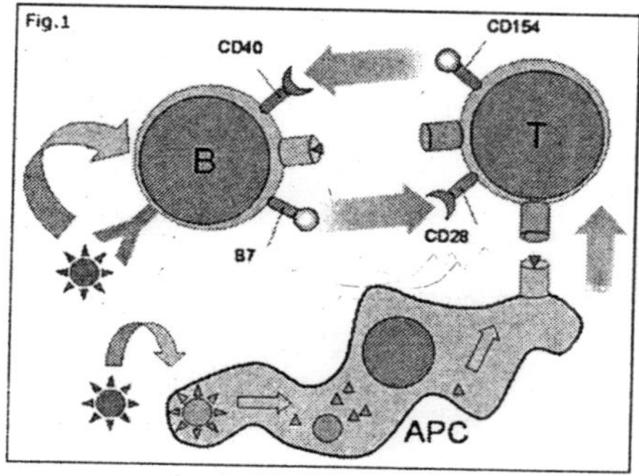

Fig. Antigen

The body's immune defenses normally coexist peacefully with cells that carry distinctive "self" marker molecules. But when immune defenders encounter foreign cells or organisms carrying markers that say "nonself," they quickly launch an attack. Anything that can trigger this immune response is called an antigen.

An antigen can be a microbe such as a virus, or a part of a microbe such as a molecule. Tissues or cells from another person (except an identical twin) also carry nonself markers and act as foreign antigens. This explains why tissue transplants may be rejected.

In abnormal situations, the immune system can mistake self for nonself and launch an attack against the body's own cells or tissues. The result is called an autoimmune disease. Some forms of arthritis and diabetes are autoimmune diseases. In other cases, the immune system responds to a seemingly harmless foreign substance such as ragweed pollen. The result is allergy, and this kind of antigen is called an allergen.

THE STRUCTURE OF THE IMMUNE SYSTEM

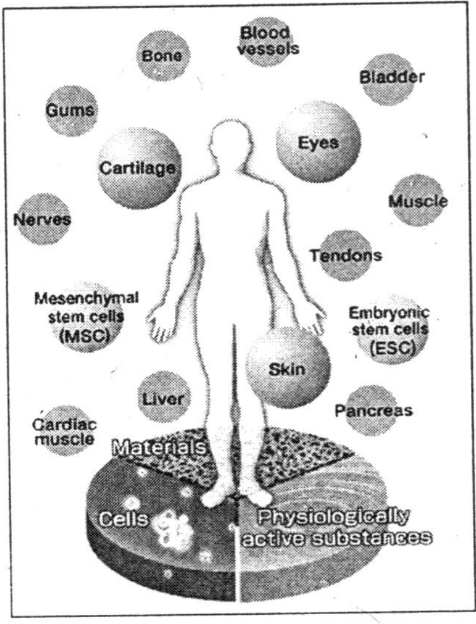

Fig. Organs of the Immune System

The organs of the immune system are positioned throughout the body. They are called lymphoid organs because they are home to lymphocytes, small white blood cells that are the key players in the immune system. Bone marrow, the soft tissue in the hollow centre of bones, is the ultimate source of all blood cells, including lymphocytes.

The thymus is a lymphoid organ that lies behind the breastbone. Lymphocytes known as *T* lymphocytes or *T* cells ("*T*" stands for "thymus") mature in the thymus and then migrate to other tissues. B lymphocytes, also known as B cells, become activated and mature into plasma cells, which make and release antibodies.

- *T* cells from the thymus concentrate in the paracortex.
- *B* cells develop in and around the germinal centers.
- Plasma cells occur in the medulla.

Lymphocytes can travel throughout the body using the blood vessels. The cells can also travel through a system of lymphatic vessels that closely parallels the body's veins and arteries.

Cells and fluids are exchanged between blood and lymphatic vessels, enabling the lymphatic system to monitor the body for invading microbes. The lymphatic vessels carry lymph, a clear fluid that bathes the body's tissues. Small, bean-shaped lymph nodes are laced along the lymphatic vessels, with clusters in the neck, armpits, abdomen, and groin.

Each lymph node contains specialized compartments where immune cells congregate, and where they can encounter antigens. Immune cells, microbes, and foreign antigens enter the lymph nodes via incoming lymphatic vessels or the lymph nodes' tiny blood vessels.

All lymphocytes exit lymph nodes through outgoing lymphatic vessels. Once in the bloodstream, lymphocytes are transported to tissues throughout the body. They patrol everywhere for foreign antigens, then gradually drift back into the lymphatic system to begin the cycle all over again.

The spleen is a flattened organ at the upper left of the abdomen. Like the lymph nodes, the spleen contains specialized compartments where immune cells gather and work. The spleen serves as a meeting ground where immune defenses confront antigens. Other clumps of lymphoid tissue are found in many parts of the body, especially in the linings of the digestive tract, airways, and lungs—territories that serve as gateways to the body. These tissues include the tonsils, adenoids, and appendix.

IMMUNE CELLS AND THEIR PRODUCTS

The immune system stockpiles a huge arsenal of cells, not only lymphocytes but also cell-devouring phagocytes and their relatives. Some immune cells take on all intruders, whereas others are trained on highly specific targets. To work effectively, most immune cells need the cooperation of their comrades. Sometimes immune cells communicate by direct physical contact, and sometimes they communicate releasing chemical messengers.

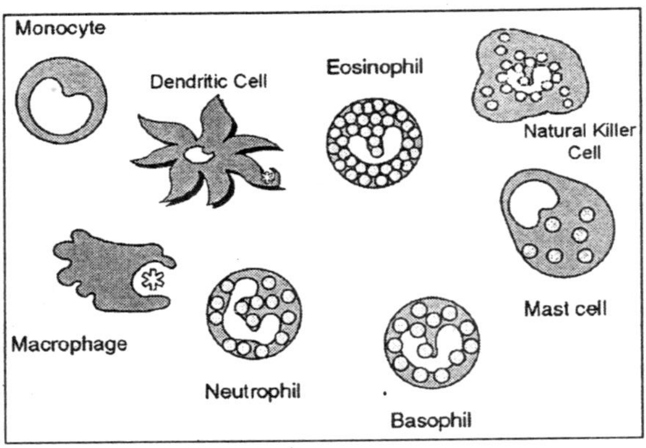

Fig. Immune Cells

The immune system stores just a few of each kind of the different cells needed to recognize millions of possible enemies. When an antigen first appears, the few immune cells that can respond to it multiply into a full-scale army of cells. After their job is done, the immune cells fade away, leaving sentries behind to watch for future attacks.

All immune cells begin as immature stem cells in the bone marrow. They respond to different cytokines and other chemical signals to grow into specific immune cell types, such as *T* cells, *B* cells, or phagocytes. Because stem cells have not yet committed to a particular future, their use presents an interesting possibility for treating some immune system disorders. Researchers currently are investigating if a person's own stem cells can be used to regenerate damaged immune responses in autoimmune diseases and in immune deficiency disorders, such as HIV infection.

B Cells

B cells and T cells are the main types of lymphocytes. B cells work chiefly by secreting substances called antibodies into the body's fluids.

Antibodies ambush foreign antigens circulating in the bloodstream. They are powerless, however, to penetrate cells. The job of attacking target cells—either cells that have been infected by viruses or cells that have been distorted by cancer—is left to T cells or other immune cells.

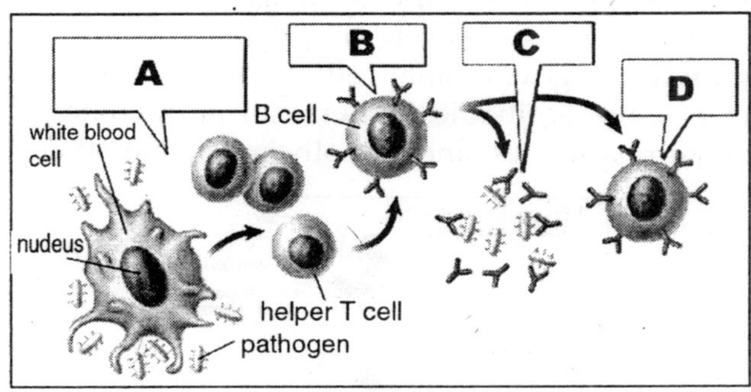

Fig. *B* Cells

Each B cell is programmed to make one specific antibody. For example, one B cell will make an antibody that blocks a virus that causes the common cold, while another produces an antibody that attacks a bacterium that causes pneumonia.

When a *B* cell encounters the kind of antigen that triggers it to become active, it gives rise to many large cells known as plasma cells, which produce antibodies.

- Immunoglobulin G, or IgG, is a kind of antibody that

works efficiently to coat microbes, speeding their uptake by other cells in the immune system.

- IgM is very effective at killing bacteria.
- IgA concentrates in body fluids—tears, saliva, and the secretions of the respiratory and digestive tracts—guarding the entrances to the body.
- IgE, whose natural job probably is to protect against parasitic infections, is responsible for the symptoms of allergy.
- IgD remains attached to *B* cells and plays a key role in initiating early *B* cell responses.

T Cells

Unlike *B* cells, *T* cells do not recognize free-floating antigens. Rather, their surfaces contain specialized antibody-like receptors that see fragments of antigens on the surfaces of infected or cancerous cells. *T* cells contribute to immune defenses in two major ways: some direct and regulate immune responses, whereas others directly attack infected or cancerous cells. Helper *T* cells, or *Th* cells, coordinate immune responses by communicating with other cells. Some stimulate nearby *B* cells to produce antibodies, others call in microbe-gobbling cells called phagocytes, and still others activate other *T* cells.

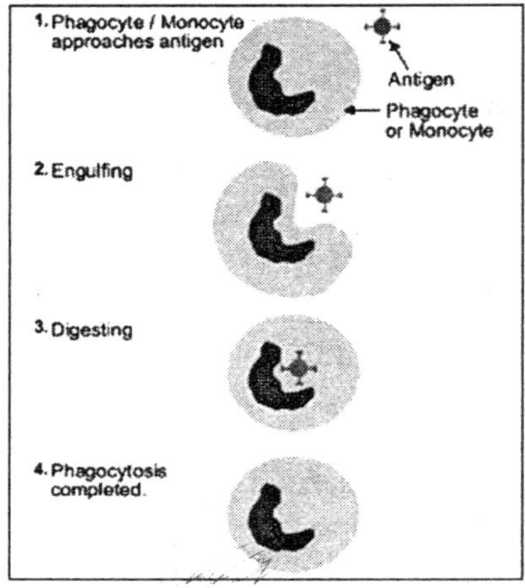

Fig. Phagocytes

Cytotoxic *T* lymphocytes (CTLs)—also called killer *T* cells—perform a different function. These cells directly attack other cells carrying certain foreign or abnormal molecules on their surfaces. CTLs are especially useful for attacking viruses because viruses often hide from other parts of the immune system while they grow inside infected cells.

CTLs recognize small fragments of these viruses peeking out from the cell membrane and launch an attack to kill the infected cell. In most cases, *T* cells only recognize an antigen if it is carried on the surface of a cell by one of the body's own major histocompatibility complex, or MHC, molecules. MHC molecules are proteins recognized by *T* cells when they distinguish between self and nonself. A self-MHC molecule provides a recognizable scaffolding to present a foreign antigen to the *T* cell. In humans, MHC antigens are called human leukocyte antigens, or HLA.

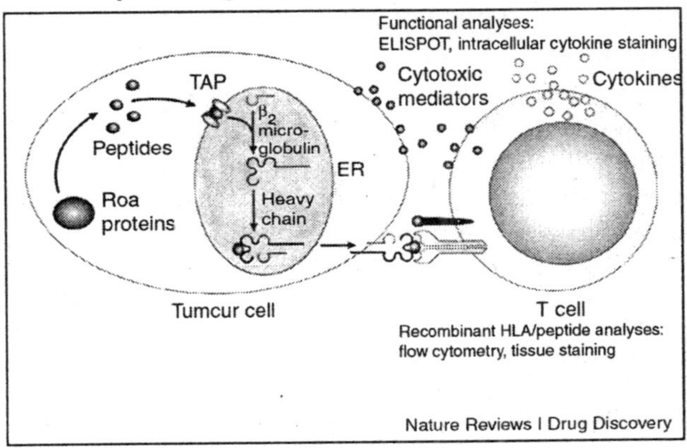

Fig. Leukocyte Antigens

Although MHC molecules are required for *T* cell responses against foreign invaders, they also create problems during organ transplantations. Virtually every cell in the body is covered with MHC proteins, but each person has a different set of these proteins on his or her cells. If a *T* cell recognizes a nonself-MHC molecule on another cell, it will destroy the cell. Therefore, doctors must match organ recipients with donors who have the closest MHC makeup. Otherwise the recipient's

T cells will likely attack the transplanted organ, leading to graft rejection.

Natural killer (NK) cells are another kind of lethal white cell, or lymphocyte. Like CTLs, NK cells are armed with granules filled with potent chemicals. But CTLs look for antigen fragments bound to self-MHC molecules, whereas NK cells recognize cells lacking self-MHC molecules. Thus, NK cells have the potential to attack many types of foreign cells. Both kinds of killer cells slay on contact.

The deadly assassins bind to their targets, aim their weapons, and then deliver a lethal burst of chemicals. *T* cells aid the normal processes of the immune system. If NK *T* cells fail to function properly, asthma, certain autoimmune diseases—including type 1 diabetes—or the growth of cancers may result. NK *T* cells get their name because they are a kind of *T* lymphocyte that carries some of the surface proteins, called "markers," typical of NK *T* cells. But these *T* cells differ from other kinds of *T* cells.

They do not recognize pieces of antigen bound to self-MHC molecules. Instead, they recognize fatty substances (lipids and glycolipids) that are bound to a different class of molecules called CD1d. Scientists are trying to discover methods to control the timing and release of chemical factors by NK *T* cells, with the hope they can modify immune responses in ways that benefit patients.

Phagocytes and Their Relatives

Phagocytes are large white cells that can swallow and digest microbes and other foreign particles. Monocytes are phagocytes that circulate in the blood. When monocytes migrate into tissues, they develop into macrophages. Specialized types of macrophages can be found in many organs, including the lungs, kidneys, brain, and liver.

Macrophages play many roles. As scavengers, they rid the body of worn-out cells and other debris. They display bits of foreign antigen in a way that draws the attention of matching lymphocytes and, in that respect, resemble dendritic cells. And they churn out an amazing variety of powerful chemical signals, known as monokines, which are vital to the immune response. Granulocytes are another kind of immune cell.

They contain granules filled with potent chemicals, which allow the granulocytes to destroy microorganisms. Some of these chemicals, such as histamine, also contribute to inflammation and allergy. One type of granulocyte, the neutrophil, is also a phagocyte.

Neutrophils use their prepackaged chemicals to break down the microbes they ingest. Eosinophils and basophils are granulocytes that "degranulate" by spraying their chemicals onto harmful cells or microbes nearby. Mast cells function much like basophils, except they are not blood cells. Rather, they are found in the lungs, skin, tongue, and linings of the nose and intestinal tract, where they contribute to the symptoms of allergy.

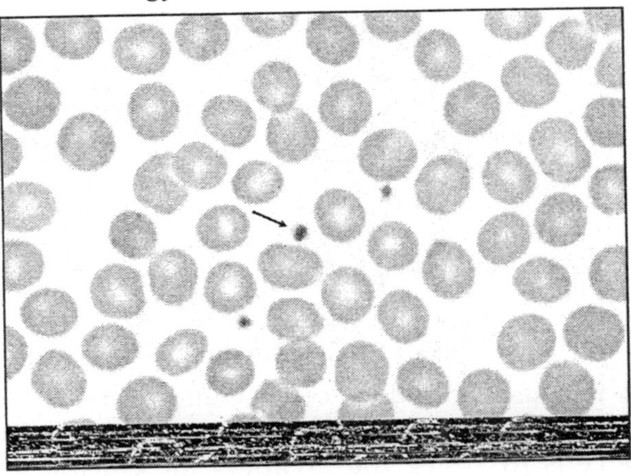

Fig. Platelets

Related structures, called blood platelets, are cell fragments. Platelets also contain granules. In addition to promoting blood clotting and wound repair, platelets activate some immune defenses.

Dendritic cells are found in the parts of lymphoid organs where *T* cells also exist. Like macrophages, dendritic cells in lymphoid tissues display antigens to *T* cells and help stimulate *T* cells during an immune response. They are called dendritic cells because they have branchlike extensions that can interlace to form a network.

T Cell Receptors

T cell receptors are complex protein molecules that peek through the surface membranes of *T* cells. The exterior part of a *T* cell receptor recognizes short pieces of foreign antigens that are bound to self-MHC molecules on other cells of the body.

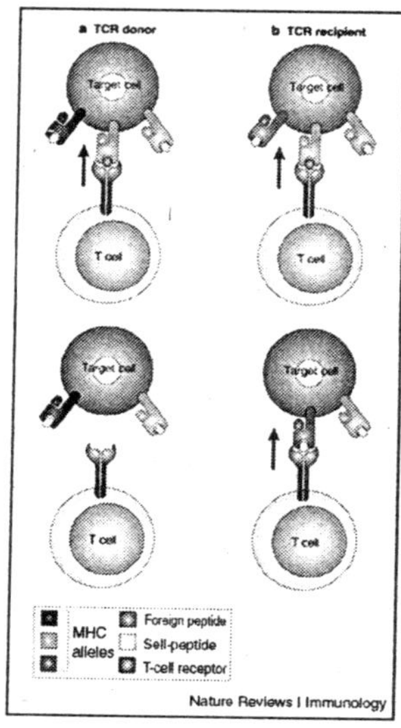

Fig. *T* Cell Receptors

It is because of their *T* cell receptors that *T* cells can recognize disease-causing microorganisms and rally other immune cells to attack the invaders, or kill the invaders themselves. Toll-like receptors (TLRs), which occur on cells throughout the immune system, are a family of proteins the body uses as a first line of defence against invading microbes.

Like *T* cell receptors, some TLRs peek through the surface membranes of immune cells, allowing them to respond to microbes in the cells' environment. Some TLRs are activated by molecules that make up viruses, whereas other TLRs respond to molecules that make up the cell walls of bacteria. Once activated, TLRs relay the alarm to other actors in the immune system. For example, some TLRs play important roles in the all-purpose "first-responder" arm of the immune system, also called the innate immune system. In short order, the innate immune system responds with a surge of chemical signals that together cause inflammation, fever, and other responses to infection or injury. Other TLRs help initiate responses from genetically identical groups of lymphocytes, called clones, that are already programmed to recognize specific antigens. Such responses are called adaptive immunity.

Overall, the cellular receptors important for the first-line responses of innate immunity are encoded by genes people inherit from their parents.

In contrast, adaptive immune responses rely on antigen receptors that are pieced together in the genomes of lymphocytes during their development in various tissues of the body. In addition to TLRs, other kinds of innate immune receptors can stimulate phagocytosis by macrophages, trigger the inflammatory responses that help control local infections, and play a range of crucial roles in defending the body against invading microbes.

Cytokines

Cells of the immune system communicate with one another by releasing and responding to chemical messengers called cytokines. These proteins are secreted by immune cells and act on other cells to coordinate appropriate immune responses.

Cytokines include a diverse assortment of interleukins, interferons, and growth factors. Some cytokines are chemical switches that turn certain immune cell types on and off. One cytokine, interleukin 2 (IL-2), triggers the immune system to produce *T* cells.

IL-2's immunity-boosting properties have traditionally made it a promising treatment for several illnesses. Clinical studies are underway to test its benefits in diseases such as cancer, hepatitis C, and HIV infection and AIDS. Scientists are studying other cytokines to see whether they can also be used to treat diseases.

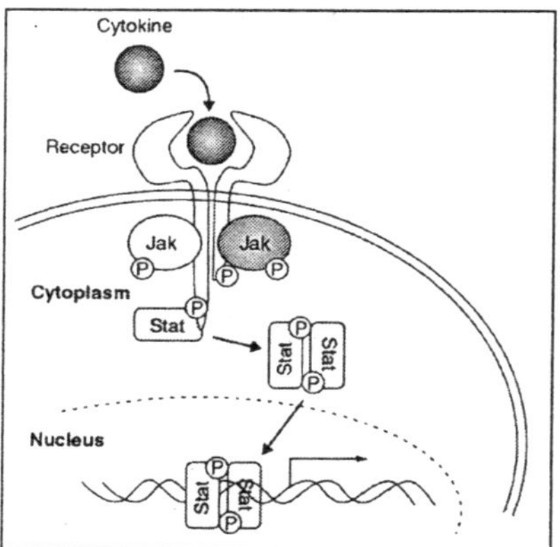

Fig. Cytokines

One group of cytokines chemically attracts specific cell types. These so-called chemokines are released by cells at a site of injury or infection and call other immune cells to the region to help repair the damage or ht off the invader. Chemokines often play a key role in inflammation and are a promising target for new drugs to help regulate immune responses.

Complement

The complement system is made up of about 25 proteins that work together to assist, or "complement," the action of antibodies in destroying bacteria. Complement also helps to rid the body of antibody-coated antigens (antigen-antibody complexes). Complement proteins, which cause blood vessels to become dilated and then leaky, contribute to the redness, warmth, swelling, pain, and loss of function that characterize an inflammatory response.

Complement proteins circulate in the blood in an inactive form. When the first protein in the complement series is activated—typically by antibody that has locked onto an antigen—it sets in motion a domino effect. Each component takes its turn in a precise chain of steps known as the complement cascade.

The end products are molecular cylinders that are inserted into— and that puncture holes in—the cell walls that surround the invading bacteria. With fluids and molecules flowing in and out, the bacterial cells swell, burst, and die. Other components of the complement system make bacteria more susceptible to phagocytosis or beckon other immune cells to the area.

MOUNTING AN IMMUNE RESPONSE

Infections are the most common cause of human disease. They range from the common cold to debilitating conditions like chronic hepatitis to life-threatening diseases such as AIDS. Disease-causing microbes (pathogens) attempting to get into the body must first move past the body's external armor, usually the skin or cells lining the body's internal passageways.

The skin provides an imposing barrier to invading microbes. It is generally penetrable only through cuts or tiny abrasions. The digestive and respiratory tracts—both portals of entry for a number of microbes—also have their own levels of protection.

Microbes entering the nose often cause the nasal surfaces to secrete more protective mucus, and attempts to enter the nose or lungs can trigger a sneeze or cough reflex to force microbial invaders out of the respiratory passageways. The stomach contains a strong acid that destroys many pathogens that are swallowed with food.

If microbes survive the body's front-line defenses, they still have to find a way through the walls of the digestive, respiratory, or urogenital passageways to the underlying cells. These passageways are lined with tightly packed epithelial cells covered in a layer of mucus, effectively blocking the transport of many pathogens into deeper cell layers.

Mucosal surfaces also secrete a special class of antibody called IgA, which in many cases is the first type of antibody to encounter an invading microbe. Underneath the epithelial layer a variety of immune cells, including macrophages, *B* cells, and *T* cells, lie in wait for any microbe that might bypass the barriers at the surface.

Next, invaders must escape a series of general defenses of the innate immune system, which are ready to attack without regard for specific antigen markers. These include patrolling phagocytes, NK *T* cells, and complement. Microbes cross the general barriers then confront specific weapons of the adaptive immune system tailored just for them. These specific weapons, which include both antibodies and *T* cells, are equipped with singular receptor structures that allow them to recognize and interact with their designated targets.

Bacteria, Viruses, and Parasites

The most common disease-causing microbes are bacteria, viruses, and parasites. Each uses a different tactic to infect a person, and, therefore, each is thwarted by different components of the immune system. Most bacteria live in the spaces between cells and are readily attacked by antibodies. When antibodies attach to a bacterium, they send signals to complement proteins and phagocytic cells to destroy the bound microbes.

Some bacteria are eaten directly by phagocytes, which signal to certain T cells to join the attack.

All viruses, plus a few types of bacteria and parasites, must enter cells of the body to survive, requiring a different kind of immune defence.

Infected cells use their MHC molecules to put pieces of the invading microbes on their surfaces, flagging down CTLs to destroy the infected cells. Antibodies also can assist in the immune response by attaching to and clearing viruses before they have a chance to enter cells.

Parasites live either inside or outside cells. Intracellular parasites such as the organism that causes malaria can trigger *T* cell responses.

Extracellular parasites are often much larger than bacteria or viruses and require a much broader immune attack. Parasitic infections often trigger an inflammatory response in which eosinophils, basophils, and other specialized granule-containing cells rush to the scene and release their stores of toxic chemicals in an attempt to destroy the invaders. Antibodies also play a role in this attack, attracting the granule-filled cells to the site of infection.

IMMUNITY: NATURAL AND ACQUIRED

Long ago, physicians realized that people who had recovered from the plague would never get it again—they had acquired immunity.

This is because some of the activated *T* and *B* cells had become memory cells. Memory cells ensure that the next time a person meets up with the same antigen, the immune system is already set to demolish it.

Immunity can be strong or weak, shortlived or long-lasting, depending on the type of antigen it encounters, the amount of antigen, and the route by which the antigen enters the body.

Immunity can also be influenced by inherited genes. When faced with the same antigen, some individuals will respond forcefully, others feebly, and some not at all. An immune response can be sparked not only by infection but also by immunization with vaccines.

Some vaccines contain microorganisms—or parts of microorganisms— that have been treated so they can provoke an immune response but not full-blown disease. Immunity can also be transferred from one individual to another by injections of serum rich in antibodies against a particular microbe (antiserum). For example, antiserum is sometimes given to protect travelers to countries where hepatitis A is widespread. The antiserum induces passive immunity against the hepatitis A virus.

Passive immunity typically lasts only a few weeks or months. Infants are born with weak immune responses but are protected for the first few months of life by antibodies they receive from their mothers before birth. Babies who are nursed can also receive some antibodies from breast milk that help to protect their digestive tracts.

Immune Tolerance

Immune tolerance is the tendency of *T* or *B* lymphocytes to ignore the body's own tissues. Maintaining tolerance is important because it prevents the immune system from attacking its fellow cells.

Scientists are hard at work trying to understand how the immune system knows when to respond and when to ignore an antigen. Tolerance occurs in at least two ways— central tolerance and peripheral tolerance. Central tolerance occurs during lymphocyte development. Very early in each immune cell's life, it is exposed to many of the self molecules in the body.

If it encounters these molecules before it has fully matured, the encounter activates an internal self-destruct pathway, and the immune cell dies.

This process, called clonal deletion, helps ensure that "self-reactive" T cells and B cells, those that could develop the ability to destroy the body's own cells, do not mature and attack healthy tissues.

Because maturing lymphocytes do not encounter every molecule in the body, they must also learn to ignore mature cells and tissues.

In peripheral tolerance, circulating lymphocytes might recognize a self molecule but cannot respond because some of the chemical signals required to activate the T or B cell are absent.

So-called clonal anergy, therefore, keeps potentially harmful lymphocytes switched off.

Peripheral tolerance may also be imposed by a special class of regulatory *T* cells that inhibits helper or cytotoxic *T*-cell activation by self antigens.

Vaccines

For many years, healthcare providers have used vaccination to help the body's immune system prepare for future attacks.

Vaccines consist of killed or modified microbes, components of microbes, or microbial DNA that trick the body into thinking an infection has occurred. A vaccinated person's immune system attacks the harmless vaccine and prepares for invasions against the kind of microbe the vaccine contained. In this way, the person becomes immunized against the microbe.

Vaccination remains one of the best ways to prevent infectious diseases, and vaccines have an excellent safety record.

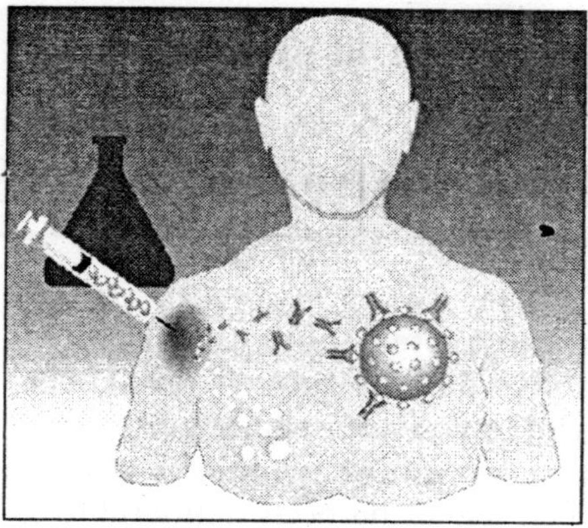

Fig. Vaccines

Previously devastating diseases such as smallpox, polio, and whooping cough have been greatly controlled or eliminated through worldwide vaccination programs.

Chapter 16

Disorders of the Immune System

The immune system helps the body defend against various microbes and pollutants. However, the immune system itself can have various failings.

An impaired immune system is called immunocompromise and can leave the body vulnerable to various viral, bacterial, or fungal opportunistic infections. Causes of immune deficiency can include various illnesses such as viruses, chronic illness, or immune system illnesses. The other type of immune disorder involves an over-active immune response. There are several different classes of diseases from an excessive response by the immune system:

- Allergies – a true allergy to a substance or food is caused by the immune response.
- Asthma – caused by an allergic reaction affecting the airway passages.
- Anaphylaxis – an extremely dangerous over-reaction that can lead to shock.
- Autoimmune diseases – a group of more than 100 diseases where the body's own immune system gets confused and starts to attack good body cells.

Allergic Diseases

The most common types of allergic diseases occur when the immune system responds to a false alarm. In an allergic person, a normally harmless material such as grass pollen, food particles, mold, or house dust mites is mistaken for a threat and attacked. Allergies such as pollen allergy are related to the antibody known as IgE. Like other antibodies, each IgE antibody is specific; one acts against oak pollen and another against ragweed, for example.

Autoimmune Diseases

Sometimes the immune system's recognition apparatus breaks down, and the body begins to manufacture *T* cells and antibodies directed against self antigens in its own cells and tissues. As a result, healthy cells and tissues are destroyed, which leaves the person's body unable to perform important functions.

Misguided *T* cells and autoantibodies, as they are known, contribute to many autoimmune diseases. For instance, *T* cells that attack certain kinds of cells in the pancreas contribute to a form of diabetes, whereas an autoantibody known as rheumatoid factor is common in people with rheumatoid arthritis. People with systemic lupus erythematosus (SLE) have antibodies to many types of their own cells and cell components.

SLE patients can develop a severe rash, serious kidney inflammation, and disorders of other important tissues and organs. No one knows exactly what causes an autoimmune disease, but multiple factors are likely to be involved. These include elements in the environment, such as viruses, certain drugs, and sunlight, all of which may damage or alter normal body cells.

Hormones are suspected of playing a role because most autoimmune diseases are far more common in women than in men. Heredity, too, seems to be important. Many people with autoimmune diseases have characteristic types of self-marker molecules.

Immune Complex Diseases

Immune complexes are clusters of interlocking antigens and antibodies. Normally, immune complexes are rapidly removed from the bloodstream.

Sometimes, however, they continue to circulate and eventually become trapped in the tissues of the kidneys, lungs, skin, joints, or blood vessels.

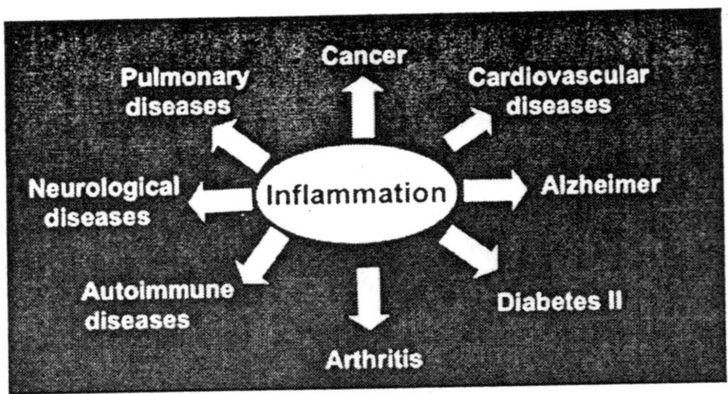

Fig. Inflammation

There, they set off reactions with complement that lead to inflammation and tissue damage. Immune complexes work their mischief in many diseases. These include malaria and viral hepatitis, as well as many autoimmune diseases.

IMMUNE DEFICIENCY DISORDERS

When the immune system is missing one or more of its parts, the result is an immune deficiency disorder. These disorders can be inherited, acquired through infection, or produced as a side effect by drugs such as those used to treat people with cancer or those who have received transplants.

Temporary immune deficiencies can develop in the wake of common virus infections, including influenza, infectious mononucleosis, and measles. Immune responses can also be depressed by blood transfusions, surgery, malnutrition, smoking, and stress. Some children are born with poorly functioning immune systems.

Some have flaws in the *B* cell system and cannot produce antibodies. Others, whose thymus is either missing or small and abnormal, lack *T* cells.

Very rarely, infants are born lacking all of the major immune defenses. This condition is known as severe combined immune deficiency disease or SCID. AIDS is an immune deficiency disorder caused by a virus (HIV) that infects immune cells.

HIV can destroy or disable vital *T* cells, paving the way

for a variety of immunologic shortcomings. The virus also can hide out for long periods in immune cells. As the immune defenses falter, a person develops AIDS and falls prey to unusual, often life-threatening infections and rare cancers.

IMMUNOLOGY AND TRANSPLANTS

Each year thousands of lives in the United States are prolonged by transplanted organs including the kidneys, heart, lung, liver, and pancreas. For a transplant to "take," however, the body's natural tendency to rid itself of foreign tissue must be overridden.

One way to avoid the rejection of transplanted tissue is tissue typing, which ensures that markers of self on the donor's tissue are as similar as possible to those of the recipient. Every cell in the body has a double set of six major tissue antigens, and each of the antigens exists, in different individuals, in as many as 20 varieties. The chance of two people having identical transplant antigens is about one in 100,000.

A second way to avoid transplant rejection is to lull the recipient's immune system into a less active state. This can be done with powerful immunosuppressive drugs such as cyclosporine A, or by using laboratory-manufactured antibodies that attack mature T cells.

Bone Marrow Transplants

When the immune response is severely depressed—in infants born with immune disorders or in people with cancer, for example—one possible remedy is a transfer of healthy bone marrow. Once introduced into the circulation, transplanted bone marrow cells can develop into functioning *B* and *T* cells.

In bone marrow transplants, a close match is extremely important. Not only is there a danger that the body will reject the transplanted bone marrow cells, but mature *T* cells from the bone marrow transplant may counterattack and destroy the recipient's tissues.

To prevent this situation, known as graft-versus-host disease, scientists use drugs or antibodies to "cleanse" the donor marrow of potentially dangerous mature *T* cells.

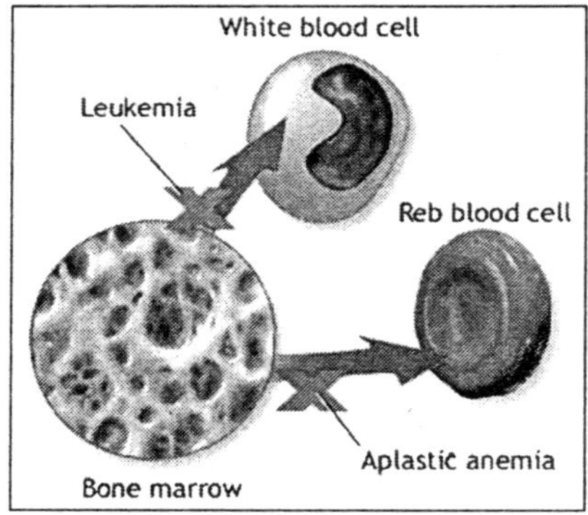

Fig. Bone Marrow Transplants

THE IMMUNE SYSTEM AND THE NERVOUS SYSTEM

Evidence is mounting that the immune system and the nervous system are linked in several ways.

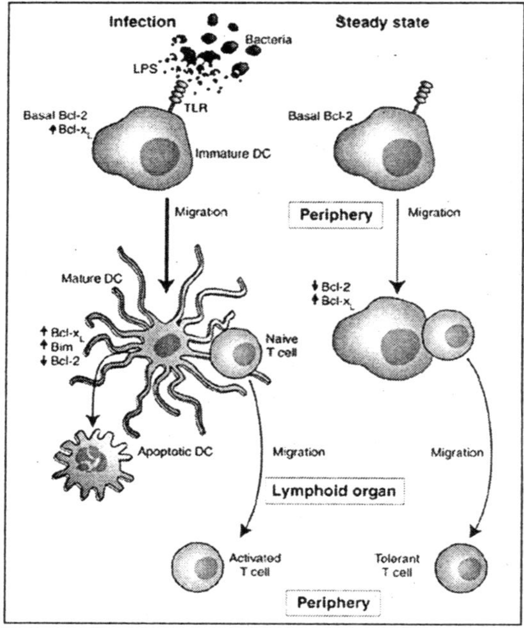

Fig. Lymphoid Organs

One well-known connection involves the adrenal glands. In response to stress messages from the brain, the adrenal glands release hormones into the blood.

In addition to helping a person respond to emergencies by mobilizing the body's energy reserves, these "stress hormones" can stifle the protective effects of antibodies and lymphocytes. Another link between the immune system and the nervous system is that the hormones and other chemicals that convey messages among nerve cells also "speak" to cells of the immune system.

Indeed, some immune cells are able to manufacture typical nerve cell products, and some lymphokines can transmit information to the nervous system. Moreover, the brain may send messages directly down nerve cells to the immune system. Networks of nerve fibers have been found connecting to the lymphoid organs.

FRONTIERS IN IMMUNOLOGY

Scientists are now able to mass-produce immune cell secretions, both antibodies and lymphokines, as well as specialized immune cells. The ready supply of these materials not only has revolutionized the study of the immune system itself but also has had an enormous impact on medicine, agriculture, and industry. Monoclonal antibodies are identical antibodies made by the many clones of a single *B* cell.

Because of their unique specificity for different antigens, monoclonal antibodies are promising treatments for a range of diseases. Researchers make monoclonal antibodies by injecting a mouse with a target antigen and then fusing *B* cells from the mouse with other long-lived cells.

The resulting hybrid cell becomes a type of antibody factory, turning out identical copies of antibody molecules specific for the target antigen.

Mouse antibodies are "foreign" to people, however, and might trigger an immune response when injected into a human. Therefore, researchers have developed "humanized" monoclonal antibodies. To construct these molecules, scientists take the antigen-binding portion of a mouse antibody and attach it to a human antibody scaffolding, greatly reducing the foreign portion of the molecule.

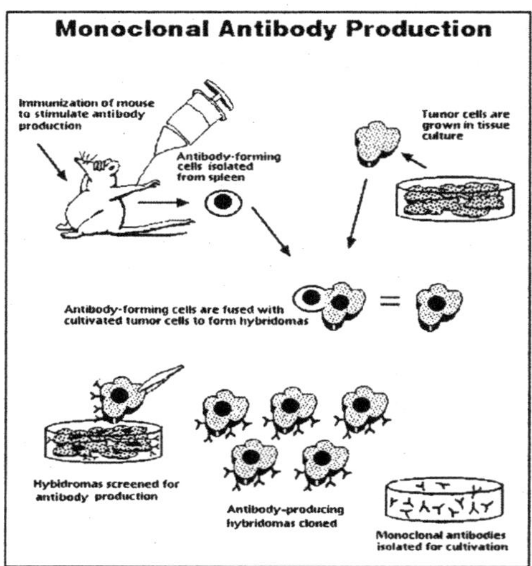

Fig. Monoclonal Antibodies

Because they recognize very specific molecules, monoclonal antibodies are used in diagnostic tests to identify invading pathogens or changes in the body's proteins. In medicine, monoclonal antibodies can attach to cancer cells, blocking the chemical growth signals that cause the cells to divide out of control.

In other cases, monoclonal antibodies can carry potent toxins into certain cells, killing the dangerous cells while leaving their neighbors untouched.

Chapter 17

Gene Therapy

Genetic engineering also holds promise for gene therapy—replacing altered or missing genes or adding helpful genes. One disease in which gene therapy has been successful is SCID, or severe combined immune deficiency disease.

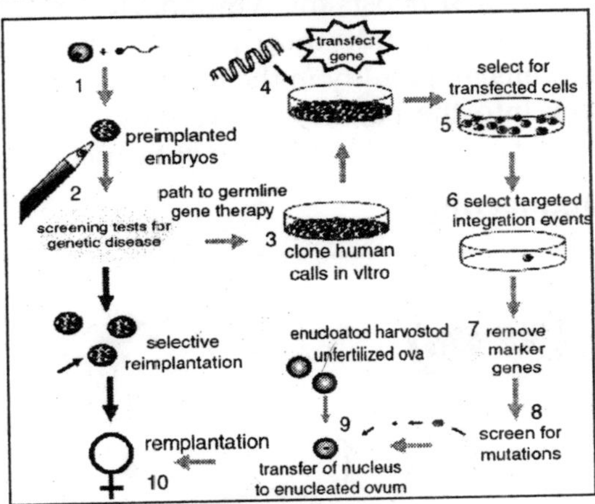

Fig. Gene Therapy

SCID is a rare genetic disease that disables a person's immune system and leaves the person unable to fight off infections. It is caused by mutations in one of several genes that code for important components of the immune system.

Until recently, the most effective treatment for SCID was transplantation of blood-forming stem cells from the bone marrow of a healthy person who is closely related to the patient. However, doctors have also been able to treat SCID by giving the patient a genetically engineered version of the missing gene. Using gene therapy to treat SCID is generally accomplished by taking blood-forming cells from a person's own bone marrow, introducing into the cells a genetically changed virus that carries the corrective gene, and growing the modified cells outside the person's body.

After the genetically changed bone marrow cells begin to produce the enzyme or other protein that was missing, the modified blood-forming marrow cells can be injected back into the person. Once back inside the body, the genetically modified cells can produce the missing immune system component and begin to restore the person's ability to fight off infections. Cancer is another target for gene therapy.

In pioneering experiments, scientists are removing cancer-fighting lymphocytes from the cancer patient's tumor, inserting a gene that boosts the lymphocytes' ability to make quantities of a natural anticancer product, then growing the restructured cells in quantity in the laboratory. These cells are injected back into the person, where they can seek out the tumor and deliver large doses of the anticancer chemical.

Genetic Engineering

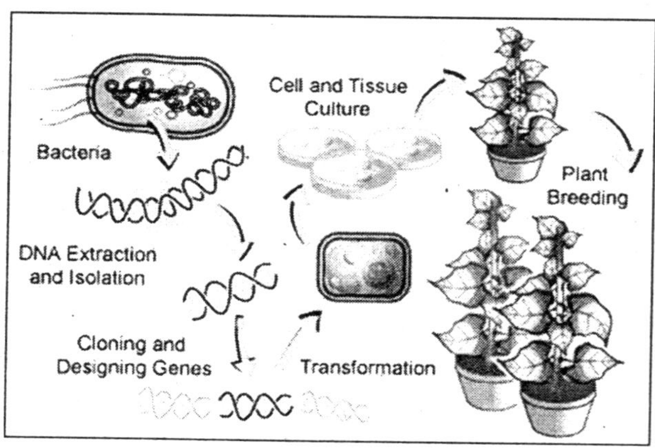

Fig. Genetic Engineering

Genetic engineering allows scientists to pluck genes—segments of DNA—from one type of organism and combine them with genes of a second organism.

In this way, relatively simple organisms such as bacteria or yeast (a type of fungus) can be induced to make quantities of human proteins, including hormones such as insulin as well as lymphokines and monokines. They can also manufacture proteins from infectious agents, such as the hepatitis virus or HIV, for use in vaccines.

Immunoregulation

Research into the delicate checks and balances that control the immune response is increasing knowledge of normal and abnormal immune system functions. Someday it may be possible to treat autoimmune diseases such as systemic lupus erythematosus by suppressing parts of the immune system that are overactive.

Scientists are also devising ways to better understand the human immune system and diseases that affect it. For example, by transplanting immature human immune tissues or immune cells into SCID mice, scientists have created "humanized" mice, a living model of the human immune system. Scientists are manipulating the immune system of humanized SCID mice to discover ways to benefit human health.

Humanized mice are also being used in research on transplantation and autoimmune and allergic diseases, and to manufacture molecules that help regulate immune system function and immune tolerance.

ANTIGEN PRESENTING CELLS (APC)

Antigen presenting cells are a functionally defined group of cells which are able to take up antigens and present them to T lymphocytes in a recognizable form (in the groove of an MHC class II molecule). Although many cells can do this the cells which are most efficient, the so-called "professional antigen presenting cells", are macrophages and dendritic cells. These are professional cells because they are highly effective at producing the "second signal" required for *T* cell activation.

The APC first internalizes the antigen (maybe in the form of a bacteria or bacterial product), processes it (breaks it down into antigenic peptides [epitopes] by digesting it with lysosyme) and then expresses the antigen fragment on its surface in the groove of an MHC class II molecule. Now the antigen is in the form recognizable by *T* cells. Antigen presentation by APC is a required first step in Th cell activation.

MAJOR HISTOCOMPATABILITY COMPLEX (MHC)

MHC stands for major histocompatability complex, but we'll just call it MHC. In humans the same proteins are called human leukocyte antigens or HLA. In basic immunology we talk about MHC (the proteins first identified in laboratory animals) because we know how they work in experimental systems. However, clinical immunology may refer to HLA. There are two classes of MHC molecules; class I and class II. The difference in the two classes is important.

MHC Class I

All cells in the body have MHC class I protein on the surface. Our finger cells, our liver cells, our B cells, our gut cells. They all have MHC class I on the surface. Everyone has MHC class I proteins but only identical twins will have identical MHC class I. The genes for the MHC proteins are part of the immunoglobulin superfamily of genes and therefore show diversity between people. Since everyone has a different MHC, this allows our body to tell what cells belong to it and what cells are foreign. The function of MHC-I is to sample the internal contents of the cell and show them to the immune system.

If the contents of the cell are purely self then no attraction of killer cells occurs. If, however, the cell is virally infected, then the killer cells can recognize the viral antigens presented by the Class-I MHC and the cells are killed. Unfortunately foreign MHC is seen by the immune system so that the cells in transplants activate the killer cells in a similar manner as virally infected cells. The end result is the same, cell death (and transplant rejection).

MHC Class II

Class II MHC are proteins only expressed by antigen

presenting cells (APC). The function of these proteins is to present an antigen to T helper cells to activate an immune response which will provide both humoral (antibody) and cell mediated immunity.

The class II MHC consists of an alpha and a beta chain with a transmembrane segment to hold them on the surface of the cell. At the end farthest from the cell is a cleft where the processed antigen sits. The processed antigen consists of a small peptide of about 13-16 amino acids. Every MHC class II on the surface of a cell contains an antigen fragment because the MHC II is not expressed on the surface until it has antigen in the cleft. MHC-II picks up the antigen that has been ingested (phagocytosed) by the APC. It does not present self antigen like MHC-I does.

TH CELL ACTIVATION

T cell activation is, in general, similar for both Th and Tc cells but each are considered seperately for clarity. Th cell activation is initiated by the interaction of TcR-CD3 complex with antigen-MHC class II molecules on the surface of an antigen presenting cell. This interaction initiates a cascade of biochemical events in the *T* cell that eventually results in growth and proliferation of the T cell. This occurs primarily through an increase in IL-2 secretion by the *T* cell and an increase in IL-2 receptors on the T cell surface.

IL-2 is a potent T cell growth cytokine (IL-2) which, in *T* cell activation, acts in an autocrine fashion to promote the growth, proliferation and differentiation of the *T* cell recently stimulated by antigen. The *T* cell receptor is an antigen recognition molecule and therefore the *T* cell that best responds to the antigen presented is the one that gets turned on. Naive, bystander *T* cells have no IL-2 receptors and thus cannot be involved.

The activation takes place through the *T* cell receptor complex and is aided by the CD4 molecule on Th cells and possibly other accessory molecules, such as CD45, CD28 and CD2. These molecules are accessory, but the APC also provides a co-stimulatory signal. The primary signal is the antigen-MHC

II and TcR-CD^3 interaction (and accessory molecules) while the co-stimulatory signal occurs with cytokines (IL-1 or IL-12). Activated Th cells then continue to become effector cells whose role includes B cell help and cytokine production.

T CELL HELP TH CELLS

The generation of an immune response both humoral by B cells and cell-mediated by Tc cells (CD^{8+}) depends on the activation of Th cells. The importance of these CD^{4+} cells has become obvious as these are the cells affected in AIDS (aquired immune deficiency syndrome).

Mature *B* cells that have already seen antigen require contact with a *T* cell in order to become plasma cells or memory cells (*T–B* cell interaction). The *T* cells provide signals to the B cell through contact of the TcR complex and MHC-antigen, In addition, the activated T cell produces cytokines such as IL-2 and IL4,5,6 or IFNg which stimulate *B* cell proliferation and differentiation into antibody secreting *B* cells.

The type of cytokines produced by the *T* cells helps the plasma cells to produce different classes of antibodies. So far it is unclear what causes this to happen, but we do know that there are two subtypes of Th cells that produce different cytokines and promote a different type of immune response.

T Helper Cell Subsets

Our understanding of how and why *T* cells behave in an immune response has markedly increased with the disovery that there are at least two subsets of *T* helper cells. The two subsets look the same and have the same *T* cell markers and receptors. However, they secrete very different cytokines upon activation. In addition to other cytokines, the Th1 subset produces great amounts of IL-2 and IFN gamma; cytokines particularly important in cell mediated immunity.

The Th2 subset is very good at providing *B* cell help by secreting IL-4,-5 and 6. Because the cytokines have different effects in an immune response, sometimes activation of one subset may be preferable to the other. For example, virus infection is best combatted with interferon's anti-viral activity and cytotoxic *T* cells. This response is more likely to be stimulated by Th1 cytokines. The cytokines produced by the two subsets also have a cross-regulatory role. In other words, an activated Th2 cell secreting IL-4,-5,and 6 will downregulate the Th1 cells in the neighborhood. Likewise, Th1 cytokines downregulate the Th2 responses.

Cytotoxic *T* cell Activation

Cytotoxic *T* cells (Tc) are derived from a lymphocyte stem cell matured in the thymus. These cells are characterized by the presence of the CD8 marker on their surface, and an antigen-specific *T* cell receptor which recognizes antigens in the context of MHC class I.

The main role of the cytotoxic *T* cell, as the name suggests, is to kill other cells. The requirement for MHC class I on the target cell means that the Tc cell is very important in recognizing and destroying self-cells that have been altered or infected. The Tc cell must first become activated and mature into a cytotoxic *T* lymphocyte (CTL).

Activation of the resting Tc cell is a two step process. First, the TcR on the CD^{8+} cell must interact with an antigen-MHC (class I) on the surface of a target cell. Secondly, the CD^{8+} Tc cell must be stimulated by cytokines. The IL-2 is probably supplied by activated Th cells. Resting Tc do not express IL-2 receptors, but antigen stimulation increases the expression of IL-2 receptors on the surface on the Tc cell. This ensures that only the cells recognizing the antigen will become activated.

Cytotoxic *T* Lymphocyte Activity

Activated CD^{8+} *T* cells (CTL's) are very effective at destroying target cells, especially virus-infected cells and tumor cells. The killing happens in three steps:

- Conjugate formation between the CTL and the target cell
- Membrane attack on the target cell
- Dissociation of the two cells and target cell death

Conjugate formation

Through the interaction of the TcR-CD^3 complex and the

MHC I -antigen, the target cell and the CTL form a conjugate. After antigen stimulation, the CTL receives other signals from the target cell through binding of accessory molecules and receptors.

Membrane Attack

Immediately following the formation of the conjugate (within 2 minutes!), granules in the CTL start to move through the cytoplasm towards the end nearest the target cell. The CTL changes shape and becomes flattened towards the target allowing an area of contact between the two cells. Contents of the granules are released from the CTL (exocytosis) and are responsible for initiated membrane damage to the target cell. The granules release cytokines, enzymes and a molecule called perforin that is able to poke holes in the target cell membrane.

Target Cell Death

In addition to the perforin-mediated damage to the target cells, the CTL's have other mechanisms for attacking the cells. Soon after the CTL contact, the target cells are found to undergo apoptosis (programmed cell death). Exactly how the CTL's induce apoptosis is an area still being investigated. It has been proposed that the CTL's may release a factor (posibly TNF beta) that induces the cell death.

PRIMARY LYMPHOID ORGANS

Bone Marrow

All cells of the blood orginate from pluripotent stem cells in the bone marrow. Vertebrate animals (including humans) will die when given high dose radiation because stem cells will be destroyed. Theoretically, 1 stem cell can reconstitute the animal. In addition: Bone Marrow serves as site of maturation of B lymphocytes in most mammals studied to date. Because there is not a discrete site, the process is very difficult to study.

Stromal cells in the bone marrow very important in this process. Soluble factors are not as well characterized in the bone marrow but IL-7 is known to be an important signal. There also is known to be a negative selection process. Many pre-*B* cells die in the bone marrow. If they do not form functional Ig molecules or if they possess Ig molecules which recognize and bind to self antigens they will undergo death by Apoptosis. As many as 90 per cent of differentiating B cells are believed to have this fate. B cells that survive this selection process leave the bone marrow through efferent blood vessels.

Thymus

Flat, bilobed organ situated above the heart and below the thyroid gland. Each lobe is surrounded by a capsule and is divided into lobules, which are separated from each other by strands of connective tissue called trabeculae. Each lobule is organized into two compartments: the cortex (outer compartment) the medulla (inner compartment).

Cortex and Medulla are both crisscrossed by a 3D network of stromal cells composed of:

- Loosely packed thymic epithelial cells.
- Interdigitating dendritic cells.
- Macrophages.

These cells physically interact with the developing *T* cells (thymocytes) as well as secreting soluble factors which influence *T* cell maturation. In the cortex, the network is densely packed with thymocytes. Cells are less dense in medulla. In medulla, the epithelial cells are more visible and we also find HASSAL's Corpuscles.

The thymus is at its largest relative size at birth and its largest actual size is at puberty. Following puberty the thymus begins to shrink. In elderly individuals it is usually less than 3 grams in weight.

Relationship between aging and a decline in immune responsiveness. Stress can also result in shrinkage of the thymus.

Precursor *T* cells enter thymus from the blood (there are no afferent lymphatic vessels] and mature into functional *T* lymphocytes. Precursor *T* cells first enter the cortex which is densely packed with cells due to tremendous degree of proliferation. However, the vast majority of these cells are destined to die due to (APOPTOSIS) - programmed cell death. The T cells (thymocytes) go through a selection process in the thymus based upon the TcR that they possess. The complete process of thymic education is a two-step process in which Thymic cortical epithelial cells function as the effector cells in a process known as *Positive Selection.*

In positive selection, T cells which bear a TcR which can bind SELF-MHC are selected to survive and proliferate. Cells which are not positively selected are triggered to undergo APOPTOSIS (*T* cells which lack a functional TcR, or *T* cells which possess a TcR without affinity for self-MHC)

Positively selected thymocytes must go through a second phase of selection known as *Negative Selection.* Functionally, during negative selection any *T* cell that is presented antigen +MHC within the thymus is triggered to undergo APOPTOSIS.

The self peptides encountered in the thymus are derived from proteins expressed in thymus + other proteins brought to the thymus via the blood stream. Of course not all potentially auto-reactive *T* cells can be deleted and peripheral deletion of autoreactive *T* cells is also important. Negative selection can apparently be mediated by a variety of different cell types. Most importantly: thymic macrophages and dendritic cells serve as APCs in the process of negative selection The surviving *T* cells migrate to the medulla where they continue maturation and finally leave the thymus through the postcapillary venules or efferent lymphatics. Stromal cells secrete soluble factors which are important in *T* cell maturation.

Some examples include:

- Alpha-1-thymosin
- Beta-4-thymosin
- Thymopoietin
- Thymulin

Peyer's Patches

Peyer's patches are areas of lymphoid tissue located in the wall of the intestine, and in some mammalian species such as sheep, cattle and rabbits, Peyer's patches have a function similar to the bursa of fabricius of birds and bone marrow of other mammals: *B* cell differentiation and maturation.

SECONDARY LYMPHOID ORGANS

Secondary lymphatic Tissues control the quality of immune responses. Differences among the various lymphatic tissues significantly affect the form of immunity and relate to how antigens are acquired by these organs. Lymph nodes are filters of lymph, the spleen is a filter of blood and mucosal associated lymphatic tissues acquire antigens by transcytosis to lymphoid tissue from the "external" environment across specialized follicle-associated epithelial cells.

SKIN AND DIFFUSE LYMPHATIC TISSUES

Skin and epithelial surfaces are the first line of defence against antigens in the environment which threaten an organism's integrity. Epithelium with underlying loose connective tissue, blood vessels and lymphatics may be considered part of the diffuse lymphatic system, especially for contact sensitizing antigens that fix to epithelial cells. The primary function of epithelium used to be regarded as a physical barrier but recently it has been shown to permit passage of drugs and antigens at a slow rate.

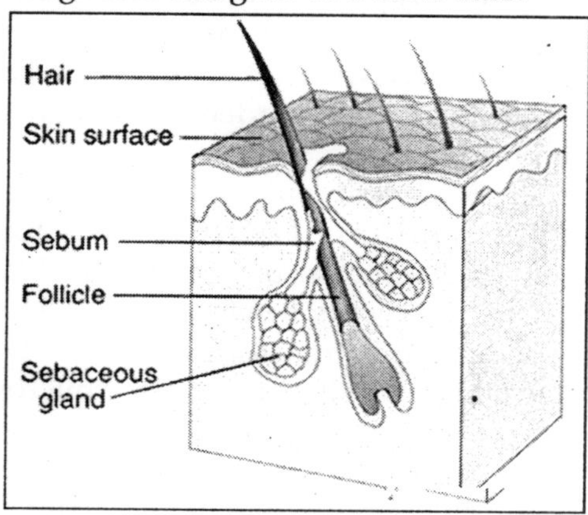

Fig. Keratinocytes

Keratinocytes, mononuclear cells and nerve endings in the skin secrete numerous factors that may attract, activate and effect differentiation of *T*-cells locally or in regional lymph nodes. Lymphocytes and Thy-l+ dendritic epidermal cells are infrequently seen in the dermis or in intra-epidermal locations of normal skin. Because of the vast surface area the total number of lymphocytes in the skin may be great despite the small numbers in any single location. Although there is considerable speculation about homing of lymphocytes to the skin, few are present in afferent lymph.

Langerhans cells populate the epithelial layer above basal cells and keratinocytes in a uniformally distributed pattern presumably limited by contacts between extended dendritic processes. Taken together it appears that the skin has become a transducer and gatherer of environmental information for transmission to regional lymphatic tissues. Mononuclear phagocytes are usually found in perivenular locations in the superficial dermis.

Precursors for Langerhans cells and monocytes enter the skin from the blood to replace cells that regularly exit into the afferent lymph of regional lymph nodes. Secondary Lymphoid-Tissue Chemokine (SLC) and CC Chemokine Receptor 7 (CCR7) appear to participate in the emigration pathway of mature dendritic cells from the skin to regional lymph nodes. The mean turnover time of Langerhans cells in mouse skin is about three weeks. Langerhans cells in afferent lymph draining the skin are regarded as the sentinels of skin-associated lymph tissue.

The so-called passenger leukocyte that initiates allograft rejection via antigen/Class I MHC expression is also a Langerhans cell, and removal of the cell or ablation of the afferent lymphatics prevents sensitization. The underlying connective tissue of skin is transformed into a "lymph node-like" microenvironment by alterations in the microvasculature and reticulum to accomodate local recirculation, lodging and proliferation of lymphocytes in situation where deposited antigen persists. Therefore, the skin has a barrier and sentinel function which depend upon a constant traffic of lymphocytes

Langerhans cells and mononuclear cells through the skin and into regional lymph nodes via afferent lymph.

Lymph and Lymphatics

Lymph is a clear body fluid which clots like blood. Lymph forms when dissolved proteins and solutes filter out of venules and capillaries because of local differences in luminal hydrostatic and osmotic pressure. Whenever the epithelial barrier is broached, the neuropeptide "substance P" is released by axons and increases lymph production through hyperemic effects. Hyperemia increases fluid transudate.

This aids fluid transport to the regional lymph node of cells, antigens and cytokines. Afferent lymphatic vessels have their origins in the reticular connective tissue beneath the epithelium of the skin, gut, and urogenital tract and within the mesenchymal tissues of all organs. The endothelial cells of lymphatics are anchored to the surrounding reticular fibre meshwork, causing the lymphatics to dilate rather than collapse when tissue hydrostatic pressure exceeds the pressure of fluid in the lymphatic.

This opens valve-like junctions between endothelial cells and allows intravasation of tissue fluids. Lymph flows unidirectionally toward lymph nodes because valves prevent back flow under normal physiological conditions. Lymph capillaries merge into larger lymphatics which drain into lymph nodes. Efferent lymph from regional lymph nodes may drain into one or more additional nodes before flowing into major efferent lymphatics. The thoracic duct carries lymph draining from the gut and the lower half of the body. For additional information on the thoracic duct.

LYMPH NODES

The main function of lymph nodes is to trap antigens and cells containing antigen that flow into them via afferent lymphatics and to provide a site for clonal expansion of lymphoid cells recruited from the millions of cells that enter and leave via various routes. However, lymph nodes have different ways of handling cells, solutes and particles as sources of antigens that will be presented. Lymph nodes are connective tissue bags filled with mobile cells organized into functional compartments by a meshwork of reticulin fibers ensheathed by fibroblastic reticular cells; and, supplied by a system of specialized blood vessels and nerves.

A "cortex" and "medulla" is distinguishable in histological preparations by the relative density of small lymphoid cells contained in the reticular matrix. The superficial cortex contains a lymphatic sinus, macrophage rich zone and *B*-cell follicles.

The deep cortex is a high traffic zone where migrant or recirculating *T* and *B* lymphocytes enter from the blood. Lymphocytes in the deep cortex are directed toward specific *B* or *T*-cell microenvironments by fibroblastic reticular cell corridors where the lymphocytes encounter antigen presenting cells. If they are not activated by antigen displayed in Class II MHC along with costimulatory signals the *T* or *B*-cells crawl into lymphatic channels and out of the lymph node.

The medullary cords contain sessile *B*-lymphoblasts and plasma cells which accumulate there after immune reactions and the medullary lymph sinuses are the highways out of the lymph node. The cortex is organized into hemispheric lobules where the flat surfaces face the afferent lymph supply and the round central borders merge with stromal chords and sinuses to form the medulla.

Cortical lobules in lymph nodes of all mammalian species have a constant vertical dimension. The lateral dimensions of cortical lobules are highly variable which accounts for the rapid 2-4 fold node enlargement that occurs 6-18 hours after antigen inoculation.

Antigen Acquisition and Presentation

Afferent lymph drains into a flat lymphatic antechamber, (the subcapsular sinus) which distributes the lymph over the superior surface of cortical lobules packed with lymphocytes, scattered macrophages and interdigitating dendritic cells. Motile cells (DC and Monocytes), transporting presentable antigens, crawl directly into the superficial cortex by passing through "pores" in the floor of the subcapsular sinus. Larger particles and dead cells pass into lymphatic capillaries (cortical sinuses) that form an extensive perivenular plexus in the deep cortex before emptying into the medullary sinuses.

These cortical sinuses define the boundaries of "cortical cords." There is considerable phagocytic activity at all portals of free antigen access to lymph nodes. This limits the amount of particulate antigen that could access the cortical *T* and *B* cells. Soluble materials should "diffuse down" into the cortex but this is not obvious. Instead, one sees soluble labels staining the fibroblastic reticulum that courses directly from lymph sinus wall to the perivenular channels of HEV.

Macrophages are found in the subcapsular sinus near the pores and suspended from reticular fibers that cross the lymphatic channels. Macrophages are also prevalent in the large lymphatic sinuses in the medulla. Antigens that escape phagocytosis within a lymph node face phagocytosis in other lymph nodes through which the lymph must pass before entering major efferent collecting ducts. Antigens that succeed in eluding lymph node entrapment will ultimately be captured by blood monocytes or macrophages in the spleen, liver or bone marrow.

Mononuclear cells spaced throughout the deep cortex participate in antigen processing and presentation. Not all macrophages in these sites express surface Ia-antigen or perform antigen presentation functions. Interdigitating dendritic cells (IDC) constitutively express Ia-antigen and exhibit physical contacts with macrophages and numerous cortical *T*-cells. This capacity to distribute its cell surface over numerous lymphocytes may explain the IDC's relative effectiveness in stimulating mixed lymphocyte culture reactions.

Although the full repertoir of cytokines secreted by, or affecting IDC are not known, the macrophages located near IDC may have a positive effect by secreting cytokines that enhance interaction and stimulation of T cells by IDC. It is now believed that interdigitating dendritic cells are derived from Langerhans cells which migrate into lymph node cortex from the skin. It is also possible that local differentiation of a migrant precursor cell in presence of GM-CSF, or gene switching of daughter cells produced through local division of resident mononuclear cells may be involved. Fibroblastic reticular cells in lymph nodes also bind antigen without phagocytosis.

Three to four days after inoculation, labeled antigens are found on the surfaces of dendritic cells within B-cell areas of the lymph node cortex. Cytophilic antibody (Fc receptors) and complement (C3b receptors) facilitate binding of antigen by the dendritic cells within germinal centers of B-cell follicles. Antigen remains bound to follicular dendritic cells for a long time and the same antigen will bind to these cells with greater avidity on secondary exposure, presumably due to cytophilic antibody.

The molecular mechanisms of antigen presentation by accessory cells after primary exposure to *T*-dependent antigens has recently been described. Ia-antigen itself is a "receptor" for immunogenic peptides produced after the antigen is degraded in an acidic lysomal compartment. Presumably, Ia-antigen is a transport protein that collects degraded peptides in Golgiassociated lysosomes and shuttles them to the cell surface to be presened to helper T-cells. Thus the "processed" immunogenic determinants are selected by the antigen presenting cell. B-cells may also present antigen to Tcells when membrane immunoglobulin captures antigen rather than presents it.

The requirement for MHC restriction in provision of *T*-cell help for B-cell responses anticipated coordinate recognition of class II MHC and antigen by helper *T*-cells. Uptake of antigen by the *B*-cell also results in processing and presentation of antigenic epitopes via Ia-antigen just as in "antigen presenting cells", and capture and presentation of antigen is greatest when the *B*-cell membrane immunoglobulin is specific for the antigen.

Germinal Follicles

Germinal follicles are discrete lymphoid compartments where *B*-cells (that are reacting to *T*-dependent antigens) undergo division, isotype switching and memory development.

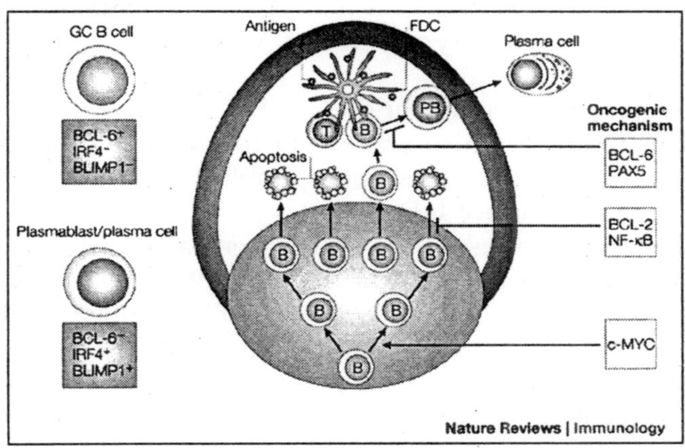

Fig. Germinal Follicles

B-cell follicles are found in all peripheral lymphatic tissues including foci of chronic inflammation. Although some germinal follicles develop oligoclonally, there are far fewer germinal follicles than there are antigen reactivities among B-cells so it is possible that follicles have a "changing of the guard" periodically when new antigens trigger development of B cell memory.

The stroma associated with the mantel zone and germinal centre of *B*-cell follicles express the chemokine BCL that is chemotactic for B-cells expressing CXCR5 receptor. "Activated" CD-4 T-cells have the CXCR5 receptor up-regulated, presumably to enable antigen specific *T*-cell help for the reactive B-cells in the germinal centre.

In lymph nodes, B-cell follicles partially interrupt the subcapsular sinus in the superficial cortex. A germinal centre develops in a follicle as proliferating *B*-cells displace the cortical reticular fibers into a basket-like enclosure that separates central lymphoblasts from the peripheral mantel of small *B*-cells. Lymphoblasts, follicular dendritic cells and "tingible-body" macrophages reside inside the enclosure. Recent morphological and genetic analyses indicate that germinal centre light zones (LZ) and dark zones (DZ) carry out somewhat different functions in B cell development.

On the outside of GC, the mantle zone is comprised almost entirely of small *B*-lymphocytes with IgM+/IgD surface phenotype.

A sprinkling of Helper (L_3T_4, CD^4) and Suppressor (Lyt-2, CD^8) *T*-cells are also present in the mantel zone. In the germinal centre the large B-lymphoblasts and intermediate cells have lost surface IgD.

They now express IgM (80-90 per cent) and another isotype such as IgG (30 per cent); IgM^+/IgD^+ lymphocytes in follicular mantle zones have not yet switched while germinal centre B-cells have. Therefore, it is likely that rearrangement of the immunoglobulin heavy chain genes occurs in (or en route to) germinal centers with mantle zone Bcells serving as a pre-switch pool.

Switched *B*-cells initiate low level transcription of messenger RNA for the "secondary" isotype while still expressing membrane IgM. IgM is deleted and the "secondary" isotype is expressed usually after a second exposure to antigen. Inside germinal centers the activated *B*-cell sees antigen on follicular dendritic cells and receives help from CD^{4+} *T*-cells and/or their factors.

The frequency of CD^{4+} *T*-cells in germinal follicles is small relative to B-cells but they outnumber CD^{8+} *T*-cells 3:1. *B*-cells divide and migrate rapidly out of the follicle as lymphoblasts which leave the node to lodge in the spleen, inflammatory foci, and other lymphatic tissues. Some large IgM+ *B*-cells lodge in the marginal zone of the spleen and switch after they re-encounter antigen.

Labeled *B*-lymphoblasts "home" to germinal centers without accumulating in the small cell mantle after intravenous infusion. This may replenish germinal centers with varieties of antigen specific clonal precursors, permitting follicular dendritic cells to present a single antigen to B-cells responsive to different epitopes, or the B-blasts may provide additional antigen presenting capacity if the B-cell happens to be binding and processing nominal antigen.

Lymph Node Vasculature

The vasculature of the lymph node cortical lobules is specialized to meet the demands of a tissue populated by transient cells. Arterial vessels enter lymph nodes at the hilus and branch once or twice as they cross cortical lobules to supply a network of capillaries and arteriovenous communications running beneath the floor of the subcapsular sinus in the superficial cortex. Arteriovenus communications and short segments postcapillary venules connect with HEV lined by plump polygonal endothelial cells.

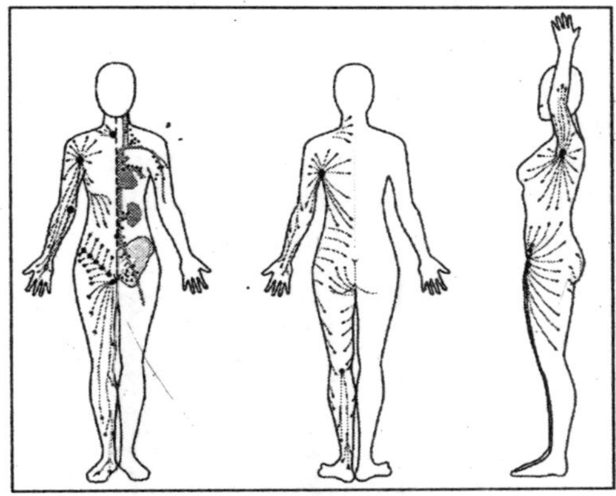

Fig. Vasculature

Distal segments of HEV open into elastic capacitance vessels draining segments of cortical lobules. These segmental veins merge to form lobular veins near the hilum. Circumferential smooth muscle bundles form contractile structures at sites where segmental and lobular veins join before connecting with the systemic circulation. These contractile vascular structures may prevent dramatic intranodal changes in blood flow from affecting systemic circulation.

Studies of microvascular changes in stimulated lymph nodes demonstrated a biphasic shift in blood flow after antigen inoculation. An early transient increase in flow up to 30 times normal was attributed to hyperemia. The increased flow, caused by increased arteriovenous shunting resulted in increased lymphocyte traffic rates. This was followed by a second, more gradual wave of increased blood flow due to angiogenesis and enlargement of the network of HEV between 3 and 6 days after alloantigenic stimulation or after inoculation with adjuvants. The relationship of this angiogenesis response to dendritic cells and secretion of VEGF.

These antigen-induced alterations in the microcirculation directly affected transvascular lymphocyte migration into lymph nodes; initially, by increasing the rate of immigration; and, subsequently, by enlarging the surface area of HEV endothelium available for migration. Myelinated and unmyelinated nerve fibers are found coursing through the stroma of lymph nodes where they are associated with arteriovenous communications and venous sphincters. Recent studies clearly indicate that noradrenergic and peptidergic nerve fibers supply diffuse areas populated with lymphoid cells in addition to the vasculature.

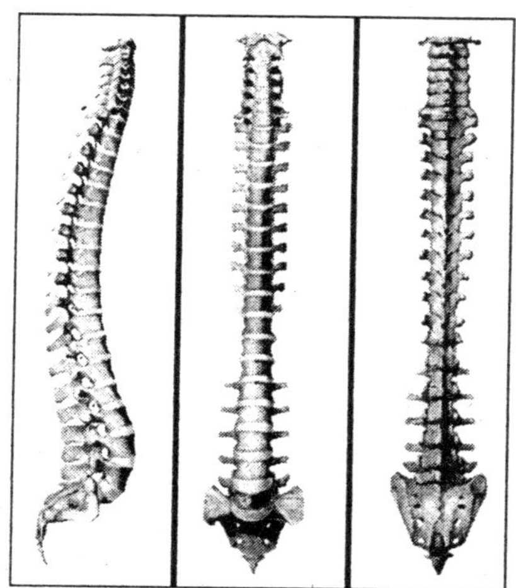

Fig. Bundles of nerve fibers

Bundles of nerve fibers enter lymph nodes at the hilus and follow arteries and veins to the floor of the subcapsular sinus where the nerves form a plexus.

Noradrenergic fibers extend into the paracortex near lymphoid aggregates adjacent to HEV. Furthermore, norepinephrine which is depletable by 6hydroxydopamine is found in lymph nodes. Depletion of norepinephrine with that agent resulted in a diminished immune response in draining lymph nodes after subcutaneous injection of antigen, suggesting an immunoregulatory role for the products of neural secretion in lymph nodes.

LYMPHOCYTE MIGRATION INTO LYMPH NODES

Gowans and collaborators demonstrated that large-scale emigration of recirculating lymphocytes from blood to lymph occurs across the walls of HEV. Accumulation of lymphocytes in HEV lumens was due to selective adhesion of circulating lymphocytes to endothelial surfaces via receptor ligand-like interactions.

Polymorphonuclear leukocytes never adhered to HEV in vivo unless the lymph node was involved in acute inflammation. Lymphocytes adhered to HEV luminal surfaces via microvilli.

This "tethering" gained strength as more microvilli interdigitated with pits on luminal surface of the endothelial cell. This kind of cell-cell interaction is effective in resisting hydrodynamic shear forces, and it involves receptors associated with the cytoskeleton. The existence of a glycoprotein "receptor" for recognition in lymphocyte homing was hypothesized a long while ago by Gesner and Ginsburg, and Marchesi and Gowans.

Adhesion inhibition studies with monoclonal antibodies identified several lymphocyte membrane glycoproteins that may be responsible for organ - selective homing of lymphocytes in mucosal or peripheral lymphatic tissues. Continuation in this line of investigation has lead to a consensus hypothesis that is based sequential use of overlapping receptor - coreceptor pairs to form progressively stronger attachments to endothelium.

The specificity of organ specific homing could result from specific combinations of complimentary adhesive molecules and a cascade of cell interactions. The original "Homing Receptors" that enabled "organ specific" lymphocyte homing are part of the consensus. Both MadCAM-1 (Mucosal Vascular Addressin) and PNad-1 (peripheral node adressin) are bound by alpha 4 beta 7 integrin and/or L-selectin because of specificity for sulfated carbohydrates. Specific oligosaccharides inhibited in vitro binding of lymphocytes to HEV thereby demonstrating that the lymphocyte receptor recognizes a carbohydrate moiety on the HEV.

Luminal lymphocytes polarize and squeeze between endothelial cells to enter a perivascular channel formed by the sheath of reticular cells. The unique permeability of the HEV from lymphatic tissue toward the blood permits cytokines or other factors in adjacent lymphatic tissue to diffuse along spaces between HEV endothelial cells to yield chemotactic gradients.

Fibroblastic reticular cells ensheathing reticular fibers serve as conduits that carry "chemokines" and other factors directly to the HEV perivenular channels. This unique anatomic system provides for rapid and discrete communication between tissue and blood interface. When antigen inoculation causes altered blood flow, cytokine release, and increased tissue fluid pressures in regional lymph nodes, the resultant interendothelial gradient in HEV accelerates lymphocyte migration into the lymphatic tissue.

After squeezing out of perivenular channels into corridors and entering the deep cortical interstitium, lymphocytes crawl along cell processes and through a latticework of reticular fibers until they reach reactive foci in *T* or *B*-cell areas. They also cross lymphatic endothelium to enter intermediary sinuses that deliver them into the efferent lymph.

Contact guidance along fibroblastic reticular cells was proposed as a mechanism for intranodal redistribution of migrants because lymphocytes bind to these cells and migrate rapidly on their surfaces in vitro.

Aggregation or repulsion of lymphocyte sets was also proposed to explain compartmentalization of lymphoid cells within lymphatic tissues. The precise mechanisms that govern accumulation and distribution of lymphocyte subsets within specific lymphoid compartments are not known.

LYMPHOCYTE TRAFFIC DEPENDENT CORTEX

The deep cortex of lymph nodes, which appears in histological sections as a diffusely cellular area containing specialized vascular structures called high endothelial venules (HEV), has also been called the *T* dependent cortex *T/B* ANIMATION because of the adverse effect of neonatal thymectomy upon maturation of recirculating lymphocytes, and because a preponderance of *T* cells are present in immunohistological preparations.

B cells also migrate into the deep cortex. In "resting" lymph nodes the reticular connective tissue surrounding HEV contains numerous small lymphocytes in cords of reticular cell matrix between the HEV and lymphatic vessels of the cortical "intermediary" sinuses.

The size of the deep cortex depends upon the rate of lymphocyte immigration via HEV, the transit time of cells traversing the reticulum, and the rate of egress of immigrsnts vis efferent lymphatic channels. Indeed, the paracortex is capable of rapid 3 to 5 fold enlargement 6 to 24 hours after a strong stimulus induced by viral infection, injection or immunologic adjuvants, or inoculation of alloantigens. Accelerated lymphocyte traffic usually results in logjams of small lymphocytes in cortical sinuses.

While most of the acute enlargement of the deep cortex during the first 24 hours after inoculation is due to increased lymphocyte traffic into the node, some of it is due to sequestration and activation of antigen reactive cells, causing them to remain to proliferate or release lymphokines. The outcome of this selection by antigen and cytokines is the subsequent outpouring of antigen reactive cells in the efferent lymph between 72 hours and 100 hours after inoculation.

The two alternating images linked are of lymph nodes from a normal rat and from a rat whose recirculating lymphocytes were depleted by continuous thoracic duct drainage for up to 14 days. Depletion of recirculating lymphocytes reduces a lymph node to its stromal and vascular components.

The Paracortex is collapsed into a small mass of vessels

and connective tissue. Germinal centers (follicle small lymphocyte mantle zones are absent) are all that remain of the B cell compartment. Both images are shown at exactly the same magnification.

The deep cortex shrinks rapidly after chronic thoracic duct drainage or neonatal thymectomy. Both procedures are experimental manipulations that selectively deplete recirculating lymphocytes. Although the deep cortex is commonly known as the *T*-dependent cortex it is not comprised solely of *T* cells.

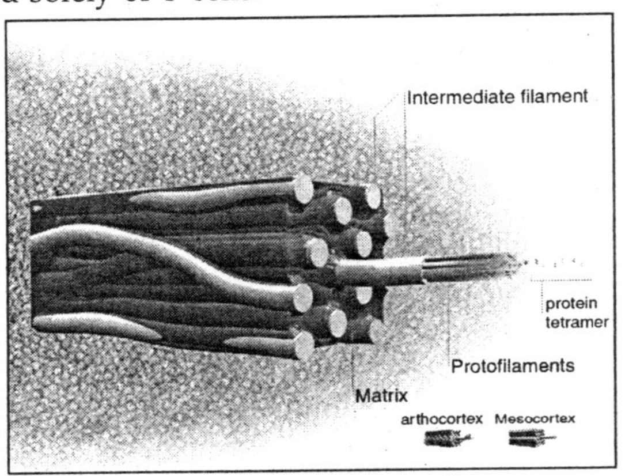

Fig. Paracortex

The lymphocytes populating the deep cortex are about 75 per cent *T*-cells and 25 per cent B-cells, 90 per cent of them are small lymphocytes and virtually all belong to the recirculating pool. Small *B* cells appear to migrate up to the superficial cortex after briefly accumulating around HEV.

Small *T* cells move out through the perivenular channels into corridors of reticulum comprising the paracortex. There are.slightly more CD^{8+} *T*-cells compared with CD^{4+} *T*-cells in resting lymph nodes but the ratio can vary considerably. This also applies to the ratio of *T* cells to *B*-cells. Normal *T* and *B*-cell recirculation through lymph nodes seems to be under control of constitutively expressed "Secondary Lymphoid Tissue Chemokine (SLC) and "B-Lymphocyte Chemokine" (BLC).

After alloantigenic stimulation lymph nodes fill up with *T*-cells. In contrast, after inoculation of an antigen which selectively induces an antibody response, the deep cortex contains nearly 60 per cent B-cells. Therefore, the deep cortex is a dynamic lymphatic compartment which responds to stimulation by cytokines, antigens, and other biological response modifiers.

THE SPLEEN

Structure of the Spleen

The spleen filters blood and is the largest single lymphoid organ in the body. It has a dense fibrous capsule with muscular trabeculae extending inward to subdivide the spleen into lobules and provide anchors for the reticular meshwork supporting the cellular parenchyma.

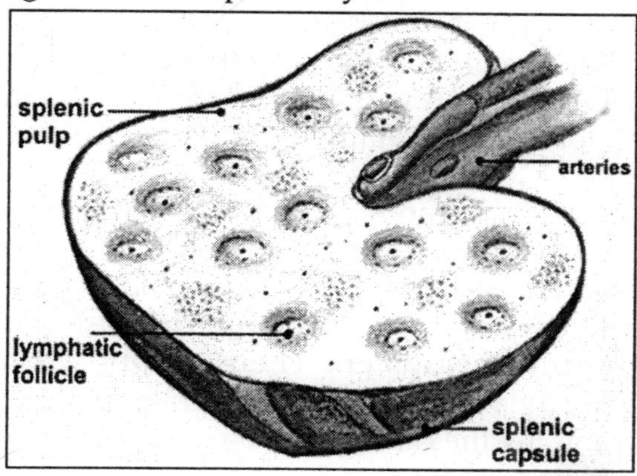

Fig. Structure of the Spleen

The spleen is grossly divisible into white pulp, formed of cylindrical collections of lymphocytes around the arteries, and red pulp, which contains erythrocyte-rich blood in cords of reticulum. The splenic cords are a continuous sponge-like reticular tissue which crisscross between fenestrated walls of splenic venous sinuses. The cords contain erythrocytes, lymphocytes, macrophages, granulocytes, and plasma cells.

The splenic venous sinuses are lined by elongated

endothelial cells that resemble barrel staves. The venous sinus wall is ringed by hoops of reticular fibers. All blood cells liberated into the splenic cords must squeeze through slits between these endothelial cells to enter the sinuses and return to the blood circulation.

Splenic sinus lining cells express antigens associated with monocytes, macrophages, endothelial cells, and *T*-lymphocytes. The spleen monitors for abnormal cells in blood by mechanical seiving in venous sinuses and exploration of surface antigens by mononuclear phagocytes.

Blood and Lymph Vasculature of the Spleen

The spleen is supplied by the splenic artery which enters through the hilus and branches to pass through trabeculae into medium sized arteries surrounded by a sheath of lymphatic tissue (periarteriolar lymphatic sheath, PALS). These vessels terminate in small arterioles which pass through PALS and empty into the reticulum of the red pulp cords or pulp sinuses. Within the PALS narrow perpendicular branches of the central arteriole supply a ring of sinuses within the marginal zone of the PALS.

These sinuses fill readily with radio opaque perfusates and materials used to make vascular corrosion casts for scanning electron microscopy; otherwise they would not be noticed. Marginal sinuses are critically important to lymphocyte recirculation in the spleen. Their supply vessels arise perpendicularly which may concentrate lymphoid cells through skimming the cells from the peripheral stream of the central arteriole.

Blood from red pulp cords and marginal sinuses empties into the venous sinusoidal system which connects by gradual transition to pulp veins and spenic veins. The spleen has not afferent lymphatic vessels, but efferent lymphatics arise within the white pulp near central arterioles. The efferent lymphatics cross perpendicularly between PALS areas to form an interconnecting network in cleared preparations of spleens whose lymphatics are labeled with carbon. The structure of the lymphatic network provides for cellular communication among the anatomically separated white pulp areas. Lymphatic trunks pass through trabeculae, merge with each other, and exit at the hilus. Major efferent lymphatics of the spleen wind down the hilar blood vessels to join the thoracic duct near the thoracic duct near the pancreas. Parts of the intrasplenic lymphatic network have anatomical relationships resembling the "bridging zones" of Mitchell because they appear to link white pulp with red pulp in histological sections. Noradrenergic fibers enter the hilus of the spleen with the vasculature, follow the trabeculae and branching vasculature, and ramify mainly within the white pulp along the central arterioles and periarteriolar lymphatic sheath.

Fibers branch from a dense periarterial plexus to permeate and terminate among diffuse small lymphoid cells and accessory cells in the PALS. Immunocytochemical data indicate that nerve fibers with neuropeptide-Y, Met-enkephalin, cholecystokinin 8, and neurotensin immunoreactivity are present along the central artery with sparse fibers entering lymphoid tissue. Cholecystokinin-8 bearing fibers are more prevalent in the PALS, however.

Functions of the Spleen

The spleen receives a high proportion of the cardiac output which contributes to its effectiveness as a filter of blood and a site for lymphocyte recirculation and lodging. Filtration of efete cells, debris and microorganisms occurs primarily in the reticulum of the red pulp chords and in the seive-like endothelium of venous sinuses.

Reticular cells provide anchoring sites for mononuclear phagocytes that perform reticuloendothelial functions in the red pulp and marginal zone. The white pulp of the spleen contains a peripheral lymphatic tissue microenvironment for antigen trapping, cellular collaboration, lymphocyte proliferation and antibody production in addition to providing young bone marrow emigrant B-cells a place to complete maturation. The spleen is the primary site for initiation of immune responses to antigens and pathogens that have invaded the blood stream in addition to being a partner in every other immune response in the body. Antigen-laden mononuclear cells and lymphoblasts, released into efferent lymph from other lymphatic tissues 48 to 100 hours after antigen priming, lodge in the spleen and set up satellite zones of *T*- and *B*-cell proliferation. During active immune responses lymphoblastic B-cells committed to plasma cell differentiation lodge in the red pulp cords and sinuses where they mature and commence secreting antibody.

Lymphocyte Migration in the Spleen

Lymphocyte traffic in the spleen moves across a blood tissue interface into sites of antigen presentation in the marginal zone and periarteriolar lymphatic sheath. Antibodies, enzymes, and drugs that are known to block lymphocyte homing to lymph nodes do not interfere with entry of the PALS by recirculating lymphocytes. Many of the lymphocytes entering the spleen in the arterial blood bypass the white pulp and flow into the reticulum of the red pulp cords and sinuses to exit via the splenic vein.

These cells are returned to the blood within 2 to 3 hours after entering the spleen. However, some small arterioles terminate in sinuses adjacent to the white pulp; others empty directly into the marginal zone. Lymphocytes in the marginal sinuses enter the marginal zone, leaving other leukocytes in the lumen to be carried away in the blood. Lymphocytes arriving in the marginal zone move across a vascular interface that is populated by antigen-binding macrophages and B-cells to enter the PALS. Once in the PALS, lymphocytes may remain within the spleen for up to 12 hours.

The PALS is analogous to the lymph node paracortex because it contains recirculating *T*- and *B*-cells in addition to interdigitating dendritic cells. Segregation of *T*- and *B*-cells into discrete zones within the white pulp is not rigorous. Follicles located in the border between the PALS and the marginal zone contain predominantly *B*-cells, and more recirculating *T*- than *B*-cells are found in the PALS.

Recent molecular data indicate that fibroblastic reticular cell stroma associated with PALS and *B*-cell follicles selectively express SLC (*T*-cell attracting chemokine) and BLC (*B*-cell attracting chemokine. Follicles in the spleen are identical to those described in the section on the lymph node. After variable periods of residence in the PALS, some recirculating cells exit through the efferent lymphatics, while others move into the red pulp sinuses. The magnitude of lymphocyte recirculation via efferent lymph versus venous return has not been determined, but traffic studies with viable lymphocytes whose motility has been arrested indicate that over 80 per cent of the lymphocytes which leave the spleen in the venous blood never migrated into white pulp.

If immunocompetant cells encounter an appropriate antigenic stimulus, they proliferate in the same manner as that seen in peripheral nodes; B-cells form large germinal centers and antibody secreting cells can be found within the splenic cords after migration through the blood or direct movement out from the white pulp.

The marginal zone of the spleen contains a heterogenous assortment of mononuclear cells with specialized functions. A principal function of the marginal zone is antigen trapping. Intravenously injected tracer antigens reproducibly accumulate in marginal zones regardless of the nature of the antigen. A subset of macrophages in the marginal zone appear to be important in presenting type-2 thymus independent antigens to *B*-cells. These cells are capable of selective uptake and retention of iv injected RITC-ficoll. They are negative for surface Ia-antigen and expression of the macrophage-related F4/80 antigen.

A splenic dendritic cell (Den-l), which is a potent accessory cell for some *B*-cell responses, has been cloned. This clone produces a unique cytokine that induces polyclonal *B*-cell proliferation in the absence of costimulators. The antibody forming cells generated by interaction of B-cells with these cells do not necessarily localize in the immediate proximity, however. The marginal zone contains specialized metallophilic macrophages that specifically label with a new monclonal antibody (MOMA-l).

These cells form a ring around the periarteriolar sheath

and follicular areas on the inner side of the marginal zone. The same cells show high nonspecific esterase activity and can be distinguished from the marginal zone macrophages by MOMA-l staining and the lack of selective FITC-ficoll uptake.

The marginal zone contains numerous *B*-cells which express membrane IgM^+, IgD^-, IL–2 receptors, have alkaline phosphatase on their surfaces and appear to be relatively sessile components of the marginal zone; i.e. they do not move into PALS unless stimulated by antigen to enter *B* cell follicles. Lipopolysaccharide (LPS) infusion rapidly depletes them from the marginal zone followed by expansion of the IgM+, IgD– cells in germinal centers.

Depletion and regeneration of these cels following anti-IgM or IgD treatment permitted MacLennan to determine that the marginal zone IgM^+, IgD^- *B*-cells are post follicular IgM^+, IgD^+ cells because the follicle cells regenerated before any marginal zone B-cells appeared during recovery from *B*-cell depletion.

This population of *B*-cells appears to be enriched in those that selectively respond to type-2 thymus independent antigens while the follicular Bcells respond to thymic dependent antigens.

MUCOSAL LYMPHATIC TISSUES

The mucosal lymphatic tissues are nonencapsulated submucosal lymphoid nodules and diffuse lymphocytic infiltrates in the submucosa of intestinal and respiratory tracts. These organs work collectively with regional lymph nodes and spleen to produce *B*- and *T*-effector cells which lodge in lamina propria and in intraepithelial locations wherever there is mucosa.

Nasopharyngeal lymphatic tissues (i.e., tonsils and adenoids in man), bronchus associated lymphatic tissues, Peyer's patches, appendix and isolated follicles in intestitinal mucosae vary with regard to type of surface epithelium (statified squamous, ciliated columnar, or absorptive columnar) and relative proportions of *T*- and *B*-cells but the similarity of these tissues to Peyer's patches is greater than the difference, especially since all have "*M*" cells in their follicle associate epithelium.

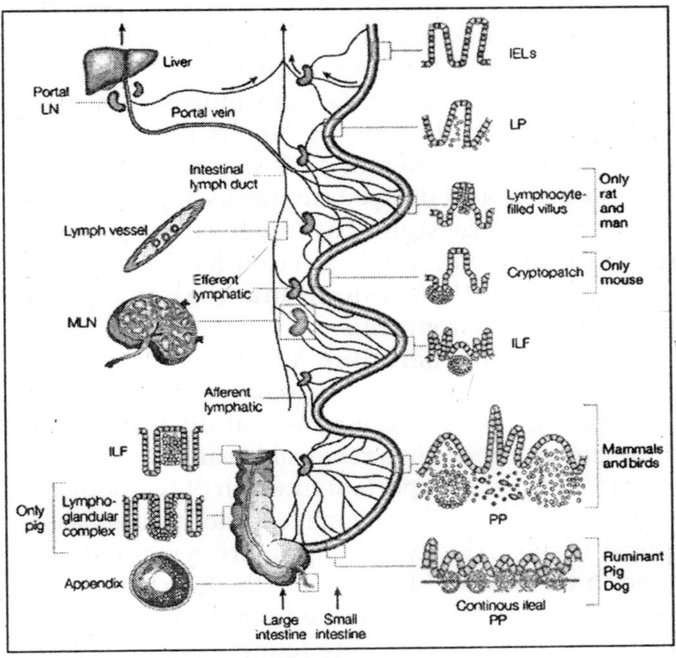

Fig. Lymphatic Tissues

The prototypical Mucosal Lymphatic Tissue is the Peyer's patch which has a unique dome epithelium that is specialized to sample environmentai antigens. Peyer's patches contain lymphoid compartments that are analogous to the deep cortex and follicles of lymph nodes, but there are no afferent lymphatics and no medullary cords for local accumulation of plasma cells. Each Peyer's patch contains multiple individual B-cell follicles separated by diffuse lymphoid tissue in interfollicular areas.

Functions of Mucosal Lymphatic Tissues

The mucosal immune system appears to have two paradoxically opposite immune purposes while also participating in continued diversification of the immunoglobulin repertoire. Mucosal lymphatic tissues amplify development of committed *B*-cells for secretory IgA responses to enviromental antigens IgA responses to environmental antigens and programs certain environmental antigens for systemic tolerance induction.

The tolerance that results from mucosal immunization does not effect production of *B*-cells committed to IgA secretion. "Mucosal" tolerance is manifested by antigen-specific suppression of delayed cutaneous hypersensitivity and reduced IgG expression. The mucosal immune system therefore exerts a Yin Yang effect on local versus systemic immunity initiated in unique mucosal and peripheral lymphoid microenvironments.

Cells retaining germline VH/VL and Fc genes are consistently recovered from mucosal (and splenic) lymphoid follicles, giving rise to suggestions that diversification of the immunoglobulin genes in a renewable early *B*-cell population may be an additional important function of lymphoid follicles in peripheral tissues like the spleen and mucosa associated lymphoid tissues. This type of "primary lymphoid" function is found in the Bursa of Fabricius in the Chicken.

However, recently described similarities in structure, development and function were found among the Bursa, the Ileal Peyers Patch in the Sheep, and the appendix in the Rabbit and Human.

Each of these structures become populated with lymphoid cells only after the respective species enters the "outside" world and comes in contact with environmental microbes and antigens. There has even been a suggestion that environmental microbes are essential stimulants for diversification of immunoglobulin genes.

M-Cells

The dome epithelium covering each follicle is composed of cuboidal absorptive epithelial cells interrupted by delicate membraneous cells which have luminal microfolds instead of microvillus borders.

The "*M*" cells endocytose and transport various materials

without lysomal degradation. Antigen is deposited into small lymphocytes, small mononuclear phagocytes and dendritic cells immediately beneath *M*-cells above the B-cell mantle of Peyer's patch germinal follicles.

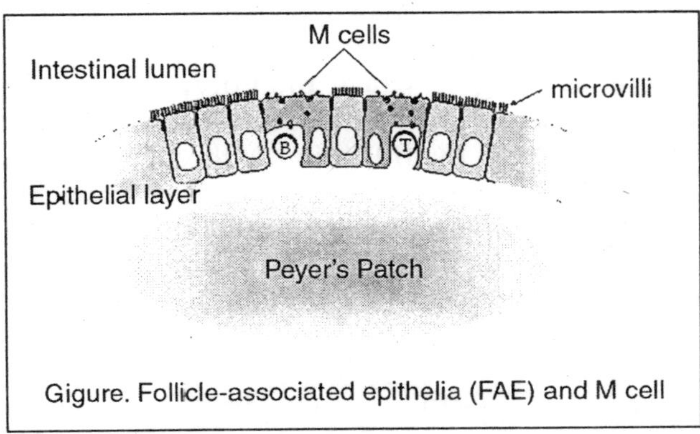

Gigure. Follicle-associated epithelia (FAE) and M cell

Fig. *M*-Cells

Minute quantities of intact antigen and products of digestion are transported to the lamina propria and lacteals by ordinary absorptive epithelial cells anywhere in the small bowel. It is important to point out that these products enter interfollicular areas and do not have access to the dome area. Between the dome epithelium and the follicles there is a thin region of reticulum, containing a delicate plexus of blood vessels and plasma cells.

Follicles

Peyer's patch follicles, located beneath the dome epithelium, have a mantle of small *B*-cells surrounding germinal centers that is thicker facing the dome. The germinal centers contain large and intermediate sized Blymphoblasts, follicular dendritic cells, macrophages and rare *T*-cells. Regardless how many follicles a patch contains, each is only one follicle thick, providing an intimate association with the overlying epithelium. Precursor B-cells with surface IgM and IgD enter Peyer's patches via HEV and migrate into the mantle zone above the germinal centre where they may come in contact with dendritic cells and *T*-helper cells. Membrane IgD is lost when the *B*-cells enter the germinal centre to begin proliferating. A small number of cells will express surface IgM with IgA or IgA alone but most will express only membrane IgM. Recent data indicate that switching and committment to IgA expression occurs in nearly 90Z of IgM only germinal centre cells although surface phenotypes and antibody produced in short term culture does not reflect this committment. In situ hybridization with c DNA probes (for the rearranged RNA message of IgA) does not label any cells in the small cell mantel, lightly label all of the cells expressing surface IgM-only and heavily label a minority population which synthesize and secrete dimeric IgA.

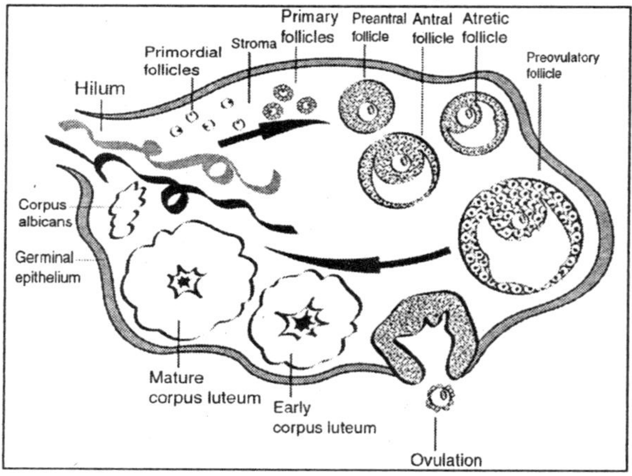

Fig. Follicles

Immature "mucosal" B-cells exit Peyer's patches down mesenteric lymphatics to regional lymph node where they contact additional recirculating helper or immunoregulatory cells. Most B-cells pass out of the mesenteric node via efferent lymph to the blood.

These cells temporarily lodge in the spleen where they proliferate and mature 5-7 days before returning to the lamina propria of the intestine and other mucosal sites. Once the B-cells reach the mucosa they divide once more and mature into IgA secreting plasma cells.

Lymphocyte Traffic Areas

The interfollicular areas are diffusely infiltrated with small and medium sized lymphocytes, usually of *T*-cell origin but numerous IgM^+/IgD^+ *B*-cells are also present. Recirculating lymphocytes emigrate from the blood into Peyer's patches via a network of HEV surrounding follicles.

Lymphocyte recirculation across HEV in patches is very rapid. Accumulation of labeled lymphocytes peaks 90 minutes after infusion, which is the shortest transit time of any organized lymphatic tissue. Lymphocyte traffic to the patch is augmented by recirculation across venules which are not recognizable as HEV.

Some lymphoid cells enter the lamina propria via thin-walled venules adjacent to intestinal crypts far away from the Peyer's patch and flow along submucosal lymphatics until they reach lymphatic plexuses in interfollicular areas. *T*-Lymphocyte traffic patterns related to mucosal and peripheral lymphatic tissues are segregated with some crossing over occurring in mesenteric lymph node and the spleen. Griscelli demonstrated the existence of two populations of *T*-lymphocytes which selectively migrated to the gut lamina propria when the cells came from mesenteric lymph node or the spleen when the cells came from a peripheral lymph node.

These distinctive traffic circuits are also found with murine *T*- and *B*-lymphoblasts but not small *T*-cells which migrated randomly to mucosal or peripheral lymphatic tissues regardless of their origin.

In sheep, small *T*-cells exhibited a pronounced asymetry of recirculation. Labeled small *T*-cells from intestinal lymph are twice as likely to be recovered in intestinal lymph than in nodal lymph, and vice versa. Furthermore, intestinal recirculating *T*-cells migrated through the small intesting while nodal *T*-cells did not. Similar lung-, gut- and peripheral lymph node associated traffic patterns were demonstrated in sheep by Spencer and Hall.

Small *B*-cells and B-lymphoblasts show preference for migration into Peyer's patches and gut lamina propria, respectively.

There is some preference for "secondary" B-cells to return to the mucosa where they first encountered antigen. However, B-cells committed to IgA expression lodge in laminae propriae of other mucosae anywhere in the body. This ability of Peyer's patch B-lymphoblasts to populate mucosal sites in the conjunctive, upper respiratory tract, bronchi, mammary glands, and gastrointestinal tracts is regarded as "The common mucosal immune system" where gut associated lymphatic tissues supply the bulk of the *B*-cells.

The immunoregulatory *T*-cell component of this sytem is segregated into traffic patterns which permit crossregulation between peripheral and mucosal immune systems within the spleen and mesenteric lymph nodes. The innervation of Peyer's patches is extensive and the diversity of adrenergic and peptidergic nerve endings in patches is greater than for any other peripheral lymphatic tissue.

Noradrenergic fibers enter at the serosal surface, course longitudinally with the muscularis, and form interfollicular plexuses which ramify through the diffuse *T*-dependent areas near HEV.

Fibers extend into the mushroom villi between the domes to supply enterochromaffin cells and plasma cells. Blood flow appears to be the limiting factor that determines lymphocyte extravasation in mucosal sites.

Immunoglobulin Isotype Committment

Cebra proposed that the special feature of mucosal lymphatic tissues that promotes the generation of IgA-only clonal precursors is an environment that allows antigen-driven division without maturation to plasma cells. The feature of the Peyer's patch environment that might promote this situation could include inaccessibility of T helper cells, excessive presence of T suppressor cells or humoral factors.

The tempo of *T*-cell recirculation through mucosal lymphatic tissues is consistent with the segregation of antigentriggering and subsequent maturation of IgA precursors, compelled to migrate to various lymphatic tissues to complete differentiation.

With a mean transit time through the Peyer's patches of under two hours *T*-cells do not remain long enough to participate in local maturation of IgM-expressing IgA-precursor cells. *T*-cells, carrier-primed by the same antigens as the *B*-cells in the Peyer's patches, catch up with IgA precursors in lymph sinuses of mesenteric nodes or in the spleen.

T-helper cells that secrete IL-4 may provide isotype specific growth and maturation advantages to *B*-cells committed to IgA expression; but, they do not cause immunoglobulin heavy chain gene switching. Recent studies by Koltoff indicate that *T*-cells are effective in selectively enhancing growth of IgA-committed B-cells in bulk cultures but cannot cause IgM-only *B*-cells to switch to any specific Ig-isotype under stringent clonal conditions.

Thus, the microenvironmental contribution of Peyer's patches to *B*-cell commitment for IgA expression is not as simple as providing a selective lymphokine. There may be other cell types in the mucosal microenvironment that participate in the programming *B*-cell switching. Mucosal dendritic cells are candidates for this role. Two recent in vitro studies indicate that IgA is expressed in high frequency when B-cells are cultured in the presence of dendritic cells from thoracic duct lymph or dendritic cells isolated by enzymatic disruption of Peyer's patches.

Intraepithelial Lymphocytes

Intraepithelial lymphocytes (IEL) are a large heterogeneous population of immune cells in the intestinal epithelium. Two basic characteristics distinguish IEL from peripheral lymphoid cells. First, 80–90X of these cells are CD^{8+}/CD^{4-}, of which only half are $Thy-^{1+}$. In contrast, the pre-dominant phenotype of lamina propria *T*-cells is CD^{4-}/CD^{8-}. Secondly, isolated IEL exhibit in vitro effector functions which identify them as natural killer-like or Cytotoxic cells that are spontaneous or "natural" but without the identifying characteristics of NK cells.

Further studies by Klein indicate that 40-70X of IEL bind

a monoclonal that recognizes immature thymocytes and *B*-cells, but these cells also expressed the cytotoxic activation antigen despite being Thy–l–. At least some of the cytotoxic cells in the intraepithelial compartment may therefore be thymic independent.

In mucosae, cytotoxicity directed toward virus-infected or otherwise parasitized epithelial cells would be prophylactic or protective because the lysed cell would be rapidly eliminated from the body in the mucosal stream. If IgA is also present, it would prevent reinfection of mucosal epithelium.

The mucosal immune system is the most dispersed, the most diverse and the most complicated lymphocytic system in the body.

Like the thymus, it plays a role in generating antigen reactive lymphoid cells which will become specific effectors upon further maturation and on the other hand the mucosal lymphatic tissues are responsible for inducing tolerance to antigens that are commonly experienced in the enteric canal. There is a great deal remaining to be learned about this important and enigmatic lymphoid system.

Chapter 18

Antibody Interactions

Interactions between antigen and antibody involve non-covalent binding of an antigenic determinant (epitope) to the variable region (complementarity determining region, CDR) of both the heavy and light immunoglobulin chains.

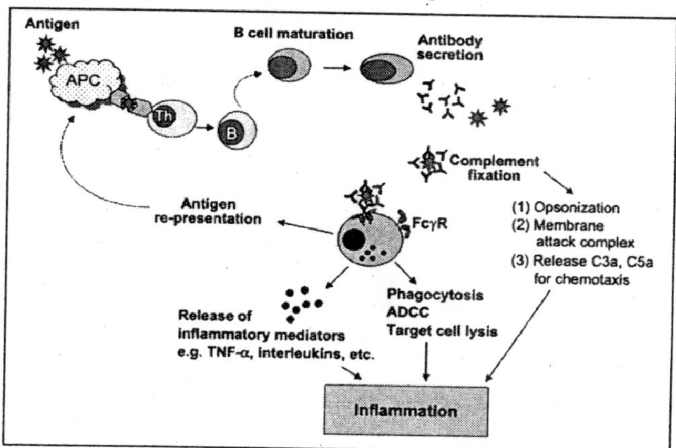

Fig. Antibody Interactions

These interactions are analogous to those observed in enzyme-substrate interactions and they can be defined similarly.

To describe the strength of the antigen-antibody interaction, one can define the affinity constant (K) as shown:

Affinity $K = [Ab - Ag]$

$[Ab] \times [Ag] = 10^4$ to 10^{12} L/mol

If the interaction between antigen and antibody were totally random, one would expect the concentrations of free antigen, free antibody and bound Ag – Ab complex to all be equivalent. In other words,

Affinity $K = 1$

$1 \times 1 = 10^0$ L/mol

Therefore, the greater the K, the stronger the affinity between antigen and antibody. These interactions are the result of complementarity in shapes, hydrophobic interactions, hydrogen bonds and Van der Waals forces.

ANTIGEN-ANTIBODY RATIOS

Experimentally, if one adds a known concentration of antibody to a tube and then adds increasing amounts of the specific antigen, the Ag – Ab complexes will begin to precipitate. If one continues to add increasing amounts of antigen, the complexes will begin to dissolve and return to solution. The following graph illustrates this process.

Tube #	1	2	3	4	5	6	7
Amount of precipitate	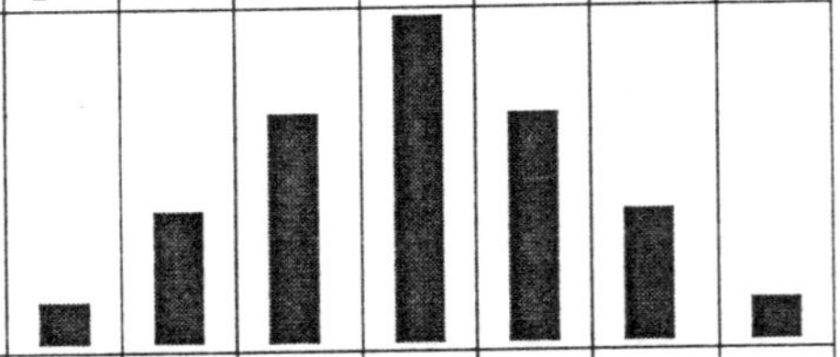						
Amount of Ag							
(arbitrary units)	1	2	3	4	5	6	7

If one then measures the amount of antigen and antibody remaining in the supernatant, one sees the following:

Excess Ab	+	+	+	–	–	–	–
Excess Ag	–	–	–	–	+	+	+

The left portion of the graph (tubes 1–3) illustrates "Antibody Excess", since not all of the antibody that is available to bind to antigen has actually bound antigen. The right portion of the graph (tubes 5–7) illustrates "Antigen Excess", where there is not enough antibody to bind to all of the available antigen. In the middle (tube 4) is a region known as "Equivalence". Here, the ration of antigen to antibody is perfect, so that all the antigen molecules and all of the antibody molecules are part of a complex. These are interesting experimental observations that do have relevance to situations occurring in the human body. For example:

- Antibody excess might occur when a person is exposed to a virus from which they have recently recovered. Hence, their body would contain a relatively large concentration of antiviral antibodies. These antibodies could quickly act to block cell receptors on the viral surface and prevent adsorption to host cells, thereby preventing disease.
- Antigen excess might occur early in the first infection by a microorganism. A person would have relatively few antibodies and these would form complexes but they would be very small. Such small complexes probably would not be phagocytosed or removed by the kidneys and could become lodged near tissue surfaces. Later, when antibody becomes available, the size of the complexes can increase leading to effective elimination by phagocytes or tissue damage where the smaller complexes had become lodged.
- Equivalence would occur when a person is exposed to an agent to which they have circulating antibodies. The correct ratio of antigen to antibody would produce extensive lattice formation, leading to enhanced phagocytosis, opsonization or agglutination, effectively eliminating the foreign agent.

CROSSREACTIVITY

Crossreactivity can occur when two (or more) antigens share similar structural features. Consider three different antigens, as shown on the right. Antibody produced in response to Ag_1 is very specific and would, therefore, have a large affinity constant (K) when combining with Ag1. However, Ag_2 is similar in shape to Ag_1 and is capable of interacting with anti-Ag1 antibody via two of three sites.

The interaction between Ab and Ag_2 is not as strong as the interaction between Ab and Ag1 (i.e., K is much smaller) but is still strong enough to allow binding. Hence, Ag_1 and

Ag_2 are said to *cross-react*. Ag_3, in contrast, cannot interact very well with anti-Ag_1 antibody and would have a K value so low that significant binding would not occur. Ag_3, therefore, would not cross-react with Ag_1. Would antibody produced in response to Ag_2 bind Ag_3? Would antibody produced in response to Ag_2 bind Ag_1?

Crossreactivity also forms the basis for several diagnostic tests. For example, infection with *Treponema pallidum* (syphilis) causes the production of antibodies that cross-react with a substance found in cardiac muscle, cardiolipin. Since it is much easier to obtain pure cardiolipin than pure *Treponemal* antigens, this cross-reaction is used to test for syphilis (Wassermann test). Likewise, antibodies produced against certain *Rickettsia* cross-react with antigens from *Proteus*. Since the latter are much easier to obtain, they can be used to test for the former.

CELL MEDIATED IMMUNITY

The second arm of the immune response is refered to as Cell Mediated Immunity (CMIR). As the name implies, the functional "effectors" of this response are various immune cells. These functions include:

- Phagocytosis and killing of intracellular pathogens
- Direct cell killing by cytotoxic T cells
- Direct cell killing by NK and K cells

These responses are especially important for destroying intracellular bacteria, eliminating viral infections and destroying tumor cells.

MACROPHAGE ACTIVATION

While the production of antibody through the humoral immune response can effectively lead to the elimination of a variety of pathogens, bacteria that have evolved to invade and multiply within phagocytic cells of the immune response pose a different threat. The following graphics illustrate this dilemma: This process can be further illustrated by considering the following experiment known as "Koch's phenomenon":

- Inoculation of an unimmunized guinea pig with a lethal dose of the intracellular pathogen *Mycobacterium tuberculosis* (MT) results in death of the animal. Inoculation with a sub-lethal dose induces immunity.

- Inoculation of an MT-immunized guinea pig with a lethal dose of MT causes a local reaction ("delayed hypersensitivity") one to two days later.
- Inoculation of an MT-immunized guinea pig with a lethal dose of a different intracellular pathogen, *Listeria monocytogenes* (LM) again results in death of the animal.
- Inoculation of an MT-immunized guinea pig with a lethal dose of LM and MT causes a delayed hypersensitivity reaction.

These results demonstrate the specific (*T*-cell mediated) and non-specific (macrophage mediated) aspects of this type of cell mediated immunity.

CELL MEDIATED CYTOTOXICITY

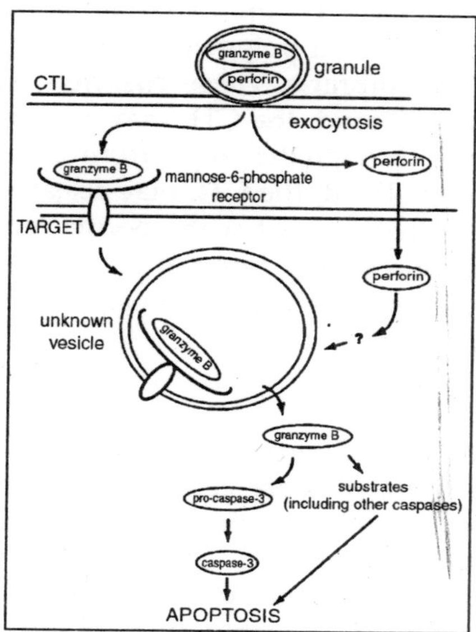

Fig. Cell Mediated Cytotoxicity

The second half of the cell-mediated immune response is involved in rejection of foreign grafts and the elimination of tumors and virus-infected cells. The effector cells involved in these processes are cytotoxic *T*-lymphocytes (CTLs), NK-cells and *K*-cells. Each of these effector cells recognizes their target by different means, described below.

Cytotoxic *T*-Lymphocytes

CTLs, like other *T*-cells are both antigen and MHC-restricted. That is, CTLs require i) recognition of a specific antigenic determinant and ii) recognition of "self" MHC. Briefly, CTLs recognize antigen via their T-cell receptor. This receptor makes specific contacts with the antigenic determinant and the target cell's class I MHC molecule. CTLs also express CD^8, which may assist the antigen recognition process. Once recognition is successful, the CTL "programs" the target cell for self-destruction. This process is thought to occur in one of several possible ways.

First, CTLs may release a substance known as perforin in the space between the CTL and its target. In the presence of calcium ions, the perforin polymerizes, forming channels in the target cell's membrane. These channels may cause the target cell to lyse. Second, the CTL may also release various enzymes that pass through the polyperforin channels, causing target cell damage. Third, the CTL may release lymphokines and/or cytokines that interact with specific receptors on the target cell surface, causing internal responses that lead to destruction of the target cell. CTLs principally act to eliminate endogenous antigens.

NK Cells

NK cells are part of a group know as the "large granular lymphocytes". These cells are generally non-specific, MHC-unrestricted cells involved primarily in the elimination of neoplastic or tumor cells.

The precise mechanism by which they recognize their target cells is not clear. Probably, there is some type of NK-determinant expressed by the target cells that is recognized by an NK-receptor on the NK cell surface. Once the target cell is recognized, killing occurs in a manner similar to that produced by the CTL.

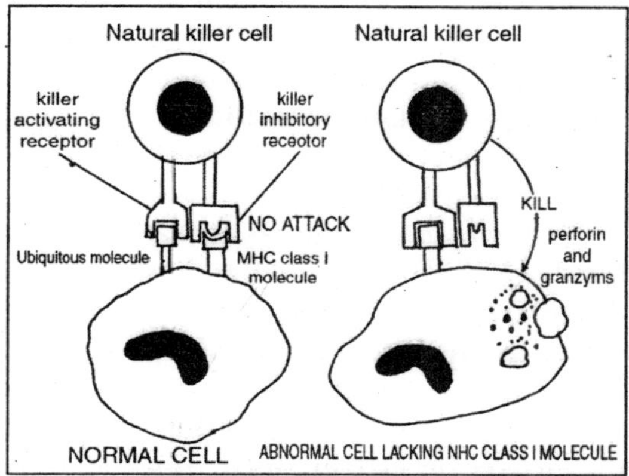

Fig. NK Cells

K-Cells

K-cells are probably not a separate cell type but rather a separate function of the NK group. *K*-cells contain immunoglobulin Fc receptors on their surface and are involved in a process known as Antibody-dependent Cell-mediated Cytotoxicity (ADCC).

ADCC occurs as a consequence of antibody being bound to a target cell surface via specific antigenic determinants expressed by the target cell. Once bound, the Fc portion of the immunoglobulin can be recognized by the *K*-cell. Killing then ensues by a mechanism similar to that employed by CTLs. This type of CMIR can also result in Type II hypersensitivities.

COMPLEMENT

Components and Functions of the Complement System

The complement system found in the blood of mammals is composed of heat labile substances (proteins) that combine with antibodies or cell surfaces. This complex, multicomponent system is composed of about 26 proteins. The "complement cascade" is constitutive and non-specific but it must be activated in order to function. The functions of complement include:

- Making bacteria more susceptible to phagocytosis

- Directly lysing some bacteria and foreign cells
- Producing chemotactic substances
- Increasing vascular permeability
- Causing smooth muscle contraction promoting mast cell degranulation.

The complement system can be activated via two distinct pathways; the classical pathway and the alternate pathway. Once initiated, a cascade of events (the "complement cascade") ensues, providing the functions listed above. Most of the complement components are numbered (e.g., C_1, C_2, C_3, etc.) but some are simply refered to as "Factors". Some of the components must be enzymatically cleaved to activate their function; others simply combine to form complexes that are active.

ACTIVATION OF THE COMPLEMENT CASCADE

Classical Pathway

The classical pathway starts with C_1; C_1 binds to immunoglobulin Fc (primarily IgM and IgG); C_1 is recognition complex composed of 22 polypeptide chains in 3 subunits; C_1q, C_1r, C_1s. C_1q is the actual recognition portion, a glycoprotein containing hydroxyproline and hydroxylysine that looks like a tulip flower. Upon binding via C_1q, C_1r is activated to become a protease that cleaves C1s to a form that activates (cleaves) both C_2 and C_4 to C_2a/b and C_4a/b. C_2b and C_4b combine to produce C_3 convertase (C_3 activating enzyme). C_4a has anaphylactic activity (inflammatory response).

C_3 is central to both the classical and alternative pathways. In classical, C_4b_2b convertase cleaves C_3 into C_3a/b. C_3a is a potent anaphylatoxin.

C_3b combines with C_4b_2b to form $C_4b_2b_3b$ complex that is a C_5 convertase. C_3b can also bind directly to cells making them susceptible to phagocytosis. C_5 is converted by C_5 convertase (i.e. $C_4b_2b_3b$) to C_5a/b. C_5a has potent anaphylatoxic and chemotaxic activities.

C_5b functions as an anchor on the target cell surface to which the lytic membrane-attack complex (MAC) forms. MAC includes C_5b, C_6, C_7, C_8 and C_9. Once C_9 polymerizes to form a hole in the cell wall, lysis ensues.

Alternate Pathway

The alternate pathway may be initiated by immunologic (e.g. IgA or IgE) or non-immunologic (e.g. LPS) means. The cascade begins with C_3. A small amount of C_3b is always found in circulation as a result of spontaneous cleavage of C_3 but the concentrations are generally kept very low.

However, when C_3b binds covalently to sugars on a cell surface, it can become protected. Then Factor B binds to C_3b. In the presence of Factor D, bound Factor B is cleaved to Ba and Bb; Bb contains the active site for a C_3 convertase. Next. properdin binds to C_3bBb to stabilize the C_3bBb convertase on cell surface leading to cleavage of C_3. Finally, a C_3bBb_3b complex forms and this is a C_5 convertase, cleaving C_5 to C_5a/b. Once formed, C_5b initiates formation of the membrane attack complex as described above.

Generally, only Gram-negative cells can be directly lysed by antibody plus complement; Gram-positive cells are mostly resistant. However, phagocytosis is greatly enhanced by C_3b binding (phagocytes have C_3b receptors on their surface) and antibody is not always required. In addition, complement can neutralize virus particles either by direct lysis or by preventing viral penetration of host cells.

Regulation Of The Complement Cascade

Because both the classical and alternate pathways depend upon C_3b, regulation of the complement cascade is mediated via 3 proteins that affect the levels and activities of this component.

- C_1 Inhibitor inhibits the production of C_3b by combining with and inactivating C_1r and C_1s. This prevents formation of the C_3 convertase, C_4b_2b.
- Protein H inhibits the production of C3b by inhibiting the binding of Factor B to membrane-bound C_3b, thereby preventing cleavage of B to Bb and production of the C_3 convertase, C_3bBb.
- Factor I inhibits the production of C_3b by cleaving C_3b into C_3c and C_3d, which are inactive. Factor I only works on cell membrane bound C_3b, mostly on red blood cells (i.e. non-activator surfaces).

Chapter 19

Hypersensitivity

Occasionally, the immune system responds inappropriately to the presence of antigen. These responses are refered to as *hypersensitivities*. There are four different types of hypersensitivities that result from different alterations of the immune system. These types are classified as:

- Type I: Immediate Hypersensitivity
- Type II: Cytotoxic Hypersensitivity
- Type III: Immune Complex Hypersensitivity
- Type IV: Delayed Hypersensitivity

REGULATION OF THE COMPLEMENT CASCADE

Type I or Immediate Hypersensitivity

Type I or Immediate Hypersensitivity can be illustrated by considering the following experiment: First, a guinea pig is injected intravenously with an antigen. For this example, bovine serum albumin (BSA, a protein) will be used. After two weeks, the same antigen will be reinjected into the same animal. Within a few minutes, the animal begins to suffocate and dies by a process called *anaphylactic shock.*

Instead of reinjecting the immunized guinea pig, serum is transferred from this pig to a "naive" (unimmunized) pig. When this second guinea pig is now injected with BSA, it also dies of anaphylactic shock. However, if the second pig is injected with a different antigen (e.g. egg white albumin), the pig shows no reaction.

If immune cells (T-cells and macrophages instead of serum) are transfered from the immunized pig to a second pig, the result is very different; injection of the second pig with BSA has no effect.

These results tell us that:

The reaction elicited by antigen occurs very rapidly (hence the name "immediate hypersensitivity").

The hypersensitivity is mediated via serum-derived components (i.e., antibody).

The hypersensitivity is antigen-specific (as one might expect for an antibody-mediated reaction).

The details of this reaction can be summarized as follows:

Initial introduction of antigen produces an antibody response. More specifically, the type of antigen and the way in which it is administered induce the synthesis of IgE antibody in particular. Immunoglobulin IgE binds very specifically to receptors on the surface of mast cells, which remain circulating.

Reintroduced antigen interacts with IgE on mast cells causing the cells to degranulate and release large amounts of histamine, lipid mediators and chemotactic factors that cause smooth muscle contraction, vasodilation, increased vascular permeability, broncoconstriction and edema. These reactions occur very suddenly, causing death.

Examples of Type I hypersensitivities include allergies to penicillin, insect bites, molds, etc.

A person's sensitivity to these allergens can be tested by a cutaneous reaction. If the specific antigen in question is injected intradermally and the patient is sensitive, a specific reaction known as *wheal and flare* can be observed within 15 minutes. Individuals who are hypersensitive to such allergens must avoid contact with large inocula to prevent anaphylactic shock.

Type II Hypersensitivity

Type II or Cytotoxic Hypersensitivity also involves antibody-mediated reactions. However, the immunoglobulin class (isotype) is generally IgG. In addition, this process involves K-cells rather than mast cells. K-cells are, of course, involved in antibody-dependent cell-mediated cytotoxicity (ADCC). Type II hypersensitivity may also involve complement that binds to cell-bound antibody. The difference here is that the antibodies are specific for (or able to cross-react with) "self" antigens. When these circulating antibodies react with a host cell surface, tissue damage may result.

There are many examples of Type II hypersensitivity.

These Include: Pemphigus: IgG antibodies that react with the intracellular substance found between epidermal cells.

Autoimmune hemolytic anemia (AHA): This disease is generally inspired by a drug such as penicillin that becomes attached to the surface of red blood cells (RBC) and acts as hapten for the production of antibody which then binds the RBC surface leading to lysis of RBCs.

Goodpasture's syndrome: Generally manifested as a glomerulonephritis, IgG antibodies that react against glomerular basement membrane surfaces can lead to kidney destruction.

Type III Hypersensitivity

Type III or Immune Complex hypersensitivity involves circulating antibody that reacts with free antigen. These circulating complexes can then become deposited on tissues. Tissue deposition may lead to reaction with complement, causing tissue damage. this type of hypersensitivity develops as a result of systematic exposure to an antigen and is dependent on:

- The type of antigen and antibody.
- The size of the resulting complex.

More specifically, complexes that are too small remain in circulation; complexes too large are removed by the glomerulus; intermediate complexes may become lodged in the glomerulus leading to kidney damage.

One example of a Type III hypersensitivity is serum sickness, a condition that may develop when a patient is injected with a large amount of e.g., antitoxin that was produced in an animal. After about 10 days, anti-antitoxin antibodies react with the antitoxin forming immune complexes that deposit in tissues. Type III hypersensitivities can be ascertained by intradermal injection of the antigen, followed by the observance of an "*Arthus*" reaction (swelling and redness at site of injection) after a few hours.

Type IV Hypersensitivity

Type IV or Delayed Hypersensitivity can be illustrated by considering the following experiment:

First, a guinea pig is injected with a sub-lethal dose of *Mycobacterium tuberculosis* (MT). Following recovery of the animal, injection of a lethal dose of MT under the skin produces only erythema (redness) and induration (hard spot) at the site of injection 1-2 days later.

Instead of reinjecting the immunized guinea pig, serum is transfered from this pig to a "naive" (unimmunized) pig. When this second guinea pig is now injected with MT, it dies of the infection.

If immune cells (T-cells and macrophages instead of serum) are transfered from the immunized pig to a second pig, the result is very different; injection of the second pig with MT causes only erythema and induration at the site of injection 1-2 days later.

In a separate experiment, if the immunized guinea pig is injected with a lethal dose of *Listeria monocytogenes* (LM) instead of MT, it dies of the infection. However, if the pig is simultaneously injected with both LM and MT, it survives.

These results tell us that: The reaction elicited by antigen occurs relatively slowly (hence the name "delayed hypersensitivity"). The hypersensitivity is mediated via T-cells and macrophages. The hypersensitivity illustrates both antigen-specific (T-cell) and antigen non-specific (macrophage) characteristics.

The details of this reaction can be summarized as follows: Initial introduction of antigen produces a cell-mediated response. *Mycobacterium tuberculosis* is an intracellular pathogen and recovery requires induction of specific T-cell clones with subsequent activation of macrophages.

Memory T-cells respond upon secondary injection of the specific (i.e. MT) antigen, but not the non-specific (i.e., LM) antigen. Induction of the memory T-cells causes activation of macrophages and destruction of both specific (MT) and non-specific (LM) microorganisms.

IMMUNOGLOBULINS

Immunoglobulins generally assume one of two roles: immunoglobulins may act as *i*) plasma membrane bound antigen receptors on the surface of a B-cell or ii) as antibodies free in cellular fluids functioning to intercept and eliminate antigenic determinants.

Basic Immunoglobulin Structure

Immunoglobulins are composed of four polypeptide chains: two "light" chains (lambda or kappa), and two "heavy" chains (alpha, delta, gamma, epsilon or mu). The type of heavy chain determines the immunoglobulin isotype (IgA, IgD, IgG, IgE, IgM, respectively).

Light chains are composed of 220 amino acid residues while heavy chains are composed of 440-550 amino acids. Each chain has "constant" and "variable". Variable regions are contained within the amino (NH_2) terminal end of the polypeptide chain (amino acids 1-110).

When comparing one antibody to another, these amino acid sequences are quite distinct. Constant regions, comprising amino acids 111-220 (or 440-550), are rather uniform, in comparison, from one antibody to another, within the same isotype. "Hypervariable" regions, or "Complementarity Determining Regions" (CDRs) are found within the variable regions of both the heavy and light chains. These regions serve to recognize and bind specifically to antigen. The four polypeptide chains are held together by covalent disulfide (–S–S–) bonds.

Structural differences between immunoglobulins are used for their classification. As stated above, the type of heavy chain an immunoglobulin possesses determines the immunoglobulin "isotype". More specifically, an isotype is determined by the primary sequence of amino acids in the constant region of the heavy chain, which in turn determines the three-dimensional structure of the molecule.

Since immunoglobulins are proteins, they can act as an antigen, eliciting an immune response that generates anti-immunoglobulin antibodies. However, the structural (three-dimensional) features that define isotypes are not immunogenic in an animal of the same species, since they are not seen as "foreign".

For example, the five human isotypes, IgA, IgD, IgG, IgE and IgM are found in all humans and a result, injection of human IgG into another human would not generate antibodies directed against the structural features (determinants) that define the IgG isotype. However, injection of human IgG into a rabbit *would* generate antibodies directed against those same structural features.

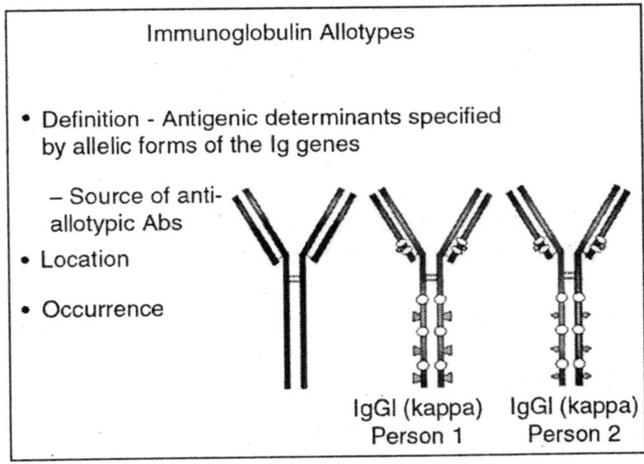

Fig. Immunoglobulin Allotypes

Another means of classifying immunoglobulins is defined by the term "allotype". Like isotypes, allotypes are determined by the amino acid sequence and corresponding three-dimensional structure of the constant region of the immunoglobulin molecule. Unlike isotypes, allotypes reflect genetic differences between members of the same species.

This means that not all members of the species will possess any particular allotype. Therefore, injection of any specific human allotype into another human could possibly generate antibodies directed against the structural features that define that particular allotypic variation.

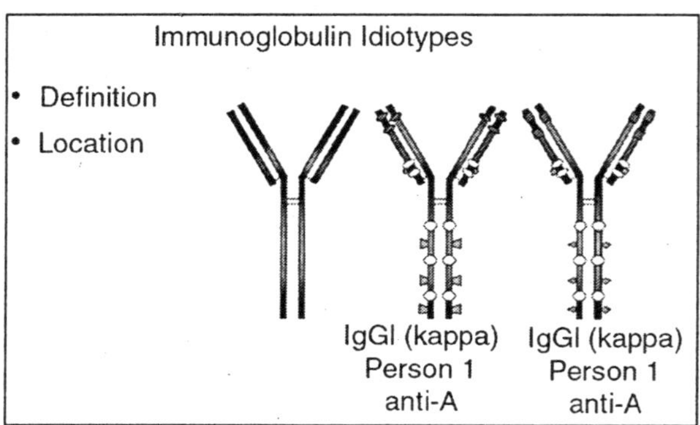

Fig. Immunoglobulin Idiotype

A third means of classifying immunoglobulins is defined by the term "idiotype". Unlike isotypes and allotypes, idiotypes are determined by the amino acid sequence and corresponding three-dimensional structure of the variable region of the immunoglobulin molecule.

In this regard, idiotypes reflect the antigen binding specificity of any particular antibody molecule. Idiotypes are so unique that an individual person is probably capable of generating antibodies directed against their own idiotypic determinants. This probability forms the basis of the Idiotypic Network Hypothesis to be described later.

BASIC IMMUNOGLOBULIN FUNCTION

Antibodies function in a variety of ways designed to eliminate the antigen that elicited their production. Some of these functions are independent of the particular class (isotype) of immunoglobulin. These functions reflect the antigen binding capacity of the molecule as defined by the variable and hypervariable (idiotypic) regions.

For example, an antibody might bind to a toxin and prevent that toxin from entering host cells where its biological effects would be activated. Similarly, a different antibody might bind to the surface of a virus and prevent that virus from entering its host cell. In contrast, other antibody functions are dependent upon the immunoglobulin class (isotype). These functions are contained within the constant regions of the molecule. For example, only IgG and IgM antibodies have the ability to interact with and initiate the complement cascade.

Likewise, only IgG molecules can bind to the surface of macrophages via Fc receptors to promote and enhance phagocytosis.

Generation of Antibody Diversity

The immune system has the capacity to recognize and respond to about 10^7 different antigens. This extreme diversity can be generated in at least three possible ways:

- Multiple genes in the germ line DNA.
- Variable recombination during the differentiation of germ line cells into B-cells.
- Mutation during the differentiation of germ line cells into B-cells.

The genetic makeup of a germ line cell and a mature B-cell at the loci controlling heavy chain production. Germ line DNA has many (up to 200) different variable (V) region genes, in addition to 12 diversity (D) region genes and four joining (J) region genes. During differentiation of this cell into the B-cell, rearrangement of the DNA occurs.

This rearrangement aligns one of the many V genes with one of the D genes and one of the J genes, producing a functional VDJ recombinant gene. Since any of the genes may recombine with any others, this rearrangement has the potential to generate 200 × 12 × 4 = 9600 different possible combinations. The same type of event occurs in the genes encoding the immmunoglobulin light chains where about 200 different V regions may recombine with about 5 different J regions giving rise to 200 × 5 = 1000 possible light chains. Since in any particular B-cell, any light chain combination can occur along with any heavy chain combination, the total possible immunoglobulin combinations approaches 10^7 (9600 × 1000).

A second way that diversity can result is through a process of variable or "inaccurate" recombination. Three possible recombination events between the variable (V) and joining (J) regions of an immunoglobulin light chain. In the first event, a proline-tryptophan dipeptide sequence is produced in the resulting protein.

However, in the second and third events, differential recombination places proline-arginine or proline-proline sequences into the resulting immunoglobulin. These types of events may also occur between the V and D regions and the D and J regions of the heavy chain DNA sequence.

A third way that diversity can result is through a process of mutation. This process simply involves changes in DNA sequence that occur during differentiation of the B-cell. An A: T to G:C transition mutation could change a serine residue into a glycine residue in the resulting immunoglobulin. This process may, in part, explain the diversity observed in hypervariable (CDR) regions.

Immunoglobulin Production

The production of immunoglobulins by B-cells or plasma cells occurs in different stages. During differentiation of the B-cells from precursor stem cells, rearrangement, recombination and mutation of the immunoglobulin V, D, and J regions occurs to produce functional VJ (light chain) and VDJ (heavy chain) genes.

At this point, the antigen specificity of the mature B-cell has been determined.

Each cell can make only one heavy chain and one light chain, although the isotype of the heavy chain may change. Initially, a mature B-cell will produce primarily IgD (and some membrane IgM) that will migrate to the cell surface to act as the antigen receptor.

Upon stimulation by antigen, the B-cell will differentiate into a plasma cell expressing large amounts of secreted IgM. Some cells will undergo a "class switch" during which a rearrangement of the DNA will occur, placing the VDJ gene next to the genes encoding the IgG, IgE or IgA constant regions.

Upon secondary induction (i.e. the secondary response), these B-cells will differentiate into plasma cells expressing the new isotype. Most commonly, this results in a switch from IgM (primary response) to IgG (secondary response).

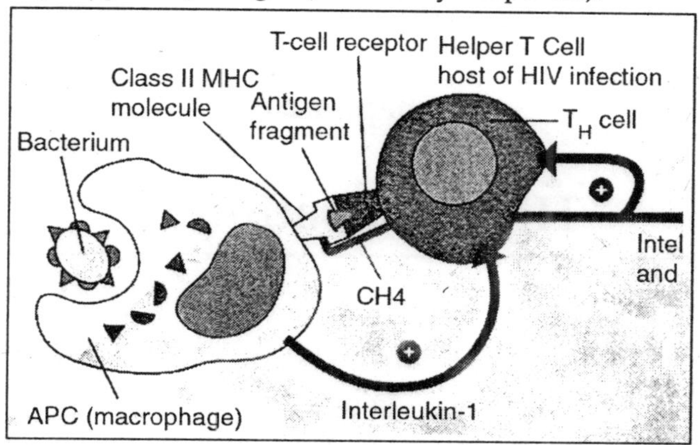

Fig. Antigen

The factors that lead to production of IgE or IgA instead of IgG are not well understood.

Chapter 20

Histocompatibility

MAJOR HISTOCOMPATIBILITY COMPLEX

The Major Histocompatibility Complex (MHC) is a set of molecules displayed on cell surfaces that are responsible for lymphocyte recognition and "antigen presentation". The MHC molecules control the immune response through recognition of "self" and "non-self" and, consequently, serve as targets in transplantation rejection.

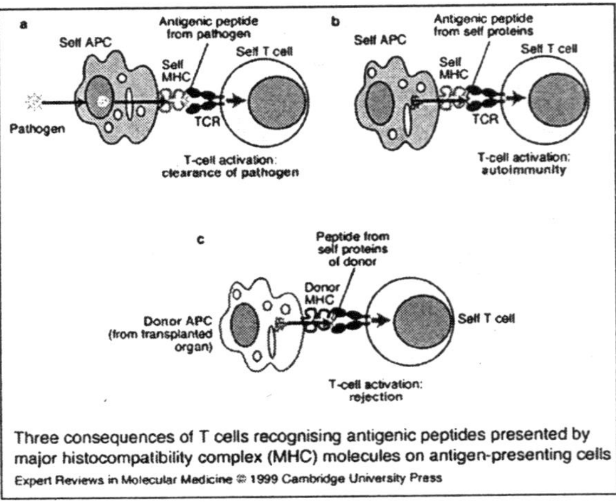

Fig. Histocompatibility

The Class I and Class II MHC molecules belong to a group of molecules known as the Immunoglobulin Supergene Family, which includes immunoglobulins, T-cell receptors, CD_4, CD_8, and others.

The major histocompatibility complex is encoded by several genes located on human chromosome 6. Class I molecules are encoded by the BCA region while class II molecules are encoded by the D region. A region between these two on chromosome 6 encodes class III molecules, including some complement components.

Class I Molecules

Class I molecules are composed of two polypeptide chains; one encoded by the BCA region and another (β2-microglobulin) that is encoded elsewhere.

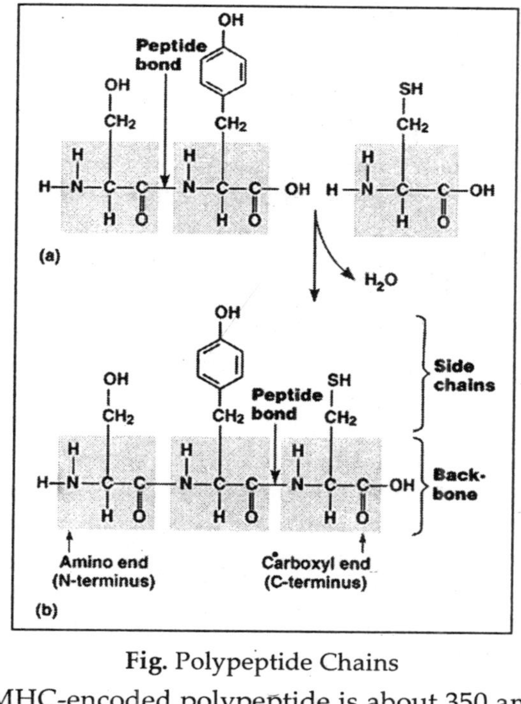

Fig. Polypeptide Chains

The MHC-encoded polypeptide is about 350 amino acids long and glycosylated, giving a total molecular weight of about 45 kDa. This polypeptide folds into three separate domains called alpha-1, alpha-2 and alpha-3. β_2-microglobulin is a 12 kDa polypeptide that is non-covalently associated with the alpha-3 domain. Between the alpha-1 and alpha-2 domains lies a region bounded by a beta-pleated sheet on the bottom and two alpha helices on the sides. This region is capable of binding (via non-covalent interactions) a small peptide of about 10 amino acids. This small peptide is "presented" to a T-cell and defines the antigen "epitope" that the T-cell recognizes.

The following images illustrate the structure of the class I MHC as seen schematically, and three dimensionally from the side and from the top (T-cell perspective). The MHC-encoded polypeptide is shown in blue, the β_2-microglobulin is green and the peptide antigen is red.

Class II Molecules

Class II molecules are composed of two polypeptide chains, both encoded by the D region. These polypeptides (alpha and beta) are about 230 and 240 amino acids long, respectively, and are glycosylated, giving molecular weights of about 33 kDa and 28 kDa.

These polypeptides fold into two separate domains; alpha-1 and alpha-2 for the alpha polypeptide, and beta-1 and beta-2 for the beta polypeptide.

Between the alpha-1 and beta-1 domains lies a region very similar to that seen on the class I molecule.

This region, bounded by a beta-pleated sheet on the bottom and two alpha helices on the sides, is capable of binding (via non-covalent interactions) a small peptide of about 10 amino acids.

This small peptide is "presented" to a T-cell and defines the antigen "epitope" that the T-cell recognizes. The structure of the class II MHC as seen schematically, and three dimensionally from the side and from the top (T-cell perspective).

Class I *vs.* Class II Molecules

While class I and class II molecules appear somewhat structurally similar and both present antigen to T-cells, their functions are really quite distinct. First, class I molecules are found on virtually every cell in the human body. Class II molecules, in contrast, are only found on B-cells, macrophages and other "antigen-presenting cells" (APCs). Second, class I molecules present antigen to cytotoxic T-cells (CTLs) while class II molecules present antigen to helper T-cells (TH-cells).

This specificity reflects the third difference, the type of antigen presented. Class I molecules present "endogenous" antigen while class II molecules present "exogenous" antigens. An endogenous antigen might be fragments of viral proteins or tumor proteins.

Presentation of such antigens would indicate internal cellular alterations that if not contained could spread throughout the body. Hence, destruction of these cells by CTLs is advantageous to the body as a whole. Exogenous antigens, in contrast, might be fragments of bacterial cells or viruses that are engulfed and processed by e.g. a macrophage and then presented to helper T-cells. The TH-cells, in turn, could activate B-cells to produce antibody that would lead to the destruction of the pathogen.

T-Cell Receptor (TCR) Molecules

The T-cell receptor molecule (TCR) is structurally and functionally similar to the B-cell immunoglobulin receptor. TCR is composed of two, disulfide-linked polypeptide chains, alpha and beta, each having separate constant and variable domains much like immunoglobulins.

The variable domain contains three hypervariable regions that are responsible for antigen recognition.

Genetic diversity is ensured in a manner analogous to that for immunoglobulins. Thus, just like the B-cell surface immunoglobulin provides antigen specificity to its B-cell, the TCR allows T-cells to recognize their particular antigenic moiety.

However, T-cells cannot recognize antigen without help; the antigenic determinant must be presented by an appropriate (i.e., self) MHC molecule. Upon recognition of a specific antigen, the signal is passed to the CD_3 molecule and then into the T-cell, prompting T-cell activation and the release of lymphokines. The following images illustrate the structure of the TCR as seen schematically, and three dimensionally from the side.

Antigen Recognition By T-Cells

The TCR provides the specificity for an individual T-cell to recognize its particular antigen. However, this recognition is "MHC-restricted" because the TCR also requires interactions with MHC. Also, interactions between the CD4 molecule (found on helper T-cells) and class II MHC or the CD8 molecule (found on cytotoxic T-cells) and class I MHC stabilize and consummate the antigen recognition process, allowing helper T-cells to respond to "exogenous" antigens (leading to B-cell activation and the production of antibody) or cytotoxic T-cells to respond to "endogenous" antigens (leading to target cell destruction).

HUMORAL IMMUNITY

The production of antibody involves three distinct phases:

Induction phase: Ag reacts with specific T and B cells

Expansion and Differentiation phase: Induced lymphocyte clones proliferate and mature to a functional stage (i.e., Ag receptor cells mature to Ag effector cells)

Effector phase: Abs or T cells exert biological effects either:

- Independently or
- Through the action of macrophages, complement, other non-specific agents

Antigen Presenting Cells(APCs)

Induction of the humoral immune response begins with the recognition of antigen. Through a process of clonal selection, specific B-cells are stimulated to proliferate and differentiate. However, this process requires the intervention of specific T-cells that are themselves stimulated to produce lymphokines that are responsible for activation of the antigen-induced B-cells. In other words, B cells recognize antigen via immunoglobulin receptors on their surface but are unable to proliferate and differentiate unless prompted by the action of T-cell lymphokines.

In order for the T-cells to become stimulated to release lymphokines, they must also recognize specific antigen. However, while T-cells recognize antigen via their T-cell receptors, they can only do so in the context of the MHC molecules. This "antigen-presentation" is the responsibility of the antigen-presenting cells (APCs).

Several types of cells may serve the APC function. Perhaps the best APC is, in fact, the B-cell itself. When B-cells bind antigen, the antigen becomes internalized, processed and expressed on the surface of the B-cell. Expression occurs within the class II MHC molecule, which can then be recognized by T-helper cells ($CD4^+$).

Other types of antigen-presenting cells include the macrophage and dendritic cells. These cells either actively phagocytose or pinocytose foreign antigens. The antigens are then processed in a manner similar to that observed for the B-cells. Next, specific antigen epitopes are expressed on the macrophage or dendritic cell surface. Again, this expression occurs within the class II MHC molecule, where T-cell recognition occurs. The stimulated T-cells then release lymphokines that act upon "primed" B-cells (B-cells that have already encountered antigen), inducing B-cell proliferation and differentiation.

Differentiation of B-Lymphocytes

B-cells begin their lives in the bone marrow as multipotential stem cells. These completely undifferentiated cells serve as the source for all of the cellular components of the blood and lymphoid system. The initial differentiation step that ultimately leads to the mature B-cell involves DNA rearrangements joining the D and J segments of the immunoglobulin heavy chain genes.

Next, DNA rearrangements joining the variable (V) region to the DJ segments of the immunoglobulin heavy chain, as well as similar rearrangements within the light chain genes gives rise to the pre-B-cell. Establishment of the B-cell specificity and consequent expression of surface immunoglobulin gives rise to the "virgin", fully functional B-cell. Each of these steps is entirely independent of antigen. The antigen-dependent stages of B-lymphocyte differentiation occur in the spleen, lymph nodes and other peripheral tissue. These stages are, of course, initiated upon encounter with antigen and activation by T-cell lymphokines. The activated B-cell first develops into a B-lymphoblast, becoming much larger and shedding all surface immunoglobulin. The B-lymphoblast then develops into a plasma cell, which is, in essence, an antibody factory.

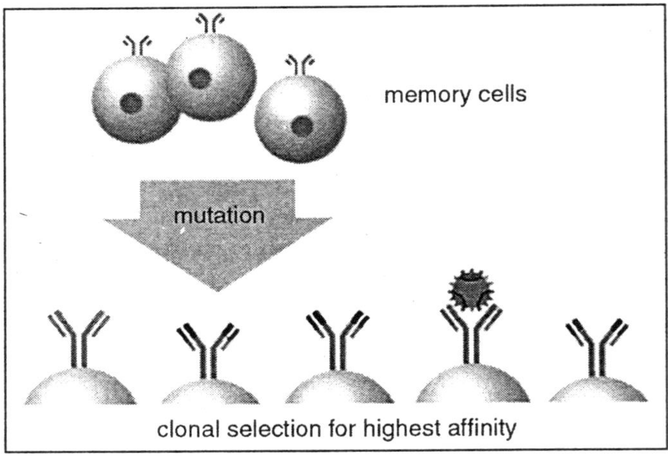

Fig. Memory Cells

This terminal differentiation stage is responsible for production of primarily IgM antibody during the "primary response". Some B-cells, however, do not differentiate into plasma cells. Instead, these cells undergo secondary DNA rearrangements that place the constant region of the IgG, IgA or IgE genes in conjunction with the VDJ genes. This "class switch" establishes the phenotype of these newly differentiated B-cells; these cells remain as long-lived "memory cells". Upon subsequent encounter with antigen, these cells respond very quickly to produce large amounts of IgG, IgA or IgE antibody, generating the "secondary response".

Regulation of the Humoral Response

Regulation of the immune response is possibly mediated

in several ways. First, a specific group of T-cells, suppressor T-cells, are thought to be involved in turning down the immune response. Like helper T-cells, suppressor T-cells are stimulated by antigen but instead of releasing lymphokines that activate B-cells (and other cells), suppressor T-cells release factors that suppress the B-cell response.

Other means of regulation involve interactions between antibody and B-cells.

One mechanism, "antigen blocking", occurs when high doses of antibody interact with all of the antigen's epitopes, thereby inhibiting interactions with B-cell receptors. A second mechanism, "receptor cross linking", results when antibody, bound to a B-cell via its Fc receptor, *and* the B-cell receptor both combine with antigen. This "cross-linking" inhibits the B-cell from producing further antibody.

Another means of regulation that has been proposed is the idiotypic network hypothesis. This theory suggests that the idiotypic determinants of antibody molecules are so unique that they appear foreign to the immune system and are, therefore, antigenic.

Thus, production of antibody in response to antigen leads to the production of anti-antibody in response, and anti-anti-antibody and so on. Eventually, however, the level of $anti_n$-antibody is not sufficient to induce another round and the cascade ends.

Chapter 21

Cell Structure

ANIMAL CELL

Animal cells are typical of the eukaryotic cell, enclosed by a plasma membrane and containing a membrane-bound nucleus and organelles. Unlike the cells of the two other eukaryotic kingdoms, plants and fungi, animal cells don't have a cell wall. This feature was lost in the distant past by the single-celled organisms that gave rise to the kingdom Animalia. The lack of a rigid cell wall allowed animals to develop a greater diversity of cell types, tissues, and organs. Specialized cells that formed nerves and muscles — tissues impossible for plants to evolve — gave these organisms mobility.

The ability to move about by the use of specialized muscle tissues is the hallmark of the animal world. (Protozoans locomote, but by nonmuscular means, i.e. cilia, flagella, pseudopodia.) The animal kingdom is unique amongst eukaryotic organisms because animal tissues are bound together by a triple helix of protein, called collagen. Plant and fungal cells are bound together in tissues or aggregations by other molecules, such as pectin. The fact that no other organisms utilize collagen in this manner is one of the indications that all animals arose from a common unicellular ancestor.

Animals are a large and incredibly diverse group of organisms. Making up about three-quarters of the species on Earth, they run the gamut from sponges and jellyfish to ants, whales, elephants, and — of course — human beings.

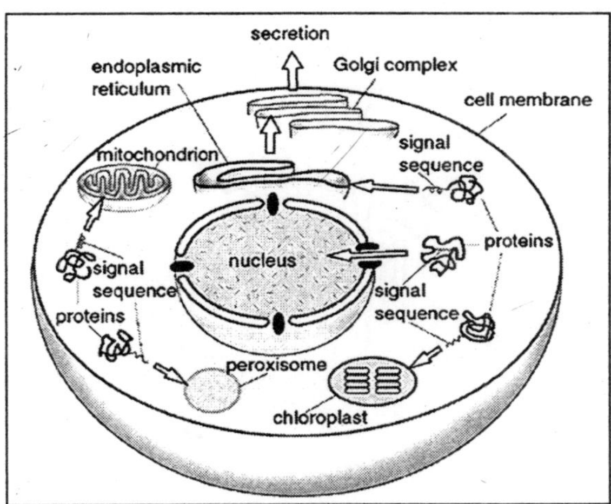

Fig. Animal cells

Being mobile has given animals the flexibility to adopt many different modes of feeding, defence, and reproduction.

The earliest fossil evidence of animals dates from the Vendian Period, with coelenterate-type creatures that left traces of their soft bodies in shallow-water sediments. The first mass extinction ended that period, but during the Cambrian Period which followed, an explosion of new forms began the evolutionary radiation that produced most of the major groups, or phyla, known today. Vertebrates (animals with backbones) are not known to have occurred until the Ordovician Period.

Centrioles

Found only in animal cells, these paired organelles are found together near the nucleus, located at right angles to each other. Each centriole is made of nine bundles of microtubules (three per bundle) arranged in a ring. They have a role in building cilia and flagella, during which time they are referred to as basal bodies.

Centrioles also play a role in cell division, although not as significant a role as once thought. Plant cells reproduce without centrioles and in experiments that have removed centrioles from animal cells, the cells were able to reproduce successfully without the organelles.

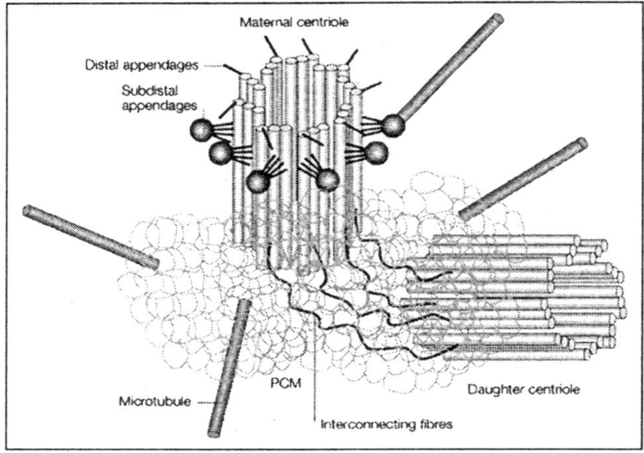

Fig. Centrioles

Apparently they organize the microtubules in the mitotic spindles during mitosis and meiosis. The mitotic spindles in plant cells are less tightly organized. These structures are self-replicating and make copies of themselves just before cell division begins. As the cell prepares to divide, the centrioles separate and move toward opposite poles of the cell. As they're moving apart, they radiate microtubules in a spindle-shaped formation that spans the cell from pole to pole. The spindle fibers act as guides for the alignment of the chromosomes as they separate.

Cilia and Flagella

Cilia and flagella are made up of microtubules, which are composed of linear polymers of globular proteins called tubulin. The core (axoneme) contains two central fibers that are surrounded by an outer ring of nine double fibers and covered by the cellular membrane.

These motile appendages are constructed by basal bodies (kinetostomes), which also function as centrioles. The basal body is located at the base of each filament, anchoring it to the cell and controlling its movement. Cilia and flagella have the same structure. The only difference is that the flagella are longer. For single-celled eukaryotes, cilia and flagella are essential for the locomotion of individual organisms. Protozoans belonging to the phylum Ciliophora are covered with cilia. Flagella are a characteristic of the protozoan group Mastigophora.

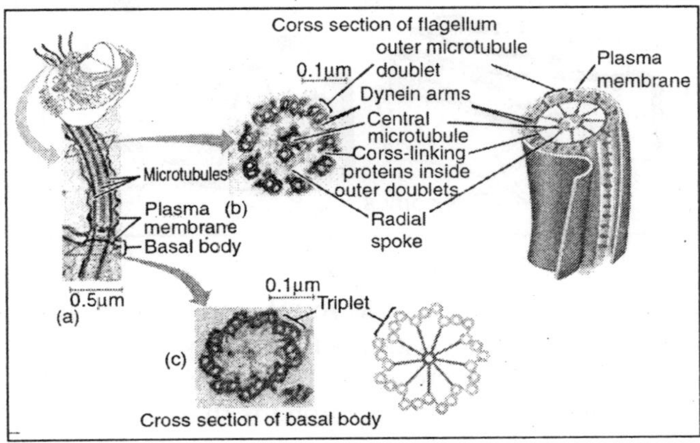

Fig. Cilia and Flagella

In multicellular organisms, cilia function to move fluid or materials past an immobile cell as well as moving a cell or group of cells. The respiratory tract in humans is lined with cilia that keep inhaled dust, smog, and potentially harmful microorganisms from entering the lungs.

Cilia generate water currents to carry food and oxygen past the gills of clams and transport food through the digestive systems of snails. Flagella are found primarily on gametes, but also create the water currents necessary for respiration and circulation in sponges and coelenterates.

Endoplasmic Reticulum

The endoplasmic reticulum (ER) is an network of sacs that manufactures, processes, and transports chemical compounds for use inside and outside of the cell. The ER is a continuous membrane with branching tubules and flattened sacs that extend throughout the cytoplasm. It is connected to the double-layered nuclear envelope, providing a connection between the nucleus and the cytoplasm.

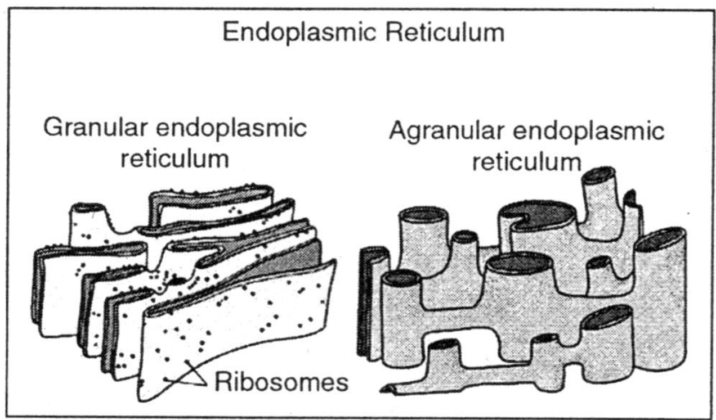

Fig. Endoplasmic Reticulum

There are two kinds of ER, rough and smooth. Rough ER is covered with ribosomes, giving it a bumpy appearance when viewed through the microscope. This type of ER is involved mainly with the production of proteins that will be exported, or secreted, from the cell.

The ribosomes assemble amino acids into units of proteins, which are transported into the rough ER for further processing.

Once inside, the proteins are folded into the correct three-dimensional conformation, as a flattened cardboard box might be opened up and folded into its proper shape in order to become a useful box. Chemicals, such as carbohydrates or sugars, are added, then the ER either transports the completed proteins to areas of the cell where they are needed, or they are sent to the Golgi apparatus for export.

Smooth ER has a smoother appearance than rough ER when viewed through the microscope because it does not have ribosomes attached to it. This portion of the ER is involved with the production of lipids (fats), carbohydrate metabolism, and detoxification of drugs and poisons. Smooth ER is also involved with metabolizing calcium to mediate some cell activities. In muscle cells, smooth ER releases calcium to trigger muscle contractions. Cells specializing in lipid and carbohydrate metabolism (brain, muscle) or detoxification (liver) usually have more of this type of ER.

Golgi Apparatus

The Golgi apparatus (GA), also called Golgi body or Golgi complex, is a series of five to eight cup-shaped, membrane-covered sacs that look something like a stack of deflated balloons. The GA is the distribution and shipping department for the cell's chemical products.

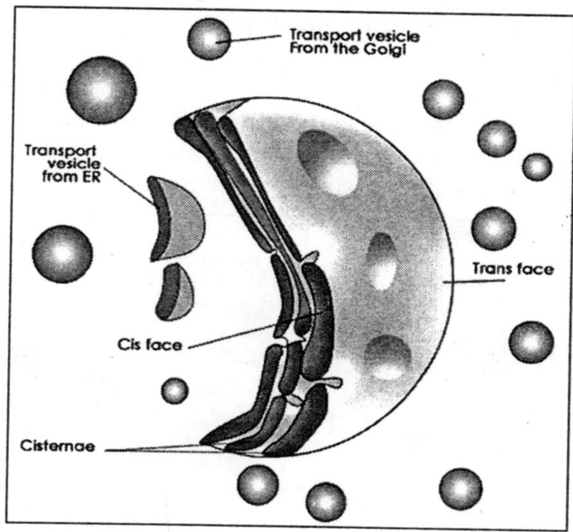

Fig. Golgi Apparatus

It modifies proteins and lipids (fats) that have been built in the endoplasmic reticulum and prepares them for export as outside of the cell. The number of GAs in each cell varies according to its function, but animal cells generally contain between ten and twenty per cell.

Proteins and lipids built in the smooth and rough endoplasmic reticulum bud off in tiny bubble-like vesicles that move through the cytoplasm until they reach the GA. The vesicles fuse with the GA membrane and release the molecules into the organelle.

Once inside, the compounds are further processed by the GA, which adds molecules or chops tiny pieces off the ends. Once completed, the product is extruded from the GA in a vesicle and directed to its final destination inside or outside the cell. The exported products are secretions of proteins or glycoproteins that are part of the cell's function in the organism. Other products are returned to the endoplasmic reticulum or become lysosomes.

Lysosomes

The main function of these microbodies is digestion. Lysosomes break down cellular waste products and debris from outside the cell into simple compounds, which are transferred out into the cytoplasm as new cell-building materials. Like other microbodies, lysosomes are spherical organelles contained by a single layer membrane.

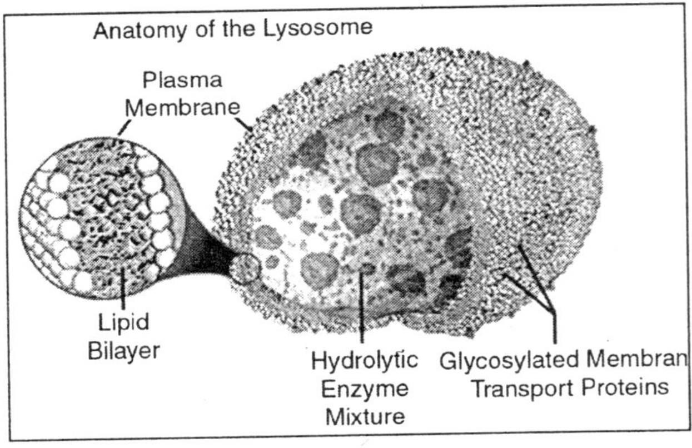

Fig. Lysosomes

This membrane protects the rest of the cell from the lysosomes' harsh digestive enzymes that would otherwise damage it. Lysosomes originate in the Golgi apparatus, but the digestive enzymes are manufactured in the rough endoplasmic reticulum. Lysosomes are found in all eukaryotic cells, but are most numerous in disease-fighting cells, such as white blood cells.

Some human diseases are caused by lysosome enzyme disorders. Tay-sachs disease is caused by a genetic defect that prevents the formation of an essential enzyme that breaks down a complex lipid called ganglioside. An accumulation of this lipid damages the nervous system, causes mental retardation and death in early childhood. Arthritis inflammation and pain are related to the escape of lysosome enzymes.

Microfilaments

Microfilaments are solid rods made of globular proteins called actin and are common to all eukaryotic cells. Long chains of the molecules are intertwined in a helix to form individual microfilaments. These filaments are primarily structural in function and are an important component of the cytoskeleton, along with microtubules. In association with myosin, microfilaments help to generate the forces used in cellular contraction and basic cell movements.

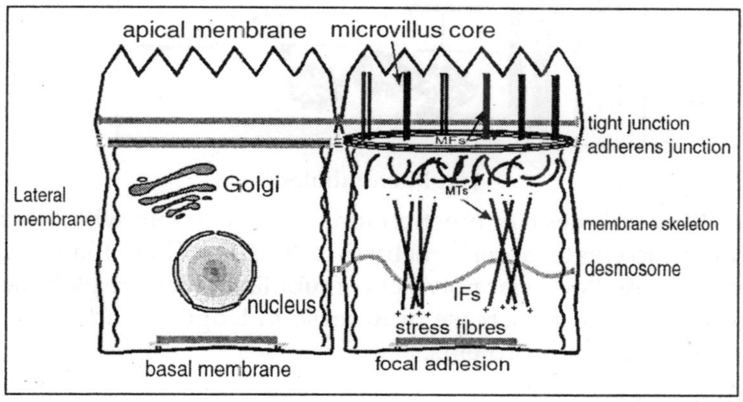

Fig. Microfilaments

They enable a dividing cell to pinch off into two cells and are involved in amoeboid movements of certain types of cells. They also enable the contractions of muscle cells.

Microtubules

These straight, hollow cylinders are found throughout the cytoplasm of all eukaryotic cells (prokaryotes don't have them) and perform a number of functions. Microtubules form part of the cytoskeleton that gives structure and shape to a cell, serve as conveyor belts moving other organelles through the cytoplasm, are the major components of cilia and flagella, and participate in the formation of spindle fibers during cell division (mitosis).

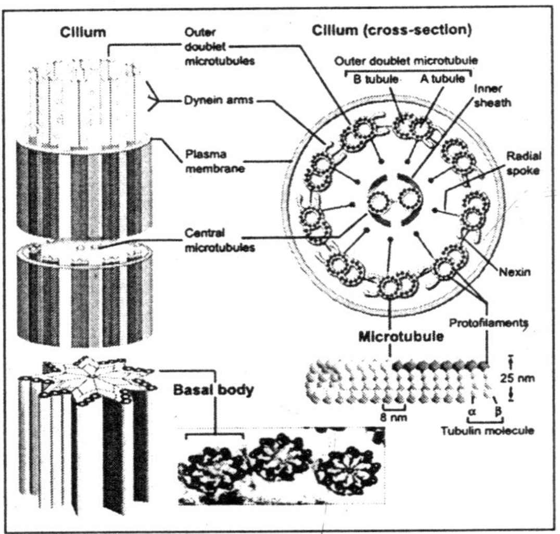

Fig. Microtubules

Microtubules can function individually or join with other proteins to create larger structures (e.g., cilia). These filaments are composed of linear polymers of tubulin, which are globular proteins, and can increase or decease in length by adding or removing tubulin proteins.

Mitochondria

Mitochondria (singular, mitochondrion) are oblong shaped organelles that are found in the cytoplasm of every eukaryotic cell. They occur in varying numbers, depending on the cell and its function. These organelles are the power generators of the cell, converting oxygen and nutrients into ATP (adenosine triphosphate).

ATP is the chemical energy "currency" of the cell that powers the cell's metabolic activities. This process is called aerobic respiration and is the reason animals breathe oxygen. The mitochondrion is different from other organelles because it has its own DNA and reproduces independently of the cell in which it is found; an apparent case of endosymbiosis. Scientists hypothesize that millions of years ago small, free-living prokaryotes were engulfed, but not consumed, by larger prokaryotes; perhaps because they were able to resist the digestive enzymes of the engulfing organism.

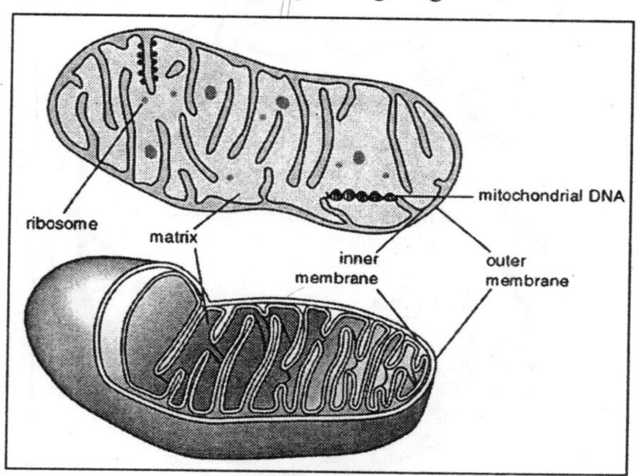

Fig. Mitochondria

The two organisms developed a symbiotic relationship over time, the larger organism providing the smaller with ample nutrients and the smaller organism providing ATP molecules to the larger one. Eventually, the larger organism developed into the eukaryotic cell, the smaller organism into the mitochondrion. Nonetheless, there are a number of prokaryotic traits that mitochondria continue to exhibit.

Their DNA is circular, as it is in the prokaryotes, and their ribosomes and reproductive methods (binary fission) are more like those of the prokaryotes. Mitochondrial DNA can be used study different aspects of inheritance. In most animal species, mitochondria are inherited through the maternal lineage.

A sperm carries mitochondria in its tail as an energy source for its long journey to the egg.

When it attaches to the egg during fertilization, the tail falls off. Consequently, the only mitochondria the new organism gets are from the egg its mother provided. Unlike nuclear DNA, mitochondrial DNA doesn't get shuffled every generation, so it is presumed to change at a slower rate. That fact is being used to study human evolution and suggests that modern humans descended from a small group of hominids in Africa around 200,000 years ago.

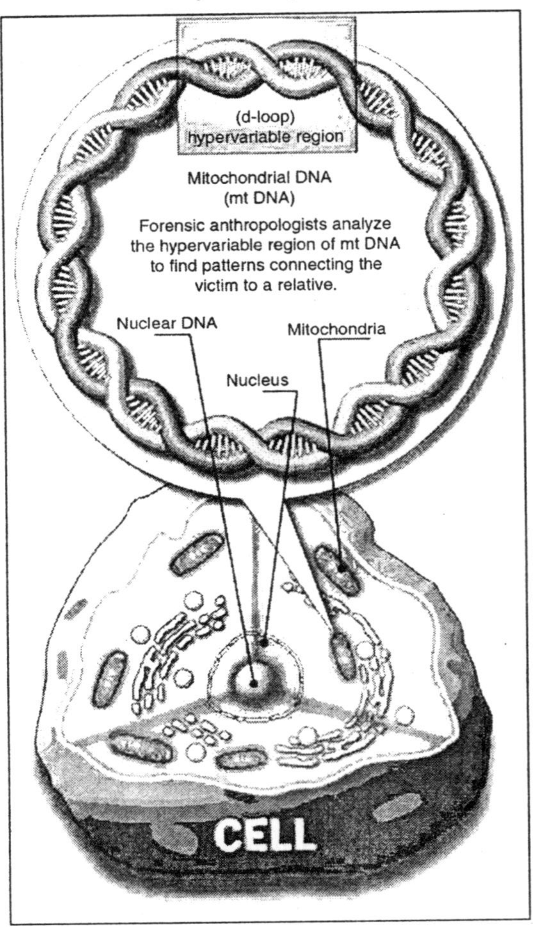

Fig. Mitochondrial DNA

Mitochondrial DNA is also being used in forensic science, as a tool for identifying corpses or body parts, and has been implicated in a number of genetic diseases such as Alzheimer's disease and diabetes.

NUCLEUS

The nucleus is a highly specialized organelle that serves as the information and administrative centre of the cell. This organelle has two major functions.

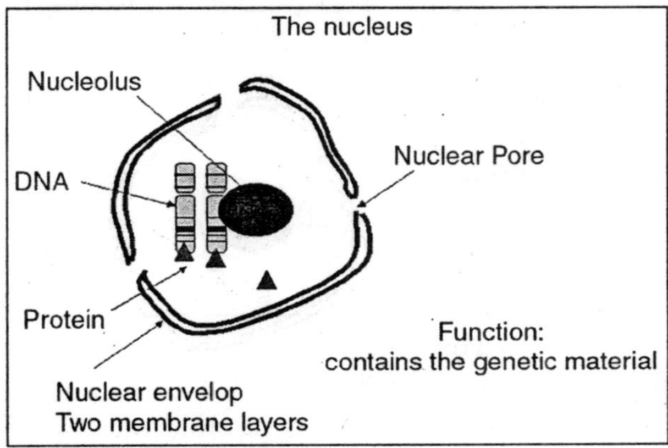

Fig. Nucleus

It stores the cell's hereditary material, or DNA, and it coordinates the cell's activities, which include intermediary metabolism, growth, protein synthesis, and reproduction (cell division). Only the cells of advanced organisms, known as eukaryotes, have a nucleus.

Generally there is only one nucleus per cell, but there are exceptions such as slime molds and the Siphonales group of algae. Simpler one-celled organisms (prokaryotes), like the bacteria and cyanobacteria, don't have a nucleus. In these organisms, all the cell's information and administrative functions are dispersed throughout the cytoplasm. The spherical nucleus occupies about 10 per cent of a cell's volume, making it the cell's most prominent feature.

Most of the nuclear material consists of chromatin, the unstructured form of the cell's DNA that will organize to form chromosomes during mitosis or cell division. Also inside the nucleus is the nucleolus, an organelle that synthesizes protein-producing macromolecular assemblies called ribosomes. A double-layered membrane, the nuclear envelope, separates contents of the nucleus from the cellular cytoplasm.

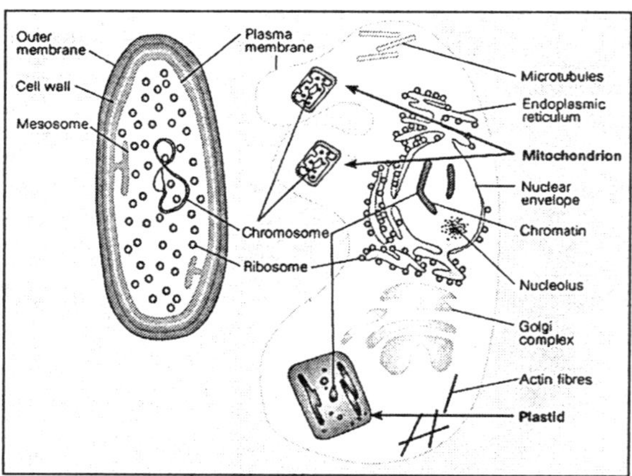

Fig. Cellular Cytoplasm

The envelope is riddled with holes called nuclear pores that allow specific types and sizes of molecules to pass back and forth between the nucleus and the cytoplasm. It is also attached to a network of tubules, called the endoplasmic reticulum, where protein synthesis occurs. These tubules extend throughout the cell and manufacture the biochemical products that a particular cell type is genetically coded to produce.

Chromatin/Chromosomes

Packed inside the nucleus of every human cell is nearly 6 feet of DNA, which is divided into 46 individual molecules, one for each chromosome and each about 1.5 inches long. Packing all this material into a microscopic cell nucleus is an extraordinary feat of packaging.

For DNA to function, it can't be crammed into the nucleus like a ball of string. Instead, it is combined with proteins and organized into a precise, compact structure, a dense string-like fibre called chromatin. Each DNA strand wraps around groups of small protein molecules called histones, forming a series of bead-like structures, called nucleosomes, connected by the DNA strand.

Under the microscope, uncondensed chromatin has a "beads on a string" appearance. The string of nucleosomes, already compacted by a factor of six, is then coiled into an even denser structure, compacting the DNA by a factor of 40. This compression and structuring of DNA serves several functions. The overall negative charge of the DNA is neutralized by the positive charge of the histone molecules, the DNA takes up much less space, and inactive DNA can be folded into inaccessible locations until it is needed.

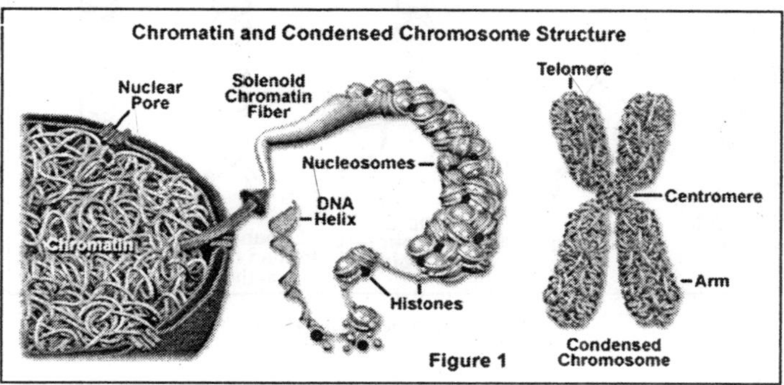

Fig. Chromosomes

There are two types of chromatin. Euchromatin is the genetically active portion and is involved in transcribing RNA to produce proteins used in cell function and growth. Heterochromatin contains inactive DNA and is the portion of chromatin that is most condensed, since it not being used. Throughout the life of a cell, chromatin fibers take on different forms inside the nucleus.

During interphase, when the cell is carrying out its normal functions, the chromatin is dispersed throughout the nucleus in what appears to be a tangle of fibers.

This exposes the euchromatin and makes it available for the transcription process. When the cell enters metaphase and prepares to divide, the chromatin changes dramatically. First, all the chromatin strands make copies of themselves through the process of DNA replication.

Then they are compressed to an even greater degree than at interphase, a 10,000-fold compaction, into specialized structures for reproduction, termed chromosomes. As the cell divides to become two cells, the chromosomes separate, giving each cell a complete copy of the genetic information contained in the chromatin.

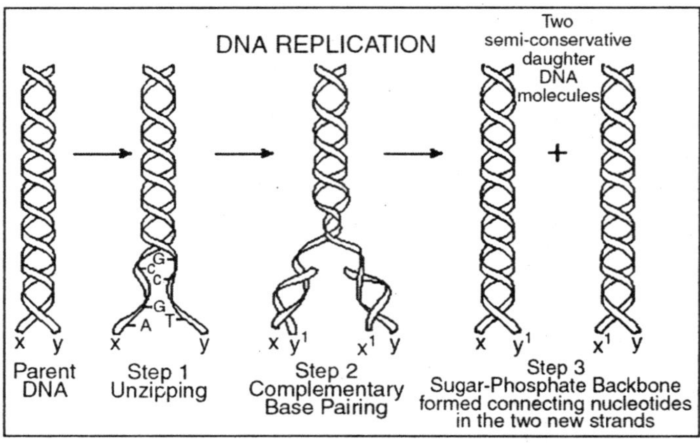

Fig. DNA Replication

Nucleolus

The nucleolus is a membrane-less organelle within the nucleus that manufactures ribosomes, the cell's protein-producing structures. Through the microscope, the nucleolus looks like a large dark spot within the nucleus. A nucleus may contain up to four nucleoli, but within each species the number of nucleoli is fixed.

After a cell divides, a nucleolus is formed when chromosomes are brought together into nucleolar organizing regions. During cell division, the nucleolus disappears. Some studies suggest that the nucleolus may be involved with cellular aging and, therefore, may affect the aging of an organism.

Nuclear Envelope

The nuclear envelope is a double-layered membrane that encloses the contents of the nucleus during most of the cell's lifecycle. The space between the layers is called the perinuclear space and appears to connect with the rough endoplasmic reticulum.

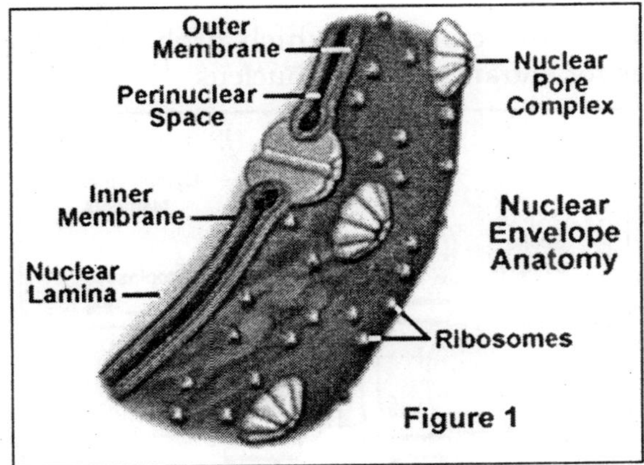

Fig. Perinuclear Space

The envelope is perforated with tiny holes called nuclear pores. These pores regulate the passage of molecules between the nucleus and cytoplasm, permitting some to pass through the membrane, but not others.

The inner surface has a protein lining called the nuclear lamina, which binds to chromatin and other nuclear components. During mitosis, or cell division, the nuclear envelope disintegrates, but reforms as the two cells complete their formation and the chromatin begins to unravel and disperse.

Nuclear Pores

The nuclear envelope is perforated with holes called nuclear pores. These pores regulate the passage of molecules between the nucleus and cytoplasm, permitting some to pass through the membrane, but not others. Building blocks for building DNA and RNA are allowed into the nucleus as well as molecules that provide the energy for constructing genetic material.

The pores are fully permeable to small molecules up to the size of the smallest proteins, but form a barrier keeping most large molecules out of the nucleus. Some larger proteins, such as histones, are given admittance into the nucleus. Each pore is surrounded by an elaborate protein structure called the nuclear pore complex, which probably selects large molecules for entrance into the nucleus.

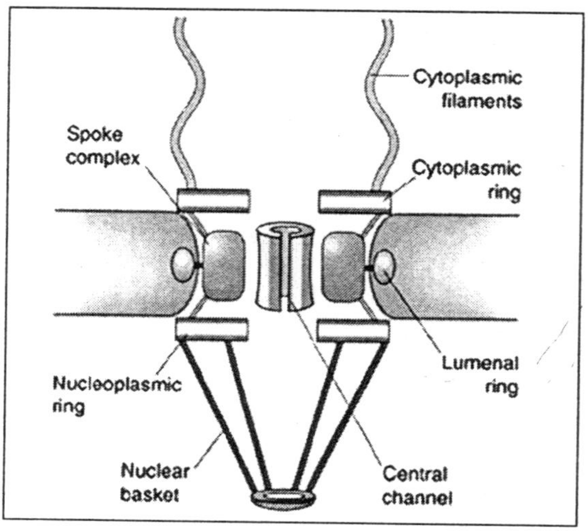

Fig. Nuclear Pores

PEROXISOMES

Microbodies are a diverse group of organelles that are found in the cytoplasm, roughly spherical and bound by a single membrane.

There are several types of microbodies but peroxisomes are the most common. Peroxisomes function to rid the cell of toxic substances, in particular, hydrogen peroxide — a common byproduct of cellular metabolism.

These organelles contain enzymes that convert the hydrogen peroxide to water, rendering the potentially toxic substance safe for release back into the cell. Some types of peroxisomes, such as those in liver cells, detoxify alcohol and other harmful compounds by transferring hydrogen from the poisons to molecules of oxygen.

Peroxisomes are similar in appearance to lysosomes, another type of microbody, but the two have very different origins. Lysosomes are formed in the Golgi complex while peroxisomes are self-replicating. Unlike mitochondria, however, peroxisomes and lysosomes do not have their own internal DNA molecules.

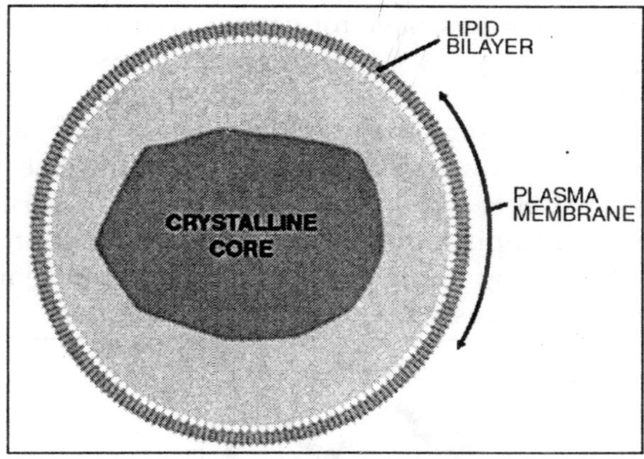

Fig. Peroxisomes

Except for mature red blood cells, all human cells have peroxisomes. Since the early 1980s, a number of metabolic disorders have been found to be caused by molecular defects in the peroxisomes. Two major categories have been described so far. The first category is Disorders of Peroxisome Biogenesis (PBD) in which the organelle fails to develop normally, causing defects in numerous peroxisomal proteins.

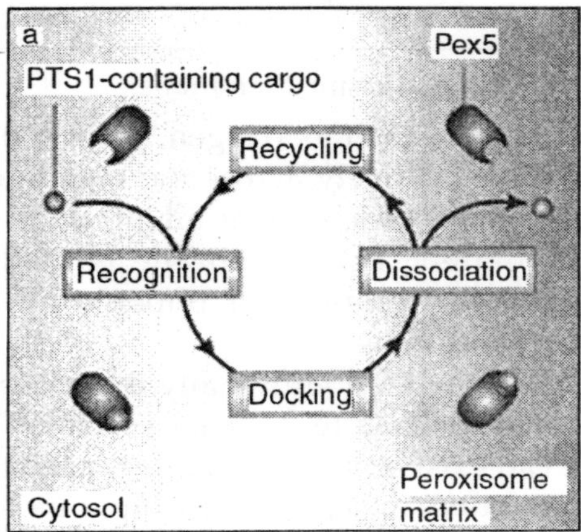

Fig. Peroxisomal Proteins

The second category includes involves defects of single peroxisomal enzymes. At present, there are no treatments for these genetic disorders, save for genetic counseling.

Plasma Membrane

All living cells, prokaryotic and eukaryotic, have a plasma membrane that encloses their contents. The membrane has two functions. First, it is a boundary holding the cell constituents together and keeping other substances out. Second, it is permeable, allowing nutrients and other essential elements to enter the cell and waste materials to leave the cell.

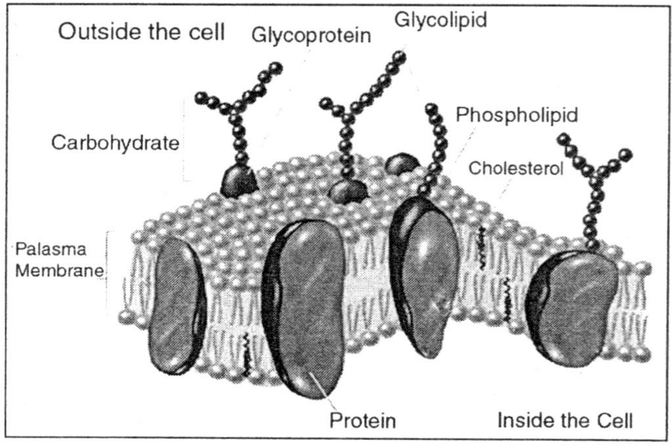

Fig. Plasma Membrane

Small molecules, such as oxygen, carbon dioxide, and water, are able to pass freely across the membrane, but the passage of larger molecules, such as amino acids and sugars, is carefully regulated. The membrane is made of a two molecule thick layer (bilayer) of phospholipids, an oily substance found in all cells.

This layer is embedded with many diverse proteins and has carbohydrates attached to its outer surface. The lipids in the membrane can exist either in a gel-like, nearly solid, state or in a liquid-like state, which gives the lipid molecules more mobility. In living cells, the membrane seems to be in a transition between the two states, depending on physical conditions and what lipids and proteins are present in the membrane layer.

In prokaryotes and plants, the plasma membrane is the inner layer of protection. A rigid cell wall forms the outside boundary for the cell. While it has pores that allow materials to enter and leave the cell, they are not as selective about what passes through. The membrane, which lines the cell wall, provides the final filter between the cell interior and the environment.

Eukaryotic animal cells probably descended from prokaryotes that lost their cell walls. With only the flexible membrane left to enclose them, they were able to expand in size and complexity. Eukaryotic cells are generally ten times larger than prokaryotic cells and have membranes enclosing interior components, the organelles.

Like the exterior plasma membrane, these membranes also regulate the flow of materials, allowing the cell to segregate its chemical functions into discrete internal compartments. Although plants have evolved another version of the cell wall, in animal cells the plasma membrane is the only barrier between the cell and its environment.

Ribosomes

All living cells contain ribosomes, tiny organelles composed of approximately 60 per cent RNA and 40 per cent protein. In eukaryotes, ribosomes are made of four strands of RNA. In prokaryotes, they consist of three strands of RNA.

Ribosomes are scattered throughout the cytoplasm and are the protein production sites for the cell. Some of the proteins are synthesized for the cell's own use, particularly in single-celled organisms.

In multicellular organisms, many of the proteins produced by a specialized cell, e.g. antibodies, will be transported and used elsewhere in the organism.

Eukaryote ribosomes are produced and assembled in the nucleolus. Three of the four strands are produced there, but one is produced outside the nucleolus and transported inside to complete the ribosome assembly.

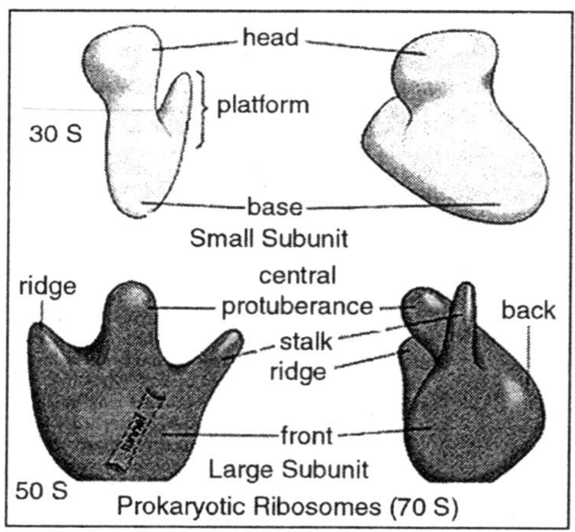

Fig. Ribosomes

Ribosomal proteins enter the nucleolus and combine with the four strands to create the two subunits that will make up the completed ribosome.

The ribosome units leave the nucleus through the nuclear pores and unite once in the cytoplasm. Some ribosomes will remain free-floating in the cytoplasm, creating proteins for the cell's use.

Others will attach to the endoplasmic reticulum and produce the proteins that will be "exported" from the cell. Protein synthesis requires the assistance of two other RNA molecules. Messenger RNA (mRNA) provides instructions from the cellular DNA for building a specific protein.

Transfer RNA (tRNA) brings the protein building blocks, amino acids, to the ribosome. Once the protein backbone amino acids are polymerized, the ribosome releases the protein and it is transported to the Golgi apparatus.

There, the proteins are completed and released inside or outside the cell.

CELL CULTURE

Cell culture is the process by which prokaryotic, eukaryotic or plant cells are grown under controlled conditions. In practice the term "cell culture" has come to refer to the culturing of cells derived from multicellular eukaryotes, especially animal cells.

The historical development and methods of cell culture are closely interrelated to those of tissue culture and organ culture.

Animal cell culture became a routine laboratory technique in the 1950s, but the concept of maintaining live cell lines separated from their original tissue source was discovered in the 19th century.

The 19th-century English physiologist Sydney Ringer developed salt solutions containing the chlorides of sodium, potassium, calcium and magnesium suitable for maintaining the beating of an isolated animal heart outside of the body.

In 1885 Wilhelm Roux removed a portion of the medullary plate of an embryonic chicken and maintained it in a warm saline solution for several days, establishing the principle of tissue culture.

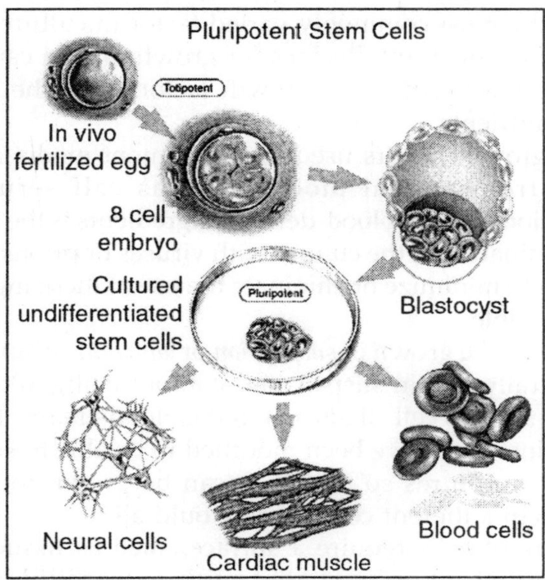

Fig. Cell Culture

Cell culture techniques were advanced significantly in the

1940s and 1950s to support research in virology. Growing viruses in cell cultures allowed preparation of purified viruses for the manufacture of vaccines. The Salk polio vaccine was one of the first products mass-produced using cell culture techniques.

This vaccine was made possible by the cell culture research of John Franklin Enders, Thomas Huckle Weller, and Frederick Chapman Robbins, who were awarded a Nobel Prize for their discovery of a method of growing the virus in monkey kidney cell cultures.

Maintaining Cells in Culture

Cells are grown and maintained at an appropriate temperature and gas mixture in a cell incubator. Culture conditions vary widely for each cell type, and variation of conditions for a particular cell type can result in different phenotypes being expressed. Aside from temperature and gas mixture, the most commonly varied factor in culture systems is the growth medium. Recipes for growth media can vary in pH, glucose concentration, growth factors, and the presence of other nutrients.

The growth factors used to supplement media are often derived from animal blood, such as calf serum. One complication of these blood-derived ingredients is the potential for contamination of the culture with viruses or prions. Current practice is to minimize or eliminate the use of these ingredients where possible.

Cells can be grown in *suspension* or *adherent* cultures. Some cells naturally live in suspension, without being attaching to a surface, such as cells that exist in the bloodstream. There are also cell lines that have been modified to be able to survive in suspension cultures so that they can be grown to a higher density than adherent conditions would allow.

Adherent cells require a surface, such as tissue culture plastic, which may be coated with extracellular matrix components to increase adhesion properties and provide other signals needed for growth and differentiation. Most cells derived from solid tissues are adherent.

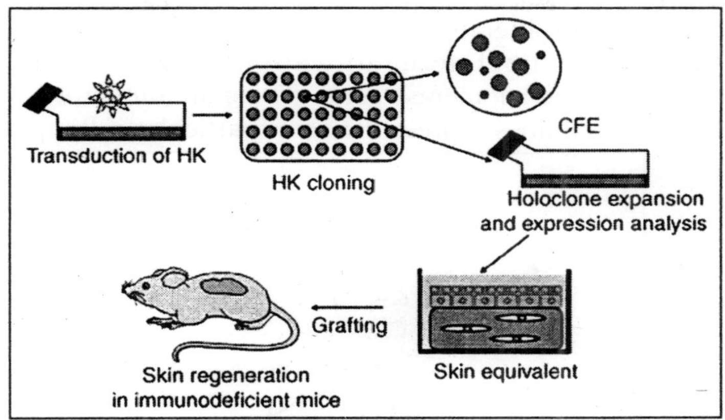

Fig. Organotypic Culture

Another type of adherent culture is *organotypic culture* which involves growing cells in a three-dimensional environment as opposed to two-dimensional culture dishes. This 3D culture system is biochemically and physiologically more similar to *in vivo* tissue, but is technically challenging to maintain because of many factors (e.g. diffusion).

CONCEPTS IN MAMMALIAN CELL CULTURE

Isolation of Cells

Cells can be isolated from tissues for *ex vivo* culture in several ways. Cells can be easily purified from blood, however only the white cells are capable of growth in culture.

Mononuclear cells can be released from soft tissues by enzymatic digestion with enzymes such as collagenase, trypsin, or pronase, which break down the extracellular matrix. Alternatively, pieces of tissue can be placed in growth media, and the cells that grow out are available for culture. This method is known as explant culture. Cells that are cultured directly from a subject are known as primary cells.

With the exception of some derived from tumours, most primary cell cultures have limited lifespan. After a certain number of population doublings cells undergo the process of senescence and stop dividing, while generally retaining viability. An established or immortalised *cell line* has acquired the ability to proliferate indefinitely either through random mutation or deliberate modification, such as artificial expression of the telomerase gene. There are numerous well established cell lines representative of particular cell types.

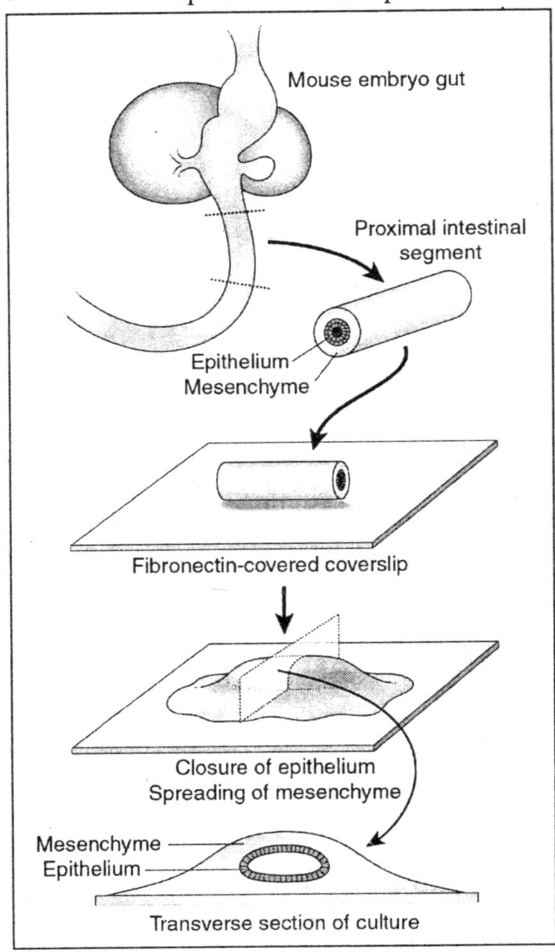

Fig. Explant Culture

MANIPULATION OF CULTURED CELLS

As cells generally continue to divide in culture, they

generally grow to fill the available area or volume. This can generate several issues:

- Nutrient depletion in the growth media
- Accumulation of apoptotic/necrotic (dead) cells.
- Cell-to-cell contact can stimulate cell cycle arrest, causing cells to stop dividing known as contact inhibition or senescence.
- Cell-to-cell contact can stimulate cellular differentiation.

Among the common manipulations carried out on culture cells are media changes, passaging cells, and transfecting cells. These are generally performed using tissue culture methods that rely on sterile technique. Sterile technique aims to avoid contamination with bacteria, yeast, or other cell lines. Manipulations are typically carried out in a biosafety hood or laminar flow cabinet to exclude contaminating micro-organisms. Antibiotics can also be added to the growth media.

Media Changes

Media changes replenish nutrients and avoid the build up of potentially harmful metabolic byproducts and dead cells. In the case of suspension cultures, cells can be separated from the media by centrifugation and resuspension in fresh media. In the case of adherent cultures, the media can be removed directly by aspiration and replaced.

Passaging Cells

Passaging (also known as subculture or splitting cells) involves transferring a small number of cells into a new vessel. Cells can be cultured for a longer time if they are split regularly, as it avoids the senescence associated with prolonged high cell density. Suspension cultures are easily passaged with a small amount of culture containing a few cells diluted in a larger volume of fresh media. For adherent cultures, cells first need to be detached; this is commonly done with a mixture of trypsin-EDTA, however other enzyme mixes are now available for this purpose. A small number of detached cells can then be used to seed a new culture.

Transfection and Transduction

Another common method for manipulating cells involves the introduction of foreign DNA by transfection. This is often performed to cause cells to express a protein of interest. More recently, the transfection of RNAi constructs have been realized as a convenient mechanism for suppressing the expression of a particular gene/protein.

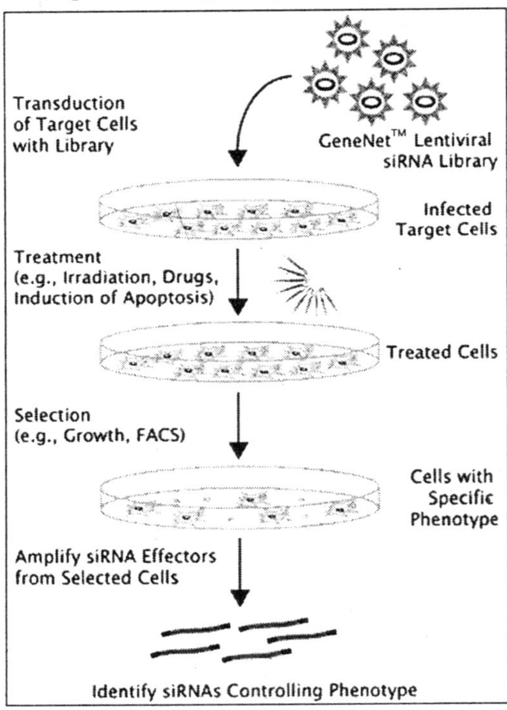

Fig. Transduction

DNA can also be inserted into cells using viruses, in methods referred to as transduction, infection or transformation. Viruses, as parasitic agents, are well suited to introducing DNA into cells, as this is a part of their normal course of reproduction.

GENERATION OF HYBRIDOMAS

It is possible to fuse normal cells with an immortalised

cell line. This method is used to produce monoclonal antibodies. In brief, lymphocytes isolated from the spleen (or possibly blood) of an immunised animal are combined with an immortal myeloma cell line (B cell lineage) to produce a hybridoma which has the antibody specifity of the primary lymphoctye and the immortality of the myleoma.

Selective growth medium (HA or HAT) is used to select against unfused myeloma cells; primary lymphoctyes die quickly in culture and only the fused cells survive. These are screened for production of the required antibody, generally in pools to start with and then after single cloning.

APPLICATIONS OF CELL CULTURE

Mass culture of animal cell lines is fundamental to the manufacture of viral vaccines and many products of biotechnology. Biological products produced by recombinant DNA (rDNA) technology in animal cell cultures include enzymes, hormones, immunobiologicals (monoclonal antibodies, interleukins, lymphokines), and anticancer agents. Although many simpler proteins can be produced using rDNA in bacterial cultures, more complex proteins that are glycosylated (carbohydrate-modified), currently must be made in animal cells. An important example of such a complex protein is the hormone erythropoietin. The cost of growing mammalian cell cultures is high, so research is underway to produce such complex proteins in insect cells or in higher plants.

Tissue Culture and Engineering

Cell culture is a fundamental component of tissue culture and tissue engineering, as it establishes the basics of growing and maintaining cells.

Vaccines

Vaccines for polio, measles, mumps, rubella, and chickenpox are currently made in cell cultures. Due to the H5N1 pandemic threat, research into using cell culture for influenza vaccines is being funded by the United States government. Novel ideas in the field include recombinant DNA-based vaccines, such as one made using human adenovirus (a common cold virus) as a vector, or the use of adjuvants.

CULTURE OF NON-MAMMALIAN CELLS

Plant Cell Culture Methods

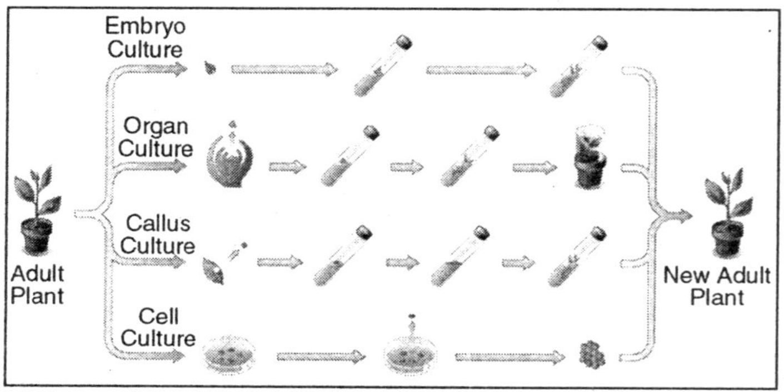

Fig. Plant Cell Culture

Plant cell cultures are typically grown as cell suspension cultures in liquid medium or as callus cultures on solid medium. The culturing of undifferentiated plant cells and calli requires the proper balance of the plant growth hormones auxin and cytokinin.

Bacterial/Yeast Culture Methods

For bacteria and yeast, small quantities of cells are usually grown on a solid support that contains nutrients embedded in it, usually a gel such as agar, while large-scale cultures are grown with the cells suspended in a nutrient broth.

Viral Culture Methods

The culture of viruses requires the culture of cells of mammalian, plant, fungal or bacterial origin as hosts for the growth and replication of the virus. Whole wild type viruses, recombinant viruses or viral products may be generated in cell types other than their natural hosts under the right conditions. Depending on the species of the virus, infection and viral replication may result in host cell lysis and formation of a viral plaque.

Chapter 22

Biology of Cultured Cells

CELL ADHESION MOLECULES

The first suggestion that cell adhesion molecules exist came from Moscona in 1952 who noted that in experiments with developing chick embryos the cells of completely disrupted tissues could reassemble and reform the original tissue structure. It was suggested that the ability of cells to re-aggregate was due to the presence of adhesion molecules present within the cell membrane.

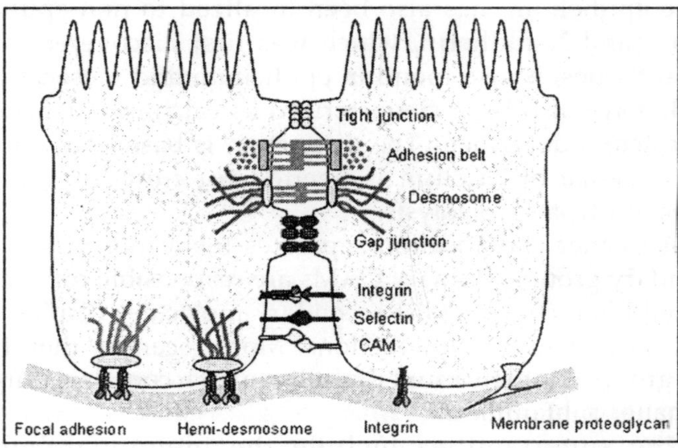

Fig. Cell Adhesion Molecules

Since the pioneering work of Moscona, Townes and Holtfreter in the 1950s, significant steps have been made in the identification, characterization and classification of several distinct cell adhesion systems in developing embryos. These systems allow cells to interact dynamically with adjacent cells and the extra-cellular matrix. They are of importance in the development, morphogenesis, maintenance and regeneration of the form, structure and organization of organisms. In addition, they appear to have a role to play in an organism's ability to resist tumour invasion and metastasis.

The functional units of cell adhesion systems are multiprotein complexes. Three classes exist:

- Cell adhesion molecules (CAMs);
- Extra-cellular matrix proteins;
- Cytoplasmic linking proteins.

The cell adhesion molecules (CAMs) are glycoproteins and are present on the external surface of the cell membrane. They recognize and interact either with other CAMs on adjacent cells or with proteins of the extra-cellular matrix. Confusingly, there are a number of classifications of CAMs. At first they were described on the basis of the tissue in which they were initially identified. It is now clear that they are ubiquitous and not limited to single tissues.

For example, E-cadherin, which was initially discovered in the epithelium, has also been localized to non-epithelial tissues; and N-cadherin, which was first discovered in the neural tissues, is also found in epithelium and mesenchyme. CAMs have also been characterized as calcium-independent and calcium-dependent The distinction is based on whether the presence of calcium is needed for function and for protection from proteolysis.

A further classification separates them into primary or secondary groups. Primary CAMs are responsible for specific cell-cell interactions whilst secondary molecules display cell-extra-cellular matrix interactions. More recently, they have been grouped into five super-families which contain a number of smaller subfamilies.

The superfamilies include the immunoglobulins, cadherins, integrins, selectins and proteoglycans (including the syndecan subfamily of adhesion molecules). The extra-cellular matrix (ECM) proteins are large glycoproteins and can form complex macromolecular arrays. They include the collagens, fibronectins, laminins and proteoglycans and have the ability to bind to the cell adhesion molecules and a wide variety of ECM proteins.

The fibronectins are a family of multi-adhesive matrix proteins which can regulate the shape of cells and the organisation of the cell cytoskeleton by their attachments to the cell adhesion molecules and the extra cellular matrices. They arc essential for migration and cellular differentiation of many cell types during embryogenesis.

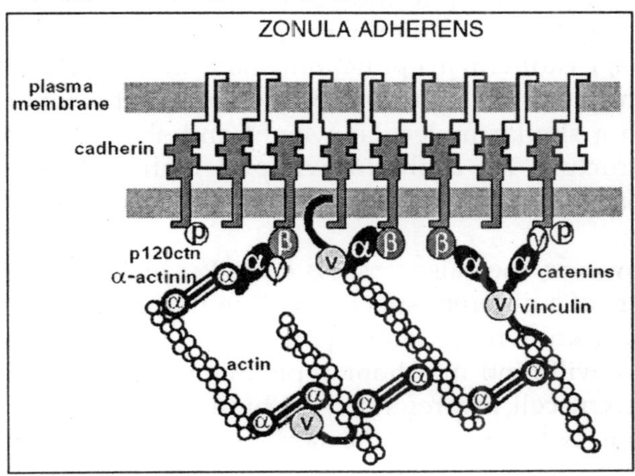

Fig. Catenins

They also have an important role in wound healing as they facilitate migration of macrophages and other cells of the immune system into the affected area. The cytoplasmic plaque proteins (known as catenins) link the cell adhesion system to the cytoskeleton of the cell.

They can regulate the functions of the CAMs and activate a number of second messenger systems within the cell. This review will concentrate on the cell adhesion molecule component of the cell adhesion system.

Cell adhesion Molecules

The Cadherins

The cadherins are a family of calcium-dependent CAMs

which originally comprised a small number of transmembrane glycoproteins Each member was found to regulate cell adhesion of particular cell types and thus was thought to be fundamental for the organization of the multicellular organism. It has been shown that an error in expression of N-cadherin, either spatially or temporally, results in embryos displaying neural tube defects.

After the initial discovery of N-cadherin in neural tissue it was subsequently found to be expressed in non-neural tissues. Similarly, P-cadherin was initially discovered in mouse placenta, E cadherin in epithelium and L-CAM in chicken liver cells. There are now more than 30 members of this superfamily, all with a similar molecular weight and sharing a common basic structure. The cadherins will preferentially bind to other receptors within their subclass, a process known as homotypic binding.

However, they also possess a weaker binding affinity for members of a different subclass within the same superfamily, a process known as heterotypic binding. For example, E-cadherin will bind in a homotypic manner to E-cadherin on an adjacent cell in preference to binding in an heterotypic manner to N-cadherin on the same cell.

The Integrins

In recent years it has become apparent that there are a family of transmembrane glycoproteins present in cells involved in processes as diverse as cell migration during embryo-genesis, thrombosis, haemostasis, wound healing, immune and non-immune defence mechanisms and oncogenic transformations. These transmembrane glycoproteins, the integrins, provide a link between the cytoskeleton of the cell and the extra-cellular matrix proteins.

They are composed of structurally distinct alpha and beta subunits, and pairs of beta subunits combine with different alpha subunits to generate distinct receptors with unique binding properties.' Unlike the immunoglobulin and cadherin superfamilies, the integrins bind to a wide variety of extra-cellular matrix molecules and other cell surface proteins.

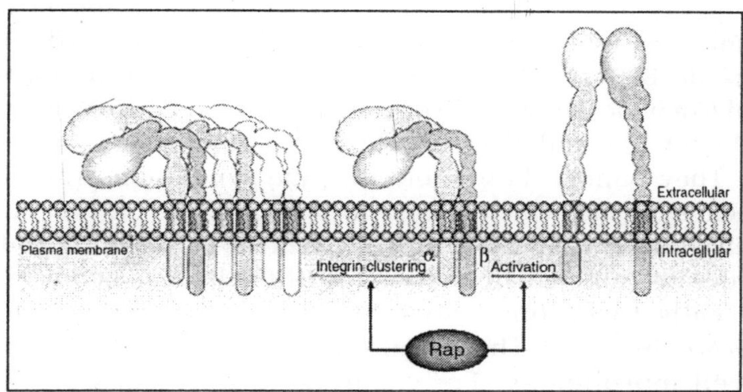

Fig. Integrins

They participate in both cell-matrix and cell-cell adhesion in a wide variety of physiologically important processes such as haemostasis and wound healing.

The Syndecans

In the early 1990s a group of scientists discovered that a number of components of the extra-cellular microenvironment had a high affinity for binding with heparin. These extra-cellular components included growth factor peptides, proteases, antiproteases and ECM molecules which, by binding to a cell, could produce changes in cell shape, motility, adhesion, proliferation and differentiation.

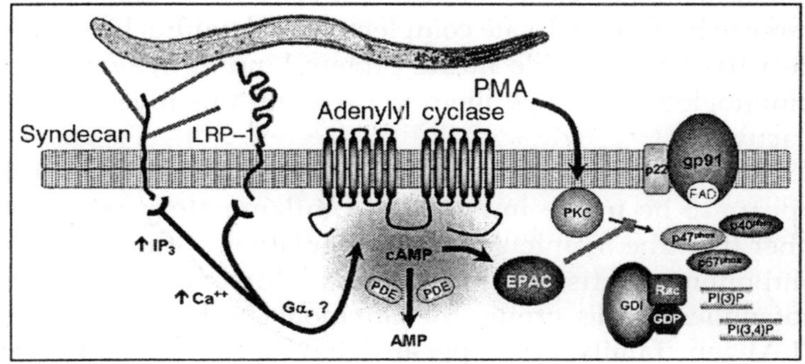

Fig. Syndecans

They discovered that there is a family of integral

membrane proteoglycans made up of heparan and chondroitin sulphate, both structural analogues of heparin, which can also bind to a wide variety of structural proteins and growth factors in the internal and external cellular environment.

They named these molecules the syndecans, from the Greek 'syndein' which means 'to bind together'. It has since been shown that by binding to extra-cellular ligands, syndecans can mediate the activity of the ligands and enable the cells to become more or less responsive to their microenvironment. They are also involved in the maintenance of cell morphology. For example, if syndecans are not expressed in epithelial cells, then the cells become rounded in shape. At present, four members of this family have been identified. Syndecan-1, the most prevalent of the group, is expressed predominantly in epithelial tissues. Syndecan-2 predominates in tissues rich in endothelial cells, Syndecan-3 is found primarily in neural tissues and Syndecan-4 mainly in the liver and kidney.

The Immunoglobulins

The immunoglobulins have been studied extensively because of their role in the inflammatory and immune responses. The members of this polypeptide group differ widely in structure and are able to bind to and neutralize protein toxins, block the attachment of some viruses to cells, opsonize bacteria, activate complement and natural killer cells. The variations possible in the protein binding region of the immunoglobulins allow them to carry out this diverse range of activities.

Although they are able to bind to cells, their main role appears to be in the immune and inflammatory responses rather than the maintenance and regulation of the structural components of tissues and organs. However, one of the subfamilies of this group, the carcinoma-embryonic antigen (CEA)-gene family, does appear to have a role in craniofacial development.

The cell adhesion molecule C-CAM is a member of this subfamily and can mediate calcium-independent cell-cell adhesion in a homotypic manner. The neural cell adhesion molecule N-CAM is another member of the immunoglobulin superfamily of proteins. Like C-CAM, it also binds cells together in a calcium-independent homotypic manner.

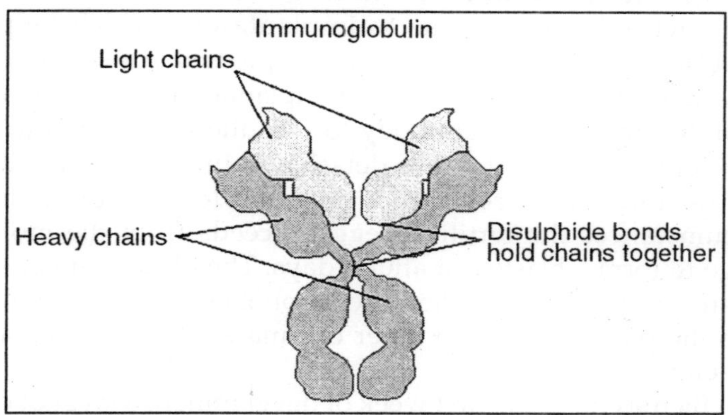

Fig. Immunoglobulins

There are at least 20 forms of N-CAM, each of which are encoded by several separate genes. N-CAM is expressed in a variety of cell types but has a prominent role in development of the nervous system.

Morphogenesis and CAMS

The ability of cells to detach from a formerly cohesive structure and to migrate and contribute to the formation of new tissues is a key event during morphogenesis. A shift from an intercellular mode of adhesion towards a preferential adhesion to an external substrate, via the components of the ECM, is required if active cell migration is to take place. The various cell adhesion molecules referred to in the preceding sections all have a role to play in these processes from a very early stage in development.

Early Development

After fertilization and fusion of the sperm and egg pronuclei, the egg divides several times into a mass of cells called a morula. Even at the one cell stage of embryogenesis, there is evidence of E-cadherin expression, and its influence increases with cell compaction at the 8—16 cell stage.

Implantation of the embryo into the uterine epithelium involves both E- and P-cadherin. Studies in mice show expression of P cadherin within the decidual cells of the endometrium in response to implantation, which help to establish the placenta. E-cadherin is expressed in all embryonic cells at this time, possibly helping implantation into the uterine wall. If a monoclonal antibody to E-cadherin is added to the developing morula, the cells detach and further development of the morula is rendered impossible. Under normal circumstances the fertilized egg proceeds from the morula stage to form a blastocyst after 4 days. The blastocyst consists of an outer cell layer called the trophoblast, which develops into the placenta, and an inner cell mass which gives rise to the embryo.

By the end of the first week of development a layer of cells known as the hypoblast has formed on the ventral surface of the inner cell mass. The remaining cells of the inner cell mass form the epiblast by the end of the second week of development and the hypoblast and epiblast constitute the bilaminar embryonic disc. These cells maintain the form of the embryonic disc by means of cell—cell and cell—extra—cellular matrix interactions, primarily due to the cell adhesion molecule E cadherin.

Transformation

The migration of cells is essential for progression of development. Epithelial cells, however, are not noted for their mobility. This is ingeniously corrected by transformation of epithelial cells into mesenchymal cells in a phenomenon known as epithelial mesenchymal transformation (EMT). At an early stage of transformation the epithelial cell down regulates its expression of cell adhesion molecules (in this instance E-cadherin) which frees the attachment of the cells from one another.

The cell then enlarges and becomes mobile via its interaction with the ECM. Cell adhesion molecules, particularly integrin, may also be involved at this stage. Their involvement has been documented in specific cell migrations, such as that of neural crest cells. After migration has been completed the phenotype can again change, with cells reverting back to their epithelial origin in a process predictably called mesenchymal epithelial transformation (MET). The close relationship between EMT and cell adhesion molecules has lead some workers to consider the gene controlling E-cadherin as the 'master gene' which switches EMT on and off.

The role of EMI in development is considerable. An early example of EMT is the formation of the three germ layers at the gastrulation phase.

Gastrulation and Embryogenesis

At the beginning of the third week of development, the cells of the epiblast lose their rigid cell-cell connections due to a reduced expression of E-cadherin in their membranes, an event coinciding with the beginnings of EMT. They interact more strongly with the extra-cellular matrix and are able to proliferate and migrate to the centre of the embryonic disc to form the 'primitive streak' on the dorsal aspect of the embryonic disc.

Here cells leave the deep surface of the primitive streak and migrate laterally, cranially and caudally to form mesenchyme: a loose embryonic connective tissue. Some of these mesenchymal cells aggregate to form a layer between the epiblast and the hypoblast known as the embryonic mesoderm. The loss of E-cadherin is associated with the ingress of mesoderm. As the mesodermal cell layer is forming the amount of N-cadherin expression in their membranes increases. Hence, the cells have changed from expression of mainly E-cadherin, when they were part of the epiblast cell layer, to predominantly N-cadherin, as part of the newly formed mesoderm layer.

Also, mesenchymal cells displace the hypoblast and form the embryonic endoderm. The epiblastic cells that remain on the surface of the embryonic disc form the layer called the embryonic ectoderm. The original bilaminar embryonic disc is now a trilaminar embryonic disc composed of three primary germ layers: ectoderm, mesoderm and endoderm. This process of germ layer formation is called gastrulation and is the beginning of embryogenesis.

As the embryo develops, these germ layers give rise to the tissues and organs of the embryo. At this stage, the mesenchymal cells in the median plane of the trilaminar disc form a midline cellular cord, known as the notochordal process. The ventral wall of the notochordal process degenerates to form a notochordal plate.

This plate then folds into a tube-like structure, the notochord. As the notochord develops it interacts with the mesenchyme adjacent to it to induce the overlying embryonic ectoderm to form the neural plate. This enlarges and invaginates along its central axis to form a neural groove which has neural folds on either side. By the end of the third week of development these folds have approached each other in the median plane and fused, converting the neural plate into a neural tube. Prior to the formation of the neural tube, c-cad 6B, ('novel' cadherin) is expressed at the neural fold, localizing at the future neural crest cell area.

Expression of c-cad 6B continues until the migration of a select group of neuroectodermal cells which ultimately develop into neural crest cells. This suggests that c-cad 6B has a role in neural fold fusion and/or in the maintenance of the presumptive neural crest cell area. As the tube is formed the associated ectoderm reduces its expression of E-cadherin and instead increases its expression of N cadherin. The overlying ectoderm, however, continues to express E-cadherin.

Neural Crest Cells and their Migration

Migration of neuroectodermal cells lying along the crest of each neural fold occurs ventrolaterally on each side of the neural tube, forming an irregular elongated mass called the neural crest which is located between the neural tube and the overlying surface ectoderm. The neural crest then divides into the right and left parts that migrate to the dorsolateral aspects of the neural tube.

It is from this position that these cells migrate further, eventually producing the majority of the orofacial mesenchyme. From this mesenchyme develops skeletal structures, facial muscle, connective tissue and smooth muscle of the major vessels of the head. It has been suggested that at least some of the neural crest cells are pre-specified before migration. This has been illustrated in rafting experiments where first branchial arch neural crest cells have been implanted into the second branchial arch with resultant formation of first arch structures.

The mechanism dictating this positional code has yet to be elucidated but may involve changes in expression of cell adhesion molecules. Numerous cell adhesion molecules are expressed and down regulated in neural crest cells during their pre-migratory and migratory stages. To test the activity of cadherin molecules, Nakagawa and Takeichi reported that migrating neural crest cells expressed alpha and beta catenins, the intracellular binding proteins for cadherins, at their cell-cell junctions suggesting that cadherin activity in general was ongoing during this period. Many of the traditional cadherins are down regulated during migration, particularly N-cadherin and N-CAM, which presumably alters the attachment of cells to one another allowing them to migrate.

Increasingly, the activity of new 'novel' cadherins are being examined. Nakagawa and Takeichi described the activity of one such 'novel' cadherin, c-cad 7,which is expressed at the initiation of migration in neural crest cells and continues to do so as migration continues. C-cad 7 promotes adhesion of a group of neural crest cells to form a small sub-population of clustered cells which eventually populates the dorsal and ventral roots.

Similarly, the selective expression of T-cadherin indicates that it too may influence the pattern and destination of migrating neural crest cells. These examples suggest that similar patterns of clustered cells may migrate as sub-populations to specific areas, with expression of specific cadherins mediating the process. It may be that particular cadherin—mediated interactions between crest cells play a role in signalling between homotypic cells, as well as influencing the sorting of heterotypic cells.

Extra-cellular cell adhesion molecules are also important in neural crest cell migration. Increased amounts of the ECM protein fibronectin have been reported which may enhance neural crest cell attachment to the substrate and promote crest cell motility by providing a framework for crest cell migration.

The importance of integrin can be illustrated by the reduction of migration and cell adhesion noted after introduction of an antiserum against chick alpha 1 integrin. In contrast, neural crest migration can still occur in beta1 integrin deficient chimeric mice embryos. The role of integrins in neural crest cell migration is thought to be associated with crest cell interactions with laminins and fibronectin.

Migration of neural crest cells is an example of EMT, with both intra- and extra-cellular cell adhesion molecules important in its initiation. The onset of EMT is strongly associated with a decrease in the expression of certain cell adhesion molecules, in this case N cadherin. The down regulation of N-CAM occurs within the same time frame, but at a rate which is unrelated to EMT progression. Changes in integrin activity have also been recorded in avian crest cells.

Secondary Palate Formation

The normal development of the secondary palate commences with the migration of ectomesenchymal cells from the neural crest to the oral cavity to give rise to the maxillary processes. On day 45 each maxillary process develops two medially directed processes, the superiorly positioned tectoseptal process and, below this, the palatal process. The mesenchyme of the tectoseptal processes contributes to the growth of the nasal septum whilst that of the two palatal processes contributes to the secondary palate.

Fusion of both processes are coincident and occur over a very short period, usually 24 h. This is normally complete in humans by the 12th week of interuterine life. The palatal shelves consist of a mass of mesenchyme surrounded by a layer two or three cells thick of undifferentiated epithelial cells.

Initially, the shelves grow vertically on the lateral sides of the tongue. The mechanisms of shelf elevation are probably multifactorial. Changes occur in the dimensions of the palatal processes, with the vertical dimension of the palatal shelves increasing but with no coincident change in width.

In addition, there is a marked concave curvature of the palatal surface. Shelf elevation may be attributed to an increase in the 'intrinsic' force of the palatal shelves produced by an accumulation and hydration of hyaluronic acid, the presence of stout collagen type I fibres and the possible contractile nature of mesenchymal cells which make up the palatal shelves.

The mechanism behind fusion itself is still unresolved. It is known that the outer layer of epithelium of the medial epithelium edge is lost leaving only the inner basal layer. It is the fate of these cells which is uncertain. There is strong evidence to suggest that programmed cell death is important. However, workers have recently considered other mechanisms, including migration of the epithelial cells to other areas of the palatal shelf and EMT.

The interest in the latter can be attributed to the finding of residual epithelial cells trapped within the mesenchyme of the palatal process after fusion. If EMT is implicated in palatal fusion the role of cell adhesion molecules in developmental anomalies, such as cleft lip and palate, should be examined.

A number of cell adhesion molecules are actively expressed prior to and after shelf fusion. The expression of E-cadherin in the epithelium of the palate, tongue and dental lamina increases with age. However, its expression is not disrupted by retinoic acid and therefore its role in the development of cleft lip and palate is unremarkable. The role of Syndecan-1 in palate development was investigated by Brinkley *etal.*

Its expression increased as the palatal shelves moved from a vertical to horizontal position. This is coincident with condensation of shelf mesenchyme. During fusion there is a decrease in syndecan expression. The explanation for changes in expression are attributed to epithelial mesenchymal interactions. Recently, N-cadherin and N-CAM have been shown to be expressed at the medial edge epithelium at the time of fusion.

Tooth Development

When the palate forms the oral epithelium proliferates downwards into the underlying ectomesenchyme forming the dental lamina. This is the first stage of tooth development. The C-CAM molecule, a member of the immunoglobulin superfamily of cell adhesion molecules, is first detected after the palatal shelves have elevated and fused.

It has an increased expression in the oral tissues and a decreased expression in the nasal epithelium, and its expression slowly increases with the age of the embryo. Increased amounts of Syndecan-1 and the ECM glycoprotein tenascin are detectable in the mesenchyme condensing around the forming epithelial bud of the dental lamina. Tenascin and Syndecan- 1 bind together and interact with the cells forming the tooth bud. Their interaction is thought to be important in the formation of the ECM. This interaction continues until the late bud stage and at the cap stage, mesenchymal expression of renascin decreases while that of Syndecan- 1 persists briefly.

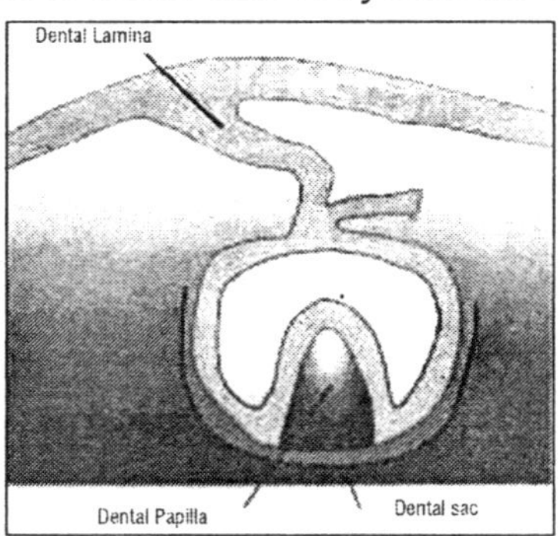

Fig: Late Bud Stage of a Developing Tooth.

During formation of the dental lamina there is increased expression of E-cadherin. In the late cap stage of tooth development, E-cadherin is not detectable in the external or internal dental epithelium surrounding the dental organ. Once the ameloblasts have been formed, E-cadherin reappears between the ameloblasts.

Chapter 23

Cell Proliferation

CELL DEATH

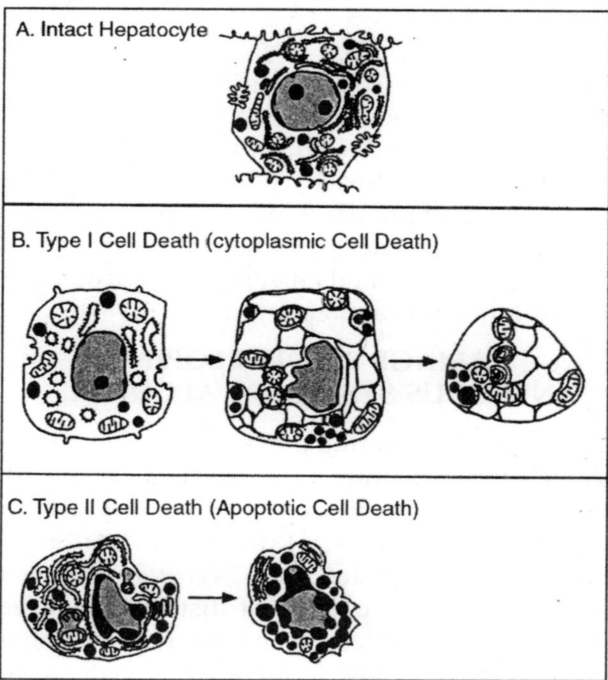

Fig. Cell Death

Terminology of Cell Death

Cell death can occur by either of two distinct1,2 mechanisms, necrosis or apoptosis. In addition, certain chemical compounds and cells are said to be cytotoxic to the cell, that is, to cause its death. Someone new to the field might ask, what's the difference between these terms? To clear up any possible confusion, we start with some basic definitions.

Necrosis and Apoptosis

The two mechanisms of cell death may briefly be defined: Necrosis ("accidental" cell death) is the pathological process which occurs when cells are exposed to a serious physical or chemical insult. Apoptosis ("normal" or "programmed" cell death) is the physiological process by which unwanted or useless cells are eliminated during development and other normal biological processes.

Cytotoxicity

Cytotoxicity is the cell-killing property of a chemical compound (such as a food, cosmetic, or pharmaceutical) or a mediator cell (cytotoxic T cell). In contrast to necrosis and apoptosis, the term cytotoxicity does not indicate a specific cellular death mechanism. For example, cell-mediated cytotoxicity (that is, cell death mediated by either cytotoxic T lymphocytes [CTL] or natural killer [NK] cells) combines some aspects of both necrosis and apoptosis.

DAMAGE TO DEVELOPING NERVOUS SYSTEM BY ALCOHOL

Maternal drinking during pregnancy can adversely affect the outcome of the offspring, with effects ranging from mild cognitive impairment, characterized by impaired mental activities, to full-blown fetal alcohol syndrome (FAS), characterized by growth deficiency, central nervous system (CNS) disorders, and a pattern of distinct facial features. Alcohol can exert these effects both directly, by acting on fetal tissue, and indirectly, by interfering with the maternal support of the growing fetus.

Such indirect mechanisms include altering the placenta' s ability to provide the necessary nutrients to the developing fetus. Alcohol also may indirectly harm the fetus by impairing the mother' s physiology. For example, alcoholism may lead to malnutrition or be combined with other drug use. This article concentrates on the mechanisms underlying alcohol' s direct effects on the fetus. Numerous mechanisms have been suggested as contributing to alcohol-induced fetal damage, particularly deficits in brain function, although none of these mechanisms has been established with certainty.

Furthermore, although alcohol itself generally is considered the primary birth-defect-inducing substance (i. e., teratogen), products resulting from alcohol s break-down (i. e., metabolism) also may play a role. For example, acetaldehyde - a toxic chemical formed by the break-down of alcohol in the liver and other tissues–can accumulate in the fetal brain after prenatal alcohol exposure and may contribute to the development of FAS. Recent literature reviews confirm that no single putative mechanism can account for all the components and variations of the anatomical and behavioral characteristics (i.e., phenotypes) found in children prenatally exposed to alcohol.

This article reviews some general challenges researchers face when trying to elucidate multiple disease mechanisms and explains the role that animal and tissue culture (i.e., in vitro) studies can play in this research. The article then explores some of the mechanisms that have been implicated in the development of alcohol-induced CNS deficits, which represent the most serious consequences of prenatal alcohol exposure.

Challenges Associated

Identifying the mechanisms contributing to alcohol-induced fetal damage is complicated by numerous factors. For example, scientists have not determined the exact cellular and molecular processes involved in normal CNS development, making it difficult to tease apart the effect that alcohol has on this system.

In addition, alcohol is known to interact with tissues in a multitude of ways, and those interactions may have both short-term and long-term effects. Finally, each person exhibits a different combination of alcohol-related effects, which is determined by the timing, level, pattern, and duration of the mother's drinking as well as by genetic factors. This variability makes it difficult to compare alcohol' s effects from one person to the next. Alcohol' s effects on the developing brain are particularly complex.

For cer tain groups of brain cells, alcohol can lead to cell death, whereas for other cell groups it interferes with cellular functions. Alcohol may even deplete cells through different mechanisms in a given cell population, depending on the developmental stage of the cells.

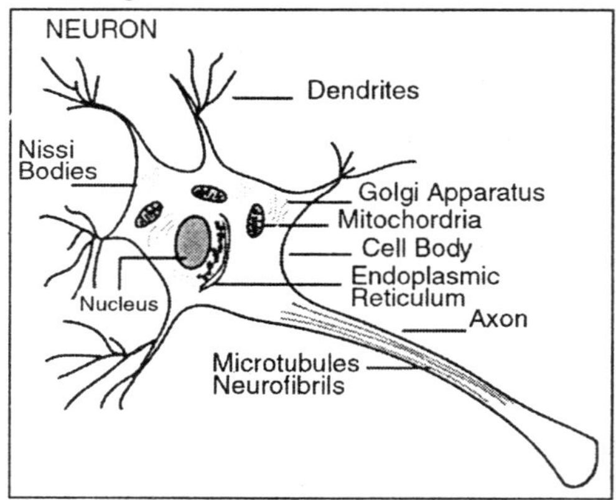

Fig. Nerve Cells

For example, nerve cells (i.e., neurons) in the fetal brain multiply through a process of cell division and then migrate during development to an appropriate location where they mature to their full form and function.

If some groups of those cells are exposed to alcohol during cell division, the generation of new cells may be reduced by an altering of the cell division rate.

If exposure occurs at a later stage of development, however, when the cells are no longer dividing, this same population of neurons can be depleted as a result of alcohol-induced cell death. As brain cells develop, they may change with respect to their susceptibility to alcohol s effects. This changing susceptibility to alcohol can be illustrated using a neuronal cell type called Purkinje cells that are located in the cerebellum, a brain region involved in motor coordination and motor learning.

During cell division, alcohol exposure appears to have minimal effects on the generation of new Purkinje cells. However, at a later stage of development when the cells begin to develop all the features of a mature neuron, Purkinje cells and perhaps other neurons) are especially vulnerable to cell death. Neurons may also die because alcohol exposure during one stage of development (e.g., before neurons migrate to their final location) interferes with subsequent developmental stages (e.g., migration or differentiation).

For example, cells from an embryonic structure called the cranial neural crest migrate to appropriate locations to form facial cartilage and bone as well as the peripheral nerves that innervate the head and face. In chick embryos, alcohol exposure before the cranial neural crest cells begin to migrate results in excessive cell death of these cells during the period of migration, leading to abnormal facial features modeling effects in humans with FAS.

These examples demonstrate that at least some consequences of alcohol exposure at different stages of CNS development are caused by different mechanisms (e.g., effects on cell division, on the survival of cells that are migrating after cell division has ended, and on the establishment of mature cell structures and functions). Even at a given developmental stage, different mechanisms of cell death may operate in response to different levels of maternal alcohol consumption and, consequently, fetal blood alcohol concentrations (BACs).

Considering these factors, multiple mechanisms leading to alcohol-related fetal damage may operate both simultaneously and sequentially over time. The presence of such a broad spectrum of potential mechanisms associated with prenatal alcohol exposure makes it difficult to identify alcohol' s effects on specific cells, as well as the consequences of these effects.

To facilitate these analyses and to be able to specifically assess alcohol' s effects on specific brain cells, researchers frequently have turned to animal models of FAS or to the study of cells grown in culture (i.e., in vitro models). The benefits and disadvantages of such model systems are discussed in the following section.

The Role of Animal

Researchers have recreated each of the major characteristics of human FAS facial abnormalities, CNS abnormalities, and growth deficiency in one or more animal models of developmental alcohol exposure (e. g., chicks, mice, rats, or primates). Such animal models allow investigators to study alcohol' s effects on fetal development in the context of an intact organism while enabling them to control several key factors that influence the type and extent of alcohol-induced structural, functional, and behavioral abnormalities in CNS development.

These factors include the pattern of alcohol exposure (e.g., constant exposure versus bingelike episodes), the BACs produced, the developmental timing and duration of exposure, the cell types or brain regions studied, and maternal and fetal genetic factors. Because animal studies can be used to mimic the conditions of maternal alcohol consumption during pregnancy and because laboratory mammals have physiological, biochemical, and genetic features in common with humans, the findings from such studies can often provide key information for inferences about effects in humans.

Nevertheless, the results should be interpreted with some caution because species differences in vulnerability to alcohol can exist. Furthermore, animal models still do not allow a detailed analysis of alcohol' s actions on individual cells. Such detailed experimental analysis of alcohol' s molecular mechanisms of action is possible, however, using in vitro approaches, in which individual cell types or tissues are grown in tissue culture.

These models allow researchers to control and manipulate both the cell and its environment, features that are essential to understanding the mechanisms underlying alcohol' s developmental effects. Even in relatively simple cell culture models, however, multiple harmful effects can occur via a variety of pathways. The main advantage of in vitro models is that experimental techniques are now available to visualize molecular events during the initial alcohol-tissue interactions and the dynamic course of the pathogenic process in living cells. As a result, in vitro studies are a necessary and powerful tool for discovering alcohol' s molecular actions. At the same time, however, it can be difficult to extrapolate the findings from the molecular actions observed in vitro to the complex mechanisms that simultaneously occur and interact in developing mammalian brains.

The gap between the molecular in vitro data and the analyses of animal models may at least in part be bridged by new techniques for conducting functional molecular analyses of alcohol-related mechanisms in whole organisms.

These new techniques include embryo culture methods, as well as manipulation of the animal' s genome (i.e., the total genetic material), such as the generation of mice that lack specific genes null mutant mice or that carry foreign genes - transgenic mice. Other emerging technologies, such as genomic and proteomic technology that can be applied both in vivo and in vitro will likely enhance the understanding of mechanisms that damage the developing nervous system. Together, these techniques, which have emerged in tandem with the recent molecular revolution in neuroscience, should yield new insights into the mechanisms of alcohol-induced damage to the developing brain.

Candidate Mechanisms

The experimental approaches described in the previous section – animal models, in vitro models, and new technologies - have helped researchers identify numerous mechanisms that may con-tribute to alcohol s detrimental effects on fetal development.

CELL DEATH MODES: NECROSIS AND APOPTOSIS

Many actions of alcohol on the developing organism, including the brain, result in cell death. Two general processes of cell death exist, called necrosis and apoptosis. These two processes can be distinguished through different patterns of morphological and biochemical changes during cell death.

Necrosis occurs when neurons are damaged by a trauma or metabolic injury and typically involves the concurrent death of groups of adjacent cells. Cells undergoing necrosis initially swell and their internal components, or organelles, break down. The cells eventually rupture and spill debris that leads to local inflammation. This inflammation can then result in the death of adjacent cells.

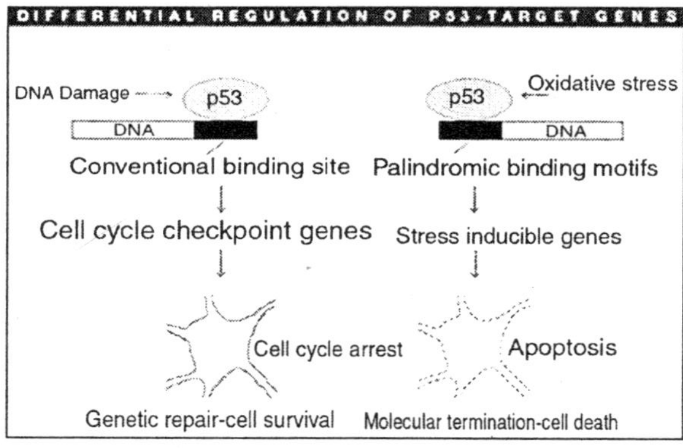

Fig. Cell Suicide

Conversely, apoptosis is a form of "cell suicide" that affects only individual cells, leaving adjacent cells intact. (Although apoptosis has been implicated in a broad range of CNS disorders, including Alzheimer s disease and amyotrophic lateral sclerosis (ALS, or Lou Gehrig' s disease), it is important to note that apoptosis is not always bad for the organism. In fact, elaborately regulated apoptotic cell death is also critical for normal CNS formation during development).

During apoptosis, the cell body shrinks, and the DNA in the nucleus becomes condensed before breaking apart into small fragments The cell' s organelles remain intact, however. Eventually, the cell breaks up into several smaller bodies that are still surrounded by a membrane. These "apoptotic bodies" then are engulfed and destroyed by scavenging cells. Although apoptosis can progress rapidly once it has been initiated, its onset may be delayed for a time after a toxic insult. At least in some cases, apoptotic cell death appears to involve the activation of a gene-directed programme for cellular self-destruction.

Numerous factors can induce apoptosis of CNS cells, including insufficient blood supply to the brain; dysfunction of the cell's energy-generating organelles, called the mitochondria; disruption of the normal calcium levels in the cells; and oxidative stress. Some of these factors can induce both apoptosis and necrosis. Alcohol can also induce apoptosis. This has been demonstrated both in animal models of early alcohol exposure (e.g., the cranial neural crest in embryos), and in isolated CNS cells grown in culture, including cells from the hypothalamus. A key event occurring in cells undergoing apoptosis is the activation of certain death-promoting enzymes called caspases. Certain caspases act as executioners during apoptosis, cutting apart and functionally destroying important proteins in the cell.

Although diverse molecular signals can initiate apoptosis, the activation of caspases is a crucial step in the progression to cell death. Accordingly, specific inhibitors of caspases have been shown to prevent apoptotic cell death in several experimental models. Not all cells exposed to one of the cell death-inducing factors mentioned above actually die. Whether a cell lives or dies is determined by the balance between certain proteins that can activate or block apoptosis.

Disruption of this balance in favour of proteins encoded by cell death genes might be involved in alcohol-induced apoptosis. One well-studied group of cell-death genes is called the bcl-2 family, a group of genes that encode related proteins. Some of these proteins promote apoptosis, whereas others can prevent apoptosis. An increase of apoptosis-preventing members of the bcl-2 protein family may be able to protect a cell against death under various conditions, including exposure to alcohol.

Mitochondrial Dysfunction

A key factor that can induce apoptosis (as well as necrosis) is oxidative stress. This term refers to the consequences of having excess levels of free radicals in the cells. Free radicals are highly reactive molecules that may be formed during various biochemical reactions in the cell. Many of these free radicals contain oxygen and are called reactive oxygen species (ROS). Typically, the levels of ROS and other free radicals are controlled by various scavenger molecules, known as antioxidants, that are normally found within the cell and which eliminate free radicals.

If ROS levels exceed the cell' s ability to eliminate them, however, or if the normal antioxidant levels within the cell are reduced due to a toxic insult such as alcohol, then oxidative stress can occur. This oxidative stress can cause damage to cellular components, such as membranes, DNA, and proteins. Moreover, oxidative stress can induce cell death processes through several mechanisms, including the release of apoptosis-inducing factors.

Alcohol can induce oxidative stress through several mechanisms. For example, certain pathways of alcohol metabolism result in the generation of ROS. Moreover, alcohol may reduce antioxidant levels.

Experimental evidence suggests that these factors may contribute to alcohol-induced cell damage and cell death in the fetus. Treatment with antioxidants appears to ameliorate alcohol-induced damage in animal models. Therefore, this line of investigation may have important implications for clinical intervention.

The alcohol-induced formation of excess levels of ROS also can damage cells and induce cell death by interfering with the function of the mitochondria - organelles surrounded by a membrane that are found in all cells and which generate most of the cell' s energy. The mitochondria serve an additional crucial function because they store calcium and regulate the calcium levels in the cell, which is particularly critical in neurons.

The controlled flow of calcium from the fluid surrounding the neuron into the neuron s interior is one of the key steps in the process of chemical communication between neurons. To ensure accurate neuronal function, calcium levels inside the neuron must be tightly regulated. Furthermore, excessive internal calcium concentrations can be toxic to neurons. Therefore, the ability of the mitochondria to actively sequester calcium is vital for maintaining neuronal function and survival.

Oxidative stress, such as the alcohol-induced formation of excess ROS levels, can also be associated with disturbed mitochondrial function, including the mitochondria' s ability to regulate internal calcium levels. Mitochondrial dysfunction can lead to both necrosis and apoptosis. When mitochondria become dysfunctional, they can undergo a process called mitochondrial permeability transition (MPT).

During this process, large holes open in the mitochondrial membrane through which the mitochondria release their contents, including calcium and a molecule called cytochrome c, into the fluid that fills the neuron. Both calcium and cytochrome c can activate caspases, which, as mentioned in the previous section, play a role in apoptosis. In addition, the MPT process plays a pivotal role in necrosis, further contributing to the deleterious consequences of alcohol-induced oxidative stress.

Cell Proliferation and Survival

Alcohol also can damage the fetal brain by a mechanism in which the generation of new cells in the cerebral cortex is hindered during development. New neurons are formed in two specific areas of the developing brain; from those cell proliferation zones, the new cells migrate to their final locations in the mature brain. Alcohol can alter the speed at which the cells divide.

Alcohol can also interfere with the activity of growth factors that regulate cell proliferation and survival. Loss of normal growth factor signaling can also interfere with or prevent normal growth and development.

Numerous growth factors are needed for cell division to proceed normally, including two factors called insulinlike growth factors (IGF) I and II. Both IGF-I and IGF-II exert their effects by binding to protein molecules called IGF-I receptors on the cell surface. Alcohol can interfere with the activity of the IGF-I receptor. As a result, IGF-I still binds to its receptor, but the receptor signaling function is blocked, and IGF-I-mediated cell division cannot proceed. This example demonstrates that alcohol can prevent the normal production of CNS cells by interfering with the growth factors that regulate cell division.

Alcohol also may induce cell death by inhibiting several growth factors that support cells that have attained their final function (i.e., that are differentiated) and no longer divide.

For example, IGF-I and the IGF-I receptor also play a role in the survival of nondividing cells and can prevent apoptosis in several models of cell death. Similar to the situation in dividing cells described above, alcohol can inhibit the IGF-I receptor in nondividing cells, thereby preventing the survival of those cells.

EFFECTS ON GLIA CELLS

Normal brain development and function require not only neurons, but also non-neuronal cells, called glia, that support the growth and development of the neurons. Various types of glial cells with specialized functions exist.

For example, migration of newly formed neurons to their final location in the developing brain requires the presence of cells called radial glia, which serve as elongated cellular tracks that direct neurons to their appropriate destinations.

Once all neurons have migrated to their final locations, the radial glial cells normally change into another type of glial cell star-shaped astrocytes. (Astrocytes provide structural support for neurons) After prenatal alcohol exposure, however, radial glia may become astrocytes prematurely.

As a result, neurons generated toward the end of the neuronal migration period, which normally would migrate to the outer layers of the cerebral cortex, lose their radial glial guides, stop migrating, and end up in abnormal positions. This model could help explain the abnormal positioning of neurons in the cortex observed after developmental alcohol exposure.

Both animal models and in vitro studies found that alcohol

exposure alters several aspects of astrocyte structure and function. Thus, alcohol exposure can reduce the overall number of astrocytes in the cortex, reduce or delay the production of the proteins that give the astrocytes their characteristic shape, and interfere with the cells production of or response to specific growth factors.

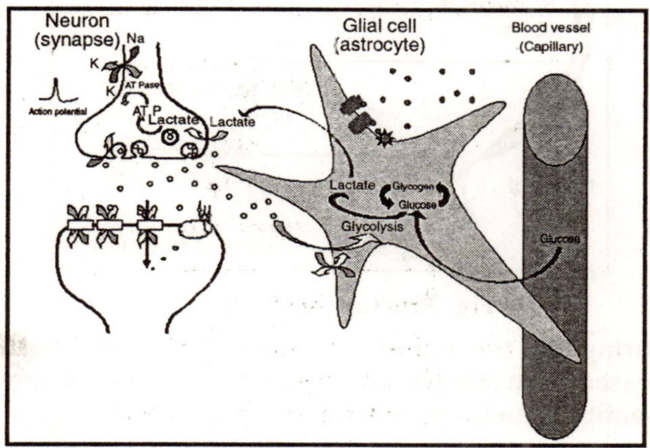

Fig. Glial Cell

Such alcohol-induced changes in astrocyte development and function could have serious consequences on neuronal migration and survival and on the correct formation of connections among neurons.

NEUROTRANSMITTER SYSTEMS

Another important mechanism through which alcohol adversely affects the structure and function of the developing brain is by interfering with the activity of neurotransmitters - brain chemicals that allow the transmission of nerve signals from one neuron to the next.

This transmission occurs at the junction between two neurons called a synapse.

At this juncture, the part of the neuron that conducts nerve signals away from the neuron s body (i.e., the axon) interacts with the branching extensions (i.e., dendrites) of a neighboring neuron that receives the nerve signal.

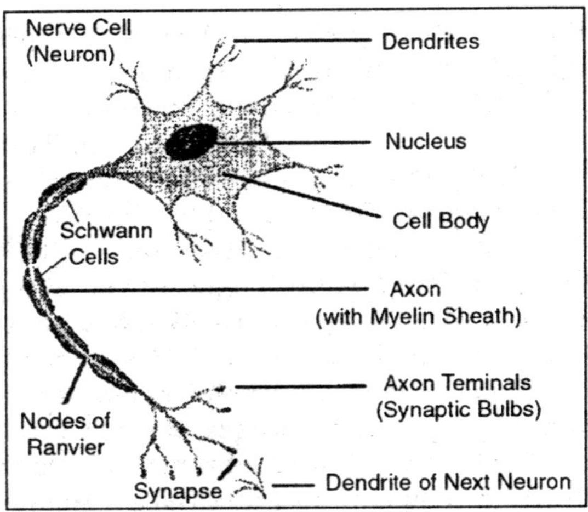

Fig. Structure of Nerve Cell

During this transmission process, the neurotrans-mitters are released from storage vesicles at the end of the axon of the signal-emitting neuron and travel across a small gap to the signal-receiving neuron.

There, the neurotransmitters interact with specific receptors to induce biochemical reactions in the signal-receiving neuron that promote or prevent the generation of a new nerve signal.

Numerous neurotransmitters exist, and some of these (e.g., glutamate, serotonin, and gamma-aminobutyric acid [GABA]) also help organize the CNS during fetal development. Prenatal alcohol exposure can alter the functions of these neurotransmitter systems, particularly the glutamate and serotonin systems.

Glutamate

To exert its actions, glutamate interacts with several receptors, including one called the NMDA receptor. During brain development, the interaction of glutamate with the NMDA receptor appears to be critical for stabilizing synapses that have been formed during sensory or other behavioral experiences.

Developmental alcohol exposure can reduce the number and/or function of NMDA receptors both during early development and during subsequent developmental stages.

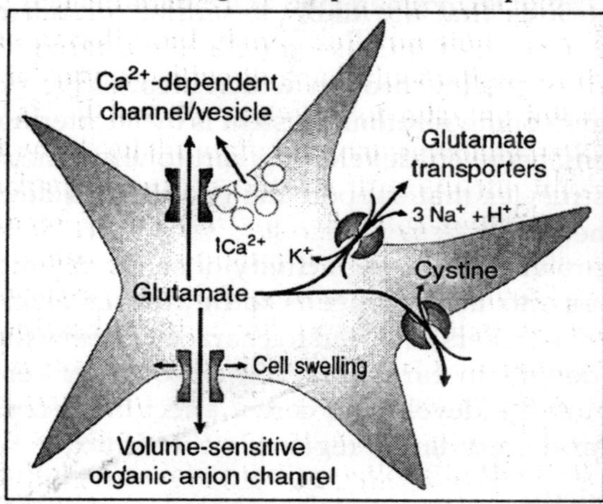

Fig. Glutamate

Through this mechanism, alcohol could affect the function of numerous neurotransmitter systems. This could play a major role in the cognitive and behavioral deficits associated with FAS.

Serotonin

Another important neurotransmitter that helps regulate brain development early in life is serotonin. The growth of serotonin-releasing (i.e., serotonergic) neurons into the brain area that eventually develops into the cortex appears to be a critical step in cortical development.

NH_2

HO

N
H

Fig. Serotonin

Developmental alcohol exposure in rats significantly delays the development of the serotonin system and alters the normal interaction between serotonin and its target sites during periods that are likely essential for normal brain development.

One mechanism through which alcohol may delay the development of the serotonin system is by interfering with the interactions between developing serotonergic neurons and nearby astrocytes that support the growth and development of those neurons.

The growth-promoting activity of serotonin and alcohol' s effects on this activity needs to be further elucidated. In addition, researchers still need to characterize how the alcohol-induced deficits in early embryonic growth of serotonergic neurons into the developing cortex affect brain organization and the function of target regions in the cortex.

Excitotoxicity

Neuronal death also can be induced by excess activity of certain neurotransmitters, including glutamate. This phenomenon, which is called excitotoxicity, may also contribute to alcohol-related damage to the developing brain. Under certain conditions, when glutamate interacts with the NMDA receptor, it causes calcium to flow into the signal-receiving neuron.

As mentioned earlier in this article, such a calcium influx is a powerful regulator of the activity and function of a neuron. In the fetus, the calcium influx generated at the NMDA receptor is an important signal in neuron development, synapse formation, and mechanisms of learning, all of which are crucial to the brain s ability to adapt to its environment. Excessive activation of the NMDA glutamate receptor, however, can lead to dangerously high calcium accumulation inside the neuron. If sufficiently severe or prolonged, the rise in intracellular calcium can lead to cell death by either apoptosis or necrosis.

Conditions of excitotoxicity can occur during withdrawal from high levels of alcohol and may thereby contribute to alcohol-induced damage to the fetal brain, particularly when the mother binge drinks. In these cases, the fetus experiences periods of heavy alcohol exposure, followed by withdrawal episodes. High levels of alcohol acutely inhibit NMDA receptor function.

During withdrawal after a binge-drinking episode, however, glutamate stimulation of NMDA receptor activity increases temporarily and may lead to excitotoxicity. Although some experimental support exists for the potential contribution of withdrawal-related events to alcohol-induced fetal brain damage, including the potential role of excitotoxicity, this hypothesis requires more research.

Glucose Transport and Uptake

Some of the harmful effects of prenatal alcohol exposure also may be associated with alcohol-induced disruption of the brain's utilization of the sugar, glucose. Glucose has several crucial functions in the body, including the brain. First, it serves as an energy source in all cells. Second, it is used in the production of various important types of molecules, including DNA and RNA building blocks (i.e., nucleic acids), fat molecules (i.e., lipids), certain hormones (i.e., steroids), and certain neurotransmitters.

To enter cells from the blood and fulfill its functions, glucose must cross the cell membrane. To this end, most mammalian cells contain specific glucose transporter proteins designated GLUT1 through GLUT7. The principal glucose transporter proteins of the brain are GLUT1 and GLUT3. In cultured rat neurons and astrocytes, short-term alcohol exposure reduced cellular glucose uptake as well as the levels of glucose transporter proteins.

Similarly, prolonged prenatal exposure of rats to alcohol reduced both glucose uptake and GLUT1 gene expression. Because of the central role that glucose plays in the body, alcohol-induced changes in glucose transport have broad implications and must be considered as an important potential contributor to both growth deficiency and CNS damage associated with prenatal alcohol exposure.

Effects on Cell Adhesion

Yet another mechanism through which alcohol may interfere with normal brain development is by reducing cell adhesion. Neurons must establish cell-to-cell contact during growth and development in order to survive, migrate to their final destination, and develop appropriate connections with neighboring cells.

Numerous cell adhesion molecules (CAMs) assist in various aspects of this process. Defects in one particular CAM called L1 can lead to abnormal brain development in humans, characterized by mental retardation, complete absence of the corpus callosum, and abnormal development of the cerebellum.

These brain abnormalities are similar to those found in patients with FAS, suggesting that prenatal alcohol exposure also may affect the L1 molecule and thereby contribute to several aspects of the FAS phenotype. This hypothesis is supported by findings that when cultured brain cells are exposed to low levels of alcohol less - than 0.05 per cent - the L1-mediated clumping together of the cells is inhibited.

In an important recent extension of the analysis of alcohol' s cell adhesion effects, researchers demonstrated that this inhibitory effect was specific to certain types of alcohol molecules. The alcohol in alcoholic beverages is chemically known as ethanol. Researchers found that only certain alcohol molecules, such as ethanol, interfere with L1-mediated cell adhesion.

Conversely, types of alcohol molecules, such as a molecule called octanol, actually block ethanol' s effect on cell adhesion in tissue cultures. Octanol even prevented the harmful effects of ethanol on mouse fetuses grown in culture, suggesting that ethanol' s effect on cell adhesion is an important contributor to the harmful consequences of prenatal alcohol exposure.

Regulation of Gene Expression

Another candidate mechanism through which prenatal alcohol exposure could damage the CNS and lead to such devastating consequences as FAS is through interfering with the normal regulation of the genes that control brain development. Researchers have not yet been able to elucidate these processes, leaving a major gap in their understanding of candidate mechanisms underlying FAS.

Although investigators have identified genes whose expression is altered by alcohol in vitro, studies of alcohol s effects on gene expression as it relates to the development of various body structures and the CNS are still in their infancy. Detailed studies of alcohol-induced changes in gene expression during critical periods of development constitute one of the highest priorities for new research.

Cutting-edge technologies, such as the gene microchip array and proteomic technologies, may provide the means to make rapid advances on this frontier in the near future.

DIFFERENCES BETWEEN NECROSIS AND APOPTOSIS

There are many observable morphological and biochemical differences between necrosis and apoptosis. Necrosis occurs when cells are exposed to extreme variance from physiological conditions (e.g., hypothermia, hypoxia) which may result in damage to the plasma membrane.

Under physiological conditions direct damage to the plasma membrane is evoked by agents like complement and lytic viruses.

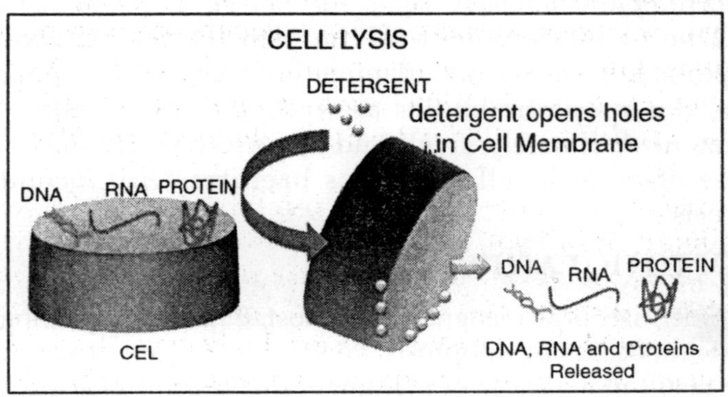

Fig. Cell Lysis

Necrosis begins with an impairment of the cell's ability to maintain homeostasis, leading to an influx of water and extracellular ions.

Intracellular organelles, most notably the mitochondria, and the entire cell swell and rupture (cell lysis).

Due to the ultimate breakdown of the plasma membrane, the cytoplasmic contents including lysosomal enzymes are released into the extracellular fluid. Therefore, *in vivo,* necrotic cell death is often associated with extensive tissue damage resulting in an intense inflammatory response. Apoptosis, in contrast, is a mode of cell death that occurs under normal physiological conditions and the cell is an active participant in its own demise ("cellular suicide").

It is most often found during normal cell turnover and tissue homeostasis, embryogenesis, induction and maintenance of immune tolerance, development of the nervous system and endocrine-dependent tissue atrophy. Cells undergoing apoptosis show characteristic morphological and biochemical features.

These features include chromatin aggregation, nuclear and cytoplasmic condensation, partition of cytoplasm and nucleus into membrane bound-vesicles (apoptotic bodies) which contain ribosomes, morphologically intact mitochondria and nuclear material. *In vivo,* these apoptotic bodies are rapidly recognized and phagocytized by either macrophages or adjacent epithelial cells.

Due to this efficient mechanism for the removal of apoptotic cells *in vivo* no inflammatory response is elicited. *In vitro,* the apoptotic bodies as well as the remaining cell fragments ultimately swell and finally lyse. This terminal phase of *in vitro* cell death has been termed "secondary necrosis".

APOPTOTIC PATHWAYS

Scientists now recognize that most, if not all, physiological cell death occurs by apoptosis, and that alteration of apoptosis may result in a variety of malignant disorders. Consequently, in the last few years, interest in apoptosis has increased greatly.

Great progress has been made in the understanding of the basic mechanisms of apoptosis and the gene products involved.

Death Receptors

Apoptosis has been found to be induced via the stimulation of several different cell surface receptors in association with caspase activation. For example, the CD95 (APO-1, Fas) receptor ligand system is a critical mediator of several physiological and pathophysiological processes, including homeostasis of the peripheral lymphoid compartment and CTLmediated target cell killing. Upon cross-linking by ligand or agonist antibody, the Fas receptor initiates a signal transduction cascade which leads to caspase-dependent programmed cell death.

Membrane Alterations

In the early stages of apoptosis, changes occur at the cell surface and plasma membrane. One of these plasma membrane alterations is the translocation of phosphatidylserine (PS) from the inner side of the plasma membrane to the outer layer, by which PS becomes exposed at the external surface of the cell.

Protease Cascade

Signals leading to the activation of a family of intracellular cysteine proteases, the caspases, (Cysteinyl-aspartate-specific proteinases) play a pivotal role in the initiation and execution of apoptosis induced by various stimuli. Different members of caspases in mammalian cells have been identified. Among the best-characterized caspases is caspase-1 or ICE (Interleukin-1-Converting Enzyme), which was originally identified as a cysteine protease responsible for the processing of interleukin 1.

Mitochondrial Changes

Mitochondrial physiology is disrupted in cells undergoing either apoptosis or necrosis. During apoptosis mitochondrial permeability is altered and apoptosis specific protease activators are released from mitochondria. Specifically, the discontinuity of the outer mitochondrial membrane results in the redistribution of cytochrome C to the cytosol followed by subsequent depolarization of the inner mitochondrial membrane.

Cytochrome C (Apaf-2) release further promotes caspase activation by binding to Apaf-1 and therefore activating Apaf-3. AIF (apoptosis inducing factor), released in the cytoplasm, has proteolytic activity and is by itself sufficient to induce apoptosis.

DNA Fragmentation

The biochemical hallmark of apoptosis is the fragmentation of the genomic DNA, an irreversible event that commits the cell to die and occurs before changes in plasma membrane permeability (prelytic DNA fragmentation). In many systems, this DNA fragmentation has been shown to result from activation of an endogenous Ca^{2+} and Mg^{2+} dependent nuclear endonuclease. This enzyme selectively cleaves DNA at sites located between nucleosomal units (linker DNA) generating mono- and oligonucleosomal DNA fragments.

CELL PROLIFERATION AND VIABILITY

Rapid and accurate assessment of viable cell number and cell proliferation is an important requirement in many experimental situations involving *in vitro* and *in vivo* studies. Examples of where determination of cell number is useful include the analysis of growth factor activity, serum batch testing, drug screening, and the determination of the cytostatic potential of anti-cancer compounds in toxicology testing. In such toxicological studies, *in vitro* testing techniques are very useful to evaluate the cytotoxic, mutagenic, and carcinogenic effects of chemical compounds on human cells.

TERMINOLOGY OF CELL PROLIFERATION

Usually, one of two parameters is used to measure the health of cells: cell viability or cell proliferation. In almost all cases, these parameters are measured by assaying for "vital functions" that are characteristic of healthy cells.

Cell Viability

Cell viability can be defined as the number of healthy cells in a sample. Whether the cells are actively dividing or are quiescent is not distinguished. Cell viability assays are often useful when non-dividing cells (such as primary cells) are isolated and maintained in culture to determine optimal culture conditions for cell populations. The most straightforward method for determining viable cell number is a direct counting of the cells in a hemocytometer.

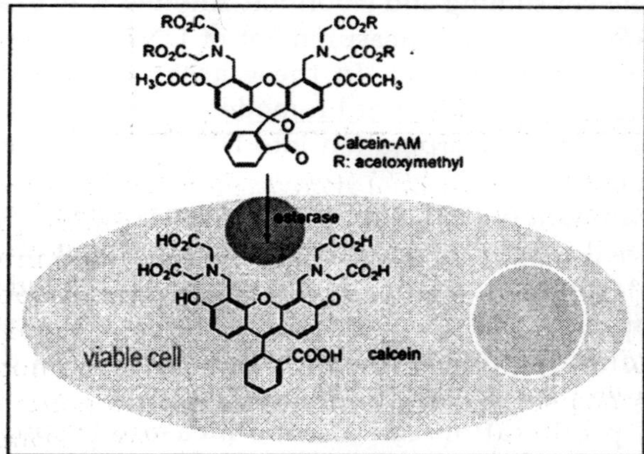

Fig. Cell Viability

Sometimes viable cells are scored based on morphology alone; however, it is more helpful to stain the cells with a dye such as trypan blue. In this case, viability is measured by the ability of cells with uncompromised membrane integrity to exclude the dye. Alternatively, metabolic activity can be assayed as an indication of cell viability. Usually metabolic activity is measured in populations of cells by incubating the cells with a tetrazolium salt (MTT, XTT, WST-1) that is cleaved into a colored formazan product by metabolic activity.

Cell Proliferation

Cell proliferation is the measurement of the number of cells that are dividing in a culture. One way of measuring this parameter is by performing clonogenic assays. In these assays, a defined number of cells are plated onto the appropriate matrix and the number of colonies that are formed after a period of growth are enumerated.

Drawbacks to this type of technique are that it is tedious and it is not practical for large numbers of samples. In addition, if cells divide only a few times and then become quiescent, colonies may be too small to be counted and the number of dividing cells may be underestimated.

Alternatively, growth curves could be established, which is also time-consuming and laborious. Another way to analyse cell proliferation is the measurement of DNA synthesis as a marker for proliferation. In these assays, labeled DNA precursors (3H-thymidine or bromodeoxyuridine) are added to cells and their incorporation into DNA is quantified after incubation.

The amount of labeled precursor incorporated into DNA is quantified either by measuring the total amount of labeled DNA in a population, or by detecting the labeled nuclei microscopically. Incorporation of the labeled precursor into DNA is directly proportional to the amount of cell division occurring in the culture.

Cell proliferation can also be measured using more indirect parameters. In these techniques, molecules that regulate the cell cycle are measured either by their activity (e.g., CDK kinase assays) or by quantifying their amounts (e.g., Western blots, ELISA, or immunohisto-chemistry).

CELL CYCLE

In an organism, the rate of cell division is a tightly regulated process that is intimately associated with growth, differentiation and tissue turnover.

Generally, cells do not undergo division unless they receive signals that instruct them to enter the active segments of the cell cycle. Resting cells are said to be in the G0 phase (quiescence) of the cell cycle. The signals that induce cells to divide are diverse and trigger a large number of signal transduction cascades.

A thorough discussion of the types of signals and the

variety of responses they can elicit are beyond the scope of this guide. Generally, signals that direct cells to enter the cell cycle are called growth factors, cytokines, or mitogens.

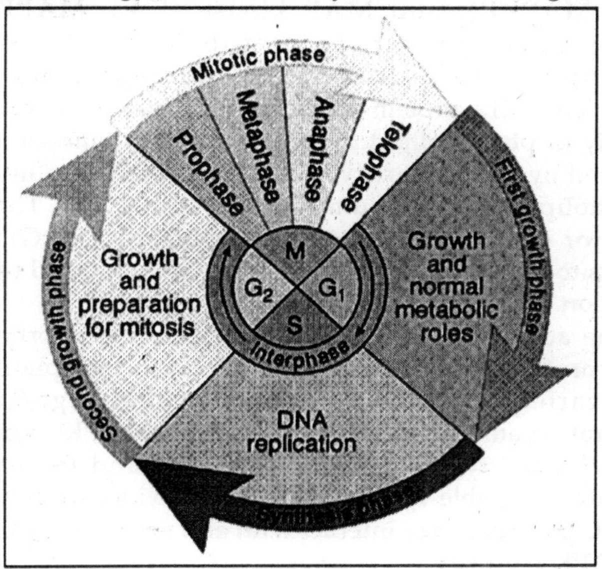

Fig. Cell Cycle

Signal Transduction Pathways

Three major types of signal transduction pathways are activated in cells in response to growth factors or mitogenic stimuli. The response to these stimuli varies from cell type to cell type and the pathways continue to grow more and more complex. These types of pathways continue to be the focus of a great deal of research and, considering the importance of cell cycle regulation in biology, the pathways will continue to grow in complexity for some time to come.

The MAP kinase (MAPK) type of pathways are triggered through a cascade of phosphorylation events that begins with a growth factor binding to a tyrosine kinase receptor at the cell surface. This causes dimerization of the receptor and an intermolecular cross-phosphorylation of the two receptor molecules. The phosphorylated receptors then interact with adaptor molecules that trigger downstream events in the cascade. The cascade works through the GTP exchange protein RAS, the protein kinase RAF (MAPKKK), the protein kinase MEK (MAPKK), and MAP kinase (Erk). MAPK then phosphorylates a variety of substrates that control transcription, the cell cycle, or rearrangements of the cytoskeleton. The protein kinase C (PKC) pathways consist of a family of phospholipid dependent protein kinases. PKC is regulated by a large variety of metabolic pathways involving phospholipids and calcium levels within a cell. The main regulator of the pathway is diacylglycerol (DAG) which appears to recruit PKC to the plasma membrane and cause its activation.

The activity of DAG is mimicked by the phorbol-ester tumor promoters. Once activated, PKC can phosphorylate a wide variety of cellular substrates that regulate cell proliferation and differentiation. Responses to PKC appear to vary with the types of PKCs expressed and the types of substrates available within a cell. Some evidence shows that the PKC pathway may interact with and exert effects through the MAPK pathway.

The JAK/STAT pathway is activated by cytokine interaction with a family of receptors called the cytokine receptor superfamily. These receptors do not contain a protein kinase domain themselves, but they associate with and activate a family of protein kinases called the JAK (Just Another Kinase or JAnus Kinase) family.

JAK family members are recruited to receptor complexes that are formed as a result of ligand binding. The high concentration of JAK in the complex leads to a cross-phosphorylation of JAK and thus activation. JAK then phosphorylates members of another protein family called STAT (signal transducers and activators of transcription). These proteins then translocate to the nucleus and directly modulate transcription.

Control of the Cell Cycle

Once the cell is instructed to divide, it enters the active

phase of the cell cycle, which can be broken down into four segments:

- During G_1 (G = gap), the cell prepares to synthesize DNA. In the latter stages of G_1, the cell passes through a restriction point (R) and is then committed to complete the cycle.
- During S phase the cell undergoes DNA synthesis and replicates its genome.
- During G_2 the cell prepares to undergo division and checks its replication using DNA repair enzymes.
- During M phase, the cell undergoes division by mitosis or meiosis and then re-enters G_1 or G_0.

In most instances, the decision for a cell to undergo division is regulated by the passage of a cell from G_1 to S phase. Progression through the cell cycle is controlled by a group of kinases called cyclin-dependent kinases (CDKs). CDKs are thought to phosphorylate cellular substrates, such as the retinoblastoma gene, that are responsible for progression into each of the phases of the cell cycle. CDKs are activated by associating with proteins whose levels of expression change during different phases of the cell cycle. These proteins are called cyclins. Once associated with cyclins, CDKs are activated by phosphorylation via CDK-activating kinase (CAKs) or by dephosphorylation via a phosphatase called CDC_{25}.

D-types cyclins are the primary cyclins that respond to external cellular factors. Their levels start off low during G_1 and increase towards the G_1/S boundary. Cyclin D regulates CDK_4 and CDK_6. Cyclin E is expressed transiently during the G_1/S transition and is rapidly degraded once the cell enters S. Cyclin E regulates CDK_2 and perhaps CDK_3. When S phase begins, levels of cyclin A increase and activate CDK_2. The cyclin A/CDK_2 complex is thought to have a direct role in DNA replication. The progression through mitosis is regulated by the presence of cyclin B. Cyclin B associates with CDC_2 and forms the primary kinase present during mitosis (MPF = "M-phase/maturation promoting factor"). During anaphase cyclin B is degraded. This degradation of cyclin B appears to regulate the cell's progression out of mitosis and into G_1.

Chapter 24

Cellular Differentiation

In developmental biology, cellular differentiation is the process by which a less specialized cell becomes a more specialized cell type. Differentiation occurs numerous times during the development of a multicellular organism as the organism changes from a single zygote to a complex system of tissues and cell types. Differentiation is a common process in adults as well: adult stem cells divide and create fully-differentiated daughter cells during tissue repair and during normal cell turnover.

Cell differentiation causes its size, shape, polarity, metabolic activity, and responsiveness to signals to change dramatically. These changes are largely due to highly-controlled modifications in gene expression.

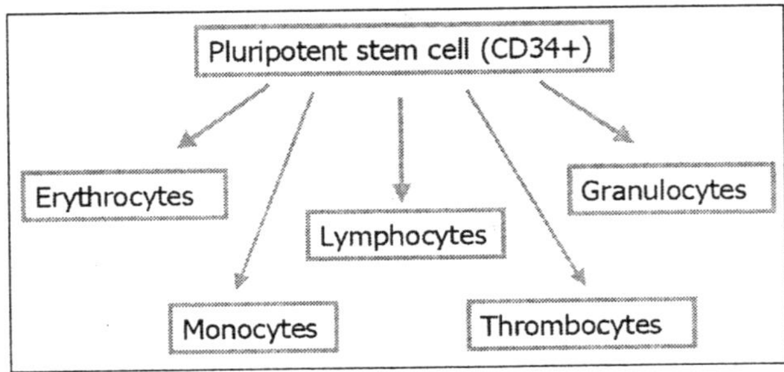

Fig. Pluripotent

With a few exceptions, cellular differentiation almost never involves a change in the DNA sequence itself. Thus, different cells can have very different physical characteristics despite having the same genome. A cell that is able to differentiate into many cell types is known as pluripotent. These cells are called stem cells in animals and meristematic cells in higher plants. A cell that is able to differentiate into all cell types is known as *totipotent*. In mammals, only the zygote and early embryonic cells are *totipotent,* while in plants, many differentiated cells can become totipotent with simple laboratory techniques.

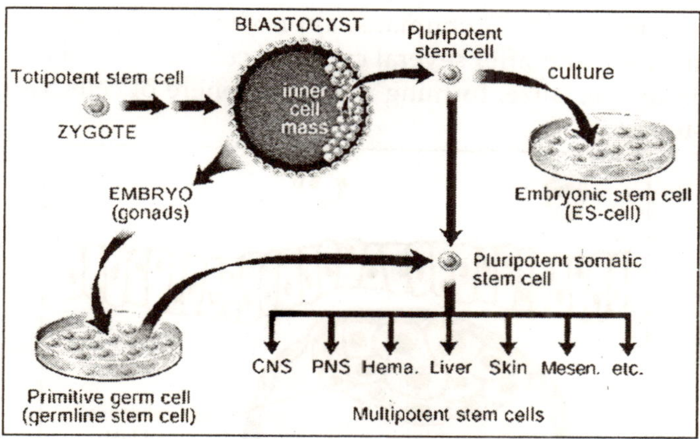

Fig. Totipotent

In cytopathology the level of cellular differentiation is used as a measure of cancer progression. "Grade" is a marker of how differentiated a cell in a tumor is.

Mammalian Cell Types

Three basic categories of cells make up the mammalian body: germ cells, somatic cells, and stem cells. Each of the approximately 100,000,000,000,000 (10^{14}) cells in an adult human has its own copy or copies of the genome except certain cell types, such as red blood cells, that lack nuclei in their fully differentiated state. Most cells are diploid; they have two copies of each chromosome. Such cells, called somatic cells, make up most of the human body, such as skin and muscle cells.

Germ line cells are any line of cells that give rise to gametes—eggs and sperm—and thus are continuous through the generations. Stem cells, on the other hand, have the ability to divide for indefinite periods and to give rise to specialized cells. They are best described in the context of normal human development.

Development begins when a sperm fertilizes an egg and creates a single cell that has the potential to form an entire organism. In the first hours after fertilization, this cell divides into identical cells. In humans, approximately four days after fertilization and after several cycles of cell division, these cells begin to specialize, forming a hollow sphere of cells, called a blastocyst.

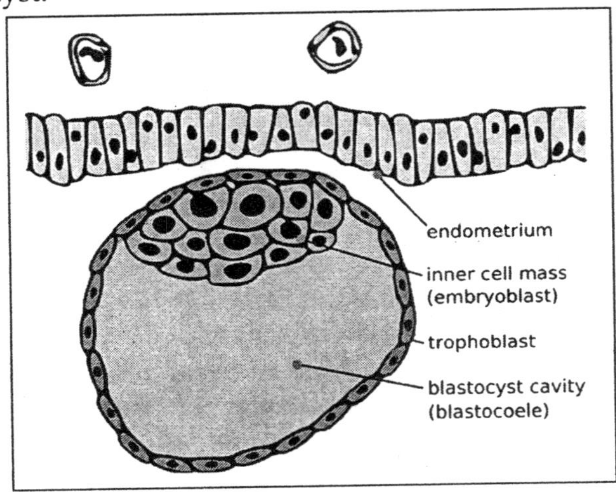

Fig. Blastocyst

The blastocyst has an outer layer of cells, and inside this hollow sphere, there is a cluster of cells called the inner cell mass. The cells of the inner cell mass will go on to form virtually all of the tissues of the human body. Although the cells of the inner cell mass can form virtually every type of cell found in the human body, they cannot form an organism. These cells are referred to as pluripotent.

Pluripotent stem cells undergo further specialization into multipotent progenitor cells that then give rise to functional cells. Examples of stem and progenitor cells include:

- *Hematopoietic stem cells* (adult stem cells) from the

bone marrow that give rise to red blood cells, white blood cells, and platelets.

- *Mesenchymal stem cells* (adult stem cells) from the bone marrow that give rise to stromal cells, fat cells, and types of bone cells.
- *Epithelial stem cells* (progenitor cells) that give rise to the various types of skin cells.
- *Muscle satellite cells* (progenitor cells) that contribute to differentiated muscle tissue.

DEDIFFERENTIATION

Dedifferentiation is a cellular process often seen in lower life forms such as worms and amphibians in which a partially or terminally differentiated cell reverts to an earlier developmental stage, usually as part of a regenerative process. Dedifferentiation also occurs in plants· Cells in cell culture can lose properties they originally had, such as protein expression, or change shape. This process is also termed dedifferentiation.

Some believe dedifferentiation is an aberration of the normal development cycle that results in cancer, whereas others believe it to be a natural part of the immune response lost by humans at some point as a result of evolution. A small molecule dubbed reversine, a purine analog, has been discovered that has proven to induce dedifferentiation in myotubes. These dedifferentiated cells were then able to redifferentiate into osteoblasts and adipocytes.

MECHANISMS

Each specialized cell type in an organism expresses a subset of all the genes that constitute the genome of that species. Each cell type is defined by its particular pattern of regulated gene expression. Cell differentiation is thus a transition of a cell from one cell type to another and it involves a switch from one pattern of gene expression to another. Cellular differentiation during development can be understood as the result of a gene regulatory network.

A regulatory gene and its cis-regulatory modules are nodes in a gene regulatory network; they receive input and create output elsewhere in the network. The systems biology approach to developmental biology emphasizes the importance of investigating how developmental mechanisms interact to produce predictable patterns (morphogenesis).

A few evolutionarily conserved types of molecular processes are often involved in the cellular mechanisms that control these switches. The major types of molecular processes that control cellular differentiation involve cell signaling. Many of the signal molecules that convey information from cell to cell during the control of cellular differentiation are called growth factors. Although the details of specific signal transduction pathways vary, these pathways often share the following general steps.

A ligand produced by one cell binds to a receptor in the extracellular region of another cell, inducing a conformational change in the receptor. The shape of the cytoplasmic domain of the receptor changes, and the receptor acquires enzymatic activity. The receptor then catalyzes reactions that phosphorylate other proteins, activating them.

A cascade of phosphorylation reactions eventually activates a dormant transcription factor or cytoskeletal protein, thus contributing to the differentiation process in the target cell. Cells and tissues can vary in competence, their ability to respond to external signals.

Induction refers to cascades of signaling events, during which a cell or tissue signals to another cell or tissue to influence its developmental fate. Yamamoto and Jeffery investigated the role of the lens in eye formation in cave and surface-dwelling fish, a striking example of induction· Through reciprocal transplants, Yamamoto and Jeffery found that the lens vesicle of surface fish can induce other parts of the eye to develop in cave and surface-dwelling fish, while the lens vesicle of the cave-dwelling fish cannot·

Other important mechanisms fall under the category of asymmetric cell divisions, divisions which give rise to daughter cells with distinct developmental fates. Asymmetric cell divisions can occur because of segregation of cytoplasmic determinants or because of signaling. In the former mechanism, distinct daughter cells are created during cytokinesis because of an uneven distribution of regulatory molecules in the parent cell; the distinct cytoplasm that each daughter cell inherits results in a distinct pattern of differentiation for each daughter cell.

A well-studied example of pattern formation by asymmetric divisions is body axis patterning in Drosophila. RNA molecules are an important type of intracellular differentiation control signal. The molecular and genetic basis of asymmetric cell divisions has also been studied in green algae of the genus *Volvox,* a model system for studying how unicellular organisms can evolve into multicellular organisms. In *Volvox carteri,* the 16 cells in the anterior hemisphere of a 32-celled embryo divide asymmetrically, each producing one large and one small daughter cell. The size of the cell at the end of all cell divisions determines whether it will become a specialized germ or somatic cell.

CELL SIGNALING PATHWAYS

Cells use a large number of clearly defined signalling pathways to regulate their activity. In this module, attention is focused on the ON mechanisms responsible for transmitting information into the cell. These signalling pathways fall into two main groups depending on how they are activated. Most of them are activated by external stimuli and function to transfer information from the cell surface to internal effector systems. However, some of the signalling systems respond to information generated from within the cell, usually in the form of metabolic messengers.

For all of these signalling pathways, information is conveyed either through protein–protein interactions or it is transmitted by diffusible elements usually referred to as second messengers. Cells often employ a number of these signalling pathways, and cross-talk between them is an important feature. Attention is focused on the properties of the major intracellular signalling pathways operating in cells to regulate their cellular activity.

During the processes of development, specific cell types

select out those signalling systems that are suitable to control their particular functions. One of the aims of this website is to understand how these unique cell-specific signalsomes function to regulate different mammalian cell types.

Communication through Chemical Signals

Cells are enclosed within a lipophilic plasma membrane, which represents a formidable barrier that has to be crossed by all incoming signals. Hydrophobic hormones, such as the steroid hormones, can simply diffuse across this cell-surface barrier to gain access to protein receptors located in either the cytoplasm or the nucleus. More elaborate mechanisms are required for the water-soluble stimuli (e.g. hormones, neurotransmitters and growth factors) that cannot cross the plasma membrane.

The basic concept of a cell signalling pathway, therefore, concerns the mechanisms responsible for receiving this external information and relaying it through internal cell signalling pathways to activate the sensor and effector mechanisms that bring about a change in cellular responses.

Cell signalling is a dynamic process in that there are on mechanisms during which information flows down the signalling pathway in response to external stimuli opposed by the off mechanisms that are responsible for switching off the signalling system once external stimuli are withdrawn. Some of these basic principles of cell signalling are explored in this introductory module, which will also briefly outline the contents of the other modules.

Most signalling pathways begin with the arrival of external cell stimuli usually in the form of a chemical signal, which is received by receptors at the cell periphery that function as molecular antennae embedded in the plasma membrane. These receptors then function to transfer information to a variety of transducers and amplifiers to produce intracellular messengers. These messengers stimulate the sensors and effectors responsible for activating cellular responses. These ON mechanisms responsible for transmitting information into the cell are counteracted by the OFF mechanisms that switch off this flow of information once stimuli are withdrawn. These cell signalling pathways utilize a variety of information transfer mechanisms such as diffusion, direct protein–protein interactions or covalent modifications, such as protein phosphorylation, acetylation and nitrosylation.

The effectiveness of information transfer is greatly enhanced through the spatial and temporal aspects of cell signalling pathways. Each cell type has a unique repertoire of cell signalling components that will be referred to here as the cellular signalsome.

During the final stages of development, cells express a particular phenotype, and this process of differentiation includes the expression of a distinctive set of signalling components (a cell-type-specific signalsome) required to control their particular functions. These signalsomes have a high degree of plasticity and are constantly being remodelled to cope with changing demands. Abnormal remodelling of cellular signalsomes creates signalling defects that have great significance for the onset of many diseases.

Signalling pathways do not operate in isolation, and a key element of cellular control mechanisms is the extensive cross-talk between signalling pathways. These highly integrated signalling mechanisms act through different effectors (e.g. muscle proteins, secretory vesicles, transcription factors, ion channels and metabolic pathways) to control the activity of cellular processes such as development, proliferation, neural signalling, stress responses and apoptosis.

CELL STIMULI

Cells are sensitive to an enormous variety of stimuli. Many of these are chemical in nature, such as hormones, neurotransmitters and growth factors. However, cells can also detect other modalities, and this is particularly evident in sensory cells that are tuned to stimuli such as mechanical deformation (sound and touch), temperature, light and oxygen tension. One way of describing this diversity is to consider the nature of the stimuli that feed into different signalling pathways. This approach reveals that some signalling mechanisms are used by many different signalling pathways, whereas other pathways respond to a specific set of stimuli. An example of the former is the cyclic AMP signalling pathway, which was the first signalling pathway to be clearly defined.

The major stimuli for this signalling pathway fall into two main classes: neurotransmitters and hormones. They all act by engaging G protein-coupled receptors (GPCRs), which use heterotrimeric GTP-binding proteins (G proteins) to activate the amplifier adenylyl cyclase that converts ATP into the second messenger cyclic AMP. Some of the stimuli that activate this signalling pathway belong to a group of lipid-derived stimuli known as the eicosanoids that include the prostaglandins, thromboxanes and leukotrienes.

The inositol 1,4,5-trisphosphate ($InsP_3$)/diacylglycerol (DAG) signalling pathway is also used by a very large number of stimuli that are mainly neurotransmitters and hormones. These external stimuli bind to GPCRs, which are coupled to G proteins to activate the amplifier phospholipase C (PLC). PLC hydrolyses an inositol lipid to generate the two second messengers, $InsP_3$ and DAG. This signalling pathway is also used by other groups of stimuli such as the growth and survival factors and some of the Wnt stimuli that control development.

The cytokines are a diverse group of stimuli that function mainly in the control of haematopoiesis and immune responses, particularly during inflammation. There are over 40 members of the family that seem to function through two main receptor types.

There are a number of peptides such as atrial natriuretic peptide (ANP), brain-type natriuretic peptide (BNP), C-type natriuretic peptide (CNP) and guanylin that act through the particulate guanylyl cyclases (pGCs).There are a number of stimuli capable of opening ion channels. In this case, the ion channel carries out all the signalling functions.

Most of the stimuli described above arrive at the cell surface through a process of diffusion. In some cases, however, stimuli can be presented to receptors by special molecules. A classic example is the role of MHCII on the antigen-presenting cell that binds fragments of antigen that are then presented to the T cell receptor. Another example is the role of CD14 in presenting lipopolysaccharide (LPS) to the Toll-like receptor (TLR).

Eicosanoids

The eicosanoids are a group of stimuli that are derived from the metabolism of arachidonic acid (AA). AA is a fatty acid that is located on the *sn*-2 position of many phospholipids. The enzyme phospholipase A_2 (PLA_2), which is activated by Ca^{2+} and by the mitogen-activated protein kinase (MAPK) signalling pathway, releases AA, which is then metabolized via two pathways to produce the prostanoids (prostaglandins and thromboxanes) and the leukotrienes.

The first step in the formation of the prostanoids is the enzyme cyclooxygenase (COX) that converts AA into the unstable cyclic endoperoxides prostaglandins PGG_2 and PGH_2. The latter is a precursor that is converted by various isomerases into the prostaglandins PGI_2, PGD_2, PGE_2 and PGF_{2a}, and thromboxane A_2 (TXA_2).

The formation of the leukotrienes begins with the enzyme 5-lipoxygenase (5-LO) that converts AA into the hydroxyperoxide 5-hydroperoxyeicosatetraenoic acid (5-HPETE). The activity of 5-LO depends upon a 5-lipoxygenase-activating protein (FLAP) that may function by presenting AA to 5-LO.

The 5-HPETE is converted into leukotriene A_4 (LTA_4), which is the precursor for two enzymes. An LTC_4 hydrolase converts LTA_4 into LTB_4, whereas an LTC_4 synthase converts LTA_4 into LTC_4. This synthetic step depends upon the conjugation of LTA_4 with glutathione to form LTC_4, which is thus referred to as a cysteinyl-containing leukotriene, as are its derivatives LTD_4 and LTE_4.

These eicosanoids are lipophilic and can thus diffuse out from their cell of origin to act on neighbouring cells. Eicosanoids are the family of G protein-coupled receptors (GPCRs) that detect these eicosanoids. These receptors are connected to either the cyclic AMP signalling pathway or the inositol 1,4,5-trisphosphate ($InsP_3$)/diacylglycerol (DAG) signalling pathway.

Inflammatory cells such as macrophages, mast cells and blood platelets produce large amounts of these eicosanoids during an inflammatory response. In macrophages, there is an increase in the expression of COX that results in an increase in the release of mediators such as platelet-activating factor (PAF) and PGE_2. In blood platelets, PGI_2 acting on its receptor activates cyclic AMP formation, which then inhibits platelet activation.

Phospholipase A_2 (PLA_2)

Phospholipase A_2 (PLA_2) functions to cleave the *sn*-2 ester bond of phospholipids to release the fatty acid to leave behind a lysophospholipid. Both of these products have a role in forming stimuli for cell signalling. If the lysophospholipid is derived from phosphatidylcholine with an alkyl linkage in the *sn*-1 position, it functions as a precursor for an acetyltransferase that converts it into platelet-activating factor (PAF). The free fatty acid that is released from the *sn*-2 position is often arachidonic acid (AA), which is a precursor for the synthesis of the eicosanoids, such as the prostaglandins, thromboxanes and leukotrienes.

There is a large family of PLA_2 enzymes that function to provide these two signalling precursors. Humans have about 15 enzymes that differ with regard to their cellular distribution and how they are activated. Some of the enzymes are secreted ($sPLA_2$).

The cytosolic forms fall into two main groups: the Ca^{2+}–sensitive ($cPLA_2$) and Ca^{2+}–insensitive ($iPLA_2$) groups. In the case of $cPLA_2$, enzyme activity is activated by Ca^{2+} that acts through Ca^{2+}/calmodulin-dependent protein kinase II (CaMKII), which phosphorylates Ser-515 causing the enzyme to translocate to internal membranes such as the endoplasmic reticulum.

Full activation of the enzyme is also dependent upon the mitogen-activated protein kinase (MAPK) signalling pathway.

Mast cells provide an example of how PLA_2 functions to generate cell signalling stimuli.

Steroids

Steroid hormones are hydrophobic stimuli responsible for regulating a number of physiological processes such as reproduction (oestrogen and testosterone), glucose metabolism and stress responses (cortisol) and salt balance (aldosterone). Steroids have two main modes of action, genomic and non-genomic.

The genomic action depends on the fact that the steroids are hydrophobic and thus can pass through the plasma membrane to bind to intracellular receptors, which are transcription factors capable of activating gene transcription. The non-genomic action depends on steroids binding to cell-surface receptors that are then coupled to conventional intracellular signalling pathways.

Aldosterone and cortisol biosynthesis illustrates how steroids are synthesized from cholesterol through a series of reactions carried out by enzymes associated with either the mitochondrion or the smooth endoplasmic reticulum.

Aldosterone and Cortisol Biosynthesis

Aldosterone and cortisol are synthesized in the zona glomerulosa and the zona fasciculata/reticularis cells of the adrenal gland respectively. Like other steroids, they are synthesized in a series of steps carried out by a number of different enzymes:

- CYP11A1 is a side chain cleavage enzyme that initiates the process by cleaving off the side chain of cholesterol to produce pregnenolone.
- 3b-HSD is a short-chain hydroxysteroid dehydrogenase that converts pregnenolone into progesterone. The two synthetic pathways now diverge.
- CYP21A is a steroid 21-hydroxylase that converts progesterone into 11-deoxycorticosterone.
- CYP11B2 (11b-hydroxylase), which is highly expressed in the zona glomerulosa cells, is an

aldosterone synthase that carries out the last three steps of converting 11-deoxycorticosterone into corticosterone, 18-OH-corticosterone and finally aldosterone.

- CYP17 is a steroid 17a-hydroxylase and 17,20-lyase that converts progesterone into 17a-progesterone that begins the synthetic sequence that results in the formation of cortisol.
- CYP21A converts 17a-progesterone into 11-deoxycortisol.
- CYP11B1, which is strongly expressed in the zona fasciculata/reticularis carries out the final step of cortisol synthesis by converting 11-deoxycortisol into cortisol.

Aldosterone

Aldosterone is mainly synthesized and released by the zona glomerulosa cells in the cortical region of the adrenal gland. However, it can also be produced by other cells located in the nervous system, heart, kidney and blood vessels. This extra-adrenal production of aldosterone may be particularly important for tissue repair.

The enzymes responsible for aldosterone and cortisol biosynthesis are located on the mitochondrion and smooth endoplasmic reticulum. One of the primary genomic actions of aldosterone is to regulate Na^+ reabsorption by the distal convoluted tubule (DCT) and the colon.

RECEPTORS

In order to respond to the myriad cell stimuli outlined in the previous section, cells have evolved an equally impressive battery of cell-surface receptors. These receptors have two main functions: they have to detect incoming stimuli and then transmit this information to the internal transducers that initiate the signalling pathway.

Receptors vary enormously in the way they carry out this transfer of information across the plasma membrane. The diverse receptor types can be separated into two main groups depending on the number of membrane-spanning regions they have to embed them into the membrane:

- Single membrane-spanning receptors
  - Protein tyrosine kinase-linked receptors (PTKRs)
  - Serine/threonine kinase-linked receptors (S/TKRs)
  - Particulate guanylyl cyclases (pGCs)
  - Non-enzyme-containing receptors
- Multi-membrane-spanning receptors
  - G protein-coupled receptors (GPCRs)
  - Ion channel receptors

PROTEIN TYROSINE KINASE-LINKED RECEPTORS

The protein tyrosine kinase-linked receptors (PTKRs) are the classical example of single-membrane-spanning receptors that contain an enzymatic activity. There are about 20 subfamilies of these PTKRs. The extracellular domain is responsible for binding the growth or survival factors, whereas the cytosolic region contains the tyrosine kinase domain that is the transducer responsible for initiating the process of signal transduction. One of the characteristics of these receptors is that they can transmit information down a number of cell signalling pathways by assembling a number of transducers and amplifiers. A classical example is the platelet-derived growth factor receptor (PDGFR), which can generate multiple output signals.

Ephrin (Eph) Receptor Signaling

The ephrin (Eph) receptors belong to one of the largest family of protein tyrosine kinase-linked receptors (PTKRs). These receptors function to transfer information between cells that come into contact with each other and are particularly important for spatial patterning during a variety of developmental processes such as axonal guidance, cell morphogenesis and bone cell differentiation. The Eph receptor family is divided into ten A types (EphA1–EphA10) and six B types (EphB1–EphB6).

The stimuli for these Eph receptors are the ephrins, which are expressed on the surface of cells. These ephrins are divided into A and B types determined by their ability to bind to either the EphA or EphB receptors. There are six ephrin-A ligands (ephrin-A1–ephrin-A6), which are attached to the membrane through a glycosyl phosphati-dylinositol (GPI) anchor, and three transmembrane ephrin-B ligands (ephrin-B1–ephrin-B3). The ephrin-B ligands, which have a cytoplasmic domain, are unusual in that they have a dual function.

Not only do they function as a ligand to activate the EphB receptors, but also they function as a receptor in that the cytoplasmic domain can convey information in the reverse direction. The Eph receptor/ephrin complex is a bidirectional signalling system and, through the forward and reverse signalling modes, information can be conveyed to both interacting cells.

Craniofrontonasal syndrome (CFNS) is caused by a loss-of-function mutation of the gene that encodes ephrin-B1. The Eph receptors have an N-terminal ephrin-binding domain followed by a cysteine-rich region and then two fibronectin type III domains. These ectodomains are connected through a typical transmembrane domain to the cytoplasmic domains. There is a relatively long juxtamembrane segment that connects to the kinase domain. At the C-terminal region there is a sterile a-motif (SAM) domain that may participate in receptor dimerization.

Finally there is a PDZ domain-binding motif. When cells approach each other, the ephrin-A ligands bind to the EphA receptors and the ephrin-B ligands bind to the EphB receptors and the resulting interactions induce dimerization as is typical of other PTKRs. In addition to dimers, the Eph receptor/ephrin complex can form higher-order aggregates, and this clustering may be facilitated by interactions between the fibronectin type III domains and the SAM domain. As the receptors are brought together, transphosphorylation by the kinase domains results in the phosphorylation of multiple sites, which then provides the binding motifs to recruit a range of signalling transducers.

One of the main functions of Eph receptor signalling is to modulate the dynamics of cell movement by altering both actin remodelling and cell adhesion. A number of the downstream signalling pathways are thus directed towards the control of actin assembly.

There are subtle differences in the action of EphA and EphB receptors that appear to be adapted to control different cellular processes. One of the functions of EphA receptors in retinal ganglion neurons is to induce growth cone collapse by activating the Rho signalling mechanism. The activated EphA receptor binds to the Rho guanine nucleotide exchange factor (GEF) ephexin, which is responsible for stimulating Rho.

The Rho GTP then activates the Rho kinase (ROK) that stimulates the actin–myosin contractions responsible for collapsing the growth cone. Aggregating platelets communicate with each other through the bidirectional Eph signalling system. Most information is available for the forward signalling pathways initiated by the EphB receptors.

While the EphA receptors are mainly linked to Rho activation, the EphB receptors are coupled to the Rho GEFs kalirin and intersectin that activate Rac and Cdc42 respectively. Cdc42 acts through Wiskott–Aldrich syndrome protein (WASP) and the actin-related protein complex (Arp2/3 complex) to regulate actin assembly. Kalirin, which is found in neuronal dendrites as part of the postsynaptic density (PSD) signalling elements, acts through Rac and the p21-activated kinase (PAK) to control spine morphogenesis.

Serine/Threonine-Kinase Linked Receptors (S/TKRs)

The serine/threonine kinase-linked receptors (S/TKRs) are typical single membrane-spanning receptors. They have an extracellular domain that binds members of the transforming growth factor superfamily. The cytoplasmic domain has a serine/threonine protein kinase region that functions as both a transducer and amplifier in that it is activated by the receptor to phosphorylate the Smads, thus producing many copies of these messengers.

Particulate Guanylyl Cyclases (pGCs)

The particulate guanylyl cyclases (pGCs) are a family of seven single membrane-spanning receptors (pGC-A–pGC-G).

These receptors consist of an extracelllar domain, a short transmembrane segment and an intracellular domain that contains the catalytic guanylyl cyclase (GC) region that converts GTP into cyclic GMP.

The extracellular domain binds a range of peptides such as atrial natriuretic peptide (ANP), brain-type natriuretic peptide (BNP), C-type natriuretic peptide (CNP) and guanylin. These peptides act through different pGC receptor types. For example, ANP and BNP act through pGC-A, whereas guanylin and uroguanylin, which are released within the intestine, act on pGC-C to stimulate intestinal secretion. For these receptors, the guanylyl cyclase region of the cytoplasmic domain functions both as a transducer and as an amplifier.

The remaining receptor isoforms pGC-B, -D, -E, -F and -G are orphan receptors, as there are no known stimuli. pGC-E and pGC-F are expressed in photoreceptors in the eye, where they function to produce cyclic GMP in phototransduction.

NON-ENZYME-CONTAINING RECEPTORS

There are a number of membrane-spanning receptors that lack any enzyme activity, and the role of signal transduction is carried out by binding various transducers and amplifiers during receptor activation as occurs during cytokine signalling. The large number of Type I and Type II cytokines act through a relatively simple signalling pathway. When they bind to the receptor, information is transferred to the associated Janus kinases (JAKs), which are tyrosine kinases that double up as transducers and amplifiers.

They phosphorylate the signal transducers and activators of transcription (STATs), which are transcription factors that act as the messenger to carry information into the nucleus.

Tumour necrosis factor (TNF) and related cytokines act on a trimeric receptor which is more versatile in that it recruits a number of transducers and amplifiers to relay information to different signalling pathways:

- Pro-caspase 8 is converted into caspase 8, which functions as an initiator caspase in the extrinsic pathway of apoptosis

- The sphingomyelinase (SMase) enzymes are recruited to the active receptor to initiate the sphingomyelin signalling pathway by producing the second messenger ceramide
- The nuclear factor kB (NF-kB) signalling pathway is initiated by binding a complex of scaffolding proteins and enzymes [e.g. inhibitory kB kinase (IKK)], which constitute a transducing complex that activates the transcription factor NF-kB

Integrin signalling is another example of a signalling system based on membrane-spanning receptors that lack enzyme activity.

Integrin Signaling

The integrins are cell-surface receptors that function in cell adhesion either to the extracellular matrix (ECM) or to specific cell-surface ligands during cell–cell interactions. Integrins have two main functions.

Firstly, they provide a link between the internal cytoskeleton and the ECM as occurs at focal adhesion complexes and podosomes. Secondly, they have a very important signalling function that is unusual because they can signal in both directions.

In the conventional *outside-in* mode, external stimuli activate the integrin receptor, which then engages various transducing elements to transmit signals into the cell. In the *inside-out* mode, intracellular signals coming from other receptors, often growth factor receptors, induce a conformational change in the integrin receptors that greatly enhance their affinity for their external ligands.

An example of inside-out signalling is found in osteoclasts, where the colony-stimulating factor-1 receptor (CSF-1R) functions to sensitize the integrin receptors.

Integrins are composed of transmembrane a and b-subunits that come together to form a functional heterodimer. There are 18 a-subunits and eight b-subunits that can combine to form 24 different heterodimers that have their own binding specificities and characteristic expression patterns. The following examples illustrate the expression of different combinations in specific cell types:

- a_2/b_1 and $aIIb/b_3$ are expressed in blood platelets
- a_v/b_3 is expressed in osteoclasts and functions in bone resorption
- a_L/b_2, which is also known as lymphocyte function-associated antigen-1 (LFA1), is expressed on lymphocytes and contributes to the immunological synapse
- a_4/b_7, which is also known as very late antigen 4 (VLA4), is expressed on lymphocytes
- a_7/b_1 is expressed in cardiac cells, where it functions in the link between the sarcolemma and the sarcomeres

The integrins have large extracellular regions that interact with specific sequences on the extracellular matrix proteins (cell–matrix interactions) or specific cell-surface ligands (cell–cell interactions). The intracellular cytoplasmic domain is relatively short (40–70 amino acids) and lacks enzyme activity. There is considerable information on integrin structure and the dramatic conformational changes that occur during the formation of adhesion complexes.

The structure of the a_L subunit illustrates the large cytoplasmic head that consists of a series of domains beginning with an I domain followed by the b-propeller region and then three domains that end in the transmembrane domain. One of the linkers between these three domains is known as *genu* (knee), because it is the region where the molecule bends over when it assumes the low-affinity state. Finally, there is a short cytoplasmic domain, which has a GFFKR hinge motif and the KK motif that binds to RAPL.

The b_2 begins with an I-like domain followed by a hybrid domain, which is linked through a *genu* to four integrin epidermal growth factor (I-EGF) repeats. The latter are connected to a b-tail, which connects to the transmembrane region and the cytoplasmic tail region. The latter has a hinge region and a NPXF motif that binds to talin. External stimuli such as intercellular adhesion molecule (ICAM) in the case of the a_L/b_2 integrin in lymphocytes, bind to the I domain of the a_L subunit through a reaction that depends upon Mg^{2+} occupying the metal-ion-dependent adhesion site (MIDAS). The molecule then undergoes the large conformational changes that bring about the changes in affinity.

In order to transmit information into the cell, the relatively short cytoplasmic domains of the integrin receptors function as a signalling platform to assemble various transducing elements. An important component of the transducing mechanism are kinases such as the integrin-linked kinase (ILK) and focal adhesion kinase (FAK), which can relay information down conventional signalling pathways such as the mitogen-activated protein kinase (MAPK) signalling pathway, the PtdIns 3-kinase signalling pathway, the inositol 1,4,5-trisphosphate ($InsP_3$)/Ca^{2+} signalling cassette and the signalling pathways activated by the monomeric G proteins (Rho, Rac and Cdc42) that are critical for remodelling the actin cytoskeleton.

More detailed information on the signalling and skeletal functions of integrins is presented in the following sections:

- Blood platelet aggregation
- Cardiac cell contractile and cytoskeletal elements
- Osteoclast activation in bone resorption
- Focal adhesion complex formation

Glanzmann's thrombasthenia is a bleeding disorder that has been linked to mutations in the b_3 integrin subunit.

G PROTEIN-COUPLED RECEPTORS (GPCRS)

The G protein-coupled receptors (GPCRs) represent a very large superfamily of receptors that are capable of responding to an enormous number and variety of extracellular stimuli (light, odorants, neurotransmitters, hormones and proteases). As their name implies, they are coupled to the heterotrimeric G proteins that function as the transducers to relay information to different signalling pathways such as the cyclic AMP signalling pathway and the inositol 1,4,5-trisphosphate ($InsP_3$)/diacylglycerol (DAG) signalling pathway.

The GPCRs are characterized by having seven-membrane-spanning regions with the N-terminus facing the outside and the C-terminus lying in the cytoplasm. The external ligands, which usually bind to a pocket formed by the external regions of some of the transmembrane domains, induce a conformational change in the receptor that is then transmitted through the membrane to activate the GTP-binding proteins (G proteins).

These G proteins fall into two main groups: the heterotrimeric G proteins and the monomeric G proteins. It is the heterotrimeric G proteins that are the main transducers responsible for transferring information from the GPCRs to a number of signalling pathways.

The receptor activates the G proteins by functioning as a guanine nucleotide exchange factor (GEF) to induce the exchange of GDP for GTP. When the G protein is bound to GTP it activates a variety of downstream effectors including adenylyl cyclase and phospholipase C.

Ca^{2+}-Sensing Receptor (CaR)

The Ca^{2+}-sensing receptor (CaR) belongs to the family of G protein-coupled receptors (GPCRs). The primary function of the CaR is to regulate parathyroid hormone (PTH) synthesis and release, but it is also expressed in many other cell types, such as bone cells, neurons, intestine, kidney, skin, pancreas and heart. It is a typical GPCR with the usual seven transmembrane domains, a large extracellular N-terminal domain and a C-terminal domain of 216 amino acids, some of which have potential phosphorylation sites for protein kinase C (PKC) and protein kinase A (PKA).

The CaR operates as a dimer, with the two subunits linked together by two disulfide bonds. In the parathyroid gland, the CaR dimers are located on caveolae where they appear to be linked to both caveolin and the scaffolding protein filamin.

In addition to responding to Ca^{2+}, CaR is also sensitive to a number of other agonists that fall into three main groups. The first group consists of related inorganic ions (Mg^{2+} and Gd^{3+}) or organic polycations (neomycin and spermine) that act in much the same way as Ca^{2+} to directly activate the receptor. The other two groups function indirectly as allosteric regulators that alter the affinity of the receptor, either positively (calcimimetics) or negatively (calcilytics).

In the parathyroid gland, CaR functions as a 'calciostat' in that it is very sensitive to small fluctuations in the plasma level of Ca^{2+}. It has the potential of relaying information to the parathyroid cell through different signalling pathways. The main mechanism appears to be through the inositol 1,4,5-trisphosphate ($InsP_3$)/Ca^{2+} signalling cassette.

It may also act to inhibit the Ca^{2+}-inhibitable isoform of adenylyl cyclase (AC) via a pertussis-insensitive G protein (G_i). When expressed in other cell types, CaR has also been found to activate the mitogen-activated protein kinase (MAPK) signalling pathway and phospholipase A_2 (PLA_2). Various diseases characterized by an alteration in Ca^{2+} homoeostasis have been linked to inherited mutations in the CaR:

- Familial hypocalciuretic hypercalcaemia (FHH)
- Neonatal severe hyperparathyroidism (NSHPT)
- Autosomal dominant hypocalcaemia (ADH)

ION CHANNEL RECEPTORS

Ion channels play a crucial role in many aspects of cell signalling. One of their most obvious roles is to function as receptors for a number of external stimuli. These receptors are multifunctional in that they detect the incoming stimulus, they transduce the information into channel opening and, by virtue of conducting large amounts of charge, they markedly amplify the signal. Such amplification is the reason such channel receptors are such effective transducers of sensory information.

Ion Channels describes the properties of these ion channel receptors and it also describes how some ion channels are important effectors for different intracellular messengers. Their sensitivity to cyclic nucleotides such as cyclic AMP and cyclic GMP is a critical component of sensory transduction.

TRANSDUCERS AND AMPLIFIERS

Transducers and amplifiers are considered together

because their activities are intimately connected and sometimes the two functions reside in the same molecule. The transducers and amplifiers are connected to the receptors, where they receive information coming in from the outside and transform it into the internal messengers.

There are many different mechanisms of information transduction. One of the classical mechanisms is found for the G protein-coupled receptors (GPCRs) that use heterotrimeric G proteins to relay information to amplifiers such as adenylyl cyclase or phospholipase C (PLC).

The monomeric G proteins also function as transducers for the tyrosine kinase-linked receptors. In this case, the tyrosine kinase associated with the receptor functions together with G proteins such as Son-of-sevenless (SoS) to carry out the transduction event. For receptors that have serine/threonine kinase activity, the enzyme functions as both a transducer and an amplifier. In the case of those receptors that lack enzymatic activity, the transducers and amplifiers are drawn on to the receptor as occurs for the cytokine receptors and for the Hedgehog and Frizzled receptors.

Finally, the ion channel receptors are an example where a single protein combines the function of receptor, transducer and amplifier.These different transducers and amplifiers produce many different intracellular messengers that carry information to the internal sensors and effectors.

INTRACELLULAR MESSENGERS

The role of intracellular messengers is to carry information generated at the cell surface to the internal sensors and effectors. These messengers can take many forms. The concept of an internal messenger first emerged in the cyclic AMP signalling pathway where the external stimulus was considered to be the first messenger, whereas the cyclic AMP formed during information transduction was referred to as the second messenger.

However, the term 'second messenger' can be confusing because there are examples where there are additional messengers within a signalling pathway. Therefore, to avoid confusion, the term 'intracellular messenger' will be used to refer to the agents that carry information within the cell.

Intracellular messengers can take many different forms:

- Ca^{2+} is one of the major intracellular messengers that transmits information for the Ca^{2+} signalling pathways.
- Cyclic AMP that functions in the cyclic AMP signalling pathway.
- Ca^{2+}-mobilizing second messengers:
  - Inositol 1,4,5-trisphosphate ($InsP_3$) that functions as a messenger linking receptors to the Ca^{2+} signalling pathway.
  - Cyclic ADP-ribose (cADPR) that releases Ca^{2+} from internal stores.
  - Nicotinic acid–adenine dinucleotide phosphate (NAADP).
- Lipid messengers:
  - Diacylglycerol (DAG) is a messenger in the phosphoinositide signalling pathway that links receptor activation to protein kinase C and protein phosphorylation.
  - $PtdIns3,4,5P_3$ is the lipid messenger that functions in the PtdIns 3-kinase signalling pathway.
  - $PtdIns4,5P_2$ is the lipid messenger that functions as a membrane messenger to activate many cellular processes.
  - Phosphatidic acid (PA) is a messenger for the phospholipase D signalling pathway.
  - Ceramide functions in the sphingomyelin signalling pathway.
- Protein kinase messengers. Certain protein kinases can carry information to different regions of the cell:
  - The mitogen-activated protein kinase (MAPK) signalling pathway generates active kinases such as extracellular-signal-regulated kinase 1/2 (ERK1/2) and c-Jun N-terminal kinase (JNK) that carry information into the cell and very often into the nucleus.

- Transcription factors that are activated at the cell surface or within the cytoplasm function as messengers carrying information into the nucleus:
  - Nuclear factor kB (NF-kB) carries information from the cell surface into the cell.
  - The signal transducers and activators of transcription (STATs) transfer information from cell-surface receptors into the nucleus.
  - Smads transfer information from the transforming growth factor b (TGF-b) receptor superfamily into the nucleus.
  - b-Catenin functions as a messenger in the canonical Wnt pathway.
  - GLI1 functions in the Hedgehog signalling pathway

The function of all of these intracellular messengers is to transmit information to the sensors and effectors that are responsible for the final function of the cell signalling pathways to activate a whole host of cellular processes.

Sensors and Effectors

The intracellular messengers that are produced by the different signalling pathways function to regulate cellular processes. Just as the cell has receptors to detect external stimuli, the cell contains internal sensors to detect these intracellular messengers. Typical examples of such sensors are the Ca^{2+}-binding proteins that detect increases in Ca^{2+} and relay this information to different effectors to control processes such as contraction and secretion.

Some sensors are also effectors. For example, there are enzymes that respond to messengers such as cyclic AMP and cyclic GMP that not only detect the messenger, but also carry out various effector functions. While some of these effectors might be relatively simple, consisting of a single downstream effector system, there are more complicated effectors made up of multiple components such as those driving processes such as exocytosis, phagocytosis, actin remodelling and gene transcription.

The activation of these sensors and effectors completes the flow of information down the cell signalling pathways. The operation of such sensors and effectors is described in more detail in Module 4: Sensors and Effectors.

Signalling pathways are composed of the ON mechanisms that generate a flow of information into the cell and the OFF mechanisms that switch off this internal flow of information, enabling cells to recover from stimulation. Module 5: OFF Mechanisms describes how the intracellular messengers and their downstream effectors are inactivated.

The second messengers cyclic AMP and cyclic GMP are inactivated by phosphodiesterases. Inositol 1,4,5-trisphosphate ($InsP_3$) metabolism is carried out by both inositol trisphosphatase and inositol phosphatases. Diacylglycerol (DAG) metabolism also occurs through two enzyme systems, DAG kinase and DAG lipase.

In the case of Ca^{2+} signalling, recovery is carried out by the Ca^{2+} pumps and exchangers that remove Ca^{2+} from the cytoplasm. Many of these second messengers activate downstream effectors through protein phosphorylation, and these activation events are reversed by corresponding protein phosphatases.

Ion Channels

Ion channels have two main signalling functions: either they can generate second messengers or they can function as effectors by responding to such messengers. Their role in signal generation is mainly centred on the Ca^{2+} signalling pathway, which has a large number of Ca^{2+} entry channels and internal Ca^{2+} release channels, both of which contribute to the generation of Ca^{2+} signals.

Ion channels are also important effectors in that they mediate the action of different intracellular signalling pathways. For example, Ca^{2+} carries out some of its signalling functions by acting on Ca^{2+}-sensitive K^+ channels and Ca^{2+}-sensitive Cl^- channels.

The cystic fibrosis transmembrane conductance regulator (CFTR), which conducts anions (Cl^- and HCO_3^-), contributes to the osmotic gradient for the parallel flow of water in various transporting epithelia.

Many of these epithelia also express aquaporins, which are water channels that increase the flux of water during fluid secretion or absorption. The metabolic messenger ATP acts on the ATP-sensitive K^+ (K_{ATP}) channels to modulate membrane potential in many cell types. The cation channel of sperm (CatSper) is sensitive to intracellular pH and has a vital role to play in maintaining sperm hyperactivity prior to fertilization. Finally, there are hyperpolarization-activated cyclic nucleotide-gated (HCN) channels that play an important role in various pacemaker mechanisms both in the heart and nervous system.

A large number of genetic diseases have been traced to mutations in various channels. For example, mutations in the cystic fibrosis transmembrane conductance regulator (CFTR) channel is the cause of cystic fibrosis. Many of the other channelopathies have been linked to changes in the channels that gate either the entry or the release of Ca^{2+}.

ENERGY METABOLISM IN CELLS

Living organisms depend on a steady supply of energy available to do work. It takes energy to produce new cellular materials, to maintain the organization of membranes and organelles, and to fuel movement and active transport. The common energy for use in living systems is chemical energy released when chemical bonds are broken.

Metabolism encompasses all of the chemical reactions within the cell. Catabolic reactions are breakdown reactions that reduce large molecules to smaller components, releasing energy. Digestion and cell respiration are primarily catabolic processes. Anabolic reactions are synthetic reactions that produce larger molecules from small ones, storing energy in chemical bonds.

Photosynthesis and the manufacture of proteins are examples of anabolic processes. Metabolic pathways, both anabolic and catabolic, are chains of reactions involving multiple steps.

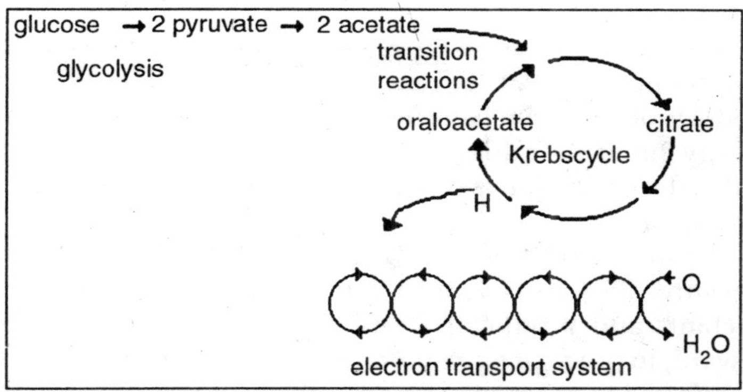

Fig. Energy Metabolism in Cells

Among the most important of metabolic pathways are those involved in glycolysis and cellular respiration — the chain of reactions that allows basic nutrients, such as the simple sugar glucose, to be metabolized to release energy in a form that can be easily used by the cell.

Potential energy is stored in the chemical bonds of molecules. Carbon-carbon and carbon-hydrogen bonds are relatively rich in energy. This energy is released when the chemical bond is broken. Some compounds contain high energy bonds. Notable among theses is adenosine triphosphate [ATP] with its high-energy phosphate bond. The high-energy bond is the one between the last two phosphate groups.

Adenosine Phosphate

This bond has high energy because both of the phosphate groups contain several oxygen atoms, and the electronegative oxygen atoms tend to pull electrons toward themselves. The resulting tension produces a high energy condition. When this bond is broken as the phosphate group is removed, energy is released. Free energy is the energy that can be made available to do work.

Essentially, this is the amount of energy that can be released by breaking chemical bonds in a molecule. When the products of a reaction have less free energy [symbolized by G] than the reactants, energy is released by the reaction. These reactions that release energy are exergonic. oxidation reactions and dephosphorylation reactions are exergonic. When the products of a reaction have more free energy than the reactants, energy must be added. These reactions are endergonic. Reduction reactions and phosphorylation reactions are endergonic.

Under standard conditions [sea-level atmospheric pressure, 25°C, pH 7, and 1 molar concentrations of all reactants and products], exergonic reactions tend to move forward, to produce more product, and to release energy. They are referred to as spontaneous reactions. Endergonic reactions do not progress under standard conditions. They are non-spontaneous.

The change in standard free energy in an exergonic reaction is graphed below. Initially, the energy of the reactants, labeled $G_{reactants}$, is high and the final $G_{products}$, the energy of the products of the reaction is low.

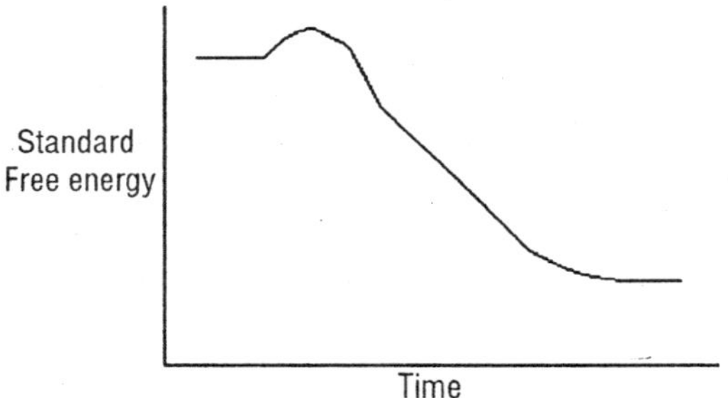

In an endergonic reaction, the graph would look like the one below; the initial $G_{reactants}$ is low and the final $G_{products}$ is high.

Endergonic reactions may be described either by the change in G value [[[Delta]]G] or by the equilibrium constant. [[Delta]]G is the difference between $G_{products}$ and $G_{reactants.}$ Exergonic reactions will have a -[[Delta]]G value; exergonic reactions will have a +[[Delta]]G.

$$[[Delta]]G = G_{products} - G_{reactants}$$

The equilibrium constant [K_{eq}] is the ratio of the concentration of products to the concentration or reactants at equilibrium. [The equilibrium point of a reaction is the point at which it is equally likely to progress forward or to go in reverse. At equilibrium, there is no net reaction.]

K_{eq} = [products] [reactants]

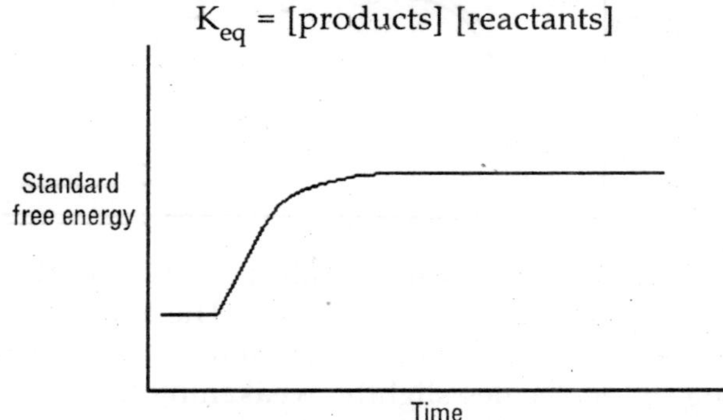

Exergonic reactions, which tend to move forward, will have K_{eq} values greater than 1 [since the reaction will reach equilibrium only when relatively large amounts of product have been formed, and when little reactant remains. Endergonic reactions have K_{eq} values less than one. Endergonic reactions tend to reach equilibrium when the concentration of reactants is still high and the concentration of products is low.

Endergonic reactions are important to many cell processes. Endergonic reactions may be 'pushed' forward by coupling them with exergonic reactions. The energy released from the exergonic reaction can be used to drive the endergonic reaction.

Endergonic reactions may also be 'pulled' forward by removing the products of the reaction as fast as they form, so that the reaction is prevented from reaching its equilibrium point. Chemical reactions, both spontaneous and non-spontaneous, require an initial input of energy, the energy of activation, to begin the reaction. This energy is needed to bring reactant molecules into position and condition for reaction.

In non-living systems, this energy can be supplied by heat. Extreme elevated temperatures are not compatible with living systems; an alternative mechanism for the initiation of chemical reactions in living systems is needed.

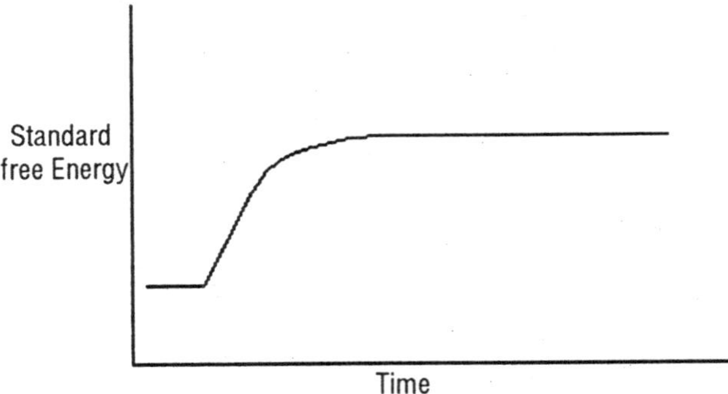

Enzymes serve as organic catalysts that lower the energy of activation, thus facilitating reactions. The enzyme is able to interact with the reactants of a reaction to 'set them up' by distorting the molecules slightly, weakening their bonds and making it more likely that the molecules undergo a chemical change. Enzymes can also act to 'collect' molecules in a local area and to position them in such a way that bonds are easily formed between them.

The convenient 'energy package' that recaptures energy for the cell's use is the molecule ATP [adenosine triphosphate]. The bonds between the phosphate groups in this molecule are high energy bonds that release a standard amount [7.3 kcal/ mole]] of energy when the bond is broken. A basic goal of the respiratory pathways is to release the chemical energy of glucose or other nutrients and recapture that energy in the high energy bonds of ATP molecules.

When glucose is fully oxidized, it is broken down to carbon dioxide and water. The chemical equation for this reaction is:

$$C_6H_{12}O_6 + 6\ O_2 \rightarrow 6\ CO_2 + 6\ H_2O$$

In the process, energy is released. It is possible to measure the energy released by a reaction in an instrument called a calorimeter. A calorimeter measures the amount of heat released when the sugar [or any other substance] is oxidized. When glucose is fully oxidized, 686 kilocalories of energy are released from each mole of glucose. When oxidation occurs in a single step — as it does during combustion, when a substance is burned — the energy is lost as heat and light, and is no longer available to do biological work.

The slow step-wise controlled release of energy during glycolysis and respiration allows a significant proportion of the energy of the sugar to be converted to energy stored in ATP molecules.The process can be broken into four groups of reactions: glycolysis, the transition reactions, the Krebs cycle, and the electron transport system.

Glycolysis is a universal metabolic pathway. All organisms — plant, animal, and bacterial cells — use the same glycolytic pathway. Note that all of the steps in the second column [after the formation of phosphoglyceraldehyde [PGAL] are doubled, since there are two molecules of PGAL formed from one original glucose molecule.

If oxygen is available [and if an organism is aerobic, and able to use oxygen in its respiration], the pyruvate formed during glycolysis continues on through the transition reactions, the Krebs cycle, and the electron transport system. Although oxygen is not involved in the reactions until the end of the electron transport system, the entire series [beginning from pyruvate] is considered to be aerobic or oxygen-dependent.

Aerobic respiration extracts far more usable energy from the glucose molecule than glycolysis alone. If oxygen is not available, the pyruvate formed during glycolysis is sent along a different metabolic pathway. Anaerobic respiration requires no oxygen and produces no additional usable energy, but it does allow glycolysis to continue. Some organisms, such as yeast cells, use alcoholic fermentation as the anaerobic pathway; others, such as human muscle cells, use lactic acid fermentation.

Chapter 25

Basic Principles of Cell Culture

TYPES OF CELL CULTURE

Primary Explantation Versus Disaggregation

When cells are isolated from donor tissue, they may be maintained in a number of different ways. A simple small fragment of tissue that adheres to the growth surface, either spontaneously or aided by mechanical means, a plasma clot, or an extracellular matrix constituent, such as collagen, will usually give rise to an outgrowth of cells. This type of culture is known as a *primary explant*, and the cells migrating out are known as the *outgrowth*.

Cells in the outgrowth are selected, in the first instance, by their ability to migrate from the explant and subsequently, if subcultured, by their ability to proliferate. When a tissue sample is disaggregated, either mechanically or enzymatically the suspension of cells and small aggregates that is generated will contain a proportion of cells capable of attachment to a solid substrate, forming a monolayer.

Those cells within the monolayer that are capable of proliferation will then be selected at the first subculture and, as with the outgrowth from a primary explant, may give rise to a cell line.

Tissue disaggregation is capable of generating larger cultures more rapidly than explant culture, but explant culture may still be preferable where only small fragments of tissue are available or the fragility of the cells precludes survival after disaggregation.

Proliferation Versus Differentiation

Generally, the differentiated cells in a tissue have limited ability to proliferate. Therefore, differentiated cells do not contribute to the formation of a primary culture, unless special conditions are used to promote their attachment and preserve their differentiated status.

Usually it is the proliferating committed precursor compartment of a tissue such as fibroblasts of the dermis or the basal epithelial layer of the epidermis, that gives rise to the bulk of the cells in a primary culture, as, numerically, these cells represent the largest compartment of proliferating, or potentially proliferating, cells.

However, it is now clear that many tissues contain a small population of regenerative cells which, given the correct selective conditions, will also provide a satisfactory primary culture, which may be propagated as stem cells or mature down one of several pathways toward differentiation.

This implies that not only must the correct population of cells be isolated, but the correct conditions must be defined to maintain the cells at an appropriate stage in maturation to retain their proliferative capacity if expansion of the population is required. This was achieved fortuitously in early culture of fibroblasts by the inclusion of serum that contained growth factors, such as plateletderived growth factor (PDGF), that helped to maintain the proliferative precursor phenotype.

However, this was not true of epithelial cells in general, where serum growth factors such as transforming growth factor (TGF) inhibited epithelial proliferation and favored differentiation. It was not until serum-free media were developed that this effect could be minimized and factors positive to epithelial proliferation, such as epidermal growth factor and cholera toxin, used to maximum effect.

Although undifferentiated precursors may give the best opportunity for expansion in vitro, transplantation may require that the cells be differentiated or carry the potential to differentiate. Hence, two sets of conditions may need to be used, one for expansion and one for differentiation. In general, it can be said that differentiation will probably require a selective medium for the cell type, supplemented with factors that favour differentiation, such as retinoids, hydrocortisone, and planar-polar compounds, such as sodium butyrate (NaBt).

In addition, the correct matrix interaction, homotypic and heterotypic cell interaction, and, for epithelial cells, the correct cellular polarity will need to be established, usually by using an organotypic culture. This assumes, of course, that tissue replacement will require the graft to be completely or almost completely differentiated, as is likely to be the case where extensive tissue repair is carried out. However, there is also the option that cell culture will only be required to expand a precursor cell type and the process of implantation itself will then induce differentiation, as appears to be the case with stem cell transplantation.

Organotypic Culture

Dispersed cell cultures clearly lose their histologic characteristics after disaggregation and, although cells within a primary explant may retain some of the histology of the tissue, this will soon be lost because of flattening of the explant with cell migration and some degree of central necrosis due to poor oxygenation. Retention of histologic structure, and its associated differentiated properties, may be enhanced at the air/medium interface, where gas exchange is optimized and cell migration minimized, as distinct from the substrate/medium interface, where dispersed cell cultures and primary outgrowths are maintained.

This so-called *organ culture* will survive for up to 3 weeks, normally, but cannot be propagated. An alternative approach, with particular relevance to tissue engineering, is the amplification of the cell stock by generation of cell lines from specific cell types and their subsequent recombination in *organotypic culture.*

This allows the synthesis of a tissue equivalent or construct on demand for basic studies on cell-cell and cell-matrix interaction and for in vivo implantation. The fidelity of the construct in terms of its real tissue equivalence naturally depends on identification of all the participating cell types in the tissue in vivo and the ability to culture and recombine them in the correct proportions with the correct matrix and juxtaposition.

So far this has worked best for skin but even then, melanocytes have only recently been added to the construct, and islet of Langerhans cells are still absent, as are sweat glands and hair follicles, although some progress has been made in this area.

There are a great many ways in which cells have been recombined to try to simulate tissue, ranging from simply allowing the cells to multilayer by perfusing a monolayer to highly complex perfused membrane or capillary beds. These are termed *histotypic cultures* and aim to attain the density of cells found in the tissue from which the cells were derived.

It is possible, using selective media, cloning, or physical separation methods to isolate purified cell strains from disaggregated tissue or primary culture or at first subculture. These purified cell populations can then be combined in organotypic culture to recreate both the tissue cell density and, hopefully, the cell interactions and matrix generation found in the tissue. Filter well inserts provide the simplest model system to test such recombinants, but there are many other possibilities including porous matrices, perfused membranes, and concentric double microcapillaries.

Substrates and Matrices

Initially, cultures were prepared on glass for ease of observation, but cells may be made to grow on many different charged surfaces including metals and many polymers. Traditionally, a net negative charge was preferred, such as found on acidwashed glass or polystyrene treated by electric ion discharge, but some plastics are also available with a net positive charge (e.g., Falcon Primaria), which is claimed to add some cell selectivity.

In either case, it is unlikely that the cell attaches directly to synthetic substrates and more likely that the cell secretes matrix products that adhere to the substrate and provide ligands for the interaction of matrix receptors such as integrins. Hence it is a logical step to treat the substrate with a matrix product, such as collagen type IV, fibronectin, or laminin, to promote the adhesion of cells that would otherwise not attach.

Suffice it to say at this stage that scaffolds have the same requirements as conventional substrates in terms of low toxicity and ability to promote cell adhesion, often with the additional requirement of a three-dimensional geometry. If the polymer or other material does not have these properties, derivatization and/or matrix coating will be required. Most studies suggest that cell cultivation on a three-dimensional scaffold is essential for promoting orderly regeneration of engineered tissues in vivo and in vitro.

Scaffolds investigated to date vary with respect to material chemistry (e.g., collagen, synthetic polymers), geometry (e.g., gels, fibrous meshes, porous sponges, tubes), structure (e.g., porosity, distribution, orientation, and connectivity of the pores), physical properties (e.g., compressive stiffness, elasticity, conductivity, hydraulic permeability), and degradation (rate, pattern, products).

In general, scaffolds should be made of biocompatible materials, preferentially those already approved for clinical use. Scaffold structure determines the transport of nutrients, metabolites, and regulatory molecules to and from the cells, whereas the scaffold chemistry may have an important role in cell attachment and differentiation. The scaffold should biodegrade at the same rate as the rate of tissue assembly and without toxic or inhibitory products.

Mechanical properties of the scaffold should ideally match those of the native tissue being replaced, and the mechanical integrity should be maintained as long as necessary for the new tissue to mature and integrate.

ISOLATION OF CELLS FOR CULTURE

Tissue Collection and Transportation

The first, and most important, element in the collection of tissue is the cooperation and collaboration of the clinical staff. This is best achieved if a member of the surgical team is also a member of the culture project, but even in the absence of this, time and care must be spent to ensure the sympathy and understanding of those who will provide the clinical material. It is worth preparing a short handout explaining the objectives of the project and spending some time with the person likely to be most closely involved with obtaining samples.

This may be the chief surgeon (who will need to be informed anyway), or it may be a more junior member of the team willing to set up a collaboration, one of the nursing staff, or the pathologist, who may also require part of the tissue. Whoever fulfils this role should be identified and provided with labeled containers of culture medium containing antibiotics, bearing a contact name and phone number for the cell culture laboratory.

A refrigerator should be identified where the containers can be stored, and the label should also state clearly The next step is best carried out by someone from the laboratory collecting the sample personally, but it is also possible to leave instructions for transportation by taxi or courier.

If a third party is involved, it is important to ensure that the container is well protected preferably double wrapped in a sealed polythene bag and an outer padded envelope provided with the name, address, and phone number of the recipient at the laboratory. Refrigeration during transport is not usually necessary, as long as the sample is not allowed to get too warm, but if delivery will take more than an hour or two, then one or two refrigeration packs, such as used in picnic chillers, should be included but kept out of direct contact.

If the tissue sample is quite small, a further tissue sample (any tissue) or a blood sample should be obtained for freezing. This will be used ultimately to corroborate the origin of any cell line that is derived from the sample by DNA profiling. A cell line is the culture that is produced from subculture of the primary, and every additional subculture after this increases the possibility of cross-contamination, so verification of origin is important.

In addition, the possibility of misidentification arises

during routine subculture and after recovery from cryopreservation.

Biosafety and Ethics

All procedures involved in the collection of human material for culture must be passed by the relevant hospital ethics committee. A form will be required for the patient to sign authorizing research use of the tissue, and preferably disclaiming any ownership of any materials derived from the tissue. The form should have a brief layman's description of the objectives of the work and the name of the lead scientist on the project.

The donor should be provided with a copy. All human material should be regarded as potentially infected and treated with caution. Samples should be transported securely in double-wrapped waterproof containers; they and derived cultures should be handled in a Class II biosafety cabinet and all discarded materials autoclaved, incinerated, or chemically disinfected.

Each laboratory will its own biosafety regulations that should be adhered to, and anyone in any doubt about handling procedures should contact the local safety committee (and if there is not one, create it!). Rules and regulations vary among institutions and countries, so it is difficult to generalize, but a good review can be obtained in Caputo.

Record Keeping

When the sample arrives at the laboratory, it should be entered into a record system and assigned a number. This record should contain the details of the donor, identified by hospital number rather than by name, tissue site, and all information regarding collection medium, time in transit, treatment on arrival, primary disaggregation, and culture details, etc.

This information will be important in the comparison of the success of individual cultures, and if a long-term cell line is derived from the culture, this will be the first element in the cell line's provenance, which will be supplemented with each successive manipulation or experimental procedure.

Such records are best maintained in a computer database where each record can be derived from duplication of the previous record with appropriate modifications. There may be issues of data protection and patient confidentiality to be dealt with when obtaining ethical consent.

Disaggregation and Primary Culture

Briefly, the tissue will go through stages of rinsing, dissection, and either mechanical disaggregation or enzymatic digestion in trypsin and/or collagenase. It is often desirable not to have a complete single-cell suspension, and many primary cells survive better in small clusters.

Disaggregated tissue will contain a variety of different cell types, and it may be necessary to go through a separation technique such as density gradient separation or immunosorting by magnetizable beads (MACS), using a positive sort to select cells of interest or a negative sort to eliminate those that are not required or by using fluorescence-activated cell sorting (FACS).

The cell population can then be further enriched by selection of the correct medium (e.g., keratinocyte growth medium (KGM) or MCDB 153 for keratinocytes), many of which are now available commercially and supplementing this with growth factors. Survival and enrichment may be improved in some cases by coating the substrate with gelatin, collagen, laminin, or fibronectin

SUBCULTURE

Frequently, the number of cells obtained at primary culture may be insufficient to create constructs suitable for grafting. Subculture gives the opportunity to expand the cell population, apply further selective pressure with a selective medium, and achieve a higher growth fraction and allows the generation of replicate cultures for characterization, preservation by freezing, and experimentation.

Briefly, subculture involves the dissociation of the cells from each other and the substrate to generate a single-cell suspension that can be quantified. Reseeding this cell suspension at a reduced concentration into a flask or dish generates a secondary culture, which can be grown up and subcultured again to give a tertiary culture, and so on.

In most cases, cultures dedifferentiate during serial passaging but can be induced to redifferentiate by cultivation on a 3D scaffold in the presence of tissue-specific differentiation factors (e.g., growth factors, physical stimuli). However, the cell's ability to redifferentiate decreases with passaging. It is thus essential to determine, for each cell type, source, and application, a suitable number of passages during subculture.

Life Span

Most normal cell lines will undergo a limited number of subcultures, or passages, and are referred to as *finite cell lines*. The limit is determined by the number of doublings that the cell population can go through before it stops growing because of senescence. Senescence is determined by a number of intrinsic factors regulating cell cycle, such as Rb and p53 and is accompanied by shortening of the telomeres on the chromosomes. Once the telomeres reach a critical minimum length, the cell can no longer divide.

Telomere length is maintained by telomerase, which is downregulated in most normal cells except germ cells. It can also be higher in stem cells, allowing them to go through a much greater number of doublings and avoid senescence. Transfection of the telomerase gene hTRT into normal cells with a finite life span allows a small proportion of the cells to become immortal although this probably involves deletion or inactivation of other genes.

Growth Cycle

Each time that a cell line is subcultured it will grow back to the cell density that existed before subculture (within the limits of its finite life span). This process can be described by plotting a growth curve from samples taken at intervals throughout the growth cycle which shows that the cells enter a latent period of no growth, called the *lag period*, immediately after reseeding. This period lasts from a few hours up to 48 h, but is usually around 12–24 h, and allows the cells to recover from trypsinization, reconstruct their cytoskeleton, secrete matrix to aid attachment, and spread out on the substrate, enabling them to reenter cell cycle. They then enter exponential growth in what is known as the *log phase,* during which the cell population doubles over a definable period, known as the *doubling time* and characteristic for each cell line.

As the cell population becomes crowded when all of the substrate is occupied, the cells become packed, spread less on the substrate, and eventually withdraw from the cell cycle. They then enter the *plateau* or *stationary phase,* where the growth fraction drops to close to zero. Some cells may differentiate in this phase; others simply exit the cell cycle into G0 but retain viability.

Cells may be subcultured from plateau, but it is preferable to subculture before plateau is reached, as the growth fraction will be higher and the recovery time (lag period) will be shorter if the cells are harvested from the top end of the log phase. Reduced proliferation in the stationary phase is due partly to reduced spreading at high *cell density* and partly to exhaustion of growth factors in the medium at high *cell concentration.* These two terms are not interchangeable.

Density implies that the cells are attached, and may relate to monolayer density (two-dimensional) or multilayer density (three-dimensional). In each case there are major changes in cell shape, cell surface, and extracellular matrix, all of which will have significant effects on cell proliferation and differentiation. A high density will also limit nutrient perfusion and create local exhaustion of peptide growth factors.

In normal cell populations this leads to a withdrawal from the cycle, whereas in transformed cells, cell cycle arrest is much less effective and the cells tend to enter apoptosis. Cell concentration, as opposed to cell density, will exert its main effect through nutrient and growth factor depletion, but in stirred suspensions cell contactmediated effects are minimal, except where cells are grown as aggregates.

Cell concentration per se, without cell interaction, will not influence proliferation, other than by the effect of nutrient and growth factor depletion. High cell concentrations can also lead to apoptosis in transformed cells in suspension, notably in myelomas and hybridomas, but in the absence of cell contact signaling this is presumably a reflection of nutrient deprivation.

Serial Subculture

Each time the culture is subcultured the growth cycle is repeated. The number of doublings should be recorded with each subculture, simplified by reducing the cell concentration at subculture by a power or two, the so-called *split ratio.* A split ratio of two allows one doubling per passage, four, two doublings, eight, three doublings, and so on. The number of elapsed doublings should be recorded so that the time to senescence can be predicted and new stock prepared from the freezer before the senescence of the existing culture occurs.

CRYOPRESERVATION

If a cell line can be expanded sufficiently, preservation of cells by freezing will allow secure stocks to be maintained without aging and protect them from problems of contamination, incubator failure, or medium and serum crises. Ideally, $1 \times 10^6 \times 10^7$ cells should be frozen in 10 ampoules, but smaller stocks can be used if a surplus is not available.

The normal procedure is to freeze a token stock of one to three ampoules as soon as surplus cells are available, then to expand remaining cultures to confirm the identity of the cells and absence of contamination, and freeze down a seed stock of 10^{-20} ampoules. One ampoule, thawed from this stock, can then be used to generate a using stock. In many cases, there may not be sufficient doublings available to expand the stock as much as this, but it is worth saving some as frozen stock, no matter how little, although survival will tend to decrease below 1×10^6 cells/ml and may not be possible below 1×10^5 cells/ml. Factors favoring good survival after freezing and thawing are:

- High cell density at freezing ($1 \times 10^6 \times 10^7$ cells/ml).
- Presence of a preservative, such as glycerol or dimethyl sulfoxide (DMSO) at 5–10 per cent.

- Slow cooling, 1°C/min, down to 70°C and then rapid transfer to a liquid nitrogen freezer.
- Rapid thawing.
- Slow dilution, ~20-fold, in medium to dilute out the preservative.
- Reseeding at 2- to 5-fold the normal seeding concentration. For example, if cells are frozen at 5×10^6 cells in 1 ml of freezing medium with 10 per cent DMSO and then thawed and diluted 1:20, the cell concentration will still be 2.5×10^5 cells/ml at seeding, higher than the normal seeding concentration for most cell lines, and the DMSO concentration will be reduced to 0.5 per cent, which most cells will tolerate for 24 h.
- Changing medium the following day (or as soon as all the cells have attached) to remove preservative. Where cells are more sensitive to the preservative, they may be centrifuged after slow dilution and resuspended in fresh medium, but this step should be avoided if possible as centrifugation itself may be damaging to freshly thawed cells.

There are differences of opinion regarding some of the conditions for freezing and thawing, for example, whether cells should be chilled when DMSO is added or diluted rapidly on thawing, both to avoid potential DMSO toxicity. In the author's experience, chilling diminishes the effect of the preservative, particularly with glycerol, and rapid dilution reduces survival, probably due to osmotic shock.

Culturing in diluted DMSO after thawing can be a problem for some cell lines if they respond to the differentiating effects of DMSO, for example, myeloid leukemia cells, neuroblastoma cells, and embryonal stem cells; in these cases it is preferable to centrifuge after slow dilution at thawing or use glycerol as a preservative.

CHARACTERIZATION AND VALIDATION

Cross-Contamination

There has been much publicity about the very real risks

of cross-contamination when handling cell lines particularly continuous cell lines.

This is less of a problem with short-term cultures, but the risk remains that if there are other cell lines in use in the laboratory, they can cross-contaminate even a primary culture, or misidentification can arise during subculture or recovery from the freezer. If a laboratory focuses on one particular human cell type, superficial observation of lineage characteristics will be inadequate to ensure the identity of each line cultured. Precautions must be taken to avoid cross-contamination:

- Do not handle more than one cell line at a time, or, if this is impractical, do not have culture vessels and medium bottles for more than one cell line open at one time, and never be tempted to use the same pipette or other device for different cell lines.
- Do not share media or other reagents among different cell lines.
- Do not share media or reagents with other people.
- Ensure that any spillage is mopped up immediately and the area swabbed with 70 per cent alcohol.
- Retain a tissue or blood sample from each donor and confirm the identity of each cell line by DNA profiling:
  - When seed stocks are frozen,
  - Before the cell line is used for experimental work or transplantation.
- Keep a panel of photographs of each cell line, at low and high densities, above the microscope, and consult this regularly when examining cells during maintenance. This is particularly important if cells are handled over an extended period, and by more than one operator.
- If continuous cell lines are in use in the laboratory, handle them after handing other, slower-growing, finite cell lines.

Microbial Contamination

Antibiotics are often used during collection, transportation, and dissection of biopsy samples because of the intrinsic contamination risk of these operations. However, once the primary culture is established, it is desirable to eliminate antibiotics as soon as possible. If the culture grows well, then antibiotics can be removed from the bulk of the stocks at first subculture, retaining one culture in antibiotics as a precaution if necessary.

Antibiotics can lead to lax aseptic technique, can inhibit some eukaryotic cellular processes, and can hide the presence of a microbial contamination.

If a culture is contaminated this must become apparent as soon as possible, either to indicate that the culture should be discarded before it can spread the contamination to other cultures or to indicate that decontamination should be attempted. The latter should only be used as a last resort; decontamination is not always successful and can lead to the development of antibiotic-resistant organisms.

Most bacterial, fungal, and yeast infections are readily detected by regular careful examination with the naked eye (e.g., by a change in the colour of culture medium) and on the microscope.

However, one of the most serious contaminations is mycoplasma, which is not visible by routine microscopy. Any cell culture laboratory should have a mycoplasma screening programme in operation, but those collecting tissue for primary culture are particularly at risk. The precautions that should be observed are as follows:

Treat any new material entering the laboratory from donors or from other laboratories as potentially infected and keep it in quarantine.

Ideally, a separate room should be set aside for receiving samples and imported cultures. If this is not practicable, handle separately from other cultures, preferably last in the day and in a designated hood, and swab the hood down after use with 2 per cent phenolic disinfectant in 70 per cent alcohol. Use a separate incubator, and adhere strictly to the rules given above regarding medium sharing.

Screen new cultures as they arrive, and existing stocks at regular intervals, e.g., once a month. There are a number of tests available, but the most reliable and sensitive are fluorescence microscopy after staining with Hoechst 33258 or PCR with a panmycoplasma primer.

The latter is more sensitive but depends on the availability of PCR technology, whereas the former is easier, will detect any DNA-containing contamination, but requires a fluorescence microscope.

Both techniques are best performed with so-called *indicator cultures*. The test culture is refreshed with antibiotic-free medium and, after 3–5 days, the medium is transferred to an antibiotic-free, 10 per cent confluent indicator culture of a cell line such as 3T6 or A549 cells, which are well spread and known to support mycoplasma growth.

After a further 3 days (the indicator cells must not be allowed to reach confluence) the indicator culture is fixed in acetic methanol and stained with Hoechst 33258, or harvested by scraping for PCR (trypsinization may remove the mycoplasma from the cell surface).

Discard all contaminated cultures. If the culture is irreplaceable, decontamination may be attempted (under strict quarantine conditions) with agents such as Mycoplasma Removal Agent (MRA), ciprofloxacin, or BM-cycline. Briefly, the culture is rinsed thoroughly, trypsinized (wash by centrifuging three times after trypsinization), and subcultured into antibiotic-containing medium.

This procedure should be repeated for three subcultures and then the culture should be grown up antibioticfree and tested after one, two, and four further antibiotic-free subcultures, whereupon the culture reenters the routine mycoplasma screening programme.

Characterization

Most laboratories will have, as an integral part of the research programme, procedures in place for the characterization of new cultures.

Species identification (in case the cells are misidentified or cross-contaminated) will be unnecessary if DNA profiling is being used to confirm cell line identity; otherwise, chromosome analysis or isoenzyme electrophoresis can be used. The lineage or tissue or origin can be determined by using antibodies to intermediate filament proteins, for example, cytokeratins for epithelial cells, vimentin for mesodermal cells such as fibroblasts, endothelium, and myoblasts, desmin for myocytes, neurofilament protein for neuronal and some neuroendocrine cells, and glial fibrillary acidic protein for astrocytes.

Some cell surface markers are also lineage specific, for example, EMA in epithelial cells, A2B5 in glial cells, PECAM-1 in endothelial cells, and N-CAM in neural cells, and have the additional advantage that they can be used in cell sorting by magnetic sorting or flow cytometry.

Morphology can also be used, but can be ambivalent as similarities can exist between cells of very different origins. Spontaneous transformation is unlikely in normal cells of human origin, but indicators are a more refractile appearance under phase contrast with a lower cytoplasmic/nuclear ratio, piling up of the cells and loss of contact inhibition and density limitation of growth, increased clonogenicity in agar, and the ability to form tumors in immune-deprived hosts, such as the Nude or SCID mouse.

Where transformation is detected, it is more likely to be due to cross-contamination, although it is possible that the tissue sample may have contained some preneoplastic cells that have then progressed in culture.

Differentiation

As stated above, a prerequisite for sustained growth in culture is the ability for the cells to proliferate, and this may preclude differentiation. If differentiation is required, then it is generally necessary for the cells to withdraw from the cell cycle. This can be achieved by removing, or changing, the growth factor supplementation.

The O_2 A common precursor of astrocytes and oligodendrocytes remains as a proliferating precursor cell in PDGF and bFGF, whereas combining bFGF with ciliary neurotropic factor (CNTF) results in differentiation into a type

2 astrocyte and embryonal stem cells, which remain as proliferating primitive cells in the presence of leukemia inhibitory factor (LIF), will differentiate in the absence of LIF and in the presence of a positively acting factor such as phorbol myristate acetate (PMA, also known as TPA). There are four main parameters governing the entry of cells into differentiation:

Soluble factors such as growth factors (e.g., EGF, KGF, TGF-^{2} and HGF, NGF), cytokines (IL-6, oncostatin-M, GM-CSF, interferons), vitamins (e.g., retinoids, vitamin D3, and vitamin K) and calcium and planar polar compounds (e.g., DMSO and NaBt).

Interaction with matrix constituents such as collagen IV, laminin, and proteoglycans. Heparan sulfate proteoglycans (HSPGs), in particular, have a significant role not only in binding to cell surface receptors but also in binding and translocating growth factors and cytokines to high-affinity cell surface receptors.

Enhanced cell-cell interaction will also promote differentiation. Homotypic contact interactions can act via gap junctions, which tend to coordinate the response among many like cells in a population by allowing free intercellular flow of second messenger molecules such as cyclic adenosine monophosphate (cAMP) and via cell adhesion molecules such as E-cadherin or NCAM, which signal via anchorage to the cytoskeleton.

Heterotypic interactions, in solid tissues at least, will tend to act across a basal lamina and are less likely to involve direct cell-cell contact.

Signaling is achieved by, on the one hand, modification of the matrix by the mutual contribution of both cell types, and, on the other, by reciprocal transmission of cytokines and growth factors across the basal lamina, such as the transfer of KGF and GM-CSF from dermal fibroblasts to the basal layer of the epidermis in response to IL-1± and -^{2} diffusing from the epidermis to the dermal fibroblasts.

The position, shape, and polarity of the cells may induce, or at least make the cells permissive for the induction of, differentiation. Epidermal keratinocytes and bronchial epithelial cells require to be close to the air/liquid interface, presumably to enhance oxygen availability, and secretory cells, such as thyroid epithelium, need the equivalent of the acinar space, that is, no direct access to nutrient or hormones, above them in a thin fluid space.

When cells are grown on collagen at this location, and the collagen gel is allowed to retract, a shape change can occur, for example, from a flat squamous or cuboidal cell into a more columnar morphology, and this, combined with matrix interaction, allows the establishment of polarity in the cells, such that secretory products are released apically and signaling receptors and nutrient transporters locate basally.

Combining these effects in vitro may require strict attention to culture geometry, for example, by growing cells in a filter well insert on a matrix incorporating stromal fibroblasts, and providing differentiation inducers basally in a defined, nonmitogenic medium. Similar conditions may be created in a perfused capillary bed or scaffold.

Chapter 26

Monoclonal Antibodies

Antibodies are proteins produced by the B lymphocytes of the immune system in response to foreign proteins, called antigens. Antibodies function as markers, binding to the antigen so that the antigen molecules can be recognized and destroyed by phagocytes. The part of the antigen that the antibody binds to is called the epitope. The epitope is thus a short amino acid sequence that the antibody is able to recognize.

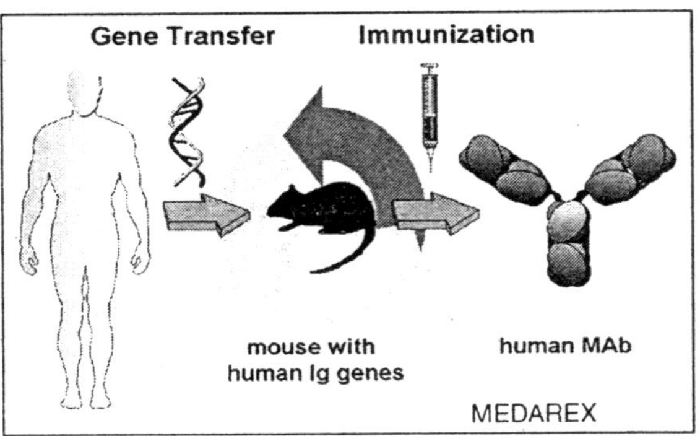

Fig. Monoclonal Antibodies

Structurally antibodies are proteins consisting of four polypeptide chains. These four chains form a quaternary structure somewhat resembling a Y shape. Each B cell in an organism synthesizes only one kind of antibody. In an organism, there is an entire population of different types of B cells and their respective antibodies that were produced in response to the various antigens that the organism had been exposed to. However to be useful as a tool, molecular biologists need substantial amounts of a single antibody (and that antibody alone).

Therefore we need a method to culture a population of B cells derived from a single ancestral B cell, so that this population of B cells would allow us to harvest a single kind of antibody.

This population of cells would be correctly described as monoclonal, and the antibodies produced by this population of B cells are called monoclonal antibodies. In contrast, antibodies obtained from the blood of an immunized animal are called polyclonal antibodies.

Antibodies or immunoglobulins are a crucial component of the immune system, circulating in the blood and lymphatic system, and binding to foreign antigens expressed on cells. Once bound, the foreign cells are marked for destruction by macrophages and complement. In the context of cancer immunotherapy, monoclonal antibodies have brought to light a wide array of human tumor antigens. In addition to targeting cancer cells, antibodies can be designed to act on other cell types and molecules necessary for tumor growth. For example, antibodies can neutralize growth factors and thereby inhibit tumor expansion.

Monoclonal antibodies are made by injecting human cancer cells, or proteins from cancer cells, into mice so that their immune systems create antibodies against foreign antigens. The murine cells producing the antibodies are then removed and fused with laboratory-grown cells to create hybrid cells called hybridomas. Hybridomas can indefinitely produce large quantities of these pure antibodies.

PRODUCTION OF MONOCLONAL ANTIBODIES

The production of monoclonal antibodies was pioneered by Georges Kohler and Cesar Milstein in 1975. Let us see how their method, now tried and tested for over 20 years, would be applied in a particular case.In order for us to isolate a B lymphocyte producing a certain antibody, we first have to induce the production of such a B cell in an organism.

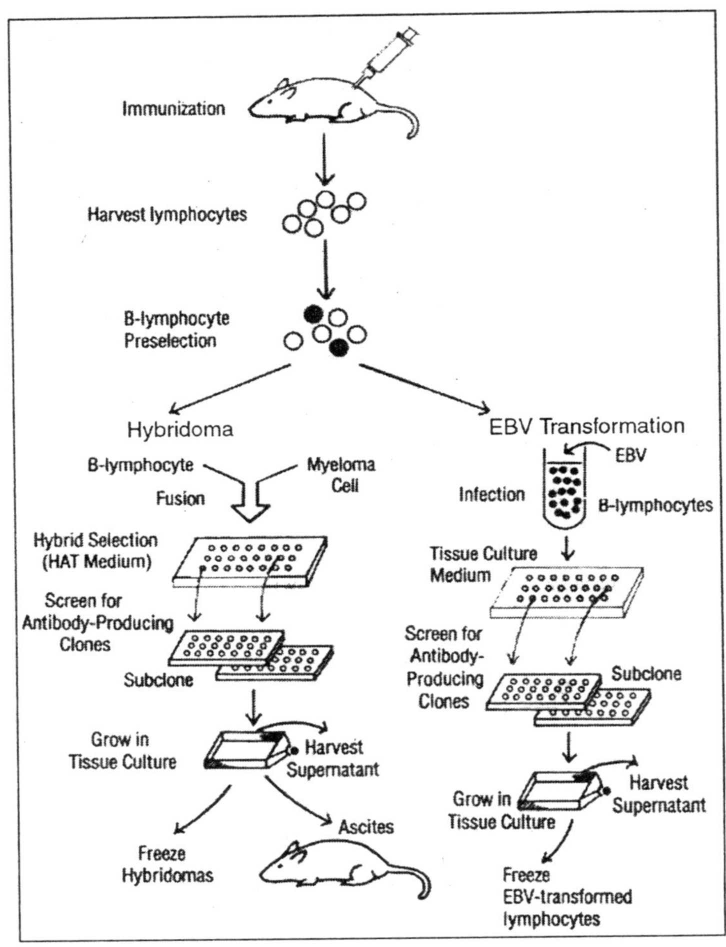

Fig. Monoclonal Antibodies Production

For example, if we need an antibody for avian SERCA2 protein, we would inject the protein into a mouse. This is typically done in two doses, an initial "priming" dose and a second "booster" dose 10 days later. Since the protein is of foreign origin, the mouse immune system recognizes it as such and soon some of the B cells in the mouse would begin production of the antibody to avian SERCA2.

A sample of B cells is extracted from the spleen of the

mouse and added to a culture of myeloma cells (cancer cells). The intended result is the formation of hybridomas, cells formed by the fusion of a B cell and a myeloma cell. The fusion is done by using polyethylene glycol, a virus or by electroporation.

The next step is to selct for the hybridomas. The myeloma cells are HGPRT- and the B cells are HGPRT. HGPRT is hypoxanthine-guanine phosphoribosyl transferase, an enzyme involved in the synthesis of nucleotides from hypoxanthine, an amino acid. The culture is grown in HAT (hypoxanthine-aminopterin-thymine) medium, which can sustain only HGPRT cells. The myeloma cells that fuse with another myeloma cell or do not fuse at all die in the HAT medium since they are HGPRT-. The B cells that fuse with another B cell or do not fuse at all die because they do not have the capacity to divide indefinitely. Only hybridomas between B cells and myeloma cells survive, being both HGPRT+ and cancerous.

The initial collection of B cells used is heterogenous, i.e. they do not all produce the same antibody. Therefore the hybridoma population too does not produce a single antibody. There is also another complication.

A hybridoma cell is initially tetraploid, having been formed by the fusion of two diploid cells. However the extra chromosomes are somehow lost in subsequent divisions in a random manner.

This means that we cannot be certain that the hybridomas will all produce the desired antibody or even any antibody at all. Screening is required to decide which hybridoma cells are actually producing the desired antibody.

Each hybridoma is cultured and screened after doing SDS (sodium dodecyl sulfate - polyacrylamide gel electrophoresis) and Western blots. The probe used is the epitope of the antibody that is desired, which may be labeled by radioactivity or immunofluorescence. Once we are sure that a certain hybridoma is producing the right antibody, we can culture that hybridoma indefinitely and harvest monoclonal antibodies from it.

Uses of Monoclonal Antibodies

Monoclonal bodies have a variety of academic, medical and commercial uses. It would be impossible to list all of these here. But the following list should indicate how ubiquitous monoclonal antibody technology has become in biotechnology.

- Antibodies are used in several diagnostic tests to detect small amounts of drugs, toxins or hormones, e.g., monoclonal antibodies to human chorionic gonadotropin (HCG) are used in pregnancy test kits. Another diagnostic uses of antibodies is the diagnosis of AIDS by the ELISA test.
- Antibodies are used in the radioimmunodetection and radioimmunotherapy of cancer, and some new methods can even target only the cell membranes of cancerous cells. A new cancer drug based on monoclonal antibody technology is Ritoxin.
- Monoclonal antibodies can be used to treat viral diseases, traditionally considered "untreatable". In fact, there is some evidence to suggest that antibodies may lead to a cure for AIDS.
- Monoclonal antibodies can be used to classify strains of a single pathogen, e.g. Neisseria gonorrhoeae can be typed using monoclonal antibodies.
- Researchers use monoclonal antibodies to identify and to trace specific cells or molecules in an organism, e.g., developmental biologists at tyhe University of Oregon use monoclonal antibodies to find out which proteins are responsible for cell differentiation in the respiratory system.
- OKT3, an antibody to the T3 antigen of T cells, is used to alleviate the problem of organ rejection in patients who have had organ transplants.

DEVELOPMENT OF MONOCLONAL ANTIBODIES

Monoclonal antibodies have several roles in cancer therapy. Monoclonal antibodies have been used in a variety of ways in the management of cancer including diagnosis, monitoring, and treatment of disease. They aid in diagnosis, such as the application of flow cytometry in the identification of different subsets of non-Hodgkin's lymphoma. We can use monoclonal antibodies to monitor disease progression, such as the measurement of carcinoembryonic antigen in colon cancer. Most importantly, we can utilize monoclonal antibodies directly as therapy.

Relative to treatment, monoclonal antibodies can react against specific antigens on cancer cells and may enhance the patient's immune response. Monoclonal antibodies can be programmed to act against cell growth factors, thus blocking cancer cell growth. We can conjugate or link monoclonal antibodies to anticancer drugs, radioisotopes, other biologic response modifiers, or other toxins.

When the antibodies bind with antigen-bearing cells, they deliver their load of toxin directly to the tumor. Monoclonal antibodies may also be used to preferentially select normal stem cells from bone marrow or blood in preparation for a hematopoietic stem cell transplant in patients with cancer. There are a number of considerations when using monoclonal antibodies for therapy. First, a target antigen must be selected. It is important that this antigen is presented uniquely by the tumor cells and not on normal tissues.

The immunogenicity of the monoclonal antibody itself is a concern because of how they are derived. As they are often derived from non-human monoclonal antibodies, they are capable of eliciting an immune response themselves. Half-life is another factor. Will it be long enough to have the desired effect? There are also logistical problems such as cost and availability.

Anti-idiotype monoclonal antibodies are a good example of this, as their development has been prohibited by cost. Finally, a decision as to whether or not the monoclonal antibody will be used alone or if it will be conjugated (i.e., attach radioisotopes, toxins, or chemotherapy) in order to get the desired therapeutic effect.

Mechanism of Action

Monoclonal antibodies achieve their therapeutic effect

through various mechanisms. They can have direct effects in producing apoptosis or programmed cell death. They can block growth factor receptors, effectively arresting proliferation of tumor cells. In cells that express monoclonal antibodies, they can bring about anti-idiotype antibody formation.

In this situation Rituximab (IDEC-C2B8), a chimeric antibody, targets the CD20 antigen. This antigen is expressed on a significant number of B cell malignancies. Rituximab is an IgG monoclonal antibody, and has an Fc receptor. The Fc fragment of the monoclonal antibody binds the Fc receptors found on monocytes, macrophages, and natural killer cells. These cells in turn engulf the bound tumor cell and destroy it. Natural killer cells secrete cytokines that lead to cell death, and they also recruit B cells.

Indirect effects include recruiting cells that have cytotoxicity, such as monocytes and macrophages. This type of antibody-mediated cell kill is called antibody-dependent cell mediated cytotoxicity (ADCC). Monoclonal antibodies also bind complement, leading to direct cell toxicity, known as complement dependent cytotoxicity (CDC).

ANTIBODY THERAPY

Antibody therapy can be used in a variety of ways to treat cancer.

As described above, they may act through ADCC or CDC. An alternative approach is to conjugate the monoclonal antibody to a toxin, a cytotoxic agent, or a radioisotope. With conjugated monoclonal antibodies a toxin is actually bound to the antibody, which then attaches to the antigen. The antibody conjugate is absorbed into the cell itself, resulting in cell death.

We can attach a radioisotope such as iodide-131 to directly infuse the cancer cell with radiotherapy, and also mitigate the effects to normal surrounding tissue. Finally, we can attach chemotherapy. The chemotherapeutic agent is taken directly into the targeted malignant cell, rather than being systemically absorbed.

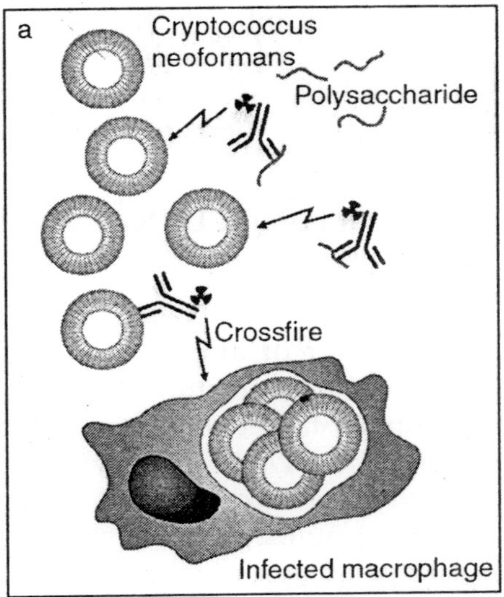

Fig. Antibody Therapy

Obstacles to Successful Therapy

There are a number of obstacles to successful therapy with monoclonal antibodies. The antigen distribution of malignant cells is highly heterogeneous, so some cells may express tumor antigens while others do not. Antigen density can vary as well, with antigens expressed in concentrations too low for monoclonal antibodies to be effective.

Tumor blood flow is not always optimal. If monoclonal antibodies need to be delivered via the blood, it may be difficult to reliably get the therapy to the site. High interstitial pressure within the tumor can prevent the passive monoclonal antibodies from binding. Sometimes the tumor antigen is even released, so the antibody binds to a free-floating antigen and not the tumor cell. Since monoclonal antibodies are derived from mouse cell lines, the possibility of an immune response to the antibodies exists. This response not only decreases the efficacy of monoclonal antibody therapy, but also eliminates the possibility of re-treatment.

Very rarely do we see cross-reactivity with normal tissue

antigens – in general target antigens that are not cross reactive with normal tissue antigens are chosen. Despite these obstacles, there has been tremendous success in the clinical application of monoclonal antibodies in hematologic malignancies and solid tumors.

APPLICATIONS OF MONOCLONAL ANTIBODIES

Hematologic Malignancies

There are a number of antigens and corresponding monoclonal antibodies for the treatment of B cell malignancies. One of the most popular target antigens is CD20, found on B cell malignancies. The CD52 antigen is targeted by the monoclonal antibody alemtuzumab, which is indicated for treatment of chronic lymphocytic leukemia.

The CD22 is targeted by a number of antibodies, and has recently demonstrated efficacy combined with toxin in chemotherapy-resistant hairy cell leukemia. Two new monoclonal antibodies targeting CD20, tositumomab and ibritumomab, have been submitted to the Food and Drug Administration (FDA). These antibodies are conjugated with radioisotopes.

The first monoclonal antibody to receive FDA approval was rituximab. Rituximab is a chimeric unconjugated monoclonal antibody directed at the CD20 antigen, a signature B cell antigen. CD20 has an important functional role in B cell activation, proliferation, and differentiation. This antigen is a transmembrane protein composed of 297 amino acids. The intracellular portion contains phosphorylation sequences for protein kinase C, calmodulin, and casein kinase.

CD20 is thought to act as a calcium channel as well, given the great structural homology between the CD20 protein and the calcium channels. When CD20 was introduced into cell lines by transfection, an increase in intracellular calcium was observed within the transfected cells. With monoclonal antibody stimulation, we see calcium influx within the cells. Calcium chelators blocked apoptosis induced by CD20 stimulation by monoclonal antibodies.

When monoclonal antibodies attach and particularly cross-link CD20 antigen, an increase in intracellular calcium is again observed. This increase appears to activate the SER family of tyrosine kinases, resulting in further phosphorylation of the CD20 inner cytoplasmic chain and also phospholipase C-gamma. At the same time there is an upregulation of C-myc and myb messenger ribonucleic acid (RNA), an increase in adhesion molecule expression and an upregulation of MHC class II proteins. The ultimate result is caspase 3 activation, causing cell apoptosis.

As previously mentioned, CD20 is a natural focus for monoclonal antibody therapy because of its relatively high degree of expression in B cell malignancies, perhaps as high as 95 per cent in follicular lymphomas, even with the heterogeneity discussed earlier. The monoclonal antibody rituximab was designed specifically to target CD20. Rituximab is predominantly human (95 per cent).

The variable light and heavy chain portion of rituximab is murine, but the remainder is humanized so the formation of human anti-mouse antibody is not significant. Rituximab is thought to induce cell apoptosis by inducing calcium influx, releasing caspase activity. In addition, evidence of indirect effects through ADCC and CDC has been observed.

Rituximab is indicated for treatment of low-grade lymphomas refractory to conventional chemotherapy. Based upon this work it has been evaluated for first-line and combination therapy. Results of studies using rituximab as first-line treatment of low-grade non-Hodgkin lymphoma have been encouraging.

Patients who had not received any prior therapy were treated with rituximab 375 mg/m^2 on a weekly basis for 4 weeks and then re-evaluated 2 weeks post-therapy. The patients who achieved a complete or partial response, or who had stable disease received rituximab maintenance therapy (weekly for 4 weeks every 6 months). Patients who showed evidence of progression were taken off maintenance therapy.

At the time of initial re-evaluation at 6 weeks, 54 per cent of the patients showed objective response to treatment. An additional 36 per cent had stable disease or minor response. At the time of publication 13 patients had undergone a second course of treatment, and 4 additional responses were documented. Four patients improved from partial to complete response. Treatment with rituximab was well tolerated, with only 1 of the 39 patients experiencing grade 3-4 infusion related toxicity.

These responses were durable as well. For patients who achieved partial or complete response, one-year follow-up showed no evidence of disease progression. One-year survival was 69 per cent; survival at two years 67 per cent. While overall survival is not an unusual finding in low-grade lymphoma, the duration of response remains relatively impressive. Rituximab has been combined with conventional chemotherapy for patients with intermediate grade or diffuse large cell non-Hodgkin lymphoma.

CHOP (cyclophosphamide, doxorubicin, vincristine, and prednisone) is standard therapy for this type and stage of disease. In a multi-institutional study, 33 patients with newly diagnosed large cell lymphoma received six infusions of rituximab 375 mg/m^2 on day 1 of each cycle combined with six doses of CHOP on day 3 of each cycle.

The overall response rate was 94 per cent, with 20 patients (61 per cent) achieving a complete response. Eleven patients (33 per cent) experienced partial response, and 2 patients were found to have progressive disease. Median duration of response and time to progression had not been reached after a median follow-up time of 26 months. Twenty-nine patients remained in remission during this observation period. The most frequent adverse events associated with rituximab were fever and chills, primarily during the first infusion.

The investigators concluded that rituximab did not appear to increase the toxicity of therapy. hese results were confirmed in a phase III randomized trial of CHOP and rituximab in elderly patients conducted by the French Lymphoma Cooperative Group (GELA). These patients had stage II to IV diffuse large cell lymphoma, were newly diagnosed and therapy-naïve. This study focused on elderly patients (60 to

80 years), because the efficacy of CHOP is decreased in the elderly. Patients had Eastern Cooperative Oncology Group (ECOG) performance status of 0 to 2. Patients were randomized to receive CHOP alone with cytokine support or CHOP with rituximab, given on day 1. These regimens were administered every 21 days for 6 to 8 cycles.

Interim results were reported at the 2000 annual meeting of the American Society of Hematology. Patients who received CHOP plus rituximab showed a statistically significant ($P < 0.0005$) advantage in event-free survival over CHOP alone. This advantage translated into an improvement in overall survival of 83 per cent with rituximab ($P < 0.01$) versus 68 per cent in these 400 patients. A similar study is now in progress, with an additional arm to evaluate the value of rituximab as maintenance therapy for responders.

Solid Tumors

Compared to hematologic malignancies, solid tumors do not have as many specific targets for monoclonal antibodies that are not cross-reactive with antigens on normal tissues. Two significant monoclonal antibodies have been used in solid tumors: edrecolomab and trastuzumab.

Edrecolomab targets the 17-1A antigen seen in colon and rectal cancer, and has been approved for use in Europe for these indications. Its antitumor effects are mediated through ADCC, CDC, and the induction of an anti-idiotypic network. In an initial study of 189 patients with resected stage II colorectal cancer, treatment with edrecolomab reduced the relative risk of mortality by 32 per cent compared with observation alone ($P < 0.01$). Edrecolomab is undergoing investigation in two large phase III trials in patients with stage III colon cancer, either as monotherapy or in combination with fluorouracil-based chemotherapy.

In the US, the most commonly used monoclonal antibody for the treatment of solid tumors is trastuzumab, which targets the HER-2/neu antigen. This antigen is seen on 25 per cent to 35 per cent of breast cancers. Trastuzumab is thought to work in a variety of ways: downregulation of HER-2 receptor expression, inhibition of proliferation of human tumor cells that overexpress HER-2 protein, enhancing immune recruitment and ADCC against tumor cells that overexpress HER-2 protein, and downregulation of angiogenesis factors. This last mechanism may be very important in terms of metastatic disease.

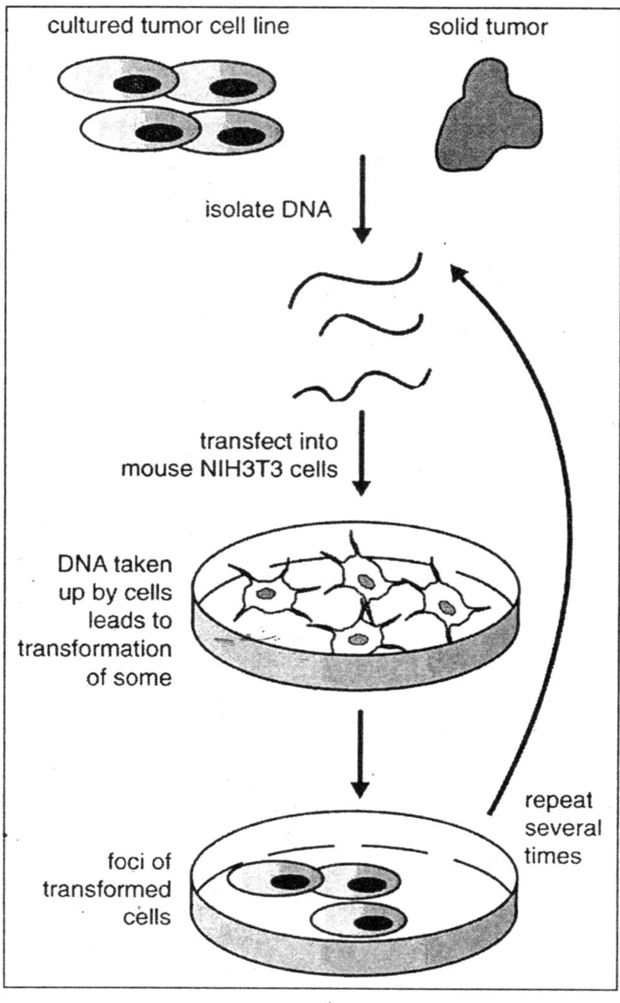

Fig. Solid Tumors

In phase I and II trials of patients with metastatic breast cancer, treatment with a combination of trastuzumab and cisplatin resulted in prolongation of survival and higher response rates than that seen with cisplatin alone.

Trastuzumab plus chemotherapy when compared to chemotherapy alone was associated with longer time to disease progression (median 7.4 months vs. 4.6 months, P<0.001), a higher rate of objective response (50 per cent vs. 32 per cent, P<0.01), a longer duration of response (median 9.1 vs. 6.1 months, P<0.001), a lower rate of death at 1 year (22 per cent vs. 33 per cent, P=0.008), longer survival (median survival 25.1 vs. 20.3 months, P=0.01) and a 20 per cent reduction in the risk of death.

This trial evaluated 469 patients with metastatic breast cancer that overexpressed HER-2/neu. Patients were randomly assigned to receive chemotherapy alone (n=234) or chemotherapy plus trastuzumab (n=235). The only significant adverse event was cardiac dysfunction, which was managed with standard medical treatment.

Trastuzumab has also been studied as monotherapy. This study involved 214 patients with relapsed metastatic breast cancer who had been heavily pre-treated: 90 per cent had received prior anthracyclines and 65 per cent had received taxane therapy. The majority had either lung or liver metastasis. All patients received a loading dose of trastuzumab (4 mg/kg), followed by weekly maintenance therapy of 2 mg/kg until evidence of disease progression.

Primary endpoints were tumor assessment relative to response. Secondary endpoints were duration of response, time to tumor progression, time to treatment failure, and quality of life.

Eight complete and 26 partial responses were identified for an objective response rate of 15 per cent in the intent-to-treat population (95 per cent CI; 11 per cent to 21 per cent). Median duration of response was 9.1 months; median duration of survival was 13 months. Toxicity was minimal, although cardiac dysfunction occurred in 4.7 per cent of patients. Patients who had higher overexpression of HER-2/neu had a better overall response rate. As first-line monotherapy, trastuzumab has demonstrated efficacy and safety in patients with metastatic breast cancer. This study included 114 women with HER-2-overexpressing metastatic breast cancer with no prior chemotherapy. Patients were randomized to receive a loading dose of trastuzumab 4 mg/kg followed by 2 mg/kg weekly, or an 8 mg/kg loading dose followed by 4 mg/kg weekly.

Primary endpoint was overall response rate. Secondary endpoints were disease relapse, time to tumor progression, and overall survival.

Complete response rates were relatively low (7/114) regardless of loading and maintenance doses. Partial response rates were nearly identical at 19 per cent vs. 21 per cent, for an overall response rate of 24 per cent vs. 28 per cent, which was not statistically significant. Seventeen (57 per cent) of 30 patients with an objective response and 22 (51 per cent) of 43 patients with clinical benefit had not experienced disease progression at 1-year follow-up.

These investigators found no benefit to the higher versus lower dose of trastuzumab. However, they noted a difference in response that correlated to HER-2 overexpression. Overexpression was measured by immunohistochemistry (IHC) and then rechecked with fluorescent *in situ* hybridization (FISH). Interestingly, patients who were positive for overexpression by IHC were not necessarily positive by FISH. Overexpression as confirmed by FISH was strongly correlated to response, and these patients appeared to have garnered the most clinical benefit from treatment with trastuzumab.

When we look at these data together first-line monotherapy with trastuzumab appears to have better overall response. However, median time to disease progression remains disappointingly short. The incidence of cardiac toxicity cannot be minimized, particularly in patients with prior anthracycline therapy or cardiac disease. These patients had a 10 per cent incidence of severe myocardial toxicity and one death from ventricular arrhythmia. Still, these data suggest the possibility of significant response rates, including improvement in overall survival. While early reports on the use of monoclonal antibodies may have been overenthusiastic, the results of these studies show there is still cause for cautious optimism as we go forward. Cure may yet elude us, but stable disease is an attainable, highly desirable goal.

MONOCLONAL ANTIBODY TECHNOLOGY

Substances foreign to the body, such as disease-causing bacteria and viruses and other infectious agents, known as antigens, are recognized by the body's immune system as invaders. Our natural defenses against these infectious agents are antibodies, proteins that seek out the antigens and help destroy them.

Antibodies have two very useful characteristics. First, they are extremely specific; that is, each antibody binds to and attacks one particular antigen. Second, some antibodies, once activated by the occurrence of a disease, continue to confer resistance against that disease; classic examples are the antibodies to the childhood diseases chickenpox and measles.

The second characteristic of antibodies makes it possible to develop vaccines. A vaccine is a preparation of killed or weakened bacteria or viruses that, when introduced into the body, stimulates the production of antibodies against the antigens it contains.

It is the first trait of antibodies, their specificity, that makes monoclonal antibody technology so valuable. Not only can antibodies be used therapeutically, to protect against disease; they can also help to diagnose a wide variety of illnesses, and can detect the presence of drugs, viral and bacterial products, and other unusual or abnormal substances in the blood.

Given such a diversity of uses for these disease-fighting substances, their production in pure quantities has long been the focus of scientific investigation. The conventional method was to inject a laboratory animal with an antigen and then, after antibodies had been formed, collect those antibodies from the blood serum (antibody-containing blood serum is called antiserum). There are two problems with this method: It yields antiserum that contains undesired substances, and it provides a very small amount of usable antibody. Monoclonal antibody technology allows us to produce large amounts of pure antibodies in the following way: We can obtain cells that produce antibodies naturally; we also have available a class of cells that can grow continually in cell culture. If we form a hybrid that combines the characteristic of "immortality" with the ability to produce the desired substance, we would have, in effect, a factory to produce antibodies that worked around the clock.

In monoclonal antibody technology, tumor cells that can replicate endlessly are fused with mammalian cells that produce an antibody.

The result of this cell fusion is a "hybridoma," which will continually produce antibodies. These antibodies are called monoclonal because they come from only one type of cell, the hybridoma cell; antibodies produced by conventional methods, on the other hand, are derived from preparations containing many kinds of cells, and hence are called polyclonal. An example of how monoclonal antibodies are derived is described below.

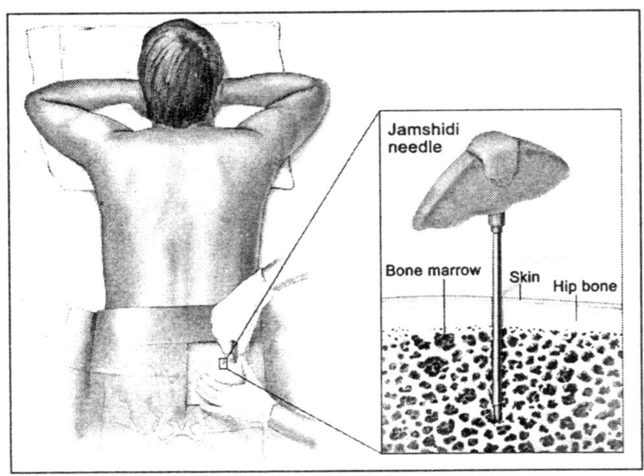

Fig. Myeloma

A myeloma is a tumor of the bone marrow that can be adapted to grow permanently in cell culture. When myeloma cells were fused with antibody-producing mammalian spleen cells, it was found that the resulting hybrid cells, or hybridomas, produced large amounts of monoclonal antibody. This product of cell fusion combined the desired qualities of the two different types of cells: the ability to grow continually, and the ability to produce large amounts of pure antibody.

Because selected hybrid cells produce only one specific antibody, they are more pure than the polyclonal antibodies produced by conventional techniques.

They are potentially more effective than conventional drugs in fighting disease, since drugs attack not only the foreign substance but the body's own cells as well, sometimes producing undesirable side effects such as nausea and allergic reactions. Monoclonal antibodies attack the target molecule and only the target molecule, with no or greatly diminished side effects.

Chapter 27

Cytokines and Cancer Therapy

Cytokines are the messengers of the immune system. Cytokines are substances, either proteins or glycoproteins, secreted by immune cells. They have autocrine and paracrine functions, so that they function locally or at a distance to enhance or suppress immunity. In cancer therapy, we generally use cytokines to enhance immunity. Cytokines regulate the innate immune system: natural killer (NK) cells, macrophages, and neutrophils.

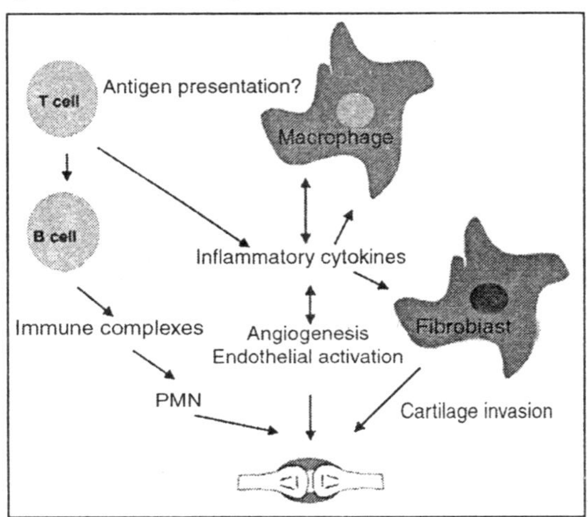

Fig. Cytokines

They also regulate the adaptive immune system, the T and B cell immune responses. In the immune system, cytokines function in cascades. Thus clinical trials of individual cytokines are rarely useful, since cytokines tend not to work individually.

Some of the individual cytokines that have been tested and found ineffective for cancer treatment include interleukin 1 beta (IL-1 beta), although it may be useful because it helps to mediate the severe toxicity of interleukin 2 (IL-2).

Tumor necrosis factor (TNF) certainly sounded promising, but in fact caused severe hypotension when used systemically. Interleukin 4 (IL-4) showed minimal anti-cancer activity and was toxic. Interleukin 6 (IL-6) had some activity against cancer cells, but turned out to be a growth factor for myeloma cells. Granulocyte-macrophage colony-stimulating factor (GM-CSF), used primarily in stem cell transplant to reconstitute the myeloid series, has been studied for melanoma with controversial results. Which cytokines are important for cancer? IL-2 and interferon-alfa 2b are two cytokines approved by the FDA for treatment of cancer.

IL-2 has demonstrated activity against renal cell, melanoma, lymphoma, and leukemia. Interferon has activity in the same histologies but also in Kaposi's sarcoma, chronic myelogenous leukemia, and hairy cell leukemia. Overall, cytokines are substances that appear to have application in the treatment of hematologic malignancies or immunogenic tumors. Here we will examine the major cytokines currently in use or under evaluation for cancer therapy: interferon alfa, IL-2, GM-CSF, and interleukin-12 (IL-12).

Interferon Alfa

Interferon alfa is actually a family of molecules rather than a single molecule. They are encoded by closely related genes on chromosome 9, encoding proteins that are variably glycosylated. These are comprised of about 150 amino acids, and they bind to certain receptors on the surface of immune cells. They are known to have profound and diverse effects on gene expression.

Interferon alfa has many roles. It upregulates genes like MHC class I, tumor antigens, and adhesion molecules. It is also an anti-angiogenic agent. It is extremely active in the immune system, promoting B and T cell activity. Interferon alfa can stimulate macrophages, even dendritic cells, and upregulates Fc receptors. Interferon was isolated in 1970 by investigators looking for antiviral substances. The substance we now know as interferon was isolated from white cells, and called interferon because it interfered with viral infection. It turns out type 1 and 2 interferons do have some antitumor activity, but not the hoped-for level.

Interferon's activity in cancer has been well documented, however. In kidney cancer we have seen small but consistent response rates in a number of studies. In a randomized controlled study of interferon alfa versus megestrol acetate, investigators found a minimal advantage to interferon in terms of survival (8.5 months vs. 6 months). However disease-free survival at 2 years was poor. With a response rate in the 5 per cent to 10 per cent range, it is somewhat surprising that interferon is now considered standard practice in the treatment of kidney cancer, especially in patients who are relatively sick and cannot tolerate IL-2. Optimal dose, route, and schedule are unknown.

Most investigators would agree that for any therapeutic effect in kidney cancer, interferon doses of 5 million to 10 million units 3 times a week subcutaneously are reasonable. Still, in looking through most of the established literature on the uses of interferon alone in phase 2 trials, the median response rate is only about 11 per cent with a range of 8 per cent to 29 per cent. Clearly, interferon alfa is not very active in kidney cancer and is in fact not FDA-approved for this use.

It is considered second-line therapy for renal cell carcinoma, and then as a palliative maneuver. Eligible patients should be offered interleukin-2, which is currently the only FDA-approved drug for renal cell carcinoma. Interferon has a long history in metastatic melanoma. A series of small phase 2 trials with high-dose interferon (up to 100 million units IV daily) reported response rates of 20 per cent to 22 per cent.

While not markedly different than any single chemotherapy in metastatic melanoma, this response rate was significant and spurred continued interest. Results in combination with cimetidine were also disappointing. Five trials of interferon alfa with cimetidine showed no advantage.

The definitive test of whether adding any agent to interferon enhances its antitumor effect was an ECOG trial with dacarbazine alone, dacarbazine with tamoxifen, dacarbazine with interferon, or interferon with tamoxifen. In the approximately 260 patients, median time to progression was a brief 3.7 months. Overall survival in the interferon plus dacarbazine arm was 9.1 months, which was identical to the dacarbazine only arm.

Interferon was then evaluated in the adjuvant setting.45 In this setting, interferon's activity in melanoma is high. Patients with stage 3 and 2B resected high-risk melanoma were randomly allocated to observation or high-dose interferon for one year (N = 287). Interferon was administered intravenously every day for 4 weeks, at a dose of 20 million units/m^2, followed by a maintenance regimen of 10 million units/m^2 3 times a week subcutaneously.

The group treated with interferon showed definite disease-free and overall survival benefits, which ultimately led to FDA approval of this regimen. Survival with interferon was 3.8 years versus 2.78 with observation – essentially a year more with interferon. Relapse-free survival also showed a 1-year advantage with interferon. The real advantage was 27 per cent augmentation of survival at 7 years of follow-up, and at 9 years of follow-up the P value was between 0.03 and 0.04. It may not be a huge increment, but it is significant.

The follow-up confirmatory trial included 642 patients who received either a year of high-dose interferon, 2 years of lower dose interferon given subcutaneously at 3 million units a week, or observation. No differences in survival were shown. The E1690 trial did not confirm the results of the earlier trial. The hazard ratio for survival for high-dose interferon was only 1.0, meaning it was not different from observation. Amazingly, the median survival in the observation arm was 6 years.

The median survival in the observation arm before was, as we said, a little less than 3 years. How could the survival change? A third trial was initiated, comparing high-dose interferon versus a vaccine. One year of interferon was compared to 18 months of vaccination with ganglioside vaccine. The vaccine arm was not terribly toxic, while interferon had a fair degree of toxicity. This trial accrued 774 patients, but was terminated early. A clear disease-free and survival advantage for the interferon arm was evident early in the study. Relapse-free survival was now again prolonged with a hazard ratio of 1.6.

Overall survival was prolonged with interferon with a hazard ratio of 1:3; stage 2B patients benefited and the P value for overall survival with a little over 2 years of follow-up was 0.04. At 3 years of follow-up the P value was even better at 0.03.

The third trial essentially replicated the results of the first. The appearance of improvement in relapse-free survival with the same treatment regimen over time is probably a function of better supportive care, better staging — CT scans, PET scans, and MRIs lead to more accurate diagnosis.

The treatment has not changed, but the patient population has. These studies laid the foundation for interferon as standard adjuvant therapy for stage III resected melanoma. The toxicity remains an issue. When we treat patients with this regimen, we prescribe an antidepressant to decrease some of the toxicity, the depression, and decreased energy level. Vigorous hydration is essential as well with high-dose interferon, as patients tend to become dehydrated.

Interferon has been approved as treatment in other histologies as well. Interferon is definitely active in hairy cell leukemia: Nine complete and 17 partial responses were documented by bone marrow core biopsies. Peripheral blood hematologic indices improved or normalized in all patients. Previously untreated patients showed higher complete remission ($P = 0.02$) than in those who had undergone splenectomy.

Therapy was well tolerated, with most patients experiencing tumor remission reporting improved quality of life. Another study found similar results in a small population of patients, with an overall response rate of 93 per cent. Peripheral blood counts returned to normal levels. This study also assessed natural killer cell activity and immunologic surface markers, and noted normalization of both measures after therapy. A significant survival benefit of more than 89 months in a phase 2 trial in patients with chronic myelogenous leukemia suggests interferon has activity here as well. This survival advantage was independent of cytogenetic improvement with interferon, which was also noted.

Interleukin

Interleukin-2 (IL-2) is a T cell growth factor that binds to a specific tripartite receptor on T cells. In dose escalation studies, patients treated with high doses of IL-2 showed clinical responses, although severe toxicity was seen. Response rates were as high as 24 per cent at the highest IL-2 dose in patients with renal cell carcinoma. However, the toxicities of treatment were limiting. These toxicities stem from what is known as a capillary leak syndrome. Giving IL-2 in high doses is comparable to inducing a controlled state of septic shock.

Low blood pressure, low systemic vascular resistance, high cardiac output, grade 3/4 hematologic toxicity, hepatic toxicity, renal toxicity, and pulmonary edema have all been documented. Toxicity is nearly always reversible. The typical IL-2 regimen – 600,000 to 720,000 IU/kg, an average of 50 million units of IL-2 per dose given three times a day as a bolus over 15 minutes.

The maximum most patients can tolerate is 14 doses. A rest period of 5 to 9 days between cycles is recommended, and patients must be treated in a step-down situation or ICU. This regimen was evaluated in 255 patients with renal cell carcinoma. Objective responses were initially reported in 15 per cent of patients, with 17 (7 per cent) complete responses.

These figures were confirmed in follow-up. These numbers were not particularly impressive, but the median duration of response (54 months) and median duration for complete responses (80 months) demonstrated the potential benefit of IL-2 in these patients.

Median survival for all patients was 16.3 months, with 10 per cent to 20 per cent estimated to be alive 5 to 10 years after treatment. The toxicity associated with IL-2 therapy is a major consideration. These patients require intensive management. Dopamine is administered prophylactically and therapeutically for low-urine output, neosynephrine is given for hypotension, acetaminophen for fever and chills, cimetidine for gastric acidity, and oxacillin for neutrophil dysfunction that leads to sepsis. Concomitant antibiotic therapy is essential, as the rate of sepsis without antibiotic therapy is as high as 27 per cent. Low-dose IL-2 has shown activity in renal cell cancer as well. Objective response rates of 18 per cent to 23 per cent were reported, without the toxicity of high dose IL-2. A larger study of 425 patients evaluated the activity of low-dose IL-2 in combination with interferon-alfa, as well as each agent alone. Interferon-alfa or IL-2 alone had low response rates, but the response rate for the combination was significantly higher ($P<0.01$), with significantly improved 1-year event-free survival ($P=0.01$).

IL-2 is indicated for use in renal cell carcinoma and melanoma on the basis of duration of response, rather than achievement of response. A substantial number of patients with renal cancer who had an objective response to treatment are still alive, and stable or progression-free, 1 year post-treatment. With melanoma, we see patients who have responded alive and progression-free 5 to 15 years later. In metastatic melanoma this duration of survival is uncommon when chemotherapy is used.

Clearly, patient selection is critical for IL-2 therapy. How can we predict who will respond to high-dose IL-2. As expected, melanoma patients with high performance status do well, as do patients who develop vitiglio or autoimmune thyroiditis. These findings support the idea that IL-2 generates a T cell response that reacts with melanocytes. Another important determinant of response was the amount of IL-2 given during the first cycle. The immunologic effect is manifested by the height of rebound leukocytosis.

Lymphocyte levels plunge during initial treatment with IL-2, and rebound when treatment stops. This rebound is an indication of the response to IL-2. IL-2 has shown activity in non-Hodgkin's lymphoma and leukemia and lymphoma post-stem cell transplant. Low-dose IL-2 was also evaluated in combination with histamine, but no differences in response were observed compared with IL-2 alone. An interesting modification of the IL-2 molecule, known as BAY 50-4798, has shown potential in animal models. This agent is the same as IL-2 with two modified amino acids. It appears to have the therapeutic effect of IL-2 without the toxicity. Phase I dose escalation studies are underway. What we have seen so far is potential for clinical response with very little toxicity compared to IL-2.

GRANULOCYTE-MONOCYTE COLONY STIMULATING FACTOR

Granulocyte-monocyte colony stimulating factor (GM-CSF) is a well-known cytokine, approved for use in stem cell and bone marrow transplant to reconstitute the myeloid series. We have data suggesting it also might have application as monotherapy in melanoma. In this trial, 48 patients with stage III and IV melanoma were treated with long-term, chronic, intermittent GM-CSF after surgical resection.

The theory is that GM-CSF would reconstitute antigen-presenting cells, and thus the ability to mount an immune response. Overall and disease-free survival were significantly prolonged by GM-CSF therapy in patients who were clinically disease-free. Median survival was 37.5 months versus 12.2 months in matched controls ($P < 0.001$). Treatment was well tolerated. However, the strength of these observations is questionable. The matched controls (disease, stage, age, sex) came from a database of patients from 1960 to 1988, before PET and MRI scans, sentinel node biopsy, sophisticated staging and treatment. Whether the therapeutic effect was due to general improvements in treatment over the past 41 years or to GM-CSF therapy will have to be confirmed in studies with true control arms.

Interleukin-12 (IL-12) is a very exciting cytokine. It is a heterodimeric protein that promotes NK and T cell activity and is a growth factor for B cells. It has demonstrated antitumor activity in mouse models. Alone, IL-12 shows minimal potential for therapeutic effect. However, IL-12 may have value as a vaccine adjuvant. When IL-12 was paired with peptide vaccines in patients with resected stage 3 and 4 melanoma, IL-12 appeared to boost the response to the vaccine. GM-CSF is also being evaluated as an adjuvant for vaccine therapy.

Cytokines such as IL-2 have modest antitumor activity in metastatic renal cell carcinoma and melanoma. The effect is remarkable for its duration rather than strength, and its application may have greatest significance for long-term survival. IL-2 at high doses is toxic. Low doses may have promising activity post-transplant as an immune restorative agent, and that effect is under evaluation in large randomized cooperative group studies. IL-12 and GM-CSF are unlikely to be stand-alone agents for any histology, but show promise as adjuvant therapy and in combination with other cytokines.

VACCINE APPROACHES AGAINST CANCER

The antibody-based therapies we have been discussing are a form of passive immunotherapy. That is, the molecules or substances are introduced into the body, rather than the body creating its own immune response. Vaccines, on the other hand, are considered active immunotherapy because they generate an intrinsic immune response. They are also considered a form of specific immunotherapy because they attempt to stimulate an immune response that can directly target the tumor antigens, in contrast to non-specific approaches such as cytokines that broadly stimulate the immune system.

Efforts to treat cancer with vaccines date back to the origins of immunology. Patients have been injected with autologous and allogeneic malignant cells, usually irradiated to prevent further growth. However, measuring immune response was problematic. Now that we have identified several tumor antigens and the immune response they provoke, we have made progress in developing cancer vaccines.

As Dr. Disis explained, tumor cells express specific antigens on the cell surface, usually within the MHC molecules. The problem with tumor cells is that they cannot stimulate a T cell response by naïve T cells, in part because they lack necessary co-stimulatory molecules. However, dendritic cells – a type of antigen-presenting cell – can provide them. We can take a dendritic cell and import tumor antigens into it by a variety of mechanisms. The dendritic cell can then present the tumor antigens on their surface within MHC molecules, ready to activate T cells. Once the T cells are activated, they are capable of recognizing and destroying antigen-expressing tumors.

Peptide, Autologous, and Viral Vector Vaccines

Numerous approaches to stimulate the immune system to recognize tumors have been tried over the years. Vaccines consisting of peptide or protein administered with an adjuvant have been the most frequently used. These adjuvants might be compounds such as bacterial cell wall components that incite an inflammatory response or cytokines such as IL-12 or GM-CSF, as Dr. Weber described.

Monocytes, neutrophils, eosinophils, and T cells are all recruited to the site of the inflammation where adjuvant is used, but it is believed that *in situ* dendritic cells ultimately take up the proteins and peptides, process them if necessary, and present them on the cell surface as peptides capable of binding to the MHC molecules. Dendritic cells can then stimulate T cells that have the receptors to recognize those particular peptides. These studies are primarily in melanoma.

The targeted antigens are gp100, tyrosinase, and MAGE, which are found in melanoma. Dr. Disis' group worked with HER-2/neu peptides. Lymphomas may be targeted by using idiotype protein. Adjuvants include IL-12, QS-21, and GM-CSF. Immunologic response was monitored by various assays—ELISPOT and microcytotoxicity in particular, but also by delayed-type hypersensitivity (DTH) skin testing. All of these studies report at least a few patients developing some type of antigen-specific immune response, to varying degrees. This response will certainly depend on the antigen, the tumor system, the patient, and disease stage.

We see they have been modest, although there are some instances of complete or partial remissions, or prolongation of survival. An interesting approach to detecting a clinical response was used in lymphoma patients, in whom circulating tumor cells could be detected by polymerase chain reaction (PCR) prior to immunization. PCR enables us to actually search for lymphoma cells by identifying the particular chromosomal translocation abnormality they are known to possess.

In this study, tumor cells could not be detected in peripheral blood of some patients following immunization. This method of measuring clinical responses is becoming increasingly important, particularly for diseases that cannot be monitored by a CT scan.

Tumor cell-based vaccines are another vital area of research. Some studies have used autologous tumor, in which tumor cells are extracted from surgical resection or biopsy specimens. Allogeneic cell lines have also been developed for tumors such as melanoma that likely encompass many of the tumor-associated antigens expressed by the melanomas of most affected individuals. Tumor cells can also be modified to make them more immunogenic.

To accomplish that, tumor cells may be infected with various types of viruses so that viral proteins are expressed on the surface, and transduced with genes expressing cytokines such as IL-2 and GM-CSF, or genes for HLA molecules or co-stimulatory molecules. The idea is irradiate the cells so they can no longer proliferate, then inject the tumor cells back into the patient.

What we hope to see is the immune system activated by either the tumor cell or the inflammatory response that includes recruitment of dendritic cells. As the injected tumor cells undergo apoptosis or are destroyed by the inflammatory reaction, antigens are picked up by the dendritic cells and represented to the T cells.

These studies employed several strategies. One of the most popular is to transduce with a vector containing GM-CSF, so that the tumor secretes GM-CSF and sets up an inflammatory response.

In animal studies, this strategy has been the most promising in inducing a protective immune response. Newcastle virus is used to infect the cells in another study to some effect, as shown by median survival of 46 weeks. Use of Bacille Calmette-Guerin (BCG) as an inflammation inducing adjuvant along with autologous colorectal cancer cells showed increased DTH but no survival benefit. In fact, most of these tumor cell approaches show an immune response, but again limited clinical response.

Nonetheless, CancerVax, which has been heavily tested in melanoma, has been suggested in nonrandomized studies to provide a survival benefit.

Viral vectors and indeed, naked DNA in the form of plasmids encoding tumor antigens, can be used to immunize people. Initially these vaccines were administered to muscle cells, but it is likely that dendritic cells were ultimately the targets infected by the virus, or were picking up the antigen released by apoptotic muscle cells. Poxviruses are another popular way to apply this approach, and a considerable amount of work has been done with vaccinia and avian and fowl pox vectors.

One of the most interesting strategies is the prime boost approach. An example of this is when a patient is first immunized with vaccinia virus encoding the gene for CEA. Vaccinia is very immunogenic, so it can only be used once or twice – after the first injection patients develop high neutralizing antibody titers. In subsequent immunizations the antibody immediately binds to the virus, making it difficult to get true immunization against the encoded tumor antigen. However if we follow with an avian vector expressing CEA.

Dendritic Cell Vaccines

Dendritic cells (DC), as mentioned earlier, are in the final common pathway for activating naïve T cells by many of these vaccine strategies. Remember these are the professional antigen-presenting cells, and can be found in most areas of the body. They circulate in peripheral blood, and are present as Langerhans cells in the epidermis.

Variables associated with employing dendritic cell vaccines are numerous. First we must consider what is the best source or lineage of dendritic cell to use. Should we use DC directly isolated from the peripheral blood or generate them ex vivo from precursors? Next, how do we load the antigen? Maturation and/or activation is another factor to consider, as data suggest immature dendritic cells may give a diminished immune response.

Route of administration is always a question with tumor vaccines, as there are advantages and disadvantages to all of the available routes. Perhaps most importantly, we need to understand how best to evaluate the immune and clinical response to dendritic cell vaccines to permit efficient development of this strategy.

The clinical utility of dendritic cell vaccines was initially limited by the fact that they are not numerous, and they are difficult to obtain. It is now possible to obtain them in large numbers by generating them in vitro. We can directly isolate them from peripheral blood, although that is still difficult because there are so few unless they are mobilized by cytokines such as Flt3-ligand. CD34+ cells from bone marrow or peripheral blood can be stimulated with a cocktail of cytokines to generate dendritic cells.

Monocytes in the peripheral blood, either adherent to plastic or selected for CD14, and cultured in media containing GM-CSF and IL-4, or GM-CSF and IL-13, can generate dendritic cells as well. We can also convert non-dendritic cells to dendritic cells.

In this process, cells from patients with chronic myelogenous leukemia, for example, may be differentiated into dendritic cells. These cells contain the bcr-abl translocation and thus present bcr-abl peptides on the surface of the cell.

However they are generated, it is important that dendritic cells are mature. Mature cells have upregulated MHC molecules and certain chemokines and chemokine receptors, and downregulated others.

The changes in the chemokine receptors allow the dendritic cells to migrate from their peripheral site into the lymph nodes. Mature dendritic cells produce IL-12, essential not just for attraction of NK cells but also cytolytic T cells, and for shifting helper T cell responses in a Th1 direction. We have shown in vitro that matured dendritic cells, loaded with antigens, seem to stimulate greater cytolytic T cell activity against tumor antigens – CEA in this case.

The sequence of loading and maturation is important as well. For example, if we use a protein or messenger RNA to load a dendritic cell, this processing is something only immature cells are good at so the cells should be matured after they are loaded. If we load with peptide, which requires no processing, we mature first and then load to optimize the number of MHC molecules on the surface.

IL-12 is a good indicator of functional maturity, although many of the maturation strategies do not actually result in IL-12 production. For example, maturation with CD40 ligand alone results in CD80 and CD83 upregulation, but only small amounts of IL-12. What we have shown to be most effective for maturation is either CD40 ligand plus interferon gamma or adding lipopolysaccharide (LPS).

The route of administration of dendritic cells continues to be controversial.

Intradermal, subcutaneous, or in some cases intralymphatic are some of the more commonly used methods. We looked at the ability of dendritic cells to migrate to areas with high lymphocyte concentrations – after all, to boost immunity, the dendritic cells must meet the T cells somewhere. We labeled the dendritic cells with indium-111 so we could measure them with a gamma camera.

When we injected the cells intravenously, they migrated to the lungs first and then redistributed to liver and spleen over the next 6 to 24 hours. We saw no dendritic cells in the lymph nodes.

In fact, only intradermal administration demonstrated dendritic cells in the lymph nodes, and that is about 1 per cent. It is possible that greater migration to critical areas may be achieved through manipulating chemokines receptors.

As Dr. Disis has mentioned, a certain threshold of antigen-specific T cells must be achieved to have a clinical response. Depending on the type of assay used we believe that threshold needs to be 1 per cent or more of circulating cells. How to apply in vivo indicators of immune response such as DTH reactivity is unclear.

We demonstrated in a phase I study that it was safe and feasible to administer the dendritic cell vaccine, and found a single minor response in about 25 patients. We did see T cell infiltration at the injection sites where the vaccine was administered, although there was little pure DTH reactivity in terms of erythema and induration.

A variety of in vitro assays have also been used to measure T cell responses from the peripheral blood. The enzyme-linked immunoSPOT (ELISPOT) assay is one used in many investigations.

This assay involves coating plates with antibodies that can recognize a particular cytokine, such as interferon gamma production by T cells in response to antigen. T cells and antigen are added and T cell stimulation is permitted for several hours. An antibody that can recognize the cytokine at a different site is then introduced. Binding by this antibody is demonstrated by an enzymatic reaction as in a typical enzyme-linked immunoassay (ELISA).

Ultimately, we see a spot at each site where there is a cytokine-secreting T cell. This assay can be difficult to read, particularly when there are a large number of spots. Automated readers are available. This data is from a healthy volunteer who is EBV-positive, which means the volunteer had been exposed to EBV and had a specific response. This assay shows approximately 75 to 100 cells per 200,000 peripheral blood mononuclear cells recognized EBV.

Other assays used include the tetramer assay, which is essentially a phenotypic assay, and intracellular cytokine assay. The intracellular cytokine assay is a functional assay that can measure multiple cytokines, use defined antigens, and measure class I or II response.

However, it is technically demanding to perform and requires 5 hours of in vitro stimulation. Regardless of the assay used, it is important to look at a variety of time points after immunization to quantify results. In order to stimulate an immune response that can be measured by an assay, the immune system first must recognize it. Dr. Disis has introduced the importance of tumor antigens. Most of the studies conducted to date use a target cell mixed with a peptide or protein, so that the target cell expresses the antigen we want the immune system to attack.

This approach permits us to determine if the T cells recognizing the antigen of interest have been stimulated, but it does not tell us whether the T cells can destroy actual patient tumor cells.

One study found that autologous dendritic cells loaded with idiotype (ID) protein could stimulate cytolytic T cells capable of recognizing patient myeloma cells. HER-2/neu is another antigen used in dendritic cell vaccines. Currently it is under investigation in patients with no evidence of disease, as they may be more likely to mount an immune response and receive clinical benefit.

Patients received dendritic cells loaded with a protein fragment of HER-2/neu or the control antigen, KLH, every 3 weeks. In vitro, we were able to demonstrate that this approach did indeed induce T cells that lysed HER-2/neu-expressing target cells. In the first two patients we have detected induction of HER-2/neu and KLH-specific T cell responses by ELISPOT. The response to KLH demonstrates a primary immune response can be elicited, since most people have not been exposed to it.

Several generalizations can be made from these studies. First, dendritic cell-based immunizations are feasible and safe. Clinically significant adverse events are exceedingly rare. Second, low levels of antigen-specific immune responses can be induced in some patients in most of the studies. Third, clinical responses, including complete responses, can be evoked in a minority of patients and these responses are more common in studies targeting melanoma. Fourth, no one strategy appears to be significantly more effective.

This suggests the continued need to use immune response

assays to help guide further development of more potent vaccines. If it is true that antigen-specific T cells must account for 0.1 per cent to 1 per cent of circulating T cells in order for there to be clinical activity, we will need to significantly improve these vaccines before being able to make definitive statements about their clinical studies.

VITRO PRODUCTION OF MONOCLONAL ANTIBODY

A major advantage of using mAb rather than polyclonal antiserum is the potential availability of almost infinite quantities of a specific monoclonal antibody directed toward a single epitope (the part of an antigen molecule that is responsible for specific antigen-antibody interaction). In general, mAb are found either in the medium supporting the growth of a hybridoma in vitro or in ascitic fluid from a mouse inoculated with the hybridoma. mAb can be purified from either of the two sources but are often used as is in media or in ascitic fluid. In vitro methods should be used for final production of mAb when this is reasonable and practical. Many commercially available devices have been developed for in vitro cultivation. These devices vary in the facilities required for their operation, the amount of operator training required, the complexity of operating procedures, final concentration of antibody achieved, cost, and fluid volume accommodated. The cost of additional equipment should be considered in the cost of in vitro production methods.

Each hybridoma cell line responds differently to a given in vitro production environment. This section describes in vitro production methods that are available and discusses the usefulness and limitations of each method.

BATCH TISSUE-CULTURE METHODS

The simplest approach for producing mAb in vitro is to grow the hybridoma cultures in batches and purify the mAb from the culture medium. Fetal bovine serum is used in most tissue-culture media and contains bovine immunoglobulin at about 50 mg/ml. The use of such serum in hybridoma culture medium can account for a substantial fraction of the immunoglobulins present in the culture fluids. To avoid contamination with bovine immunoglobulin, several companies have developed serum-free media specifically formulated to support the growth of hybridoma cell lines. In most cases, hybridomas growing in 10 per cent fetal calf serum (FCS) can be adapted within four passages (8-12 days) to grow in less than 1 per cent FCS or in FCS-free media. However, this adaptation can take much longer and in 3-5 per cent of the cases the hybridoma will never adapt to the low FCS media.

After this adaptation, cell cultures are allowed to incubate in commonly used tissue-culture flasks under standard growth conditions for about 10 days; mAb is then harvested from the medium. The above approach yields mAb at concentrations that are typically below 20 mg/ml. Methods that increase the concentration of dissolved oxygen in the medium may increase cell viability and the density at which the cells grow and thus increase mAb concentration.

Some of those methods use spinner flasks and roller bottles that keep the culture medium in constant circulation and thus permit nutrients and gases to distribute more evenly in large volumes of cell-culture medium. The gas-permeable bag (available through Baxter and Diagnostic Chemicals), a fairly recent development, increases concentrations of dissolved gas by allowing gases to pass through the wall of the culture container. All these methods can increase productivity substantially, but antibody concentrations remain in the range of a few micrograms per milliliter.

Most research applications require mAb concentration of 0.1-10 mg/ml, much higher than mAb concentrations in batch tissue-culture media (Coligan and others). If unpurified antibodies are sufficient for the research application, low-molecular-weight cutoff filtration devices that rely on centrifugation or gas pressure can be used to increase mAb concentration.

Alternatively, tissue-culture supernatants can be purified by passage over a protein A or protein G affinity column, and mAb can then be eluted from the column at concentrations suitable for most applications. However, bovine or other immunoglobulin present in the culture medium will contaminate the monoclonal antibody preparation. Either concentration step can be performed in a day or less with minimal hands-on time.

In short, batch tissue-culture methods are technically relatively easy to perform, have relatively low startup costs, have a start-to-finish time that is similar to that of the ascites method, and make it possible to produce quantities of mAb comparable with those produced by the mouse ascites method.

The disadvantages of these methods are that large volumes of tissue-culture media must be processed, the mAb concentration achieved will be low (around a few micrograms per milliliter), and some mAb are denatured during concentration or purification.

In fact, a random screen of mAb revealed that activity was decreased in 42 per cent by one or another of the standard concentration or purification processes.

SEMIPERMEABLE-MEMBRANE-BASED SYSTEMS

As mentioned above, growth of hybridoma cells to higher densities in culture results in larger amounts of mAb that can be harvested from the media.

The use of a barrier, either a hollow fibre or a membrane, with a low-molecular-weight cutoff (10,000-30,000 kD), has been implemented in several devices to permit cells to grow at high densities. These devices are called semipermeable-membrane-based systems. The objective of these systems is to isolate the cells and mAb produced in a small chamber separated by a barrier from a larger compartment that contains the culture media.

Culture can be supplemented with numerous factors that help optimize growth of the hybridoma. Nutrient and cell waste products readily diffuse across the barrier and are at equilibrium with a large volume, but cells and mAb are retained in a smaller volume (1-15 ml in a typical membrane system or small hollow-fibre cartridge). Expended medium in the larger reservoir can be replaced without losing cells or mAb; similarly, cells and mAb can be harvested independently of the growth medium. This compartmentalization makes it possible to achieve mAb concentrations comparable with those in mouse ascites.

Two membrane-based systems are available: the mini-PERM and the CELLine(Integra Bioscience, Ijamsville, MD). The Celline has the appearance of and is handled similarly to a standard T Flask but is separated into two chambers by a semi-permeable membrane and a gas-permeable membrane is on its underside next to the cell chamber. The mini-PERM has a similar design but is cylindrical and comes with a motor unit that functions to roll the fermentor continuously to allow gas and nutrient distribution.

Startup for these units costs about $300-800 and requires a CO_2 incubator. The advantage of membrane-based systems is that high concentrations of mAb can be produced in relatively low volumes and fetal calf serum can be present in the media reservoir with only insignificant crossover of bovine immunoglobulins into the cell chamber.

A disadvantage is that the mAb may be contaminated with dead cell products. Technical difficulty is slightly more than that of the batch tissue-culture methods but should not present a problem for laboratories that are already doing tissue culture. The total mAb yield from a membrane system ranges from 10-160 mg according to Unisyn literature.

In the hollow-fibre bioreactor, medium is continuously pumped through a circuit that consists of a hollow-fibre cartridge, gas-permeable tubing that oxygenates the media, and a medium reservoir. The hollow-fibre cartridge is composed of multiple fibers that run through a chamber that contains hybridoma cells growing at high density. These fibers are semipermeable and serve a purpose similar to that of membrane-based systems.

The hollow-fibre bioreactor is technically the most difficult of in vitro systems, partly because of the susceptibility of cells grown at extremely high density to environmental changes and toxic metabolic-byproduct buildup.

The hollow-fibre bioreactor is designed to provide total yields of 500 mg mAb or more. Startup of this kind of system usually costs more than $1,200. For those reasons, hollow-fibre reactors are used only if large quantities of mAb are needed.

The hollow-fibre reactor has been successfully used in many independent laboratories. If investigators are unable to invest the time or material costs, several institutional core facilities and government and commercial contract laboratories produce mAb from a hybridoma. For example, commercial contract laboratories typically charge $11/mg to produce 1,000 mg with hollow-fibre reactors.

Recently, several workshops, forums, and publications have discussed the use of the alternative methods to replace mice for production of mAb. Their conclusions indicate that alternative methods can often provide an adequate means of generating most of the mAb needed by the research community.

In vitro methods for producing mAb are appropriate in numerous situations, and it is the responsibility of the researcher to produce scientific justification for using the mouse ascites method. It is the responsibility of the IACUC to evaluate researchers'scientific justification and to approve or disapprove the use of mouse ascites methods.

Chapter 28

Use of Hormones in Animal Production

Hormone-dependent sex differences in growth rate have been known for a long time. It has also been known that growth rate and FCE (feed conversion efficiency) are higher in intact males than in castrates. It was natural, then, that the availability of hormones and other natural or synthetic substances displaying hormonal activity led to experiments aiming at their use to increase production.

Beginning in the mid-1950s, DES (diethylstilboestrol) and hexoestrol were administered to cattle increasingly in the US and the UK respectively, either as feed additives or as implants, and other types of substances also gradually became available. In general, such treatment has resulted in 10–15 per cent increases in daily gains, similar improvements in FCE and improvement of carcass quality (increased lean/fat ratio). Thus there has been a substantial reduction in the amount of energy required per unit weight of protein produced and the economic implications of this have been great.

While the use of hormonally active substances in animal production rose, opposition to their use also increased, because of the theoretical possibility that residues in edible tissues might endanger consumers. The factors leading to the ban on DES in the US, first imposed in 1973, have been described. Several reports confirm that DES endangers the health of animals and man, when repeatedly used in large doses. However, as regards risks due to the presence of residues in meat produced according to regulations, no documented deleterious effects have ever been reported in man, either from DES or any other substance with hormonal activity.

A distinction should be made between the hormones as such, for which the metabolism in the body is relatively well known, and synthetic or other substances for whose metabolic inactivation the body may not possess the enzymes necessary. When natural hormones are used in animal production, claims of zero-tolerance residue levels are not meaningful, since these compounds occur in detectable and highly variable concentrations in body fluids as well as in the tissues of all animals, treated or not.

For other substances with hormonal activity the situation is different. However, when residue levels are extremely low, it seems reasonable to weigh the potential risks against the undisputed positive effects some of these compounds have in animal protein production. This paper will discuss types of substances with hormonal activity currently in use or under investigation, their effects, mechanism of action, metabolism/ elimination, tissue levels, risks to the consumer and their economic importance.

Finally, other avenues to increased animal production as alternatives to use of hormones will be briefly envisaged. For the sake of simplicity the term *hormone* will be used, even if incorrectly, to cover all substances with hormonal activity, whether natural or synthetic.

HORMONE PREPARATIONS USED IN ANIMAL PRODUCTION

Hormones of Endogenous Origin

These comprise the "classical" steroid sex hormones, oestradiol-17â, testosterone and progesterone. The two former are used either in the free form or as esters, mainly those of propionic or benzoic acid.

Esterification generally causes prolongation of the half-life of the compounds in the body by 40 to 50 per cent. The natural hormones having low bioavailability when administered orally, owing to rapid conjugation and metabolic transformation in the liver, they are therefore administered by subcutaneous implantation.

Hormones of Exogenous Origin

Of the *oestrogens,* the stilbene derivatives diethyl-stilboestrol (DES) and hexoestrol possess high biological activity and have been used most widely. They are active orally as well as by implantation. Other orally active oestrogens include ethynyl-oestradiol, a more slowly metabolized derivative of the true hormone, with higher activity. An oestrogen with an entirely different structure is zeranol, a derivative of a resorcylic acid lactone occurring in the fungus *Giberella zeae.*

The synthetic androgens comprise a large number of substances, most of which are steroids. Of these, trenbolone acetate (TBA) possesses strong anabolic properties and has received much attention during recent years, used alone or in combination with an oestrogen. Another anabolic steroid is methyl-testosterone.

Of synthetic gestagens, only one will be mentioned here: melengestrol acetate, which stimulates growth in heifers but not in steers, and which can also be used for the suppression of oestrus. Numerous other gestagens also exist, but at present few other than progesterone and melengestrol acetate are used to stimulate growth. In addition to these substances, numerous others exist, and some of them are used more or less frequently in clinical veterinary medicine. However, clinical applications of hormones are not considered to be of consequence to the consumer, since such treatment is much less frequent than the use of hormones to promote growth.

Hormone preparationsin current use as growth stimulants are listed, which also shows modes of application, dosages, etc. It will be noted that almost all preparations currently in use are based on implantation, the site usually being the base of the ear, or less frequently, the dewlap.

RANGE OF APPLICATION

In *cattle* the use of hormones is limited to veal calves and beef cattle. *Veal calves* are produced mainly in continental Europe, to an extent of about 8 million per year. Research has demonstrated that hormone treatment improves growth rate, nitrogen retention and FCE during the five- to six-week period before slaughter.

Beef cattle, including steers as well as heifers, were treated in large numbers, especially in the USA and the UK, with DES or hexoestrol, administered orally, until the use of these compounds was restricted. During the last several years, practice has changed dramatically in the direction of increased use of implants of natural steroids, synthetic anabolic steroids and the phyto-oestrogen zeranol.

In *sheep*, especially in wether lambs, some increase in gain has been reported but results are somewhat ambiguous.

In *swine*, hormone treatment may increase growth rate, FCE and lean/fat ratio of the carcass in male castrates.

Poultry generally do not appear to respond to oestrogens by increased gain but by changes in lipid deposition. In male and female turkeys, androgens have recently been reported to increase growth rate as well as FCE.

MODES OF APPLICATION

When DES was used as a feed additive, a usual procedure was to start treatment of steers at a body weight of 360 kg and continue administration for 120 to 170 days. Since restrictions on its use were imposed, most preparations have been administered as implants, whose effect is usually limited to 80 to 100 days. Practice varies with management systems. Animals may be implanted at live weights from 270 to 450 kg.

Depending upon the age and weight at the time of implantation, the animals are either slaughtered at the end of this first period, or fed for an additional period, either without further treatment or after a second implant to act for another 80 to 100 days. Most types of implants in use are not removable, but removable types have recently been tested and their effects described. When tested in steers, no reduction in performance was recorded when the implants were withdrawn 32 and 39 days before slaughter.

Implantation is subcutaneous, usually at the base of the ear, thus eliminating the risk that residues of the implantation site will be present in edible tissue.

EFFECTS OF HORMONES

Veal Calves

In veal calves, hormone treatment may begin at a body weight of about 65 kg, the animals being slaughtered at about 170 kg. Implants of 20 mg oestradiol-17â + 200 mg progesterone in males and 20 mg oestradiol-17â + 200 mg testosterone in females resulted in a 20 per cent increase in daily gain and 21 per cent higher nitrogen retention in the period studied. In other studies, improvements of 10 to 12 per cent in gain and 10 per cent in FCE have been reported.

Nitrogen retention is about 70 per cent in the very young veal calf, but decreases gradually to below 40 per cent at the age of about 15 weeks. For ages of 10 to 15 weeks, the average conversion of feed protein to body protein is about 40 per cent; this rate can be increased to about 60 per cent by hormone treatment. The effective preparations were DES, oestradiol-17â, and the combination of TBA + oestradiol-17â. More recently, positive effects have been reported for zeranol alone (36 mg) and for zeranol (36 mg) + TBA (140 mg), with increases in nitrogen retention of the same order as for DES and E_2 + TBA. When zeranol + TBA was implanted at the age of 56 days, the growth rate up to day 106 increased by 18 per cent.

Steers

The most extensive studies of the effects of hormones on growth and FCE have been carried out on steers, under strictly controlled conditions as well as in the field. Since 1975, most studies have involved implants of oestrogens alone, androgens alone, or combined oestrogen/androgen preparations, although many trials have also been based on oestrogen/ progesterone combinations during recent years.

Oestrogen implants have included DES, hexoestrol, oestradiol-17â and zeranol. *DES implants* have, as in previous studies, resulted in an increase of about 12 per cent in gain and in improvement in FCE of the order of 10 per cent. *Hexoestrol implants,* usually in doses of 30 to 60 mg, have been shown in numerous experiments to lead to considerable improvement in growth rate and FCE. In 19 trials carried out over the years on experimental husbandry farms in the UK, the overall average increase in gain produced by 45 or 60 mg hexoestrol implants was 0.16 kg a day, and in only 2 of the trials was it less than half.

Oestradiol - 17â implants alone (30 mg) have resulted in a 24 per cent increase in gain and a 13 per cent improvement in FCE. *Zeranol implants,* usually at 36 mg, have consistently improved gain as well as FCE. In a series of 21 UK trials over several years, the average response to zeranol implants alone was an increase in daily gain of 0.15 kg. Only in one trial was there no response. Similar results have been obtained in Ireland. Positive effects on gain in steers have been observed under a variety of experimental conditions, under controlled feeding, on *ad lib* feeding of standardized rations, and on pasture. *TBA implants* administered alone at a dose of 300 mg have also had positive effects on growth even if combination with an oestrogen has yielded better responses (*vide infra*). In a series of 8 trials in the UK, the average additional daily gain amounted to 0.09 kg, with considerable variation among trials. Similar results have been reported from Ireland.

Combined Preparations

A number of trials have been carried out with implants containing two hormones. The combination of an oestrogen with an anabolic steroid, or with progesterone, has met with the greatest responses. *Synovex-S* has consistently increased gain as well as FCE, with responses averaging about 20 per cent and 17 per cent respectively *Hexoestrol + TBA* (usually 30 or 45 mg hexoestrol + 300 mg TBA) has resulted in marked increases in gain of the order of 30 per cent and in FCE of the order of 20 per cent. *Oestradiol-17â + TBA* (20/140 mg) has given similar results as has *Zeranol + TBA* (36/300 mg), also recently tested. Hormone preparations have also been tested in combination with substances such as monensin, which increase

FCE by promoting propionic acid formation in the rumen. Results have varied from no effect to marked additional gain.

Reimplantation, tested under various forms of management with varying results has not gained general acceptance. Lamming has stated that "repeat implantation of hormone is not likely to produce the benefits obtained from its initial use, since a second dramatic change in the endocrine balance of the animal is not likely to occur. In addition, double implantation increases the possibility of exceeding the optimum dose rate and the chance of deleterious side effects occurring." The evidence for highly significant positive effects on the growth rate and FCE of steers is thus beyond dispute, the most marked effects being provoked by implants combining an oestrogen with an androgen of high anabolic activity.

Bulls

Since the entire male animal produces its own anabolic androgen, testosterone, an effect of additional hormones similar to that for steers is not to be expected. The number of trials with bulls is also limited. Positive effects on gain have been reported using DES alone and combined oestrogen/TBA implants in other studies, no effect on gain has been recorded while a certain increase in the deposition of fat in the carcass has been observed.

Heifers

Recent trials with beef-producing heifers have mostly been based on the use of an androgen, although oestrogens have been tested, alone or in combination. Thus, zeranol has been reported to increase gain while in other trials no response has been observed. TBA administered alone (300 mg) has led to increases in weight gain and FCE of the order of 36 per cent and 25 per cent respectively.

In other trials, combinations of an oestrogen with TBA or testosterone have yielded significant growth responses. In general, it appears that the effect of TBA alone in heifers corresponds closely to the effect of combined oestrogen/TBA implants in steers.

Sheep

Trials have mainly concerned wether lambs, and positive effects of hormonal treatment have been reported using DES hexoestrol and zeranol although other reports have indicated that zeranol yields no significant effects. Wether lambs implanted with TBA + oestradiol-17â have shown increases in gain, carcass weight and FCE. In general, however, the results obtained in sheep thus far do not warrant the same clear-cut conclusions as for steers and heifers.

Swine and Poultry

There is little evidence that existing hormonal preparations influence the growth rate and FCE to an extent that would be interesting from a practical point of view. The lean/fat ratio in male castrate and female pig carcasses may be increased by the use of oestrogen/androgen combinations. In poultry, redistribution of fat in the body is a known effect of oestrogens. Recent research indicates improved growth rate and FCE using androgens in young male and female turkeys.

Undesirable Side Effects in Treated Animals

Reported side effects of hormone treatment for growth stimulation are few and generally concern the use of oestrogens in steers. Changes in body conformation such as feminization and raised tail-heads were described as early as 1958. Similarly, bulling has occurred with increased frequency although in most animals it is limited to the first few days after implantation. However, it has been reported from Kansas that 2.2 per cent of all steers fed in pens have to be removed, at an estimated loss of $23 per head. In a study of the effect of reimplantation of oestrogens in steers, all animals were given a 30 mg DES implant at a live weight of 260 kg, and then reimplanted 91 days later, with either 30 mg DES or Synovex S. Following the second implant, the frequency of the steer-buller syndrome was 1.65 per cent for the DES-DES group, and 3.36 per cent for the DES-Synovex S group. The economic advantage of using DES + DES was estimated at $1.15 per head. The steer-buller syndrome is a special problem in feedlots.

Chapter 29

Action of Hormones

No reliable explanation of how the growth-promoting hormones act has yet been furnished. Some observations indicate an indirect influence through changes in the balance of endogenous hormones. Thus there have been reports of DES and TBA increasing the levels of growth hormone and/or of insulin in plasma these hormones are known to stimulate amino acid transport across the cell membrane.

However, others have found no such effect. Bulls fed DES (10 mg/day) over two years had significantly higher plasma testosterone levels than controls those levels are positively correlated with growth. Recent experiments indicate that DES reduces the rate of muscle catabolism in steers.

As regards the anabolic androgens, evidence exists indicating competition with glucocorticoids for receptor sites on the muscle cell membrane. Since glucocorticoids have a catabolic effect on tissues, their displacement from muscle cells would reduce catabolism. TBA alone, and even more when combined with oestradiol-17α, causes a marked decrease in the concentration of total thyroxin in plasma of steers. In another study, combined oestradiol-17α-progesterone implants (20 + 200 mg) in steers caused a uniform but slight increase in thyroxine binding capacity. The significance of these findings is not yet clear. For a fuller discussion of possible mechanisms of action of the hormones.

ENDOGENOUS HORMONES IN BODY FLUIDS AND TISSUES

Any discussion of possible health hazards connected with

the use of hormones in animal production must take into account the normal occurrence of hormones and their metabolites in body fluids and tissues, and the fact that the levels of these hormones vary greatly, according to the physiological state of the animal. Thus, oestrogen levels in the blood of female farm animals may vary from a few pg up to 5–6 000 pg per ml plasma.

As to males, the plasma of stallions and entire male pigs contains high levels of oestrogens, although mainly in the conjugated form. Milk also contains oestrogens in very high concentrations in the first drawings after parturition; in non-pregnant animals, levels in the range of 80–100 pg/ml have been reported. For the sake of comparison, levels of oestrogen activity normally present in products of plant origin widely used in human nutrition are included.

METABOLISM, ROUTES AND RATES OF ELIMINATION

The general patterns of metabolism and elimination of endogenous hormones in farm animals have been outlined. In ruminants, testosterone and oestradiol-17α are rapidly converted to their epimers, biologically much less active, epitestosterone and oestradiol-17α. Progesterone is partially converted to androgens before excretion. In the pig, epimerization of testosterone and oestradiol-17α does not appear to take place to a significant degree. The faecal route of elimination dominates in ruminants, while in the pig urinary excretion is more important.

Progesterone

After repeated injections of progesterone to cows and steers over 2 to 3 weeks followed by ^{14}C-progesterone for 2 to 5 days, the animals were slaughtered 2 to 3 hours after the last injections. Activity levels were 2 to 7 times higher in the fat, 3 times higher in the kidneys, and 13 times higher in the liver than in the muscle.

Excretion of radioactivity amounted to 50 per cent and 12 per cent in faeces and 2.0 per cent and 1.2 per cent in urine in cows and steers respectively. About 50 per cent of the activity in muscle and milk was associated with unchanged progesterone, most of the remaining activity being associated with a mono-hydroxy compound. Cooking or frozen storage did not affect the nature or quantity of metabolites.

Oestradiol-17β

Following daily injections of 1 mg oestradiol-17α or its benzoate to heifers and steers for 11 days, followed by the ^{14}C-compounds on days 12, 13 and 14, the animals were slaughtered 3 hours after the last injections, when residual levels were maximal. In muscle extracts, oestradiol-17α represented the major fraction of extracted activity (38 to 71 per cent), followed by oestrone (17 to 45 per cent).

Levels in muscle were 161 to 225 pg/g and 40 to 86 pg/g for oestradiol-17α and oestrone respectively. In fat the levels were 3 to 5 times higher. The authors conclude that residual levels are extremely low when these hormones are administered as growth stimulants to growing/finishing cattle. Glucosides of the 17α- and the 17α-epimers, and the glucoronide of the 17α- epimer are the major metabolites in cattle. When oestradiol-17α was administered orally to swine, plasma concentrations were very high 7 min after administration. Oestradiol was completely conjugated during absorption and its first passage through the liver. Some conversion to oestrone took place.

DES

The metabolism of DES in food-producing animals has been reviewed recently. The substance seems to be eliminated to a large extent in unaltered form. After oral administration of ^{14}C-DES to beef cattle, 99.5 per cent of the radioactivity was excreted within 5 days after withdrawal. In liver extracts, radioactivity associated with DES-conjugate and free DES was found to be 75 per cent and 25 per cent respectively.

Higher than background levels of activity were observed after withdrawal in kidney, liver, bile and urine/faeces for up to 5, 7, 9 and 11 to 12 days respectively.

The fate of 24-mg DES implants containing ^{14}C-DES and implanted in the dewlap of calves was studied over 98 days. Free radioactivity was almost completely associated with unchanged DES. At the time of slaughter, levels were less than 0.1 ppb in muscle and fat, and 1 to 1.5 ppb in liver and kidney. In a study in steers implanted with ^{14}C-DES, on day 120 after implantation radioactivity in muscle was not distinguishable from background.

It was above background in spleen, lung, adrenal glands and kidney, but less than levels corresponding to 0.5 ppb. In a similar study on steers, 120 days after implantation, levels in liver, kidney, lungs and salivary glands were in the range of 0.07 to 0.13 ppb of DES equivalent. In a recent study of DES metabolism in rhesus monkeys and chimpanzees, most of the substance was excreted with the urine.

Extracts in the organic and aqueous phase mostly contained unchanged DES in the free and conjugated form respectively. Current evidence indicates that the oxidative metabolism of DES leads to at least three compounds that may have cytotoxic or mutagenic activity but these have not been identified as DES metabolites in ruminants, but in the mouse.

Zeranol

Using a gas chromatographic method with a sensitivity limit of 20 ppb, no residues of zeranol could be detected in edible tissue from cattle slaughtered 65 days following implantation of 36 mg, or from lambs 40 days following implantation of 12 mg. In another study, tritiated zeranol was implanted in cattle as part of 36-mg doses. Skeletal muscle obtained 10, 30 and 50 days following implantation contained no detectable residual activity. This confirms previous results based on the use of ^{14}C-labelled zeranol.

Trenbolone Acetate (TBA)

Trenbolone is a 17α-OH steroid esterified in the 17 position with acetic acid. Upon release in the organism the ester is rapidly hydrolyzed to the free compound TB-17α-OH and acetate. In cattle the 17α-OH compound is rapidly transformed to its 17α-OH epimer, in the same manner as oestradiol-17â in this species.

The 17α epimer possesses only about 5 per cent of the biological activity of the 17α epimer. Another metabolite of TBA in cattle is the 17-keto compound, analogous to oestrone; quantitatively it appears to be of very little importance. Following intravenous injection of TBA, levels of TB-17α-OH and TB-17α-OH of 0.05 and 0.005, 0.10 and 1.0, 0 and 191 ppb have been recorded for muscle, liver and bile respectively. Other metabolites occurred in extremely small quantities in cattle. Similar findings have been reported in studies based on the use of implants.

The major route of excretion is by faeces. Metabolism studies of TBA thus clearly show that the substance is rapidly subjected to biological inactivation in cattle, mainly by epimerization of the free steroid to the 17α-compound, and that the major route of excretion is via the bile.

HORMONE-TREATED ANIMALS

Much work has been devoted to the development of sensitive methods of detecting hormone residues in meat from hormone-treated animals. As regards compounds given orally, it should in principle be possible to realise claims of zero-tolerance residue levels, by selecting the proper withdrawal time. During recent years, the use of implants has, however, gained in importance. While removable implants have been tested in steers, with no decrease in performance when withdrawn 32 and 39 days before slaughter the wide use of non-removable implants makes residue studies important. Determination of normal levels of endogenously produced natural hormones is also important, to enable risk evaluation to be carried out in realistic terms.

Several residue studies have been made of synthetic as well as natural compounds, mainly in cattle. When regulations governing dose, sites of implantation and timing in relation to slaughter are adhered to, residue levels of DES hexoestrol and oestradiol-17α in edible tissues have generally been in the lower ppb to the ppt range, i.e. from a few ng/g down to some hundred pg/g of tissue. In the latter case there was almost complete overlap between values for untreated and treated steers after 105 days. Zeranol implants have so far not left detectable residues in edible tissue.

Most studies of androgens have concentrated on TBA. The ester being rapidly hydrolyzed, measurements of residues have been limited to the free compound and/or its major metabolite. Results based on radio-immunoassay of extracts or on radioactivity measurements have indicated levels in edible tissue of the order of 1 ppb or below. In a recent study using implants containing tritiated TBA in heifers, it was found that when slaughter took place 60 days after implantation, the major proportion of tritium-containing residues was not extractable with organic solvents. In muscle 95.5 per cent, in liver 94.4 per cent, in kidney 98.8 per cent and in fat 59.1 per cent of the radioactivity remained in the aqueous phase, not quantifiable by radio-immunoassay. This suggests that the major part of the residues after TBA implantation occurs in a non-extractable, possible covalently bound form in tissues.

Residue levels of gestagens have been also measured, in connection with their use as growth stimulants. Residues of melengestrol acetate used as a feed additive in daily doses of 0.25 to 0.50 mg per head have consistently been below the sensitivity levels of the methods used (i.e., below 10 ppb in fat, liver, muscle and kidney), whether or not the compound was withdrawn 48 hrs before slaughter.

HORMONES IN FOOD

According to the Agricultural Research Service, United States Department of Agriculture (ARS), the average per caput consumption of beef is 157 g per day in the US. Calculations show that 157 g of beef from an animal implanted 61 days before slaughter with a combined implant containing 20 mg oestradiol-17α + 200 mg progesterone or testosterone will contain 3.43 ng oestrogen and 19.5 ng progesterone or 16 ng testosterone.

Normal levels of these hormones in certain dairy foods, shows that some foods represent hormone sources vastly richer than meat from hormonetreated animals. Based on these values, and averages for consumption of various foods, the relative contribution of meat from hormone-treated animals to the total consumption of hormones has been calculated on the assumption of proper use of the hormones.

Table: Hormones in Certain Dairy Foods

	Oestrogens (pg/ml)	*Progesterone (ng/ml)*
Milk, from non-pregnant cows	80	9.5
Milk, from pregnant cows	126	
Cream		73
Butter		133

It is clear that in most cases the contribution from meat of treated animals is insignificant when hormones have been properly used, and must be considered to be biologically without impact. This becomes even more evident when seen in relation to normal endogenous hormone production in man. It will be seen that even for oestrogens, the hormones considered the greatest risk, the maximal contribution from meat (assuming proper use of the hormones) is less than 0.01 per cent in the prepubertal boy who represents the lowest endogenous oestrogen production.

Thus far the discussion has been limited to the natural hormones. For synthetic substances the situation may be different. But again, considering the very low residue levels found *when* hormones have been properly used, the question may be raised whether the risk to the consumers is being grossly overestimated.

The figures represent effective fractions (i.e. 10 per cent of real fractions), to take into account the low bio-availability of the hormones absorbed orally.

In the production of meat for human consumption, a hormonally-induced increase in growth rate of the order of 10 per cent evidently has major economic implications. The improvement in FCE which usually accompanies the increase in gain adds to the economic benefits, and at the same time makes possible greater production of edible protein per unit energy used, and this in itself is of importance in a world lacking in protein supplies. Some of the hormones that have become available recently appear on average to increase gain as well as FCE considerably beyond the 10 per cent level, and in examining whether they should be approved for use in animal production, the risk/benefit analysis must take this fact into account.

Few analyses of the economic advantages of using hormones as growth stimulants appear to have been made. For the UK, a recent calculation is based on the estimated increased return to producers for 1 350 000 cattle treated over a 12-month period. Assuming that 1 155 000 of these were steers and 195 000 were heifers, and that the estimated daily gain was only 0.06 to 0.11 kg for steers and 0.05 to 0.06 kg for heifers, depending on the preparation used, the overall gross increased return was calculated at £21 306 000, without taking into consideration improvements in FCE.

Estimated from sales of the preparations during a year. Based on results from Meat and Livestock Commission trials using yard finishing cattle receiving only one implant. Based on 1978 data showing that 0.1 kg increase in daily gain gave an increase in gross margin of £13 per head, and that increase in slaughter weight averaged 85 p per head. From these figures are subtracted the cost of treatment.

Does not include costs of veterinary services, etc.

These calculations must be taken as an example only. Availability of the various feeds, variations in feed and product prices as well as in types of management from time to time and from place to place may play an important role. However, shortening the time required for producing a certain weight at slaughter will represent an economic advantage, especially under feedlot conditions, since non-feed costs also contribute significantly to the total cost of production (10 to 18 cents per head per day in the USA).

ALTERNATIVES TO THE USE OF HORMONES

Growth rates are influenced by many factors, especially

genetic constitution and feeding. Over time, selection as well as improvements in management systems, feed composition and feeding programmes have contributed much to increasing productivity in meat as well as milk. Although it is difficult to evaluate the exact relative contributions of these factors, the overall improvements have been dramatic.

An example is the increase in milk yield per head in US dairy cattle. In the period 1944–1975, the number of dairy cows decreased by 33 per cent, while the average yield per cow increased by 60 per cent. These gains represented a saving of about 23 billion kg of total digestible nitrogen per year, the volume of milk produced remaining relatively constant.

The saving is equivalent to about 1.1 billion bushels of maize. Data illustrating progress in beef production over the years are scarce, but increases in productivity similar to those for milk production are unlikely. In addition to the use of hormones, many avenues are still open for increasing productivity in meat and milk production including breeding programmes, regulation of rumen fermentation, optimalization of the balance between the indirect and direct feeding of the ruminant organism proper, and disease control.

Breeding Programmes

Systematic selection of high-quality sires, combined with an increase in the number of offspring from high-yielding females through embryo transfer, may bring about further improvements in beef and milk production. In many countries, development along these lines has hardly begun. However, the establishment of effective breeding associations and the strict organization of programme planning and execution are prerequisites for realizing the potentials in this sector.

Regulation of Rumen Fermentation

The microbial systems in the rumen are extremely complex, and the balance between the various strains of bacteria is susceptible to changes brought about by many factors. The recent introduction of substances such as monensin offers great promise in altering the fermentation pattern to the benefit of productivity by increasing FCE. Since the very extensive breakdown of carbohydrates and protein represents loss of much energy, research is currently being conducted in many laboratories in order to find new methods of increasing FCE.

Indirect and the Direct Feeding of the Ruminant Organism Proper

To a large extent, feeding a ruminant means feeding the rumen microbes which then themselves serve as feed for the organism proper. This is indirect feeding, expensive in energy. On the other hand, the ruminant possesses, in the postruminal part of its digestive tract, all the enzymes necessary for utilizing all types of nutrients except cellulose.

The rumen microbes are necessary for the utilization of cellulose, which globally represents an enormous source of energy. However, it is possible to sustain an adequate microbial population in the rumen even when ruminal breakdown of part of the easily digestible nutrients is prevented.

Enabling nutrients to bypass the rumen will increase the utilization of feed for production, and also create a more adequate supply of amino acids. Increased rumen bypass of nutrients can currently be brought about by several means, including formaldehyde and heat treatment of protein-rich feeds. A third method, aiming more at specific substances that may be rate-limiting for production (e.g. certain amino acids), or of significance in treatment of diseases, is protection against rumen degradation by such means as incorporation into the ration of long-chain fatty acid mixtures in the form of small pellets. In the future, new methods of increasing rumen bypass will undoubtedly contribute significantly to increased productivity of ruminants.

Disease Control

Whatever management system is adopted, effective disease control is essential for productivity. In many areas of the world, infectious and parasitic diseases inflict heavy losses on animal production. A recent study has disclosed nearly a one-to-one relationship between investment in agricultural research and annual productivity of edible protein in ruminants.

An increase of about 45 per cent in scientist/man years and a corresponding increase in funding for research and development is considered sufficient to raise productivity in this sector by 50 per cent. Investment in disease control is an important aspect of this work. Annual world mortality losses from disease exceed 50 million cattle and buffalo, and 100 million sheep and goats. Non-lethal diseases are believed to lead to an equivalent reduction in production. Thus, investment in disease control holds great promise for future augmentation of animal protein production.

In these perspectives, the significance of hormones in animal production may seem marginal, leading to the question of what priority to give to the various efforts to increase productivity and production. In the global context it is, however, at least at present, impossible to adopt one approach to the exclusion of others. As long as preparations exist that combine positive effects on yield and feed utilization with low or non-existing risk to the consumer, there will be a market for them. What is more, the use of hormonally-active substances in the future may not be limited to those currently available. Common to the present compounds, natural or synthetic, is that they are degraded in the body only to a limited extent.

An entirely different situation exists for proteid hormones, which are broken down completely to amino acids, leaving no residues whatever. An example is the growth hormone which not only stimulates growth but also milk secretion, even in high-yielding cows. This anabolic hormone is currently available only in small quantities for research.

However, a recent breakthrough in the use of recombinant DNA technique has made large-scale microbial production of species-specific peptide hormones a realistic possibility. Combined with the development of miniaturized automatic delivery systems for subcutaneous use, a new era may be visualized as regards the use of hormones in animal production.

Measuring Hormone Levels in Animal Products

The question of the use in animal production of anabolic agents, including the natural steroid hormones and their derivatives, has been — and continues to be — badly stated, and therefore misunderstood. Those who brought the problem before the public had neither the information nor the technical competence, especially in biology, to speak with perfect objectivity.

Hormones are found naturally in many animal products and, as phytohormones, in many plants. Only some of them raise public health problems, and these should be distinguished from the others. Indeed, a study of the practical and health aspects of the use of hormones in animal production should underlie all regulation of the matter, and therefore any decision on the measurement of hormone levels.

A considerable volume of documentation exists on the problem which is as important for the developed as for the developing countries. Items 1 to 11 in the Bibliography of this paper indicate a number of general studies.

Use of Hormones in Animal Productions

A reading of many popular, and even official, papers reveals that different types of anabolic agents, whatever their structure, are grouped together under the general name of hormones, among them those whose activity as oestrogens is emphasized and those which reinforce masculine traits.

The first type includes those not found in mammals and birds but obtained exclusively through organic synthesis: diethylstilboestrol (DES) or stilboestrol, dienoestrol, hexoestrol and their derivatives. These compounds, active when administered orally or as implants, are harmful for laboratory animals, whether given directly or as residues in the flesh and offal of treated animals.

The use of these substances, a potential danger to human health, is prohibited in many countries, particularly since the methods of measuring DES levels in meat are unreliable, so that measurements must be made on the excreta. The second type of anabolic agents consists of the natural steroids, normally present in all animals: testosterone, oestradiol, oestrone, progesterone. Secreted by the endocrine glands, these substances and their derivatives, although absorbed by the intestines, are deactivated in the liver. They produce measurable effects only when implanted.

The third type, found in the higher plants and moulds, consists of substances with a more or less clear oestrogenic and, at times, anabolic activity. Among these phytoestrogens, only zearalenone has been isolated for use in animal production: as zeranol, it gives good results, but has an effect on genital formation and the level of thyroxine in plasma. It should be added that Sharaf and Gomaa attribute oestrogenic effects to Vitamins E, B_6, C and perhaps A, while Nelson *et al.* have studied the same effects in O-p′-DDT. This paper discusses only the consequences for the consumer of the use in animal production of the first two types: the synthetic agents such as DES and the natural steroids oestradiol, progesterone and testosterone.

First mention should be made of cases of cancer of the vagina and hypogonadism observed in the offspring of women treated with DES. Such observations are verified by research based on the so-called relay methodology based on the fact that methods of analysis for a residue and its metabolites fail to reflect practical realities, since they may lead to findings of residues without bioavailability or vice versa.

Further, the individual biological evaluation of each metabolite fails to reflect the overall consumption of all of them. The relay, a farm animal, changes this situation. In order to determine whether this was true of the first two types of anabolic agents, rats and mice of both sexes received balanced diets containing 20 per cent of the flesh or 6 per cent of the liver of male and female calves implanted with DES (24 mg in each of two implants), oestradiol and progesterone (20 + 200 mg in two doses) or oestradiol + testosterone (20 + 200 mg in two doses). Animals were implanted at the beginning of the growth period and 60 days later, 38 days prior to slaughter. Chemical measurement of residues showed DES levels of 30 g/kg in the flesh, but none in the liver. Levels of natural steroid hormones did not exceed levels in untreated animals.

Feeding rats and mice on the flesh of DES-implanted calves led to growth in both sexes, sterilized the females and sharply atrophied the testicles, seminal ducts and penis of the males. Liver behaviour was less clear. Pokrovski *et al.* and Nesterin obtained similar results with flesh. It was also observed that DES persisted for nearly 40 days in the environment and was found in alfalfa.

Animals fed on the flesh and liver of calves implanted with oestradiol + progesterone or oestradiol + testosterone behaved in the same manner as controls fed on the flesh and liver of control calves. After 24 months, the percentage of tumors (primarily mammary adenomata, frequent in old Wistar rats) was the same in both groups.

Trenbolone acetate (TBA), a testosterone derivative, leaves little or no residue. Measured by radio-immunoassay, Gropp *et al.* also studied their biological effects by the relay toxicity method, on Sprague-Dawley rats fed on the flesh of calves implanted with a mixture of 20 mg oestradiol-17â + 140 mg TBA, as well as with doses 10 and 25 times higher. The results of this study, which continued for 110 weeks and covered two generations, showed the absence of any unfavourable influence on growth, organ weight, reproduction, teratogenicity and cancerogenicity.

Only females fed on the flesh of control calves to which TBA had been added in doses 25 times normal, had significantly lower growth($P< 0.05$). The biochemical constants in the blood were the same in both groups. These experiments show that compounds of natural steroid male and female hormones, and their derivatives, appear to be without danger for the consumer.

They are useful in animal production, obviating a number of often costly and polluting therapeutic operations.

On the contrary, the use of DES-type non-steroid anabolic agents has serious disadvantages. It is difficult to detect their residues in meat. Relay toxicity results show that analyses and biological behaviour fail to agree. The indiscriminate prohibition of all types of anabolic agents may be leading to the use of DES and similar substances.

Analyses of urine in which they can be identified most easily, have revealed them in 20 per cent of cases. In the United States, McClung estimates that 500 000 bovines are being treated illegally. In Europe, illegal use appears to be fairly general, but is tending to decrease.

The use of DES and similar substances should thus be prohibited, while that of the steroids and their derivatives should be authorized under veterinary control. This was the view of many writers as early as 1967, and their conclusions were confirmed by the FAO/WHO Joint Committee of 1975 and a recent international symposium. Any new anabolic agent should be severely tested before it is distributed widely. There is a need to develop practical, economic and rapid — but very exact — methods of detecting residues, particularly those originating in products chemically different from the natural steroid hormones and their derivatives.

MEASURING HORMONE LEVELS

It should be clearly understood that this discussion concerns only measurement methods intended to detect residues remaining in flesh and viscera. These methods are, however, not very reliable, and other biological material must be examined to determine whether the animals have been treated. Specialists agree that there are still difficulties in testing for residues in the tissues of animals intended for use as food.

Many different methods have been proposed and discussed at the EEC level. While the trend is increasingly toward the use of radio-immunoassay, it is undeniable that the techniques are highly complex and still far from being perfected. DES and similar substances should preferably be looked for in urine and faeces.

Account should also be taken of the fact that no difference may be found between animals properly implanted with anabolic agents and untreated animals. This problem has received insufficient attention. The proceedings of the FAO/WHO Joint Meeting of March 1975 contain reports emphasizing this lack of difference and the relatively high, but physiological, levels of testosterone and progesterone in the flesh of bulls and gestating cows, respectively. This being the case, how can the natural presence of these substances be distinguished from their presence due to the use of natural products?

Further, it should not be forgotten that very many animal feeds contain phytohormones which can cause modifications in the morphology and histology of certain organs. After summarizing and stating the principles of the four main groups of useful techniques enumerated by Kroes *et al.* and Richou-Bac and Pantaléon — histological, biological, physico-chemical and radioimmunoassay methods — this paper will study in detail the techniques that can be applied in practice for routine control.

Histological methods used specially in young bovines, consist in the observation of modifications in the male and female genitalia.

The prostate and the bulbo-urethral glands are particularly examined in male calves and lambs, while in females attention is focused on the glands of Bartholin, the teats, the vagina and the cervix. Modifications are obvious when DES and zeranol, as well as hexoestrol, have been used. In pigs the effects are less discernable.

It is generally agreed that the histological method makes it possible to conclude that an oestrogen has been used at some earlier time. The use of natural steroid oestrogens, in combination with testosterone or progesterone, leads only to slight histological modifications. The phyto-oestrogens may also have some effect and lead to modifications; this is the case of Fusariuminfected maize.

The biological methods used are those of Astwood which measures the increase in uterus weight of pre-puberal mice and rats, and of Allen and Doisy based on modifications of the cytology of the vagina of adult mice after ovarectomy. The latter test, applied by Stob *et al.* in 1954, is the most sensitive, but it is responsive only to compounds with oestrogenic activity. Martin reports that the use of concentrated extracts placed in the vagina of mice also gives good results.

Physico-chemical methods, which make it possible to identify with precision the anabolic agent sought for, require a fairly long extraction, followed by thin-layer or gaseous-phase chromatography.

Hans and Abraham report that the latter techniques, combined with mass spectometry, can now be contemplated. Sensitivity is high. Further off is the possible use of liquid/liquid chromatography. Thin-layer chromatography is currently used, and Verbeke gives details in an EEC document while Waldschmidt and Schuller and Stephany have described the method.

Radio-immunoassay, first described for the oestrogenic steroids by Jiang and Ryan is based on competition between tritium-labelled and unlabelled antigens for the corresponding specific antibodies.

This technique, increasingly used, is highly sensitive, making it possible to detect 0.05 to 0.005 ppb of DES and 0.1 ppb of TBA. It is, however, complex and long, although it is being improved. It has the same advantages as all uses of radioactive compounds. The technique has been well studied by Hoffmann and Hoffmann and Karg.

Two new methods may also be mentioned: high-pressure liquid chromatography which, according to Richou-Bac and Pantaléon does not appear to be sufficiently sensitive, and the so-called Elisa method, applied by Ruitenberg *et al.* in the Netherlands to detect pig infection by *Trichinella spiralis*. This method, which uses an enzyme for labelling, followed by measurement with 450 nm spectrophotometry, needs further development, but it may be possible to automatize its use. Kroes, working at Bilthoven (Netherlands) is studying how it can be used to measure hormone levels.

In practice, the most urgent problem appears to be the detection of DES. Bories *et al.* and other authors have shown that it is eliminated in the urine and faeces. It has also been demonstrated that the saliva does not contain DES but that it does contain progesterone.

DES cannot be detected in the flesh of animals treated three weeks before slaughter. It can be detected in the urine after one month by thin-layer chromatography and radio-immunoassay.

The bladder need only contain from 20 to 50 ml of urine for the thin-layer chromatographic method to be applied. The method, used in 1973 to detect oestradiol residues, was modified and improved by Verbeke in 1979. Vogt and Oehrle have employed it for detecting steroid oestrogen residues and Stilbene in calves' urine.

Ryan and Hoffmann use it, together with radio-immunoassay, to study linked trenbolone residues.

For the latter compound, according to Richou-Bac and Pantaléon Schuller and Stephany believe it possible to use thin-layer chromatography to attain a sensitivity level of 0.005 ppb, comparable to that for radioimmunoassay.

Radio-immunoassay is based on competition between labelled and unlabelled antigens, i.e. a labelled hormone (indicated below by H*) and an unlabelled hormone (H), for the same antibody molecule. The diagram below shows the process of the reaction.

The levels of AB and H* being constant, any increase in H provokes a reduction in the labelled hormone localized in the specific antibody (H* - AB). The measurement, established by calculating the relationship

$$\frac{H^* - AB}{H^*}$$

depends on the measurement of radioactivity levels between the localized labelled hormone and the free hormone.Kroes *et al.* and Richou-Bac and Pantaléon give details of extraction and measurement techniques. Hoffmann and Laschuetza have used the method to examine blood plasma and flesh for DES.

The hormone is rendered antigenic by localization on seric albumen. Labelling the hormone should in no case alter the molecule.

The specificity of immunoassay depends on the intrinsic characteristics of the antibody and the efficiency of the hormone purification method. Many techniques exist for separating the free H*, but immunoprecipitation appears to be the method most commonly used. The immunological behaviour of the hormone present in its matrix should be identical with that of the reference standard.

Flesh, fat and viscera cannot be used for DES, which can be identified only in the urine, with a sensitivity of about 6 ppb. Many analyses can be made by thin-layer chromatography and radio-immunoassay.

The French control services use only the former, estimating the cost of one analysis at 25 FFr (= approx. US $5). The possibility of techniques applicable to flesh is under study, but in this regard the techniques speak of "chemical torture".

They concede that routine control of meat for all hormones appears to be difficult.

The same services also use urine for measuring natural hormones, applying Verbeke's technique. No results are obtained from flesh, even for natural hormones, over 3 weeks following treatment, levels being too low.

Encyclopaedia

of

BIOINFORMATICS SCIENCE, TECHNOLOGY AND ENGINEERING

Encyclopaedia of Bioinformatics Science, Technology and Engineering

Volume 3

Prof. Nirmal Chandra Pradhan

ANMOL PUBLICATIONS PVT. LTD.
NEW DELHI-110 002 (INDIA)

ANMOL PUBLICATIONS PVT. LTD.

Regd. Office: 4360/4, Ansari Road, Daryaganj,
New Delhi-110 002 (India)
Ph.: 23278000, 23261597

Branch Office: No. 1015, Ist Main Road, BSK IIIrd Stage
IIIrd Phase, IIIrd Block,
Bangalore-560 085 (India)
Tel.: 080-41723429
Visit us at: www.anmolpublications.com

Encyclopaedia of Bioinformatics Science, Technology and Engineering

© Reserved

First Edition, 2009

ISBN 978-81-261-4120-3 (Set)

PRINTED IN INDIA

Printed at Mehra Offset Press, Delhi.

Contents

Volume 3

Preface

Bioinformatics is the application of computer technology to the management of biological information. Computers are used to gather, store, analyze and integrate biological and genetic information which can then be applied to gene-based drug discovery and development. The need for Bioinformatics capabilities has been precipitated by the explosion of publicly available genomic information resulting from the Human Genome Project. The goal of this project - determination of the sequence of the entire human genome.

The science of Bioinformatics, which is the melding of molecular biology with computer science, is essential to the use of genomic information in understanding human diseases and in the identification of new molecular targets for drug discovery. In recognition of this, many universities, government institutions and pharmaceutical firms have formed bioinformatics groups, consisting of computational biologists and bioinformatics computer scientists. Such groups will be key to unraveling the mass of information generated by large scale sequencing efforts underway in laboratories around the world. Recent years have seen an explosive growth in biological data. Large sequencing projects are producing increasing quantities of nucleotide sequences. The latest release of GenBank exceeded one billion base pairs. Not only the size of sequence data is rapidly increasing, but also the number of characterized genes from many organisms and protein structures doubles about every two years. To cope with this great quantity of data, a new scientific discipline has emerged: *bioinformatics, biocomputing* or *computational biology*.

Author

Chapter 30

Transgenic Animals

Prior to the development of molecular genetics, the only way of studying the regulation and function of mammalian genes was through the observation of inherited characteristics or spontaneous mutations. Long before Mendel and any molecular genetic knowledge, selective breeding was a common practice among farmers for the enhancement of chosen traits, e.g., increased milk production.

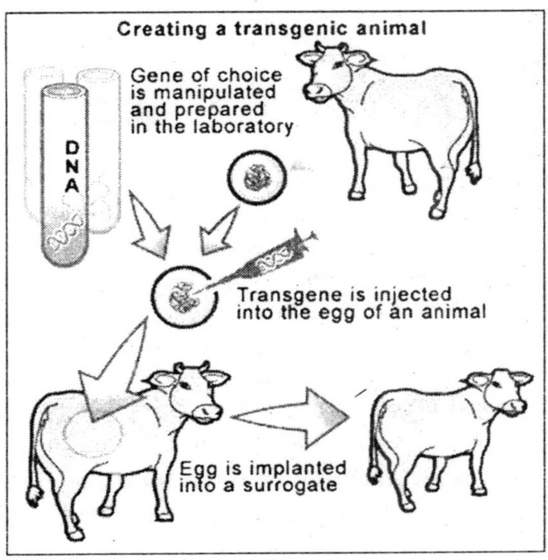

Fig. Transgenic Animals

During the 1970s, the first chimeric mice were produced. The cells of two different embryos of different strains were combined together at an early stage of development (eight cells) to form a single embryo that subsequently developed into a chimeric adult, exhibiting characteristics of each strain.

The mutual contributions of developmental biology and genetic engineering permitted rapid development of the techniques for the creation of transgenic animals. DNA microinjection, the first technique to prove successful in mammals, was first applied to mice and then to various other species such as rats, rabbits, sheep, pigs, birds, and fish.

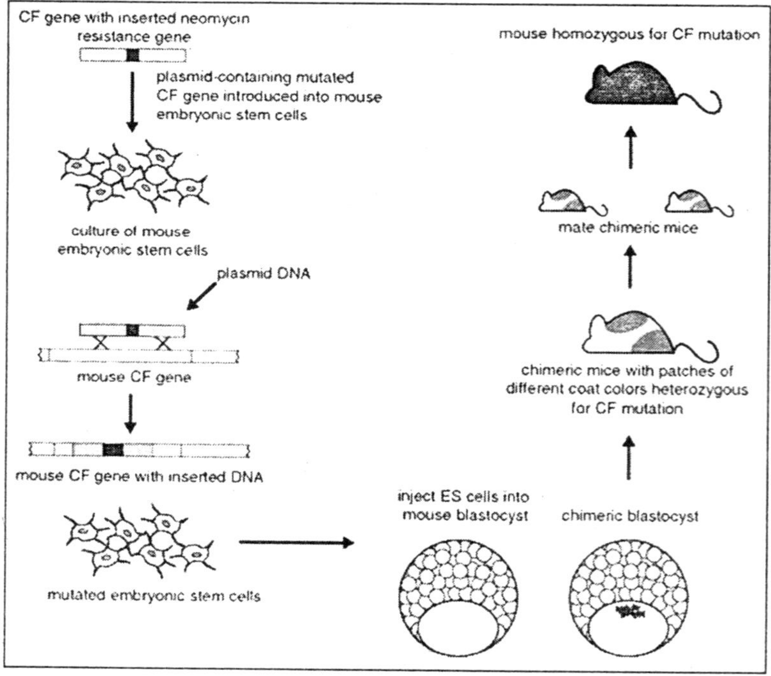

Fig. Chimeric Mice

Two other main techniques were then developed: those of retrovirus-mediated transgenesis and embryonic stem (ES) cell-mediated gene transfer.

Since 1981, when the term transgenic was first used by J.W. Gordon and F.H. Ruddle, there has been rapid development in the use of genetically engineered animals as investigators have found an increasing number of applications for the technology.

Methods of Creation of Transgenic Animals

For practical reasons, i.e., their small size and low cost of housing in comparison to that for larger vertebrates, their short generation time, and their fairly well defined genetics, mice have become the main species used in the field of transgenics.

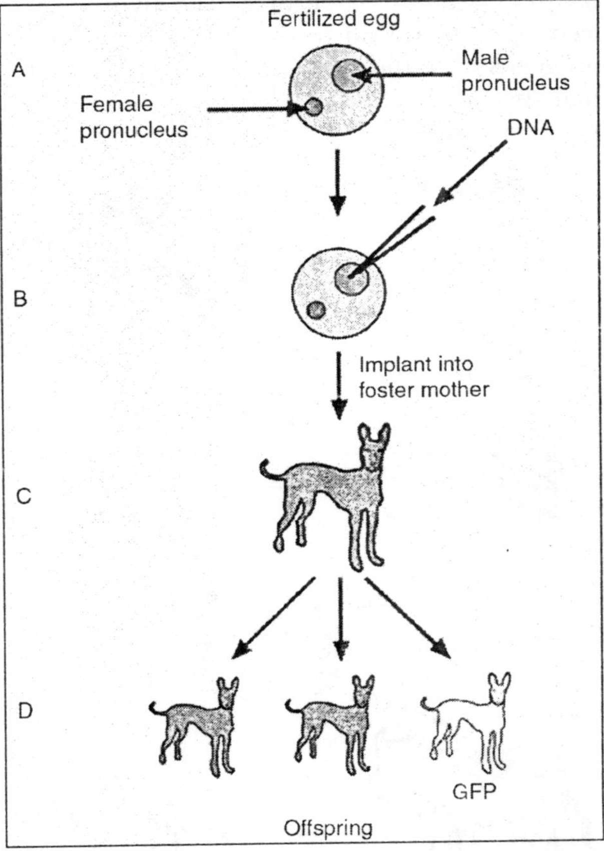

Fig. Methods of Creation of Transgenic Animals

The three principal methods used for the creation of transgenic animals are DNA microinjection, embryonic stem cell-mediated gene transfer and retrovirus-mediated gene transfer.

DNA Microinjection

This method involves the direct microinjection of a chosen gene construct (a single gene or a combination of genes) from another member of the same species or from a different species, into the pronucleus of a fertilized ovum. It is one of the first methods that proved to be effective in mammals. The introduced DNA may lead to the over- or under-expression of certain genes or to the expression of genes entirely new to the animal species.

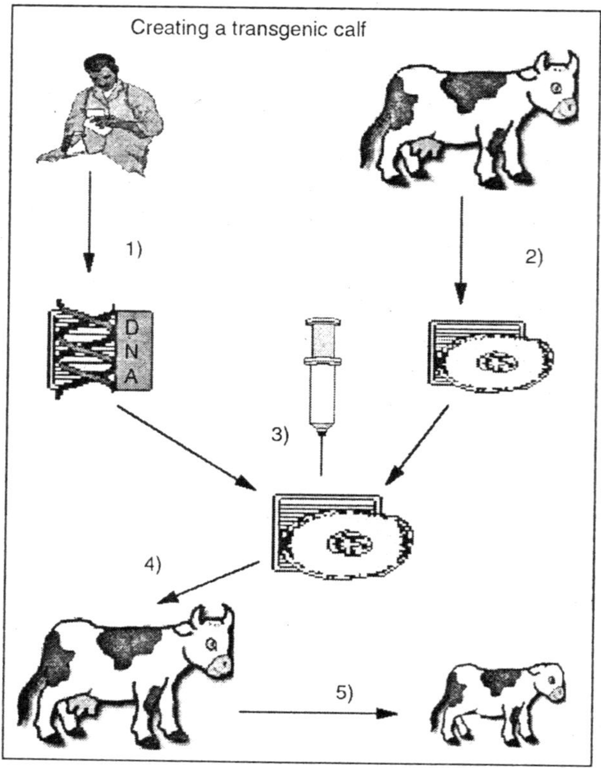

Fig. DNA Microinjection

The insertion of DNA is, however, a random process, and

there is a high probability that the introduced gene will not insert itself into a site on the host DNA that will permit its expression. The manipulated fertilized ovum is transferred into the oviduct of a recipient female, or foster mother that has been induced to act as a recipient by mating with a vasectomized male. A major advantage of this method is its applicability to a wide variety of species.

Embryonic Stem Cell-Mediated Gene Transfer

This method involves prior insertion of the desired DNA sequence by homologous recombination into an in vitro culture of embryonic stem (ES) cells. Stem cells are undifferentiated cells that have the potential to differentiate into any type of cell (somatic and germ cells) and therefore to give rise to a complete organism. These cells are then incorporated into an embryo at the blastocyst stage of development. The result is a chimeric animal. ES cell-mediated gene transfer is the method of choice for gene inactivation, the so-called knock-out method.

This technique is of particular importance for the study of the genetic control of developmental processes. This technique works particularly well in mice. It has the advantage of allowing precise targeting of defined mutations in the gene via homologous recombination.

Retrovirus-Mediated Gene Transfer

To increase the probability of expression, gene transfer is mediated by means of a carrier or vector, generally a virus or a plasmid. Retroviruses are commonly used as vectors to transfer genetic material into the cell, taking advantage of their ability to infect host cells in this way. Offspring derived from this method are chimeric, i.e., not all cells carry the retrovirus. Transmission of the transgene is possible only if the retrovirus integrates into some of the germ cells. For any of these techniques the success rate in terms of live birth of animals containing the transgene is extremely low.

Providing that the genetic manipulation does not lead to abortion, the result is a first generation (F1) of animals that need to be tested for the expression of the transgene.

Depending on the technique used, the F1 generation may result in chimeras. When the transgene has integrated into the germ cells, the so-called germ line chimeras are then inbred for 10 to 20 generations until homozygous transgenic animals are obtained and the transgene is present in every cell. At this stage embryos carrying the transgene can be frozen and stored for subsequent implantation.

TRANSGENIC ANIMALS AS BIOTECHNOLOGY

Transgenic animals are just one in a series of developments in the area of biotechnology. Biotechnology has transformed the way in which we understand processes such as engineering and manufacturing. These terms now include the use of living organisms or their parts to make or modify products, to change the characteristics of plants or animals, or to develop micro-organisms for specific uses.

The novel uses of biological techniques such as recombinant DNA techniques, cell fusion techniques, mono and polyclonal antibody technology and biological processes for commercial production have altered traditional distinctions and methods. Genetic manipulations at the level of DNA have also changed long held views as to what is considered to be animal, plant and human. In turn, these changes have made it more difficult to evaluate the ways in which animals are used and have obscured distinctions between pure and applied research.

Consideration of the acceptability of creating specific transgenic animal strains or genetic manipulation involving interchanging DNA between species and kingdoms could be a simple animal care issue or a societal decision. The following is an attempt to show what the ability to create transgenic animals or engage in other forms of DNA manipulation means in terms of traditional ACC functions, not forgetting that this impacts on wider considerations of human responsibility for the welfare of other life forms.

The creation of transgenic animals is resulting in a shift from the use of higher order species to lower order species, and is also affecting the numbers of animals used. This shift in the patterns of animal use is being monitored by the CCAC through the use of the Animal Use Data Form.

An example of the replacement of higher species by lower species is the possibility to develop disease models in mice rather than using dogs or non-human primates. In the long term, a reduction in the number of animals used, for example to study human diseases, is possible due to a greater specificity of the transgenic models developed. On the other hand, the success of the method has led to using its potential for investigating a wider range of diseases and conditions.

The actual use of some species may be increased, in addition to the numbers of animals which are sacrificed as donors during the creation process. The potential of the technology has also made it possible to consider employing cattle, swine, sheep and goats as processing units to manufacture proteins or as organ donors.

The complex interactive processes of living mammals are not reproducible in vitro. However, transgenic animals provide a means of evaluating genetic modifications in terms of anatomical and physiological changes in a complex system. Transgenic models are more precise in comparison to traditional animal models, for example the oncomouse with its increased susceptibility to tumor development enables results for carcinogenicity studies to be obtained within a shorter time-frame, thus reducing the course of tumor development in experimentally affected animals.

However, models are not strict equivalents, so as with any other system care must be taken in drawing conclusions from the data. A representative, but non-inclusive, list of purposes for which transgenic animals have been used indicates the wide ranging application of this biotechnology:

- In medical research, transgenic animals are used to identify the functions of specific factors in complex homeostatic systems through over- or under-expression of a modified gene (the inserted transgene);
- In toxicology: as responsive test animals (detection of toxicants);

- In mammalian developmental genetics;
- In molecular biology, the analysis of the regulation of gene expression makes use of the evaluation of a specific genetic change at the level of the whole animal;
- In the pharmaceutical industry, targeted production of pharmaceutical proteins, drug production and product efficacy testing;
- In biotechnology: as producers of specific proteins;
- Genetically engineered hormones to increase milk yield, meat production; genetic engineering of livestock and in aquaculture affecting modification of animal physiology and/or anatomy; cloning procedures to reproduce specific blood lines; and
- Developing animals specially created for use in xenografting.

Important general considerations include the extent to which experience acquired in the laboratory with regard to husbandry should influence industry standards for keeping animals created specifically as living machines for the production of proteins, antibodies, etc. What words are appropriate to describe and evaluate the condition of animals now used as production units? The successful cloning of Dolly underlines the fact that innovative developments in animal science are part of the mainstream of biotechnology. In addition, the use of xenografts, at least at the public health level makes animal and human welfare inseparable.

ANIMAL WELFARE AND ETHICS

The CCAC Guidelines on Transgenic Animals

In addition to the Guide to the Care and Use of Experimental Animals, the CCAC has recently developed guidelines on transgenic animals that will be subject to regular review due to the fast rate of evolution of this field.

The intention behind the production of these guidelines is to assist the ACC members and investigators in evaluating the ethics of technological aspects of the proposed creation, care and use of transgenic animals; to ensure that transgenic animals are used in accordance with the CCAC statement Ethics of Animal Investigation; and to ensure that the well-being of Canadians and the environment are protected.

ACC Protocol Review

The creation and use of transgenic animals are subject to all of the considerations raised by the CCAC guidelines on animal use protocol review. However, special consideration must be given to the procedures involved and in particular to possible welfare concerns for both the parent generation and the progeny.

Evaluation of proposals for the creation of transgenic animals may be divided into two interrelated parts: first, the justification for creation of the particular transgenic animal; and, secondly the welfare issues underlying the creation process itself. Special attention must be devoted to new protocols that use, for example, previously uncharacterized vectors or new transgenes, and/or are being performed by investigators who are new to the techniques.

As in all animal experimentation, justification for the use of the transgenic animal involves weighing the possible benefits of the experiment (e.g., advances in biomedical knowledge, the understanding and treatment of disease, improvements in production of foodstuffs or pharmaceuticals) versus the consideration of the ethical cost of the experiment in terms of the potential suffering of the animal.

This is particularly difficult for novel transgenic animals as it is not possible to predict with absolute certainty what the effect of a novel transgenic manipulation will be on the animals. For this reason, the protocol must include a strategy to address unanticipated suffering and to establish endpoints for the termination of the experiment. For these reasons, the guidelines require a transgenic information sheet to be completed with the protocol submission. In addition, a separate protocol is required for the creation of a novel transgenic animal, and for its subsequent use.

Category D level of invasiveness must be assigned to each

creation protocol, until the effects on the progeny are known. Any harmful effects observed must be reported to the ACC. The review of transgenic animal use protocols must take into consideration the effects of the transgenic modification on the animal itself, in addition to the subsequent effects incurred by the procedures.

As with any other laboratory animals, transgenic animals must be accorded high standards of care and use. Therefore, all standard operating procedures for laboratory management, human health (investigators and technicians involved with transgenic animals), and animal welfare (as part of the creation protocol, during the subsequent development of the progeny and as part of the animal use protocol) have to be evaluated accordingly.

LABORATORY ANIMAL MANAGEMENT

The implications of transgenic techniques for animal welfare have been discussed by Moore and Mepham. Specific welfare aspects have to be taken into consideration with transgenic animals. They include: the extent of discomfort experienced by the parents during the experimental procedures; the effect of the expression of the transgene (the modified gene inserted) on the created transgenic animal; and the effects on their progeny.

Physical and biological containment for transgenic animals should be adequate to assure the biosafety of the animal care staff which work with the animals, to prevent any possibility of the transfer of the gene within the non-transgenic colonies maintained in the same facilities, and to protect potentially immuno-compromised transgenic animals from pathogens.

The purpose of any breeding operation is to preserve the traits of interest and to restrict causes of genetic variability. Some problems associated with the breeding of transgenic mice may arise. For example, during the creation of transgenic mice, contamination of the media used in collecting eggs and blastocysts for microinjection can occur.

Establishing and maintaining lines requires careful

management. As part of the general process of transgenic animal creation, each animal used must be carefully identified. Cage cards and good records with details of breeding information are necessary to be able to identify with certainty the genetic characteristics and modifications of the animal. The data recorded should include the identity, breeding, pedigree and any other pertinent data such as any dates, observations or laboratory analysis information. Since transgenic animals are not easily replaceable, the cost of containment is an important factor in transgenic experimental design.

Embryo freezing is used for the preservation of transgenic strains. To protect colonies against disease, contamination or any other cause of loss, a large number of preimplanted embryos are kept, by cryopreservation. This also reduces the cost of maintaining a transgenic mouse line when it is not needed for experimentation.

With the development of transgenic animals, just a small genomic change can induce unpredictable and quite drastic changes at the level of the whole animal. This is the main challenge for transgenic animal management. It is therefore important to have a clear procedure for monitoring the animals and for dealing with unanticipated suffering.

THE NEED FOR ETHICAL REVIEW

The evaluation of animal and human welfare as it may be affected by biotechnology is a complex issue. One of the elements most notable in this process is the absence of an informed sense of the processes involved. ACCs share the responsibility for educating members on relevant aspects of animal care and use. Education concerning transgenic animal care and use is of particular importance, involving the careful consideration of the reasons for manipulating the genome of any organism as genetic engineering is a sensitive social issue.

The ethical review process is complicated by the fact that many techniques and developments in biotechnology are eligible for patent. The reluctance of biotechnologists to reveal proprietary information is understandable. Nonetheless, current work in fish culture, genetically altered animals for organ replacement, and cloning underlines the fact that transgenic research is wide ranging. To be able to provide a competent review, ACC members need to develop a similar broad understanding of the underlying principles.

A thorough discussion of biotechnology issues, including transgenic animals is needed, particularly to develop some consensus as to the relative value of benefits to be obtained from the use of transgenic animals. One of the more challenging questions is how to account for the interests of the animals involved. The field of transgenic animal biotechnology is likely to become of increasing importance as the techniques develop further and are applied to many more animal species. Welfare and ethical concerns will also continue to evolve. Consequently, education together with thoughtful ethical decision-making will remain the keystone of the review of transgenic protocols.

Antibody Production in Transgenic Animals

Numerous monoclonal antibodies are being produced in the mammary gland of transgenic goats. Cloned transgenic cattle produce a recombinant bispecific antibody in their blood. Purified from serum, the antibody is stable and mediates target cell-restricted T cell stimulation and tumour cell killing. An interesting new development is the generation of trans-chromosomal animals.

A human artificial chromosome containing the complete sequences of the human immunoglobulin heavy and light chain loci was introduced into bovine fibroblasts, which were then used in nuclear transfer. Transchromosomal bovine offspring were obtained that expressed human immunoglobulin in their blood. This system could be a significant step forward in the production of human therapeutic polyclonal antibodies. Further studies will show whether the additional chromosome will be maintained over future generations and how stable expression will be.

Blood Replacement

Functional human haemoglobin has been produced in

transgenic swine. The transgenic protein could be purified from the porcine blood and showed oxygen-binding characteristics similar to natural human haemoglobin. The main obstacle was that only a small proportion of porcine red blood cells contained the human form of haemoglobin. Alternative approaches to produce human blood substitutes have focused on the chemical cross-linking of haemoglobin to the superoxide-dismutase system.

XENOTRANSPLANTATION OF PORCINE ORGANS

Today more than 250,000 people are alive only because of the successful transplantation of an appropriate human organ (allotransplantation). However, progress in organ transplantation technology has led to an acute shortage of appropriate organs, and cadaveric or live organ donation does not meet the demand in Western societies.

To close the growing gap between demand and availability of appropriate human organs, porcine xenografts from domesticated pigs are considered to be the best alternative. Essential prerequisites for successful xenotransplantation are: – overcoming the immunological hurdles – preventing the transmission of pathogens from the donor animal to the human recipient – ensuring the compatibility of the donor organs with human anatomy and physiology.

The major immunological obstacles in porcine-to-human xenotransplantation are:

- Hyperacute rejection (HAR)
- Acute vascular rejection (AVR)
- Cellular rejection, and eventually
- Chronic rejection.

Hyperacute rejection occurs within seconds or minutes, when, in the case of a discordant organ (e.g. in transplanting from pig to human), pre-existing antibodies react with antigenic structures on the surface of the porcine cells and activate the complement cascade; in other words, the antigen-antibody complex triggers formation of the membrane attack complex. Induced xenoreactive antibodies are thought to be responsible for AVR, which occurs within days after transplantation of a xenograft; disseminated intravascular coagulation (DIC) is a predominant feature of AVR.

Human thrombomodulin and heme-oxygenase 1 are crucially involved in the etiology of DIC and are targets for future transgenic modifications to improve the long-term survival of porcine xenografts by creating multi-transgenic pigs. When using a discordant donor species such as the pig, overcoming HAR and AVR are the pre-eminent goals.

Two main strategies have been successfully explored for long-term suppression of HAR: synthesis of human regulators of complement activity (RCAs) in transgenic pigs and the knockout of á-gal epitopes, the antigenic structures on the surface of the porcine cells that cause HAR. Survival rates, after the transplantation of porcine hearts or kidneys expressing transgenic RCA proteins to immunosuppressed nonhuman primates, reached 23 to 135 days, indicating that HAR can be overcome in a clinically acceptable manner.

The successful xenotransplantation of porcine organs with a knockout of the 1,3⟨galactosyltransferase gene, eliminating the 1,3-⟨-gal-epitopes produced by the 1,3-⟨-galactosyltransferase enzyme, has recently been demonstrated. Upon transplantation of porcine hearts or kidneys from these ⟨-gal-knockout pigs to baboons, survival rates reached two to six months.

A particularly promising strategy to enhance long-term graft tolerance is the induction of permanent chimerism via intraportal injection of embryonic stem (ES) cells or the co-transplantation of vascularised thymic tissue. Recent findings have revealed that the risk of porcine endogenous retrovirus transmission is negligible and show that this critical danger could be eliminated, paving the way for preclinical testing of xenografts.

Despite further challenges, appropriate lines of transgenic pigs are likely to be available as organ donors within the next five to ten years. Guidelines for the clinical application of porcine xenotransplants are already available in the USA and are currently being developed in other countries. The general consensus of a worldwide debate is that the technology is ethically acceptable provided that the individual's well-being does not compromise public health. Economically, xenotransplantation will be viable, as the enormous costs of maintaining patients suffering from severe kidney disease using dialysis or supporting those suffering from chronic heart disease could be reduced by a functional kidney or heart xenograft.

FARM ANIMALS AND HUMAN DISEASES

Mouse physiology, anatomy and life span differ significantly from those of humans, making the rodent model inappropriate for many human diseases. Farm animals, such as pigs, sheep or even cattle, may be more appropriate models in which to study potential therapies for human diseases that require longer observation periods than those possible in mice, e.g. atherosclerosis, non-insulin-dependent diabetes, cystic fibrosis, cancer and neuro-degenerative disorders.

Cardiovascular disease is an increasing health problem in aging Western societies, where coronary artery diseases account for the majority of deaths. Because genetically modified mice do not manifest myocardial infarction or stroke as a result of atherosclerosis, new animal models, such as swine that exhibit these pathologies, are needed to develop effective therapeutic strategies.

An important porcine model has been developed for the rare human eye disease retinitis pigmentosa (PR). Patients with PR suffer from night blindness early in life, a condition attributed to a loss of photoreceptors. Transgenic pigs that express a mutated rhodopsin gene show great similarity to the human phenotype and effective treatments are being developed. The pig could be a useful model for studying defects of growth-hormone releasing hormone (GHRH), which are implicated in a variety of conditions such as Turner syndrome, hypochrondroplasia, Crohn's disease, intrauterine growth retardation or renal insufficiency.

Application of recombinant GHRH and its myogenic expression has been shown to alleviate these problems in a porcine model. Development of further ovine and porcine models of human diseases is underway.

An important aspect of nuclear-transfer-derived large animal models for human diseases and the development of regenerative therapies is that somatic cloning per se does not result in shortening of the telomeres and thus does not necessarily lead to premature ageing. Telomeres are highly repetitive DNA sequences at the ends of the chromosomes that are crucial for their structural integrity and function and are thought to be related to lifespan. Telomere shortening is usually correlated with severe limitation of the regenerative capacity of cells, the onset of cancer, ageing and chronic disease with significant impacts on human lifespans.

Expression of the enzyme telomerase, which is primarily responsible for the formation and rebuilding of telomeres, is suppressed in most somatic tissues postnatally. Recent studies have revealed that telomere length is established early in pre-implantation development by a specific genetic programme and correlates with telomerase activity.

Transgenic Animals in Agriculture Carcass Composition

Transgenic pigs bearing a human metallothionein promoter/porcine growth-hormone gene construct showed significant improvements in economically important traits such as growth rate, feed conversion and body fat/muscle ratio without the pathological phenotype known from previous growth hormone constructs.

Similarly, pigs transgenic for the human insulin-like growth factor-I had ~30 per cent larger loin mass, ~10 per cent more carcass lean tissue and ~20 per cent less total carcass fat. The commercialisation of these pigs has been postponed due to the current lack of public acceptance of genetically modified foods.

Recently, an important step towards the production of more healthful pork has been made by the creation of the first pigs transgenic for a spinach desaturase gene that produces increased amounts of non-saturated fatty acids.

These pigs have a higher ratio of unsaturated to saturated

fatty acids in striated muscle, which means more healthful meat since a diet rich in non-saturated fatty acids is known to be correlated with a reduced risk of stroke and coronary diseases.

Lactation

The physicochemical properties of milk are mainly due to the ratio of casein variants, making these a prime target for the improvement of milk composition. Transgenic mice have been developed with various modifications demonstrating the feasibility of obtaining significant alterations in milk composition in larger animals, but at the same time, showing that unwanted side effects can occur. Dairy production is an attractive field for targeted genetic modification.

It may be possible to produce milk with a modified lipid composition by modulating the enzymes involved in lipid metabolism, or to increase curd and cheese production by enhancing expression of the casein gene family in the mammary gland. The bovine casein ratio has been altered by over-expression of betaand kappa-casein, clearly underpinning the potential for improvements in the functional properties of bovine milk.

Furthermore, it may be possible to create 'hypoallergenic' milk by knockout or knockdown of the â-lactoglobulin gene; to generate lactose-free milk by a knockout or knockdown of the á-lactalbumin locus, which is the key molecule in milk sugar synthesis; to produce 'infant milk' in which human lactoferrin is abundantly available; or to produce milk with a highly improved hygienic standard by increasing the amount of lysozyme.

Lactose-reduced or lactose-free milk would render dairy products suitable for consumption by the large numbers of adult humans who do not possess an active intestinal lactase enzyme system. Although lactose is the main osmotically active substance in milk and a lack thereof could interfere with milk secretion, a lactase construct has been tested in the mammary gland of transgenic mice.

In hemizygous mice, this reduced lactose contents by 50 per cent to 85 per cent without altering milk secretion.

Unfortunately, mice with a homozygous knockout for ⟨-lactalbumin could not nurse their offspring because of the high viscosity of their milk.

These findings demonstrate the feasibility of obtaining significant alterations in milk composition by applying the appropriate strategy. In the pig, transgenic expression of a bovine lactalbumin construct in sow milk has been shown to result in higher lactose contents and greater milk yields, which correlated with improved survival and development of piglets. Any increased survival of piglets at weaning would provide significant commercial advantages to the producer.

Wool Production

Transgenic sheep carrying a keratin-IGF-I construct showed that expression in the skin and the amount of clear fleece was about 6.2 per cent greater in transgenic than in nontransgenic animals. No adverse effects on health or reproduction were observed. Approaches designed to alter wool production by transgenic modification of the cystein pathway have met with only limited success, although cystein is known to be the ratelimiting biochemical factor for wool growth.

Environmentally Friendly Farm Animals

Phytase transgenic pigs have been developed to address the problem of manure-related environmental pollution. These pigs carry a bacterial phytase gene under the transcriptional control of a salivary-gland-specific promoter, which allows the pigs to digest plant phytate.

Without the bacterial enzyme, phytate phosphorus passes undigested into manure and pollutes the environment. With the bacterial enzyme, fecal phosphorus output was reduced by up to 75 per cent. Developers expect these environmentally friendly pigs to enter commercial production in Canada within the next few years.

TRANSGENIC ANIMALS AND DISEASE RESISTANCE

Transgenic strategies to increase disease resistance In most cases, susceptibility to pathogens originates from the interplay of numerous genes; in other words, susceptibility to pathogens is polygenic in nature. Only a few loci are known that confer resistance against a specific disease. Transgenic strategies to enhance disease resistance include the transfer of major histocompatibility-complex genes, T-cell-receptor genes, immunoglobulin genes, genes that affect lymphokines or specific disease-resistance genes.

A prominent example for a specific disease resistance gene is the murine Mx-gene. Production of the Mx1-protein is induced by interferon and was discovered in inbred mouse strains that are resistant to infection with influenza viruses. Microinjection of an interferon- and virus-inducible Mx-construct into porcine zygotes resulted in two transgenic pig lines that expressed the Mx-messenger RNA (mRNA); unfortunately, no Mx protein could be detected.

Recently the bovine MxI gene was identified and shown to confer antiviral activity as a transgenic construct in Vero cells. Transgenic constructs bearing the immunoglobulin-A (IgA) gene have been successfully introduced into pigs, sheep and mice in an attempt to increase resistance against infections.

The murine IgA gene was expressed in two transgenic pig lines, but only the light chains were detected and the IgA-molecules showed only marginal binding to phosphorylcholine. High levels of monoclonal murine antibodies with a high binding affinity for their specific antigen have been produced in transgenic pigs. Attempts to increase ovine resistance to Visna virus infection via transgenic production of Visna envelope protein have been reported. The transgenic sheep developed normally and expressed the viral gene without pathological side effects.

However, the transgene was not expressed in monocytes, the target cells of the viral infection. Antibodies were detected after an artificial infection of the transgenic animals. It has also proved possible to induce passive immunity against an economically important porcine disease in a transgenic mouse model. The transgenic mice secreted a recombinant antibody in milk that neutralised the corona virus responsible for transmissible gastroenteritis (TGEV) and conferred resistance against TGEV. Strong mammarygland- specific expression was achieved for the entire duration of lactation. Verification of this work in transgenic pigs is anticipated in the near future. Knockout of the prion protein is the only secure way to prevent infection and transmission of spongiform encephalopathies like scrapie or bovine spongiform encephalopathy. The first successful targeting of the ovine prion locus has been reported; however, the cloned lambs carrying the knockout locus died shortly after birth. Cloned cattle with a knockout for the prion locus have also been generated.

Transgenic animals with modified prion genes will be an appropriate model for studying the epidemiology of spongiform encephalopathies in humans and are crucial for developing strategies to eliminate prion carriers from farm animal populations.

Transgenic Approaches to Increase Disease Resistance

The levels of the anti-microbial peptides lysozyme and lactoferrin in human milk is many times higher than in bovine milk. Transgenic expression of the human lysozyme gene in mice was associated with a significant reduction of bacteria and reduced the frequency of mammary gland infections. Lactoferrin has bactericidal and bacteriostatic effects, in addition to being the main iron source in milk.

These properties make an increase in lactoferrin levels in bovine transgenics a practical way to improve milk quality. Human lactoferrin has been expressed in the milk of transgenic mice and cattle at high levels and is associated with an increased resistance against mammary gland diseases.

Lycostaphin has been shown to confer specific resistance against mastitis caused by *Staphylococcus aureus*. A recent report indicates that transgenic technology has been used to produce cows that express a lycostaphin gene construct in the mammary gland, thus making them mastitis-resistant. Emerging transgenic technologies

Lentiviral Transfection of Oocytes and Zygotes

Recent research has shown that lentiviruses can overcome previous limitations of viral-mediated gene transfer, which included the silencing of the transgenic locus and low expression levels. Injection of lentiviruses into the perivitelline space of porcine zygotes resulted in a very high proportion of piglets that carried and expressed the transgene. Stable transgenic lines have been established by this method.

The generation of transgenic cattle by lentiviruses requires microinjection into the perivitelline space of oocytes and has a lower efficiency than that obtained in pigs. Lentiviral gene transfer in livestock promises unprecedented efficiency of transgenic animal production. Whether the multiple integration of lentiviruses into the genome is associated with unwanted side-effects like oncogene activation or insertional mutagenesis remains to be investigated.

Chimera Generation

Embryonic stem cells with pluripotent characteristics have the ability to participate in organ and germ cell development after injection into blastocysts or by aggregation with morulae. True ES cells (that is, those able to contribute to the germ line) are currently only available from inbred mouse strains. In mouse genetics, ES cells have become an important tool for generating gene knockouts, gene knockins and large chromosomal rearrangements.

Embryonic stem-like cells and primordial germ cell cultures have been reported for several farm animal species, and chimeric animals without germ line contribution have been reported in swine and cattle. Recent data indicate that somatic stem cells may have a much greater potency than previously assumed. Whether these cells will improve the efficiency of chimera generation or somatic nuclear transfer in farm animals has yet to be shown.

Culture of Spermatogonia

Transplantation of genetically altered primordial germ cells into the testes of host male animals is an alternative approach to generating transgenic animals. Initial experiments in mice showed that the depletion of endogenous spermatogenesis by treatment with busulfan prior to transplantation is effective and compatible with recolonisation by donor cells.

Recently, researchers succeeded in transmitting the donor haplotype to the next generation after germ-cell transplantation. The major obstacles to this strategy are the lack of efficient *in vitro* culture methods for primordial germ/ prospermatogonial cells, and the limitations this imposes on gene transfer techniques for these cells.

Ribonucleic acid Interference

Ribonucleic acid interference is a conserved posttranscriptional gene regulatory process in most biological systems, including fungi, plants and animals. Common mechanistic elements include siRNAs with 21 to 23 nucleotides, which specifically bind complementary sequences on their target mRNAs and shut down expression. The target mRNAs are degraded by exonucleases and no protein is translated. The RNA interference seems to be involved in gene regulation by controlling/suppressing the translation of mRNAs from endogenous viral elements.

The relative simplicity of active siRNAs has facilitated the adoption of this mechanism to generate transient or permanent knockdowns for specific genes. For transient gene knockdown, synthetic siRNAs can be transfected into cells or embryonic stages. For stable gene repression, the siRNA sequences must be incorporated into a gene construct. The conjunction of siRNA and lentiviral vector technology may soon provide a method with high gene transfer efficiency and highly specific gene knockdown for livestock.

Chapter 31

Tissue Culture

Most methods of plant transformation applied to GM crops require that a whole plant is regenerated from isolated plant cells or tissue which have been genetically transformed. This regeneration is conducted *in vitro* so that the environment and growth medium can be manipulated to ensure a high frequency of regeneration.

In addition to a high frequency of regeneration, the regenerable cells must be accessible to gene transfer by whatever technique is chosen.

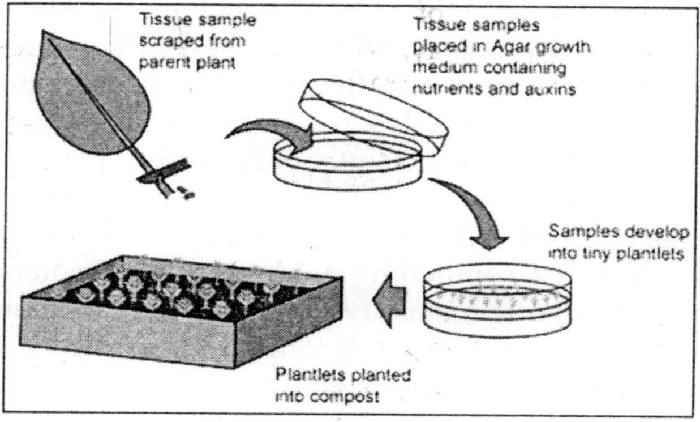

Fig. Plant Tissue Culture

The primary aim is therefore to produce, as easily and as quickly as possible, a large number of regenerable cells that are accessible to gene transfer.

The subsequent regeneration step is often the most difficult step in plant transformation studies. However, it is important to remember that a high frequency of regeneration does not necessarily correlate with high transformation efficiency.

PLASTICITY AND TOTIPOTENCY

Two concepts, plasticity and totipotency, are central to understanding plant cell culture and regeneration. Plants, due to their sessile nature and long life span, have developed a greater ability to endure extreme conditions and predation than have animals.

Many of the processes involved in plant growth and development adapt to environmental conditions. This plasticity allows plants to alter their metabolism, growth and development to best suit their environment. Particularly important aspects of this adaptation, as far as plant tissue culture and regeneration are concerned, are the abilities to initiate cell division from almost any tissue of the plant and to regenerate lost organs or undergo different developmental pathways in response to particular stimuli.

When plant cells and tissues are cultured *in vitro* they generally exhibit a very high degree of plasticity, which allows one type of tissue or organ to be initiated from another type. In this way, whole plants can be subsequently regenerated. This regeneration of whole organisms depends upon the concept that all plant cells can, given the correct stimuli, express the total genetic potential of the parent plant.

This maintenance of genetic potential is called 'totipotency'. Plant cell culture and regeneration do, in fact, provide the most compelling evidence for totipotency. In practical terms though, identifying the culture conditions and stimuli required to manifest this totipotency can be extremely difficult and it is still a largely empirical process.

When cultured *in vitro*, all the needs, both chemical and physical, of the plant cells have to met by the culture vessel, the growth medium and the external environment (light, temperature, etc.).

The growth medium has to supply all the essential mineral ions required for growth and development. In many cases (as the biosynthetic capability of cells cultured *in vitro* may not replicate that of the parent plant), it must also supply additional organic supplements such as amino acids and vitamins.

Many plant cell cultures, as they are not photosynthetic, also require the addition of a fixed carbon source in the form of a sugar (most often sucrose).

One other vital component that must also be supplied is water, the principal biological solvent. Physical factors, such as temperature, pH, the gaseous environment, light (quality and duration) and osmotic pressure, also have to be maintained within acceptable limits.

Culture media used for the *in vitro* cultivation of plant cells are composed of three basic components:

- Essential elements, or mineral ions, supplied as a complex mixture of salts
- An organic supplement supplying vitamins and/or amino acids
- A source of fixed carbon; usually supplied as the sugar sucrose.

Macroelements

As is implied by the name, the stock solution supplies those elements required in large amounts for plant growth and development. Nitrogen, phosphorus, potassium, magnesium, calcium and sulphur (and carbon, which is added separately) are usually regarded as macroelements. These elements usually comprise at least 0.1 per cent of the dry weight of plants. Nitrogen is most commonly supplied as a mixture of nitrate ions (from the KNO_3) and ammonium ions (from the NH_4NO_3).

Theoretically, there is an advantage in supplying nitrogen in the form of ammonium ions, as nitrogen must be in the reduced form to be incorporated into macromolecules. Nitrate ions therefore need to be reduced before incorporation. However, at high concentrations, ammonium ions can be toxic to plant cell cultures and uptake of ammonium ions from the medium causes acidification of the medium. In order to use ammonium ions as the sole nitrogen source, the medium needs to be buffered. High concentrations of ammonium ions can also cause culture problems by increasing the frequency of vitrification (the culture appears pale and 'glassy' and is usually unsuitable for further culture). Using a mixture of nitrate and ammonium ions has the advantage of weakly buffering the medium as the uptake of nitrate ions causes OH– ions to be excreted. Phosphorus is usually supplied as the phosphate ion of ammonium, sodium or potassium salts. High concentrations of phosphate can lead to the precipitation of medium elements as insoluble phosphates.

Microelements

These elements are required in trace amounts for plant growth and development, and have many and diverse roles. Manganese, iodine, copper, cobalt, boron, molybdenum, iron and zinc usually comprise the microelements, although other elements such as nickel and aluminium are frequently found in some formulations.

Iron is usually added as iron sulphate, although iron citrate can also be used. Ethylenediaminetetraacetic acid (EDTA) is usually used in conjunction with the iron sulphate. The EDTA complexes with the iron so as to allow the slow and continuous release of iron into the medium. Uncomplexed iron can precipitate out of the medium as ferric oxide.

Organic Supplements

Only two vitamins, thiamine (vitamin B1) and myoinositol (considered a B vitamin) are considered essential for the culture of plant cells *in vitro*. However, other vitamins are often added to plant cell culture media for historical reasons. Amino acids are also commonly included in the organic supplement.

The most frequently used is glycine (arginine, asparagine, aspartic acid, alanine, glutamic acid, glutamine and proline are also used), but in many cases its inclusion is not essential. Amino acids provide a source of reduced nitrogen and, like ammonium ions, uptake causes acidification of the medium. Casein hydrolysate can be used as a relatively cheap source of a mix of amino acids.

Carbon Source

Sucrose is cheap, easily available, readily assimilated and relatively stable and is therefore the most commonly used carbon source. Other carbohydrates (such as glucose, maltose, galactose and sorbitol) can also be used, and in specialised circumstances may prove superior to sucrose.

Gelling Agents

Media for plant cell culture *in vitro* can be used in either liquid or 'solid' forms, depending on the type of culture being grown. For any culture types that require the plant cells or tissues to be grown on the surface of the medium, it must be solidified (more correctly termed 'gelled'). Agar, produced from seaweed, is the most common type of gelling agent, and is ideal for routine applications.

However, because it is a natural product, the agar quality can vary from supplier to supplier and from batch to batch. For more demanding applications gelling agents are available. Purified agar or agarose can be used, as can a variety of gellan gums. These components, then, are the basic 'chemical' necessities for plant cell culture media. However, other additions are made in order to manipulate the pattern of growth and development of the plant cell culture.

PLANT GROWTH REGULATORS

We have already briefly considered the concepts of plasticity and totipotency. The essential point as far as plant cell culture is concerned is that, due to this plasticity and totipotency, specific media manipulations can be used to direct the development of plant cells in culture.

Plant growth regulators are the critical media components in determining the developmental pathway of the plant cells. The plant growth regulators used most commonly are plant hormones or their synthetic analogues.

Classes of Plant Growth Regulators

There are five main classes of plant growth regulator used in plant cell culture, namely:

- Auxins;
- Cytokinins;
- Gibberellins;
- Abscisic acid;
- Ethylene.

Each class of plant growth regulator will be briefly looked at.

Auxins

Auxins promote both cell division and cell growth The most important naturally occurring auxin is IAA (indole-3-acetic acid), but its use in plant cell culture media is limited because it is unstable to both heat and light. Occasionally, amino acid conjugates of IAA (such as indole-acetyl-L-alanine and indole-acetyl-L-glycine), which are more stable, are used to partially alleviate the problems associated with the use of IAA.

It is more common, though, to use stable chemical analogues of IAA as a source of auxin in plant cell culture media. 2,4-Dichlorophenoxyacetic acid (2,4-D) is the most commonly used auxin and is extremely effective in most circumstances. Other auxins are available, and some may be more effective or 'potent' than 2,4- D in some instances.

Cytokinins

Cytokinins promote cell division. Naturally occurring cytokinins are a large group of structurally related (they are purine derivatives) compounds. Of the naturally occurring cytokinins, two have some use in plant tissue culture media. These are zeatin and 2iP (2-isopentyl adenine). Their use is not widespread as they are expensive (particularly zeatin) and relatively unstable. The synthetic analogues, kinetin and BAP (benzylaminopurine), are therefore used more frequently. Non-purine-based chemicals, such as substituted phenylureas, are also used as cytokinins in plant cell culture media. These substituted phenylureas can also substitute for auxin in some culture systems.

Gibberellins

There are numerous, naturally occurring, structurally

related compounds termed 'gibberellins'. They are involved in regulating cell elongation, and are agronomically important in determining plant height and fruit-set. Only a few of the gibberellins are used in plant tissue culture media, GA3 being the most common.

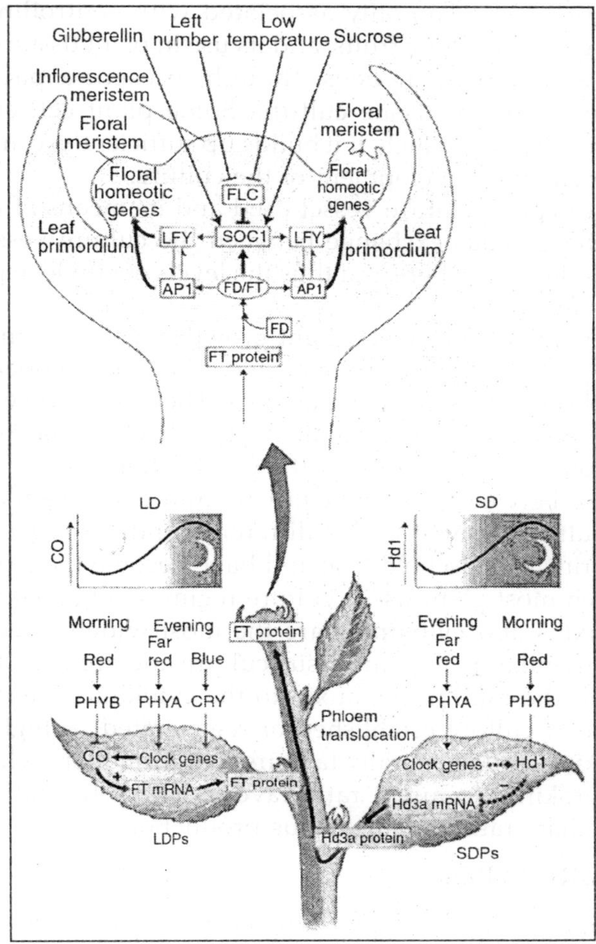

Fig. Gibberellins

Abscisic Acid

Abscisic acid (ABA) inhibits cell division. It is most

commonly used in plant tissue culture to promote distinct developmental pathways such as somatic embryogenesis.

Ethylene

Ethylene is a gaseous, naturally occurring, plant growth regulator most commonly associated with controlling fruit ripening in climacteric fruits, and its use in plant tissue culture is not widespread. It does, though, present a particular problem for plant tissue culture. Some plant cell cultures produce ethylene, which, if it builds up sufficiently, can inhibit the growth and development of the culture.

The type of culture vessel used and its means of closure affect the gaseous exchange between the culture vessel and the outside atmosphere and thus the levels of ethylene present in the culture.

Generalisations about plant growth regulators and their use in plant cell culture media have been developed from initial observations made in the 1950s. There is, however, some considerable difficulty in predicting the effects of plant growth regulators: this is because of the great differences in culture response between species, cultivars and even plants of the same cultivar grown under different conditions. However, some principles do hold true and have become the paradigm on which most plant tissue culture regimes are based.

Auxins and cytokinins are the most widely used plant growth regulators in plant tissue culture and are usually used together, the ratio of the auxin to the cytokinin determining the type of culture established or regenerated. A high auxin to cytokinin ratio generally favours root formation, whereas a high cytokinin to auxin ratio favours shoot formation. An intermediate ratio favours callus production.

CULTURE TYPES

Cultures are generally initiated from sterile pieces of a whole plant. These pieces are termed 'explants', and may consist of pieces of organs, such as leaves or roots, or may be specific cell types, such as pollen or endosperm. Many features of the explant are known to affect the efficiency of culture initiation. Generally, younger, more rapidly growing tissue (or tissue at an early stage of development) is most effective. Several different culture types most commonly used in plant transformation studies will now be examined in more detail.

Callus

Explants, when cultured on the appropriate medium, usually with both an auxin and a cytokinin, can give rise to an unorganised, growing and dividing mass of cells. It is thought that any plant tissue can be used as an explant, if the correct conditions are found.

In culture, this proliferation can be maintained more or less indefinitely, provided that the callus is subcultured on to fresh medium periodically.

During callus formation there is some degree of dedifferentiation (i.e. the changes that occur during development and specialisation are, to some extent, reversed), both in morphology (callus is usually composed of unspecialised parenchyma cells) and metabolism. One major consequence of this dedifferentiation is that most plant cultures lose the ability to photosynthesise. This has important consequences for the culture of callus tissue, as the metabolic profile will probably not match that of the donor plant.

This necessitates the addition of other components—such as vitamins and, most importantly, a carbon source—to the culture medium, in addition to the usual mineral nutrients. Callus culture is often performed in the dark (the lack of photosynthetic capability being no drawback) as light can encourage differentiation of the callus.

During long-term culture, the culture may lose the requirement for auxin and/or cytokinin. This process, known as 'habituation', is common in callus cultures from some plant species (such as sugar beet).

Callus cultures are extremely important in plant biotechnology. Manipulation of the auxin to cytokinin ratio in the medium can lead to the development of shoots, roots or somatic embryos from which whole plants can subsequently be produced. Callus cultures can also be used to initiate cell suspensions, which are used in a variety of ways in plant transformation studies.

Cell-Suspension Cultures

Callus cultures, broadly speaking, fall into one of two categories: compact or friable. In compact callus the cells are densely aggregated, whereas in friable callus the cells are only loosely associated with each other and the callus becomes soft and breaks apart easily. Friable callus provides the inoculum to form cell-suspension cultures. Explants from some plant species or particular cell types tend not to form friable callus, making cell-suspension initiation a difficult task. The friability of callus can sometimes be improved by manipulating the medium components or by repeated subculturing.

The friability of the callus can also sometimes be improved by culturing it on 'semi-solid' medium (medium with a low concentration of gelling agent). When friable callus is placed into a liquid medium (usually the same composition as the solid medium used for the callus culture) and then agitated, single cells and/or small clumps of cells are released into the medium. Under the correct conditions, these released cells continue to grow and divide, eventually producing a cell-suspension culture.

A relatively large inoculum should be used when initiating cell suspensions so that the released cell numbers build up quickly. The inoculum should not be too large though, as toxic products released from damaged or stressed cells can build up to lethal levels. Large cell clumps can be removed during subculture of the cell suspension. Cell suspensions can be maintained relatively simply as batch cultures in conical flasks. They are continually cultured by repeated subculturing into fresh medium.

This results in dilution of the suspension and the initiation of another batch growth cycle. The degree of dilution during subculture should be determined empirically for each culture. Too great a degree of dilution will result in a greatly extended lag period or, in extreme cases, death of the transferred cells. After subculture, the cells divide and the biomass of the culture increases in a characteristic fashion, until nutrients in the medium are exhausted and/or toxic by-products build up to inhibitory levels—this is called the 'stationary phase'. If cells are left in the stationary phase for too long, they will die and the culture will be lost.

Therefore, cells should be transferred as they enter the stationary phase. It is therefore important that the batch growth-cycle parameters are determined for each cell-suspension culture.

Protoplasts

Protoplasts are plant cells with the cell wall removed. Protoplasts are most commonly isolated from either leaf mesophyll cells or cell suspensions, although other sources can be used to advantage.

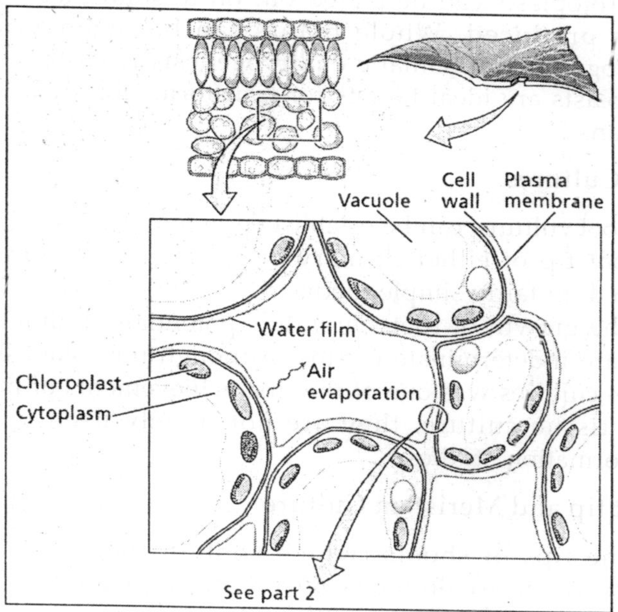

Fig. Protoplasts

Two general approaches to removing the cell wall (a difficult task without damaging the protoplast) can be taken—

mechanical or enzymatic isolation. Mechanical isolation, although possible, often results in low yields, poor quality and poor performance in culture due to substances released from damaged cells.

Enzymatic isolation is usually carried out in a simple salt solution with a high osmoticum, plus the cell wall degrading enzymes.

It is usual to use a mix of both cellulase and pectinase enzymes, which must be of high quality and purity. Protoplasts are fragile and easily damaged, and therefore must be cultured carefully.

Liquid medium is not agitated and a high osmotic potential is maintained, at least in the initial stages. The liquid medium must be shallow enough to allow aeration in the absence of agitation.

Protoplasts can be plated out on to solid medium and callus produced. Whole plants can be regenerated by organogenesis or somatic embryogenesis from this callus. Protoplasts are ideal targets for transformation by a variety of means.

Root Cultures

Root cultures can be established *in vitro* from explants of the root tip of either primary or lateral roots and can be cultured on fairly simple media.

The growth of roots *in vitro* is potentially unlimited, as roots are indeterminate organs. Although the establishment of root cultures was one of the first achievements of modern plant tissue culture, they are not widely used in plant transformation studies.

Shoot tip and Meristem Culture

The tips of shoots (which contain the shoot apical meristem) can be cultured *in vitro*, producing clumps of shoots from either axillary or adventitious buds. This method can be used for clonal propagation.

Shoot meristem cultures are potential alternatives to the more commonly used methods for cereal regeneration as they are less genotype-dependent and more efficient (seedlings can be used as donor material).

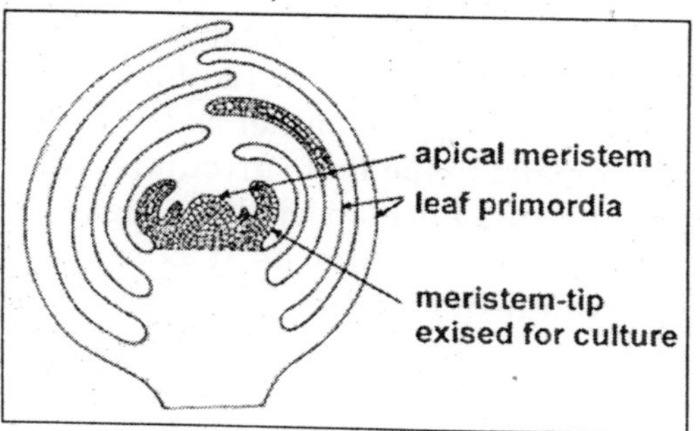

Fig. Meristem Culture

Embryo Culture

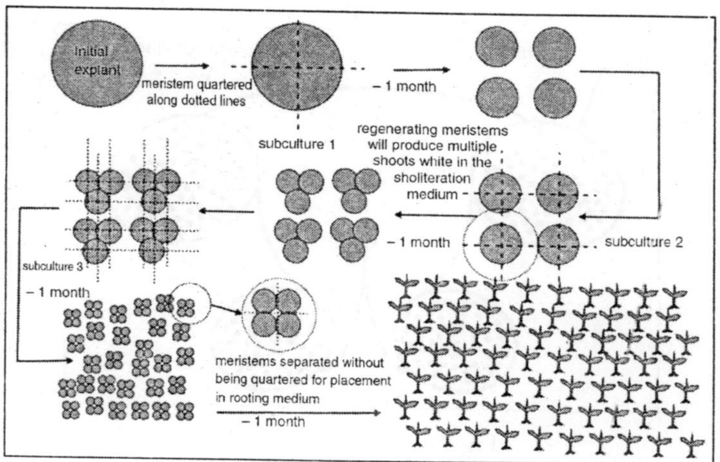

Fig. Embryo Culture

Embryos can be used as explants to generate callus cultures or somatic embryos. Both immature and mature embryos can be used as explants. Immature, embryo-derived embryogenic callus is the most popular method of monocot plant regeneration.

Microspore Culture

Haploid tissue can be cultured *in vitro* by using pollen or anthers as an explant. Pollen contains the male gametophyte, which is termed the 'microspore'. Both callus and embryos can be produced from pollen. Two main approaches can be taken to produce *in vitro* cultures from haploid tissue. The first method depends on using the anther as the explant.

Anthers (somatic tissue that surrounds and contains the pollen) can be cultured on solid medium (agar should not be used to solidify the medium as it contains inhibitory substances). Pollen-derived embryos are subsequently produced via dehiscence of the mature anthers. The dehiscence of the anther depends both on its isolation at the correct stage and on the correct culture conditions. In some species, the reliance on natural dehiscence can be circumvented by cutting the wall of the anther, although this does, of course, take a considerable amount of time.

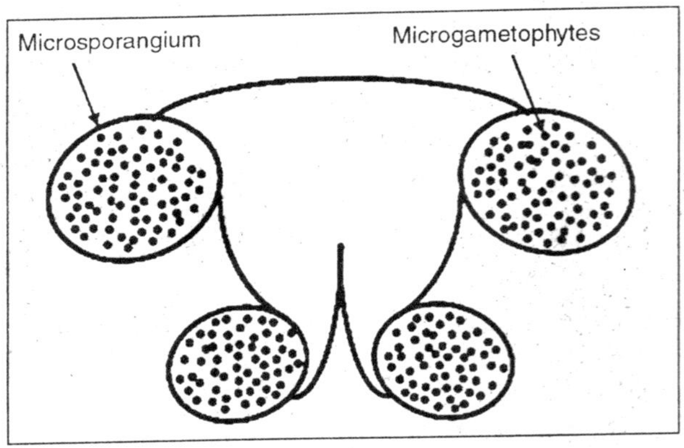

Fig. Microspore Culture

Anthers can also be cultured in liquid medium, and pollen released from the anthers can be induced to form embryos, although the efficiency of plant regeneration is often very low. Immature pollen can also be extracted from developing anthers and cultured directly, although this is a very time-consuming process. Both methods have advantages and disadvantages. Some beneficial effects to the culture are observed when anthers are used as the explant material. There is, however, the danger that some of the embryos produced from anther culture will originate from the somatic anther tissue rather than the haploid microspore cells.

If isolated pollen is used there is no danger of mixed embryo formation, but the efficiency is low and the process is time-consuming. In microspore culture, the condition of the donor plant is of critical importance, as is the timing of isolation. Pretreatments, such as a cold treatment, are often found to increase the efficiency. These pretreatments can be applied before culture, or, in some species, after placing the anthers in culture.

Plant species can be divided into two groups, depending on whether they require the addition of plant growth regulators to the medium for pollen/anther culture; those that do also often require organic supplements, e.g. amino acids. Many of the cereals (rice, wheat, barley and maize) require medium supplemented with plant growth regulators for pollen/anther culture. Regeneration from microspore explants can be obtained by direct embryogenesis, or via a callus stage and subsequent embryogenesis.

Haploid tissue cultures can also be initiated from the female gametophyte (the ovule). In some cases, this is a more efficient method than using pollen or anthers. The ploidy of the plants obtained from haploid cultures may not be haploid. This can be a consequence of chromosome doubling during the culture period. Chromosome doubling (which often has to be induced by treatment with chemicals such as colchicine) may be an advantage, as in many cases haploid plants are not the desired outcome of regeneration from haploid tissues. Such plants are often referred to as 'di-haploids', because they contain two copies of the same haploid genome.

Chapter 32

Plant Regeneration

Having looked at the main types of plant culture that can be established *in vitro*, we can now look at how whole plants can be regenerated from these cultures. In broad terms, two methods of plant regeneration are widely used in plant transformation studies, i.e. somatic embryogenesis and organogenesis.

Somatic Embryogenesis

In somatic (asexual) embryogenesis, embryo-like structures, which can develop into whole plants in a way analogous to zygotic embryos, are formed from somatic tissues. These somatic embryos can be produced either directly or indirectly. In direct somatic embryogenesis, the embryo is formed directly from a cell or small group of cells without the production of an intervening callus.

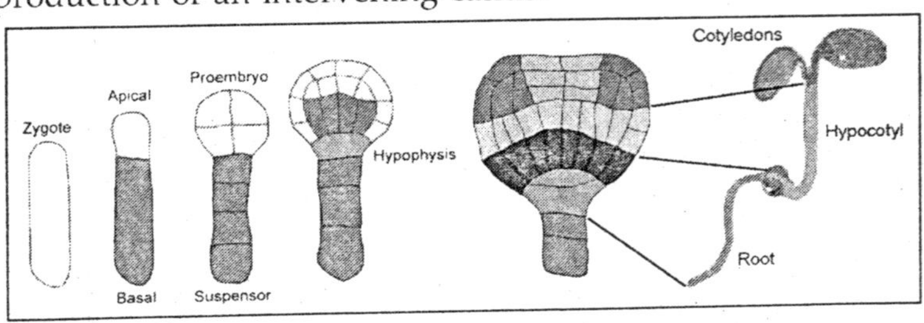

Fig. Embryogenesis

Though common from some tissues (usually reproductive tissues such as the nucellus, styles or pollen), direct somatic embryogenesis is generally rare in comparison with indirect somatic embryogenesis. In indirect somatic embryogenesis, callus is first produced from the explant. Embryos can then be produced from the callus tissue or from a cell suspension produced from that callus.

Somatic embryogenesis from carrot is the classical example of indirect somatic embryogenesis. Somatic embryogenesis usually proceeds in two distinct stages. In the initial stage (embryo initiation), a high concentration of 2,4-D is used. In the second stage (embryo production) embryos are produced in a medium with no or very low levels of 2,4-D.

IMPORTANCE OF GENOTYPE

The major influence on tissue-culture response appears to be genetic, with culture requirements varying between species and cultivars.

Model genotypes that responded well to culture *in vitro* were initially used in plant transformation studies. However, most of the model genotypes used were not elite, commercial cultivars.The commercial cultivars tended to respond poorly to culture *in vitro*. One of the main aims is therefore to identify the components that make up a widely applicable, optimal culture regime.

Many factors have been investigated for their ability to improve the culture response from elite cultivars, including media components (such as alternative carbon sources, macro- and microelement concentrations and composition), media preparation method and donor plant condition and growth conditions.

ORGANOGENESIS

Somatic embryogenesis relies on plant regeneration through a process analogous to zygotic embryo germination. Organogenesis relies on the production of organs, either directly from an explant or from a callus culture. There are three methods of plant regeneration via organogenesis. The first two methods depend on adventitious organs arising either from a callus culture or directly from an explant.

Alternatively, axillary bud formation and growth can also be used to regenerate whole plants from some types of tissue culture. Organogenesis relies on the inherent plasticity of plant tissues, and is regulated by altering the components of the medium.

In particular, it is the auxin to cytokinin ratio of the medium that determines which developmental pathway the regenerating tissue will take.

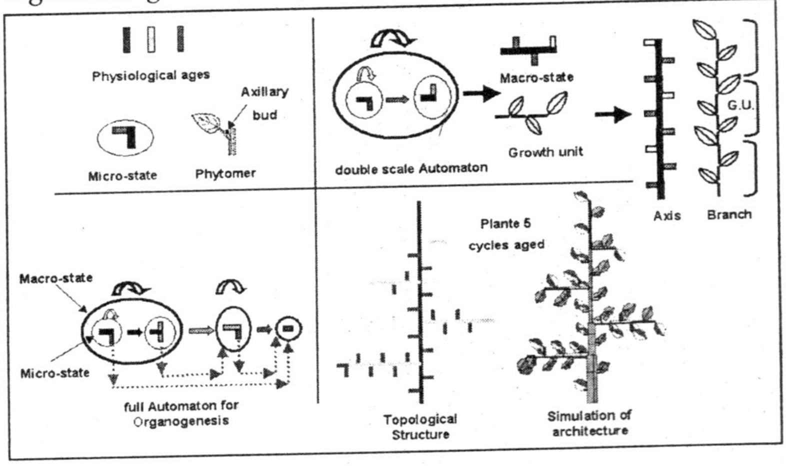

Fig. Organogenesis

It is usual to induce shoot formation by increasing the cytokinin to auxin ratio of the culture medium. These shoots can then be rooted relatively simply.

Plant Transformation Protocols

Various methods of plant regeneration are available to the plant biotechnologist. Some plant species may be amenable to regeneration by a variety of methods, but some may only be regenerated by one method.

The various methods that can be used to transform plants will be considered, but it is worthwhile briefly considering the interaction of plant regeneration methodology and transformation methodology here.

Not all plant tissue is suited to every plant transformation method, and not all plant species can be regenerated by every method.

There is therefore a need to find both a suitable plant tissue culture/regeneration regime and a compatible plant transformation methodology.

ART AND SCIENCE OF MICROPROPAGATION

Of all the terms which have been applied to the process, "micropropagation" is the term which best conveys the message of the tissue culture technique most widely in use today.

The prefix "micro" generally refers to the small size of the tissue taken for propagation, but could equally refer to the size of the plants which are produced as a result.

Micropropagation allows the production of large numbers of plants from small pieces of the stock plant in relatively short periods of time.

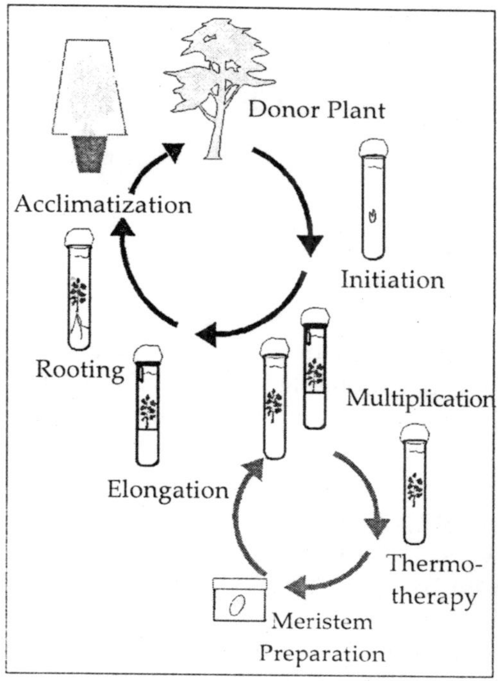

Fig. Micropropagation

Depending on the species in question, the original tissue

piece may be taken from shoot tip, leaf, lateral bud, stem or root tissue. In most cases, the original plant is not destroyed in the process — a factor of considerable importance to the owner of a rare or unusual plant.

Once the plant is placed in tissue culture, proliferation of lateral buds and adventitious shoots or the differentiation of shoots directly from callus results in tremendous increases in the number of shoots available for rooting. Rooted "microcuttings" or "plantlets" of many species have been established in production situations and have been successfully grown on either in containers or in field plantings.

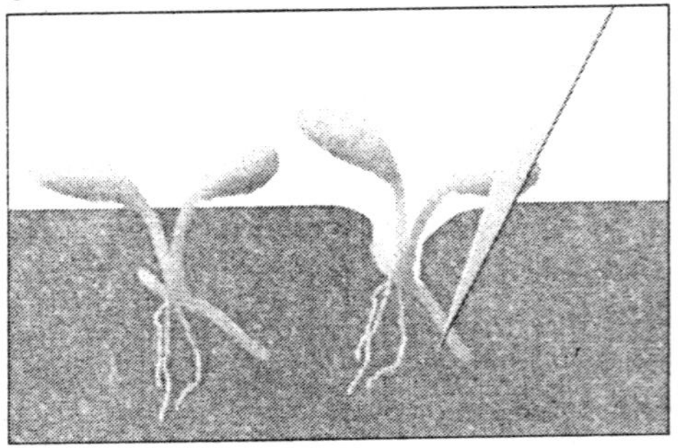

Fig. Plantlets

The two most important lessons learned from these trials are that this methodology is a means of accelerated asexual propagation and that plants produced by these techniques respond similarly to any own-rooted vegetatively propagated plant. Micropropagation offers several distinct advantages not possible with conventional propagation techniques. A single explant can be multiplied into several thousand plants in less than one year. With most species, the taking of the original tissue explant does not destroy the parent plant.

Once established, actively dividing cultures are a continuous source of microcuttings which can result in plant production under greenhouse conditions without seasonal interruption. Using methods of micropropagation, the nurseryman can rapidly introduce selected superior clones of ornamental plants in sufficient quantities to have an impact on the landscape plant market.

PLANT IMPROVEMENT THROUGH TISSUE CULTURE

In introducing this research update, it was mentioned that the major impact of tissue culture technology would not be in the area of micropropagation, but rather in the area of controlled manipulations of plant germplasm at the cellular level. The ability to unorganize, rearrange, and reorganize the constituents of higher plants has been demonstrated with a few model systems to date, but such basic research is already being conducted on ornamental trees and shrubs with the intent of obtaining new and better landscape plants.

Plats With Pest Resistance

Perhaps the most heavily researched area of tissue culture today is the concept of selecting disease, insect, or stress resistant plants through tissue culture. Just as significant gains in the adaptability of many species have been obtained by selecting and propagating superior individuals, so the search for these superior individuals can be tremendously accelerated using in vitro systems.

Such systems can attempt to exploit the natural variability known to occur in plants or variability can be induced by chemical or physical agents known to cause mutations. All who are familiar with bud sports, variegated foliage and other types of chimeras have an appreciation for the natural variability in the genetic makeup or expression in plants. Chimeras are the altered cellular expressions which are visible, but for each of these which are observed many more differences probably exist but are masked by the overall organization of the plant as a whole.

For example, even in frost-tender species, certain cells or groups of cells may be frost hardy. However, because most of the organism is killed by frost, the tolerant cells eventually die because they are unable to support themselves without the remainder of the organized plant. Plant tissues grown in vitro can be released from the organization of the whole plant through callus formation.

If these groups of cells are then subjected to a selection agent such as freezing, then those tolerant ones can survive while all those which are susceptible will be killed. This concept can be applied to many types of stress as well as resistance to fungal and bacterial pathogens and various types of phytotoxic chemical agents. The goal of selecting such resistant cell lines would be to reorganize whole plants from them which would retain the selected resistance Current research in this area extends across many interests including attempts to select salt tolerant lines of tomato, freezing resistant tobacco plants, herbicide resistant agronomic crops, and various species of plants with enhanced pathogen resistance.

PATHOGEN FREE PLANTS

Another purpose for which plant tissue culture is uniquely suited is in the obtaining, maintaining, and mass propagating of specific pathogen-free plants. The concept behind indexing plants free of pests is closely allied to the concept of using tissue culture as a selection system. Plant tissues known to be free of the pathogen under consideration (viral, bacterial, or fungal) are physically selected as the explant for tissue culture. In most cases, the apical domes of rapidly elongating shoot tips are chosen. These are allowed to enlarge and proliferate under the sterile conditions of in vitro culture with the resulting plantlets tested for presence of the pathogen (a procedure called indexing).

Cultures which reveal the presence of the pathogen are destroyed, while those which are indexed free of pathogen are maintained as a stock of pathogen-free material. Procedures similar to these have been used successfully to obtain virus-free plants of a number of species and bacteria-free plants of species known to have certain leaf spot diseases. The impact of obtaining pathogen-free nursery stock can only be speculative, since little research documenting viral, bacterial, or fungal diseases transmitted through propagation of woody ornamentals is available.

SOMATIC HYBRIDIZATION

The ability to fuse plant cells from species which may be incompatible as sexual crosses and the ability of plant cells to take up and incorporate foreign genetic codes extend the realm of plant modifications through tissue culture to the limits of the imagination. Most such manipulations are carried out using plant "protoplasts". Protoplasts are single cells which have been stripped of their cell walls by enzymatic treatment.

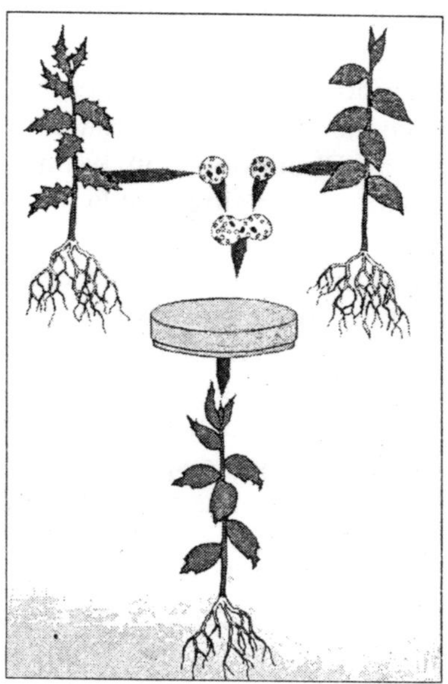

Fig. Somatic Hybridization

A single leaf treated under these conditions may yield tens of millions of single cells, each theoretically capable of eventually producing a whole plant. This concept has fueled speculation as diverse as the possibilities of obtaining nitrogen-fixing corn plants on the one extreme to discovering a yellow-flowered African violet on the other extreme.

The observation that has provided the impetus for most of this research is that when cells are stripped of their cell walls and brought into close contact, they tend to fuse with each other. This "somatic hybridization" is not subject to the same incompatibility problems that limit traditional plant breeding strategies. It is conceivable then that one could hybridize a Juneberry with a crabapple or a plum, but the fundamental research required to demonstrate such an event has yet to be conducted.

The potential use of somatic hybridization to bring about novel combinations of genetic material has been demonstrated in the genera Petunia and Nicotiana. Research funded in part by the Horticultural Research Institute at the University of Wisconsin is investigating the feasibility of using such techniques with woody species. Brent McGown and co-workers have succeeded in obtaining naked cells from tissue cultures of Betula and Rhododendron, but as of yet, they have neither obtained plants from single cells not achieved cellular fusion.

However, further research in this area promises to have a tremendous impact on our concepts of woody plant diversity. Just as remarkable as the idea of fusing plant protoplasts is the idea of incorporating foreign genetic material into the genetic code of plant cells. Such transformations have been carried out in the so-called "gene-splicing" experiments where the information for making insulin was incorporated into bacteria.

Not only is the desired information transmitted to succeeding generations of bacteria, but the bacterial cultures become synthesizers of insulin as well. Plant cells can be made to take up foreign genetic codes, but evidence that this can be transmitted into the daughter cells and serve the intended function is lacking. Plant tissue culture research is multi-dimensional. While most nurserymen have been introduced to the techniques and advantages of micropropagation, few have ventured to use it as a propagation tool.

The applicability of micropropagation for woody trees has been demonstrated as feasible since all aspects of the technology have confirmed the fact that trees produced by this method look like and grow like their counterparts produced by traditional methods of cloning.Other dimensions of tissue culture research have been less well publicized.

The potential for selecting pathogen free plants, for selecting stress-tolerant and pathogen-resistant clones of plants, and the novel genetic combinations to be achieved through somatic hybridization are all lines of research which can have a profound impact on the nursery industry.

ANTHER CULTURE

Crosses between distantly related species can bring together novel gene combinations. However, the hybrid offspring can be few in number, genetically unstable and require years of further selection and screening before any advantageous characteristics can be brought near to commercial use.

Anther culture (androgenesis), to generate haploid plants from pollen microspores, is one way to shorten this process. It allows novel allele combinations, particularly ones involving recessive characters, to be assessed in intact plants. Useful individuals can then be developed into homozygous and fertile plants through chromosome doubling techniques, and brought into a breeding programme.

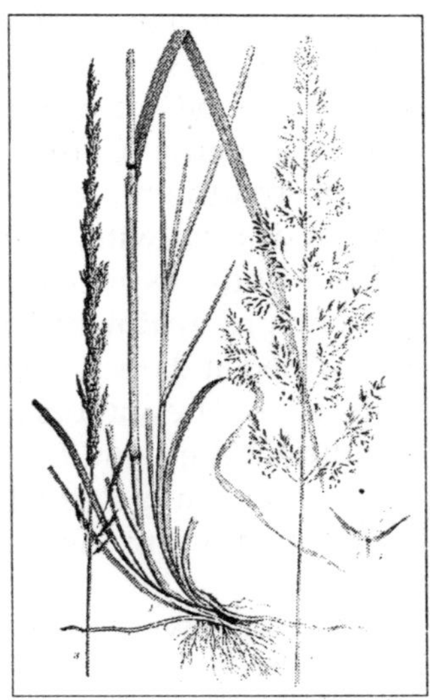

Fig. Festulolium

We have recently been involved in a collaborative project with the Institute of Grassland and Environmental Research (IGER) to use this approach to improve cold-tolerance and fodder quality in grazing grasses. Crosses between *Lolium multiflorum* (Italian ryegrass) and *Festuca arundinacea* (tall fescue) should offer valuable combinations of characteristics. The *Lolium* species should provide good growth characteristics, while the *Festuca* provides stress-tolerance.

One hybrid individual (*Festulolium*) resulting such a cross had already shown drought-tolerance characteristics. However, the out-breeding nature of these grass species, along with the hexaploid genome of *F. arundinacea* and autotetraploid *L. multiflorum* indicated that a lengthy breeding programme might be necessary.The research project therefore aimed to produce androgenic plants from the existing pentaploid *Festulolium* plant and assess them for cold tolerance.

This exploited the expertise in tissue culture at the University of Liverpool together with experience in breeding for stress-tolerance at IGER.Anthers containing immature pollen (microspores) are the starting material for androgenesis. Flowers have to be selected at the correct developmental stage, which varies from species to species. In addition, some individual genotypes may not be amenable to anther culture, or require specific pretreatments. Careful microscopy and testing of successful pre-treatments of related species are therefore necessary when dealing with a new species.

For the Graminae, microspores must be at the mononucleate stage and no pre-treatment is necessary. The cut flowers were surface sterilised and opened in sterile conditions under a binocular microscope. The anthers were dissected and transferred to a solid nutrient medium. Large numbers could be placed on each petri dish. Callus developed, which was transferred to a different medium to initiate embryos and so generate haploid plants.

Over 200 androgenic plants were produced at Liverpool, each originating from a different microspore. Each therefore represented a genetically different individual. Testing for phosphoglucosisomerase, where a different isozyme was contributed by each of the five chromosome groups within the *Festulolium* plant, indicated that the pollen-derived plants had a wide variety of chromosome combinations from each of the parents of the hybrid. The freezing-tolerance of these plants varied considerably, with three individuals able to survive the extreme cold of -14 degrees Celsius.

When the chromosome complement of two of these plants was examined using genomic *in situ* hybridisation (GISH), they carried virtually the whole genome of *F. pratensis*, a parent of *F. arundinacea* noted for its freezing-tolerance. Unfortunately, the fertility of these two plants was not restored by chromosome doubling, so that they could not be used for further breeding. However, they demonstrated the potential of androgenesis for rapid assessment of the genetic potential available from a difficult breeding combination, indicating that this type of wide cross revealed characters of cold and drought tolerance that were worth pursuing.

Somaclonal Variation for Disease Resistance

Plant tissue cultures isolated from even a single cell can show variation after repeated subculture. Distinct lines can be selected with their own particular morphology and physiology. It suggests that the tissue culture contains a population of genotypes whose proportion can be altered by imposing an appropriate selection pressure. This variation can be transmitted to plants regenerated from the tissue cultures, and is called somaclonal variation.

It provides an additional source of novel variation for exploitation by plant breeders. The carrot cultivar Fancy was used in our laboratory to generate a series of 197 regenerant progeny lines. These plants showed considerable morphological variation. They were tested for resistance to the leaf spot pathogen *Alternaria dauci*, which can cause total necrosis of mature leaves. They had a greater degree of variation in response than the parental cultivar.

DEMONSTRATION OF TISSUE CULTURE

The starting point for all tissue cultures is plant tissue, called an explant. It can be initiated from any part of a plant - root, stem, petiole, leaf or flower - although the success of any one of these varies between species. It is essential that the surface of the explant is sterilised to remove all microbial contamination. Plant cell division is slow compared to the growth of bacteria and fungi, and even minor contaminants will easily over-grow the plant tissue culture.

The explant is then incubated on a sterile nutrient medium to initiate the tissue culture. The composition of the growth medium is designed to both sustain the plant cells, encourage cell division, and control development of either an undifferentiated cell mass, or particular plant organs. The concentration of the growth regulators in the medium, namely auxin and cytokinin, seems to be the critical factor for determining whether a tissue culture is initiated, and how it subsequently develops.

The explant should initially form a callus, from which it is possible to generate multiple embryos and then shoots, forming the basis for plant regeneration and thus the technology of micropropagation. The first stage of tissue culture initiation is vital for information on what combination of media components will give a friable, fast-growing callus, or a green chlorophyllous callus, or embryo, root or shoot formation. There is at present no way to predict the exact growth medium, and growth protocol, to generate a particular type of callus.

These characteristics have to be determined through a carefully designed and observed experiment for each new plant species, and frequently also for each new variety of the species which is taken into tissue culture. The basis of the experiment will be media and protocols that give the desired effect in other plant species, and experience.

The Demonstration

The strategy for designing a medium to initiate tissue culture, showing how growth regulators and other factors modulate development, can be demonstrated using the African Violet, a popular house plant. Leaf sections are the source of explants. This demonstration is regularly carried out by a student class, and gives reliable results. Sterile supplies are provided from central facilities, and provision of sterile working areas (for example, in laminar flow hoods) is an advantage, although cultures can be initiated in an open laboratory with careful aseptic technique.

The standard precautions used during any laboratory work involving chemicals or microbes should be adopted. If we are in any doubt about safety hazards associated with this demonstration, we should conslt our local safety adviser.

Step 1: Selection of the Leaves

Leaves are cut from healthy plants, leaving a short length of petiole attached. They should be selected to each yield several explants of leaf squares with approximately 1 cm sides. The youngest and oldest leaves should be avoided.

Wash the dust off the leaves in a beaker of distilled water, holding the leaf stalk with forceps.

Step 2: Surface Sterilisation and Preparation of the Explants

This part of the procedure should be carried out in a sterile working area, or with meticulous aseptic technique.

The leaf, with the petiole still attached, should be immersed in 70 per cent ethanol for 30 seconds, then transferred to a sterile petri dish. Sterile scissors and forceps are then used to cut squares from the leaf as explants, each with approximately 1 cm sides.

The explants are transferred into a 10 per cent hypochlorite bleach solution for 5 minutes, gently agitating once or twice during this time. They are then washed free of bleach by immersing in four successive beakers of sterile distilled water, leaving them for 2-3 minutes in each.

Three explants are placed on each petri dish of growth medium, with the upper epidermis pressed gently against the surface of the agar to make good contact.

The petri dishes are sealed with plastic film to prevent moisture loss, and incubated at 25°C in 16h light/8h dark.

Step 3: Assessment of Tissue Culture Development

The explants are incubated for 4 - 6 weeks, and inspected

at weekly or fortnightly intervals. The growth of obvious bacterial or fungal colonies indicates contamination, and data from such cultures is obviously suspect. The development of dark brown tissue cultures can also be a consequence of contamination.The media used in the demonstration are designed to show the effects of auxin, cytokinin, sucrose and mineral salts on development. The media were based on the well-known Murashige and Skoog inorganic medium.

Typical Results

Absence of sucrose inhibits this production. Shoot production is also limited on the low sucrose concentration, but comparable with the control at high sucrose.At zero and low levels of cytokinin, callus forms where the leaf surface is in contact with the medium, while at high levels, shoot formation is stimulated.At zero and low levels of auxins there is a stimulus to shoot formation, but at high concentrations, large numbers of roots are formed.At low and zero levels of MS salts, there is no growth at all.

These very obvious variations demonstrate the importance of a carbon and inorganic salt source for plant growth, as well as the effect of the auxin:cytokinin ration on the control of plant development.

Chapter 33

Pollen Culture

A culture of plant cells derived from pollen in a synthetic medium: the progeny generated will have a single set of chromosomes. Clapham reported the first successful culturing of barley pollen. In order to use barley pollen culture in plant breeding or in mutation and other genetic studies, the techniques must be somewhat improved.

Attempts were thus made to increase the efficiency of plantlets from barley pollen. Knowing that the pollen of various varieties have a rather wide range of response to the culturing process, 16 spring barley varieties and 1 winter barley variety were tested using a modified Linsmaier and Skoog's medium with 12 per cent sucrose and 1 p.p.m. each of IAA and BAP. Between 300 and 1,200 anthers were plated for each spring variety.

All the spring varieties produced callus and several, including Akka, Zephyr, Unitan, Traill, and Trebi produced at least one plantlet with shoot and or roots. The winter variety, on the other hand did not respond—not even producing callus, though more than 2,000 anthers were plated. Since none of the varieties appeared superior to Akka, this variety was used in most of the subsequent work.

A principle objective was to induce a large percentage of the pollen grains to divide and thus hopefully start a large number along the developmental course of forming plantlets.

Anther and Pollen Culture

Anther and pollen culture is tissue culture method of production involving culturing of anthers. Haploids develop directly from pollen grains in culture, either through direct formation of embryos from pollen grains or formation of callus and subsequent plant regeneration.

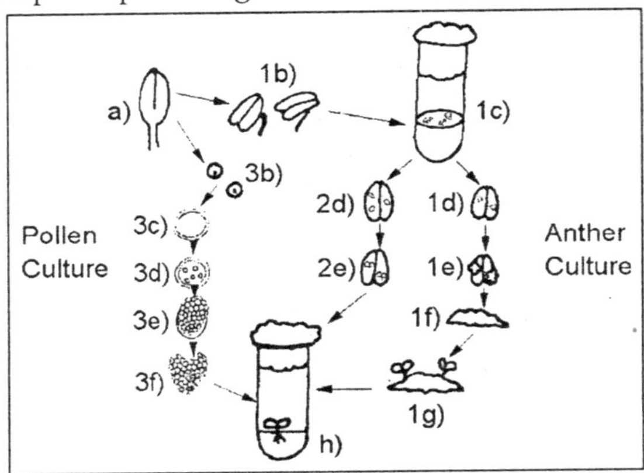

Fig. Anther and Pollen Culture

This technique requires optimizing culture conditions which vary between species and between genotypes. Induction frequency can be high and the production time to mature plants of three to four months with direct development of homozygous individuals.

POLLEN GRAIN CULTURE

The culture of immature anthers and pollen grains is done so as to induce the pollen grains to develop into multicellular forms, particularly into embryos, with half the normal number of chromosomes for the species. When such haploid embryos are treated with chromosome doubling agents, e.g. colchicine, their normal chromosome number is restored (and thus their fertility) and, in addition, the plants are pure lines. Pure line formation is a natural tendency for self fertilizing species and can be obtained with cross fertilizing species with repeated inbreeding for 10 or more generations.

So far the induction of haploid plant formation from anther and pollen grain cultures has been successful mainly with naturally self fertilizing species and thus, on chromosome doubling, are in theory very similar if not identical with the parents. However, by first crossing many lines from a self fertilizing species, new combinations of genes are formed, and haploid plants produced by anther culture from such crosses could be an extremely valuable and quick way of obtaining pure lines of these new combinations.

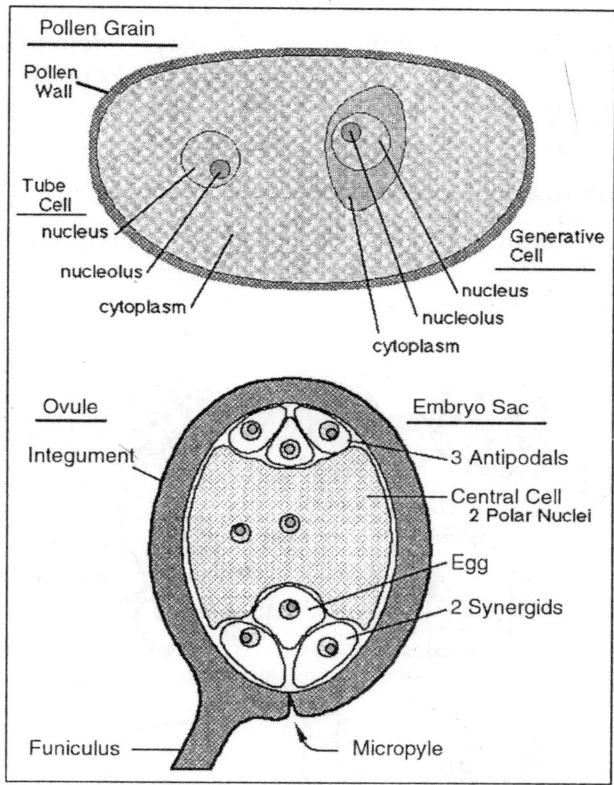

Fig. Pollen Grain Culture

If we can find out how to obtain haploid plants from anther culture of cross fertilizing species, they also could be extremely valuable relative to breeding programmes and the selection of improved strains. Pollen is produced within the anthers (microsporangia or pollen sacs) of the flower. During its development from an undifferentiated mound of cells (anther primordium) the anther forms two general groups of cells. The reproductive or sporogenous cells give rise to the microspores and are formed from cells located centrally within the developing anther.

The nonreproductive cells form discrete anther tissues layers and include the epidermal, cortical and tapetal cell layers surrounding the sporogenous cells. The tapetum which is the innermost layer of the pollen sac plays a dominant role particularly during the microspore stage. For example, many male sterile mutations affect tapetal cell functions and development is often arrested during the microspore stage.

Microsporogenesis and Microgametogenesis

Two distinct and successive developmental phases, microsporogenesis and microgametogenesis, lead to the production of the mature microgametophytes.

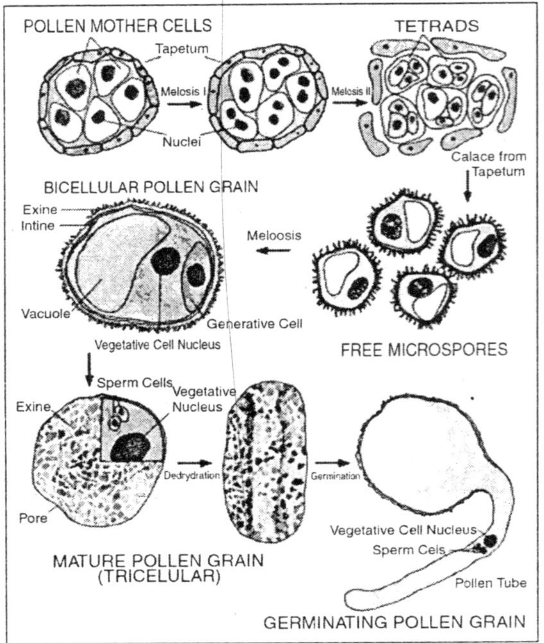

Fig. Microsporogenesis

Microsporogenesis comprises the events which lead to the

formation of the haploid unicellular microspores. During microsporogenesis the diploid sporogenous cells differentiate as microsporocytes (pollen mother cells or meiocytes) which divide by meiosis to form four haploid microspores.

Each diploid meiocyte gives rise to a tetrad of four haploid microspores and microsporogenesis is complete with the formation of distinct single-celled haploid microspores.

Microgametogenesis

It comprises events which lead to the progressive development of the unicellular microspores into mature microgametophytes containing the gametes. This phase begins with the expansion of the microspore which is commonly associated with the formation of a single large vacuole. Vacuolation is accompanied by the displacement of the microspore nucleus to an eccentric position against the microspore wall.

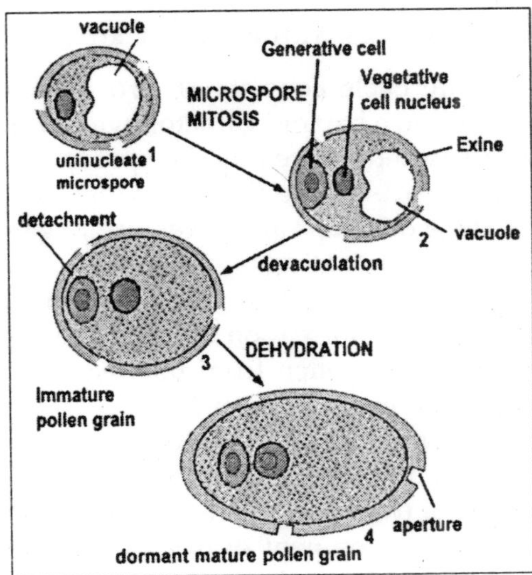

Fig. Microgametogenesis

In this position the nucleus undergoes first pollen mitosis which results in the formation of two unequal cells, a large vegetative cell and a small generative cell each containing a haploid nucleus. The generative cell subsequently detaches from the pollen grain wall and is engulfed by the vegetative cell forming a unique 'cell within a cell' structure.

The engulfed generative cell divides once more by mitosis to form the two sperm cells completely enclosed within the vegetative cell cytoplasm either before pollen is shed (tricellular pollen) or within the pollen tube (bicellular pollen).

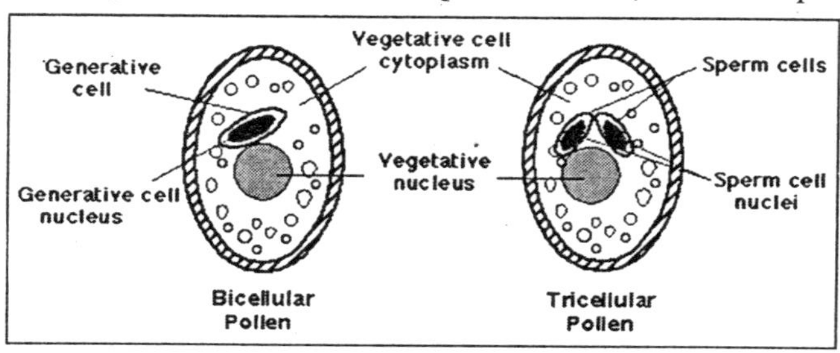

Fig. Bicellular Pollen and Tricellular Pollen

Reproduction Involves Anthers and Carpels

The plant life cycle consists of two phases: a diploid sporophyte stage and a haploid gametophyte stage. The flower is a diploid structure produced by the sporophyte. However, within the fertile parts of the flower (anther and carpel) the gametophyte stage develops and gives rise to gametes which recombine during fertilization.

The key events that take place in the anther are microsporogenesis, the formation of microspores by meiosis, and the production of pollen from microspores. The pollen grain represents the microgametophyte.

The key events that take place in the carpel are megasporogenesis, the formation of megaspores by meiosis, the development of the megagametophyte (also known as the embryo sac), the development of ovules which mature into seeds after fertilization, and the development of fruits to disperse the seeds.

Microsporogenesis and Pollen Formation

Pollen forms in the anthers, the enlarged tips of stamens.

Each anther usually contains four pollen sacs, which can also be considered microsporangia.

Each pollen sac contains numerous microsporocytes, diploid cells (2N) that undergo meiosis to produce four haploid microspores. Each microspore undergoes one round of mitosis and matures into a pollen grain.

Although it only consists of two cells, the pollen grain represents the microgametophyte stage of the angiosperm life cycle. Thus, a pollen grain consists of two haploid cells surrounded by a tough outer coat.

- Exine - outer wall of pollen grain - contains sporopollenin, a tough polymer containing carotenoids (which is why pollen is yellow). The exine makes the pollen grain resistant to decay.
- Tube cell - one of the haploid cells, will form a pollen tube that burrows through the style after pollination.
- Generative Cell - the other haploid cell, will undergo mitosis to form two sperm as the pollen tube grows.

Megasporogenesis and Ovule Formation

The megagametophyte, also called the embryo sac, is a a 7- celled haploid structure that forms inside each ovule. One of these cells is the egg, will undergo fertilization to form the zygote.

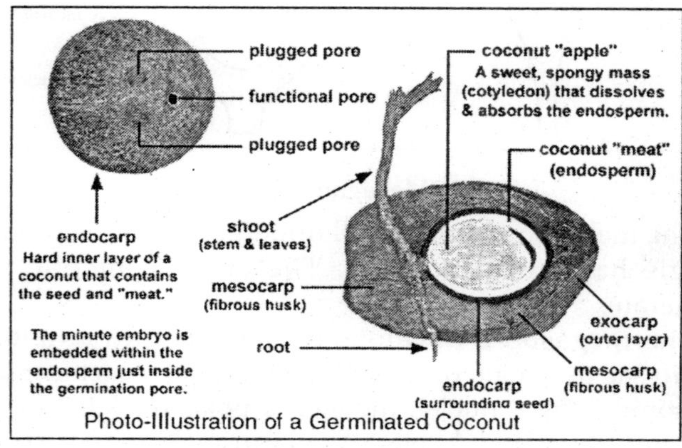

Fig. Megasporogenesis

A second fertilization event will also take place. Thus, fertilization occurs in the ovule. For sexual reproduction, anthers usually split open, releasing mature pollen that is picked up by pollinating agents and transferred to other flowers.

MEGASPOROGENEIS AND MEGAGAMETOGENESIS

Megasporogeneis and megagametogenesis occur within the ovules found in the ovaries of carpels. An ovule is attached to the ovary wall by a structure called the funiculus. The funiculus will also be a seed's attachment to the ovary tissue after fertilization and development of an embryo. It can be seen in some fruits. Each ovule contains layers of sterile tissue, called the integuments and a nucellus layer that surround a single megaspore mother cell. A small opening between the integuments is the micropyle.

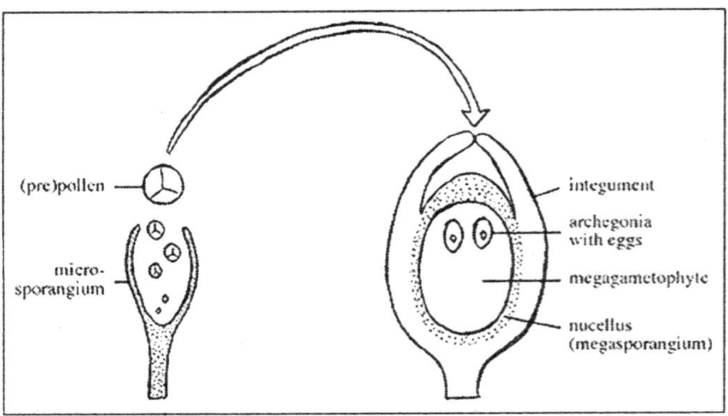

Fig. Megagametophyte

The megaspore mother cell undergoes meiosis producing a single haploid megaspore. The other 3 meiotic nuclei degenerate.

The megaspore develops by mitosis into an elongate embryo sac (a female or megagametophyte) which contains 7 cells comprising 8 nuclei: the egg, 2 polar nuclei, 2 synergid cells, and 3 antipodal cells. At maturity, the embryo sac is contained within the ovule.

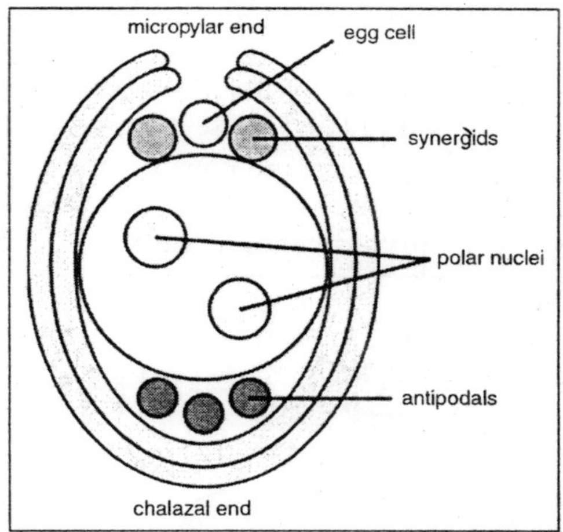

Fig. Ovules Contain Mature Eggs

The cells at the micropyle end of the embryo sac form the egg apparatus, and include the egg and the two synergids. The three antipodal cells are located at the opposite end of the embryo sac, the chalazal end. A central cell, containing the two polar nuclei is located in the centre of the embryo sac. The stigma becomes receptive (sticky and shiny) when ovules contain mature eggs.

POLLEN GERMINATION

Successful pollination is just one step in the process of reproduction and development in flowering plants. If pollen of the correct type is deposited on the receptive stigma of a carpel (one which contains at least one ovule with a mature egg), chemicals secreted by the surface stigmatic tissue of the stigma will stimulate the pollen grain to "germinate". A receptive stigma can often be discerned visually, because its surface will be coated with a sugary, sticky secretion.

Pollen germination itself involves uptake of water from surrounding stigma cells by the pollen grain, swelling it until it splits open. As pollen germinates, it grows a pollen tube that grows through the length of the style, through a special tissue called the transmitting tissue that forms a path to the ovary with its ovule(s). The style may have mechanisms to facilitate the growth of the tube. Each ovule must be fertilized by separate sperm, so flowers that have many ovules must also have many germinated pollen grains. For example, each corn silk is a single style. An average corn on the cob may have 400-500 grains. That is the result of 400-500 pollinations (and fertilizations and embryo developments).

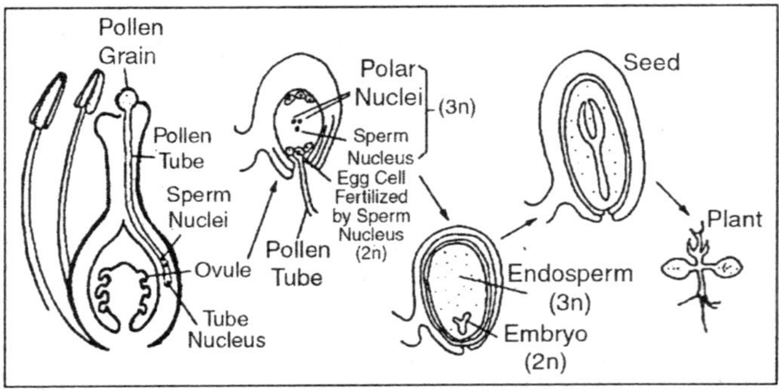

Fig. Pollen Germination

Each pollen grain contains two sperm that migrate down the pollen tube. Ultimately, the sperm will migrate to the embryo sac at a region called the micropyle, which is at the base of the synergids and egg.

POLLEN MORPHOLOGY

Microsporogenesis, microgametogenesis and pollen morphology of six species of genus *Passiflora* L. belonging to three subgenera were studied with light and scanning microscopy; *P. caerulea* was also examined with transmission microscopy. The tapetum is secretory, microspore tetrads are tetrahedral and pollen grains are two-celled when shed.

Small Ubisch bodies are attached to a peritapetal membrane; they are a product of tapetal activity and the rough endoplasmic reticulum (ERr) appears to be involved in their origin. The pollen grains of all the species are subspheroidal, zonocolpate, geminicolpate. Each pair of colpi anastomoses at the poles. The exine is semitectate, reticulate, heterobrochate. The muri are simplibaculate, wavy.

The lumina have clavate bacula of varying height. The colpus structure is similar to that of the lumina but generally with fewer and smaller bacula.

Lumina size and amount of bacula inside the lumina vary between subgenera. The grains from subgenera *Passiflora* and *Dysosmia* differ from those of *Decaloba* in their size and number of colpi. The pollen and microsporangium morphology of the species of subgenera *Passiflora* and *Dysosmia* are more similar than those of subgenus *Decaloba*.

Pollination and Pollinating Agents

Pollination is the transfer of pollen (containing sperm) from one flower to a second. Since plants cannot move, and sperm in flowering plants is not motile (no flagella), sperm transfer requires a pollinating agent. This dependency poses some problems for the typical plant:

- We must attract an agent
- We must have adequate pollen for transfer
- We must convince the agent to take our pollen to another flower of our species (usually by providing a reward)
- In fact, we want to be so special that the pollinating agent will visit our species exclusively, for at least the time that we are in flower. (Because pollen production takes energy, which should be conserved whenever possible.)
- We do not want to produce too much of an attractant or reward because our pollinator must leave we and go visit another flower of our species in order to achieve the goal of pollen transfer.

  Filling up us or taking our pollen to a different kind of flower is not useful

Chapter 34

Ovule Culture

DEVELOPMENTAL STAGES

Generally, gynogenesis has two or many stages and each stage may have distinct requirements. In rice, two stages, viz, induction and regeneration, are recognized. During induction, ovaries are floated on a liquid medium having low auxin and kept in dark, while for regeneration they are transferred on to an agar medium with higher auxin concentration and incubated in light. Haploid plants generally originate from egg cell in most of the species (in vitro parthenogenesis) but in some species, e.g., rice, they arise chiefly from synergids; in atleast Allium tuberosum even antipodals produce haploid plants (in vitro apogamy).

As in anther culture, gynogenesis may occur either via embryogenesis or through plantlet regeneration from callus. In rice MCPA generally leads to a small amount of protocorm like callus formation from which shoots and roots regenerate, while picloram promotes embryo regeneration. In contrast, sugarbeet usually shows embryo development, while in sunflower embryos regenerate following a callus phase. In general, regeneration from a callus phase appears, at least for the present, to be easier than direct embryogenesis.

Limitations of Ovary Culture

Ovary culture has mainly two limitations: i. so far it has been successful only in less than two dozen species, and ii. the frequency of responding ovaries (1-5 per cent) and the number of plantlets/ovary (1-2) is quite low. iii. Therefore, anther culture is preferred over ovary culture. Only in those cases where anther culture fails, e.g., sugarbeet, and for male sterile lines, ovary culture assumes significance.

EMBRYO CULTURE

What do we understand by Embryo culture? Embryo culture is the sterile isolation and growth of an immature or mature embryo in vitro, with the goal of obtaining a viable plant. The first attempt to grow the embryos of angiosperms was made by Hannig in 1904 who obtained viable plants from in vitro isolated embryos of two crucifers Cochleria and Raphanus.

In 1924, Dietrich grew embryos of different plant species and established that mature embryos grew normally but those excised from immature seeds failed to achieve the organization of a mature embryo.

They grew directly into seedlings, skipping the stages of normal embryogenesis and without the completion of dormancy period. Laibach demonstrated the practical application of this technique by isolating and growing the embryos of interspecific cross Linum perenne and L. austriacum that aborted in vivo.

This led Laibach to suggest that in all crosses where viable seeds are not formed, it may be appropriate to excise their embryos and grow them in an artificial nutrient medium. Embryo Culture is now a well-established branch of plant tissue culture.

TYPES OF EMBRYO CULTURE

Mature embryo culture

It is the culture of mature embryos derived from ripe seeds. This type of culture is done when embryos do not survive in vivo or become dormant for long periods of time or is done to eliminate the inhibition of seed germination. Seed dormancy of many species is due to chemical inhibitors or acids, mechanical resistance present in the structures covering the embryo, rather than dormancy of the embryonic tissue.

Immature Embryo Culture/embryo Rescue

It is the culture of immature embryos to rescue the embryos of wide crosses. This is mainly used to avoid embryo abortion with the purpose of producing a viable plant.

The underlying principle of embryo rescue technique is the aseptic isolation of embryo and its transfer to a suitable medium for development under optimum culture conditions. Florets are removed at the proper time and either florets or ovaries are sterilized. Ovules can then be removed from the ovaries. The tissue within the ovule, in which the embryo is embedded, is already sterile.

For mature embryo culture, either single mature seeds are disinfected or if the seeds are still unripe then the still closed fruit is disinfected. The embryos can then be aseptically removed from the ovules. Utilization of embryo culture to overcome seed dormancy requires a different procedure. Seeds that have hard coats are sterilized and soaked in water for few hours to few days. Sterile seeds are then split and the embryos excised. The most important aspect of embryo culture work is the selection of medium necessary to sustain continued growth of the embryo. In most cases, a standard basal plant growth medium with major salts and trace elements may be utilized.

Technique

Mature embryos can be grown in a basal salt medium with a carbon energy source such as sucrose. But young embryos, in addition require different vitamins, amino acids, and growth regulators and in some cases natural endosperm extracts. Young embryos should be transferred to a medium with high sucrose concentration (8-12 per cent); which approximate the high osmotic potential of the intracellular environment of the young embryosac, and a combination of hormones which supports the growth of heart-stage embryos (a moderate level of auxin and a low level of cytokinin).

Reduced organic nitrogen as aspargine, glutamine or casein hydrolysate is always beneficial for embryo culture. Malic acid is often added to the embryo culture medium. After one or two weeks when embryo ceases to grow, it must be transferred, to a second medium with a normal sucrose concentration, low level of auxin and a moderate level of cytokinin which allows for renewed embryo growth with direct shoot germination in many cases.

In some cases where embryo does not show shoot formation directly, it can be transferred to a medium for callus induction followed by shoot induction. After the embryos have grown into plantlets in vitro, they are generally transferred to sterile soil and grown to maturity.

ORGANOGENIC POTENTIAL OF EMBRYO CALLUS

An embryo callus is reported to possess a high regenerative capacity compared to those derived from mature organs, such as the leaf, stem and root. This is especially true of crucifers, cereals and millets. The age of the embryo considerably influences the regenerative ability of maize calluses. Green and Phillips obtained a differentiating callus from maize embryos excised 18 DAP. No differentiation occurred if the callus originated from a mature embryo.

Immature embryos of oats, barley, sorghum, wheat, rice, rye, rye-grass and triticale are good explants for initiating callus capable of plant regeneration. The suspension cultures of Pennisetum americanum, known to differentiate whole plants from isolated protoplasts, were derived from immature embryos. To obtain a callus with morphogenetic potential excised immature embryos are placed on an agar medium with the scutellum facing up in the presence of 2,4-D alone or in combination with cytokinin.

An embryo callus of Picea abies regenerated numerous shoot buds but only in one instance did the root differentiate.

APPLICATIONS

Prevention of Embryo Abortion in Wide Crosses

Successful interspecific hybrids have been seen in cotton, barley, tomato, rice, legume, flax and well known intergeneric hybrids include wheat x barley, wheat x rye, barley x rye, maize x Tripsacum, Raphanus sativus x Brassica napus. Distant hybrids have also been obtained via embryo rescue in Carica and Citrus species. Embryo rescue technique has been successfully used for raising hybrid embryos between Actidinia deliciosa x A. eriantha and A. deliciosa x A. arguata.

Production of Haploids

Embryo culture can be utilized in the production haploids or monoploids. Kasha and Kao have developed a technique to produce barley monoploids. Interspecific crosses are made with Horeum bulbosum as the pollen parent, and the resulting hybrid embryos are cultured but they exhibit H. bulbosum chromosome elimination resulting in monoploids of the female parent H. vulgare.

Overcoming Seed Dormancy

Embryo culture technique is applied to break dormancy. Seed dormancy can be caused by numerous factors including endogenous inhibitors, specific light requirements, low temperature, storage requirements and embryo immaturity. These factors can be circumvented by embryo excision and culture.

Shortening of Breeding Cycle

There are many species that exhibit seed dormancy that is often localized in the seed coat and/or in the endosperm. By removing these inhibitions, seeds germinate immediately. Seeds sometimes take up and give O_2 very slowly or not at all through the seed coat, and so germinate slowly if at all, e.g. Brussels sprouts, rose, apple, oil palm and iris. These (Ilex) are important plants for Christmas decorations. Ilex embryos remain in the immature heart-shaped stage though the fruits have reached maturity.

Prevention of Embryo Abortion with Early Ripening Stone Fruits

Some species produce sterile seeds that will not germinate under appropriate conditions and eventually decay in soil e.g., early ripening varieties of peach, cherry, apple, plum. Seed sterility may be due to incomplete embryo development. which results in the death of the germinating embryo. In crosses of early ripening stone fruits, the transport of water and nutrients to the yet immature embryo is sometimes cut off too soon resulting in abortion of the embryo. Eg: Macapuno coconuts.

These are priced for their characteristic soft endosperm which fills the whole nut. These nuts always fail to germinate because the endosperm invariably rots before germinating embryo comes out of the shell. Embryo culture has been practised as a general method in horticultural crops include avocado, peach, nectarine and plum. Two cultivars 'Goldcrest peach' and 'Mayfire nectarine' have resulted from embryo culture and commercially grown.

Clonal Micropropagation

The regenerative potential is an essential prerequisite in nonconventional methods of plant genetic manipulations. Because of their juvenile nature, embryos have a high potential for regeneration and hence may be used for in vitro clonal propagation. This is essentially true of conifers and graminaceous members. Complete plantlets in vitro from conifers were achieved in long-leaved pine (Pinus palustris) through embryo culture.

This was followed by plantlet formation and their clonal micropropagation from embryos of P. elliottii, P. radiata, P. regida, P. monticola, P. taeda, P. virginiana and P. sabiniana. Cotyledonary buds were also induced in P. coulleri and P. strobus. Both organogenesis and somatic embryogenesis have been induced in major cereals and forage grasses from embryonic tissues. Generally, callus derived from immature embryos of cereals has the desired morphogenetic potential for regeneration and clonal propagation.

As an exception, plant regeneration of orchid grass can be achieved in subcultured tissues derived from mature embryos. Probably these embryos are very small and the degree of development as well as organisation may be similar to that of immature embryos of large cereal grains.

Germination of seeds of obligatory parasites without the host is impossible in vivo, but is achievable with embryo culture. Plant embryo culture is an invaluable breeding technique as far as it is possible to synthesise hybrids from incompatible crosses.

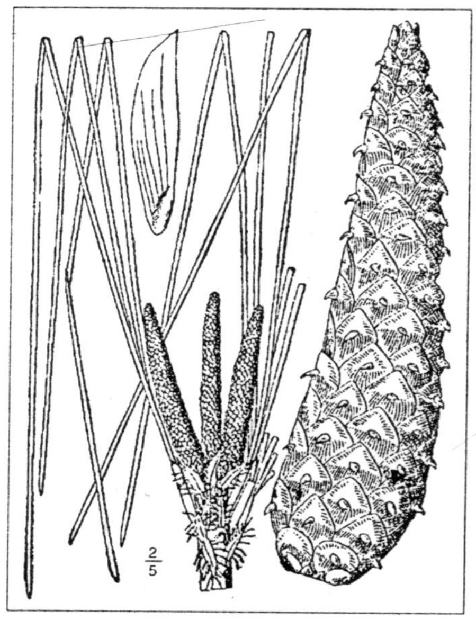

Fig. Pinus Palustris

However, the number of hybrid seedlings rescued in many instances is extremely low due to the difficulty in growing very young embryos. Further viability decreases with age of the embryo in most of the incompatible crosses. Efforts are needed to identify the requirements for embryos of progressively younger stages in major crop species.

PROTOPLAST ISOLATION, CULTURE AND FUSION

Exploitation of genetic variability is an essential strategy in plant breeding programme for the genetic improvement of crop plants. Improvements of crop plants by conventional breeding is time consuming, expensive and labour intensive. Unconventional techniques can shorten the process. One of these techniques is somatic hybridization. Somatic hybridization through protoplast fusion has been envisaged for the production of parasexual hybrids to overcome sexual incompatibility. The ultimate objective to transfer desirable traits from distantly related taxa.

This para-sexual hybridization is a means of increasing genetic variability of several crops gene pool, not only by overcoming sexual incompatibility but it can produce nuclear – cytoplasmic combinations that are difficult or impossible to obtain by conventional sexual crossing. Klercker first isolated plant protoplasts of *Stratiotes aloides* by plasmolysing and subsequently slicing the tissue.

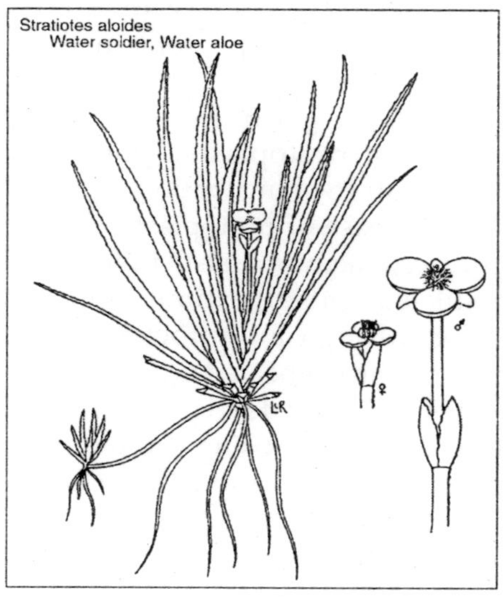

Fig. Stratiotes Aloides

Protoplasts obtained by this procedure were used mainly in physiological studies. However, limited yields, excessive damage, and restriction of the technique to specific tissues, were the major drawbacks. Large scale isolation of plant protoplasts, using cell wall degrading enzymes, gave impetus to research in protoplasts.

Initially, a two step procedure was adopted, first producing single cells and then protoplasts. However, with further experimentation, a mixed enzyme procedure was developed, involving maceration and protoplast isolation in a single step from intact tissues.

The major activity of the cell wall degrading enzymes is in the digestion of pectins, cellulase and hemicellulases (xylans) that largely constitute the plant cell wall. Pectinases degrade the galacturonic acid residues of pectins that confer the cell to cell adhesion, and apparently macerate the tissue to single cells; cellulases digest the cellulose component, conferring the spherical shape to the protoplasts, while hemicellulases assist in the breakdown of xylans.

Incubation in a mixture of the three types of enzymes in suitable ratios facilitates release of a large number of protoplasts. The impurities present in the commercially supplied crude enzymes reduce the protoplast viability and enzyme purification by gel filtration can be beneficial for sustained division of protoplasts. The pH of the enzyme mixture and temperature of incubation are some of the factors that influence the release of protoplasts.

In certain cases, vacuum assists initial penetration of the enzyme and subsequent digestion. Following protoplast release, filtering through an autoclavable mesh removes undigested tissue, while repeated washing by centrifugation, removes enzymes and finer debris. Separation of protoplasts from debris is also achieved using sucrose floatation. A two-phase system using dextran and PEG or sodium metrizoate or a discontinuous density gradient system incorporating percoll are some of the other methods used to purify plant protoplasts from debris.

PROTOPLAST CULTURE

A range of basal media for protoplast culture have been used with several modifications of hormones within each formulation. These media have been based mainly on those devised by Murashige and Skoog and Gamborg, with an addition of an osmoticum, generally a sugar, such as glucose or a sucrose or a sugar alcohol, for example mannitol or sorbitol. These sugar alcohol are used for their nearly metabolically inactive property. Kao and Michayluk developed a complex medium, wherein single protoplasts of *Vicia hajastana* were capable of colony formation.

This medium, together with minor modifications, has proved successful in obtaining sustained divisions in many of the plant systems. Caboche constituted a fully defined medium for growth of pre - cultured protoplasts at low densities. Efficient growth of pre-cultured protoplasts has also been achieved by the use of "nurse" cultures of albino protoplasts, irradiated protoplasts or protoplasts from an auxotrophic mutant. A multiple drop array (MDA) technique has been useful in analyzing the effects of different combinations and concentrations of growth regulators in a particular medium on protoplast development.

Protoplasts have been cultured in sitting or hanging droplets, thin, liquid layers embedded in agar medium, and liquid over an agar layer. Partanen et al. and Santos et al. employed a filter paper layer at the agar liquid interface of liquid over agar media for successful culture of fern and Medicago mesophyll protoplasts, respectively. Other than agar gelling agent, alginate, which enables protoplasts to be plated without a temperature shock.

The aligmate can be liquified by addition of a chelating agent, making possible the recovery of developing colonies. In addition, other factors that influence division and subsequent colony formation, include the temperature of incubation, the initial light intensity and the type of culture vessel.

Methods of Protoplast Culture

Consequent to isolation and adjusting of culture density, protoplasts are plated by any one of the following methods that have been usually employed.

- Liquid layers
- Embedded in agar/agarose
- Liquid over agar/agarose
- Alginate encapsulation
- Hanging/sitting drop culture

- Filter paper substratum placed on agar
- Microdrop array (M.D.A.) technique

In certain instances agarose has been used in place of agar and this has improved the culture response.

- Liquid layers: 10 ml of the protoplast suspension at requisite density is plated in 90 mm petridish. Proportionately 3 ml are plated in a 50 mm dish and 1.5 ml in a 35 mm dish. Seal with parafilm and incubate in moist environment.
- Embedded in Agar/agarose:
  - Melt double strength agar (1.4 per cent) medium with 9 per cent mannitol and bring to 45°C.
  - First pour 5 ml of protoplast suspension into a 90 mm petri dish (at double the required density).
  - Pour 5 ml of molten double strength agar medium into 5 ml of protoplast suspension in the 90 mm petri dish.
  - Mix thoroughly and allow agar to set. Seal dishes with parafilm.

Alternatively, agarose at a final concentration of 1-1.2 per cent can be employed and blocks of embedded protoplasts can be transferred to liquid media in petridishes or flasks for slow shaking.

Maintenance of Protoplast Cultures:

- Maintain the protoplast cultures initially for 10-15 days under dark conditions at temperatures ranging from 25-30°C depending upon a particular species or variety.
- Provide with continuous illumination first at a low intensity of 1,000 lux and gradually increased to 3,000 lux.
- Maintain relative humidity in containers during this period to avoid desiccation.

Cell wall formation is evident by the change of shape of the protoplasts within 24-48 hrs of plating. Divisions occur by 3-7 days of culture initiation. For sustained growth of protoplasts the osmoticum of the culture medium has to be lowered by sequential dilutions periodically. For protoplasts that are embedded in agar or agarose and alginate beads, small blocks containing the dividing protoplasts are transferred to the surface of the same medium but with lower osmoticum. The same can further be shifted again to a medium with a lower osmotic pressure. Once the protoplasts have grown up to a callus stage, the procedure for regeneration of plants there from is essentially the same as that employed in regular tissue culture techniques.

PROTOPLAST FUSION

Mechanism of Protoplast Fusion

Major constituents of the cell membrane are phospholipids and proteins. Lipid molecules consist of a polar head group that are hydrophilic, to which is attached one or two long hydrocarbon chains of tails which are hydrophobic.

The phospholipids are arranged in a bilayer into which peripheral or integral structural proteins are embedded in a mosaic-like fashion.

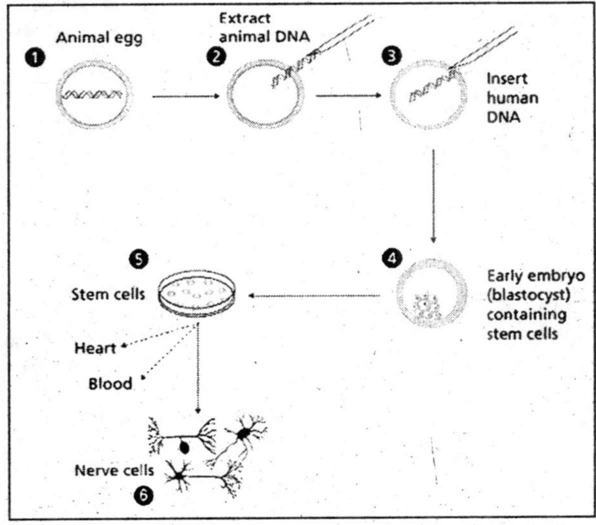

Fig. Protoplast Fusion

Exogenous chemical stimuli cause disturbances in the

intra-membranous proteins and glyco-proteins, increasing thereby the membrane fluidity and resulting in fusion of membranes at the point of contact. Such fluidity is related to the proportion of saturated phospholipids in the membrane and is influenced by temperature. The intra-membranous phosphate groups are postulated to induce a negative surface charge on the protoplasts. This tends to repel the protoplasts from one another.

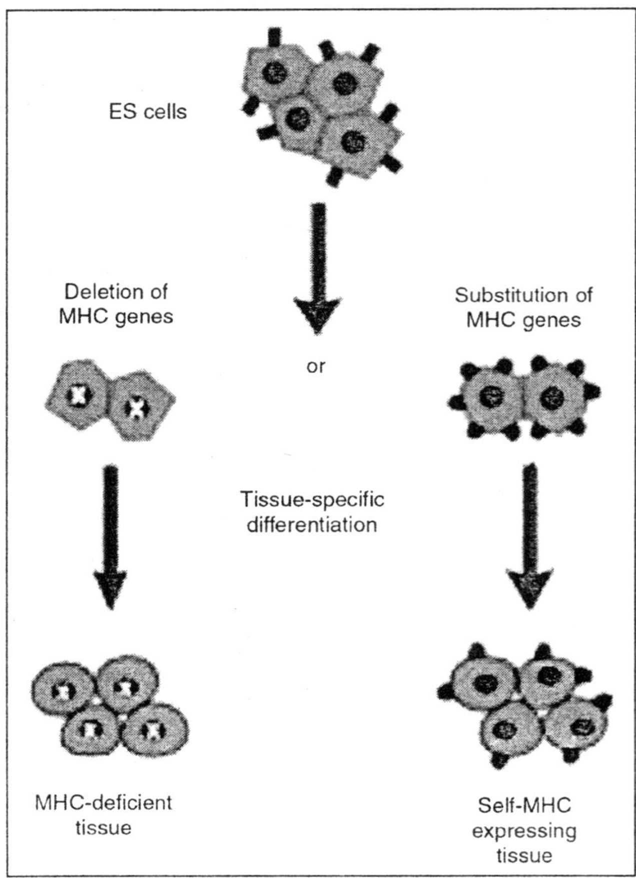

Fig. Genetic Manipulation

The underlying mechanism of fusion by chemical methods

and also electric field induced fusion of plant protoplasts, remains similar to that suggested by Poste and Allison, 1971 for animal cells. According to these workers, membrane proximity of the fusion partners is an essential pre-requisite followed by an induction phase, whereby the charges induced in electrostatic potential of the membrane result in fusion.

It was the early part of the century, that Kuster demonstrated the ability of onion epidermal protoplasts to fuse upon de-plasmolysis.

Also Michel showed that protoplasts could be fused by using potassium nitrate as the plasmolyticum. However, fusions were rare and the culture of fusion products was difficult. In the light of enzymatic procedures for protoplast isolation, interest developed in protoplast fusion with an aim to genetic manipulation.

Power et al., reported sodium nitrate induced fusion of root protoplasts from various species. This method met with success in the production of the first somatic hybrid. However, fusion frequency following this method was low and the sodium nitrate treatment also affected the division capability of the protoplasts. This stimulated interest in other fusogens, and as a consequence rapid advances in the technique of protoplast fusion have been made in the last decade.

Fusion methods that have gained maximum attention are the use of high pH combined with high Ca^{++} in solution, polyethylene glycol (PEG) and more recently the electric field induced electrofusion. Several other methods have also been reported to induce protoplast fusion. These include manipulation with a perfusion micropipette, sea water, lysozyme and inactivated viruses.

Action of high pH and Ca^{++}

A calcium solution buffered at high pH induces aggregation of the protoplasts and their fusion. The addition of Ca^{++} causes the potential of the surface negative charge on protoplasts to be reduced, facilitating protoplast adhesion. The high alkalinity (pH 9.5 - 10.4) induces the formation of intramembranous lysophospholipids such as lysolecithin and lysophosphatidyl -ethanolamine that increase membrane fluidity that results in fusion.

This method has been successfully used in the production of numerous somatic hybrid plants. Protoplasts derived from cell suspension cultures generally are not as tolerant to the use of high pH Ca^{++} method as the mesophyll derived protoplasts.

Action of Polyethyleneglycol (PEG)

PEG has been most widely used to fuse plant protoplasts. Both, the concentration and molecular weights of PEG are important in relation to fusion. PEG, whose general formula is $HOCH_2$-(CH_2-O-CH_2) -CH_2OH is a water soluble compound whose other linkages make the molecule slightly negative in charge. Thus, addition of Ca^{++} ($CaCl_2$ or Ca $(NO_3)_2$ links the compound with membrane surfaces. The high molecular weight of the polymer acts as a bridge connecting the protoplasts together.

A strong affinity of PEG for water causes local membrane dehydration and increased fluidity. This in combination with the reduction of an exclusion volume between adjacent protoplasts causes diminishing mutual membrane electrostatic repulsion. The redistribution of glycoprotein and glycocalyx macromolecules, causes fusion. PEG of molecular weight 400 to 6,000 was found to be active in fusion, whereas PEG 200 and 20,000 was almost inactive.

Thus factors, other than purely chemical structure determine fusogenicity. Compounds structurally related to PEG, namely, polyvinyl alcohol, polyvinyl pyrrolidone and polyglycerol are known to induce fusion. Gelatin, and dextran sulphate also induce fusion to varying degrees. PEG has been used in combination with other treatments to enhance fusion frequency. An increase in the frequency of heterokaryon formation was observed when protoplasts were pre-incubated in lysozyme.

Also, combination of PEG with high pH Ca^{++} solution, or the addition of DMSO or concanavalin A gave rise to higher fusion frequencies in comparison to treatments with PEG alone. It is generally observed that in fusion experiments involving PEG as the fusogen, the elution of PEG subsequent to its application is a critical step in determining the frequency of stable fusion products.

A rapid change of osmotic pressure subsequent to fusion in the process of transferring to a culture medium results in excessive damage, particularly when mesophyll protoplasts are involved in the fusion process.

ISOLATION OF PROTOPLASTS

Protoplasts can be released from plant tissue by one of two methods. The first method involves mechanically isolating the naked cells by dissection or rupture of the cell walls. The more common method involves treating the plant tissue with enzymes that digest the cell wall material. For our work we used a combination of three enzymes: 0.5 per cent macerozyme, a pectinase, to dissolve the tissue; and, 2 per cent cellulase R10 and 0.1 per cent driselase, two cellulases, to dissolve the cell walls. The starting plant material employed for a source of protoplasts proved to be very important for successful regeneration.

Only when young shoots from tissue culture were used as the starting material were plants able to be regenerated from protoplasts. After removing the enzymes by centrifugation, the protoplasts were embedded in alginate. Protoplasts plated in liquid or in a medium solidified with agarose did not develop. The successful medium contained macro- and micronutrients, organic acids, vitamins, and high concentrations of different sugars to stabilize the naked protoplasts until the cell walls reformed.

Cell walls were formed and the cells began to divide after 8 to 10 days of growth in the dark. The medium also contained two plant growth regulators; 1 mg/liter naphthaleneacetic acid (an auxin); and, 1 mg/liter benzyladenine (a cytokinin). The complete details of the protoplast culture procedure can be found in the reference by Winkelmann and Grunewald.

After 14 days growth on the initial culture medium, the osmotic strength and the concentration of growth regulators was reduced. The osmotic strength was reduced again 10 days later. Then, after about 4 weeks of culture, small clumps of unorganized cells, or calli, could be removed and plated on a medium solidified with agarose. These calli were grown in the dark until they reached a size of 3 to 4 mm in diameter, and then, they were transferred to a medium containing 2 mg/liter benzyladenine to induce plant formation.

As soon as the young plants were visible under a stereomicroscope, the cultures were moved to the light and placed on a shoot elongation medium. The number of plants per callus varied, ranging between 5 and 50. These plants rooted easily and could be grown on in the greenhouse with few losses. The scheme presented summarizes the process from protoplast isolation to growth of the plants in the greenhouse. Different cultivars of African violet responded differently in this system.

Four of five of the cultivars we tested produced plants from protoplasts. Plants were regenerated from protoplasts of the cultivars 'Heidrun hell', 'Sarosa', 'Gracia', and 'Rokoko rosa'. Protoplasts of the cultivar 'Blanca' produced callus, but the callus died before shoots were formed. More than 2,000 plants have been transferred to the greenhouse. These are growing vigorously, and appear to be uniform.

A few plants showed chlorophyll deficiencies (either albino or with variegated leaves), and some appear to be polyploid with thick, succulent leaves and peduncles. Chromosome counts will determine whether these are really polyploid. In total, about 95 per cent of the plants appeared to be true to cultivar, indicating that this regeneration technique is a stable one.

Applications for African Violet Improvement

Protoplasts are useful in genetic manipulations because they do not have cell walls. They are ideal targets for taking up naked DNA and for fusing with protoplasts of other related species. In the related genus, Episcia, some species and selections have true yellow and red flowers.

We have applied successfully our procedure for plant regeneration from African violet protoplasts to protoplasts of

Episcia cupreata 'Tropical Topaz'. As was suggested by Bilkey and McCown, protoplast fusion between African violet and Episcia may lead to the production of new flower colors which have not been possible because of genetic barriers. Research toward this goal is now in progress.

PLANT CELL BIOTECHNOLOGY

The past two decades plant cell biotechnology has evolved as a promising new area within the field of biotechnology, focusing on the production of plant secondary metabolites. For most compounds of interest, e.g. morphine, quinine, vinblastine, atropine, scopolamine and digoxin, one has so far not been able to come to a commercially feasible process.

In contrast with the production of antibiotics by micro-organisms, the plant is an already existing, although not always a reliable, source of these compounds.

This has limited the amount of research put into developing alternative biotechnological production for each of the products mentioned. Fifty years of extensive research on the production of penicilline has led to yields of 50 *g/l* and more. Although in plant cell biotechnology the research efforts have been diluted over numerous plants, still for some products one has been able to considerably increase the production levels, e.g. shikonin (3.5 *gA)* and berberine (7 *gA)*.

Presently research particularly focuses on the possibilities to apply metabolic engineering to improve yields to commercially interesting levels. The technological and economical feasibility of plant cell cultures for the production of secondary metabolites.

Technological and Economical Feasibility

In literature plant cells are described as extremely sensitive for shear forces, necessitating the use of special low-shear bioreactors, e.g. air-lift bioreactors. However, in industry such bioreactors are not common, most processes are runned in stirred-tanks.

As a consequence, such a bioreactor is preferable for plant cell cultures, it is the lowest cost process-unit.

More recent studies on the shear sensitivity of plant cells, among others in our own laboratories, have shown that in fact plant cells in general are quite shear-stress tolerant. This is supported by the fact that a series of large scale processes have been reported with plant cell cultures, e.g. shikonin production. Plant cells have even been cultured in a *60* m3 stirred tank. The technology being,feasible, how about the economy? A number of papers has appeared on this.

Assuming a yearly production of 3000 kg/year of a compound produced by a cell culture at a level of 0.3 *g/l,* resulted in a calculated price of 1500 US $/kg. An increase of productivity with a factor (i.e. 3 *g/l)* results in a price of 430 $/kg. In both cases a fed-batch type of process was applied.

These prices are high, but a number of natural products have even much higher prices (e.g. taxol, vinblastine and vincristine). However, most of the high-value specialty chemicals are produced at too low levels in the plant cell cultures. Their production must thus be increased to make an indusmal process possible.

IMPROVING PRODUCTION

Several strategies are being followed to improve yields of secondary metabolites in plant cell cultures. First of all the screening and selection of high producing cell lines and the optimization of growth and production media can be mentioned as common approaches. In case of shikonin and berberine with success in many others with limited success. In the past years new approaches have been developed: the culturing of differentiated cells (e.g. shoots, roots and hairy roots), induction by elicitors and metabolic engineering.

With the culture of differentiated cells one has in most cases been able to get production of the desired compounds in levels comparable to that of the plant, however the culture of such differentiated tissues on a large-scale in bioreactors is a major constraint. For studies of the biosynthesis, such systems are very useful. The second approach mentioned, the use of elicitors has been successful in several cases. However, it remains limited to a certain type of compound for each plant, compounds which most likely act as phytoalexins in these plants. Therefore attention is more and more focused on metabolic engineering. How could metabolic engineering be used to increase yields? Several possibilities can be envisaged:

- Increase activity of enzymes which are limiting in a pathway;
- Induce expression of regulatory genes;
- Block competitive pathways;
- Block catabolism.

The first two possibilities require the expression of genes yielding active enzymes, the latter two approaches blocking of genes by antisense genes. In all cases the respective biosynthetic pathway has to be known on the level of products, enzymes and genes, as well as the regulation on all these levels, including aspects as compartmentation and transport.

In our studies of the production of terpenoid indole alkaloids we have cloned genes coding for tryptophan decarboxylase (TDC), smctosidine synthase (SSS) and NADPH:cytochrome P-450 reductase. The latter enzyme serves all cytochrome P-450 enzymes in the plant, among others geraniol-10-hydroxylase (GlOH), a key enzyme in the terpenoid part of the alkaloid biosynthesis. The gene encoding for GlOH has not yet been cloned, despite the fact we have been able to purify the enzyme to homogeneity.

The cloning of the gene is hampered by the fact that the plant contains many, very similar, P-450 genes. By means of PCR, 18 closely related genes coding for P-450 enzymes were detected in *Carharunthus roseus*. The fdc-gene has been expressed in tobacco plants, resulting in an active enzyme. These plants produce upto 1 per cent of DW in tryptamine, i.e. similar levels as found for the tryptamine-derived indole alkaloids in plants producing such compounds.

The activity of anthranilate synthase, the first committed enzyme of the tryptophan-pathway from chorismate, was not increased in the tryptamine producing transgenic tobacco plants. Plants can apparently make about 1 per cent of its DW in tryptophan derived compounds with its normal primary metabolic machinary.

Introduction of the rdc-gene into C. *roseus* resulted in callus cultures showing up to 10-fold increased levels of TDC activity and a concomminant increase of tryptamine levels, but no significant increase of strictosidine or other alkaloids. This supports other observations that the terpenoid part of the pathway is also a limiting factor. The *tdc*- and sss-genes are both single copy genes. Both are repressed by auxins and induced by elicitation.

Because of this similar regulation, they might be controlled by one trans-acting factor. Identification of this factor and the encoding gene is of great interest, as such a regulatory gene might also control further steps in the biosynthetic pathway. The cloning of such a regulatory gene would avoid the need to clone a large number of genes coding for the individual steps in the biosynthetic pathway.

Fig. Cinchona

In C. *roseus* upon elicitation a 2550 per cent increase is observed in the activity of AS, TDC and *SSS* (Moreno et al. in preparation), whereas the activities of PAL decreases and chalcone synthase is not affected. However, the major change is observed in the content of phenolics, 2,3-dihydroxybenzoic acid (DHBA) being the major compound formed (Moreno et al. submitted). At the same time a strong induction is seen of the enzyme isochorismate synthase, which convert chorismate into isochorismate.

In bacteria this pathway is known to lead to DHBA. This compound has antimicrobial activity, particularly in combination with UV-light and has also antifeedant properties. It is thus likely to be a major defence compound. Phytoalexin-biosynthesis in Cinchona cell cultures (anthraquinones) also use chorismate as precursor as well as mevalonic acid. In Tabernaemontana species mevalonate is used for the mterpene biosynthesis after elicitation, channelling away this precursor from the alkaloid pathway.

Blocking such pathways by means of antisense genes might be of interest to increase the availablity of these precursors for the alkaloid biosynthesis. Catabolism is another important factor in the accumulation of the alkaloids. Studies feeding "N or 14C labeled alkaloids to *Tabernaemntana divaricata* and *C. roseus* cell cultures showed that at a certain point of the growth phase the rate of catabolism equals the de-novo biosynthesis.

Identification of the enzymes involved and the cloning of the encoding genes is thus of interest to eventually use antisense gene technology to block catabolism and thus increase the production.

ISOPRENOIDS

Polymeric isoprene derivatives are a large family of substances of little functional and structural common ground: steroids, carotenoids, gibberelic acid are just some of its members. Several thousand different types of molecules from very different plant groups have been isolated and characterized. Despite their varied structures, all of them are synthesized by only a few pathways.

The starting product of all the different groups of compounds shown in the illustration above is mevalonic acid that is transformed into a phosphorylated isoprene upon phosphorylation.

This isoprene polymerizes subsequently. In the course of polymerization, the number and position of the double bonds are fixed. All green plants are able to generate linear isoprenoids in this way. While terpenes with more than five isoprene units are quite universal, many of the simpler terpenes are restricted to certain plant groups.

Sequestiterpenes, for example, are common in mosses but occur in higher plants, too. They can be found with Magnoliales, but not with Ranunculales. This example shows, why the presence of certain secondary plant products has proven to be a useful taxonomical feature. The same is also true for monoterpenes (iridoid compounds, iridians), more about this topic can be found in the section about systematics. Among the diterpenes are the gibberellins, a group of phytohormones.

Steroids are triterpenes or triterpenoids. Triterpenes are a group of molecules that contain 30 C-atoms and are generated by the polymerization of six isoprene units although a number of derivatives, some with more but most with less C-atoms are also counted among this group. Steroid molecules consist of four rings marked A, B, C and D that have a number of additional residues R. It is of some importance whether two of these are in a cis- (i.e. at the same side of the cyclic system) or in a trans-position (at opposite sites). Steroids have been shown to occur both in gymnosperms and in angiosperms.

Carotenoids are very common both in the plant and the animal kingdom though they are always of plant origin. All of them are tetraterpenes, i.e. they contain 40 C-atoms in eight isoprene residues. The pictures below show that they all have a centre of symmetry.

They are formally derived by the subsequent hydration, dehydration, ring formation, shifting of double bonds and/or methyl groups, chain elongation or shortening and the incorporation of oxygen into a non-cyclic $C_{40}H_{56}$ compound. Carotenoids can be further classified into carotenes (pure carbohydrates without additional groups) and the xanthophylls (carotenoids containing oxygen). Members of both groups are components of the pigment systems (the light traps) of chloroplasts and are involved in the primary light absorption and the photon canalization of photosynthesis. Moreover, they also function as light receptors in a number of further light-induced plant processes.

The yellow colour of many flowers is caused by carotenoid-containing chromoplasts that are usually devoid of chlorophyll. Carotenoids are also common in fruits. The red colour of ripe tomatoes and of pepper is caused by the presence of lycopene. It is a linear molecule with 13 double bonds, 11 of which are conjugated.

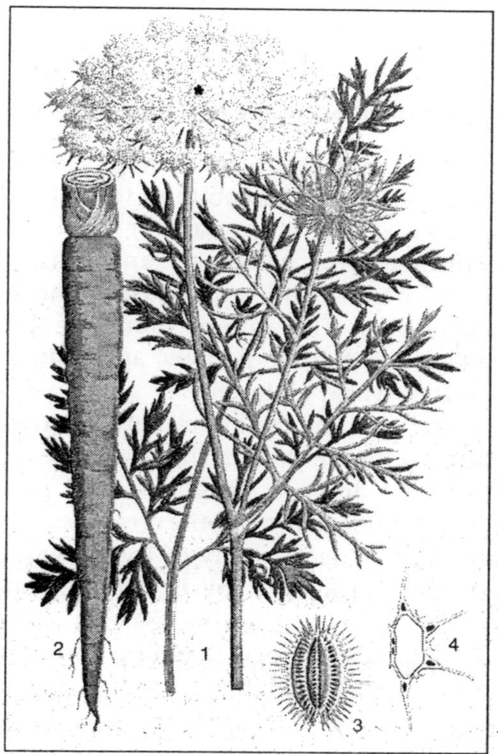

Fig. Daucus Carota

Many carotenes (and xanthophylls) are cyclic at their termini and loose as a consequence the terminal double bond(s). If such a ring is formed at just one terminus, then the result is *gamma*-carotene, if they occur at both, either *alpha*- or *beta* - carotene is formed depending on the type of ring generated. *Beta*-carotene is best-known as the pigment of the carrot (*Daucus carota*). It occurs mostly as a crystal.

One of its most important derivatives is vitamin A (a precursor of visual purple).

Xanthophylls like the common xanthophyll of green leaves or lutein are derived from carotenes. Violaxanthin, for example, is a derivative of *alpha*-carotene. The yellow pigment of corn, zeaxanthin is a *beta*-carotene derivative. Fucoxanthin, the brownish pigment of brown algae and diatoms, is another xanthophyll. Among the catabolic products of xanthophyll is the pigment of saffron, the crocetin.

PRIMARY VS. SECONDARY METABOLISMS

While primary metabolim consists of biochemical pathways that are in general common to all cells, secondary metabolisms consist of a large number of diverse processes that are specific to certain cell types. Plant pigments, alkaloids, isoprenoids, terpenes, and waxes are some examples of secondary products. The role of many of the secondary products has been rather ambiguous, and initially they were thought to be just waste materials.

However, considering their non-motile nature and the lack of sophisticated immune system that we have, plants had to develop their own defence system against pathogens and predators, and systems to lure motile creatures for fertilization and dissemination. Indeed, many of the secondary products are bacteriocidal, repellent (by bad tastes, etc), or even poisonous to pests and hervibores. Pigments of flowers would give attractive colors for insects that help with fertilization, or warning colors against predators.

Plant pigments also provide protection against environmental harms, such as free radiacls and UV irradiation. Some of the secondary products perform signalling function as plant hormones. Many of these secondary products are originally meant for defence against herbivores such as insects which would soon come up with metabolic pathways to detoxify and even utilize these defence compounds. During eveolutionary processes, animals developed a variety of dependencies to phytochemicals, including the secondary products that are, with or without modification, used as procursors for the synthesis of vital or benefitial molecules in animal body. Secondary plant products have for thousands of years played an essential role in medicine. Traditionally, they have been directly used as food and herbs.

Nowadays, they are used either directly or after chemical modification. Plant secondary metabolites represent a tremendous resources for scientific and clinical researches and new drug development.

Overall, their pharmacological value not only remains undeminished until today, but is increasing due to constant discoveries of their pontential roles in healthcare and as lead chemicals for new drug development.

PRODUCTS OF SECONDARY METABOLISMS

There is one rule: Secondary metablic products are more complex than primary metabolites. This is because secondary metabolites are derived from the primary products, such as amino acids or nucleotides, by modifications, such as methylation, hydroxylation, and glycosylation. The products of plant secondary metabolisms, despite their enormous diversity, are grouped into the following categories:

ALKALOIDS

Alkaloids consist of a host of chemicals that are derivatives of nitrogen-containing bases. The large majority of alkaloids are known as "mind altering drugs" and are synthesized from amino acids, and a few of them from purine and pyrimidine bases (psuedo alkaloids), the major components of nucleotides.

Ornithine Derivatives

Nicotine and nornicotine ("tobacco alkaloids" found in plants species in Solanaceae family), cocaine and hyoscamine that are produced from the metabolic intermediates in the biosynthesis of tropane.

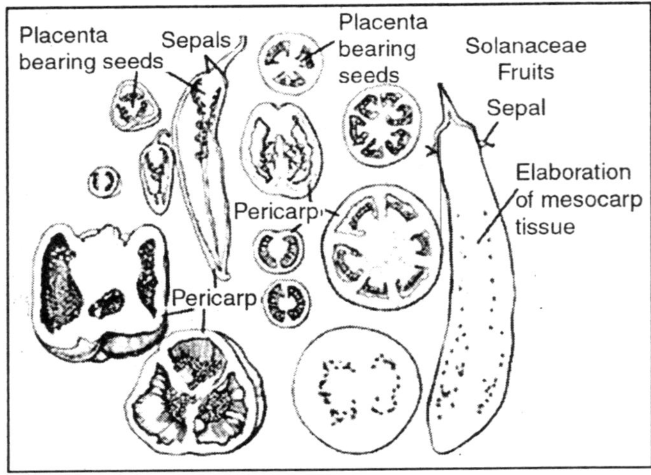

Fig. Solanaceae Family

Lysine Derivatives

Lysine is the precursor of piperidine, which provides a backbone for a number of alkaloids including bitter-tasting chemicals lupine, lupinine, and lupanine. Lycopodine found in Lycopodium (clubmoss) also belongs to this group.

Phenylalanine Derivatives

Taxine (the alkaloid in Taxus), lunarine and lunaridine (the alkaloids in Lunaria), alkaloids in Lythraceae family, cytochalasine B and D (microbial alkaloids), and ephedrine and pseudoephedrine (found in Ephedra) are some important members in this group.

Tyrosine Derivatives

Dopamine, atropine, berberine, colchicine, morphine, and benzyl-isoquinone are typical members in this group.

Tryptophane Derivatives

About 1,200 unique compounds representing about 25 per cent of all known alkaloids have been identified including indole alkaloids. d-tubocurarine, an active ingredient of curare, the arrow poison used by native South Americans, a host of other indole alkaloids found in Apocynaceae, Loganiaceae, and Rubiaceae families, and a variety of known fungal poisons.

Glycosides

Glycosides are compounds containing a carbohydrate (sugar) residue called "glycone", and a non-carbohydrate (non-sugar) residue called "aglycone" in the same molecule.

Some of the compound groups belonging to glycosides are: flavonol, phenol, cyanophore, isothiocyanate, lactone, anthraquinone, saponine, tannin, alcohol, and aldehyde glycoside groups.

Glycyrrhizinic acid found in licorice and ginseng saponins are glycosides.

Isoprenoids and Terpenes

Carotenoids, steroids, and gibberellins are well known members of polymeric isoprene derivatives, which consist of at least thousands of different phytochemicals found in a wide variety of plant species.

Isoprene, the five carbon (C5) unit molecule assembles to build up carbon skeletons for a remarkable array of isoprenoid compounds through isopentenyl pyrophosphate, the activated form of isoprene, and the major C5 building block.

The fragrances of many plants arises from C10 and C15 compounds called terpenes, which are made up of two or three isoprene units, as exemplified by myrcene, a C10 compound found in bay, limonene, another C10 compound found in lemon, and a C15 compound zingiberene found in ginger.

Successive addition of C5 gives rise to a C40 compound phytoene, which converts to lycopene by dehydrogenation, and by the cyclization at both ends, the linear molecule lycopene turns into beta-carotene.

Carotenoids are essential light-harvesting molecules for photosynthesis, and also for vision, as exemplified by beta-carotene, which is the precursor of retinal, the chromophore in all known visual pigments. Natural rubber is a linear polymer of *cis*-isoprene units.

Plant Amines

In general, plant amines are derivatives of ammonia, and generated by decarboxylation of amino acids, or by transamination of aldehydes. By definition, some amines, such as mescaline, are also alkaloids. Putrescine (diamine), spermidine, and spermine (polyamines) interact with double stranded DNA, and found in most of the eukaryotic cells. A phytohormone indole-3-acetic acid (IAA), and serotonin are the amines derived from tryptophan (tryptamines). A variety of aliphatic or aromatic amines are produced by plants as insect attractants.

Phenolic Compounds

This group consists of thousands of diverse molecules with heterogeneous structure, with common feature of having one or more phenol rings. Most of the phenolic compounds belong to flavonoids, a group of polyphenolic compounds with 15 carbon atoms (C15) based on a skeletal structure of two benzene rings joined by a linear C3 chain (C6-C3-C6-system). Flavonoids compounds are highly characteristic to plants: many of flavonoids are easily recognizable as the pigments in flowers and fruits, and occur in all parts of plants.

Important subgroups in flavonoid compounds are: anthocyanins (red and blue pigments, 250 members), aurones (yellow pigments, 20 members), biflavonoids (65 members), catechines (40 members), chalcons (yellow pigments, 60 members), dihydrochalcones (bitter-tasting principles, 10 members), flavones (cream-colored pigments of flowers, generally found in herbaceous families such as Labiatae, Umbelliferae, Compositae, 350 members, e.g. apigenin, luteolin), flavonols (repellents in leaves, generaly found in woody angiosperms, 350 members, e.g. quercitol, kaempferol, myricetin), isoflavonoids (colorless, estrogenic effect, fungicide, 15 members, e.g. rotenone), proanthocyanidins (50 members)

Polyisoprenes and Rubber-like Polymers

More than 1,800 plant polyisoprenes have been identified. Natural rubber is a linear polymer of 1,4-polyisoprenes in *cis* configuration (caoutchouc), the main source of which is Hevea brasiliensis. Palaquium gutta is the main source of gutta-percha, which consists of 1,4-polyisoprene residues in *trans* configuration and has molecuar weight far lower than that of rubber. Balata is a substance similar to gutta-percha and produced in Mimosops balata.

Chicle, the basic substance of bubble gums, is a polymer containing both *cis* and *trans* configurations (in 1:2 ratio), and produced by Achras sapota.

Rare Amino Acids

Unlike well-known 20 amino acids, rare amino acids, which are derivatives of regular amino acids, do not incorporate themselves into proteins. About 220 such derivatives have been found to be synthesized from almost all 20 regular amino acids. These derivatives are found only in one or a few specific species. In some fungi, the rare amino acids are polymerized to form small, and often cyclic polypeptides such as phalloidine or the amanitins, the toxins and inhibitors of RNA polymerases.

Octopine and nopaline are derivatives of arginine and found in the Ti plasmids of Agrobacterium tumefaciens, a pathogen causing crown gall tumors in leguminous plants. Other examples are ornithine and citrullines, the common metabloic intermediates, and canavanine, 3,4-dihydroxy phenylalanine, 5-hydroxy tryptophan, and acetidine-2-carboxylicacide, etc.

Chapter 35

Genetics

Genetics, study of the function and behaviour of genes. Genes are bits of biochemical instructions found inside the cells of every organism from bacteria to humans. Offspring receive a mixture of genetic information from both parents. This process contributes to the great variation of traits that we see in nature, such as the colour of a flower's petals, the markings on a butterfly's wings, or such human behavioral traits as personality or musical talent.

Geneticists seek to understand how the information encoded in genes is used and controlled by cells and how it is transmitted from one generation to the next. Geneticists also study how tiny variations in genes can disrupt an organism's development or cause disease. Increasingly, modern genetics involves genetic engineering, a technique used by scientists to manipulate genes.

Genetic engineering has produced many advances in medicine and industry, but the potential for abuse of this technique has also presented society with many ethical and legal controversies. Genetic information is encoded and transmitted from generation to generation in deoxyribonucleic acid (DNA). DNA is a coiled molecule organized into structures called chromosomes within cells. Segments along the length of a DNA molecule form genes.

Genes direct the synthesis of proteins, the molecular laborers that carry out all life-supporting activities in the cell. Although all humans share the same set of genes, individuals can inherit different forms of a given gene, making each person genetically unique.

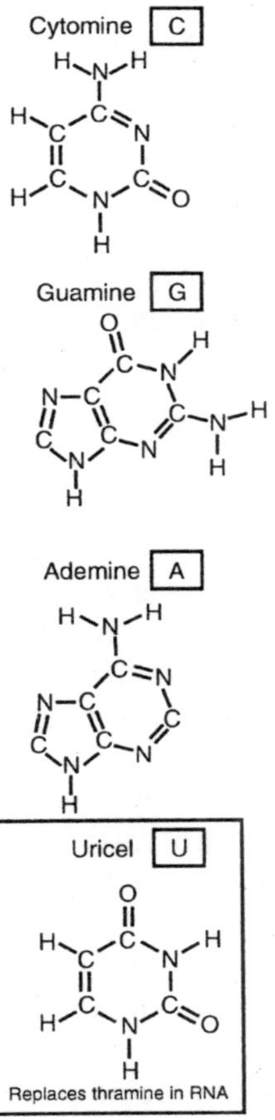

Cytomine C

Guamine G

Ademine A

Uricel U

Replaces thramine in RNA

Nitrogenous Bases

Fig. Genetics

Since the earliest days of plant and animal domestication, around 10,000 years ago, humans have understood that characteristic traits of parents could be transmitted to their offspring.

The first to speculate about how this process worked were Greek scholars around the 4th century BC, who promoted theories based on conjecture or superstition. Some of these theories remained in favor for several centuries. The scientific study of genetics did not begin until the late 19th century. In experiments with garden peas, Austrian monk Gregor Mendel described the patterns of inheritance, observing that traits were inherited as separate units. These units are now known as genes. Mendel's work formed the foundation for later scientific achievements that heralded the era of modern genetics.

The Importance of Genetics

The modern science of genetics influences many aspects of daily life, from the food we eat to how we identify criminals or treat diseases. In agriculture, genetic advances enable scientists to alter a plant or animal to make it more useful.

For instance, some food crops, such as oranges, potatoes, wheat, and rice, have been genetically altered to withstand insect pests, resulting in a higher crop yield. Tomatoes and apples have been modified so that they resist discoloration or bruising on their way to market, enhancing their appeal on supermarket shelves.

The genetic makeup of cows has been modified to increase their milk production, and cattle raised for beef have been altered so that they grow faster. Genetic technologies have also helped convict criminals.

DNA recovered from semen, blood, skin cells, or hair found at a crime scene can be analyzed in a laboratory and compared with the DNA of a suspect. An individual's DNA is as unique as a set of fingerprints, and a DNA match can be used in a courtroom as evidence connecting a person to a crime. Genetics has revolutionized the way industries produce certain substances, many of which formerly required costly and arduous manufacturing methods.

In medicine, scientists can genetically alter bacteria so that they mass-produce specific proteins, such as insulin used by people with diabetes mellitus or human growth hormone used by children who suffer from growth disorders. In other medical applications, genetic technologies have been instrumental in the development of gene therapy. In this still-experimental form of treatment, scientists try to cure disease by replacing malfunctioning genes with healthy ones. Gene therapy has shown promise in treating some devastating conditions, including some forms of cancer and cystic fibrosis.

Genetically engineered vaccines are being tested for possible use against the human immunodeficiency virus (HIV), the virus that causes acquired immunodeficiency syndrome (AIDS). The field of human genetics has been energized in recent years by the Human Genome Project, an international collaboration of scientists, governments, and drug companies from around the world. Scientists working on this project have developed detailed maps that identify the chromosomal locations of the estimated 20,000 to 25,000 human genes.

The vast databases emerging from the project help scientists study previously unknown genes as well as many genes all at once to examine how gene activity can cause disease. Scientists expect that the project will lead to the development of new drugs targeted to specific genetic disorders. Despite the benefits derived from genetic advancements, some observers have voiced concerns that genetically engineered organisms could harm people or the environment.

Others fear that new genetic technologies may enable scientists to modify genes that affect characteristics other than those responsible for disease. They warn that determining who has undesirable genetic characteristics may lead to discriminatory practices. Others are concerned about the common misperception that a person's genes determine all aspects of a person's life, including health and behaviour. This misperception leads people to blame their genetic makeup for problems, leaving no room for the influence of free will, personal responsibility, or hope for change.

These and other challenging issues place geneticists at the crossroads of science and social responsibility, where they work to promote understanding of genetic advances and prevent the abuse of them.

PRINCIPLES OF GENETICS

The site where genes work is the cell. Some organisms, such as paramecia or amoebas, are made up of a single cell.

Other organisms are made of many kinds of cells, each having a different function. For instance, a tree contains some cells that form the root system and other cells that form leaves. Each cell's function within an organism is determined by the genetic information encoded in DNA.

In animals, plants, and other eukaryotes (organisms whose cells contain a nucleus), DNA resides within membrane-bound structures in the cell. These structures include the nucleus, the energy-producing mitochondria, and, in plants, the chloroplasts (structures where photosynthesis takes place). In prokaryotes, one-celled organisms and bacteria that lack internal membrane-bound structures, DNA floats freely within the cell body.

Cell Division and Reproduction

Organisms could not grow or function properly if the genetic information encoded in DNA was not passed from cell to cell. DNA is packaged into structures called chromosomes within a cell. Every chromosome in a cell contains many genes, and each gene is located at a particular site, or locus, on the chromosome. Chromosomes vary in size and shape and usually occur in matched pairs called homologues.

The number of homologous chromosomes in a cell depends upon the organism—for example, most cells in the human body contain 23 pairs of chromosomes, while most cells of the fruit fly *Drosophila* contain 4 pairs. Within all organisms, cells divide to produce new cells, each of which requires the genetic information found in DNA. Yet simply splitting the DNA of a dividing cell between two new cells would lead to disaster—the two new cells would have different instructions and each subsequent generation of cells would have less and less genetic information to work with.

Imagine how chaotic it would be to rip an architectural blueprint in two, give each half to different contractors, and tell them to construct identical buildings. Just as each contractor would require a full copy of the blueprint to construct a complete building, each new cell needs a complete copy of an organism's genetic information to function properly.

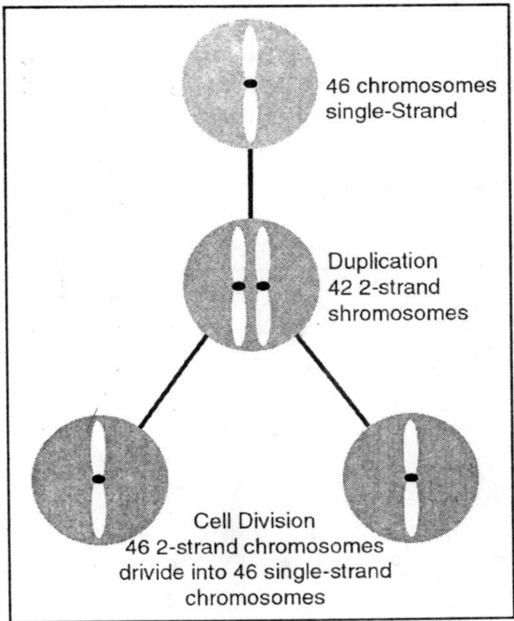

Fig. Cell Division

Organisms use two types of cell division to ensure that DNA is passed down from cell to cell during reproduction.

Simple one-celled organisms and other organisms that reproduce asexually—that is, without the joining of cells from two different organisms—reproduce by a process called mitosis. During mitosis a cell doubles its DNA before dividing into two cells and distributing the DNA evenly to each resulting cell.

Organisms that reproduce sexually use a different type of cell division. These organisms produce special cells called gametes, or egg and sperm. In the cell division known as meiosis, the chromosomes in a gamete cell are reduced by half. During sexual reproduction, an egg and sperm unite to form a zygote, in which the full number of chromosomes is restored.

Mitosis

Mitosis occurs in five stages: prophase, prometaphase,

metaphase, anaphase, and telophase. During prophase, the start of mitosis, the DNA of each chromosome replicates. Each chromosome then reorganizes into paired structures called sister chromatids, with each member of the pair containing a full copy of the DNA sequence. The sister chromatids condense, thickening until they appear joined at a single site, known as the centromere. Prometaphase is marked by the disintegration of the nuclear membrane.

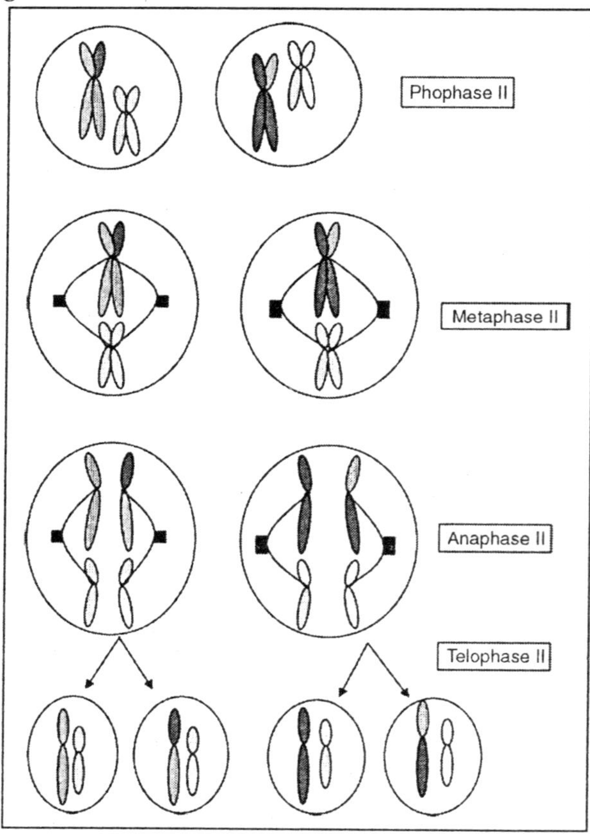

Fig. Mitosis

The sister chromatids line up in the middle of the cell halfway between the poles. In metaphase, exactly half of the chromatids face one pole, and half face the other pole. This equilibrium position is called the metaphase plate. In anaphase, the chromatid pairs split apart at the centromere, and each half of the pair then moves toward opposite poles of the cells. In telophase, the final stage of mitosis, a nuclear membrane forms around the chromosomes at each pole of the cell. Mitosis ends with the formation of two new cells, each with a matching full set of chromosomes as well as an identical complement of cellular structures.

Meiosis

During meiosis, two cell divisions occur to produce four daughter cells from the original parent cell.

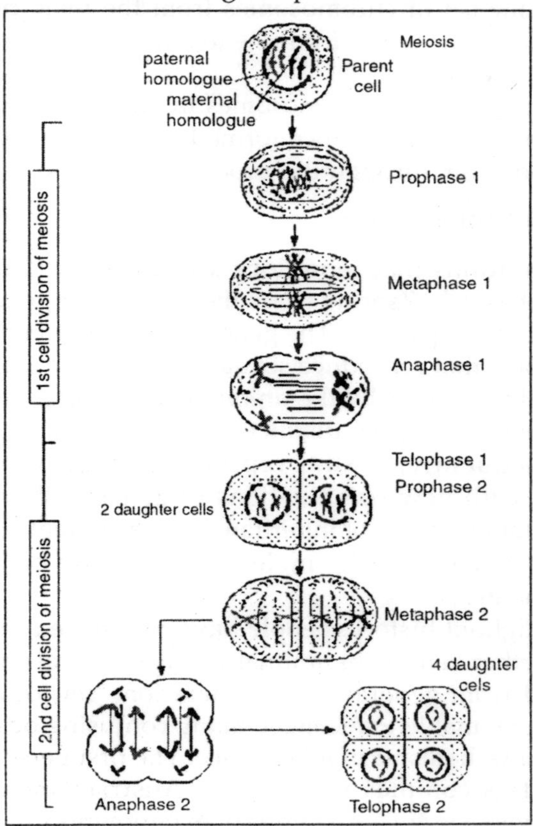

Fig. Meiosis

Each resulting cell has half the chromosomal DNA of the parent cell. A half set of chromosomes in an organism is known as the haploid number. In the first cell division of meiosis the chromosomes of a gamete cell duplicate and join in pairs. The paired chromosomes align at the equator of the cell, and then separate and move to opposite poles in the cell.

The cell then splits to form two daughter cells. As meiosis proceeds, the two daughter cells undergo another cell division to form four cells, each of which bears half of the number of chromosomes found in the other cells of the organism. Meiosis ensures that reproduction will produce a zygote that has received one set of chromosomes from the male parent and one set of chromosomes from the female parent to form a full set of chromosomes.

The entire set of chromosomes in an organism is known as the diploid number. Once formed, the zygote continues to divide and grow through the process of mitosis.

Patterns of Inheritance

In life forms that reproduce asexually, such as bacteria and amoebas, all offspring share the exact same genes and are identical to their parents. The genetic transmission that occurs in organisms that reproduce sexually is far more complex. An individual that forms by the union of two gametes inherits its chromosomes from two distinct parents.

Consequently, sexual reproduction guarantees that offspring with new combinations of genes will continually arise. Certain patterns of inheritance were evident long before scientists discovered the molecular structure of DNA and chromosomes.

Throughout history, people have recognized that certain traits, whether in humans, animals, or agricultural crops, could be passed from generation to generation. Yet for centuries, people were unable to reconcile many confusing observations about the mechanisms of inheritance. The first person to make sense of this complex subject was Austrian monk Gregor Mendel, who conducted a series of experiments on pea plants beginning in the 1850s.

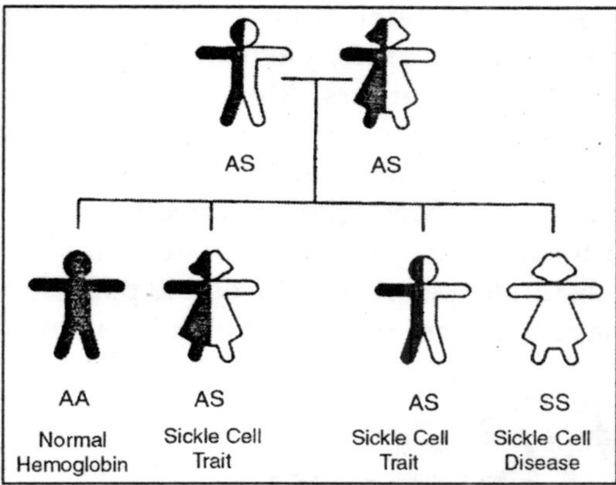

Fig. Patterns of Inheritance

Mendel observed the results of crossbreeding plants with different characteristics, such as height, flower colour, and seed shape. His conclusions from these experiments led him to develop explanations for how traits are transmitted from generation to generation. Mendel's theories form the foundation of modern genetics.

Mendel's Rules

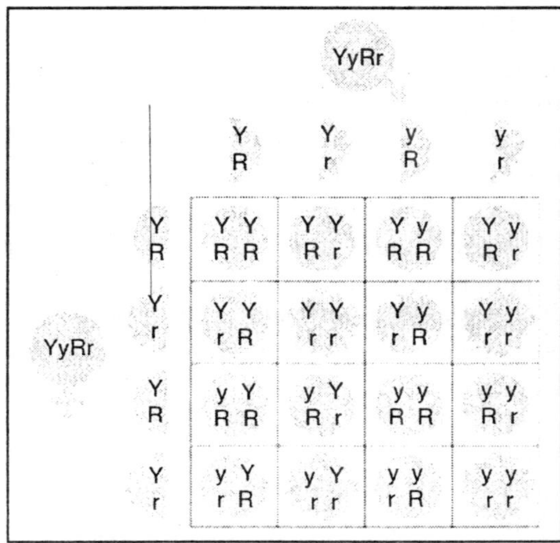

Fig. Mendel's Rules of Gene

In his research, Mendel observed that characteristics were inherited as separate units, each of which was inherited independently of the others. Mendel suggested that each parent has pairs of these units but contributes only one of each pair to offspring. The units that Mendel described were later given the name *genes*. Mendel recognized that a gene can exist in different forms. Today these alternate forms are known as alleles. For example, pea seeds, the edible part of the plant we call peas, have a texture trait controlled by a single gene.

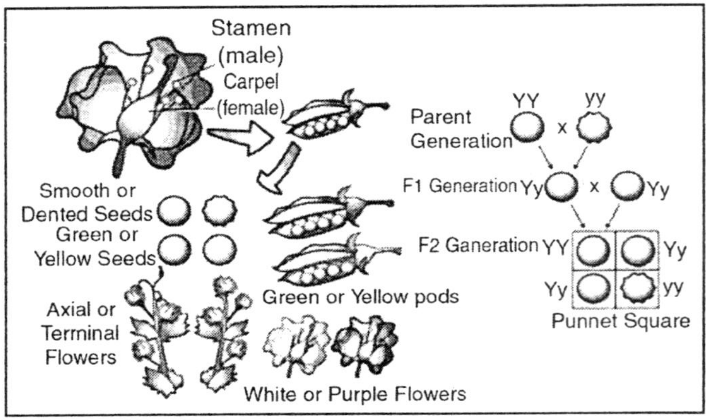

Fig. Gene Occurs in two Alleles of Pea

This gene occurs in two alleles: one corresponding to round (smooth) peas, the other to wrinkled peas. Although an individual can carry only two alleles for a particular gene, each gene may have dozens of different alleles. Mendel's experiments focused on interbreeding different strains of pea plants and then observing the traits that appeared in subsequent generations.

When he crossbred plants with round peas and those with wrinkled peas, he discovered that all of the resulting offspring had round peas.

Today we know that peas are round and smooth when they contain the right amount of sugar. If peas are missing the gene that produces a protein called starch branching enzyme 1 (SBE1), the peas make too much sugar, causing the peas to swell and then wrinkle and shrivel as they dry.

Mendel concluded that when an organism has two different alleles corresponding to the same genetic trait, one of the two may be dominant. The other allele is said to be recessive, meaning that its presence will be detectable only if an organism has inherited the recessive gene from both parents.

For convenience, geneticists designate alleles by a single letter—the dominant allele is represented by a capital letter and the recessive allele by a small letter. In the pea texture example, a plant inherits one allele for pea texture from each parent. The dominant allele that produces SBE1, resulting in round, smooth peas, is designated as *R*, while the recessive allele that does not produce SBE1 and produces wrinkled peas is designated as *r*.

To determine the set of alleles an organism has for a given trait just by visual observation can often be difficult. In the pea plant example, for instance, plants with smooth peas might be carrying two dominant alleles for that characteristic (RR) or one dominant and one recessive allele (Rr). Geneticists use the term genotype to refer to the combination of genes that code for a trait, while the term phenotype describes the physical manifestation of that trait.

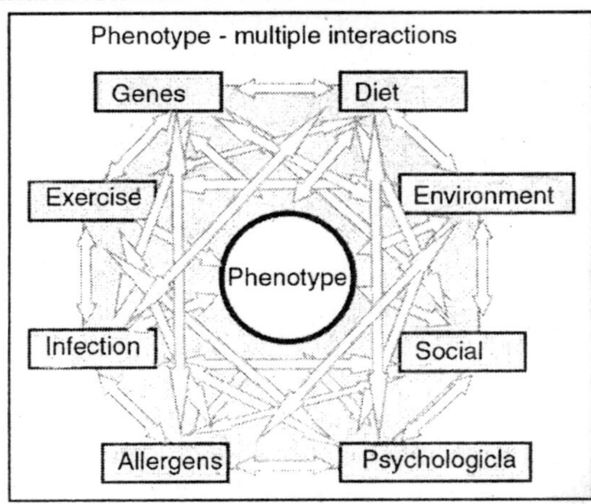

Fig. Phenotype

Therefore, the presence of two dominant alleles for pea

texture (*RR*) would reflect the genotype while a smooth pea indicates the phenotype. Mendel did not limit his experiments to testing the rules of inheritance of single traits. He also studied plant traits involving multiple pairs of genes, breeding plants that have round, yellow seeds with plants that produce wrinkled, green seeds. Such experiments demonstrated that the patterns of inheritance he observed in his experiments with single traits also apply to cases involving more complex gene combinations.

Exceptions to Mendel's Rules

Mendel published his studies in a science journal in 1865, at which time no other scientist commented on his work. Since that time, geneticists have learned that sometimes genes do not easily conform to so-called Mendelian patterns of inheritance.

Incomplete Dominance

In cases of incomplete dominance, the inheritance of a dominant and a recessive allele results in a blending of traits to produce intermediate characteristics. For example, four-o'clock paint plants may have red, white, or pink flowers. Plants with red flowers have two copies of the dominant allele *R* for red flower colour (*RR*). Plants with white flowers have two copies of the recessive allele *r* for white flower colour (*rr*). Pink flowers result in plants with one copy of each allele (*Rr*), with each allele contributing to a blending of colors.

Quantitative Inheritance

Mendel focused his studies on traits determined by a single pair of genes, and the resulting phenotype was easy to distinguish. A tall plant can be markedly different from a short one, and a green pea can easily be distinguished from a yellow one. There are some traits, however, that are not easy to distinguish. Human skin colour, for example, may be any of a wide variety of shades. Traits such as skin colour differ from the ones Mendel studied because they are determined by more than one pair of genes.

In this form of inheritance, known as quantitative inheritance, each pair of genes has only a slight effect on the trait, while the cumulative effect of all the genes determines the physical characteristics of the trait. At least four pairs of genes control human skin colour. Multiple genes also control many traits important in agriculture, such as milk production in cows and ear length in corn.

Multiple Alleles

Another exception to Mendelian genetics involves genes with multiple alleles. Certain traits are controlled by multiple alleles that have complex rules of dominance.

In humans, for example, the gene for blood type has three alleles: I_A, I_B, and i. With three alternatives for each member of a gene pair, there are six possible combinations of these genes (I_AI_A, I_BI_B, ii, I_Ai, I_Bi, I_AI_B).

Although there are six possible combinations, humans have only four major blood types: A, B, AB, and O.

This results because both I_A and I_B dominate over i, but not over each other, so a person with a gene combination of I_AI_A or I_Ai has blood type A. The gene combinations I_BI_B and I_Bi both produce blood type B. I_AI_B results in a blood type AB, and ii results in blood type O.

Gene Linkage

In his experiments, Mendel was careful to study traits in pea plants where one trait did not appear to influence another, such as the plant's height or the pea's texture. These two phenotypes (height and texture) occur randomly with respect to one another in a manner known as independent assortment. Today scientists understand that independent assortment occurs when the genes affecting the phenotypes are found on different chromosomes.

An exception to independent assortment develops when genes appear near one another on the same chromosome. When genes occur on the same chromosome, they are inherited as a single unit. Genes inherited in this way are said to be linked. For example, in fruit flies the genes affecting eye colour and wing length are inherited together because they appear on the same chromosome.

But in many cases, genes on the same chromosome that are inherited together produce offspring with unexpected allele combinations. This results from a process called crossing over. Sometimes at the beginning of meiosis, a chromosome pair (made up of a chromosome from the mother and a chromosome from the father) may intertwine and exchange sections of chromosome.

The pair then breaks apart to form two chromosomes with a new combination of genes that differs from the combination supplied by the parents. Through this process of recombining genes, organisms can produce offspring with new combinations of maternal and paternal traits that may contribute to or enhance survival.

Sex-Linked Traits

Most chromosome pairs consist of identical, or homologous, partners. In many species, including humans, there is one pair of chromosomes in which the partners noticeably differ from each other. These are called the sex chromosomes because they determine the differences between males and females.

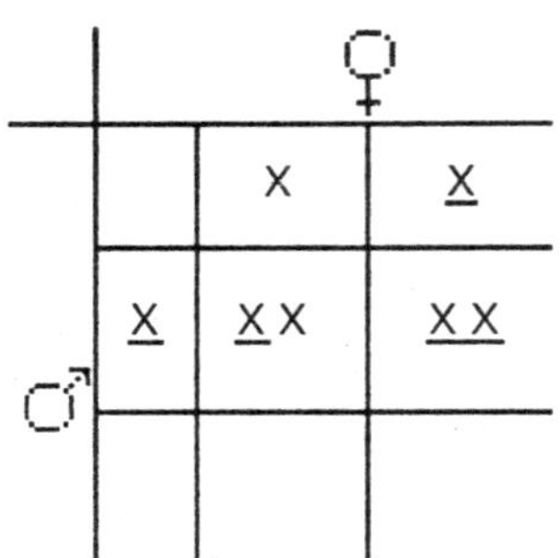

Fig. Sex-Linked Traits

Genes located on the sex chromosomes display different patterns of inheritance than genes located on other chromosomes. In human females, the sex chromosomes consist of two X chromosomes, while males have an X chromosome and a shorter Y chromosome with many fewer genes. In males the X chromosome contains many genes that have no corresponding gene on the Y chromosome.

A male's X chromosome may contain a recessive allele associated with a genetic disorder, such as hemophilia or Duchenne muscular dystrophy. In this case, males do not have a normal second copy of the gene on the Y chromosome to mask the effects of the recessive gene, and disease typically results. Additional examples of sex-linked traits include red-green colour blindness in humans and eye colour in fruit flies.

THE GENETIC CODE

The structure of DNA encodes all the information every cell needs to function and thrive. In addition, DNA carries hereditary information in a form that can be copied and passed intact from generation to generation.

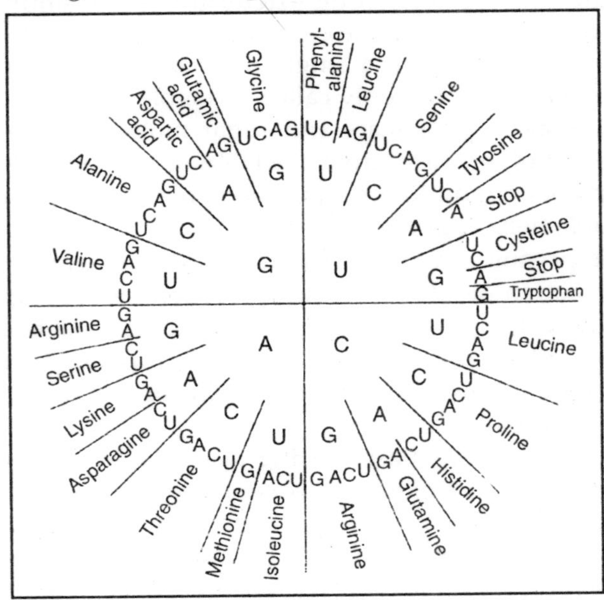

Fig. Genetic Code

A gene is a segment of DNA. The biochemical instructions found within most genes, known as the genetic code, specify the chemical structure of a particular protein. Proteins are composed of long chains of amino acids, and the specific sequence of these amino acids dictates the function of each protein. The DNA structure of a gene determines the arrangement of amino acids in a protein, ultimately determining the type and function of the protein manufactured.

DNA Structure

DNA molecules form from chains of building blocks called nucleotides. Each nucleotide consists of a sugar molecule called deoxyribose that bonds to a phosphate molecule and to a nitrogen-containing compound, known as a base. DNA uses four bases in its structure: adenine (A), cytosine (C), guanine (G), and thymine (T). The order of the bases in a DNA molecule—the genetic code—determines the amino acid sequence of a protein. In the cells of most organisms, two long strands of DNA join in a single molecule that resembles a spiraling ladder, commonly called a double helix.

Alternating phosphate and sugar molecules form each side of this ladder. Bases from one DNA strand join with bases from another strand to form the rungs of the ladder, holding the double helix together.

The pairing of bases in the DNA double helix is highly specific—adenine always joins with thymine, and guanine always links to cytosine. These base combinations, known as complementary base pairing, play a fundamental role in DNA's function by aiding in the replication and storage of genetic information.

Complementary base pairing also enables scientists to predict the sequence of bases on one strand of a DNA molecule if they know the order on the corresponding, or complementary, DNA strand. Scientists use complementary base pairing to help identify the genes on a particular chromosome and to develop methods used in genetic engineering. Genes line up in a row along the length of a DNA molecule. In humans a single gene can vary in length from 100 to over 1,000,000 bases.

Genes make up less than 2 per cent of the length of a DNA molecule. The rest of the DNA molecule is made up of long, highly repetitive nucleotide sequences. Once dismissed as "junk" DNA, scientists now believe these nucleotide sequences may play a role in the survival of cells. Identifying the function of these sequences is a thriving field of genetics research.

PROTEIN SYNTHESIS

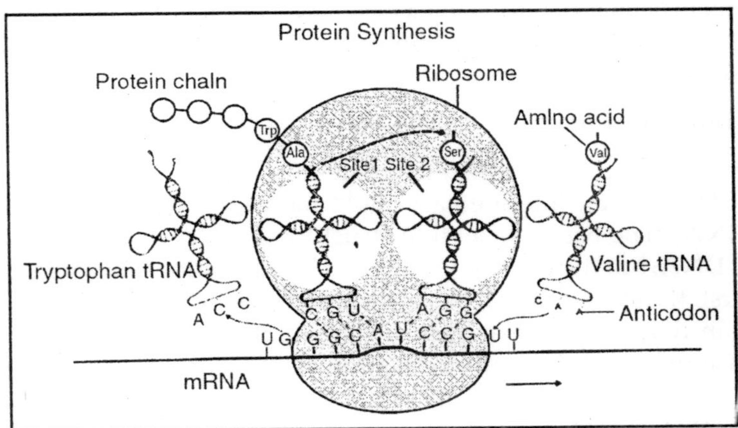

Fig. Protein Synthesis

DNA replication ensures that the genetic instructions encoded in DNA can be used continuously through generations to produce the proteins that build and operate the cells of an organism. The process of tapping the genetic code to create proteins, known as protein synthesis, has two crucial steps: transcription and translation.

DNA Replication

In order for inherited traits to be transmitted from parent to child, the genetic information encoded in DNA must be copied with great precision during cell division. The accuracy of DNA replication depends upon the complementary pairing of bases. During replication, the DNA double helix unwinds and bonds joining the base pairs break, separating the DNA molecule into two separate strands.

Each strand of DNA directs the synthesis of another complementary strand. The unpaired bases of each DNA strand attach to bases floating within the cell. But the DNA strand's unpaired bases bond only with specific, complementary bases—for example, an adenine base will bond only with a thymine base and a cytosine bases will pair only with a guanine base. Once all of the bases of a DNA strand bond to complementary bases, the complementary bases then link to each other, forming a new DNA double-helix molecule. Thus the original DNA molecule replicates into two DNA molecules that are exact duplicates.

Transcription

Transcription transfers the genetic code from a molecule of DNA to an intermediary molecule called ribonucleic acid (RNA). The basic nucleotide structure of RNA resembles that of DNA, but the two compounds have three critical differences. First, the structure of RNA incorporates the sugar ribose rather than deoxyribose, the sugar in DNA. Second, RNA uses the base uracil (U) instead of thymine (T).

In RNA uracil binds with adenine just as thymine does in DNA. Third, RNA usually exists as a single strand, unlike the double-helix structure that normally characterizes DNA. Transcription involves the production of a special kind of RNA known as messenger RNA (mRNA).

The process begins when the two strands of a DNA molecule separate, a task directed by the enzyme RNA polymerase. After the double helix splits apart, one of the strands serves as a template, or pattern, for the formation of a complementary mRNA molecule. Free-floating individual bases within the cell bind to the bases on the DNA template using complementary base pairing. The individual bases then link together to form a strand of mRNA.

In eukaryotes (organisms whose cells have a nucleus), the mRNA strand undergoes an additional step before the next stage of protein synthesis can occur. The mRNA strand consists of coding regions called exons separated by regions called introns. The introns do not contribute to protein synthesis.

Special enzymes in the nucleus remove the introns from the mRNA strand. The remaining exons then link together to form an mRNA strand that contains the entire code for making a protein.

Translation

Once transcription is complete and the genetic code has been copied onto mRNA, the genetic code must be converted into the language of proteins. That is, the information coded in the four bases found in mRNA must be translated into the instructions encoded by the 20 amino acids used in the formation of proteins. This process, called translation, takes place in cellular organelles called ribosomes.

In eukaryotes, mRNA travels out of the nucleus into the cell body to attach to a ribosome. In prokaryotes (organisms without a nucleus), the ribosome clasps mRNA and starts translation before these strands have finished transcription and separated from the DNA. In both eukaryotes and prokaryotes, the ribosome acts like a workbench and clamp that holds the mRNA strand and coordinates the activity of enzymes and other molecules essential to translation.

From Protein Structure and Function by Gregory A Petsko and Dagmar Ringe

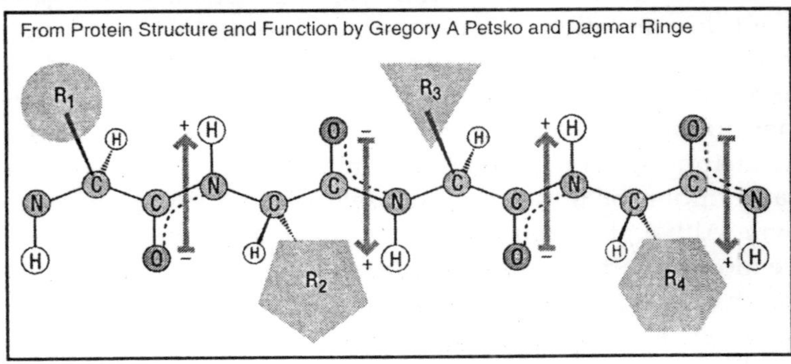

Fig. Polypeptide Chain

Another form of RNA called transfer RNA (tRNA) is found in the cytoplasm of the cell. There are many different types of tRNA, and each type binds with one of the 20 amino acids used in protein formation. One end of a tRNA binds with a specific amino acid. The other end carries three bases, known as an anticodon. The tRNA with an amino acid attached travels to the ribosome where the mRNA is stationed.

The anticodon of the tRNA undergoes complementary base pairing with a series of three bases on the mRNA, known as the codon. The mRNA codon codes for the type of amino acid carried by the tRNA. A second tRNA bonds with the next codon on the mRNA. The resident tRNA transfers its amino acid to the amino acid of the incoming tRNA and then leaves the ribosome.

This process continues repeatedly, with new tRNA receiving the growing chain of amino acids, known as a polypeptide chain, from a resident tRNA. The ribosome moves the mRNA strand one codon at a time, making new codons available to bind with tRNAs. The process ends when the entire sequence of mRNA has been translated. The polypeptide chain falls away from the ribosome as a newly formed protein, ready to go to work in the cell.

MUTATIONS

Occasionally mistakes occur during DNA replication and protein synthesis. Any alteration in the structure of a gene results in a mutation. Mutations occur during DNA replication when the chemical structure of genes undergoes random modifications. Once a change has occurred, the altered genes continue to replicate in their changed form unless another mutation occurs. Sometimes mutations occur during transcription or translation, causing protein synthesis to go awry. Although mutations may occur in any living cell, they are most important when they occur in gametes because then the change affects the traits of following generations. Most mutations harm an organism.

If a mutation occurs in a gene sequence that codes for a particular protein, the mutation may result in a change in the amino acid sequence directed by the gene. This change, in turn, may affect the function of the protein. The implications can be significant: The amino acid sequence distinguishing normal hemoglobin from the altered form of hemoglobin responsible for sickle-cell anemia differs by only a single amino acid.

Mutation

Original Individual

(÷(−√(− (* b b)(* (* 2 2)(* a c)))b)(* 2 a))

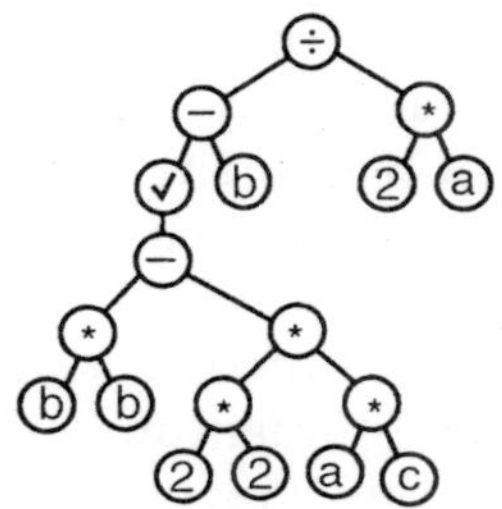

Mutated Individuals

(÷(−√(+ (* b b)(* (* 2 2)(* a c)))b)(* a a))   (÷ (−√(− (* b b)(* (* 2 a))b)(* a 2))

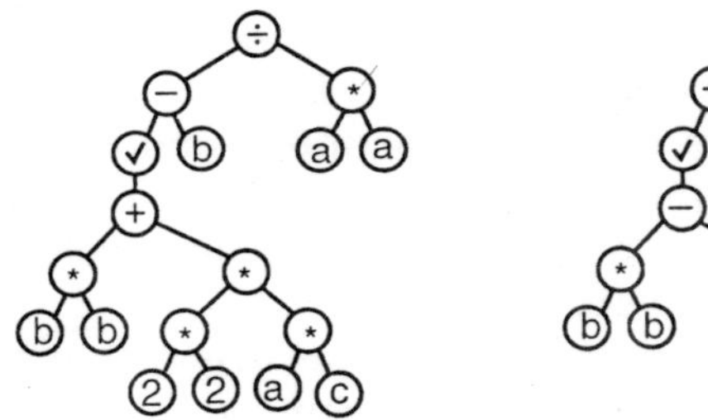

Fig. Mutations

Some mutations may be neutral or silent and do not affect the function of a protein. Occasionally a mutation benefits an organism. Over the course of evolutionary time, however, mutations serve the crucial role of providing organisms with previously nonexistent proteins.

In this way, mutations are a driving force behind genetic diversity and the rise of new or more competitive species better able to adapt to changes, such as climate variations, depletion of food sources, or the emergence of new types of disease.

Mutations can produce a change in any region of a DNA molecule. In a *point mutation*, for example, a single nucleotide replaces another nucleotide. Although a point mutation produces a small change to the DNA sequence, it may cause a change in the amino acid sequence, and thus the function, of a protein. Far more serious are mutations that involve the addition or deletion of one or more bases from a DNA molecule.

Adding or subtracting even a single base from a normal sequence during transcription can disrupt translation by shifting the "reading frame" of every subsequent codon. For example, an mRNA strand may include two codons in the following sequence: AUG UGA. The addition of a cytosine base at the beginning of this sequence shifts the "spelling" of these codons so that they read: CAU GUG.

This may result in an incorrect amino acid sequence during translation, or the protein may be truncated. Known as frameshift mutation, this type of alteration could result in the production of a protein with no real function or one with a harmful effect.

Sometimes mutations are caused by transposition, in which long stretches of DNA (containing one or more genes) move from one chromosome to another. These jumping genes, called transposons, can disrupt transcription and change the type of amino acids inserted into a protein. Transposons rearrange and interrupt genes in a way that generally improves the genetic variation of a species.

While mutations can occur spontaneously, some can be caused by exposure to physical or chemical agents in the environment called mutagens. Common environmental mutagens include ultraviolet rays from the sun and various chemicals, such as asbestos, cigarette smoke, and nitrous acid.

High-energy radiation, such as medical X rays, can cause DNA strands to break, leading to the deletion of potentially important genetic information.

Radiation damage can also affect an entire chromosome, disrupting the function of many genes. In chromosomal translocation, a piece of one chromosome breaks off and merges with another chromosome. In some cases, large sections of chromosomes may break off and be lost.

The cell has highly effective self-repair mechanisms that can correct the harmful changes made by mutations and prevent some mutations from being passed on. Some 50 specialized enzymes locate different types of faulty sequences in the DNA and clip out those flaws. Another repair mechanism scans DNA after replication and marks mismatched base pairs for repair.

GENE REGULATION

The processes that enable information to be copied from genes and then used to synthesize proteins must be regulated if an organism is to survive. Different cells within an organism share the same set of chromosomes. In each cell some genes are active while others are not.

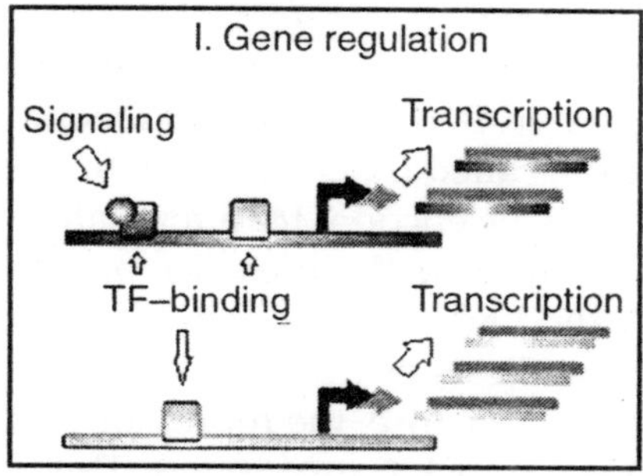

Fig. Gene Regulation

For example, in humans only red blood cells manufacture the protein hemoglobin and only pancreas cells make the digestive enzyme known as trypsin, even though both types of cells contain the genes to produce both hemoglobin and trypsin. Each cell produces different proteins according to its needs so that it does not waste energy by producing proteins that will not be used.

A variety of mechanisms regulate gene activity in cells. One method involves turning on or off gene transcription, sometimes by blocking the action of RNA polymerase, an enzyme that initiates transcription.

Gene regulation may also involve mechanisms that slow or speed the rate of transcription, using specialized regulatory proteins that bind to DNA. Depending on an organism's particular needs, one regulatory protein may spur transcription for a particular protein, and later, another regulatory protein may slow or halt transcription.

Prokaryotes

The bacterium *Escherichia coli* (commonly referred to as *E. coli*), found in the intestines of humans and other mammals, provides a good example of gene regulation. *E. coli* uses three enzymes to digest lactose, the primary sugar found in milk. The bacterium produces great quantities of lactose-digesting enzymes when lactose is present and saves energy by not synthesizing the enzymes when the sugar is not available. *E. coli* prefers consuming glucose to lactose, so the bacterium produces these enzymes in the presence of lactose only when no glucose is available.

A region of DNA known as an operon controls this gene regulation process. In *E. coli*, the operon includes at least five genes: Three genes, called lac genes, code for the enzymes that digest lactose; one gene encodes for a regulatory protein, called a repressor, that can sense the presence or absence of lactose; and one gene, called an operator, activates transcription, the first step in the synthesis of the lactose-digesting enzymes.

In the absence of lactose, the repressor protein binds with the operator gene to block transcription. This prevents the lac genes from being transcribed and halts production of the lactose-digesting enzymes. If lactose is present, the repressor protein binds to the sugar, leaving the operator gene free to trigger transcription of the lac genes. The transcription of the lac genes produces the mRNA that will direct the production of the three lactose-digesting enzymes.

Eukaryotes

Gene regulation in eukaryotes is more complex than in bacteria and other prokaryotes. Even the simplest eukaryotes have far more genes than prokaryotes, and these genes must be turned on or off as conditions dictate. Most multicellular organisms contain different types of cells that serve specialized functions. The cells of an animal's heart, blood, skin, liver, and muscles all contain the same genes. But in order to carry out their specific functions within the body, each cell must produce different proteins and respond to changing environmental stimuli, such as glucose levels in the blood or body temperature. Such specialization is possible only with sophisticated gene regulation.

Eukaryotes use a variety of mechanisms to ensure that each cell uses the exact proteins it needs at any given moment. In one method, eukaryotic cells use DNA sequences called enhancers to stimulate the transcription of genes located far away from the point on the chromosome where transcription occurs. If a specific protein binds to an enhancer site on the DNA, it causes the DNA to fold so that the enhancer site is brought closer to the site where transcription occurs.

This action can activate or speed up transcription in the genes surrounding the enhancer site, thereby affecting the type and quantity of proteins the cell will produce. Enhancers often exert their effects on large groups of related genes, such as the genes that produce the set of proteins that form a muscle cell.

Gene regulation can also take place after transcription has occurred by interfering with the steps that modify mRNA before it leaves the nucleus to take part in translation. This process typically involves removing exons (segments that code for specific proteins) and introns. These sections of the mRNA can be modified in more than one way, enabling a cell to synthesize different proteins depending on its needs.

GENES IN DEVELOPMENT

Gene regulation helps individual cells within an organism function in a specialized way. Other regulatory mechanisms coordinate the genes that determine how cells develop. All of the specialized cells in an organism, including those of the skin, muscle, bone, liver, and brain, derive from identical copies of a single fertilized egg cell. Each of these cells has the exact same DNA as the original cell, even though they have vastly different appearances and functions. Genes dictate how these cells specialize.

Early in an organism's embryonic development the overall body plan forms. Individual cells commit to a particular layer and region of the embryo, often migrating from one location to another to do so. As the organism grows, cells become part of a particular body organ or tissue, such as skin or muscle.

Ultimately, most cells become highly specialized—not only to develop into a neuron rather than a muscle cell, for example, but to become a sensory neuron instead of a motor neuron.

This process of specialization is called differentiation. At each stage of the differentiation process, specific genes known as developmental control genes actively turn on and switch off the genes that differentiate cells. One class of developmental control genes, known as homeotic genes, directs the formation of particular body parts. Activating one set of homeotic genes instructs part of an embryo to develop into a leg, for example, while another set initiates the formation of the head.

If a homeotic gene becomes altered or damaged, an organism's body development can be dramatically disrupted. A change in a single gene in some insects, for instance, can cause a leg to grow where an antenna belongs. Homeotic genes work by regulating the activity of other genes. Homeotic genes code for the production of a regulatory protein that can bind to DNA and thus affect the transcription of one or more genes. This enables homeotic genes to initiate or halt the development and specialization of characteristics in an organism.

Nearly identical homeotic genes have been identified in varied organisms, such as insects, worms, mice, birds, and humans, where they serve similar embryonic development functions. Scientists theorize that homeotic genes first appeared in a single ancestor common to all these organisms. Sometime in evolutionary history, these organisms diverged from their common ancestor, but the homeotic genes continued to be passed down through generations virtually unchanged during the evolution of these new organisms.

Chapter 36

Biochemical and Genetic Techniques

Scientists have developed a number of biochemical and genetic techniques by which DNA can be separated, rearranged, and transferred from one cell to another. Some of these laboratory methods help scientists study the properties of genes in nature—for example, by comparing DNA from different animals to find out whether those animals are closely related to each other or only distant relatives.

Other DNA techniques provide tools for genetic engineering—the alteration of genes in an organism. These tools are used in industry to develop commercial products, such as hardier crops, microbes that can break down oil slicks or decompose garbage, and improved medicines.

RECOMBINANT DNA

The DNA molecules of all life forms, from oak trees to sea horses, have the same structure and the same four bases. Scientists have made use of these similarities in a technology called recombinant DNA. In this laboratory method, one or more genes of an organism are introduced into a second organism. The new genes, sometimes known as foreign DNA, become functional in the second organism and produce a desired protein.

In this way, scientists can create changes in the genetic makeup of an organism that would be unlikely to occur through natural processes. Scientists use recombinant DNA when they want to obtain large amounts of a protein, such as insulin, produced by a gene. Insulin was once in short supply for diabetics, whose bodies lack adequate supplies. Insulin supplies were derived from cows in an expensive and time-consuming process. Today recombinant DNA techniques produce insulin cheaply and in abundance.

The first step in creating insulin using recombinant DNA is to isolate the sequence of nucleotides in the DNA of a human cell that forms the insulin gene. Scientists use *restriction enzymes,* specialized proteins that act like molecular scissors, to cut the double-stranded DNA at the point where the insulin gene occurs. The isolated DNA can then be recombined, or spliced, with a vector, a fragment of DNA that is able to transport genes from one organism to another.

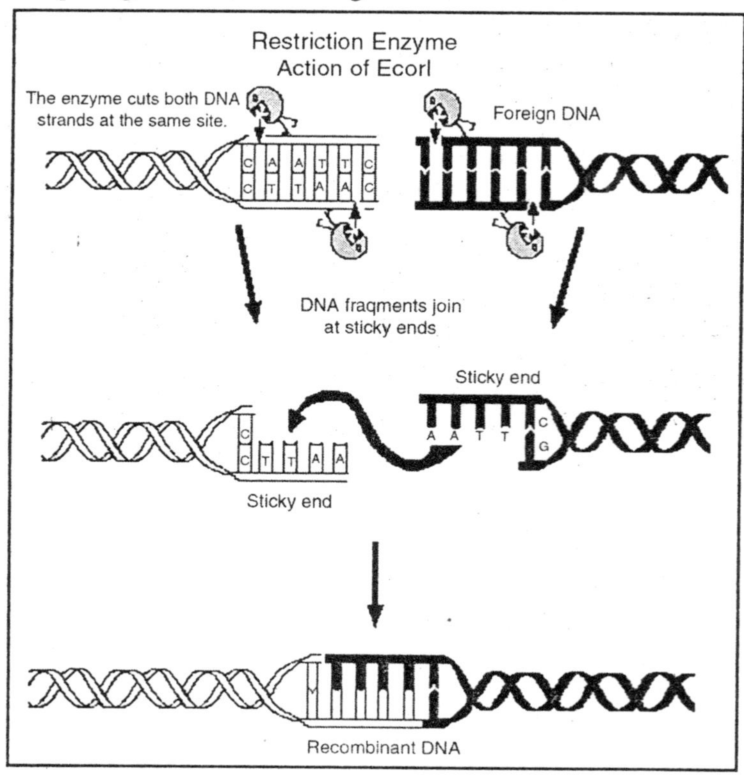

Fig. Restriction Enzymes

A vector may be a plasmid, a small, circular segment of DNA found in bacteria. Bacteriophages, viruses that are parasites of bacteria, also act as vectors. Scientists insert the vector containing the insulin gene into a bacterium, such as *E. coli*. Within just a few hours, a single *E. coli* will reproduce hundreds of times to make millions of cells, all containing exact copies of the insulin-producing gene inserted by the scientists. This process of making many cells with identical DNA is known as cloning.

DNA LIBRARIES

A DNA library is a storehouse of genetic information maintained in bacteria instead of books.

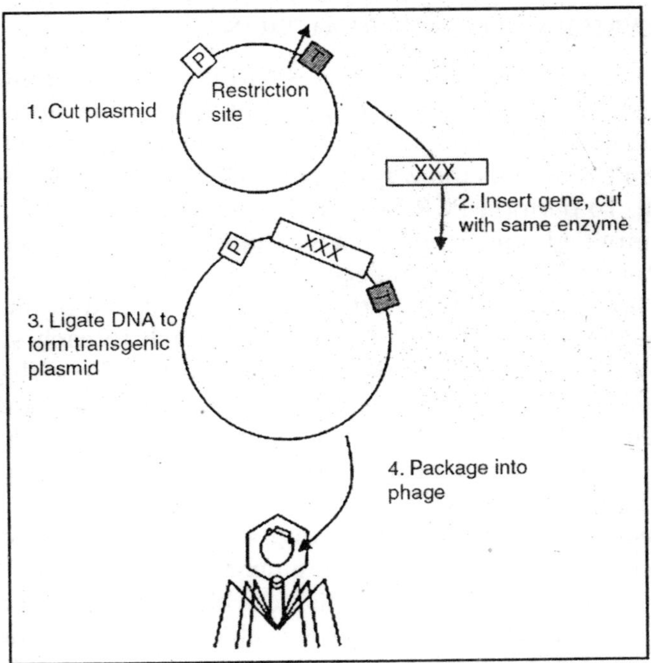

Fig. DNA Libraries

These bacteria are clones created by recombinant DNA, and the foreign DNA they hold is the library's store of information. DNA libraries are helpful to scientists who require a plentiful supply of particular DNA segments to do their work. These repositories of genetic information are stored in small tubes, which can easily be shipped to other researchers for study.

Each library has a unifying theme. For example, a library may contain the entire chromosomal DNA, or genome, of a given organism, or it may consist of genes that are active within certain types of cells, such as heart cells.

To create a library of the human genome, DNA from all the human chromosomes would be cut into many pieces. These pieces would be randomly inserted into vectors, such as plasmids, which would then be placed into a population of bacteria. Taken together, the entire population of bacteria would contain all the DNA of the human chromosomes.

POLYMERASE CHAIN REACTION

Polymerase chain reaction (PCR) offers an alternative to vector-based cloning as a means of generating numerous copies of DNA from a small initial sample. Performed in a test tube, PCR mirrors the way in which DNA is replicated within a cell. To perform PCR, scientists isolate the piece of DNA to be amplified (multiplied) in a test tube and heat it to separate the two strands of the molecule.

As cooling occurs, short pieces of DNA called primers are added to the test tube. The primers attach to each strand, marking the segment that will be cloned. Free-floating nucleotides and an enzyme called DNA polymerase are then added to the mixture. DNA polymerase uses the free-floating nucleotides to build a complementary copy of each amplified DNA segment, resulting in two new double-stranded DNA molecules. Each cycle of heating and cooling doubles the amount of the desired DNA fragment in the test tube.

In a matter of hours, scientists can obtain millions of copies of a desired piece of DNA. PCR enables scientists to amplify traces of DNA found at a crime scene or in a fossil animal to produce sufficient quantities to study.

GEL ELECTROPHORESIS

PCR and recombinant DNA techniques create large

amounts of DNA segments. To study the structure of these segments, researchers use a process known as gel electrophoresis. This technique can be used to identify genes in humans that have previously been identified in other organisms, such as fruit flies. It can also be used to compare the DNA found from blood or hair samples at a crime scene with the DNA of a suspect in the crime.

In gel electrophoresis, restriction enzymes break up the DNA under study into restriction fragments of varying lengths. Solutions containing these fragments are placed within a thick gel. An electric current is applied to the gel, causing one end of the gel to have a positive charge and the other to have a negative charge.

All of the restriction fragments begin to move from the negative end of the gel toward the positive end. The smaller fragments move faster than the larger fragments. When the current shuts off, typically after several hours, the DNA fragments have spread out across the gel, with the smaller ones closer to the positive end.

The dispersed fragments display a pattern resembling a bar code. Each bar in this pattern contains DNA fragments of a certain size. Scientists can identify specific restriction fragments by their location on the gel. A complementary sequence of DNA can be used as a probe to find a restriction fragment on the gel that has a particular nucleotide sequence.

Scientists may use DNA found in blood at a crime scene as the probe to see if it pairs up with any of the DNA fragments in the gel electrophoresis. If pairing occurs, the DNA from the crime scene is from the same person who provided the DNA sample for the gel electrophoresis.

DNA SEQUENCING

Once an interesting piece of DNA has been isolated or identified, scientists often need to determine if the sequence of nucleotides in the fragment is related to known genes and to determine what kind of protein it might make. Scientists use DNA sequencing to detect genetic mutations linked to diseases such as cystic fibrosis. Scientists have also used this method to alter the sequence of a gene and study the function of the resulting protein.

In DNA sequencing, scientists create many copies of a single-stranded DNA fragment that will be used to synthesize a new DNA strand. An equal number of copies of the fragment are placed into four different test tubes to act as the template for the synthesis of a new strand. The enzyme DNA polymerase and free nucleotides are added to each test tube. Each test tube also receives one type of dideoxy nucleotide—a nucleotide that closely resembles either adenine, guanine, thymine, or cytosine.

These nucleotides can attach to the end of the new complementary DNA strand, but they cannot bind to anything else, thus they terminate the synthesis of the new DNA strand.

DNA polymerase uses the free nucleotides to build a complementary DNA strand. If the original DNA fragment contains guanine, DNA polymerase delivers a cytosine dideoxy nucleotide to pair with the guanine base on the original strand. The cytosine links with the growing chain of nucleotides on the complementary DNA strand, but it is unable to bind with any other nucleotide. The newly formed DNA fragment terminates with the cytosine dideoxy nucleotide at the end of the chain.

The reactions in each of the four test tubes produce a series of DNA fragments in which the new strands terminate at a known base. Each test tube produces fragments that differ in length from the other test tubes. The newly formed fragments are sorted in an electrophoresis gel that can detect differences as small as one nucleotide in length. By analyzing these sorted fragments, scientists can determine the complementary base sequence for the original DNA fragment.

This sequencing method has become a routine laboratory technique, automated with specialized machines and computers that can prepare DNA samples and read nucleotide sequences far faster and more accurately than people can.

GENE CHIP

The gene chip, also known as a DNA chip or DNA

microarray, is a thumbnail-sized chip of glass or silicon that carries DNA instead of electronic circuits. Gene chips can identify the genes that are active within a cell and help identify mutated genes. In one application, scientists take a single strand of DNA that contains a defective gene and use ultraviolet light to attach the strand onto a glass or silicon chip.

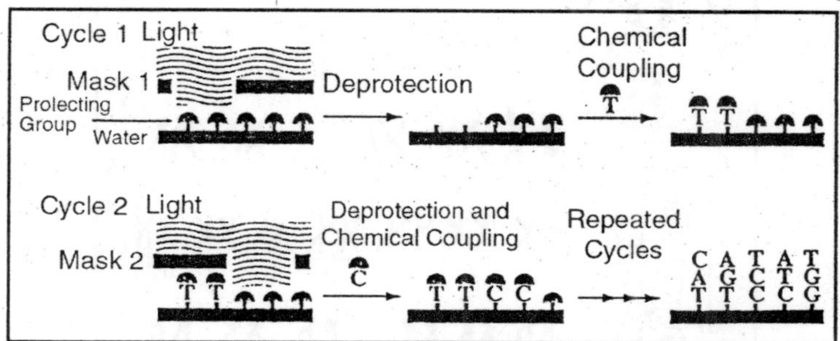

Fig. Gene Chip

A second DNA strand isolated from a patient is attached to fluorescent markers and deposited onto the chip. If the patient's DNA strand bonds with the DNA already bonded to the chip, then the individual's DNA contains the defective gene. When the DNA on the chip pairs with the fluorescent DNA, it develops a fluorescent glow that can be viewed with a microscope and interpreted by a computer.

A diagnostic gene chip may soon be manufactured to hold the DNA sequences of all the known disease-causing genes, making diagnosis for genetic disorders fast, reliable, and inexpensive. Gene chips also distinguish between active DNA—DNA that is being transcribed to produce mRNA—and inactive DNA. Researchers use these chips to learn how the transcription of a group of genes is affected when cells are exposed to a drug.

HUMAN GENETICS

Our understanding of human genetics builds on a foundation of information obtained from studying other organisms. Until the 1980s, genetic researchers focused their work on the fundamental genetic processes in simpler organisms, such as bacteria, plants, and fruit flies. Today an expanded array of tools available for the direct study of human genetics attracts scientists from around the world to collaborate to identify and study every human gene.

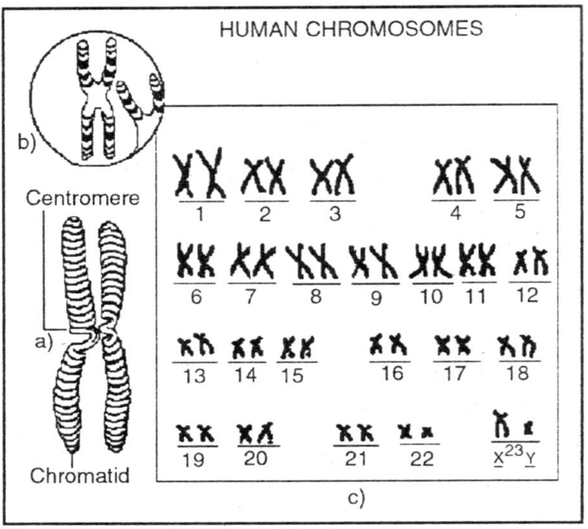

Fig. Human Genetics

The genetic principles that Mendel first discovered in plants apply to humans as well. As in all other life forms, the DNA found in human cells encodes the proteins that are essential for reproduction, survival, and growth. The unique structure and behaviour of DNA ensures that human traits are passed from generation to generation and accounts for why parents, children, and grandchildren often have similar facial features, hair colour, height, and athletic or artistic abilities.

Yet each of us inherits a unique genetic legacy from our parents and more distant ancestors. With the exception of identical twins, no two people have the exact same combination of alleles for the estimated 20,000 to 25,000 human genes. Some human traits are controlled largely by a single gene. But most inheritable characteristics are influenced by a number of genes that interact in a complex fashion.

Also, personal experiences and environmental factors

combine with genetic influences to shape certain traits, including vulnerability to disease and characteristics such as intelligence, emotions, talents, and personality.

Human Genome

Human genes reside on 23 pairs of chromosomes found in the nucleus of every body cell except gamete cells. In each pair, one of the chromosomes is inherited from the mother and the other is passed down from the father. About 2 m (7 ft) of DNA is packaged into each chromosome. All of the genes carried on chromosomes form the human genome. A lesser amount of DNA can be found in mitochondria, cellular organelles responsible for creating the energy used in cell activities.

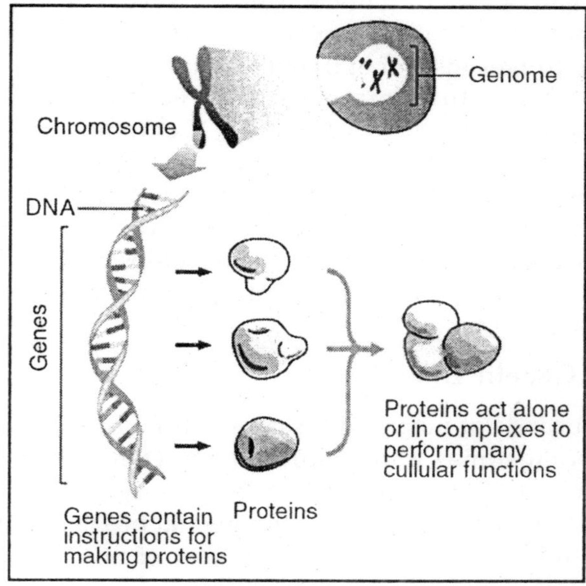

Fig. Human Genome

All but one of these 23 pairs are composed of chromosomes nearly identical in shape. Each of these 22 chromosome pairs, known as autosomes, contains the same genes (although they likely carry different alleles). The autosome pairs vary considerably in length, and scientists number them according to their relative size: Pair number 1 is the longest pair and pair number 22 is the shortest.

Rounding out the human genome is the 23rd pair of chromosomes, known as the sex chromosomes, which determine the sex of an individual. Females inherit two X chromosomes, a matched pair carrying the same genes. One X chromosome is inherited from the mother and one X chromosome is inherited from the father. Males inherit an X chromosome from their mother and a Y chromosome from their father. The Y chromosome is shorter than the X chromosome and bears far fewer genes. One gene on the Y chromosome causes an embryo to develop as a male rather than a female.

Humans produce gamete cells for sexual reproduction. These gametes contain a haploid number of chromosomes—23 chromosomes instead of the full complement of 46. Female gamete cells mature into eggs, with each egg containing chromosomes 1 through 22 and an X chromosome. Males produce gametes that mature into sperm, and each sperm cell has a single set of chromosomes 1 through 22 and either an X or a Y chromosome. During fertilization, an egg that joins with a sperm containing a Y chromosome develops into a male, and an egg fertilized by a sperm containing an X chromosome develops into a female.

Human Genetic Disorders

Thousands of inherited diseases caused by altered genes and chromosomal abnormalities affect humans. These disorders cause problems such as physical deformities, metabolic dysfunction, and developmental problems. Medical surveys indicate that roughly 1 per cent of newborns in the United States have a single-gene defect.

As many as 1 baby in 200 is born with a chromosomal abnormality serious enough to produce physical defects or mental retardation. It is misleading to say that a person "inherits the gene" for a disease, since humans are born with the same number and types of genes. We inherit allele forms of specific genes, and these alleles may be defective. Most of the known inherited genetic disorders are caused by the mutation of a single gene, resulting in alleles that produce disease.

These defects often produce disturbances in the body's biochemical processes, such as inhibiting the action of an important enzyme or stimulating the overproduction of a harmful substance.

Frequently the consequences of such problems can cause severe disability or be fatal. Many single-gene disorders follow Mendelian patterns of inheritance. A mother and father each pass an allele for a specific gene on to a child. If one of the alleles is defective and causes disease, the child will develop the disease according to a dominant-recessive pattern of inheritance.

For example, cystic fibrosis (CF), a metabolic disorder that causes a progressive loss of lung function, is caused by a mutation in the recessive allele of a gene responsible for regulating salt content in the lungs.

The recessive allele is unable to direct the production of a key protein, resulting in a salt imbalance that causes thick, suffocating mucus to build up in the lungs. If a baby inherits the defective allele from just one parent, no disease results. But the infant who inherits the defective allele from both parents will be born with the disease.

In other cases, a single dominant allele causes genetic disease. Huntington's disease, a condition characterized by involuntary movements, dementia, and eventually death, is caused by the inheritance of a pair of alleles in which a defective allele dominates the normal allele for the gene.

An affected parent has a 50 per cent chance of passing the defective allele to a child.

A child who inherits the dominant defective allele from just one parent will develop the disease. Other inherited genetic diseases are caused by defects in the genes found on the X chromosome.

Hemophilia, the inability of the blood to clot and heal a wound, is caused by a defect in an allele located on the X chromosome that helps produce proteins involved in the clotting process.

Women who inherit this defective allele usually have the normal allele on their second X chromosome, which produces enough of these clotting proteins for the body to remain healthy. Women who inherit this faulty allele have a 50 per cent chance of passing the defective allele on to their children.

Males who inherit this defective allele do not have a normal version of the allele on their Y chromosome and so cannot produce clotting proteins to heal wounds. Hemophiliacs are almost always males who have inherited an X chromosome with the faulty allele from their mother. Other genetic disorders arise due to the inheritance of an abnormal number of chromosomes or a defective chromosome structure. These chromosomal abnormalities have a devastating impact: Many fetuses with such defects, particularly those with missing chromosomes, will die prenatally, resulting in miscarriage (spontaneous abortion).

In other cases, newborns with chromosomal abnormalities suffer from physical problems or varying degrees of mental retardation.

Down syndrome occurs when an individual's cells carry an extra copy of chromosome 21. People born with this condition have characteristic facial features, short stature, severe developmental disabilities, and a shortened life expectancy.

GENETICS AND CANCER

Cancer is a common name for many diseases that affect different body tissues, including the skin and the liver. All cancers involve alterations in genes that control cell division. These alterations cause cells to replicate abnormally and form tumors.

Cancers generally arise from mutations that occur directly in the somatic cells, any cells of the body with the exception of the gametes (sperm and egg cells). Since the genetic mutations have not occurred in gametes, the mutations are not inherited by the next generation.

While cancer is not a traditional inherited genetic disorder, scientists have determined that a genetic component plays a strong role in the development of the disease.

Geneticists have identified many different genes with certain alleles that appear to increase an individual's susceptibility to cancer. A notable example involves two genes linked to breast cancer.

Researchers estimate that more than half of the women with a family history of breast cancer who inherit mutated alleles of these two genes, known as BRCA1 and BRCA2, will develop breast cancer by the age of 70.

In contrast, women who lack either of the mutated alleles have only a 13 per cent chance of developing the disease. For many cancers, researchers believe that mutations in several different genes must accumulate before cancer develops. As a person ages, errors in DNA replication may occur during cell division, or cells may be damaged by exposure to certain environmental factors, including cigarette smoke, radiation, and chemical pollutants. As a result, an accumulation of mutations may develop in two types of genes: tumor suppressor genes and oncogenes.

Tumor suppressor genes normally function to halt cell division, while oncogenes function to activate cell division. A mutation in either type of gene can stimulate nonstop cell division. These types of defects have been linked to some cases of leukemia as well as to cancers of the ovaries, lungs, colon, and other organs.

Genetics and Aging

A growing area of study focuses on the link between aging and genetics. Scientists have determined that structures called telomeres, long, repetitive sequences of nucleotides at the end of chromosomes, affect the aging process. Each time a cell divides, telomeres become shorter. When the structures shorten to a certain length, the process of cell division terminates.

The cells of these chromosomes continue to live, but they never divide again. Laboratory tests suggest that an enzyme produced in gamete cells of the human body, called telomerase, can maintain telomere length in human cells, enabling them to continue dividing, perhaps indefinitely. Scientists hoping to lasso the life-extending properties of telomerase have been confounded by research indicating that telomerase is also active in rapidly dividing cancer cells. Before telomerase can be used to slow or halt aging, scientists must learn how to manipulate the enzyme so that it does not promote cancer growth.

Genes and Behaviour

Scientists actively explore the links between genes and behaviour to determine both the patterns and the limits of genetic influence. Such studies continue to be controversial because behaviour or mental processes can be difficult to measure objectively. Furthermore, many behavioral traits, both normal and abnormal, are complex, influenced by many genes as well as by personal experiences. Studies of the possible genetic components of psychiatric disorders have yielded mixed results.

Geneticists have identified at least two genes linked to schizophrenia, a condition characterized by hallucinations, delusion, paranoia, and other symptoms. Other studies that reported the discovery of genes that influence bipolar disorder (also known as manic-depressive illness) and alcoholism have been reversed or questioned.

Though attempts to identify genes linked to these disorders have been flawed, scientists have little doubt that the conditions do have a genetic component. Scientists have established links between genes and certain antisocial or violent behaviors.

For instance, researchers have identified a gene on the X chromosome that has been tied to extremely violent behaviour in men. They identified the gene in members of several families with a multigenerational history of violent, criminal behaviour.

They identified gene codes for monoamine oxidase inhibitor (MAO), an enzyme that helps nerve cells in the brain communicate with each other. Males in an affected family who inherit a defective allele for MAO do not produce enough of the enzyme.

As a result, low levels of MAO change the activity of certain brain nerve cells, possibly contributing to socially unacceptable behaviour. While research suggests that men who have a defective allele for the MAO gene are more prone to aggressive behaviour, experts cite numerous reasons for concern and doubt.

Few scientists believe that a single gene could have a leading role in influencing complex behaviors. Others charge that these kinds of investigations promote an unreasonably simplistic view of genetic determinism, in which genes can be blamed for certain behaviors.

Critics note that studies have identified many men who carried the defective allele for MAO production and never committed a violent act.

Clearly, nongenetic influences—as varied as an individual's family life, work circumstances, attitudes, diet, and emotional state—affect complex behaviors.

Chapter 37

Identifying Genetic Disorders

Health-care professionals who specialize in genetic disorders use a variety of methods to identify inherited conditions. Analysis of a family medical history, known as pedigree analysis, is used to track the transmission of a condition through generations. Blood tests that identify specific DNA sequences can reveal carriers of a disease-causing gene who have no symptoms of the disease. Geneticists collect a person's medical family history to trace the inheritance of a genetic trait among multiple generations.

The information is placed in a pedigree, which resembles a traditional multigenerational family tree but includes information about individuals who were diagnosed with a particular disorder or who suffered from certain medical symptoms.

A pedigree can help researchers recognize diseases that express themselves in dominant or recessive alleles. Dominant disorders affect every generation. Recessive disorders may cluster in a single generation, reflecting when two parents who both carry a recessive allele for a disease have one or more children who develop the disease.

A pedigree can also identify diseases that show X-linked inheritance. Pedigree analysis can be useful when combined with certain genetic tests. A blood sample taken from a person who is at risk for a genetic disorder can be compared with a DNA sequence known to cause the disorder in question.

Other genetic tests can reveal if a person has extra chromosomes, missing chromosomes, or chromosomes that have attached to one another in unusual ways. In some cases, these chromosomal abnormalities may produce genetic disorders in children or they may affect a person's ability to conceive a child. Genetic testing can also identify disorders in a fetus, enabling parents to learn early in a pregnancy if a fetus will likely be born with health problems or develop them later in life.

Presymptomatic testing can identify DNA abnormalities in a person before health problems develop. In the case of certain inherited heart conditions, for example, these tests enable a person to make healthy lifestyle changes or take other preventative measures, such as medications, to lower the risk of illness or death.

Medical genetic testing raises challenging issues because such tests typically provide statistical possibilities rather than a definite prediction of whether a person will develop a given genetic disease. A test result may indicate, for example, that a person has a 75 per cent risk of developing colon cancer by the age of 65.

Such results enable a physician to perform appropriate screening tests on the at-risk person in order to identify the disease at its earliest stages, when it is most treatable. At the same time, however, physicians must decide at what age the person's screening should begin and whether the benefits of early screening are worth the drawbacks of frequent screening. These drawbacks include expense, patient anxiety and discomfort, and exposure to radioactivity or other harmful substances used in testing.

Different problems are posed by genetic screening tests that diagnose conditions for which no preventive measures exist, such as Alzheimer's disease, a progressive brain disorder that causes the loss of mental function. People may find it devastating to learn that they are at risk for a deadly disease that cannot be prevented by medical measures or lifestyle choices.

GENE THERAPY

A recent development in genetic technology known as gene therapy focuses on curing inherited disorders. In experiments using gene therapy, researchers have replaced defective genes with normal alleles, inactivated a mutated gene, or inserted a normal form of a gene into a chromosome. The earliest success in human gene therapy involved the treatment of infants who cannot produce adenosine deaminase (ADA), an enzyme important to normal function of the immune system. Scientists have successfully inserted the normal allele for the gene that codes for the enzyme into cells in ADA-deficient children.

Preliminary evidence indicates that this gene therapy leads to better immune function in recipients. Researchers are also exploring gene therapy's potential to help treat people with many other conditions, including certain cancers, hemophilia, heart disease, and cystic fibrosis. Although the United States Food and Drug Administration (FDA) has approved more than 400 clinical trials in gene therapy, this method of treating disease remains far from an unqualified medical success.

Treatments usually produce some improvement in the underlying condition, but not enough to consider the therapy suitable for large-scale use. The death of a patient involved in a gene therapy experiment in 1999 caused the National Institutes of Health (NIH), a federal agency that monitors gene therapy studies, to reevaluate the safety and effectiveness of gene therapy clinical trials.

HUMAN GENOME PROJECT

The Human Genome Project is the most ambitious project in the history of biology. The program's challenging goal was to identify and sequence all of the DNA in human chromosomes. The project was initiated in 1990 in the United States with government funding, and it rapidly grew into an international consortium of academic centers and drug companies in China, France, Germany, Japan, the United Kingdom, and the United States.

The consortium initially hoped to reach its goal by the year 2005. In 1998 Celera Genomics, a privately funded biotechnology firm, announced that it would sequence the human genome by the year 2000 using different sequencing strategies than those used by the public consortium.

This announcement triggered a heated race between Celera Genomics and the public consortium to complete the genome project. In June 2000 both teams declared victory when they jointly announced that they had separately completed a rough draft of the genome. The two teams published their findings simultaneously, although in two different journals, in February 2001.

The draft provided a basic outline of 90 per cent of the human genome. Scientists from the public consortium completed the final sequencing of the human genome in April 2003, two years earlier than planned. The completed human genome has provided scientists with a detailed blueprint of our complex genetic code. Large computer databases of genetic information enable scientists to look for patterns and relationships among the actions of different genes.

Among the findings about the human genome was that the number of genes in the human genome is much lower than was predicted—only about 20,000 to 25,000 genes compared to the expected 100,000 genes.

This number is a little more than twice the number of genes found in the fruit fly. Scientists are now turning their attention to studying how the relatively low number of genes in the human genome can produce the complex structures found in humans.

Scientists have long known that a single gene produces a single protein and that this single protein subsequently may be processed into several different proteins. In a new science known as proteomics, scientists seek to identify and understand the function of all the proteins in the human body. They theorize that there may be many more proteins than there are genes—that is, more than 25,000.

Among other advances, the database of proteins derived from proteomics is expected to help scientists better understand the regulation of gene expression in the body and how it leads to the complexity of cellular structures and functions. In addition, proteomics may lead to the development of breakthrough drugs for a variety of genetic disorders.

GENES AND OUR WORLD

Breakthroughs in decoding and manipulating the genetic information stored in DNA promise a world of benefits. These scientific advances already help to diagnose and treat disease, develop new medicines, bring criminals to justice, improve our food supply, and clean up the environment. At the same time, however, genetic technologies also present society with the potential for new and serious social or environmental problems. Many of the developments that worry critics of genetic technologies remain on the horizon, but the debate over their inevitable arrival is already in full swing.

Genes and the Environment

Humans have tampered with the genetic composition of other organisms for thousands of years. Most of this manipulation has been decidedly low-tech: domestication of animals and selective breeding of desirable food crops. The development and use of genetic engineering techniques has accelerated the pace at which humans can alter nature, creating some products that have unquestionable benefits and others that raise serious concerns. Some scientists fear that genetic engineering techniques will damage genetic diversity. Domestication and selective breeding, which aim to produce many similar organisms with particular desirable traits, reduce the natural variation of genes within a species.

Genetic engineering techniques take this effect a giant step further. Many crop species—including wheat, corn, tomatoes, and strawberries—have been manipulated to maximize yield, appearance, resistance to pests and chemicals, hardiness, and other commercially valuable traits. Once attractive varieties have been developed, agricultural techniques favor wide-scale planting of such genetically similar or identical stocks. The impact of this practice on biodiversity can be ominous because older plant varieties that carry diverse and useful alleles may be lost forever.

The forfeited genetic material could leave a species vulnerable to annihilation from a single factor in the environment, such as an insect pest or an infectious virus.

Fortunately, efforts to combat decreasing biodiversity are under way. Farmers, gardeners, government agencies, and other interested parties have collaborated to create seed banks to maintain genetic diversity. These banks catalogue, store, and distribute the seeds of rare or endangered plants, enabling gardeners and farmers to continue cultivating rare plant varieties so that the unique genetic makeup of these plants does not disappear.

In the same vein, zoos and other institutions breed endangered species of animals that may no longer be able to survive in their native habitats. Other animal programs seek to preserve or enhance the genetic variation within certain endangered animal populations.

For example, all cheetahs are almost identical genetically, most likely due to their near extinction about 12,000 years ago. Inbreeding among the few remaining individuals has resulted in a loss of genetic diversity in modern cheetahs that may have affected the cheetah's immune system, leaving the animal vulnerable to disease.

Scientists hope to use genetic engineering techniques to introduce new genes into the cheetah population to increase the genetic diversity of the species. The broad use of genetic engineering techniques in agriculture has raised other concerns beyond issues of biodiversity. From a consumer point of view, for instance, new technologies may have compromised a food's taste or nutritional value in exchange for a plant or animal that can be grown faster or at less cost.

In addition, some critics question the safety of genetically engineered foods. They fear that plants or animals that have received new genes will produce proteins that would not be present in nonmanipulated organisms. Such changes could have a serious effect: causing allergies or toxicity in humans who eat these foods, for example, or disrupting a plant's production of key nutrients.

Though there has been much speculation about the

potential health risks of genetically engineered foods, rigorous scientific investigation into the effects of these foods on humans is just beginning. The potential environmental impact of genetically engineered agriculture is equally controversial. Critics fear that transgenic organisms, which contain DNA from other species, could give rise to populations of genetically altered life forms that could cause disease, displace native species, or otherwise harm delicate ecosystems.

Consider the example of a genetically engineered form of oilseed rape, the plant that yields canola oil. The altered form, which has been grown commercially in the United States since 1993, contains inserted genes that increase its resistance to herbicides (weed-killing chemicals). The altered plants can grow unabated even when a farmer sprays a field with enough herbicide to kill troublesome weeds.

Yet a curious problem has emerged: The transgenic rape can interbreed with weedy relatives that grow nearby. Some scientists fear that this interbreeding could create weed varieties that also have a genetic resistance to known herbicides. The biological and economic impact of such a development could be enormous.

Genes and Society

Advances in genetic technologies allow scientists to take an unprecedented glimpse into the genetic makeup of every person. The information derived from this testing can serve many valuable purposes: It can save lives, assist couples trying to decide whether or not to have children, and help law-enforcement officials solve a crime. Yet breakthroughs in genetic testing also raise some troubling social concerns about privacy and discrimination.

For example, if an individual's genetic information becomes widely available, it could give health insurers cause to deny coverage to people with certain risk factors or encourage employers to reject certain high-risk job applicants. Furthermore, many genetically linked problems are more common among certain racial and ethnic groups—for example, the BRCA1 breast cancer allele is more common in Ashkenazi

Jews, and the blood disorder sickle-cell anemia is more prevalent among blacks of African ancestry.

Many minority groups fear that the expansion of genetic testing could create whole new avenues of discrimination. Of particular concern are genetic tests that shed light on traits such as personality, intelligence, and mental health or potential abilities. Genetic tests that indicate a person is unlikely to get along with other people could be used to limit a person's professional advancement. In other cases, tests that identify a genetic risk of heart failure could discourage a person from competing in sports. New technologies that allow the manipulation of genes have raised even more disturbing possibilities.

Gene therapy advances, which allow scientists to replace defective genes with normal alleles, give people with typically fatal diseases new hope for healthy lives. To date, gene therapy has focused on manipulating the genetic material in body cells other than gametes, so the changes will not be passed on to future generations. However, the application of gene therapy techniques to gametes—the cells involved in reproduction—seems inevitable.

Such manipulation might help prevent the transmission of disease from one generation to another, but it could also produce unforeseen problems with long-lasting consequences. For instance, many people worry that new genetic techniques could be used to alter or encourage traits now viewed as part of normal human variability, such as shortness or baldness. At various times in the past century, people have advocated efforts to improve the human condition by promoting the perpetuation of certain genes.

This concept, known as eugenics, typically involves encouraging people with "positive" genes to reproduce and discouraging those with "inferior" genes from having offspring. Many people fear that new genetic technologies used to manipulate the human genome could give people previously unattainable methods to resort to extreme forms of eugenics. Advances in genetic technologies have turned some genes into valuable commercial commodities, spawning a host of controversial questions. Who owns a genetically altered organism or the genes it contains? Is it right to patent the use of a naturally occurring gene? Some people feel that genetic material should not be owned or used for profit.

Costa Rica has enacted laws to prevent foreign companies from patenting and then profiting from genes of native Costa Rican plant and animal species. Balancing the need to limit patents on genes are concerns that the profit motive of companies must be protected to maintain incentives to make new discoveries for medical products.

The citizens of Iceland, for example, are cooperating with a biotechnology company in a study of the genetic makeup of the Icelandic people. The information compiled will be used to learn about genetic diseases.

HEREDITY

Humans have had some understanding of heredity since prehistoric times, observing how similar traits pass from parent to offspring and noting that differences arise with each generation. Most of the mechanisms of heredity, however, were shrouded in mystery until early in the 20th century. Since that time, the rate of discovery has reached a feverish pace, enabling the advancement of modern molecular biology and the current Human Genome Project.

Early Views of Heredity

In ancient times, people understood some basic rules of heredity and used this knowledge to breed domestic animals and crops. By about 5000 BC, for example, people in different parts of the world had begun applying selective breeding techniques to grow new plant varieties, including types of wheat, maize, rice, and date palms, that had never existed in the wild. Ancient people understood that the rules of inheritance also applied to humans. The ancient Greeks were particularly interested in human heredity and evolution. Greek scientists and philosophers hotly debated whether a male or female parent contributed more to an offspring.

In the 4th century BC, Aristotle speculated that acquired

characteristics, such as a scar that was incurred during life, could be passed on to offspring. He also believed in a widely held theory known as pangenesis.

This theory proposed that particles in the body, called gemmules, reside in the limbs and organs. The gemmules become imprinted with any changes acquired by the body, such as muscle development from exercise. The gemmules then move to the reproductive cells and transfer information about the body's alterations to these cells.

The reproductive cells transmit the acquired traits to offspring through particles called pangenes. The theories about the inheritance of acquired characteristics and pangenesis persisted until the middle of the 19th century. French zoologist Jean-Baptiste Lamarck formalized the theory of acquired characteristics in his treatise *Philosophie Zoologique*. Lamarck proposed that organisms evolve by responding to changes in their environment. When organisms undergo a change in order to adjust to their environment, that change acts as a trait that can be passed on to offspring.

INFLUENCES OF DARWIN AND MENDEL

A surprising supporter of pangenesis was the British naturalist Charles Robert Darwin, who believed that the theory accounted for the process of heredity and the wide variety of traits seen among offspring. Despite his mistaken belief in pangenesis, Darwin nonetheless had an enormous impact on human understanding of heredity. During his years of extensive worldwide travel, Darwin collected many observations of how related species adapt to their local environments.

Darwin and British naturalist Alfred Wallace independently formulated the theory of natural selection, which holds that members of a given species born with more favorable characteristics to deal with their environment would be most likely to survive to pass on these traits to the next generation. This important theory was popularized by Darwin's publication *On the Origin of Species*. The book was an immediate sensation, but it raised many questions.

Foremost among these was the mystery of how organisms could appear with modified or entirely new traits.

At roughly the same time that Darwin published his natural selection theories, the answer to many questions about the mechanisms of heredity were being unraveled by Gregor Mendel, a reclusive Austrian monk. Mendel conducted a long series of experiments on pea plants during the 1850s and 1860s. Mendel crossbred plants that expressed differing traits, such as height and flower colour.

His conclusions from these experiments helped him formulate a comprehensive theory of how such traits pass from one generation to another. In his studies, Mendel recognized that characteristics were inherited as discrete units, and that each of these was inherited independently of the others. He speculated that each parent has pairs of these units but passes only one to an offspring. He also noted that certain forms of one trait were always dominant over others. Today the units that Mendel described are known as genes.

EMERGENCE OF THE SCIENCE OF GENETICS

Mendel published his findings in 1866, but they went largely unnoticed for more than three decades. In the year 1900, however, Dutch botanist Hugo Marie de Vries, German botanist Karl Correns, and Austrian botanist Erich Tschermak independently rediscovered the monk's works and verified his conclusions.

Advances in cytology, the science of the structure and function of cells, enabled scientists to more deeply appreciate Mendel's work. In 1902 American biologist Walter S. Sutton and German cell biologist Theodor Boveri separately noted the parallels between Mendel's units and chromosomes. The demonstration of the chromosomal basis of inheritance gave rise to the modern science of genetics. The term *genetics* itself was coined in 1905 by British biologist William Bateson. The terms *gene* and *genotype* were contributed in 1909 by German scientist Wilhelm Johannsen.

In 1905 American biologists Edmund B. Wilson and Nettie Stevens independently discovered and identified the sex chromosomes. Wilson discovered the X chromosome in a butterfly, and Stevens discovered the Y chromosome in a beetle. The discoveries of the X and Y chromosomes helped scientists begin to unravel new patterns of inheritance. Foremost among this research was the work of American biologist Thomas Hunt Morgan on fruit flies.

In 1910 Morgan identified the first proof of a sex-linked trait, an eye-colour characteristic that resides on the X chromosome of fruit flies. With this finding, Morgan became the first scientist to pin down the location of a gene to a specific chromosome.

Morgan was also the first to explain the implications of linkage, unusual patterns of inheritance that occur when multiple genes found on the same chromosome are inherited together. A student of Morgan's, American biologist Alfred Sturtevant, found early evidence of the mechanisms of crossing over, the phenomenon in which chromosomes interchange genes.

More definitive proof emerged in the 1930s with work by American geneticists Harriet Creighton and Barbara McClintock. The pair demonstrated gene recombination with experiments on seed colour in corn. McClintock later gained notice for her work on transposable elements, large genetic segments that move within a chromosome or even between chromosomes. Her research into these elements, commonly known as jumping genes, earned McClintock the 1983 Nobel Prize in physiology or medicine.

BREAKTHROUGHS IN DNA STUDIES

While cytologists and geneticists were studying the properties and location of genes on chromosomes, other scientists focused their studies on the composition of genes.

In 1928 British microbiologist Frederick Griffith ran a series of experiments on two strains of bacteria, one that kills mice and another that is harmless to them. When Griffith injected mice with killed cells of the virulent bacteria, all of the mice survived.

But in a second trial, when Griffith injected a combined

cocktail of dead virulent bacteria and live "harmless" bacteria, the mice all died. He concluded that something in the dead virulent cells "transformed" the hereditary material of normally harmless bacteria so that they became killers. Most scientists at the time theorized that the transforming factor was composed of a protein.

The real identity of the transforming factor in this experiment was not identified until 1944, when American geneticists Oswald Avery, Colin MacLeod, and Maclyn McCarty revisited Griffith's research. After isolating different molecular components from dead bacterial cells, Avery and his colleagues determined that DNA was the agent that transformed the live harmless bacteria into killers. Despite a growing body of evidence about the function of DNA, many scientists were not ready to reject proteins as the hereditary material.

The debate was largely quieted in 1952 by American geneticists Alfred Hershey and Martha Chase. Hershey and Chase showed that when a type of virus called a bacteriophage infects a bacterium, it is the virus's DNA—not protein—that enters the bacterium to cause infection.

Their studies confirmed that DNA contained the virus's genetic information, which triggered viral replication within the bacteria. The experiments of Hershey and Chase convinced most scientists that DNA was the molecule of heredity, but many questions about the structure and mechanisms of DNA remained.

In the early 1950s researchers began to apply techniques of X-ray diffraction to learn about the basic structure of DNA. X-ray diffraction can determine molecular structures by measuring patterns of scattered X rays after they pass through a crystalline substance.

British physical chemist Rosalind Franklin and British biophysicist Maurice Wilkins used X-ray diffraction to obtain DNA images of unprecedented clarity. Yet the exact three-dimensional structure of DNA remained unclear.

The groundbreaking work of American biochemist James Watson and British biophysicist Francis Crick solved that mystery. In 1953 the two proposed a model of DNA that is still accepted today: A double helix molecule formed by two chains, each composed of alternating sugar and phosphate groups, connected by nitrogenous bases. Watson and Crick were awarded the 1962 Nobel Prize in physiology or medicine for their discoveries.

Watson and Crick speculated that the structure of DNA provided some obvious clues about how the molecule could replicate itself.

They proposed a replication model in which each strand of DNA serves as a template for making exact copies. This model of replication, called semi-conservative replication, was demonstrated in 1958 by American molecular biologists Matthew Meselson and Franklin Stahl.

Their experiments demonstrated the mechanisms of replication by tracking DNA containing a heavy nitrogen isotope through a series of replications. With DNA's structure and replication mechanisms largely solved, scientists turned their attention to identifying the genetic code—learning how a gene's nucleotide sequence determines what type of protein is made. In the late 1950s, South African geneticist Sydney Brenner and other scientists confirmed that RNA acted as an intermediary between DNA and protein production.

Researchers still were uncertain how the sequence of nucleotides in DNA corresponded to the production of specific amino acids. In 1961 Crick and Brenner determined that groups of three nucleotides, now known as codons, code for the 20 amino acids that form the foundation of proteins. The exact relationship between codons and amino acids was clarified after several important discoveries. American biochemists Marshall Nirenberg and J. Heinrich Matthaei synthesized repeated nucleotide sequences that led to the production of repeated single amino acids.

They identified how certain codon combinations code for a specific amino acid. A process developed by American geneticist Har Gobind Khorana helped scientists create a "dictionary" of codons that defined specific amino acids, thus resolving the remaining ambiguities in the genetic code. Only

12 years after the structure of DNA was deduced, the genetic code was solved.

Learning to Manipulate DNA

After scientists had unraveled the structure and replication mechanisms of DNA, many felt that the major discoveries of genetic research were resolved. They predicted that the only task left in genetics was to sort out the molecular details of how genes work. But in the process of studying gene function, researchers developed powerful new molecular techniques, enabling them to analyse and manipulate genes with a speed and precision never before possible.

A number of discoveries made during the 1960s and 1970s shed light on how distinct fragments of DNA could be isolated. The work of Swiss molecular biologist Werner Arber focused on specialized enzymes that digest, or "restrict," the DNA of viruses infecting bacteria. These enzymes were subsequently dubbed restriction enzymes. In the following decade, scientists learned that restriction enzymes could also act like molecular scissors to cut DNA.

In 1970 American molecular biologist Hamilton Smith and colleagues determined that restriction enzymes could cleave DNA molecules at precise and predictable locations. Hamilton concluded that the enzymes were able to recognize specific nucleotide sequences. Scientists quickly realized that restriction enzymes could be used in the laboratory to manipulate DNA. In 1973 American biochemist Herb Boyer used restriction enzymes to produce a DNA molecule with genetic material from two different sources. This splicing technique is now known as recombinant DNA.

Boyer inserted foreign genes into plasmids and observed that the plasmids could replicate to make many copies of the inserted genes. In subsequent experiments, Boyer, American biochemist Stanley Cohen, and other researchers demonstrated that inserting a recombinant DNA molecule into a host bacteria cell would lead to extremely rapid replication and the production of many identical copies of the recombinant DNA.

This process, known as cloning, gave scientists the power

to make many copies of desired DNA for molecular study. The speed and efficiency of DNA cloning were vastly improved in the 1980s with the invention of polymerase chain reaction (PCR). Developed by American biochemist Kary Mullis, PCR enables scientists to produce large amounts of DNA sequences in a test tube.

In a matter of hours, the process can produce millions of cloned DNA molecules. Yet all of the advances in isolating and replicating DNA would not be possible or be of much use if researchers could not determine the nucleotide sequence of genetic material.

In the late 1970s and early 1980s, British biochemist Frederick Sanger and his associates developed DNA sequencing techniques. Sanger's methods, which used special compounds called dideoxy nucleotides, rapidly yielded the exact nucleotide sequence of a desired sample.

With the use of automated equipment, the new techniques transformed genetic sequencing into a speedy, routine laboratory procedure. Many of the new techniques for isolating, sequencing, and replicating DNA have been put to practical use through the field of genetic engineering.

The Human Genome Project and the new field of proteomics have both benefited from continuing technical advances and have accelerated the development of new genetic technologies. Modern genetics is poised to radically change the practice of medicine and the biotechnology industry.

Chapter 38

Tools of Genetic Engineering

Cutting and Measuring DNA

The first step in gene cloning is cutting DNA into appropriately sized pieces. The size of DNA molecules is measured by subjecting the DNA to a technique called "electrophoresis". An electrophoretic apparatus gives DNA an electrical shove (DNA is negatively charged because of its phosphate groups.

The electrophoresis apparatus puts the DNA in an electric field, and it moves toward the positive pole), forcing it to move through a porous gel. Two types of gels are commonly used.

Agarose Gels

The most popular variant of electrophoresis is carried out with agarose gels. In the laboratory, suspensions of agarose in buffer are heated to boiling (often in microwave ovens), and the hot solution is poured on plastic or glass plates to a height of a few millimeters or so. Agarose is a derivative of agar — an edible polysaccharide extracted from seaweed. When cooled from heated solutions, solutions of agarose form a nearly transparent gel that looks like a slab of gelatin dessert.

After the gel sets, DNA solutions are placed in little depressions formed by leaving a comb in the gel. The gel is then flooded with a weak salt solution, and a voltage is applied. Since DNA is negatively charged (due to its phosphate groups), it moves toward the positive pole. Large DNA molecules advance slowly in the gel because they are impeded by the gel matrix.

Smaller ones encounter fewer barriers and move more quickly. By plotting the distance migrated against the reciprocal of size (in base pairs) of a group of DNA standards, a fairly straight line can be obtained. From these data, the length of unknowns can be easily calculated. For ease of comparison, the standards are usually run in the same gel beside the unknowns.

Gels of differing porosities can be made by adjusting the concentration of agarose. With 2 and 3 per cent solutions, double-stranded DNA's as small as 50 or 100 base pairs can be resolved. More dilute gels can resolve fragments as large as about 30,000 base pairs (30 kilobase pairs or 30kb).

Polyacrylamide Gels

When very small DNA molecules need to be analyzed or when high-resolution separations are required, electrophoresis is carried out in polyacrylamide gels. They, like the gels made from agarose, are water white. Unlike agarose, polyacrylamide gels are polymers of acrylamide, a small synthetic organic compound. Polyacrylamide gel electrophoresis (PAGE) is most often used to separate fragments of DNA that are between 6 and a few thousand base pairs in length. The technique is used mostly in DNA sequencing, a topic that we'll come back to later in the course.

Detecting DNA

DNA doesn't have any colour; it only absorbs light in the ultraviolet range. How is it visualized in a clear gel after electrophoresis? DNA can be detected in both agarose and polyacrylamide gels after staining with various dyes, incuding ethidium bromide, a dye that forms a fluorescent complex upon binding to DNA. Usually, the ethidium bromide is added to the gel before electrophoresis.

After the run, the gel is examined and often photographed under an ultraviolet light. Yellow-orange zones (bands) of fluorescence appear (similar to those in the picture at the right), indicating migration of discrete pieces of DNA. Other agents — like methylene blue or silver stains — can also be used to visualize DNA, but they are either less sensitive or less convient, and aren't used as often. By the way, many dyes that bind to DNA are mutagenic; i.e, they cause mutations.

Separating Large Fragments of DNA

Contrary to what we might expect, very large fragments of DNA readily move in agarose gels when an electrical field is applied. But they don't separate according to size. In fact, very big pieces of DNA all migrate at about the same velocity regardless of whether they are 20kb in length or 2000kb.

PFE

Since many techniques, particulary recently derived ones, depend on working with very large DNA fragments, even pieces the size of some chromosomes, several newly developed electrophoretic techniques have been developed. In one of these, called pulsed field electrophoresis (PFE), DNA molecules are analyzed on agarose gels but with a modified apparatus. Instead of having only two sets of electrodes situated at opposite ends of the electrophoresis device as in the picture shown above, the pulsed field apparatus bears four sets of electrodes.

Two are designated A^- and A^+ and the other two are called B^- and B^+. The two A's and the two B's are set at an angle of 120° with respect to one another. During electrophoresis, the DNA molecules are subjected to alternating bursts of current from the two A and B pairs, thereby pulling the DNA alternately to the right and to the left as it advances (from south to north in this case) through the gel.

The DNA seems to reorient itself each time the current is switched, and the time that it takes to turn itself around is dependent on its length. The result is that very large pieces of DNA can be separated from another on the basis of their size. The technique also allows one to estimate the length of an unknown when run beside standards of known size. One of the most impressive accomplishments of this technique is the separation of the 16 chromosomes of yeast in a single electrophoretic run.

FIE

Another procedure, field inversion electrophoresis, is also effective at separating large molecules of DNA. In it, the electrodes are set up as in ordinary electrophoresis, but the current is switched so that the DNA first moves forward and then backward. Of course, if the switching were to be done in equal time intervals — say, 1 second forward and 1 second back — the DNA wouldn't move at all.

To get it to move at all, electrophoresis is conducted forward for a longer time than backward (or at a greater voltage forward versus backward). One might expect that moving ahead three steps and back two would be equivalent to moving forward one step at a time, but that's not what happens. Apparently the DNA reorients itself during the switching cycles, just as it does in pulsed field electrophoresis. Again, the ability to change directions seems to be dependent on size. This allows different lengths of DNA to be separated from one another.

Restriction Endonucleases

One critically important advance that has greatly stimulated the rapid progress in molecular biology and genetic engineering was the discovery of a set of enzymes that are capable of cutting DNA at defined sequences. These enzymes are found in a variety of microorganisms and are called restriction endonucleases, or more simply, restriction enzymes.

The first specific restriction enzyme was discovered by Hamilton Smith in 1970, an accomplishment for which he (and two others) were awarded the Nobel Prize. Since then, some 3,059 similar enzymes have been reported, many of which have different specificities.

Nomenclature

Smith and another Nobel laureate, Daniel Nathans, devised a nomenclature for these enzymes. In brief, the name of each restriction enzyme derives from the organism from which it is isolated. The first letter of the genus name plus the first two letters of the species name form the first three letters of the restriction enzyme's name. If necessary, a letter indicating strain designation is added, and finally a number is appended that stands for the order in which the enzyme was discovered in each organism.

For example, BamHI is the name of an enzyme that is isolated from the bacterium *Bacillus amyloliquifaciens,* strain H, and it was, presumably, the first restriction endonuclease identified from that source. The first three letters of the name should be italicized, but because italicized letters are often hard to read on a computer.

What they Do

The restriction enzymes owe their usefulness to the fact that they bind to DNA at specific DNA sequences, four to eight nucleotides in size, called recognition sites. Once bound, restriction enzymes cut the DNA at or near this site. With a little thought, it should be clear that an enzyme that has a six base pair recognition site will, on the average, produce larger pieces of DNA than one that recognizes a four base site. Expressed quantitatively, the approximate size of the fragments produced by a particular enzyme, given that it is cutting a DNA containing an equal proportion of all four nucleotides, can be calculated from the formula:

Average size of fragment = 4^N

where N is the number of bases that the enzyme recognizes. Hence, a four-cutter (the shorthand name for an enzyme that recognizes a site containing four base pairs) is expected to cleave random DNA into fragments of about 4^4 (256) base pairs while an enzyme with a recognition site of six bases will produce pieces (on the average) of about 4^6 (4096) base pairs.

PROPERTIES OF RESTRICTION ENZYMES

Ends

Another interesting property of restriction enzymes is that while they often recognize a symmetrical site, they do not always cut at the axis of symmetry. For instance, the enzyme

EcoRI. Similarly, the enzyme BglII (from the microorganisms, *Bacillus globiggi* and universally and irreverently pronounced BAGEL TWO) recognizes the sequence AGATCT and cuts between the first A and G residues.

In fact, the four nucleotide single-stranded ends are the same for both BglII and BamHI. Moreover, there are at least two other six cutting enzymes that have been discovered that leave the same four nucleotide overhang: BclI and XhoII. These overhanging ends are very useful because — under the proper conditions — they may base pair with each other. In fact, because of their affinity for one another, they are often called cohesive or sticky ends.

Moreover, if molecules with these ends are treated with the appropriate enzyme — DNA ligase — their phosphodiester bonds may be rejoined (ligated). When two ends that originate from digestion by a single enzyme are ligated, the resulting molecule can be cut by the same enzyme again. But if the ends of a DNA molecule that originated with a BamHI cut and a BglII cut are joined together, the new sequence will not be cut with either enzyme. All restriction endonucleases do not generate 5' single-strand overhangs. In fact, some don't even produce an overhang at all. Several enzymes — like SacI — produce 3' single-stranded sticky ends.

And some enzymes — like PvuII — cut at the axis of symmetry, leaving perfectly aligned ends. DNA molecules without overhangs are said to have blunt ends.In addition there are restriction enzymes that cleave DNA some distance away from the sequence that they recognize. For example the enzyme HgaI makes staggered cuts that lie 5 and 10 nucleotides away from a 5 base pair sequence, GACGC. This leaves 5' overhanging ends, but, in contrast to the enzymes described above, these will be different almost every time the enzyme cuts.

The Utility of the Restriction Enzymes

The discovery of these many restriction endonucleases have allowed genetic engineers to cut pieces of DNA at specific sites and into defined sizes. The result has been that a scientist can work with a collection of molecules all of the same size and with ends of known sequence. Restriction enzymes have proved to be valuable analytical and diagnostic tools as well.

RESTRICTION ENZYMES

Restriction enzymes are DNA-cutting enzymes found in bacteria (and harvested from them for use). Because they cut within the molecule, they are often called restriction endonucleases. A restriction enzyme recognizes and cuts DNA only at a particular sequence of nucleotides. For example, the bacterium Hemophilus aegypticus produces an enzyme named HaeIII that cuts DNA wherever it encounters the sequence

5'GGCC3'
3'CCGG5'

The cut is made between the adjacent G and C. This particular sequence occurs at 11 places in the circular DNA molecule of the virus phiX174. Thus treatment of this DNA with the enzyme produces 11 fragments, each with a precise length and nucleotide sequence. These fragments can be separated from one another and the sequence of each determined.

HaeIII and AluI cut straight across the double helix producing "blunt" ends. However, many restriction enzymes cut in an offset fashion. The ends of the cut have an overhanging piece of single-stranded DNA. These are called "sticky ends" because they are able to form base pairs with any DNA molecule that contains the complementary sticky end. Any other source of DNA treated with the same enzyme will produce such molecules.

Mixed together, these molecules can join with each other by the base pairing between their sticky ends. The union can be made permanent by another enzyme, DNA ligase, that forms covalent bonds along the backbone of each strand. The result is a molecule of recombinant DNA (rDNA).

The ability to produce recombinant DNA molecules has not only revolutionized the study of genetics, but has laid the foundation for much of the biotechnology industry. The availability of human insulin (for diabetics), human factor VIII (for males with hemophilia A), and other proteins used in human therapy all were made possible by recombinant DNA.

RECOMBINANT DNA AND GENE CLONING

Recombinant DNA is DNA that has been created artificially. DNA from two or more sources is incorporated into a single recombinant molecule.

Making Recombinant DNA (rDNA)

- Treat DNA from both sources with the same restriction endonuclease (BamHI in this case).
- BamHI cuts the same site on both molecules
  5' GGATCC 3'
  3' CCTAGG 5'
- The ends of the cut have an overhanging piece of single-stranded DNA.
- These are called "sticky ends" because they are able to base pair with any DNA molecule containing the complementary sticky end.
- In this case, both DNA preparations have complementary sticky ends and thus can pair with each other when mixed.
- DNA ligase covalently links the two into a molecule of recombinant DNA.

To be useful, the recombinant molecule must be replicated many times to provide material for analysis, sequencing, etc. Producing many identical copies of the same recombinant molecule is called cloning. Cloning can be done in vitro, by a process called the polymerase chain reaction (PCR). Here, however, we shall examine how cloning is done in vivo.

Cloning in vivo can be done in

- Unicellular microbes like E. coli
- Unicellular eukaryotes like yeast and
- in mammalian cells grown in tissue culture.

In every case, the recombinant DNA must be taken up by the cell in a form in which it can be replicated and expressed. This is achieved by incorporating the DNA in a vector. A number of viruses (both bacterial and of mammalian cells) can serve as vectors. But here let us examine an example of cloning using E. coli as the host and a plasmid as the vector.

PLASMID

A plasmid is an extra-chromosomal DNA molecule separate from the chromosomal DNA which is capable of replicating independently of the chromosomal DNA. In many cases, it is circular and double-stranded. Plasmids usually occur naturally in bacteria, but are sometimes found in eukaryotic organisms (e.g., the 2-micrometre-ring in Saccharomyces cerevisiae). Plasmid size varies from 1 to over 200 kilobase pairs (kbp).

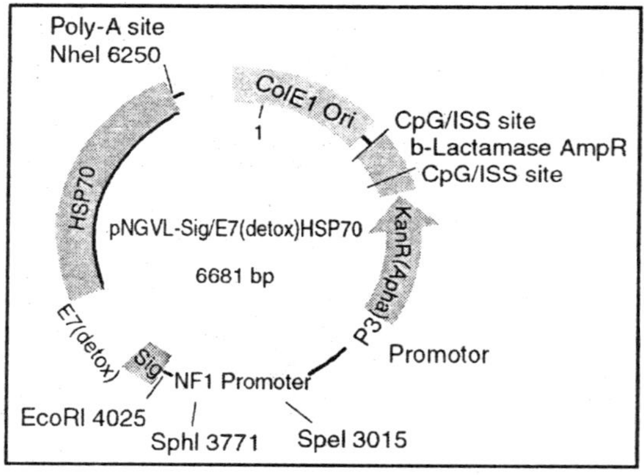

Fig. Plasmid

The number of identical plasmids within a single cell can be zero, one, or even thousands under some circumstances. Plasmids can be considered to be part of the mobilome, since they are often associated with conjugation, a mechanism of horizontal gene transfer.The term plasmid was first introduced by the American molecular biologist Joshua Lederberg in 1952.

Plasmids can be considered to be independent life-forms similar to viruses, since both are capable of autonomous replication in suitable (host) environments. However the plasmid-host relationship tends to be more symbiotic than parasitic (although this can also occur for viruses, for example with Endoviruses) since plasmids can endow their hosts with useful packages of DNA to assist mutual survival in times of severe stress. For example, plasmids can convey antibiotic resistance to host bacteria, who may then survive along with their life-saving guests who are carried along into future host generations.

Vectors

There are two types of plasmid integration into a host bacteria: Non-integrating plasmids replicate as in the top instance; whereas episomes, the lower example, integrate into the host chromosome.

Plasmids used in genetic engineering are called vectors. Plasmids serve as important tools in genetics and biotechnology labs, where they are commonly used to multiply (make many copies of) or express particular genes. Many plasmids are commercially available for such uses.

The gene to be replicated is inserted into copies of a plasmid containing genes that make cells resistant to particular antibiotics and a multiple cloning site (MCS, or polylinker), which is a short region containing several commonly used restriction sites allowing the easy insertion of DNA fragments at this location.

Next, the plasmids are inserted into bacteria by a process called transformation. Then, the bacteria are exposed to the particular antibiotics. Only bacteria which take up copies of the plasmid survive the antibiotic, since the plasmid makes them resistant. In particular, the protecting genes are expressed (used to make a protein) and the expressed protein breaks down the antibiotics.

In this way the antibiotics act as a filter to select only the modified bacteria. Now these bacteria can be grown in large amounts, harvested and lysed (often using the alkaline lysis method) to isolate the plasmid of interest.

Another major use of plasmids is to make large amounts of proteins. In this case, researchers grow bacteria containing a plasmid harboring the gene of interest. Just as the bacteria produces proteins to confer its antibiotic resistance, it can also be induced to produce large amounts of proteins from the inserted gene.

This is a cheap and easy way of mass-producing a gene or the protein it then codes for, for example, insulin or even antibiotics. However, a plasmid can only contain inserts of about 1-10 kbp. To clone longer lengths of DNA, lambda phage with lysogeny genes deleted, cosmids, bacterial artificial chromosomes or yeast artificial chromosomes could be used.

TYPES

One way of grouping plasmids is by their ability to transfer to other bacteria. Conjugative plasmids contain so-called tra-genes, which perform the complex process of conjugation, the transfer of plasmids to another bacterium. Non-conjugative plasmids are incapable of initiating conjugation, hence they can only be transferred with the assistance of conjugative plasmids, by 'accident'.

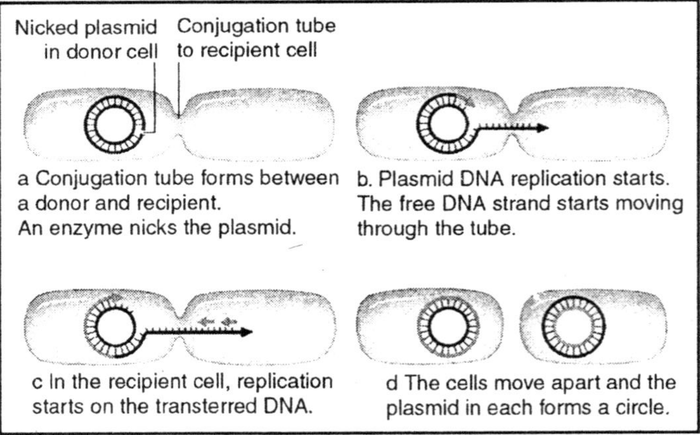

Fig. Bacterial Conjugation

An intermediate class of plasmids are mobilizable, and carry only a subset of the genes required for transfer. They can 'parasitize' a conjugative plasmid, transferring at high frequency only in its presence. Plasmids are now being used to manipulate DNA and may possibly be a tool for curing many diseases.

It is possible for plasmids of different types to coexist in a single cell. Seven different plasmids have been found in E. coli. But related plasmids are often incompatible, in the sense that only one of them survives in the cell line, due to the regulation of vital plasmid functions.

Therefore, plasmids can be assigned into compatibility groups. Another way to classify plasmids is by function. There are five main classes:

- Fertility-F-plasmids, which contain tra-genes. They are capable of conjugation.
- Resistance-(R)plasmids, which contain genes that can build a resistance against antibiotics or poisons. Historically known as R-factors, before the nature of plasmids was understood.
- Col-plasmids, which contain genes that code for (determine the production of) bacteriocins, proteins that can kill other bacteria.
- Degradative plasmids, which enable the digestion of unusual substances, e.g., toluene or salicylic acid.
- Virulence plasmids, which turn the bacterium into a pathogen.

Plasmids can belong to more than one of these functional groups. Plasmids that exist only as one or a few copies in each bacterium are, upon cell division, in danger of being lost in one of the segregating bacteria. Such single-copy plasmids have systems which attempt to actively distribute a copy to both daughter cells.

Some plasmids include an addiction system or "postsegregational killing system (PSK)", such as the hok/sok (host killing/suppressor of killing) system of plasmid R1 in Escherichia coli. They produce both a long-lived poison and a short-lived antidote. Daughter cells that retain a copy of the plasmid survive, while a daughter cell that fails to inherit the plasmid dies or suffers a reduced growth-rate because of the lingering poison from the parent cell.

Plasmid DNA Extraction

As alluded to above, plasmids are often used to purify a

specific sequence, since they can easily be purified away from the rest of the genome. For their use as vectors, and for molecular cloning, plasmids often need to be isolated.

There are several methods to isolate plasmid DNA from bacteria, the archetypes of which are the miniprep and the maxiprep/bulkprep. The former can be used to quickly find out whether the plasmid is correct in any of several bacterial clones. The yield is a small amount of impure plasmid DNA, which is sufficient for analysis by restriction digest and for some cloning techniques.

In the latter, much larger volumes of bacterial suspension are grown from which a maxi-prep can be performed. Essentially this is a scaled-up miniprep followed by additional purification. This results in relatively large amounts (several micrograms) of very pure plasmid DNA

DNA LIGASES

DNA ligases are vital enzymes required for important cellular processes such as DNA replication, repair of damaged DNA and recombination. The enzyme mediates the formation of phosphodiester bonds between adjacent 3'-OH and 5'-phosphate termini, thereby joining the nicks in double stranded DNA.

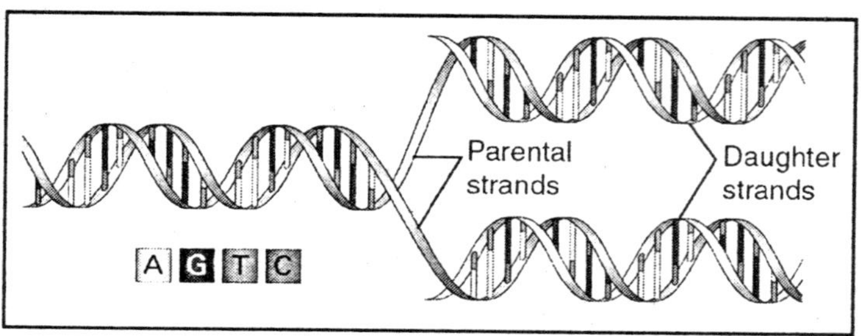

Fig. DNA Ligases

Ligases can be classified into two groups depending on their requirement for ATP or NAD+ as the cofactor. All eukaryotic and virally encoded enzymes are ATP-dependent, whereas most prokaryotic enzymes require NAD^+ for their activity.

ATP-Dependent DNA Ligase

DNA ligase from bacteriophage T7 is a monomer with a molecular weight of 41 kDa. Crystal structures of the apo form of this enzyme and its complex with ATP have been determined at 2.7 Å and 2.6 Å respectively.

The structure consists of two distinct domains, a larger N-terminal domain and a C-terminal domain. The ATP-binding site is situated in the N-terminal domain in a pocket beneath one of the beta-sheets. This pocket is lined by a number of motifs that are conserved across a wide family of nucleotidyltransferases.

The structure of the C-terminal domain is remarkably similar to the oligonucleotide binding fold, observed in a number of proteins. The DNA-binding site is proposed to be in a groove running between the two domains. The structure has provided the basis for biochemical experiments. We have made the two domains separately and looked at their biochemical properties. The larger N-terminal domain is an active ligase but with greatly reduced activity. Both domains are able to bind to DNA. The N-terminal domain binds both single and double stranded DNA, while the smaller C-terminal domain is only able to bind double stranded DNA despite having a fold that is similar to single strand DNA binding proteins.

The affinity for nicked DNA comes from a combination of these two DNA-binding affinities at the active site. Interestingly, the N-terminal domain is poor at the adenylation reaction, but this activity is stimulated greatly by addition of the C-terminal domain. Furthermore, we can demonstrate an association of the domains on gel filtration. These data suggest that a conformational change occurs during the adenylation reaction. This would be similar to that observed directly in our crystal structure of an mRNA capping enzyme, a group of enzymes that are closely related to ligases.

NAD-Dependent DNA Ligase

An NAD-dependent DNA ligase from B.stearothermophilus. The enzyme has been overexpressed and biochemical studies are in progress. Limited proteolysis has been used to investigate the domain structure of the enzyme. There are two domains, with the larger N-terminal domain retaining full self adenylation activity and the smaller C-terminal domain having all of the DNA-binding activity of the full length enzyme.

The situation is therefore different to that we found in the ATP-dependent enzyme, where both domains contribute to the activities of the enzyme. Structural studies are underway with these domains. The structure of the large N-terminal domain has been solved and work is in progress to determine the structure of the small domain.

RIBONUCLEASE A

Ribonuclease A (RNase A) is an endonuclease that cleaves single-stranded RNA. Bovine pancreatic RNase A is one of the classic model systems of protein science. The importance of bovine pancreatic RNase A was secured when the Armour purified a *kilogram* of it, and gave 10 mg samples away free to any interested scientists. The ability to have a single lot of purified enzyme instantly made RNase *the* model system for protein studies.

RNase A was the model protein used to work out many spectroscopic methods for assaying protein structure, including absorbance, circular dichroism/optical rotary dispersion, Raman, EPR and NMR spectroscopy. RNase A was also the first model protein for the development of several chemical structural methods, such as limited proteolysis of disordered segments, chemical modification of exposed side chains, and antigenic recognition.

Ribonuclease-S, which is RNase A that has been treated with subtilisin, was the third protein to have its structure solved, in 1967.Studies of the oxidative folding of RNase A led Chris Anfinsen to enunciate the thermodynamic hypothesis of protein folding, which states that the folded form of a protein represents the minimum of its free energy.

RNase A was the first protein for showing the effects of non-native isomers of X-Pro peptide bonds in protein folding.RNase A was the first protein to be studied by multiple sequence alignment and by comparing the properties of evolutionarily related proteins.

STRUCTURE AND PROPERTIES

Labeled ribbon diagram of bovine pancreatic ribonuclease A (PDB accesion code 7RSA). The backbone ribbon is colored from blue (N-terminus) to red (C-terminus). The side chains of the four disulfide-bonded cysteines are shown in yellow, with their sulfur atoms highlighted as small spheres. Residues important for catalysis are shown in magenta.

RNase A is a relatively small protein (124 residues, ~13.7 kDa). It can be characterized as a two-layer á + â protein that is folded in half to resemble a taco, with a deep cleft for binding the RNA substrate. The first layer is composed of three alpha helices from the N-terminal half of the protein. The second layer consist of three â-hairpins arranged in two a-sheets. The hairpins 61-74 and 105-124 form a four-stranded, antiparallel â-sheet that lies on helix 3 (residues 50-60). The longest â-hairpin 79-104 mates with a short â-strand (residues 42-45) to form a three-stranded, antiparallel â-sheet that lies on helix 2 (residues 24-34).

RNase A has four disulfide bonds in its native state: Cys26-Cys84, Cys58-110, Cys40-95 and Cys65-72. The first two (26-84 and 58-110) are essential for conformational folding; each joins an alpha helix of the first layer to a beta sheet of the second layer, forming a small hydrophobic core in its vicinity. The latter two disulfide bonds (40-95 and 65-72) are less essential for folding; either one can be reduced (but not both) without affecting the native structure under physiological conditions. These disulfide bonds connect loop segments and are relatively exposed to solvent. Interestingly, the 65-72 disulfide bond has an extraordinarily high propensity to form, significantly more than would be expected from its loop entropy, both as a peptide and in the full-length protein. This suggests that the 61-74 â-hairpin has a high propensity to fold conformationally.

RNase A is a basic protein (pI =8.63); its many positive charges are consistent with its binding to RNA (a poly-anion).

More generally, RNase A is unusually polar or, rather, unusually lacking in hydrophobic groups, especially aliphatic ones. This may account for its need of four disulfide bonds to stabilize its structure. The low hydrophobic content may also serve to reduce the physical repulsion between highly charged groups (its own and those of its substrate RNA) and regions of low dielectric constant (the nonpolar residues).

The N-terminal a-helix of RNase A (residues 3-13) is connected to the rest of RNase A by a flexible linker (residues 16-23). As shown by F. M. Richards, this linker may be cleaved by subtilisin between residues 20 and 21 without causing the N-terminal helix to dissociate from the rest of RNase A. The peptide-protein complex is called RNase S, the peptide (residues 1-20) is called the S-peptide and the remainder (residues 21-124) is called the S-protein.

The dissociation constant of the S-peptide for the S-protein is roughly 30 pM; this tight binding can be exploited for protein purification by attaching the S-peptide to the protein of interest and passing a mixture over an affinity column with bound S-protein. The RNase S model system has also been used for studying protein folding by coupling folding and association. The S-peptide was the first peptide from a native protein shown to have (flickering) secondary structure in isolation.

Enzymatic Mechanism

The positive charges of RNase A lie mainly in a deep cleft between two lobes. The RNA substrate lies in this cleft and is cleaved by two catalytic histidines, His12 and His119, via a 2'–3' cyclic phosphate intermediate that is stabilized by nearby lysines such as Lys7. Lys41 and Lys66.

Anti-Cancer Effects

RNase A, and to a greater extent its oligomers and some homologs (such as onconase from frogs), have cytotoxic and cytostatic effects, particularly on cancer cells. This has led to the development of onconase as a cancer therapeutic, particularly for external use against skin cancers. As with many protein drugs, the internal use of non-human ribonucleases such as onconase is limited by the patient's immune response.

Chapter 39

Polymerase Chain Reaction

The polymerase chain reaction (PCR) is a technique widely used in molecular biology. It derives its name from one of its key components, a DNA polymerase used to amplify a piece of DNA by *in vitro* enzymatic replication. As PCR progresses, the DNA thus generated is itself used as template for replication. This sets in motion a chain reaction in which the DNA template is exponentially amplified. With PCR it is possible to amplify a single or few copies of a piece of DNA across several orders of magnitude, generating millions or more copies of the DNA piece.

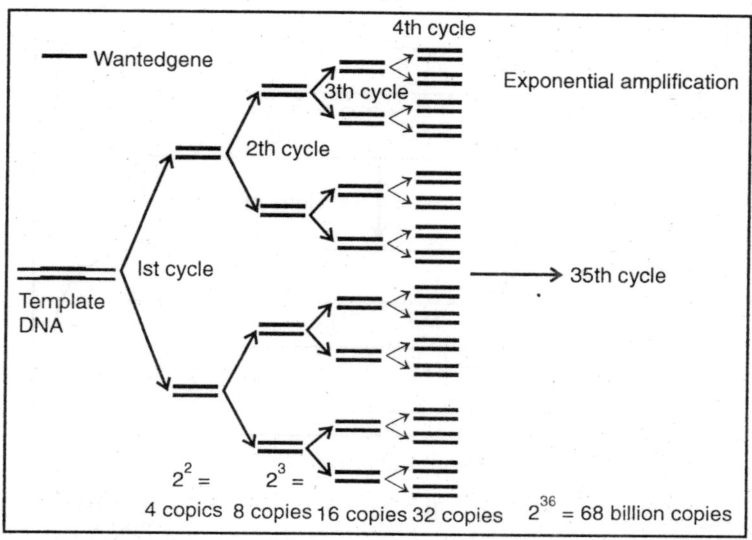

Fig. PCR Technique

PCR can be extensively modified to perform a wide array of genetic manipulations.Almost all PCR applications employ a heat-stable DNA polymerase, such as Taq polymerase, an enzyme originally isolated from the bacterium *Thermus aquaticus*. This DNA polymerase enzymatically assembles a new DNA strand from DNA building blocks, the nucleotides, using single-stranded DNA as template and DNA oligonucleotides (also called DNA primers) required for initiation of DNA synthesis.

The vast majority of PCR methods use thermal cycling, i.e., alternately heating and cooling the PCR sample to a defined series of temperature steps.

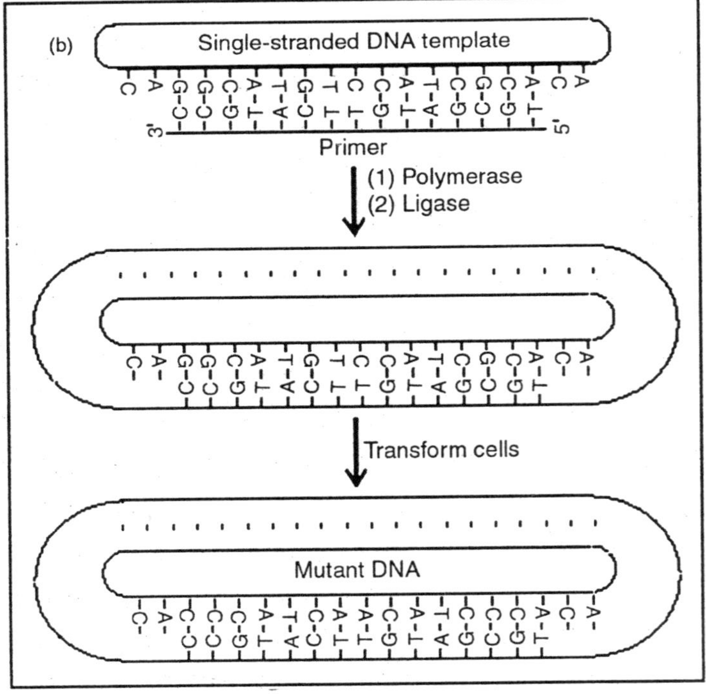

Fig. Oligonucleotide Primer

These thermal cycling steps are necessary to physically separate the strands (at high temperatures) in a DNA double helix (DNA melting) used as template during DNA synthesis (at lower temperatures) by the DNA polymerase to selectively amplify the target DNA. The selectivity of PCR results from the use of primers that are complementary to the DNA region targeted for amplification under specific thermal cycling conditions.

Developed in 1983 by Kary Mullis, PCR is now a common and often indispensable technique used in medical and biological research labs for a variety of applications. These include DNA cloning for sequencing, DNA-based phylogeny, or functional analysis of genes; the diagnosis of hereditary diseases; the identification of genetic fingerprints (used in forensics and paternity testing); and the detection and diagnosis of infectious diseases.

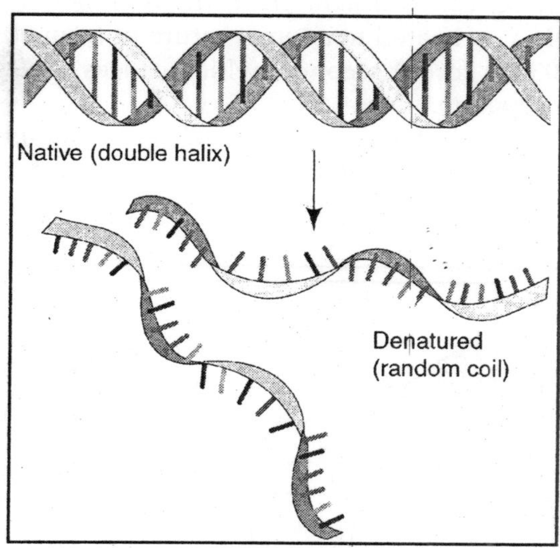

Fig. Denaturation

In 1993 Mullis won the Nobel Prize in Chemistry for his work on PCR. How PCR Works The polymerase chain reaction mimics the DNA replication, or reproduction, process that occurs naturally in living cells.

Most DNA is double-stranded—that is, each strand of DNA is paired with a complementary strand. During replication, the two strands of DNA separate and a specialized cell *enzyme* (a protein that initiates chemical reactions) called polymerase makes a copy of each strand, using the original strand as a *template,* or pattern.

Normally this copying occurs when cells divide and results in the production of one pair of daughter strands for each of the two parent strands. Polymerase requires two additional ingredients to copy DNA. The first is a supply of the four basic building blocks of DNA, called nucleotide bases. The second is a short stretch of copied DNA, called an oligonucleotide primer, consisting of several nucleotides that initiate replication. PCR uses these same ingredients to copy DNA in a vial.

There are three phases in a polymerase chain reaction. In the first phase, called *denaturation,* the template, or piece of original DNA, is heated to a temperature of from 90º to 95º C (194º to 203º F) for 30 seconds, which causes the individual strands to separate.

In the second phase, called *annealing,* the temperature of the mixture is lowered to 55º C (131º F) over a 20-second period, allowing the oligonucleotide primers to bind to the separated DNA.

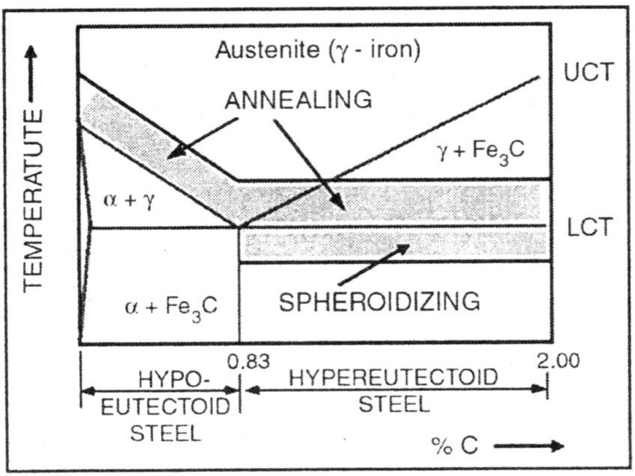

Fig. Annealing

In the third phase, called *polymerization,* the temperature of the mixture is raised to 75º C (167º F), a temperature at which the polymerase can copy the DNA molecule rapidly. These three phases are carried out in the same vial and make up one complete PCR cycle, which takes less than two minutes to complete. Theoretically, the PCR cycle can be repeated indefinitely, but the polymerase, nucleotides, and primers are usually renewed after about 30 cycles. Thirty PCR cycles can produce 1 billion DNA copies in less than three hours.

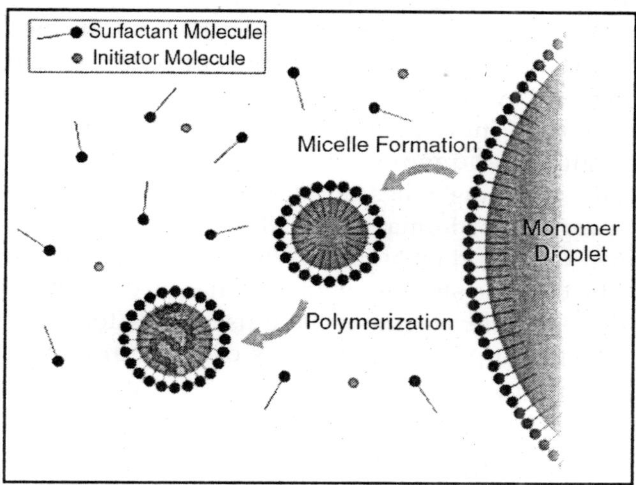

Fig. Polymerization

The polymerase used in early PCR procedures was easily destroyed by heat. Consequently, additional polymerase had to be added in each PCR cycle to replace that destroyed by the high temperatures of the first phase. In modern PCR procedures, however, the heat-stable *Taq* polymerase is usually used. It was originally extracted from *Thermus aquaticus,* a heat-loving bacterium found in the hot springs of Yellowstone National Park.

Because *Taq* polymerase is not destroyed by the high temperatures of PCR, it is only necessary to add it once, at the beginning of the reaction.

Taq polymerase is now produced for PCR by genetically engineered bacteria. The use of PCR requires great care. The chief concern is contamination of the reaction mix. PCR is so sensitive that it is possible to accidentally multiply minute amounts of contaminating DNA. Special procedures are used to ensure that such contamination is avoided.

APPLICATIONS

Because only minute amounts of relatively crude DNA samples are required for PCR, it is a valuable tool for research in biology, clinical medicine, and *forensic science* (the application of scientific investigation to law). In biological research, PCR has accelerated the study of gene function, gene mapping, and evolution. Gene-function research uses PCR to create copies of individual genes, the activities of which can then be studied and more precisely defined.

Gene mapping relies on PCR to create many copies of specific regions of human DNA. These regions can then be examined to see if they are linked to genetic diseases, such as cystic fibrosis. The study of evolution has also benefited from PCR. For example, scientists have used PCR to study DNA from insects trapped for thousands of years in amber. Even though much of the DNA has lost its structure, there is still enough intact DNA remaining to be multiplied by PCR for comparison with the DNA of present-day insects.

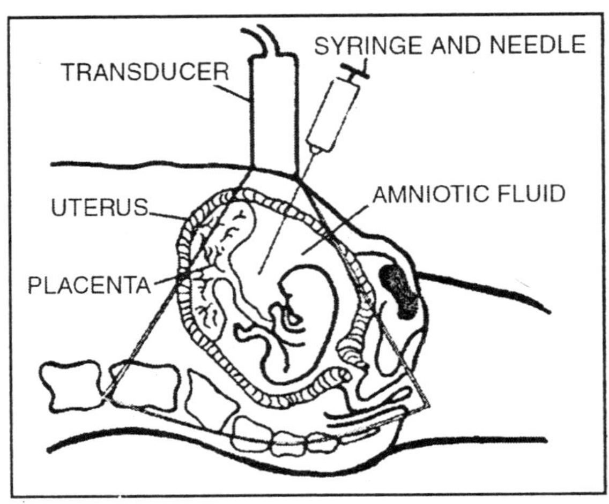

Fig. Amniocentesis

In medicine, PCR is particularly useful in prenatal testing

for genetic diseases. DNA samples obtained from a fetus by *amniocentesis,* in which a small sample of fluid is drawn from the mother's uterus, can be tested by PCR in just a few hours.

Previously, it was necessary to culture fetal cells, or grow them in a special nutrient medium, for several weeks before biochemical tests could be performed. PCR testing has also been used on cells taken from hours-old embryos fertilized *in vitro* (in a test tube) to identify an embryo that was free of disease. The embryo was then implanted in the mother's uterus, and a normal pregnancy resulted. Other medical applications of PCR include identifying viruses, bacteria, and cancerous cells in human tissues.

PCR can even be used within single cells, in a procedure called *in situ* (in-site) PCR, to identify specific cell types. In forensic science, PCR has revolutionized the process of criminal identification. PCR-based DNA-typing tests can create detailed DNA "fingerprints" that can definitively identify individuals. Such tests can also exclude or implicate suspects based on small amounts of blood, skin, hair, or semen left at a crime scene.

PCR has been used to trace industrial waste and other products. Small amounts of known DNA are inserted into batches of explosives, petroleum products, poisons, and other waste at their source to create a tag that will last indefinitely. Such DNA tags can be recovered from oil slicks or other pollutants found in public waterways for example, and then multiplied by PCR and compared with a listing of manufacturers' DNA tags. A match can provide strong evidence implicating polluters.

Human Health and the Human Genome Project

PCR has very quickly become an essential tool for improving human health and human life. Medical research and clinical medicine are profiting from PCR mainly in two areas: detection of infectious disease organisms, and detection of variations and mutations in genes, especially human genes. Because PCR can amplify unimaginably tiny amounts of DNA, even that from just one cell, physicians and researchers can examine a single sperm, or track down the elusive source of a puzzling infection. These PCR-based analyses are proving to be just as reliable as previous methods-sometimes more so-and often much faster and cheaper.

The method is especially useful for searching out disease organisms that are difficult or impossible to culture, such as many kinds of bacteria, fungi, and viruses, because it can generate analyzable quantities of the organism's genetic material for identification. It can, for example, detect the AIDS virus sooner during the first few weeks after infection than the standard ELISA test. PCR looks directly for the virus's unique DNA, instead of the method employed by the standard test, which looks for indirect evidence that the virus is present by searching for antibodies the body has made against it.

PCR can also be more accurate than standard tests. It is making a difference, for example, in a painful, serious, and often stubborn misfortune of childhood, the middle ear infection known as otitis media. The technique has detected bacterial DNA in children's middle ear fluid, signaling an active infection even when culture methods failed to detect it. Lyme disease, the painful joint inflammation caused by bacteria transmitted through tick bites, is usually diagnosed on the basis of symptom patterns. But PCR can zero in on the disease organism's DNA contained in joint fluid, permitting speedy treatment that can prevent serious complications.

PCR is the most sensitive and specific test for *Helicobacter pylori,* the disease organism now known to cause almost all stomach ulcers. Unlike previous tests, PCR can detect three different sexually transmitted disease organisms on a single swab (herpes, papillomaviruses, and chlamydia) and can even distinguish the particular strain of papillomavirus that predisposes to cancer, which other tests cannot do.

In short, if a disorder is caused by an infectious agent, PCR can, in principle, ferret out the culprit. More than 60 PCR protocols for identifying pathogens have been described to date, and at least 10 clinical products are available for detecting the evasive organisms that cause such diseases as tuberculosis, chlamydia, viral meningitis, viral hepatitis, AIDS, and cytomegalovirus.

Because PCR can easily distinguish among the tiny variations in DNA that each of us possess and that make each of us genetically unique, the method is also leading to new kinds of genetic testing. These tests diagnose not only people with inherited disorders, but also people who carry deleterious variations, known as mutations, that could be passed to their children. (These carriers are usually not themselves affected by the mutant gene, which they can lead to disease in the next generation.) Research is expected eventually to yield predictive tests: methods for finding out who is predisposed to common disorders we do not customarily consider genetic, such as heart disease, and the cancers that can arise in adulthood via mutations in body cells. This knowledge will help us take steps to prevent those diseases, which are the chief killers in the developed world.

With PCR analysis of cells shed into feces, for example, doctors have already demonstrated premalignant changes in the gastrointestinal tract, such as mutations in genes that protect against tumors. This can help them select high-risk candidates for colon cancer tests. Researchers have also detected potentially metastatic cells in the circulation of patients with newly diagnosed tumors.

PCR can provide enormous peace of mind to people who are trying to have children- for example, by reassuring anxious parents-to-be that they run no risk of having a child with a particular genetic disease. The technique even saves the lives of babies before they are born: doctors have used it for examining fetal DNA to learn whether the blood groups of mother and fetus are incompatible.

This condition often leads to severe disability and even death of the fetus, but can be treated successfully in the womb with enough advance warning-thanks to PCR. This process is also a direct way of distinguishing among the confusion of different mutations in a single gene, each of which can lead to a disorder such as Duchenne muscular dystrophy. It helps doctors track the presence or absence of DNA abnormalities characteristic of particular cancers, so that they can start and stop drug treatments and radiation therapy as soon as possible.

And it promises to greatly improve the genetic matching of donors and recipients for bone marrow transplantation.

PCR can even diagnose the diseases of the past. Former vice president and presidential candidate Hubert H. Humphrey underwent tests for bladder cancer in 1967. Although the tests were negative, he died of the disease in 1978. In 1994, researchers compared a 1976 tissue sample from his cancer-ridden bladder with his 1967 urine sample.

With the help of PCR amplification of the small amount of DNA in the 27-year-old urine, they found identical mutations in the p53 gene, well-known for suppressing tumors, in both samples. "Humphrey's examination in 1967 may have revealed the cancerous growth if the techniques of molecular biology were as well understood then as they have become," the researchers said.

Historical medical genetics has gone even further back in time with PCR. After the colour-blind British chemist John Dalton died in 1844, some tissue from his eyes was preserved. Dalton had asked for a posthumous investigation of the reason why he confused scarlet with green and pink with blue. A recent examination of DNA taken from that tissue, carefully amplified by PCR, has shown that Dalton lacked a gene for making one of the three photopigments essential for normal colour vision.

Many of the new genetic tests are the result of the Human Genome Project, the huge international effort to identify and study all human genes. Scientists expect the Human Genome Project to be finished shortly after the turn of the century. It is moving more rapidly than originally expected toward its ultimate goal, which is to sequence all the DNA in typical human cells. ("Sequence" means to determine the precise order of the four different nucleotides that make up any strand of DNA.) DNA sequencing reveals crucial variations in the nucleotides that constitute genes.

These mutational changes produce disease and even death by forcing the genes to produce abnormal proteins, or sometimes no proteins at all. DNA sequencing involves first isolating and duplicating DNA segments for nucleotide analysis. Thus PCR is an essential tool for the Human Genome

Project because it can quickly and easily generate an unlimited amount of any piece of DNA for this kind of study.

PCR and the Law

The technique's unparallelled ability to identify and copy the tiniest amounts of even old and damaged DNA has proved exceptionally valuable in the law, especially the criminal law. PCR is an indispensable adjunct to forensic DNA typing-commonly called DNA fingerprinting.

To type DNA, for example DNA extracted from blood found on a murder suspect's clothes, scientists study a handful of sites on the DNA where variation among individuals is typical. This helps them determine the likelihood that the sample matches the DNA of a specific person, for example a stabbing victim. Although in its early days DNA typing was controversial, laboratory standards have been established, and carefully done DNA typing is now accepted as strong evidence in courts throughout the world. Defendants' attorneys continue to argue about the population frequencies of certain variant stretches of DNA, but a recent major scientific commentary concluded, "the DNA fingerprinting controversy has been resolved."

DNA typing is only one of many pieces of evidence that can lead to a conviction, but it has proved invaluable in demonstrating innocence. Dozens of such cases have involved people who have spent years in jail for crimes they did not commit. One example is Kirk Bloodsworth. The Maryland waterman was wrongly imprisoned for almost nine years for the rape and murder of a 9-year-old girl, but was freed in 1993 with the aid of PCR. Even when evidence such as semen and blood stains is years old, PCR can make unlimited copies of the tiny amounts of DNA remaining in the stains for typing, as it did in Bloodsworth's case.

Ancient DNA

Archaeologists have happily seized on PCR and are applying it in an amazing variety of ways. It is helping, for example, the colorful and controversial story of the 2000-year-old Dead Sea Scrolls, which are written on parchment made out of skins from goats and gazelles. Researchers are analyzing the parchment fragments to try to identify individual animals they came from. The hope is that the genetic information will guide them in piecing together the 10,000 particles of scrolls that remain. PCR is also helping sort out relationships among vanished human groups, and tracing human migrations.

Studies on human brains that survived 8,000 years in a Florida sinkhole more or less intact indicate, for example, that the people who lived there were genetically different from today's Native Americans. Archaeologists are finding that PCR can illuminate human cultural practices as well as human biology. Analyzing pigments from 4000-year-old rock paintings in Texas, they found one of the components to be DNA, probably from bison. The animals did not live near the Pecos River at that time, so the paleo-artists must have gone to some effort to obtain such an unusual ingredient for their paint.

Taking so much trouble suggests that the paintings were not simply decorations, but had religious or magical significance. PCR can faithfully copy bits of DNA whose age numbers in the thousands-some say millions-of years. Indeed, PCR's special strengths may be best revealed in the domain that has come to be known as Ancient DNA, where minuscule amounts of archaic, badly damaged genetic material are the norm. Ancient DNA studies generally fall under either the traditional concerns of archaeology, or of evolutionary biology-even the biology of organisms that disappeared long ago.

Scientists have used PCR to correct errors in a previous analysis of DNA from the 140-year-old skin of the last quagga, an African member of the horse family. The new genetic analysis has shown that the quagga was more closely related to the zebra than to any other horselike creatures. By amplifying and analyzing DNA from bone and mummified soft tissue, scientists have also found that moas, a group of large New Zealand birds that were hunted to extinction, are not related to the still-extant New Zealand kiwi, despite the fact that both bird species could not fly.

Leaping far back in time, researchers have suggested,

however, that termites imprisoned in amber 40 million years ago differed little from the termites of today.

Ecological Studies, and Animal Behaviour

But DNA need not be ancient to provide information about evolutionary relationships. With PCR, systematists can measure differences in DNA sequences between species directly, and if they select sequences that have changed little during evolution, between major classes of organisms. The speed and automation of the process means that scientists can easily compare dozens or even hundreds of individuals, putting their conclusions on a firmer basis. With PCR, scientists can glean genetic information from the faintest traces of the shyest, rarest animal-urine, feces, scent marks, infinitesimal bits of hair or skin rubbed onto a tree as the elusive creature passes by.

In addition to information that aids classification, individuals can be identified so as to estimate population size in a particular locale, or to determine the geographic range of a single animal, or a group of them. The technique can be adapted to similar studies of plants, for analysis of patterns of seed dispersal and the relative reproductive success of specific plants. Researchers have even used PCR to study badly damaged specimens such as roadkill, or the leavings of carnivores, where little-known vertebrates have been identified among the prey. Because PCR does not require invasive samples of blood or other tissue, research need not disrupt an animal's lifestyle-a boon for behavioral studies-and should not distress people concerned about animals.

DNA extracted from feces, for example, is being explored to find out which of the approaches to mating common among olive baboons work best, by establishing which males actually are successful at fathering infants. Researchers have used the technique to aid in reducing illegal trade in endangered species, and products made from them. Because PCR is a relatively low-cost and portable technology, and likely to become more so, it is adaptable for field studies of all kinds in the developing countries. It is also a tool for monitoring the release of genetically engineered organisms into the environment.

Chapter 40

Future of PCR

The present technology for doing PCR, about the size of a microwave oven and costing several thousand dollars, seems destined for further radical improvement. By tinkering with variables such as chemical reagents and pH, researchers have already reported success at copying larger and larger pieces of DNA, including the entire genome of HIV. Extraordinary miniaturization of the hardware is also underway, as experimenters squeeze PCR onto chip-sized devices.

Crisscrossed with the tiniest of troughs to hold the reagents and the DNA, the chips are heated electrically and cool down much faster than the present generation of machines, so amplification is even speedier than today's swift process. Already researchers have reported using a handheld battery-powered gadget to copy pieces of DNA that contained eight different cystic fibrosis mutation sites. While such experimental chip-based devices are not yet ready for prime time, they are hastening the day when scientists can take them on the road, and patients will be able to get on-the-spot readouts of their DNA.

Before long it may be quite routine to diagnose an infectious or genetic disorder, or even detect an inherited predisposition to cancer or heart disease, right in the doctor's office. PCR is doing for genetic material what the invention of the printing press did for written material-making copying easy, inexpensive, and accessible. In principle, PCR can reproduce the genetic material of any organism in essentially unlimited quantities, so it can be used to analyse any cells containing that material.

Whether they are germs, rare medicinal plants, or human beings, eventually we can know whatever is recorded in their DNA. With simple organisms, to know their DNA will be to know almost everything about them. With complicated ones, like people, DNA is only part of the story, but a very big part. Thanks to PCR, we will be probing the genetic past, and peering into the genetic future, for many years to come.

Nucleic Acid Hybridization

Nucleic acid hybridization is a fundamental tool in molecular genetics which takes advantage of the ability of individual single-stranded nucleic acid molecules to form double-stranded molecules (that is, to hybridize to each other). For this to happen, the interacting single-stranded molecules must have a sufficiently high degree of base complementarity.

Standard nucleic acid hybridization assays involve using a labeled nucleic acid probe to identify related DNA or RNA molecules (that is, ones with a significantly high degree of sequence similarity) within a complex mixture of unlabeled nucleic acid molecules, the target nucleic acid.

Preparation of Nucleic acid Probes

In standard nucleic acid hybridization assays the probe is labeled in some way. Nucleic acid probes may be made as single-stranded or double-stranded molecules but the working probe must be in the form of single strands.Conventional DNA probes are isolated by cell-based DNA cloning or by PCR. In the former case, the starting DNA may range in size from 0.1 kb to hundreds of kilobases in length and is usually (but not always) originally double-stranded.

PCR-derived DNA probes have often been less than 10 kb long and are usually, but not always, originally double-stranded. Conventional DNA probes are usually labeled by incorporating labeled dNTPs during an *in vitro* DNA synthesis reaction. RNA probes can conveniently be generated from DNA which has been cloned in a specialized plasmid vector. Such vectors normally contain a phage promoter sequence immediately adjacent to the multiple cloning site. An RNA synthesis reaction is employed using the relevant phage RNA polymerase and the four rNTPs, at least one of which is labeled. Specific labeled RNA transcripts can then be generated from the cloned insert.

Oligonucleotide probes are short (typically 15–50 nucleotides) single-stranded pieces of DNA made by chemical synthesis: mononucleotides are added, one at a time, to a starting mononucleotide, conventionally the 32 end nucleotide, which is bound to a solid support. Generally, oligonucleotide probes are designed with a specific sequence chosen in response to prior information about the target DNA.

Sometimes, however, oligonucleotide probes are used which are degenerate in sequence. Typically this involves parallel syntheses of a set of oligonucleotides which are identical at certain nucleotide positions but different at others. Oligonucleotide probes are often labeled by incorporating a ^{32}P atom or other labeled group at the 52 end.

DNA and RNA can conveniently be labeled in vitro by incorporation of nucleotides (or nucleotide components) containing a labeled atom or chemical group. Although, in principle, DNA and RNA can be labeled *in vivo*, by supplying labeled deoxynucleotides to tissue culture cells, this procedure is of limited general use; it has been restricted largely to preparing labeled viral DNA from virus-infected cells, and studying RNA processing events.

A much more versatile method involves *in vitro* labeling: the purified DNA, RNA or oligonucleotide is labeled *in vitro* by using a suitable enzyme to incorporate labeled nucleotides. Two major types of procedure have been widely used:

Labeling of new strands during*in vitro*DNA or RNA synthesis. In this type of procedure, DNA or RNA polymerase is used to make labeled DNA or RNA copies of a starting DNA. The *in vitro* DNA or RNA synthesis reaction requires that at least one of the four nucleotide precursors carries a labeled group. Labeling of DNA by *in vitro* DNA synthesis is normally accomplished using one of three methods: nick-translation (this section); random primed labeling (this section); or PCR-mediated labeling. Labeling of RNA generally is carried out using an *in vitro* transcription system. End-labeling. This type of procedure involves addition of a labeled group to one or a few terminal nucleotides.

It is less widely used, but is useful for a number of procedures, including labeling of single-stranded oligonucleotides and restriction mapping. Inevitably, because only one or a very few labeled groups are incorporated, the specific activity (the amount of radioactivity incorporated divided by the total mass) of the labeled DNA is much less than that for probes in which there has been incorporation of several labeled nucleotides along the length of the DNA.

Labeling DNA by Nick Translation

The nick-translation procedure involves introducing single-strand breaks (*nicks*) in the DNA, leaving exposed 32 hydroxyl termini and 52 phosphate termini. The nicking can be achieved by adding a suitable endonuclease such as pancreatic deoxyribonuclease. The exposed nick can then serve as a start point for introducing new nucleotides at the 32 hydroxyl side of the nick using the DNA polymerase activity of *E. coli* DNA polymerase I at the same time as existing nucleotides are removed from the other side of the nick by the 52 '! 32 exonuclease activity of the same enzyme.

As a result, the nick will be moved progressively along the DNA ('translated') in the 52 '! 32 direction. If the reaction is carried out at a relatively low temperature (about 15° C), the reaction proceeds no further than one complete renewal of the existing nucleotide sequence. Although there is no net DNA synthesis at these temperatures, the synthesis reaction allows the incorporation of labeled nucleotides in place of the previously existing unlabeled ones.

RANDOM PRIMED DNA LABELING

The random primed DNA labeling method (sometimes known as oligolabeling) is based on hybridization of a mixture of all possible hexanucleotides: the starting DNA is denatured and then cooled slowly so that the individual hexanucleotides can bind to suitably complementary sequences within the DNA strands. Synthesis of new complementary DNA strands is primed by the bound hexanucleotides and is catalyzed by the Klenow subunit of DNA polymerase I (which contains the polymerase activity in the absence of associated exonuclease activities).

DNA synthesis occurs in the presence of the four dNTPs, at least one of which has a labeled group. This method produces labeled DNAs of high specific activity. Because all sequence combinations are represented in the hexanucleotide mixture, binding of primer to template DNA occurs in a random manner, and labeling is uniform across the length of the DNA.

END-LABELING OF DNA

Single-stranded oligonucleotides are usually end-labeled using polynucleotide kinase (kinase end-labeling). Typically, the label is provided in the form of a ^{32}P at the α-phosphate position of ATP and the polynucleotide kinase catalyses an exchange reaction with the 52 -terminal phosphates. The same procedure can also be used for labeling double-stranded DNA. In this case, fragments carrying label at one end only can then be generated by cleavage at an internal restriction site, generating two differently sized fragments which can be separated by gel electrophoresis and purified.

Larger DNA fragments can be end-labeled by various alternative methods. Fill-in end-labeling is one popular approach, and uses the *Klenow subunit* of *E. coli* DNA polymerase. Again, fragments carrying label at one end only can be generated by restriction cleavage and size fractionation. An alternative PCR-based method is primer-mediated 52 end-labeling.

Labeling of RNA

The preparation of labeled RNA probes (riboprobes) is most easily achieved by *in vitro* transcription of insert DNA cloned in a suitable plasmid vector. The vector is designed so that adjacent to the multiple cloning site is a phage promoter sequence, which can be recognized by the corresponding phage RNA polymerase. For example, the plasmid vector pSP64 contains the bacteriophage SP6 promoter sequence immediately adjacent to a multiple cloning site. The SP6 RNA polymerase can then be used to initiate transcription from a specific start point in the SP6 promoter sequence, transcribing through any DNA sequence that has been inserted into the multiple cloning site.

By using a mix of NTPs, at least one of which is labeled, high specific activity radiolabeled transcripts can be generated. Bacteriophage T3 and T7 promoter/RNA polymerase systems are also used commonly for generating riboprobes. Labeled sense and antisense riboprobes can be generated from any gene cloned in such vectors (the gene can be cloned in either of the two orientations) and are widely used in tissue *in situ* hybridization.

LABELLING OF NUCLEIC ACIDS

Isotopic Labeling and Detection

Traditionally, labeling of nucleic acids has been conducted by incorporating nucleotides containing radioisotopes. Such radiolabeled probes contain nucleotides with a radioisotope (often ^{32}P, ^{33}P, ^{35}S or ^{3}H), which can be detected specifically in solution or, much more commonly, within a solid specimen.

The intensity of an autoradiographic signal is dependent on the intensity of the radiation emitted by the radioisotope, and the time of exposure, which may often be long (one or more days, or even weeks in some applications). ^{32}P has been used widely in Southern blot hybridization, dot-blot hybridization, colony and plaque hybridization because it emits high energy particles which afford a high degree of sensitivity of detection. It has the disadvantage, however, that it is relatively unstable.

Additionally, its high energy α-particle emission can be a disadvantage under circumstances when fine physical resolution is required to interpret the resulting image unambiguously. For this reason, radionuclides which provide less energetic α-particle radiation have been preferred in certain procedures, for example ^{35}S-labeled and ^{33}P-labeled nucleotides for DNA sequencing and tissue *in situ* hybridization, and ^{3}H-labeled nucleotides for chromosome *in situ* hybridization. ^{35}S and ^{33}P have moderate half-lives while ^{3}H has a very long half-life.

However, the latter isotope is disadvantaged by its comparatively low energy â-particle emission which necessitates very long exposure times. ^{32}P-labeled and ^{33}P-labeled nucleotides used in DNA strand synthesis labeling reactions have the radioisotope at the á-phosphate position, because the â- and ã-phosphates from dNTP precursors are not incorporated into the growing DNA chain. Kinase-mediated end-labeling, however, uses ATP.

In the case of ^{35}S-labeled nucleotides which are incorporated during the synthesis of DNA or RNA strands, the NTP or dNTP carries a ^{35}S isotope in place of the O^- of the á-phosphate group. ^{3}H-labeled nucleotides carry the radioisotope at several positions. Specific detection of molecules carrying a radioisotope is most often performed by autoradiography.

Nonisotopic Labeling and Detection

Nonisotopic labeling systems involve the use of nonradioactive probes. Although developed only comparatively recently, they are becoming increasingly popular and are finding increasing applications in a variety of different areas. Two types of non-radioactive labeling are conducted: Direct nonisotopic labeling, where a nucleotide which contains the label that will be detected is incorporated. Often such systems involve incorporation of modified nucleotides containing a fluorophore, a chemical group which can fluoresce when exposed to light of a certain wavelength.

Indirect nonisotopic labeling, usually featuring the chemical coupling of a modified reporter molecule to a nucleotide precursor.

After incorporation into DNA, the reporter groups can be specifically bound by an affinity molecule, a protein or other ligand which has a very high affinity for the reporter group. Conjugated to the latter is a marker molecule or group which can be detected in a suitable assay.

Indirect Nonisotopic Labeling Systems

The biotin-streptavidin system utilizes the extremely high affinity of two ligands: biotin (a naturally occurring vitamin) which acts as the reporter, and the bacterial protein streptavidin, which is the affinity molecule. Biotin and streptavidin bind together extemely tightly with an affinity constant of 10^{-14}, one of the strongest known in biology. Biotinylated probes can be made easily by including a suitable biotinylated nucleotide in the labeling reaction.

Digoxigenin is a plant steroid (obtained from *Digi-talis* plants) to which a specific antibody has been raised. The digoxigenin-specific antibody permits detection of nucleic acid molecules which have incorporated nucleotides containing the digoxigenin reporter group.

The protein recognizing the reporter group is often a specific antibody, as in the digoxigenin system, or any other ligand that has a very high affinity for a specific group, such as streptavidin in the case of using biotin as the reporter. The marker can be detected in various ways. If it carries a specific fluorescent dye, it can be detected in a fluorimetric assay. Alternatively, it can be an enzyme such as alkaline phosphatase which can be coupled to an enzyme assay, yielding a product that can be measured colorimetrically.

A variety of different marker groups or molecules can be conjugated to affinity molecules such as streptavidin or the digoxigenin-specific antibody. They include various fluorophores, or enzymes such as alkaline phosphatase and peroxidase which can permit detection via colorimetric assays or chemical luminescence assays, etc.

The reporter molecules on modified nucleotides need to protrude sufficiently far from the nucleic acid backbone to facilitate their detection by the affinity molecule and so a long carbon atom *spacer* is required to separate the nucleotide from the reporter group.

PRINCIPLES OF NUCLEIC ACID HYBRIDIZATION

Nucleic acid hybridization is a method for identifying closely related nucleic acid molecules within two populations, a complex target population and a comparatively homogeneous probe population

Definition of Nucleic Acid Hybridization

Nucleic acid hybridization involves mixing single strands of two sources of nucleic acids, a probe which typically consists of a homogeneous population of *identified molecules* (e.g. cloned DNA or chemically synthesized oligonucleotides) and a target which typically consists of a *complex, heterogeneous population* of nucleic acid molecules. If either the probe or the target is initially double-stranded, the individual strands must be separated, generally by heating or by alkaline treatment. After mixing single strands of probe with single strands of target, strands with complementary base sequences can be allowed to reassociate.

Complementary probe strands can reanneal to form *homoduplexes*, as can complementary target DNA strands. However, it is the annealing of a probe DNA strand and a complementary target DNA strand to form a labeled probe-target heteroduplex that defines the usefulness of a nucleic acid hybridization assay. The rationale of the hybridization assay is to use the identified probe to query the target DNA by identifying fragments in the complex target which may be related in sequence to the probe.

Melting Temperature and Hybridization Stringency

Denaturation of double-stranded probe DNA is generally achieved by heating a solution of the labeled DNA to a temperature which disrupts the hydrogen bonds that hold the two complementary DNA strands together. The energy required to separate two perfectly complementary DNA strands is dependent on a number of factors, notably:

- Strand length - long homoduplexes contain a large number of hydrogen bonds and require more energy to separate them; because the labeling procedure typically results in short DNA probes, this effect is negligible above an original length (i.e. prior to labeling) of 500 bp;

- Base composition - because GC base pairs have one more hydrogen bond than AT base pairs strands with a high per cent GC composition are more difficult to separate than those with a low per cent GC composition;
- Chemical environment - the presence of monovalent cations (e.g. Na^+ ions) stabilizes the duplex, whereas chemical denaturants such as formamide and urea destabilize the duplex by chemically disrupting the hydrogen bonds.

A useful measure of the stability of a nucleic acid duplex is the melting temperature (T_m). This is the temperature corresponding to the midpoint in the observed transition from double-stranded to single-stranded form. Conveniently, this transition can be followed by measuring the optical density of the DNA. The bases of the nucleic acids absorb 260 nm ultraviolet (UV) light strongly.

However, the adsorption by double-stranded DNA is considerably less than that of the free nucleotides. This difference, the so-called hypochromic effect, is due to interactions between the electron systems of adjacent bases, arising from the way in which adjacent bases are stacked in parallel in a double helix. If duplex DNA is gradually heated, therefore, there will be an increase in the light absorbed at 260 nm (the optical density$_{260}$ or OD_{260}) towards the value characteristic of the free bases. The temperature at which there is a midpoint in the optical density shift is then taken as the T_m.

For mammalian genomes, with a base composition of about 40 per cent GC, the DNA denatures with a T_m of about 87°C under approximately physiological conditions. The T_m of perfect hybrids formed by DNA, RNA or oligonucleotide probes can be determined according to the formulae. Often, hybridization conditions are chosen so as to promote heteroduplex formation and the hybridization temperature is often as much as 25°C below the T_m. However, after the hybridization and removal of excess probe, hybridization washes may be conducted under more stringent conditions so as to disrupt all duplexes other than those between very closely related sequences.

Probe-target heteroduplexes are most stable thermodynamically when the region of duplex formation contains perfect base matching. Mismatches between the two strands of a heteroduplex reduce the T_m: for normal DNA probes, each 1 per cent of mismatching reduces the T_m by approximately 1°C. Although probe-target heteroduplexes are usually not as stable as reannealed probe homoduplexes, a considerable degree of mismatching can be tolerated if the overall region of base complementarity is long (>100 bp).

Increasing the concentration of NaCl and reducing the temperature reduces the hybridization stringency, and enhances the stability of mismatched heteroduplexes. This means that comparatively diverged members of a multigene family or other repetitive DNA family can be identified by hybridization using a specific family member as a probe. Additionally, a gene sequence from one species can be used as a probe to identify homologs in other comparatively diverged species, provided the sequence is reasonably conserved during evolution.

Conditions can also be chosen to maximize hybridization stringency (e.g. lowering the concentration of NaCl and increasing the temperature), so as to encourage dissociation (denaturation) of mismatched heteroduplexes. If the region of base complementarity is small, as with oligonucleotide probes (typically 15–20 nucleotides), hybridization conditions can be chosen such that a single mismatch renders a heteroduplex unstable.

KINETICS OF DNA REASSOCIATION

When double-stranded DNA is denatured, for example by heat, and the complementary single strands are then allowed to reassociate to form double-stranded DNA, the speed at which the complementary strands reassociate will depend on the starting concentration of the DNA. If there is a high concentration of the complementary DNA sequences, the time taken for any one single-stranded DNA molecule to find a complementary strand and form a duplex will be reduced.

Reassociation kinetics is the term used to measure the speed at which complementary single-stranded molecules are able to find each other and form duplexes. It is determined by two major parameters: the starting concentration (C_o) of the specific DNA sequence in moles of nucleotides per liter and the reaction time (t) in seconds. Since the rate of reassociation is proportional to C_o and to t, the C_ot value (often loosely referred to as the cot value) is a useful measure. The C_ot value will also vary depending on the temperature of reassociation and the concentration of monovalent cations. As a result, it is usual to use fixed reference values: a reassociation temperature of 65°C and a Na^+ concentration of 0.3 M NaCl.

Most hybridization assays use an excess of target nucleic acid over probe in order to encourage probe- target formation. This is so because the probe is usually *homogenous,* often consisting of a single type of cloned DNA molecule or RNA molecule, but the target nucleic acid is typically *heterogeneous,* comprising for example genomic DNA or total cellular RNA. In the latter case the concentration of any one sequence may be very low, thereby causing the rate of reassociation to be slow.

For example, if a Southern blot uses a cloned â-globin gene as a probe to identify complementary sequences in human genomic DNA, the latter will be present in very low concentration (the â-globin gene is an example of a single copy sequence and in this case represents only 0.00005 per cent of human genomic DNA). It is therefore necessary to use several micrograms of target DNA to drive the reaction. By contrast, certain other sequences are highly repeated in genomic DNA, and this greatly elevated DNA concentration results in a comparatively rapid reassociation time.

Because the amount of target nucleic acid bound by a probe depends on the copy number of the recognized sequence, hybridization signal intensity is proportional to the copy number of the recognized sequence. Single copy genes give weak hybridization signals, highly repetitive DNA sequences give very strong signals. If a particular probe is heterogeneous and contains a low copy sequence of interest, such as a specific gene, mixed with a highly abundant DNA repeat, the weak hybridization signal obtained with the former will be completely masked by the strong repetitive DNA hybridization signal. This effect can, however, be overcome by competition hybridization.

NUCLEIC ACID HYBRIDIZATION

Early experiments in nucleic acid hybridization utilized solution hybridization, involving mixing of aqueous solutions of probe and target nucleic acids. However, the very low concentration of single copy sequences in complex genomes meant that reassociation times were inevitably slow.

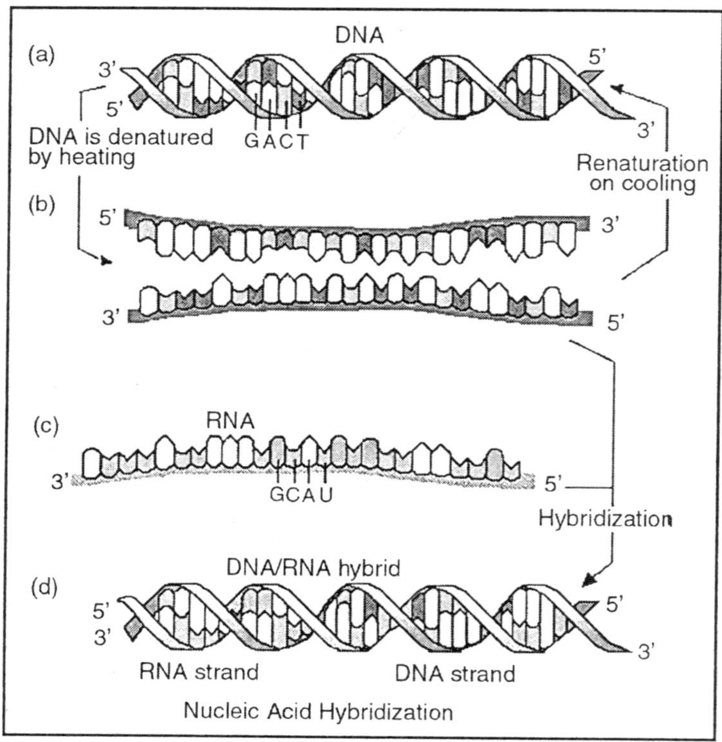

Fig. Nucleic Acid Hybridization

One widely used way of increasing the reassociation speed is to artificially increase the overall DNA concentration in aqueous solution by abstracting water molecules (e.g. by adding high concentrations of polyethylene glycol). An alternative to solution hybridization which facilitated detection of reassociated molecules involved immobilizing the target DNA on a solid support, such as a membrane made of nitrocellulose or nylon, to both of which single-stranded DNA binds readily.

Attachment of labeled probe to the immobilized target DNA can then be followed by removing the solution containing unbound probe DNA, extensive washing and drying in preparation for detection. This is the basis of the standard nucleic acid hybridization assays currently in use.

More recently, however, reverse hybridization assays have become more popular. In these cases, the probe population is unlabeled and fixed to the solid support, while the target nucleic acid is labeled and present in aqueous solution. Note, therefore, that probe and target are not primarily distinguished by which is the labeled and which is the unlabeled population.

Instead, the important consideration is that the target DNA should be the complex imperfectly understood population which the probe (whose molecular identity is known) attempts to query. Depending on the nature and form of the probe and target, a very wide variety of nucleic acid hybridization assays can be devised.

NUCLEIC ACID POPULATIONS

Numerous applications in molecular genetics involve taking an individual DNA clone and using it as a hybridization probe to screen for the presence of related sequences within a complex target of uncloned DNA or RNA. Sometimes the assay is restricted to simply checking for presence or absence of sequences related to the probe.

In other cases, useful information can be obtained regarding the size of the complementary sequences, their subchromosomal location or their locations within specific tissues or groups of cells. Dot-blot hybridization is a rapid screening method which often employs allele-specific oligonucleotide (ASO) probes to discriminate between alleles differing at a single nucleotide position.

The general procedure of dot-blotting involves taking an aqueous solution of target DNA, for example total human genomic DNA, and simply spotting it on to a nitrocellulose or nylon membrane then allowing it to dry. The variant technique of slot-blotting involves pipetting the DNA through an individual slot in a suitable template. In both methods the target DNA sequences are denatured, either by previously exposing to heat, or by exposure of the filter containing them to alkali.

The denatured target DNA sequences now immobilized on the membrane are exposed to a solution containing single-stranded labeled probe sequences. After allowing sufficient time for probe-target heteroduplex formation, the probe solution is decanted, and the membrane is washed to remove excess probe that may have become nonspecifically bound to the filter. It is then dried and exposed to an autoradiographic film. A useful application of dot-blotting involves distinguishing between alleles that differ by even a single nucleotide substitution.

To do this allele-specific oligonucleotide (ASO) probes are constructed from sequences spanning the variant nucleotide site. ASO probes are typically 15–20 nucleotides long and are normally employed under hybridization conditions at which the DNA duplex between probe and target is stable only if there is perfect base complementarity between them: a single mismatch between probe and target sequence is sufficient to render the short heteroduplex unstable.

Typically, this involves designing the oligonucleotides so that the single nucleotide difference between alleles occurs in a central segment of the oligonucleotide sequence, thereby maximizing the thermodynamic instability of a mismatched duplex. Such discrimination can be employed for a variety of research and diagnostic purposes. Although ASOs can be used in conventional Southern blot hybridization, it is more convenient to use them in dot-blot assays.

Another method of ASO dot blotting uses a reverse dot-blotting approach. This means that the oligonucleotide probes are not labeled and are fixed on a filter or membrane whereas the target DNA is labeled and provided in solution. Positive binding of labeled target DNA to a specific oligonucleotide on the membrane is taken to mean that the target has that specific sequence. This approach, and related DNA microarray methods, have many diagnostic applications.

SIZE-FRACTIONATED BY GEL ELECTROPHORESIS

Southern Blot hybridization

In this procedure, the target DNA is digested with one or more restriction endonucleases, size-fractionated by agarose gel electrophoresis, denatured and transferred to a nitrocellulose or nylon membrane for hybridization. During the electrophoresis, DNA fragments, which are negatively charged because of the phosphate groups, are repelled from the negative electrode towards the positive electrode, and sieved through the porous gel. Smaller DNA fragments move faster. For fragments between 0.1 and 20 kb long, the migration speed depends on fragment length, but scarcely at all on the base composition. Thus, fragments in this size range are fractionated by size in a conventional agarose gel electrophoresis system. To achieve efficient size-fractionation of large fragments (40 kb to several megabases), a more specialized system is required, such as a pulsed-field gel electrophoresis apparatus. Following electrophoresis, the test DNA fragments are denatured in strong alkali.

As agarose gels are fragile, and the DNA in them can diffuse within the gel, it is usual to transfer the denatured DNA fragments by blotting on to a durable nitrocellulose or nylon membrane, to which single-stranded DNA binds readily. The individual DNA fragments become immobilized on the membrane at positions which are a faithful record of the size separation achieved by agarose gel electrophoresis.

Subsequently, the immobilized single-stranded target DNA sequences are allowed to associate with labeled single-stranded probe DNA. The probe will bind only to related DNA sequences in the target DNA, and their position on the membrane can be related back to the original gel in order to estimate their size. An important application of Southern blot hybridization in mammalian genetics is the ability to identify a given DNA probe as a member of a repetitive DNA family. Many important mammalian genes belong to multigene families, and many other DNA sequences show varying degrees of repetition.

Once a newly isolated probe is demonstrated to be related to other uncharacterized sequences, attempts can then be made to isolate the other members of the family by screening genomic DNA libraries. Additionally, screening can also be conducted on genomic DNA samples from different species to identify interspecific related sequences. An important route to identifying coding DNA involves identifying sequences that are highly conserved in evolution.

Northern blot Hybridization

Northern blot hybridization is a variant of Southern blotting in which the target nucleic acid is RNA instead of DNA. A principal use of this method is to obtain information on the expression patterns of specific genes. Once a gene has been cloned, it can be used as a probe and hybridized against a Northern blot containing, in different lanes, samples of RNA isolated from a variety of different tissues. The data obtained can provide information on the range of cell types in which the gene is expressed, and the relative abundance of transcripts. Additionally, by revealing transcripts of different sizes, it may provide evidence for the use of alternative promoters, splice sites or polyadenylation sites.

Hybridization Permits Restriction Mapping

Southern blot hybridization has been used extensively in molecular genetic studies as a means of genomic restriction mapping: a labeled DNA probe from one genome can be used to infer the structure of related sequences in the same or different genomes. Because the genomic DNA samples are fractionated by separation of restriction fragments according to size, mutations which alter a restriction site, and significantly large insertions or deletions occurring between neighboring restriction sites, can be typed. Such mutations will result in altered restriction fragment lengths, that is restriction fragment length polymorphisms (RFLPs). Direct detection of pathogenic point mutations by restriction mapping Very occasionally, a pathogenic mutation directly abolishes or creates a restriction site, enabling direct screening for the pathogenic mutation. For example, the sickle cell mutation is a single nucleotide substitution at codon 6 in the â-globin gene, which causes a missense mutation, and at the same time abolishes an *Mst*II restriction site which spans codons 5 to 7.

The nearest flanking restriction sites for *Mst*II, located 1.2 kb upstream in the 52 -flanking region and 0.2 kb downstream at the 32 end of the first intron, are well conserved. Consequently, a â-globin DNA probe can differentiate the normal $â^{A}$-globin and the mutant $â^{S}$-globin alleles in *Mst*II-digested human DNA: the former exhibits 1.2 kb and 0.2 kb *Mst*II fragments, whereas the sickle cell allele exhibits a 1.4 kb *Mst*II fragment.

DETECTION OF CONVENTIONAL RFLPS

The great majority of mutations are not associated with disease; instead, they often occur within noncoding DNA sequences. As a large number of recognition sequences are known for type II restriction endonucleases, many point mutation polymorphisms will be characterized by alleles which possess or lack a recognition site for a specific restriction endonuclease and therefore display restriction site polymorphism (RSP).

Accordingly, individual RSPs normally have two detectable alleles (one lacking and one possessing the specific restriction site). RSPs can be assayed by digesting genomic DNA samples with the relevant restriction endonuclease and identifying specific restriction fragments whose lengths are characteristic of the two alleles, so-called RFLPs.

VNTR-BASED RFLPS AND DNA FINGERPRINTING

DNA probes can also be used to monitor VNTR polymorphisms where alleles differ by a variable number of tandem repeats. To do this, genomic DNA samples are digested with a restriction endonuclease which recognizes well-conserved restriction sites flanking a specific VNTR locus. The resulting restriction fragments are separated according to size on agarose gels, transferred to a suitable membrane and hybridized with a probe representing a unique sequence from the corresponding locus. The resulting pattern of locus-specific RFLPs does not reflect RSP: instead, the differences in sizes of the restriction fragments represent integral numbers of the tandemly repeated unit.

Although the term VNTR could, in theory, encompass a wide range of repeat lengths, in practice the term is usually reserved for moderately large arrays of a repeat unit which is typically in the 5–64 bp region (so-called hypervariable minisatellite DNA, distinguishing it from simple tandem repeat polymorphism (where the repeat unit length is from 1 to 4 bp; i.e. microsatellite DNA) and tandem repeat polymorphism associated with very large arrays of satellite DNA.

If the VNTR locus is a member of a repeated DNA family, the use of a VNTR repeat probe, rather than a unique flanking probe, will produce a complex polymorphic pattern. For example, hypervariable minisatellite DNA clones have been used as probes against Southern blots of appropriately digested genomic DNA. Cross-hybridization of such probes with the other members of this highly repeated DNA family results in a pattern of hybridization bands representing the summed contributions of two alleles at each of many hypervariable loci scattered throughout the genome. Consequently, the overall polymorphism of the multilocus hybridization patterns is uniquely high. Because it permits distinction between any two individuals who are not identical twins, probing with hypervariable minisatellites has been termed DNA fingerprinting.

Detection of Gene Deletions by Restriction Mapping

Certain diseases are associated with a high frequency of deletion of all or part of a gene. If a partial restriction map has been established for the gene under investigation, deletions can be screened by Southern blot hybridization using an appropriate intragenic DNA probe. If the deletion is a small one, for example a few hundred base pairs, it is often apparent as a consistent reduction in size of normal restriction fragments in the gene. An individual who is homozygous for this mutation, or is a heterozygote with one normal allele and another with a small deletion, can easily be identified by detecting the aberrant size restriction fragments.

Large deletions will lead to absence of specific restriction fragments. Homozygous deletion of large DNA segments can easily be detected as complete absence of appropriate restriction fragments associated with the gene. If, however, an individual is heterozygous for a relatively large gene deletion, the deletion may still be detected by demonstrating comparatively reduced intensity of specific gene fragments.

Chromosome in Situ Hybridization

A simple procedure for mapping genes and other DNA sequences is to hybridize a suitable labeled DNA probe against chromosomal DNA that has been denatured *in situ*.

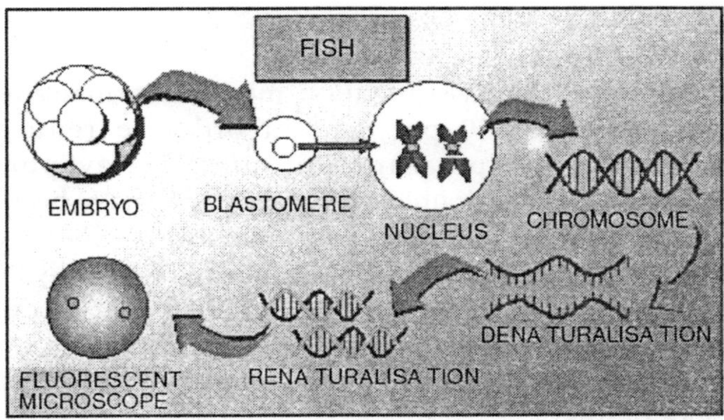

Fig. Fluorescence in Situ Hybridization

To do this, an air-dried microscope slide preparation of metaphase or prometaphase chromosomes is made, usually from peripheral blood lymphocytes or lymphoblastoid cell lines. Treatment with RNase and proteinase K results in partially purified chromosomal DNA, which is denatured by exposure to formamide.

The denatured DNA is then available for *in situ* hybridization with an added solution containing a labeled nucleic acid probe, overlaid with a coverslip. Depending on the particular technique that is used, chromosome banding of the chromosomes can be arranged either before or after the hybridization step. As a result, the signal obtained after removal of excess probe can be correlated with the chromosome band pattern in order to identify a map location for the DNA sequences recognized by the probe.

Chromosome *in situ* hybridization has been revolutionized by the use of fluorescence *in situ* hybridization (FISH) techniques.

Tissue in situ Hybridization

In this procedure, a labeled probe is hybridized against RNA in tissue sections. Tissue sections are made from either paraffin-embedded or frozen tissue using a cryostat, and then mounted on to glass slides. A hybridization mix including the probe is applied to the section on the slide and covered with a glass coverslip.

Typically, the hybridization mix has formamide at a concentration of 50 per cent in order to reduce the hybridization temperature and minimize evaporation problems. Although double-stranded cDNAs have been used as probes, single-stranded complementary RNA probes (riboprobes) are preferred: the sensitivity of initially single-stranded probes is generally higher than that of double-stranded probes, presumably because a proportion of the denatured double-stranded probe renatures to form probe homoduplexes.

cRNA riboprobes that are complementary to the mRNA of a gene are known as antisense riboprobes and can be obtained by cloning a gene in the reverse orientation in a suitable vector such as pSP64.

In such cases, the phage polymerase will synthesize

labeled transcripts from the opposite DNA strand to that which is normally transcribed *in vivo*. Useful controls for such reactions include sense riboprobes which should not hybridize to mRNA except in rare occurrences where both DNA strands of a gene are transcribed. Labeling of probes is performed using either selected radioisotopes, notably ^{35}S, or by nonisotopic labeling.

In the former case, the hybridized probe is visualized using autoradiographic procedures. The localization of the silver grains is often visualized using only dark-field microscopy (direct light is not allowed to reach the objective; instead, the illuminating rays of light are directed from the side so that only scattered light enters the microscopic lenses and the signal appears as an illuminated object against a black background).

However, bright-field microscopy (where the image is obtained by direct transmission of light through the sample) provides better signal detection. Fluorescence labeling is a popular nonisotopic labeling approach and detection is accomplished by fluorescence microscopy.

DNA FINGERPRINTING

DNA Fingerprinting, method of identification that compares fragments of deoxyribonucleic acid (DNA) It is sometimes called DNA typing. DNA is the genetic material found within the cell nuclei of all living things. In mammals the strands of DNA are grouped into structures called chromosomes. With the exception of identical twins, the complete DNA of each individual is unique. A DNA fingerprint is constructed by first extracting a DNA sample from body tissue or fluid such as hair, blood, or saliva.

The sample is then segmented using enzymes, and the segments are arranged by size using a process called electrophoresis. The segments are marked with probes and exposed on X-ray film, where they form a characteristic pattern of black bars—the DNA fingerprint. If the DNA fingerprints produced from two different samples match, the two samples probably came from the same person. DNA fingerprinting was first developed as an identification technique in 1985. Originally used to detect the presence of genetic diseases, DNA fingerprinting soon came to be used in criminal investigations and forensic science. The first criminal conviction based on DNA evidence in the United States occurred in 1988. In criminal investigations, DNA fingerprints derived from evidence collected at the crime scene are compared to the DNA fingerprints of suspects. The DNA evidence can implicate or exonerate a suspect. Generally, courts have accepted the reliability of DNA testing and admitted DNA test results into evidence.

However, DNA fingerprinting is controversial in a number of areas: the accuracy of the results, the cost of testing, and the possible misuse of the technique. The accuracy of DNA fingerprinting has been challenged for several reasons. First, because DNA segments rather than complete DNA strands are "fingerprinted," a DNA fingerprint may not be unique; large-scale research to confirm the uniqueness of DNA fingerprinting test results has not been conducted.

In addition, DNA fingerprinting is often performed in private laboratories that may not follow uniform testing standards and quality controls. Also, since human beings must interpret the test, human error could lead to false results. DNA fingerprinting is expensive. Suspects who are unable to provide their own DNA experts may not be able to adequately defend themselves against charges based on DNA evidence.

In the United States, the Federal Bureau of Investigation (FBI) has created a national database of genetic information called the National DNA Index System. The database contains DNA obtained from convicted criminals and from evidence found at crime scenes. Some experts fear that this database might be used for unauthorized purposes, such as identifying individuals with stigmatizing illnesses such as acquired immunodeficiency syndrome (AIDS).

DNA SEQUENCING

The term DNA sequencing encompasses biochemical methods for determining the order of the nucleotide bases, adenine, guanine, cytosine, and thymine, in a DNA oligonucleotide. The sequence of DNA constitutes the heritable genetic information in nuclei, plasmids, mitochondria, and chloroplasts that forms the basis for the developmental programs of all living organisms.

Determining the DNA sequence is therefore useful in basic research studying fundamental biological processes, as well as in applied fields such as diagnostic or forensic research. The advent of DNA sequencing has significantly accelerated biological research and discovery. The rapid speed of sequencing attained with modern DNA sequencing technology has been instrumental in the large-scale sequencing of the human genome, in the Human Genome Project. Related projects, often by scientific collaboration across continents, have generated the complete DNA sequences of many animal, plant, and microbial genomes.

Early Methods

For thirty years, a large proportion of DNA sequencing has been carried out with the chain-termination method developed by Frederick Sanger and coworkers in 1975. Prior to the development of rapid DNA sequencing methods in the early 1970s by Sanger in England and Walter Gilbert and Allan Maxam at Harvard, a number of laborious methods were used. For instance, in 1973 Gilbert and Maxam reported the sequence of 24 basepairs using a method known as wandering-spot analysis.

RNA sequencing, which for technical reasons is easier to perform than DNA sequencing, was one of the earliest forms of nucleotide sequencing. The major landmark of RNA sequencing, dating from the pre-recombinant DNA era, is the sequence of the first complete gene and then the complete genome of Bacteriophage MS2.

Maxam-Gilbert Sequencing

Allan Maxam and Walter Gilbert developed a DNA sequencing method based on chemical modification of DNA and subsequent cleavage at specific bases. Chemical sequencing method two years after the ground-breaking paper of Sanger and Coulson on plus-minus sequencing, Maxam-Gilbert sequencing rapidly became more popular, since purified DNA could be used directly, while the initial Sanger method required that each read start be cloned for production of single-stranded DNA.

However, with the development and improvement of the chain-termination method Maxam-Gilbert sequencing has fallen out of favour due to its technical complexity, extensive use of hazardous chemicals, and difficulties with scale-up. In addition, unlike the chain-termination method, chemicals used in the Maxam-Gilbert method cannot easily be customized for use in a standard molecular biology kit.

In brief, the method requires radioactive labelling at one end and purification of the DNA fragment to be sequenced. Chemical treatment generates breaks at a small proportion of one or two of the four nucleotide bases in each of four reactions (G, A+G, C, C+T). Thus a series of labelled fragments is generated, from the radiolabelled end to the first 'cut' site in each molecule. The fragments are then size-separated by gel electrophoresis, with the four reactions arranged side by side.

To visualize the fragments generated in each reaction, the gel is exposed to X-ray film for autoradiography, yielding an image of a series of dark 'bands' corresponding to the radiolabelled DNA fragments, from which the sequence may be inferred. Also sometimes known as 'chemical sequencing', this method originated in the study of DNA-protein interactions (footprinting), nucleic acid structure and epigenetic modifications to DNA, and within these it still has important applications.

CHAIN-TERMINATION METHODS

While the chemical sequencing method of Maxam and Gilbert, and the plus-minus method of Sanger and Coulson were orders of magnitude faster than previous methods, the chain-terminator method developed by Sanger was even more efficient, and rapidly became the method of choice. The Maxam-Gilbert technique requires the use of highly toxic chemicals, and large amounts of radiolabeled DNA, whereas the chain-terminator method uses fewer toxic chemicals and lower amounts of radioactivity. The key principle of the Sanger method was the use of dideoxynucleotides triphosphates (ddNTPs) as DNA chain terminators.

The classical chain-termination or Sanger method requires a single-stranded DNA template, a DNA primer, a DNA polymerase, radioactively or fluorescently labeled nucleotides, and modified nucleotides that terminate DNA strand elongation. The DNA sample is divided into four separate sequencing reactions, containing the four standard deoxynucleotides (dATP, dGTP, dCTP and dTTP) and the DNA polymerase. To each reaction is added only one of the four dideoxynucleotides (ddATP, ddGTP, ddCTP, or ddTTP).

These dideoxynucleotides are the chain-terminating nucleotides, lacking a 3'-OH group required for the formation of a phosphodiester bond between two nucleotides during DNA strand elongation. Incorporation of a dideoxynucleotide into the nascent (elongating) DNA strand therefore terminates DNA strand extension, resulting in various DNA fragments of varying length. The dideoxynucleotides are added at lower concentration than the standard deoxynucleotides to allow strand elongation sufficient for sequence analysis.

The newly synthesized and labeled DNA fragments are heat denatured, and separated by size (with a resolution of just one nucleotide) by gel electrophoresis on a denaturing polyacrylamide-urea gel. Each of the four DNA synthesis reactions is run in one of four individual lanes (lanes A, T, G, C); the DNA bands are then visualized by autoradiography or UV light, and the DNA sequence can be directly read off the X-ray film or gel image. In the image on the right, X-ray film was exposed to the gel, and the dark bands correspond to DNA fragments of different lengths.

A dark band in a lane indicates a DNA fragment that is the result of chain termination after incorporation of a dideoxynucleotide (ddATP, ddGTP, ddCTP, or ddTTP). The terminal nucleotide base can be identified according to which dideoxynucleotide was added in the reaction giving that band.

The relative positions of the different bands among the four lanes are then used to read (from bottom to top) the DNA sequence as indicated.There are some technical variations of chain-termination sequencing.

In one method, the DNA fragments are tagged with nucleotides containing radioactive phosphorus for radiolabelling. Alternatively, a primer labeled at the 5′ end with a fluorescent dye is used for the tagging. Four separate reactions are still required, but DNA fragments with dye labels can be read using an optical system, facilitating faster and more economical analysis and automation. This approach is known as 'dye-primer sequencing'.

The later development by L Hood and coworkers of fluorescently labeled ddNTPs and primers set the stage for automated, high-throughput DNA sequencing. The different chain-termination methods have greatly simplified the amount of work and planning needed for DNA sequencing. For example, the chain-termination-based "Sequenase" kit from USB Biochemicals contains most of the reagents needed for sequencing, prealiquoted and ready to use.

Some sequencing problems can occur with the Sanger method, such as non-specific binding of the primer to the DNA, affecting accurate read-out of the DNA sequence. In addition, secondary structures within the DNA template, or contaminating RNA randomly priming at the DNA template can also affect the fidelity of the obtained sequence. Other contaminants affecting the reaction may consist of extraneous DNA or inhibitors of the DNA polymerase.

Dye-Terminator Sequencing

An alternative to primer labelling is labelling of the chain terminators, a method commonly called 'dye-terminator sequencing'. The major advantage of this method is that the sequencing can be performed in a single reaction, rather than four reactions as in the labelled-primer method.

In dye-terminator sequencing, each of the four dideoxynucleotide chain terminators is labelled with a different fluorescent dye, each fluorescing at a different wavelength. This method is attractive because of its greater expediency and speed and is now the mainstay in automated sequencing with computer-controlled sequence analyzers. Its potential limitations include dye effects due to differences in the incorporation of the dye-labelled chain terminators into the DNA fragment, resulting in unequal peak heights and shapes in the electronic DNA sequence trace chromatogram after capillary electrophoresis.

This problem has largely been overcome with the introduction of new DNA polymerase enzyme systems and dyes that minimize incorporation variability, as well as methods for eliminating "dye blobs", caused by certain chemical characteristics of the dyes that can result in artifacts in DNA sequence traces.

The dye-terminator sequencing method, along with automated high-throughput DNA sequence analyzers, is now being used for the vast majority of sequencing projects, as it is both easier to perform and lower in cost than most previous sequencing methods.

Challenges

Modern sequencing typically produces a sequence that has poor quality in the first 15-40 bases, a high quality region of 700-900 bases, and then quickly deteriorating quality. Base calling software typically outputs an estimate of quality along with the sequence to aid in quality trimming.

Before the DNA can be sequenced, linker sequences are attached to its ends, and it is inserted into a cloning vector. The resulting sequence can therefore often contain parts of the vector or the linker sequences, which must be filtered out prior to analysis. In contrast, emerging sequencing technologies based on pyrosequencing often avoid using cloning vectors.

During PCR amplification, unrelated sequences can hybridize, and the resulting clone can be a chimaeric sequence, containing fragments from both sequences. Another problem is polymerase stuttering, where the polymerase repeatedly outputs the same fragments, giving an artificially long low-complexity part of the sequence.

Automation and Sample Preparation

Modern automated DNA sequencing instruments (DNA sequencers) can sequence up to 384 fluorescently labelled samples in a single batch (run) and perform as many as 24 runs a day. However, automated DNA sequencers carry out only DNA-size-based separation (by capillary electrophoresis), detection and recording of dye fluorescence, and data output as fluorescent peak trace chromatograms.

Sequencing reactions by thermocycling, cleanup and re-suspension in a buffer solution before loading onto the sequencer are performed separately. In the past, an operator had to trim the low quality ends of every sequence manually in order to remove the sequencing errors. However, today, software like Fast Chromatogram Viewer can automatically trim the ends at batch.

LARGE-SCALE SEQUENCING STRATEGIES

Current methods can directly sequence only relatively short (300-1000 nucleotides long) DNA fragments in a single reaction. The main obstacle to sequencing DNA fragments above this size limit is insufficient power of separation for resolving large DNA fragments that differ in length by only one nucleotide. Limitations on ddNTP incorporation were largely solved by Tabor at Harvard Medical, Carl Fuller at USB biochemicals, and their coworkers.Large-scale sequencing aims at sequencing very long DNA fragments. Even relatively small bacterial genomes contain millions of nucleotides, and the human chromosome 1 alone contains about 246 million bases.

Therefore, some approaches consist of cutting (with restriction enzymes) or shearing (with mechanical forces) large DNA fragments into shorter DNA fragments. The fragmented DNA is cloned into a DNA vector, usually a bacterial artificial chromosome (BAC), and amplified in *Escherichia coli*. The amplified DNA can then be purified from the bacterial cells (a disadvantage of bacterial clones for sequencing is that some DNA sequences may be inherently *un-clonable* in some or all available bacterial strains, due to deleterious effect of the cloned sequence on the host bacterium or other effects).

These short DNA fragments purified from individual bacterial colonies are then individually and completely sequenced and assembled electronically into one long, contiguous sequence by identifying 100 per cent-identical overlapping sequences between them (shotgun sequencing).

This method does not require any pre-existing information about the sequence of the DNA and is often referred to as *de novo* sequencing. Gaps in the assembled sequence may be filled by Primer walking, often with sub-cloning steps (or transposon-based sequencing depending on the size of the remaining region to be sequenced). These strategies all involve taking many small *reads* of the DNA by one of the above methods and subsequently assembling them into a contiguous sequence.

The different strategies have different tradeoffs in speed and accuracy; the shotgun method is the most practical for sequencing large genomes, but its assembly process is complex and potentially error-prone - particularly in the presence of sequence repeats. Because of this, the assembly of the human genome is not literally complete — the repetitive sequences of the centromeres, telomeres, and some other parts of chromosomes result in gaps in the genome assembly.

Despite having only 93 per cent of the full genome assembled, the Human Genome Project was declared complete because their definition of human genome sequencing was limited to euchromatic sequence (99 per cent complete at the time), excluding these intractable repetitive regions.The human genome is about 3 billion (3,000,000,000) bp long; if the average fragment length is 500 bases, it would take a minimum of six million (3 billion/500) to sequence the human genome (not allowing for overlap = 1-fold coverage). Keeping track of such a high number of sequences presents significant challenges, only held down by developing and coordinating several procedural and computational algorithms, such as efficient database development and management.

Resequencing or targeted sequencing is utilized for determining a change in DNA sequence from a "reference" sequence. It is often performed using PCR to amplify the region of interest (pre-existing DNA sequence is required to design the PCR primers). Resequencing uses three steps, extraction of DNA or RNA from biological tissue; amplification of the RNA or DNA (often by PCR); followed by sequencing. The resultant sequence is compared to a reference or a normal sample to detect mutations.

High-Throughput Sequencing

The high demand for low cost sequencing has given rise to a number of high-throughput sequencing technologies. These efforts have been funded by public and private institutions as well as privately researched and commercialized by biotechnology companies. High-throughput sequencing technologies are intended to lower the cost of sequencing DNA libraries beyond what is possible with the current dye-terminator method based on DNA separation by capillary electrophoresis. Many of the new high-throughput methods use methods that parallelize the sequencing process, producing thousands or millions of sequences at once.

As molecular detection methods are often not sensitive enough for single molecule sequencing, most approaches use an *in vitro* cloning step to generate many copies of each individual molecule. Emulsion PCR is one method, isolating individual DNA molecules along with primer-coated beads in aqueous bubbles within an oil phase. A polymerase chain reaction (PCR) then coats each bead with clonal copies of the isolated library molecule and these beads are subsequently immobilized for later sequencing. (commercialized by 454 Life Sciences, acquired by Roche), Shendure and Porreca et al. (also known as "polony sequencing") and Solid sequencing, (developed by Agencourt and acquired by Applied Biosystems).

Another method for *in vitro* clonal amplification is "bridge PCR", where fragments are amplified upon primers attached to a solid surface, developed and used by Solexa (now owned by Illumina). These methods both produce many physically isolated locations which each contain many copies of a single fragment. The single-molecule method developed by Stephen

Quake's laboratory (later commercialized by Helicos) skips this amplification step, directly fixing DNA molecules to a surface.

Parallelized Sequencing

Once clonal DNA sequences are physically localized to separate positions on a surface, various sequencing approaches may be used to determine the DNA sequences of all locations, in parallel. "Sequencing by synthesis", like the popular dye-termination electrophoretic sequencing, uses the process of DNA synthesis by DNA polymerase to identify the bases present in the complementary DNA molecule.

Reversible terminator methods (used by Illumina and Helicos) use reversible versions of dye-terminators, adding one nucleotide at a time, detecting fluorescence corresponding to that position, then removing the blocking group to allow the polymerization of another nucleotide. Pyrosequencing (used by 454) also uses DNA polymerization to add nucleotides, adding one type of nucleotide at a time, then detecting and quantifying the number of nucleotides added to a given location through the light emitted by the release of attached pyrophosphates.

"Sequencing by ligation" is another enzymatic method of sequencing, using a DNA ligase enzyme rather than polymerase to identify the target sequence. Used in the polony method and in the SOLiD technology offered by Applied Biosystems, this method uses a pool of all possible oligonucleotides of a fixed length, labeled according to the sequenced position. Oligonucleotides are annealed and ligated; the preferential ligation by DNA ligase for matching sequences results in a signal corresponding to the complementary sequence at that position.

Other Sequencing Technologies

Other methods of DNA sequencing may have advantages in terms of efficiency or accuracy. Like traditional dye-terminator sequencing, they are limited to sequencing single isolated DNA fragments. "Sequencing by hybridization" is a non-enzymatic method that uses a DNA microarray. In this method, a single pool of unknown DNA is fluorescently labeled and hybridized to an array of known sequences. If the unknown DNA hybridizes strongly to a given spot on the array, causing it to "light up", then that sequence is inferred to exist within the unknown DNA being sequenced.

Mass spectrometry can also be used to sequence DNA molecules; conventional chain-termination reactions produce DNA molecules of different lengths and the length of these fragments is then determined by the mass differences between them (rather than using gel separation).There are new proposals for DNA sequencing, which are in development, but remain to be proven.

These include labeling the DNA polymerase, reading the sequence as a DNA strand transits through nanopores, and microscopy-based techniques, such as AFM or electron microscopy that are used to identify the positions of individual nucleotides within long DNA fragments by nucleotide labeling with heavier elements (e.g., halogens) for visual detection and recording.

In October 2006 the NIH issued a news release describing novel sequencing techniques and announcing several grant awards.In October 2006, the X Prize Foundation established the Archon X Prize, intending to award $10 million to "the first Team that can build a device and use it to sequence 100 human genomes within 10 days or less, with an accuracy of no more than one error in every 100,000 bases sequenced, with sequences accurately covering at least 98 per cent of the genome, and at a recurring cost of no more than $10,000 (US) per genome.

Chapter 41

Plant Gene Transfer

Plant transformation mediated by the soil plant pathogen *Agrobacterium tumefaciens* has become the most used method for plant transformation. *A. tumefaciens* naturally infects the wound sites in dicotyledonous plant causing the formation of the crown gall tumours.

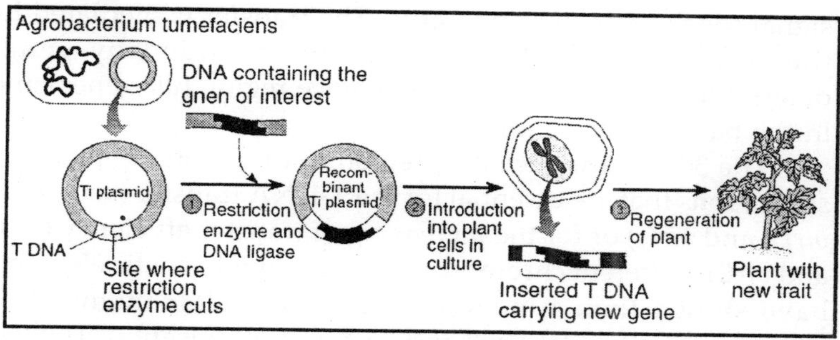

Fig. Agrobacterium Tumefaciens

Since the discovery of the bacterial origin of this neoplastic diseases large number of researches have focused on the study of this process, firstly with the hope to understand the mechanisms of oncogenesis in general and applied it to study of cancer disease in animals and humans.

Later this hypothesis was discarded and the interest on crown gall disease largely decreased until it was evident that A. tumefaciens is capable to transfer a particular DNA segment (T-DNA) of the tumour-inducing (Ti) plasmid into the nucleus of infected cells where it is subsequently stable integrated into the host genome and transcribed, causing the crown gall disease.

The T-DNA contains two types of genes: the oncogenic genes, encoding for enzymes involved in the synthesis of auxins and cytokinins and responsible for tumour formation; and the genes encoding for the synthesis of opines, a product resulted from condensation between amino acids and sugars, which are produced and excreted by the crown gall cells and consume by *A. tumefaciens* as carbon and nitrogen sources.

Outside the T-DNA, are located the genes for the opine catabolism, the genes involved in the process of T-DNA transfer from the bacterium to the plant cell and for the bacterium-bacterium plasmid conjugative transfer genes.

Virulent strains of *A. tumefaciens* and *A. rhizogenes* contain a large megaplasmid (more than 200 kb) which play a key role in tumour induction and for this reason it was named Ti plasmid, or Ri in the case of *A. rhizogenes*. The T-DNA fragment is flanked by 25-bp direct repeats, which act as a *cis* element signal for the transfer apparatus. The transfer is mediated by the co-operative action of proteins encoded by genes determined in the Ti plasmid virulence region (*vir* genes) and in the bacterial chromosome.

The 30 kb virulence (*vir*) region is a regulon organised in six operons that are essential for the T-DNA transfer (*virA, virB, virD,* and *virG*) or for the increasing of transfer efficiency (*virC* and *virE*). Different chromosomal-determined genetic elements have shown their functional role in the attachment of *A. tumefaciens* to the plant cell and bacterial colonisation. The loci *chvA* and *chvB*, involved in the synthesis and excretion of the b -1,2 glucan the *chvE* required for the sugar enhancement of *vir* genes induction and bacterial chemotaxis the *cel* locus, responsible for the synthesis of cellulose fibrils the *pscA* (*exoC*) locus, playing its role in the synthesis of both cyclic glucan and acid succinoglycan and the *att* locus, which is involved in the cell surface proteins.

The initial results of the studies on T-DNA transfer process to plant cells demonstrate three important facts for the practical use of this process in plants transformation. Firstly, the tumour formation is a transformation process of plant cells resulted from transfer and integration of T-DNA and the subsequent expression of T-DNA genes. Secondly, the T-DNA genes are transcribed only in plant cells and do not play any role during the transfer process.

Thirdly, any foreign DNA placed between the T-DNA borders can be transferred to plant cell, no matter where it comes from. These well-established facts, allowed the construction of the first vector and bacterial strain systems for plant transformation.

AGROBACTERIUM TUMEFACIENS T-DNA TRANSFER PROCESS

The process of gene transfer from *Agrobacterium tumefaciens* to plant cells implies several essential steps:

- Bacterial colonisation
- Induction of bacterial virulence system,
- Generation of T-DNA transfer complex
- T-DNA transfer
- Integration of T-DNA into plant genome.

Supported by the most recent experimental data and accepted hypothesis on T-DNA transfer we presented a hypothetical model depicting the most important stages of this process.

Bacterial Colonisation

Bacterial colonisation is an essential and the earliest step in tumour induction and it takes place when *A. tumefaciens* is attached to the plant cell surfac. Mutagenesis studies show that non-attaching mutants loss the tumour-inducing capacity. The polysaccharides of the *A. tumefaciens* (lipopolysaccharides (LPS), and capsular polysacharides (K-antigens)) are proposed to play an important role in the colonising process.

The LPS are integral part of outer membrane and capsular polysaccharides (K-antigens), lacking of lipid anchor, have strong anionic nature. There are evidences that capsular polysaccharides may play specific role during the interaction with the host plant.

The chromosomal 20kb *att* locus contains the genes required for successful bacterium attachment to the plant cell.

This locus has been extensively studied. Insertions in the left 10 kb side of this region produce avirulent mutants that could restore its attachment capacity if the culture medium is previously conditioned by the incubation of wild-type virulent bacterium with plant cells.

In contrast, mutational insertion in the 10 kb right side of the *att* locus results in the irreversible loss of attachment capacity, which can not be restored by conditioned medium. These results suggest that genes, placed at *att* left side, are involved in molecular signalling events, while the right side genes are likely to be responsible for the synthesis of fundamental components.

Induction of Bacterial Virulence System

The T-DNA transfer is mediated by products encoded by the 30-40 kb *vir* region of the Ti plasmid. This region is composed by at least six essential operons (*vir A, vir B, vir C, vir D, vir E, virG*) and two non-essential (*virF, virH*). The number of genes per operon differs, *virA, virG* and *virF* have only one gene; *virE, virC, virH* have two genes while *virD* and *virB* have four and eleven genes respectively.

The only constitutive expressed operons are *virA* and *virG*, coding for a two-component (VirA-VirG) system activating the transcription of the other *vir* genes, and has structural and functional similarities to other already described for other cellular mechanisms.

VirA is a transmembrane dimeric sensor protein that detects signal molecules, released from wounded plants. The signals for VirA activation include acidic pH, phenolic compounds, such as acetosyringone, and certain class of monosaccharides which acts sinergically with phelolic compounds.

VirA protein can be structurally defined into three domains: the periplasmic or input domain and two transmembrane domains (TM1 and TM2). The TM1 and TM2 domains act as a transmitter (signaling) and receiver (sensor). The periplasmic domain is important for monosaccharide detection. Within the periplasmic domain, adjacent to the TM2 domain is an amphipatic helix, with strong hydrophilic and hydrophobic region. This structure is characteristic for other transmembrane sensor proteins and folds the protein to be simultaneously aligned with the inner membrane and anchored in the membrane. The TM2 is the kinase domain and plays a crucial role in the activation of VirA, phosphorilating itself on a conserved His-474 residue) in response to signalling molecules from wounded plant sites. Monosaccharide detection by VirA is important amplification system to respond to low levels of phenolic compounds.

The induction of this system is only possible through the periplasmic sugar (glucose/galactose) binding protein ChvE, which interacts with VirA. Recent studies for determination of VirA regions, important for its sensing activity suggested the position, which may be involved on TM1-TM2 interaction. This interaction causes the exposure of the amphipathic helix to small phenolic compounds and suggests a putative model for the VirA-ChvE interaction.

Activated VirA has the capacity to transfer its phosphate to a conserved aspartate residue of the cytoplasmic DNA binding protein VirG. VirG functions as transcriptional factor regulating the expression of vir genes when it is phosphorilated by VirA. The C-terminal region is responsible for the DNA binding activity, while the N-terminal is the phosphorilation domain and shows homology with the VirA receiver (sensor) domain.

The activation of vir system also depends on external factors like temperature and pH. At temperatures greater than 32°C the vir genes are not expressed because of a conformational change in the folding of VirA induce the inactivation of its properties. The effect of temperature on VirA is suppressed by a mutant form of VirG ($VirG^c$), which activates the constitutive expression of the vir genes.

However, this mutant can not confer the virulence capacity at that temperature to Agrobacterium, probably because the folding of other proteins actively participate in the T-DNA transfer process are also affected at high temperature.

Generation of T-DNA Transfer Complex

The activation of *vir* genes carries out the generation of single-stranded (ss) molecules representing the copy of the bottom T-DNA strand. Any DNA placed between T-DNA borders will be transferred to the plant cell, as single strand DNA, and integrated into plant genome.

These are the only cis acting elements of the T-DNA transfer system. The proteins VirD1 and VirD2 play the key role in this step are, recognising the T-DNA border sequences and nicking (endonuclease activity) the bottom strand at each border. The nick sites are assumed as the initiation and termination sites for T-strand recovery. After endonucleotidic cleavage VirD2 remains covalently attached to the 5'-end of the ss-T-strand.

This association prevents the exonucleolitic attack to the 5'-end of the ss-T-strand and distinguishes the 5'-end as the leading end of the T-DNA transfer complex. VirD1 interacts with the region where the ss-T-strand will be originated. Experiments "in vitro" experiments evidenced that for cleavage of supercoiled stranded substrate by VirD2 is essential the presence of VirD1. The simultaneous restoration of the excised ss-T-strand is evolutionarily related to other bacterial conjugative DNA transfer processes, which include the generation of the single strand DNA.

Extensive mutation or deletion of the right T-DNA border is followed by almost completely loss of T-DNA transfer capacity, while at the left border resulted in lower transfer efficiency. This fact indicates that T-strand synthesis is in 5' to 3' direction, and it is initiated at the right border and that the termination process take place even when the left border is mutated or completely absent, although at lower efficiency. The presence of an enhancer or "overdrive" sequence next to the right border which specifically recognised by VirC1 protein could make the difference between both T-DNA borders.

Two Models for the Translocation of T-DNA-Complex

The transferring vehicle to the plant nucleus is a ssT-DNA-protein complex. According to the most accepted model, the ssT-DNA-VirD2 complex is coated by the 69 kDa VirE2 protein, a single strand DNA binding protein. This co-operative association prevents the attack of nucleases and, in addition, extends the ssT-DNA strand reducing the complex diameter to approximately 2 nm, making easier the translocation through membrane channels.

However, that association does not stabilises T-DNA complex inside *Agrobacterium*. VirE2 contains two plant nuclear location signals (NLS) and VirD2 one. This fact indicates that both proteins are presumably to play important role once the complex is in the plant cell mediating the complex uptake to the nucleus.

VirE1 is essential for the export of VirE2 to the plant cell. Bacterial strains mutated in *virE1*, cannot export VirE2 and resulted in its accumulation inside the bacterium. Such mutants can be complemented if coinfecting with strain that can export VirE2, indicating that this protein can be exported independently and that for transmission event does not necessary the transfer of VirE2 as part of the ss-T-DNA complex and that is possible the transfer of naked T-DNA to the plant cell.

From these experimental evidences an alternative model was brought. This model proposes that the transfer complex is a single-strand DNA covalently bound at its 5'-end with VirD2, but uncoated by VirE2. The independent export of VirE2 to plant cell is presented as natural process, and once the naked ssT-DNA-VirD2 complex is inside the plant cell, it is coated by VirE2.

Previous researches described the role of 9.5 kb *virB* operon in the generation of a suitable cell surface structure for the ssT-DNA complex transfer from bacterium to plant. The VirD4 protein is also required for the ss-T-DNA transport. The function of VirD4 is the ATP-dependent linkage of protein complex necessary for T-DNA translocation.

VirB are proteins that present the hydrophathy characteristics similar others membrane-associated proteins. VirD4 protein is a transmembrane protein but predominantly located at the cytoplasmatic side of the cytoplasmic membrane.

Comparative studies showed highly degree of homology between the *virB* operon and transfer regions of broad host range (BHR) plasmid in genetic organisation, nucleotide sequence and protein function.

Both systems delivery non-self transmissible DNA-protein complex to recipient host cell. In addition, they have the capacity to DNA interkingdom delivery suggesting that T-DNA transfer apparatus and conjugation systems are related and probably evolved from a common ancestral.

The majority of VirB proteins are assembled as a membrane-spanning protein channel involved both membranes. Except for VirB11, they have multiple periplasmic domains. VirB1 is the only member of VirB proteins found in the extracellular milieu although it is possible that some of the other VirB proteins may be redistributed during the process of biogenesis and functioning of the transcellular conjugal channel. That could be the case of the VirB2, which is translated as a 12 kDa proprotein, and later is later proteolically processed to its mature 7 kDa functional form.

VirB4 and VirB11 are hydrophilic ATPases necessary for active DNA transfer. Vir B11 lacks continuous sequence of hydrophobic residues, motivating of periplasmic domains. These characteristics are atypical for this type of protein and evidence the possible dinamic co-existence of different conformational forms "in vivo". VirB4 tightly associates with the cytoplasmic membrane.

It contains two putative extracellular domains conferring transmembrane topology to this protein, which is presumed to allow the ATP-dependent conformational change in the conjugation channel.

Probably, the functional forms of VirB4 and VirB11 are homo- and heterodimmers. The VirB7-VirB9 heterodimmer is assumed to stabilises other Vir proteins during assembly of functional transmembrane channel.

Recent studies identified some of the initial steps of biogenesis of ssT-DNA complex apparatus. Firstly, VirB7 and VirB9 monomers are exported to the membrane and processed. They interact each other to form covalently cross-linked homo-

and heterodimmers. Although it is widely assumed the role of both types of dimmers in the biogenesis of the transfer apparatus, it is likely that only heterodimmers are the essential ones. Subsequently the VirB7-VirB9 heterodimmer is sorted to the outer membrane.

The sorting mechanism has not been elucidated the next step implies the interaction with the other Vir proteins for assembling the transfer channel with the contribution of the transglycosidase.

Two accessory *vir* operons, present in the octopine Ti-plasmid are *virF* and *virH*. The *virF* operon encoding for a 23-kDa protein that functions once the T-DNA complex is inside the plant cells via the conjugal channel or independently, as it was assumed for VirE2 export. The role of VirF is seems to be related with the nuclear targeting of the ssT-DNA complex but its contribution is less important than in the case of VirF. The *virH* operon consists in two genes that code for VirH1 and VirH2 proteins.

These Vir proteins are no essential but could enhance the transfer efficiency, detoxifying certain plant compounds that can affect the growth of bacterial. If that is the function of VirH proteins, they play a role in the host range specificity of bacterial strain for different plant species.

Integration of T-DNA into Plant Genome

Inside the plant cell, the ssT-DNA complex is targeted to the nucleus crossing the nuclear membrane. Two Vir proteins have been found to be important in this step: VirD2 and VirE2, which are the most important; and probably VirF, which has minor contribution to this process. The two NLS of VirE2 have been considered important for the continuos nuclear import of ss-T-DNA complex, probably by keeping both sides of nuclear pore simultaneously open.

The nuclear import is probably also mediated by specific NLS-binding proteins, which are present in plant cytoplasm. The final step of T-DNA transfer is its integration into plant genome. It is considered that the integration occurs by illegitimate recombination. According to this model, paring of a few bases provides just a minimum specificity for the recombination process by positioning VirD2 for the ligation.

The 3′-end or adjacent sequences of T-DNA find some low homologies with plant DNA resulting in the first contact (synapses) between the T-strand and plant DNA forms a gap in 3'-5' strand of plant DNA.

Displaced plant DNA is subsequently cut at the 3'-end position of the gap by endonucleases, and the first nucleotide of the 5' attached to VirD2 pairs with a nucleotide in the top (5'-3') plant DNA strand.

The 3' overhanging part of T-DNA together with displaced plant DNA are digested away, either by endonucleases or by 3'-5' exonucleases. Them, the 5' attached to VirD2 end and other 3'-end of T-strand (paired with plant DNA during since the first step of integration process) joins the nicks in the bottom plant DNA strand.

Once the introduction of T-strand in the 3'-5' strand of the plant DNA is completed, a torsion followed by a nick into opposite plant DNA strand is produced. This situation activates the repair mechanism of the plant cell and the complementary strand is synthesised using the early inserted T-DNA strand as a template. VirD2 has an active role in the precise integration on T-strand in the plant chromosome. The release of VirD2 protein may provide the energy containing in its phosphodiester bond, at the Tyr29 residue, with the first nucleotide of T-strand, providing the 5'-end of the T-strand for ligation to the plant DNA.

This phosphodiester bound can serve as electrophilic substrate for nucleophilic 3'-OH from nicked plant DNA. When the mutant VirD2 protein is transferred attached to the T-strand, the integration process take place with the loss of nucleotides at the 5'-end of the T-strand.

AGRO BACTERIUM TUMEFACIENS

The first plant transformed by *Agrobacterium tumefaciens* was tobacco. Since that crucial moment in the development of plant science, a great progress in understanding the *Agrobacterium*-mediated gene transfer to plant cells has been archived. However, *Agrobacterium tumefaciens* naturally infects only dicotyledonous plants and many economically important plants, including the cereals, remained accessible for genetic manipulation during long time.

For these cases alternative direct transformation methods have been developed. However *Agrobacterium*-mediated transformation have remarkable advantages over direct transformation methods in reducing the copy number of the transgene, potentially leading to fewer problems with transgene cosuppresion and instability. In addition, It is a single-cell transformation system and avoids the obtainment of mosaic plants, which are more frequent when direct transformation is used.

The monocots have been considered to be outside the *Agrobacterium* host range and other gene-transfer methods were developed for these plants. To develop this methodologies for a monocot plant it is important to take in consideration the critical aspects in the *Agrobacterium tumefaciens*-plant interaction, the cellular and tissue culture methodologies developed for that specie. The suitable genetic materials (bacterial strains, binary vectors, reporter and marker genes, promoters) and molecular biology techniques available in the laboratory, are necessary for selection of DNA to be introduce.

This DNA must be expressible in plant making possible the identification of transformed plants in selectable medium and using molecular biology techniques test and characterise the transformation events. Transformation is currently used for genetic manipulation of more than 120 species of at least 35 families, including the most major economic crops, vegetables, ornamental, medicinal, fruit, tree and pasture plants transformation methods.

The idea, that some species can not accept the integration of foreign DNA in its genome and lack the capacity to be transformed is unacceptable under the increasing number species that have been transformed.

The optimisation of *Agrobacterium tumefaciens*-plant interaction is probably the most important aspect to be considered. It includes the integrity of bacterial strain its correct manipulation as warranty of the virulence machine integrity and the study of reaction in wounded plant tissue, which may develop necrotic process in the wounded tissue or affect the interaction and release compounds inducers or repressors of *Agrobacterium* virulence system.

The type of explant is also important fact and it must be suitable for regeneration allowing the recovering of whole transgenic plants. The establishment of method for efficient regeneration for one particular species is crucial for its transformation.

The mentioned aspects are important to establish transformation procedure for any plant but particularly for those species categorised as recalcitrant. In this category have been included cereals, legumes and woody plants, which are very difficult to transform or remain untransformed. Many species originally considered in this category has been transformed in recent years. One of these species, sugarcane, has been transformed in our laboratory.

Concluding Remarks

Agrobacterium tumefaciens is more than the causative agent of crown gall disease affecting dicotyledonous plants. It is also firstly the natural instance for the introduction of foreign gene in plants allowing its genetic manipulation.

Similarities have been found between T-DNA and conjugal transfer systems are evolutionally related and apparently evolved from a common ancestral. Although the gene transfer mechanisms remain largely unknown, great progress has been obtained in practical implementation of transformation protocols for both dicotyledonous and monocotyledonous plants. Particularly important is the extension of this single-cell transformation methodology to monocotyledonous plants.

This advance has biological and practical implications. Firstly, because of advances of *A. tumefaciens*-mediated gene transfer over the direct transformation methods, which where the only way for genetic manipulation of economically important crops as cereals and legumes. Second, it has been demonstrated that T-DNA is transferred to dicot and monocot plants by an identical molecular mechanism.

This confirmation implies that any plant can potentially be transformed by this method if suitable transformation protocol is established.

The *Agrobacterium*-mediated transformation protocols differ from one plant specie to other and, within species, from one cultivar to other.

In consequence, the optimisation of Agrobacterium-mediated transformation methodologies requires the considered of several factors that can be determined in the successful transformation of one species.

Firstly, the optimisation of *Agrobacterium*-plant interaction on competent cells from different regenerable tissues. Second, the development of suitable method for regeneration from transformed cells.

Undoubtedly, the development of transformation procedures based on A. tumefaciens-mediated gene transfer for new economically important species are advisable and the results obtained in last years evidence a promising future.

Chapter 42

Genetically Engineered Crops

Probably the most important scientific event of the 20th century was the 1953 discovery, by James Watson and Francis Crick, of the structure of the DNA molecule which is the basis of heredity. Darwin had shown how species might have changed over eons by slow, random natural processes.

Watson and Crick gave us the key to moving evolution along much faster, to suit our own purposes. A DNA molecule is like a string of letters, using a four letter alphabet, easily copied when living cells reproduce. The sequences of letters make sentences, which we call genes.

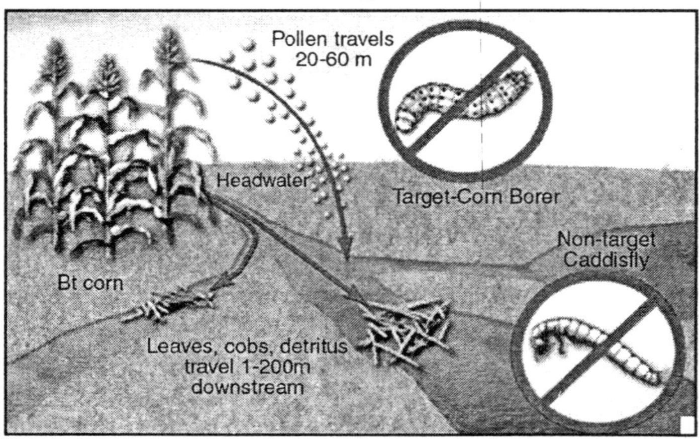

Fig. Genetically Engineered Crops

These sentences are the instructions for making and operating a living cell. There are two kinds of sequences. One kind of gene gives a cell the necessary instructions for making one of the various kinds of protein, used for structures, enzymes, signals, all the basic mechanisms of life. The other kind of sequence is used as a control mechanism so that a cell can tell when to make which proteins and when to do something else. By 1966, scientists had learned the language of protein-making gene sequences.

This language is the same for all forms of life. That means that a human gene sentence for making insulin, a kind of protein, could be transferred to, say, a yeast cell, and then the yeast cell could equally well make human insulin. The other gene sequences, the control sequences, are like switches that turn other genes on or off.

A control sequence could have different results in different organisms, just as an electrical switch can produce a different result in a car or in an oil burner.

In particular, it could control a completely different protein-making gene. Some control genes are used to turn another gene on, and others are used to turn another gene off, and some control genes turn other control genes on or off. In a simple case, suppose a cell needs protein A, but not too much. If the gene that tells the cell to make protein A is turned on, eventually the control gene will sense that there is lots of protein A available, so it will turn off the protein making gene.

Later when the supply of protein A has diminished, the control gene will relent and let the protein making gene turn back on. There are more complicated control arrangements. For example, the gene which makes insulin is turned on in pancreas cells but not in liver cells.

To understand the connection between a gene and its function requires lots of scientific work, enough to keep biologists busy for a very long time.

Even in the simplest cases, one first needs to know what sequence of letters make up the protein-making gene, and what sequences make up the control genes which turn it on or off, as well as where they are situated on the chromosome; one needs to know what signals activate the control genes; then one needs to know the chemical reactions in which the protein molecule takes part, and finally one needs to know how those chemical reactions relate to some activity of the cell. Each different organism has tens of thousands of different genes and makes a huge number of proteins. Life is enormously complex. Slowly but surely, more and more secrets of living things are being uncovered. Hundreds of genes are now understood completely. There are many more genes which have been discovered and associated with some function, but not yet understood very well.

It is now possible to transfer a gene from the DNA of one species to the DNA of another species. For cases in which scientists know exactly what a gene does and exactly how it does it, it is now possible to express that function in another species. That is genetic engineering. There are practical applications of this knowledge.

The first practical applications were in medicine, using genetically modified bacteria to make medical drugs such as interferon, human growth hormone and human insulin. The second kind of application was to modify organisms for agricultural purposes. It is this second application that will occupy us now.

FRUITS OF TRANSGENIC AGRICULTURE

Rice with Vitamin A

Rice does not contain very much vitamin A. In the poorer parts of Asia, where rice is almost the only food of the rural population, a vitamin A deficiency is common, leading to early blindness. Now Drs.

Ingo Potrykus and Peter Beyer, two genetic engineers, have transferred the genes for vitamin A from other species into rice, creating a strain of rice which is rich in vitamin A — the amount of rice in a typical third world diet could provide about fifteen per cent of the recommended daily allowance of vitamin A, sufficient to prevent blindness.

Now that a few plants with this trait have been created, they are being cross bred with other varieties of rice using conventional breeding techniques, as has been done for centuries. Such cross breeding could further increase the vitamin A content. The development of rice with vitamin A was carried out at the Swiss Federal Institute of Technology, making free use of patented technology and of the earlier research which had established the basic facts about how plants synthesize vitamins.

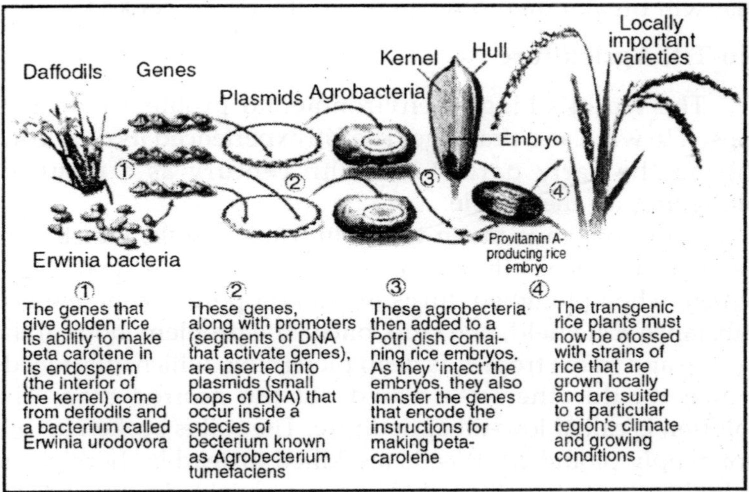

Fig. Rice with Vitamin A

The corporations holding the various patents all agreed to cost-free use of their patents as long as the rice was to be provided free to poor third-world farmers.

The new rice strain was then turned over to the International Rice Research Institute, a non-profit organization based in the Philippines, where it will be evaluated for its adaptability to various growing conditions, food safety, and environmental impacts, etc. The IRRI preserves thousands of varieties of rice with different genetic characteristics, so the new strain can be cross bred to produce varieties suitable for almost any locality.

The result is that rural Asians can soon expect to retain normal eyesight. Genetic engineers also intend to produce a rice variety rich in iron, because iron-deficiency anemia is a common problem in the same rural populations. But this is a more difficult problem than increasing rice's vitamin

A content. Rice contains a substance called phytate. Phytate prevents the body from absorbing iron, so it does little good to breed for increased iron content, and the rice plant cannot reproduce without adequate phytate in the grains. Dr. Potrykus hopes to be able to find a gene coding for a protein that will break down phytate when the rice is cooked.

No-Till Agriculture

The world's biggest environmental problem is loss of topsoil to wind and drainage. The US experienced its dust bowl during the early part of the 20th century as a result of ploughing up the prairie.

The problem is much worse in tropical soils, which may have a thin, inches thin, layer of topsoil above a type of soil which, when ploughed, turns into a non-porous concrete-like substance. One field, one crop, once. The problem, both in the prairie and in the tropics, is deep ploughing, which kills weeds which would otherwise crowd out the desired crop. The solution is called low-till agriculture. The soil is broken up but not deeply ploughed. Weeds are killed instead by herbicides. A herbicide of choice should be cheap, quickly biodegradable and non-toxic.

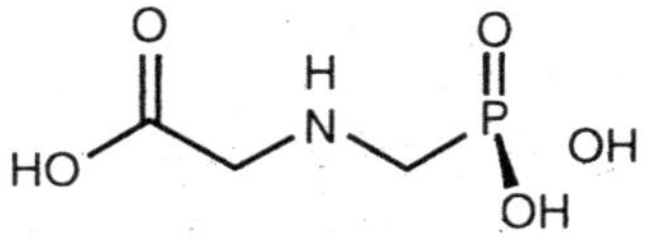

O H O HO N P OH OH

Fig. Glyphosate

An excellent choice is a chemical called glyphosate, except that glyphosate kills the crops as well as the weeds.

So genetic engineers found a gene which lets plants tolerate glyphosate, and transferred it into soybeans. Today, 63 per cent of the soybeans grown in the US are glyphosate tolerant, allowing soil saving no-till agriculture on half the US soybean acreage. There is another advantage to no-till agriculture. There are lots of plant residues beneath the ground, both root systems and humus transported by earthworms.

Ploughing brings this material to the surface, where it can

oxidize. Carbon dioxide is created, a greenhouse gas. So transgenic soybeans are a positive factor in postponing global warming. Approximately four tons of carbon dioxide are retained in the soil per acre per year.

This saving is applicable to the accounting of CO_2 reductions in the Kyoto Treaty on Climate. In fact, any technology which reduces the need to plough, spray, or till crops will reduce carbon dioxide emission. Consider a tractor pulling a ten foot wide harrow over a square mile of agricultural land. Simple arithmetic shows that the tractor will travel 528 miles, all the while burning gasoline. Perhaps we are thinking that even if ploughing has disadvantages, herbicides don't sound very good either. The very word means "plant killer".

But that is not the choice. Traditional soybeans are also grown using herbicides, most of which are far more toxic than glyphosate. On the average, the transgenic soybeans actually use 30 per cent less total herbicide than conventional soybeans. So environmentally, this is a no-brainer.

WITCHWEED CONTROL

Farmers in east Africa are plagued by a devastating parasitic weed called Striga, or witchweed. Farmers are used to dealing with weeds that grow in the soil alongside the crop and compete for nutrients. From time immemorial, they have dealt with those weeds by pulling them up by hand. Less neighbour intensive methods like spraying and ploughing are now common. But none of these methods work for the witchweed.

Striga attacks plants directly, underground, even before the weed has emerged above the soil surface. It sucks nutrients from the seeds and the roots of the crop. In some parts of Africa, the striga parasite destroys as much as 80 per cent of the crop yield. But now that a herbicide resistance trait can be transferred to a crop, scientists in Israel and Kenya, working together, have demonstrated a new strategy for striga control.

Before planting the crop, they soak its seeds in a herbicide. The seeds are unharmed, but they become poisonous to the striga parasite. The seed germinates and sprouts without interference. By the time the crop is harvested, the herbicide has decomposed and disappeared. Their demonstration used herbicide resistant transgenic corn. The same strategy would probably work with Africa's other important grains, sorghum and millet.

Soaking seeds would use far less herbicide than spraying it on the ground, and the complex spraying apparatus would not be needed. This is a significant consideration in Africa, where so many farmers are too poor to own expensive equipment.

CHEESE CHYMOSIN FROM YEASTS

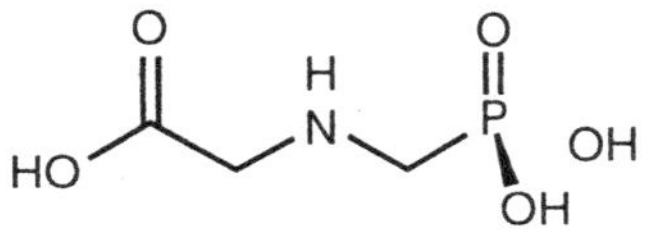

Fig. Chymosin

Hard cheeses are made from whole milk by adding an enzyme called chymosin (rennet), which was formerly extracted from the stomachs of calves, a byproduct of veal. The gene for making chymosin was transferred from cows to yeast. Yeast can be grown in vats, as any brewer knows. Although many people consider it wrong to slaughter calves, yeasts have few defenders. Besides, chymosin from yeast is cheaper and purer than chymosin from calves. So today, almost all hard cheese (over 90 per cent) is made from chymosin produced by genetic engineered yeast.

COTTON WITHOUT INSECTICIDES

Cotton farmers are plagued by various insect pests, such as the boll budworm, the tobacco budworm, and the pink bollworm. In the US south, where most of our cotton is raised, these insects were controlled using chemical insecticides. But there is a natural insecticide which has been used for almost a century by organic farmers, a bacterium called Bacillus thuringiensis, Bt for short.

The bacterium produces a toxin which is deadly to caterpillars like the three mentioned above, but harmless to almost everything else (except insects of the order lepidoptera, butterflies and moths — even the legendary boll weevil (Anthromonus grandis) is not harmed by the Bt toxin). So genetic engineers transferred the gene for Bt toxin from Bacillus thuringiensis to cotton. Then the cotton plants, which could make Bt toxin, were cross bred with other varieties in the old fashioned way.

Today, much of the US cotton crop is genetic engineered for the Bt toxin trait. The use of chemical insecticides in the cotton belt has declined dramatically, by over a million liters per year. Since the Bt toxin is inside the plant instead of sprayed onto the plant, the only insects which it can harm are those which eat the plant. The benefit of reduced spraying of cotton is overwhelming. The cotton pesticides replaced are extremely damaging to the environment.

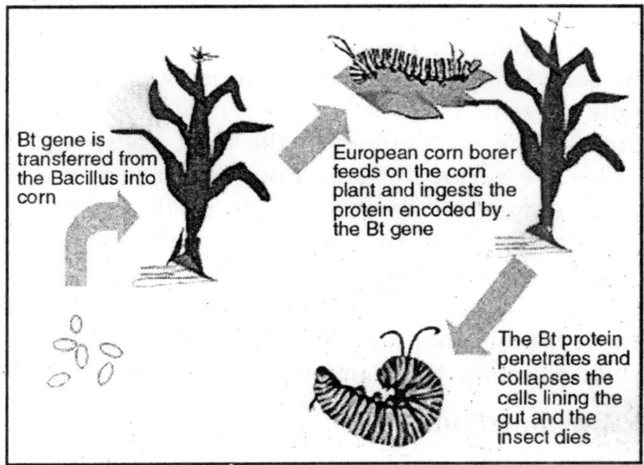

Fig. The Bt Gene

They not only kill all insects in a cotton field, harmless or not, but also nearly anything else in the field, thus depriving insectivorous birds of their food. There is no way to keep these pesticides from getting into streams and rivers, where they are a serious hazard to aquatic life. We may think of cotton as a natural material, therefore environmentally friendly, but before there was Bt protected cotton, that was just wrong. The gene for Bt toxin has been transferred into several other crops, including potatoes and corn.

Approximately 30 per cent of US corn is now transgenic, and the most popular transgenic varieties contain the Bt gene. Although we call it a toxin, to humans and other mammals and birds it's just a nutrient. The principal potato pests are not caterpillars, but beetles, and the Bt toxin that protects cotton and corn doesn't harm beetles. But there is another Bt toxin found in another variety of Bacillus thuringiensis which is deadly to beetles. The gene for that toxin was used in potatoes. Biotechnology also has overzealous advocates who exploit consumers' fears about pesticides.

Except in the case of serious accidents, there's little to worry about. Our bodies deal with many toxic substances in many foods, and as long as the amounts are small enough they give us no problems. The toxic load of pesticide residues on food is completely negligible in comparison with the toxins naturally present.

But as they are applied in the field, these agricultural pesticides are seriously hazardous. Each year many farm workers are poisoned by exposure to pesticides and farmers have nightmares about their children or pets being injured by playing near pesticides. Pesticides can also harm wild birds and small animals, and when they get into waterways they can kill fish and other aquatic life. The Bt transgenic plants reduce or eliminate this danger.

Slow Ripening Fruits

There are many fruits which ripen after picking. After they reach optimum ripeness, they begin to deteriorate. This is necessary for the life cycle of the plant, which relies on the sweet and pulpy parts to nourish the seeds. A ripe fruit literally digests itself.

When this process is rapid, it effectively means that the fruit cannot be enjoyed out of season, or far from its growing area. For example, there is a popular Malaysian papaya variety which is unavailable outside Southeast Asia because it ripens so rapidly that it cannot be shipped very far.

But it is quite easy to genetically engineer a fruit so that it does not ripen so rapidly. It doesn't even require a gene from another organism. Instead, a gene involved in the ripening process is copied with the message in reverse order. So now that plant has two genes with mirror image structure. The way an organism uses the information in a gene to make a protein involves copying the gene (DNA) onto a messenger molecule, known as messenger RNA. The modified plant copies both the original gene and the mirror image gene to produce both types of messenger RNA.

But since these messenger RNAs are exact complements of one another, they can wrap about one another just like the two strands of DNA, effectively blocking both messages. This means that the plant makes very little of the enzyme that causes ripening. This genetic engineering trick is called "antisense technology".

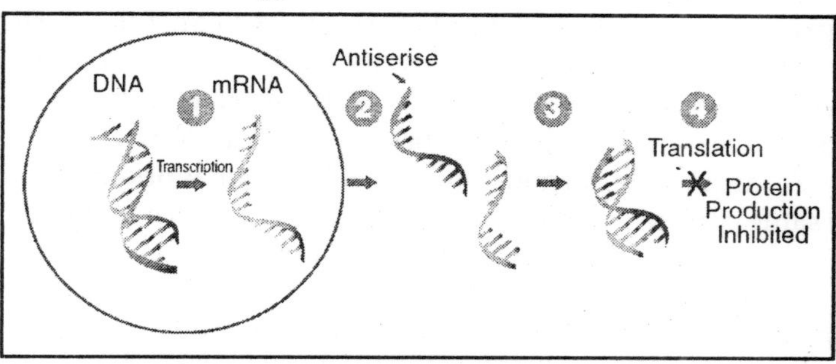

Fig. Antisense Technology

The Malaysian papaya was transformed in this way and therefore a slow ripening variety will soon be available. The very first genetic engineered plant to be commercially developed as a whole food was a slow ripening tomato, called FlavR Savr.

It was developed by Calgene, Inc. Because it could remain on store shelves for a long time, it could be left on the tomato plant until optimally ripe, and therefore the FlavR Savr tomatoes sold for a premium compared to other tomatoes. Although consumers initially liked Calgene's tomatoes, they didn't ship well and the variety was eventually dropped.

Controlled Ripening

A coffee bush ripens a few coffee beans each day for many months. The best quality beans must be picked just after ripening, so picking coffee beans is very neighbour intensive. It would obviously be preferable if the beans would all get ripe at the same time. Genetic engineering will make this possible. There is a coffee gene which turns on to initiate the last stage of ripening.

Scientists modified a control gene so that the ripening gene does not turn on until the plant is sprayed with a triggering substance (patented and sold by the company that developed the coffee variety). Therefore all the beans on a bush reach the same not quite ripe stage and stop to wait for the triggering signal. The farmer decides when to spray the bush so it can be picked completely clean a few days later. This can substantially improve the life of the small farmer. He can take a short vacation without losing part of his livelihood. He can work fewer hours per day, or he can pick all his crop in a few days and increase his income by working at another job.

A large scale farmer would need fewer workers to pick the same quantity of coffee beans, and could afford to pay them a higher wage. The control of when a crop is harvested would be valuable for other crops besides coffee. For example, the quality of grapes declines rapidly after they reach their optimum sugar content. Grape farmers now have to mobilize every available hand to harvest all their crop in a very short time. Their lives would be simpler if they could spread the harvest effort over a few weeks instead of a few days.

Large scale crops are harvested with special equipment. A farmer would not need to own a combine if he could rent it for the few days it was needed. But that wouldn't work if his neighbour needed to rent it for those same few days. If neighboring farmers could control when their crops become ready for harvest, they could share scarce and expensive equipment.

Saving the Banana

Wild bananas have seeds. They reproduce sexually, like

beans and oak trees. But they aren't easy to eat. Bananas grown on plantations have no seeds. They are reproduced by taking cuttings from older banana plants. Cultivated bananas are seedless because they have three of each type of chromosome instead of the normal two of each type. Such plants are called "triploid".

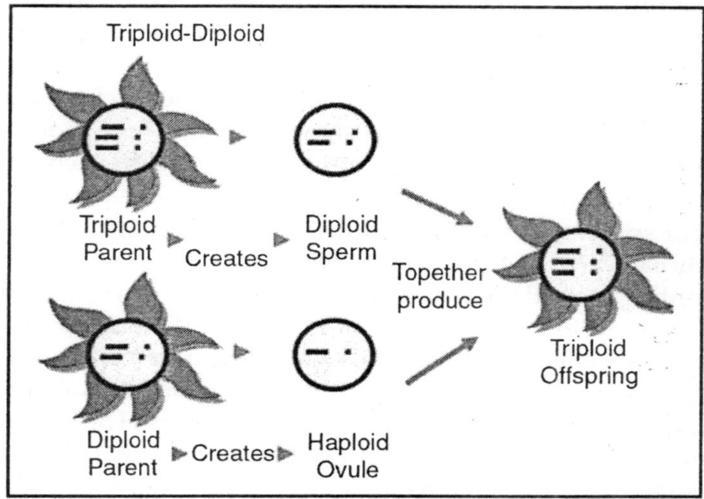

Fig. Triploid

They are always sterile. Genetic triploid freaks arise from time to time in nature, but modern breeders can also use chemicals or electric shocks to create triploid mutant cells.

Bananas have been cultivated for many thousands of years and there are about three hundred different banana varieties. Each variety was developed by crossbreeding wild bananas. Whenever a promising variety was been produced, the breeder caused it to be triploid, hence seedless. That plant was cloned by propagating cuttings, and it became the parent of its variety. All bananas plants of the same variety are genetically identical, like identical twins.

Of the three hundred varieties, only one single variety completely dominates international trade. It is called Cavendish. It is possible that we have never seen a banana other than a Cavendish banana. Certain kinds of fungus can infect and kill banana plants. In many parts of the world, Cavendish banana plants are being attacked by a fungus called black Sigatoka. Since wild bananas can reproduce sexually, they are not all identical and some wild bananas can resist black Sigatoka.

If the black Sigatoka fungus is present in a region, the resistant banana types become prevalent. But the Cavendish banana plants have no resistance. They are all identical so they all die. To grow bananas commercially, growers must spray their plants with fungicides. Year after year, the black Sigatoka fungus has been evolving resistance to these fungicides, so growers have to spray more and more fungicide each year. Approximately one third of the cost of raising a banana is the cost of spraying it with fungicides, and it gets more and more costly each year. A form of black Sigatoka banana disease, now spreading around the world, can tolerate all known fungicides.

Soon it will attack bananas in Central America and the Carribean islands, the heartland of banana culture. The Cavendish banana will become virtually extinct. Agronomists estimate that this will happen within ten years. This is not a fairy tale. It has happened before. Forty years ago, the most popular variety of banana was one called Gros Michel. But Gros Michel was susceptible to a fungus called "race 1 Panama disease". Now it is gone. Cavendish bananas, which are not susceptible to race 1 Panama disease, replaced them. (A related fungus, race 4 Panama disease, can infect Cavendish banana plants. At present it is only found in Malaysia and some nearby countries.)

In turn, some other variety of banana could replace the Cavendish. It would look different and taste different but it would still be a banana. There is only one practical way to save the Cavendish banana. It must be given some combination of genes from wild bananas which are not susceptible to the fungus. But this can't be done by any natural technique.

Cross-breeding can create other banana varieties because wild bananas reproduce sexually. But Cavendish bananas reproduce only by cloning. Conventional breeding would have to rely on rare mutated Cavendish banana plants which can produce seeds and which can therefore be crossbred, in theory. Even the mutated plants produce only a tiny number of viable seeds, as few as two or three seeds in a hundred pounds of bananas. No banana breeding experts think that they can breed fungus resistance into a Cavendish banana variety in only ten years. Biotechnology can rescue the Cavendish banana. Scientists in Belgium have used genetic engineering to transfer some fungal resistance genes from wild bananas. The transformed Cavendish plants are not susceptible to the fungus and they can then be reproduced into nursery stock by the usual method of taking cuttings.

EGGPLANT IN WINTER

The edible part of an eggplant is formed from the ovary of its flower. In this way, it is like the edible flesh of an apple, a pepper or grape. When we eat these fruits, we discard the seeds. But the plants only transform their ovaries into fruits when they start to produce seeds, although in the case of an eggplant, its seeds are so tiny that we ignore them. Eggplants will only set seeds in warm weather, so to grow them in the winter in an unheated greenhouse, the grower must use a chemical to trick the plant into beginning fruit development without setting seed.

Such fruits do not grow very large or very fast under these conditions. So eggplants are expensive in the winter. But now scientists in Italy have transferred two genes into a variety of eggplant, which not only allows the plant to set fruit in cool greenhouse conditions without chemicals, but also increases productivity of the same plant in either hot or cold weather. The eggplant variety that the Italian scientists created is seedless. One of the two transferred genes is a switching gene which is turned on only in the ovary part of a flower. That gene turns on the other transferred gene, which makes a protein involved in synthesizing a growth hormone.

The growth hormone makes the ovary grow into the fruit, just as it would have done in a traditional eggplant making seeds. Neither gene requires either seed setting or warm weather. Where does one get seeds to produce large numbers of seedless eggplants? The transformed plants produce pollen, so they can be crossed with traditional eggplant varieties and the hybrid produced by that crossing has the seedless and self-starting property. The scientists report productivity increases of 37 per cent for the new variety, and they believe that the seedless type would be more marketable.

VIRUS RESISTANT CROPS

Some viruses infect people or animals and other viruses infect plants. Plant viruses reduce the productivity of annual crops and can kill fruit trees. Some plant viruses are spread by insects. Plants can be protected from those viruses by using insecticides or other pest management methods. There is essentially nothing else that a farmer can do to protect his crop from virus damage, except to grow a different crop. But genetic engineering a plant to protect it from a particular kind of virus is quite easy.

A gene from the virus which encodes a protein in the virus' outer coat is copied into the plant's DNA. The plant then makes the coat protein, which is harmless, but which stimulates the plant's natural defenses. Virus resistance traits have been introduced into many crops, including squashes, tomatoes, potatoes, tobacco and, perhaps most dramatically papaya. Recently, cultivated Hawaiian papayas were hit by a devastating virus which essentially eradicated the commercial variety.

Only the virus resistant transgenic papayas survived. If we like papaya, we can only buy the transgenic variety. Nobody can grow any other kind.

The Potato Famine

The potato crop was devastated by a late blight fungus (Phytophthora infestans). That fungus could reappear at any time in any place and wipe out a potato crop. Some varieties of potato have previously had some resistance to late blight fungus, but now a fungal strain has appeared in Russia that destroys those previously resistant varieties.

This year a similar fungus appeared in potato fields in Prince Edward Island and 630 million pounds of potatoes, the island's principal crop, had to be destroyed. But very recently, scientists were able to transfer a gene from alfalfa to a potato plant and the resulting potato plant is able to resist the fungus and thrive.

Potatoes also rot. A principal cause of potato rot is the bacterium Erwina carotovora, which has been called the flesh eating bacteria of the plant kingdom. Now a gene that confers resistance to E. carotovora has been coupled to a control gene that turns on when a plant has been wounded, and this construct has been transferred to experimental potatoes.

As the researchers hoped, the modified potatoes, when punctured by a toothpick and exposed to E. carotovora, had almost twenty times less rot than unmodified potatoes.

Sentinel Crops

A recent innovation is a plant intended not for food but for quality control. It contains a gene derived from a luminescent jellyfish, but in all other ways it is identical to the food crop it is planted alongside. When these sentinel plants experience a lack of water, they literally glow in the dark. The farmer then knows that his crop must be watered or whether irrigation can be postponed. In the western U.S. water is scarce.

Agriculture is the biggest user of water. Wasting water is intolerable. For example, so much water is taken from the Colorado River for irrigation that the river flows into Mexico a mere trickle, and it never gets to the sea at all. So this is yet another way that transgenic crops can benefit the environment.

BUILDING WITH SILK

Silk is composed of two proteins, fibroin and sericin. The gene for fibroin has been transferred from a silkworm to a goat, and is expressed as a component of its milk. Soon we may also expect sericin to be transferred. It still remains to be seen whether technology can be developed to spin these proteins into a fibre.

That has already been accomplished with the kind of silk

spiders use to make their webs. Genes for the two spider silk proteins were transferred to cells cultured from cow udders. Those cells then made the proteins. Happily, the spider silk can be spun by forcing a solution of its two protein components through a tiny nozzle. The proteins self-assemble into spider silk strands. The same genes have since been transferred to live goats and when those goats are old enough to produce milk, it should be possible to make large quantities of spider silk very cheaply. Silk is an extraordinarily strong material, stronger than steel. In the future we may be getting our strongest building material from a farm instead of from a mine.

Safer Meat

Escherichia coli are friendly bacteria that live in our intestines and contribute to our health. But there is one strain of E. Coli that can make us sick, even kill us. We can get it from inadequately cooked meat. The E. coli infected meat comes from cattle with the virulent E. coli strain in their intestines. A cow's digestive system is adapted to digesting hay and grasses. The food first goes into a pre-stomach called a rumen.

That's why cows are called ruminants. In the rumen, microorganisms turn the indigestible cellulose into nutrients the cow can assimilate. The food is then passed to a true stomach, and finally gets to the cow's intestines, where E. coli can live. Therefore to guarantee against the virulent strain thriving in the cow's intestine, one needs to get some sort of prophylactic agent into the intestine. Antibiotics won't do. They would kill the cow's normal intestinal bacteria, and besides, it isn't a good idea to overuse antibiotics.

There are antibodies specific to the virulent strain of E. coli, but they would be destroyed by passage through the cow's stomach before reaching its intestines. Genetic engineers are working on a neat solution. They are developing a transgenic animal feed which resists complete digestion in the stomach and delivers the antibody, specific to the virulent E. coli strain, into the intestines. There are over 60,000 cases of E. coli illness in the United States each year.

There would be many more except for an extensive programme of meat inspection. Even this understates the problem, because meat, if contaminated, has to be destroyed. E. coli infections were not such a serious problem when cattle were raised exclusively on grass and hay. On that diet, there isn't much digestion going on in the cow's intestine and the E. coli populations are comparatively low.

But today's cattle spend the last weeks of their lives in feed lots, being fattened up on grain, which is digested in their intestines, leading to much higher populations of E. coli. So another way to solve the E. coli problem would be to raise leaner cattle and skip the feed lots.

Reduced Need for Fertilizers

One of the ways that farmers get better yields is by providing their plants with sources of organically bound nitrogen and phosphorus. These can be provided either by applying chemical nitrates or phosphates, or by using manures or decaying vegetation as sources of the same nutrients. Nitrogen is the largest constituent of the atmosphere, about 80 per cent.

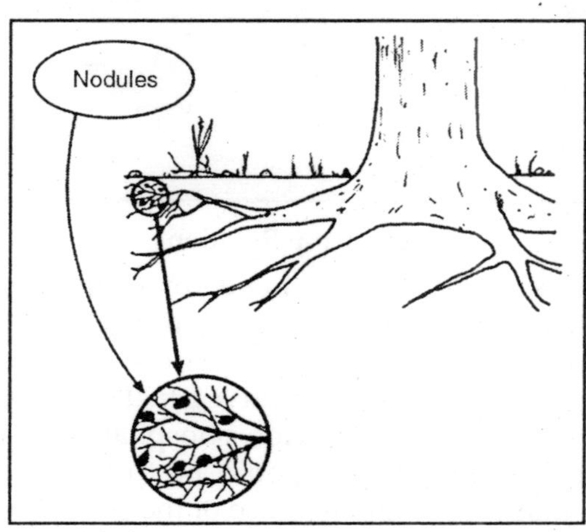

Fig. Nitrogen Fixing Bacteria

It may seem paradoxical that unfertilized plants could

suffer from a nitrogen deficiency while immersed in a sea of nitrogen gas, but it is just not available in the form they need. Of living things, only certain bacteria (and human chemists) have evolved the means to convert nitrogen from the atmosphere to a form useful to plants. But some plants, primarily legumes (peas and beans), have a symbiosis with these nitrogen fixing bacteria.

The plants provide nodules on their roots that protect the nitrogen fixing bacteria, which then enrich the soil around those roots. Not only does this permit the legumes to grow luxuriantly without nitrate fertilization, but it makes the soil fertile for other plants growing in the same soil later. The technique of *crop rotation* is one of the oldest techniques of agriculture. Scientists hope to be able to transfer the genes which direct the formation of the nodules to other crops.

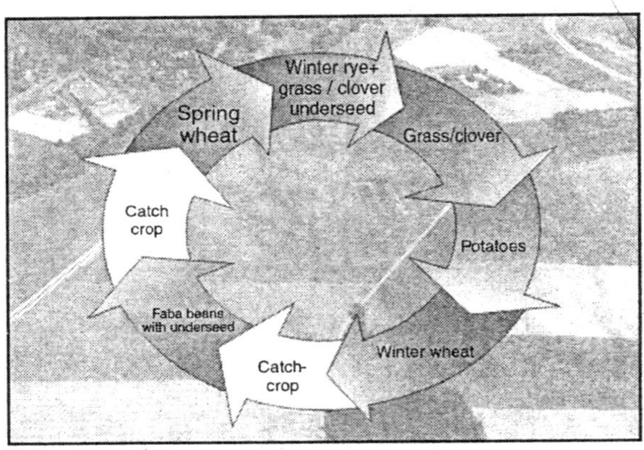

Fig. Crop Rotation

If this is successful, the need for fertilizers would be dramatically reduced. Unlike nitrogen, phosphorus is not a constituent of the atmosphere. There is no short-term likelihood that scientists will find a genetic engineering way to replace fertilizers that provide phosphates. The best hope for phosphate replacement would be to breed or engineer plants that make more efficient use of the phosphate available to them.

If it proves impossible to engineer plants for nitrogen

fixation, there are still options which can let them use fertilizers more efficiently. An enzyme called glucine dehydrogenase is involved in utilization of fertilizers.

The gene for glucine dehydrogenase is present in most crops, but it is expressed at low levels, because the control genes turn it off more than on. A genetic transformation of wheat which promoted increased synthesis of glucine dehydrogenase was 29 per cent more effective in utilizing the same amount of fertilizer as the unmodified variety. The increased efficiency can either be used to grow more crop on the same land, or to cut down on the need for fertilizer to grow the same amount of crop.

MORE FROM THE SUN

Plants derive energy from sunlight and use it to make sugar from carbon dioxide and water. This is photosynthesis. Scientists still do not have a complete understanding of how photosynthesis happens, although they know most of the steps. They know many of the genes which create the proteins needed for photosynthesis. They also know that there are differences in photosynthesis from one species to another. It happens that corn is an overachiever.

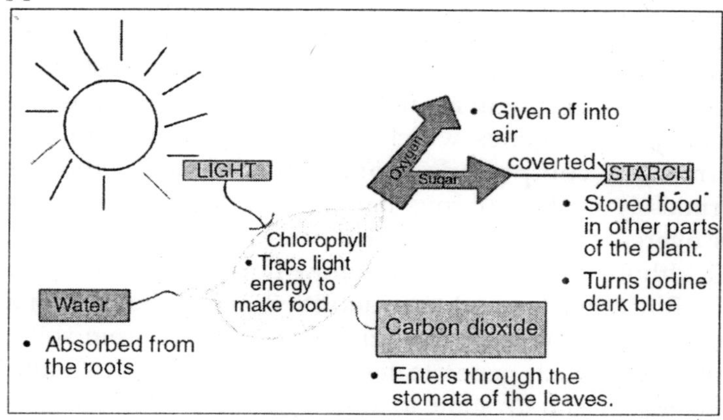

Fig. Photosynthesis

Corn plants make more sugar per unit of sunlight than any of the other grains. An international team of scientists from

Japan and from Washington State has transferred three of corn's photosynthesis genes into a rice plant. Early indications are that the transformed rice is more productive than the original rice variety. A more important potential application may be the development of very fast growing trees.

If global warming cannot be prevented by adding less carbon dioxide to the atmosphere, by burning less coal and oil, the only alternative is to depend on processes that remove it. Number one on that list is growing new trees.

Anything that makes agriculture more efficient can make more land available for growing trees. Anything that makes those trees grow faster removes more carbon dioxide from the atmosphere. For many environmentalists, preventing global warming is the highest priority. But proposed measures to restrict burning fossil fuels have encountered fierce political resistance. Opponents claim that such restrictions would be cost too much money, and would cost people their jobs, their comfort and their prosperity. By contrast, nobody loses anything if carbon dioxide is removed from the atmosphere by growing new trees.

Toxic Soils

Some soils are poor for plant growth because their mineral content is toxic. A high aluminum content is the most frequent problem, especially in acidic soils. But it has been possible to identify a few genes which enable some plants to extract aluminum compounds from soil and sequester them harmlessly in their fibrous parts. Recently, Florida scientists discovered a type of fern which can extract arsenic from the soil, although they do not yet know how the fern does this.

But other teams have identified genes that can enable plants to remove cadmium, zinc and mercury from soils. By transferring such genes to fast growing plants, it should be possible to clean up some toxic soils in much the same way as we can use bacteria to clean up oil spills. In the nearer term, there is the work of Mexican scientist Luis Herrera Estrella. He transferred into corn a gene that allows the plant to overproduce a natural chemical, citric acid, which it then excretes through the roots. Citric acid binds to aluminum and prevents the plant from taking it up from the soil. Herrera's approach is not to extract aluminum from the soil but to prevent it from passing from the soil to the plant. A much larger problem is salt-contaminated soil caused by irrigation. Rainwater is very pure, but water borrowed from rivers contains some dissolved salt. Over many years of irrigation, the salt accumulates.

But water cannot get from soil to roots if the soil water is saltier than the intracellular water.

In fact, water goes the other way, from plant to soil, and the plant dies. A gene was identified in a relative of cabbage. This gene enables the plant to pump salt from the soil into an isolated part of a cell, called a vacuole, where it is stored without harm to the plant. When salt is thus removed from the soil around the roots the plant can then take up the less salty water.

The salt-tolerance gene was experimentally transferred to a tomato plant, where a control gene keeps it turned on all the time. The resulting tomato plant is able to grow well in salty soils. Happily, the fruit is not high in salt, but the plant's stems, leaves and roots are loaded with salt, so after the growing season the plant parts could be shipped elsewhere, making the soil become less salty each year. It's one more case of an environmental problem that can be solved by gene transfer.

Biological Pest Controls

Farmers, for very obvious reasons, would prefer not to use pesticides.

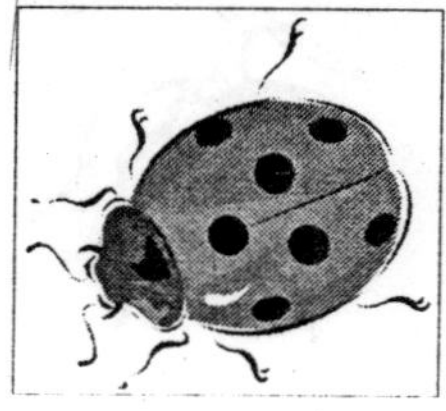

Fig. Ladybugs

They cost money and they are dangerous to use. Farmers

much prefer *Integrated Pest Management* (IPM), a system that combines many different methods of suppressing crop pests, including encouraging predatory insects.

Farmers even buy them. Agricultural distributors can supply such insects as ladybugs, praying mantids, lacewings and parasitic wasps.

Fig. Praying Mantids

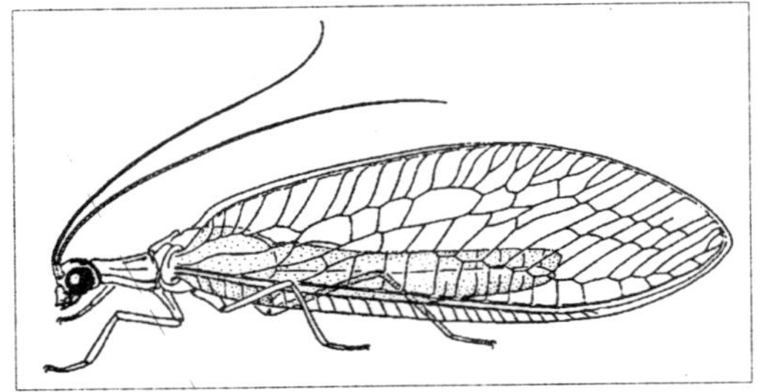

Fig. Lacewings

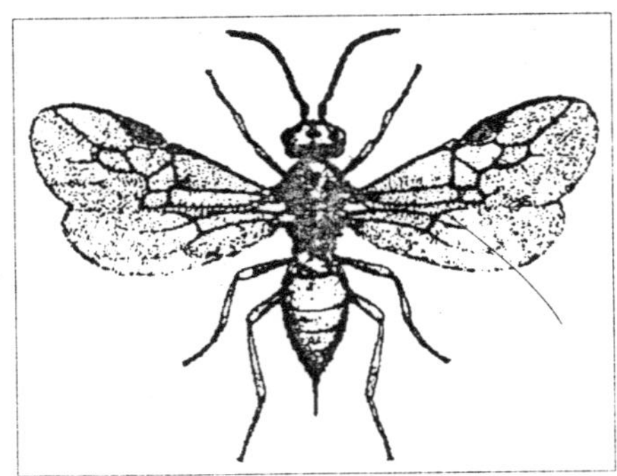

Fig. Parasitic Wasps

Integrated pest management includes using pesticides

when other means are insufficient. But when a crop is sprayed with conventional insecticide, the harmful insects are not the only victims.

Predatory insects may also be wiped out. Without any predators available, the pest populations can recover quickly, so that a second application of pesticide is required, which also kills the insect predators. This vicious cycle could be broken if genetical engineers can develop predatory insects resistant to common pesticides.

Rust Resistance

To a plant scientist, rust has nothing to do with oxidized metal. It is a plant disease caused by a fungus. It blights all the cereal crops, barley, wheat, oats, corn, millet and sorghum, *but not rice*. The rust fungus reproduces itself by forming club shaped cells called basidia.

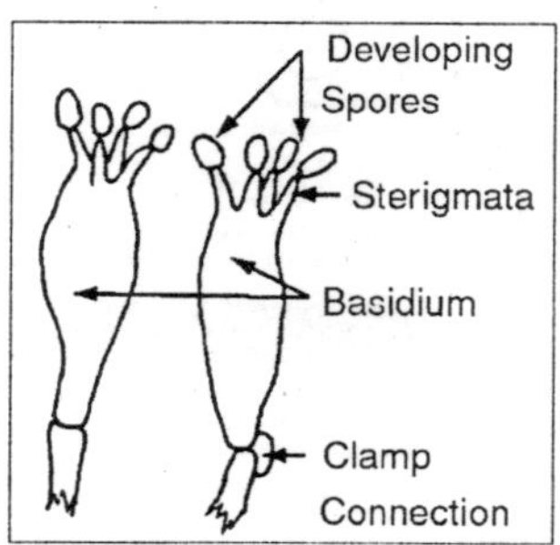

Fig. Basidia

Each basidium bears four spores. When the spores are ready, they are released and carried by the wind. The fungus infecting a single grain of wheat can easily produce millions of spores. The spores are so light that they can travel several times around the world before falling to the ground.

Although some varieties are more resistant to rust than others, no variety is immune. But rice must contain some combination of genes that confers immunity to rust. If these genes can be identified and if their function can be deciphered, it should be possible to transfer them to other cereal crops and end, once and for all, this most important cause of famine.

Fast Growing Trees

Making paper requires large amounts of natural cellulose. Some can be derived from recycling, but most of our paper is made from freshly cut trees. The best trees for paper-making are fast growing softwoods with low resin content, like aspens.

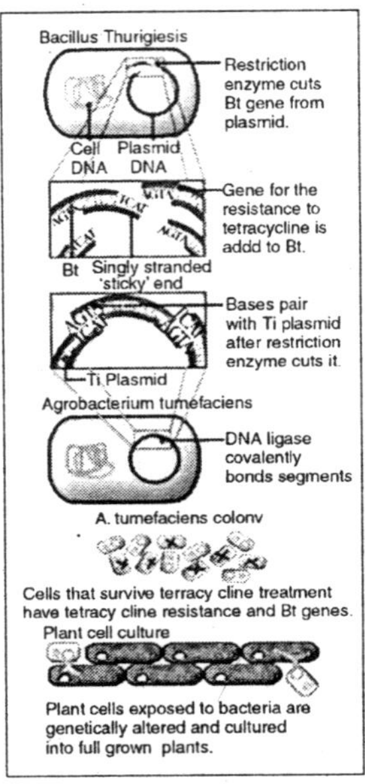

Fig. Genetic Engineering in Plants

Genetic engineers have transferred genes for pest resistance and herbicide resistance into aspen and have tinkered with the genetic switches that promote growth to create a fast growing aspen that could supply our paper needs using considerably less land. The paper-making process must bleach out the brown colour of lignin, one of the components of wood. The bleaches used to be dumped in the nearest rivers, an important and highly visible kind of pollution. This is no longer allowed, but the disposal of chemicals from paper mills is still a major headache.

FAST GROWING FISH

Most of the salmon we eat are caught wild, but some are

grown in farm ponds. It takes about three years for a salmon to grow from fingerling size to optimum marketing size.

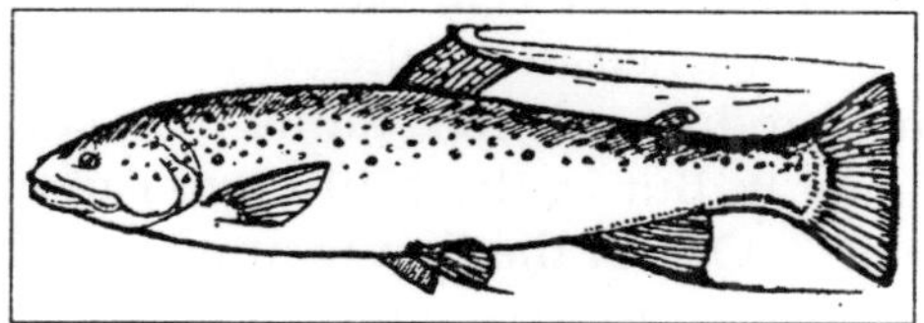

Fig. Salmon

In wild salmon a control gene turns on the gene for growth hormone, but only in the pituitary gland and primarily in warm water. So genetic engineers used a different control gene to turn on a growth hormone gene in cold water.

That control gene was transferred from an ocean pout, and it originally turned on a gene for a protein that helped the pout tolerate very cold water. The resulting creature looks and tastes just like the wild type salmon but it grows three times faster so it ought to be cheaper to produce. Wild salmon are now under environmental pressure from overfishing and because many of the streams where they lay their eggs are either polluted or inaccessible. If farmed salmon can economically replace more wild salmon, the pressure on this desirable species could be reduced dramatically. There's a need for fast growing fish in rice growing regions. Rice is planted in standing water, but it is harvested from dry ground. With a slow growing rice farmers often raised fish in the rice paddy alongside the young plants. But newer strains of rice mature almost twice as fast as traditional varieties. Although this lets farmers grow more crops per year, unfortunately the rice paddies are not flooded long enough to raise fish. If genetic engineers were able to make fish grow faster, the farmers could again exploit this valuable protein resource.

CONSUMER TRAITS

Most of the traits in the examples mentioned so far have been targeted at the producer. The no-till soybeans are cheaper to grow because ploughing costs money. The cotton is cheaper to grow because chemicals cost money. The chymosin from yeast is cheaper than chymosin from calves. The consumer doesn't know how much water or fertilizer the farmer used, or whether a salmon is one year old or three.

The vitamin enriched rice is the only one of the examples where the final product is better, rather than cheaper, for the person who eats it. But in the future we can expect to see "consumer traits". One of the first to appear will be potatoes genetically engineered to have a higher percentage of solids.

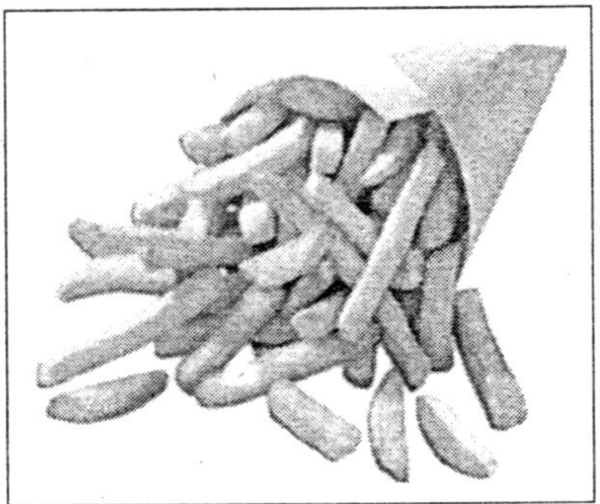

Fig. French Fried Potatoes

If we love french fried potatoes but don't like the calories, we will love the new potatoes. They will absorb less oil but stay crispier longer. Another valuable trait coming soon is coffee beans without caffeine.

Fig. Coffee Beans

Caffeine is now removed from coffee by a chemical treatment, invented by German chemist Ludwig Roselius. Many common foods are not safe for everyone. For example, peanuts cause a life threatening allergy in some people, especially children. Allergens are unusual proteins which are digested very slowly.

Someone who has an allergy to a widely used food needs to read the small print on product labels, quiz the waiter in a restaurant, etc. If a child has the allergy, the problem is that much harder to manage. But peanuts or other crops are now being genetically engineered to eliminate these allergens.

USDA scientists have identified the principal allergen in soybeans and have successfully developed modified soybeans which do not produce that allergen.

Fig. Soybeans

It is much easier to deactivate a gene, once its function is discovered, than it is to transfer a gene from one organism to another.

The same antisense trick that was used to delay ripening can be used to suppress synthesis of an allergen. Therefore, once a gene has been identified which codes for an allergenic protein, the technology to eliminate that allergen from food crops is relatively easy, unless the allergenic protein is important to the life processes of the plant.

In addition to food and fibre, genetic engineers are working to modify grass. Farmland constitutes mankind's biggest footprint on the earth, but lawns, athletic fields, golf courses, etc. have the biggest footprint in some communities. Environmentalists have many criticisms of lawns. They must be mowed regularly, which uses gasoline, creates noise pollution, and takes up people's time.

In addition, the growing grass uses a very significant amount of water, fertilizers and pesticides, to make longer leaves which are then cut off by the mower.

But genetic engineers are trying to develop a variety of grass which reaches a desired length and then dramatically slows its growth.

Combined with pest resistance genes, this new grass would be almost as simple to maintain as astroturf.

Chapter 43

Genetics of Schizophrenia

Schizophrenia is one of the more complex and less understood of the psychological disorders. For many years, studies have explored the genetic aspect of this syndrome in an effort to provide a better understanding of its etiology, symptoms, course, and development. Much of what is involved in schizophrenia still remains unclear, although great strides have been accomplished by research in the area.

In trying to understand the etiology of psychiatric disorders, one cannot avoid studying gene—environment interactions. Just as genes influence the susceptibility of an individual to the development of almost any disorder, environmental factors are also seen as important etiological contributors to disturbance. Consequently, whether or not genetic or environmental variation is involved in the development of schizophrenia is no longer considered to be an important point of contention in this area of research.

Instead, most of the current investigations appear to be concerned with acquiring a more thorough understanding of the magnitude and relevance of both environmental and genetic variation and the ways in which these interact to establish the various manifestations of schizophrenia.

Today there seems to be little doubt that a genetic basis for schizophrenia exists. However, the specific schizophrenia-producing genetic components and the method by which they are transmitted remain unclear. Genetic investigations have involved primarily three methodologies: family studies, twin studies, and adoption studies. Despite some difficulties, these methods have obtained strong evidence pointing toward the involvement of genetic factors in the development of schizophrenia. Currently, the amount of interest and speculation in this area of research is high, and on the basis of findings, sophisticated models and new, more innovative research techniques are being developed.

The major conceptual, diagnostic, and methodological difficulties inherent in this area of investigation, the evidence for the existence of genetic factors in the etiology of schizophrenia, and the models for a genetic mode of transmission that have arisen from these lines of research.

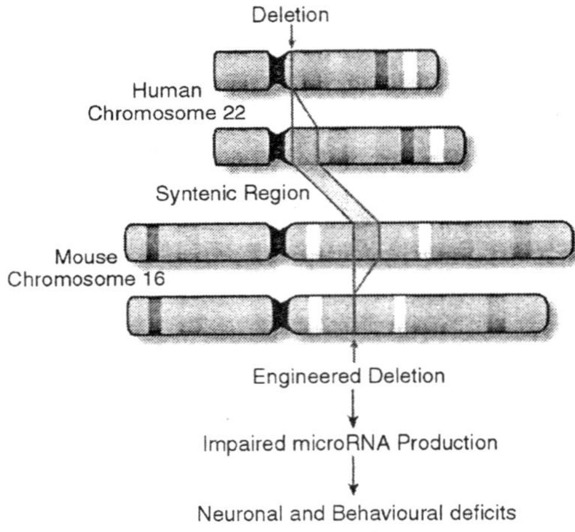

Fig. Genetics of Schizophrenia

One of the more difficult aspects of this field remains the lack of knowledge concerning the nature of schizophrenia. To date, no homogeneous group of schizophrenics has been isolated and no unitary etiology for the disorder has been found.

The possibility exists that schizophrenia is not a single, unitary disorder, but may consist of a number of different entities. Schizophrenic heterogeneity could be present at several levels. Seemingly similar phenotypic expressions may be due to quite different etiologies. As a result, the reliance upon symptomatology for the differentiation of schizophrenic subtypes may be grossly misleading. To complicate matters further, the complexity of gene operations makes it difficult to delineate the various interacting etiological components of a schizophrenic disorder.

The nature of the activating environment and the timing of that activation in addition to the gene could all combine to produce a single phenotype. Therefore, a range of possible phenotypic outcomes exists for every gene as a function of the activating environment and the time of activation. Depending on the environmental circumstances, there are a number of different genes that can produce a single phenotype. Consequently, several genetic mechanisms may underlie the expression of a single characteristic.

If schizophrenia is in fact heterogeneous, many of the present research strategies may prove inadequate because most rely on the implicit assumption of homogeneity among patients studied. A schizophrenic sample based on an overinclusive diagnostic category would result in variable schizophrenic performance. As a result, common findings among schizophrenics would be difficult to detect and valid hypotheses may be prematurely discarded on the basis that the data fail to meet statistical significance.

The failure to reach a clear consensus on the nature of schizophrenia has resulted in conflicting viewpoints. To date, genetic studies of schizophrenia have relied almost exclusively on the assumption of a model in which schizophrenia consists of a complex disorder amenable to the same type of study and displaying the same pathogenic characteristics as an inheritable disease.

Other points of view, especially those among the psychodynamic and contextualist schools of thought, hold that the conceptualization of schizophrenia as a disease entity is inadequate and unjustified. For example, Sarbin and Mancuso claim that schizophrenia is a label and the product of social processes resulting in the imposition of a moral verdict in response to certain undesirable Behaviours. Others have proposed that certain forms of familial interaction predispose an individual to schizophrenia, whereas still others have preferred to focus on the schizophrenic's intrapsychic world and dynamics. An interesting model postulated by Meehl incorporates both genetic and environmental factors to explain the occurrence of schizophrenia. Meehl suggests that all that is inherited in schizophrenia is an integrative neural defect that he labels *schizotaxia*.

The imposition of a social learning history on schizotaxic individuals results in a schizotypic personality organization. That is, all schizotaxic individuals obtain a schizotypic personality organization regardless of environmental conditions. Most of these individuals will remain compensated. However, those who encounter certain detrimental influences (e.g., schizophrenogenic mothering) will decompensate into schizophrenics. In light of this model, only a few of those individuals who have inherited the neural defect and are thereby predisposed to schizophrenia, will manifest the disorder.

Despite the wide disparity among theoretical models, most can be classified as elaborate extensions of the nature versus nurture controversy. It is apparent that this debate cannot be adequately resolved until a broader base of knowledge is developed.

Until that time, it may be wise to bear in mind that genetic research does not necessarily negate endeavors supporting contextualist and psychodynamic conceptions. Instead, genetic research addresses both environmental and genetic factors, and is interested in delineating the effects of each separately and in interaction as they pertain to schizophrenia.

DIAGNOSIS

Another research difficulty related to that of schizophrenic heterogeneity pertains to clinical diagnosis. The new DSM-III allows for considerable heterogeneity as to who will be diagnosed schizophrenic; persons need to fulfill only a certain number of criteria out of a standard set of possibilities in order to receive the diagnosis.

Schizophrenia cannot be diagnosed unless signs of the disorder have been present for at least 6 months. Illnesses of shorter duration, or illnesses that have resulted in the face of a psychosocial stressor and appear to be transient in nature are no longer included.

In fact, these disorders are not even considered to be subtypes of schizophrenia. These changes in diagnostic criteria have left schizophrenia researchers in a peculiar bind since they now must try to define the onset of a schizophrenic disorder. In the past, the onset of schizophrenia was demarcated by the appearance of psychotic symptoms. However, it is well known that the onset of schizophrenia is often quite insidious, and that prodromal features commonly predate psychotic symptoms.

Logically, it would be best if we could define onset as the point in time when one could predict with virtual certainty that a schizophrenic psychosis would occur. Unfortunately, we still lack the necessary knowledge to make such a prediction, and the definition of onset remains strictly arbitrary. Consequently, the conceptual problem of trying to determine criteria for delineating signs along a continuum marking transitions from a schizophrenia-prone personality structure to a prodomal syndrome to a full-blown psychosis still exists.

Because of these diagnostic difficulties, a standard definition for what constitutes schizophrenia remains to be determined. The absence of a standard definition remains a major hindrance to the progression of this field. The severity of the problem is due to the fact that every research project must of necessity concern itself with identifying those people who have the disorder.

Therefore, the homogeneity of the sample studied and the rigor of the investigation carried out on the sample depend on the accuracy of the diagnosis. Unfortunately, there is no independent validation and only moderate reliability for diagnosing a schizophrenic disorder. Even the combination of moderate interrater reliability and independent validation appears to be inadequate. Consequently, there is no truly homogeneous sample in schizophrenia research. Instead, investigators are forced to contend with the heterogeneous groupings that fit the schizophrenic label, the abundance of contradictory data, and the numerous disagreements over research findings. As long as this state of affairs exists, studies are limited to statistical rather than clinical significance.

At present, many diagnostic difficulties result because little is known about the unity or diversity of schizophrenia. In order to gain a clearer understanding of the nature of the disorder, many investigations have been conducted to determine the degree to which the range of phenotypic expression in schizophrenia is paralleled by genetic heterogeneity. These included certain studies that were designed to investigate whether these subtypes were fundamentally alike or different.

One such study by Winokur, Morrison, Clancy, and Crowe found that hebephrenic probands had more than three times as many schizophrenic relatives as paranoid probands. These authors determined that both children and other secondary cases of schizophrenia in the families of hebephrenics were more likely to be diagnosed hebephrenic as opposed to paranoid schizophrenic. The same type of pattern was established for paranoid schizophrenics.

The children and other secondary cases of schizophrenia in the paranoid probands were more likely to manifest paranoid rather than hebephrenic symptoms. These familial data led the au thors to conclude that process schizophrenia is not likely to be a unitary illness. Instead, they postulated the existence of two types of schizophrenia in the population studied, one of which may manifest itself mostly as hebephrenia but also occasionally as paranoid schizophrenia.

The second type would manifest itself exclusively as paranoid schizophrenia in which there are relatively few schizophrenic relatives. Together, these findings suggest that there may be only one type of hebephrenic schizophrenia, but two types of paranoid schizophrenia, one related genetically to the hebephrenic subtype.

Another study by Kety et al. in 1975 demonstrated that the diagnoses of chronic, latent or borderline, and uncertain schizophrenia tend to cluster in the biological relatives of schizophrenic adoptees. Similarly, a study by Slater and Cowie examined the classical pre-Kraepelinian categories and found a significant association between schizophrenic probands and their schizophrenic offspring with respect to type of category.

Kringlen in 1968 corroborated these results and also demonstrated that the entire set of Kraepelinian subtypes is represented in the children of every type of schizophrenic proband.

These studies are typical of investigations in this area. Their results seem to indicate that some groups of disorders are genetically related, whereas the relationships among others remain unclear. Despite some contradictory findings, the most common results from these studies provide evidence for a common genetic basis among various forms of schizophrenic disorders.

In general, the findings indicate that cases of all subtypes occur in the families of each schizophrenic subtype. They also suggest that only a modest tendency toward subtype resemblance exists within families.

In addition to considering the unity or diversity of schizophrenia, one must also consider the relationship between schizophrenia and other psychiatric disturbances when making clinical diagnoses. This is especially important if one wants to formulate diagnostic criteria that are successful at differentiating among various psychoses.

At present little is known about the specificity of the schizophrenic disorder. However, there is some evidence suggesting a genetic relationship between schizophrenia and other psychiatric disorders. In a study by Reed, Hartley, Anderson, Phillips, and Johnson, a high risk of psychoses for the siblings of psychotics was found.

When the investigators looked at specific disorders, they discovered that the risk for first degree relatives of affective probands to develop a similar disorder was less than their risk of becoming schizophrenic. There is also some indication that the risk of schizophrenia for the offspring of paranoid and hebephrenic schizophrenics is lower than the risk for children of manic—depressives. Even some of the very earliest studies discovered that schizophrenic and affectively ill individuals were about equally prevalent in the families of schizophrenics. Today as well, despite efforts to select clinically homoge neous probands, the occurrence of different types of psychoses in the same family is difficult to avoid.

The suggestion that there may be a similar genetic basis to schizophrenia and other forms of psychoses raises the possibility of overlap in the symptomatology of various psychiatric disturbances; and indeed, this has been found to be the case. Another problem exists in the possibility that symptomatology is not directly related to genetic inheritance. Bleuler speculated that secondary symptoms such as delusions and hallucinations are not inherited. He also suggested that the primary symptoms do not correspond to the transmitted anomaly itself.

As a result of these possibilies, there are some major difficulties involved in basing diagnostic criteria on clinical symptoms. However, there are now some indications that future diagnostic categories will be based on biochemical indicators rather than Behavioural symptoms in order to decrease the heterogeneity of individuals in specified subgroups of psychopathology.

Using biological markers could be advantageous in several ways. First, it could avoid the genetically heterogeneous diagnostic groupings based on standard methods of symptom classification. It is well known that even relatively well formulated clinical subtypes of schizophrenia present diverse symptomatology.

As a result, it is impossible to differentiate between genotypic and phenotypic variations under the present diagnostic systems. The use of biologically independent variables in formulating diagnoses may promote the detection of biological etiologies and the formation of more homogeneous psychiatric diagnostic categories. Individuals who exhibit divergent or less severe clinical symptoms and are thereby excluded from present investigations could be included if they met a biological criterion.

This in turn would allow the high risk population (those

individuals with the salient factor) to be identified and studied. The use of a biologically based strategy would also permit the study of non hospitalized and unmedicated individuals, thereby avoiding the inclusion of these confounds within investigations.

In addition to developing biologically based criteria, many other efforts are being made to improve the diagnostic system. For example, the United States— United Kingdom Diagnostic Project, the International Pilot Study of Schizophrenia, and the World Health Organization have provided standardized input on patients for formulating diagnoses, experts to address diagnostic issues, and cross national reference points in order to establish criteria that will improve the reliability and validity of the diagnosis of schizophrenia on an international basis. Additional reasons for optimism stem from the new trends in schizophrenia nosology.

These include an increased reliance on affective symptoms as diagnostically useful, as well as a shift toward incorporating the schizo affective disorders within the affective diagnoses. In addition, the broad use of the schizophrenic label seems to be decreasing as a function of less emphasis being placed on classic schizophrenic symptoms, and more effort being devoted to the use of operationally defined criteria tested for reliability.

Although at present it is still difficult to reach clear and definite conclusions in research dealing with schizophrenia because of the many diagnostic difficulties associated with this disorder, the increasing refinement of diagnostic issues and the development of more sophisticated research techniques makes it possible to conceive of removing the current obstacles to gathering clinically significant data and conducting therapeutic—preventive research.

FAMILY STUDIES

Despite the numerous conceptual and diagnostic difficulties inherent in schizophrenia research, much evidence has been gathered to substantiate the existence of genetic components in the transmission of the disorder. The idea that various forms of mental illness are heritable has existed for many centuries. However, it was not until early in this century that researchers were able to provide actual documentation of this fact.

The first consanguinity study that found a significant increase in the prevalence rate of schizophrenia among relatives of index cases in comparison to the rate of the general population. More specifically, he discovered an age corrected risk for schizophrenia of 4.48 per cent and 4.12 per cent for other psychoses in the siblings of schizophrenics whose parents were unaffected.

Many other studies have confirmed Rudin's basic finding of a higher rate of schizophrenia in the relatives of samples of schizophrenics than in the samples from the general population. From these studies the rate of schizophrenia in the general population has been calculated to range from.35 to 2.85per cent , with a mean around.85per cent.

The findings from the consanguinity studies also reveal a clear correlation between the degree of blood relationship in relatives of an individual affected with schizophrenia and the risk of those relatives for becoming schizophrenic. More specifically, those relatives who share fewer genes in common with an affected individual show a lower expectancy rate for the disorder than those who share more genes.

In addition, these studies have demonstrated that there are no significant differences between prevalence rates for siblings and other first degree relatives. Although the difference found in the rate of schizophrenia between parents (5.5 per cent) and siblings and children of schizophrenic probands (10.2 per cent and 13.9 per cent respectively) appears to be large and therefore contradictory to the foregoing statement, this difference has been explained by the fact that these parents are largely selected for a degree of mental health through having been married and having children.

When a correction is made for this factor, the risk in parents of schizophrenics becomes close to that of siblings and children. The somewhat higher risk in children than in siblings of schizophrenics is thought to be due to sampling error.

Additional evidence for a genetic factor in the transmission of schizophrenia is found by comparing the different rates of prevalence for the disease among offspring of matings of various psychotic combinations of parents. Estimates have been obtained indicating that the risk for schizophrenia increases as the genetic loading increases. The best estimate of the risk of schizophrenia in the sibling of a schizophrenic is 9.7 per cent when neither parent has the disorder.

This risk increases to 17.2 per cent for the siblings of the proband when one parent also has schizophrenia. When both parents are schizophrenic, the risk to their children ranges between 36.6per cent and 46.3 per cent. Although it can be argued that this raise in risk from 17.2 per cent to 46.3per cent for the siblings of a schizophrenic may be due to the extreme environmental conditions created by the mating of two schizophrenics, a study by Kallman reported that a comparable disturbed setting created by the mating of a schizophrenic and a psychopath yielded a sibling risk for schizophrenia of only 15per cent in Gottesman & Shields,. The basic findings of these studies have been further substantiated by the results of more recent investigations. For example, a study by Reed et al. reported that the risk to the offspring when neither parent was psychotic increased relative to the number of aunts and uncles who were psychiatrically disturbed.

The difference among estimates of risk is a fact significant for genetic theories. Not only does this difference demonstrate that the risk of mental illness increases as a function of increased genetic loading, but the large amount of variation in the offspring undermines the prediction of environmental theories that virtually all children from dual matings of schizophrenics would be emotionally disturbed.

Another line of evidence for the role of genetic factors in schizophrenia was provided by Manfred Bleuler, who applied the methodology of consanguinity studies to clinical symptomatology and course over time. He discovered that the biological relatives of schizophrenics who also had the disorder tended to have similar ages of onset, presenting clinical pictures, lengths of duration of illness, and outcomes. These findings can be interpreted in either an environmental or genetic fashion. However, a genetic explanation appears to carry more weight because many of the relatives in the study were geographically separated and in some cases did not even know each other. Therefore, the findings of the study indicate that genetic factors play a role in several clinical parameters of schizophrenia.

Although these family studies have contributed much toward the understanding of schizophrenia and have pointed to the role of genetic factors in the development of this disorder, some major difficulties are inherent in their methodology. The earliest family studies did not question the assumption that schizophrenia is heritable and therefore failed to consider the possibility that environmental factors such as family interactions also contribute to the etiology of the disorder.

As a result, these studies provided sources of data on the empirical risk of the occurrence of schizophrenia in different kinds of relatives, but did not elucidate the specific etiology involved in creating those risks. In other words, these studies could indicate if a trait ran in families, but could not specify what aspect of the family determined the trait. Obviously, families share more than genetic similarity, they also share a variety of social, psychological, and cultural experiences. Consequently, the results of these studies can just as validly be interpreted as supporting an environmental position.

Recently, some studies taking an environmental approach have investigated family interactions on the premise that certain conditions of family life can predispose an individual to schizophrenia. The majority of these investigations have focused on deviant role relationships and disordered communication processes among family members.

The findings of these studies generally indicate that deficits in family communication are highly correlated with schizophrenia and high risks for schizophrenia in offspring. In addition, there are some indications that families of schizophrenics contain poorly articulated and rigid role structures in the areas of dominance, autonomy, and sexuality. Despite the success of this type of research, it is unable to establish the occurrence of schizophrenia as a strictly environmental outcome.

In fact, the evidence that is presently available tends to suggest that both genetics and family process play a role in the development of schizophrenia. In light of this evidence, the current mode of investigation has changed to incorporate biological, genetic, psychological, and social factors in the context of complex, multifactorial models of causation when approaching family variables in relation to schizophrenia.

TWIN STUDIES

It was not until the advent of twin studies that the role of genetic and environmental factors in schizophrenia could be more clearly defined. The power of the twin method resides in the environmental similarities and genetic differences between monozygotic and dizygotic twins. Both types of twins share the same intrauterine environment simultaneously with their co- twin.

However, monozygotic twins share the same fertilized ovum initially, whereas dizygotic twins are no more genetically alike than ordinary siblings. As a result, one can assume the presence of a genetic factor in determining a trait if the concordance rate for that particular trait is significantly higher for monozygotic as opposed to dizygotic twins.

The methodology of twin studies also provides certain advantages. For example, through the comparison of intrapair resemblance, it becomes possible to identify subtypes or components of schizophrenia that may be more under genetic or environmental control than others. Also, by noting the variability of abnormal personality traits in the monozygotic co-twins of typical schizophrenics, it becomes feasible to identify schizophrenic "equivalents" or schizotypes.

The comparison of monozygotic twins who differ in outcome will permit the formulation of inference about the relationship between life experiences and outcome. It must be kept in mind that the twin study method relies on certain assumptions. These specify that monozygotic twins as such are not specifically predisposed to schizophrenia, and that the environments of monozygotic twins are not systematically more similar than those of dizygotic twins in such a way as to foster the disorder.

Also, when assessing the outcome of these studies it is important to be aware of the usage of statistical age corrections and their effects on research findings. Age corrections incorporate estimates of a lifetime risk for developing schizophrenia in order to compensate for a current sample while the disorder exhibits a variable age of onset. Although several methods have been developed, none is entirely satisfactory, and all present only varying degrees of approximation to the true concordance rate.

It is essential to realise that the absolute size of the concordance rate is not predictive of the relative importance of the genetic contribution to the trait in question. Concordance rates can be expressed in several fashions. The basic difference in methods depends on whether concordance is expressed in terms of the pair or the individual.

The casewise or proband method refers to the proportion of cases with an affected partner, whereas the pairwise rate considers the proportion of pairs in which both twins are affected. The individual or proband concordance method yields a higher percentage than a pairwise concordance. Therefore, only those studies that use the same method are comparable.

The most influential early twin studies were conducted by Kallman and Slater. Even though the two studies contained important differences and suffered from methodological problems, both found a significantly higher concordance rate for schizophrenia among monozygotic twins.

By the end of Kallman's study and with some diagnostic revision, but without statistical age correction, 59 per cent of 174 monozygotic pairs and 9 per cent of 517 dizygotic pairs were concordant for what Kallman labeled "definite schizophrenia." In Slater's study, 65per cent of 37 monozygotic pairs and 14 per cent of 58 same sexed dizygotic pairs were found to be concordant. Both studies used the pairwise method for calculating the concordance rates. Findings from other earlier studies were basically consistent with these results.

More recently, several studies have been conducted that are methodologically powerful, and that provide highly significant results. One of these studies, reported by Tienari, was conducted on an exclusively male sample in Finland. Initially, this investigator found a zero concordance rate for monozygotic twins.

However, by 1976 the pairwise concordance rate for the sample had risen to 16 per cent. The use of the proband method, the concordance rate for the sample was found to be between 35per cent and 53per cent depending upon which borderline cases were included. As a result, even though it took some time, a significant monozygotic—dizygotic difference was demonstrated.

A study by Kringlen was conducted in Norway and included both male and female twins. His study was especially important for several reasons. His sample was well constructed and contained the largest number of psychotic twins personally investigated among the more recent studies. His investigation was very comprehensive and included case history, pedigree, clinical analysis, rating scale data, and all functional psychoses in addition to schizophrenia. Using the more conservative pairwise method, Kringlen found a significant difference between monozygotic and dizygotic twins with concordance rates ranging from 60 per cent for monozygotic probands rated as severe to 25 per cent for the most benign.

His results indicated no difference between the concordance rates of monozygotic twins as a function of gender. He reported that whether the sexes of the dizygotic twins were the same or different, the concordance rates did not differ, but were equal to those found in ordinary siblings. These results counter findings from some earlier studies and undermine the psychodynamic thesis that concordance for schizophrenia will increase without regard to genetic similarity, but relative to the amount of likeness and resulting identity confusion among siblings. The findings of another important twin study were reported by Fischer, Harvald, and Hauge in 1969. This study was conducted in Denmark upon a sample that was very representative with respect to twin types and sexes of probands. Only those probands who met strict criteria for chronic or process schizophrenia were included.

Even so, schizophreniform, paranoid, and atypical psychoses in co-twins led to data analysis for quantitative concordance for other than process schizophrenia. As a result of this study, Fischer et al. discovered a pairwise concordance rate of 24 per cent for monozygotic twins and 10per cent for dizygotic twins.

In addition, the authors calculated the rates according to the proband method and obtained a 36 per cent concordance rate for monozygotic and 18 per cent concordance rate for dizygotic pairs. When those co-twins who exhibited a form of schizophrenic disorder different from process schizophrenia were included in their calculations, the investigators obtained proband concordance rates of 56 per cent and 26per cent for monozygotic and dizygotic twins respectively.

These figures appear to be good approximations of the true incidence rate, as the twins used in this study were followed long enough to virtually eliminate the need for any statistical age correction.

Fisher went on to report that monozygotic twin pairs concordant for schizophrenia were not at an increased risk for the disorder when compared to discordant monozygotic twins. She did not discover a relationship between severity of disturbance in the proband and the risk of developing psychosis in the co-twin.

The offspring of concordant twins were not at an increased risk for schizophrenia when compared to the offspring of discordant twins. In addition, the likelihood of a discordant twin member producing a schizophrenic child was equal to that of the schizophrenic member of the twin pair. These findings point toward a genetic hypothesis and indicate that the genetic transmission of schizophrenia is not always manifested directly.

The Maudsley Hospital based study of Gottesman and Shields conducted in England obtained similar results to those of the studies reported earlier. These investigators also obtained a significantly increased monozygotic concordance rate over that found in dizygotic twins.

Similarly, Pollin, Allen, Hoffer, Stabenau, and Hrubec, in a study based on an American sample, found a higher concordance rate for monozygotic as opposed to dizygotic twins. More specifically, the concordance rate for schizophrenia in monozygotic twins was found to be 3.3 times greater than the rate in dizygotic twins.

What is interesting about this study is that the results were based on an exclusively male sample consisting of almost 16,000 twin pairs who had successfully passed army induction exams and who had served in the armed forces during the period spanning World War II and the Korean War.

The calculations for the concordance rates were based on chart diagnoses obtained from VA records or responses to mailed questionnaires. As a result, the consistency of the findings from this study appear somewhat amazing when one considered the size of the sample, its apparent skewness toward healthy individuals, and the tendency for wartime nonpsychiatric physicians to use the diagnosis of schizophrenia more loosely.

This last factor tends to lower the homogeneity of the schizophrenic sample and the accompanying concordance rates for schizophrenia. Nevertheless, a monozygotic to dizygotic twin concordance ratio of 3.3 was obtained, and the authors concluded that the genetic loading for schizophrenia was higher than for any other psychiatric or medical diagnosis considered in their investigation.

At present, all the major twin studies carried out worldwide have demonstrated higher concordance rates for monozygotic as opposed to dizygotic twins. The monozygotic concordance rates from studies conducted before 1965 tended to average around 60 per cent or more, whereas the results from more recent studies are usually lower. These differences have been attributed to variations in methodological and diagnostic procedures. Even so, the consistency of twin studies' findings despite diversity in design and execution provides strong evidence for a hereditary factor in the etiology of schizophrenia.

From these studies it is apparent that identical twins per se are not more prone to develop the disease. However, if one twin does develop the disorder, the co-twin is at an increased risk over an ordinary sibling or dizygotic twin. In addition, a frequent finding from twin studies has been that concordance is higher for index cases with severe forms of schizophrenia.

The findings from these studies contradict the suggestion that the stress inherent in rearing twins plays a role in the development of schizophrenia by demonstrating that both monozygotic and dizygotic twin pairs are not at an increased risk for the disorder. In fact, the research literature in general suggests that life events and stress do not account for the development of a schizophrenic disorder.

Despite the usefulness of the twin method in demonstrating that genetic factors play a role in the etiology of schizophrenia, it is still important to consider the ways in which environmental factors may have contributed to the findings reported. Jackson suggested that monozygotic twins may be more prone to develop schizophrenia because of identity confusion and/or weak identity formation.

However, this idea has failed to be substantiated, as no excess of monozygotic twins has been discovered in schizophrenic samples. The possibility of one twin identifying with the other, in this case disturbed twin, has also been considered as a psychological factor that might lead to higher concordance rates in monozygotic twins. Usually this is referred to as *folie à deux*, or psychosis by association, a clinical entity that would result in clinical resemblances between the psychoses of identical twins.

Again, there appears to be a lack of evidence to support this position. There are simply too few cases of shared delusions or other symptoms of intense identification of one schizophrenic twin with the other to support the theory.

Another environmental explanation that has been put

forward postulates that both members of monozygotic twin pairs have a greater likelihood of being exposed to a schizophrenia inducing situation than do members of dizygotic twin pairs. This position would entail the assumption that monozygotic twins are treated more alike than dizygotic twins.

In fact, Kallman reported higher concordance rates in twin pairs living together for 5 or more years before the onset of the disorder than in pairs separated during that period. Even so, studies conducted on identical twins where one or both members later became schizophrenic reveal similar concordance rates for those reared apart when compared to those reared together. In addition, pairs of monozygotic twins reared apart reveal much higher concordance rates for schizophrenia than dizygotic twins reared together.

One criticism of the twin method addresses the assumption that the sharing of the intrauterine environment simultaneously by both members of the monozygotic twin pair markedly reduces the developmental and environmental variations for the members of the pair. Instead, it is often the case that one twin member is put at a disadvantage and receives less of the limited space and nutrients available than the other twin. Consequently, the disadvantaged twin may appear weaker and receive special care and attention from the parents. This parental Behaviour may in turn foster dependency and submissiveness in the individual.

In a study conducted by Pollin et al. 1966, the conclusion was reached that low birth weight was an important contributing factor to the development of the schizophrenia. However, this finding did not appear to be replicated in other studies. For example, Kringlen found no general tendency for submissive, mother dependent twins to become schizophrenic. In fact, his data suggest that birth weight and obstetric complications are not correlated with the later development of schizophrenia.

Instead, he concluded that the twin who is predisposed to become schizophrenic, is predisposed in a nonspecific manner. The conclusions in this area remain unclear however. Other more recent analyses reveal that the schizophrenic twin is often lighter at birth in discordant twin pairs. Even so, the differences in birth weights are usually small, and the schizophrenic twin's weight is usually within the normal range.

Despite the continuing controversy in this matter, it is important to remember that even if some conclusive evidence could be obtained with regard to birth weights, it would still throw little light on the etiology of schizophrenic disorders because lowered birth weights can result from numerous factors and may create any number of reactions that may or may not be etiologically linked to schizophrenia.

Taken together, the findings from the twin study method have strengthened the view that genetic factors are significantly involved in creating the higher monozygotic concordance rates reported in the literature. This is not to say that environmental variables are not important in the etiology of schizophrenia. However, the findings do point out the vital contribution twin studies have made for the genetic position in this area of research.

ADOPTION STUDIES

In the adoption strategy there are basically four approaches that can be used to separate genetic factors from other possible etiological factors related to familial Behaviour and environment the adoptees' families' method, schizophrenic adoptees are selected as index cases to be compared to a control group. The control group consists of other adoptees who are free of any psychiatric history but who match the index cases with respect to age, sex, age at transfer to adoptive parents, and social class of adoptive parents. The difference in incidence of schizophrenia is then determined for both the biological and adoptive relatives of the index and control groups.

The adoptees' study method consists of finding schizophrenics whose children were placed for adoption at an early age so as to compare the prevalence ratings of these children with either a matched control group or the population in general. A third strategy, the adoptive parents' method, starts with the adoptee schizophrenic index cases and then compares the incidence of pathology in their biological and adoptive parents.

Cross fostering is perhaps the most sophisticated research strategy among adoption studies. With this research design two different groups of adoptees are identified: those who have a biological parent who is schizophrenic, but whose adoptive parents are free of psychiatric illness, and those whose biological parents have no psychiatric history, but who are adopted by a schizophrenic. The two groups are then compared with respect to incidence of schizophrenia and other psychiatric disturbances.

One study that uses the adoptees' method was conducted by Heston. He found that the offspring of severely schizophrenic mothers who were given up for adoption within the first 3 days of life still grew up to have schizophrenia at the same rate as expected for those reared by their schizophrenic mothers.

In addition, the offspring from schizophrenic mothers showed considerably more abnormalities other than schizophrenia than would be expected for the general population. A later and more elaborate study conducted on a Danish population confirmed Heston's results.

Another Danish study conducted by Kety, Rosenthal, Wender, & Schulsinger identified adoptees who became schizophrenic and then through the use of documents determined the status of their biological and adoptive parents, siblings, and half siblings. The purpose of the study was to discover if the schizophrenia in the adoptees was a function of schizophrenia in the adoptive or biological families. Again, supporting evidence was found for a genetic factor in the transmission of schizophrenia.

It is important to note that Rosenthal used a broad diagnostic concept, the "schizophrenic spectrum" in their investigations. Rosenthal included borderline and schizoid or paranoid diagnoses along with the diagnosis of schizophrenia; and Kety included chronic, acute, or borderline schizophrenia, definite or uncertain, in addition to inadequate personality in the schizophrenic spectrum. Both studies reported that schizophrenic spectrum index cases showed an increase in the rate of schizophrenic spectrum disorders only in their biological relatives. Besides indicating that schizophrenia is heritable, these studies also offer evidence that the genetic potential for schizophrenia may manifest itself in ways other than overt schizophrenia.

Since these findings were reported, Kety has conducted interviews with the relatives of schizophrenics and the control adoptees in the field, in addition to increasing his sample size. Similarly, Rosenthal increased his sample size of index adoptees and interviewed the spouses of the schizophrenic parents. In general, the more recent data from these studies confirmed the findings of the earlier reports. More specifically, there were more schizophrenic and schizophrenia related disorders in the biological relatives of schizophrenic adoptees and in the adopted away children of schizophrenic parents than in the control groups.

The cross fostering technique, also found a greater prevalence of psychopathology among the adopted away offspring of schizophrenics, but no such increase among the cross fostered group. The investigators concluded that genetic factors play a role in the etiology of schizophrenia whereas familial psychopathology does not. An earlier study conducted by Wender, Rosenthal, and Kety looked at the parents of adult schizophrenics who had been given up for adoption as infants, and compared them with parents who had reared their own, now adult schizophrenic offspring.

These two groups were also compared to the adopting parents of normal young adults.

The findings of this research indicated that the biological parents of schizophrenics were considerably more disturbed than the adopting parents of schizophrenics, and that the adopting parents of schizophrenics were exhibiting a slightly greater degree of psychopathology than the adopting parents of normal young adults. Consequently, both the psychosocial and genetic hypotheses were supported by this research. In a later study based on the original test data, protocols were blindly analyzed, and both the adoptive and biological parents of schizophrenic patients were successfully differentiated from the adoptive parents of normal offspring. An earlier series of experiments using projective techniques in blind predictions of psychiatric patients and their families was also completed.

In addition to matching the tests of the patients with their families, test data from other members of the family were successfully employed to deduce the diagnosis, form of thinking, and degree of disorganization in the young adult psychiatric patient. These findings provided substantial support for the environmentalist position and seemed to indicate that certain types of familial communication patterns are possible etiological contributors to schizophrenia. The debate between those supporting a contextualist position and those espousing a genetic model raged on, as so often happens in this area of investigation, without arriving at a firm resolution.

Part of the difficulty in interpreting this type of research stems from the inability to avoid genetic and psychological confounds and the failure to differentiate schizophrenogenic Behaviour from the Behaviour of schizophrenics. In addition, problems arise when the findings from certain studies may be attributable to several hypotheses. For example, the Wynne et al. results may just as easily be attributed to psychological changes in the parents resulting from rearing a schizophrenic child, or genetic factors accountable for both the psychopathology and communication patterns observed.

That is, even if communication deviances are etiologically significant for schizophrenia, they do not rule out the possible important of genetic factors as influential in creating those deviances. In any case, further supportive evidence for a genetic position was obtained in a later replicative study by Wender, Rosenthal, Rainer, Greenhill, and Sarlin using the same design but including a systematic sample. The findings from this study revealed that the biological parents of schizophrenics were more disturbed than adoptive parents of either schizophrenics or normals; and that the adoptive parents did not differ on degree of psychopathology.

Despite all the advances made by the adoption research,

it is still important to realise that even the most sophisticated adoption studies are limited in their resolving power. As of yet, confounds and other research difficulties cannot be avoided. This is in part due to the many practical considerations that must be dealt with when using this type of research strategy.

For instance, only certain children are permitted to be legally adopted, and sometimes these are abnormal before placement. Adopting parents as a group usually create a restricted range of rearing environments and share a similar socioeconomic status with the biological parents. In addition, fostered children are not included in the initial sample; and the spouses of schizophrenics whose children are adopted probably include a higher proportion of socially inadequate individuals than those spouses of schizophrenics whose children are not adopted.

Although these limitations do not erode the principal conclusions obtained through adoption studies, they do limit their generalizability. In spite of the various difficulties inherent in this research method, the adoption research has made a vital contribution to the investigation of the etiology of schizophrenia. Perhaps most importantly, it has disconfirmed the hypothesis that the high rate of schizophrenia observed in the offspring of schizophrenics is necessarily due to being raised by a schizophrenic parent.

As a result of their findings, the adoption studies along with the family and twin studies have all but dispelled any doubt that a genetic factor is operating in the transmission of schizophrenia. This seems pretty remarkable when one considers that this genetic factor has surfaced in spite of all the problems in the formulation of individual diagnoses, the heterogeneous nature of the schizophrenic syndrome, and the interindividual differences that must be present in any population labeled schizophrenic.

With the knowledge that hereditary factors play an important role in the development of schizophrenia, scientists have begun to search for what exactly is transmitted in the disorder. However, so far the search for an adequate model has not yielded conclusive results. What has happened, is that investigators have proposed several theoretical models and have explored the fit between the theoretical predictions made from these models and the existing data.

The proposed models for the genetic transmission of schizophrenia can be roughly divided into two types: monogenic and polygenic. Monogenic models rest on the assumption that the action of a single or a few major genes leads to a specific metabolic deficit that in turn causes schizophrenia. Polygenic models propose that the action of many genes, each with a small effect, can, given a certain combination, produce the disorder.

Since Rudin conducted his study in 1916, it has been apparent that the mode of genetic transmission in schizophrenia does not seem to follow any simple Mendelian pattern. In all genetic studies, the number of cases of schizophrenia found in the relatives of schizophrenic probands has always been less, but never higher than would be expected with a simple Mendelian distribution.

Therefore, one can conclude that the incidence of schizophrenia was not due to random error nor could it be explained by a simple monogenic theory. Consequently, investigators have had to propose the existence of modifying factors to accompany monogenic theories in order to account for the lack of consistency with Mendelian ratios.

In 1965, Slater proposed a major partially dominant gene model that included the concept of penetrance. Penetrance refers to the percentage of cases that manifest the trait in question when the necessary genetic components are present. According to Slater's modified dominance theory, virtually all schizophrenics inherit the same abnormal gene in a single or double dose, but only one out of four will manifest the disorder, because the gene expression is only partially penetrant.

Later, modifications of this model by other investigators such as Slater and Cowie and Kidd and Cavalli Sforza produced variations in the posited penetrance of the dominant gene in the heterozygote.

Another model developed by Kallman in 1953 proposes a recessive single gene that can be modified by another variable called the "constitutional resistance to manifestation"

This variable was considered to be a polygenetically determined trait. Those individuals who were homozygous with respect to the single gene were thought to be much more likely to develop schizophrenia than those individuals who were heterozygous. However, the degree of constitutional resistance present could affect the tendency to develop schizophrenia even when the individual was homozygous.

Through this model, which allows for various combinations of homozygosity and heterozygosity interacting with high or low levels of constitutional resistance, Kallman hoped to account for different degrees of severity in the manifestation of the disorder.

One criticism that has been leveled against these monogenic models is that they cannot stand without some kind of support from such concepts as penetrance or resistance to manifestation.

These types of concepts are especially problematic because the specific mechanisms proposed have never been demonstrated, do not further the understanding of the etiology of schizophrenia, and fail to provide a means of disproving a particular model. Consequently, these types of concepts have often been labeled as ad hoc explanations whose only purpose is to bring order and consistency to a theoretical model.

When one works with polygenic models many of the difficulties inherent in the monogenic models disappear but others arise.

Mendelian distributions are no longer required and the expectancy rates for schizophrenia, which were lower than expected for monogenic positions, are not contraindicative of polygenic theory. There is good scientific precedent for this type of theory, as many traits among many different kinds of species have been shown to have complex non Mendelian, multifactorial genetic determinants.

One interesting model advanced by Falconer uses a quantitative genetic approach and makes the assumption that underlying the etiology of the disorder is a variable distributed along a continuum. This variable, called liability, includes both innate tendencies and external circumstances that may be predisposing to the development of the disorder. The variation of liability is distributed in the form of a normal, bell shaped curve.

At one end of this curve is a point called the threshold, which demarcates those individuals who are affected by the disorder from those who are not.

This point is positioned on the curve according to the prevalence of the disorder in the general population. Once the prevalence rates among different classes of relatives of affected individuals are determined, it becomes possible to use the model to calculate the heritability of the liability of the disorder in question.

In 1967, Gottesman and Shields were the first to apply Falconer's model to schizophrenia. They obtained many heritability values using different samples of subjects, and these values ranged between 45 and 106 per cent. Accordingly, the authors acknowledged that the method was subject to errors, and that the underlying assumptions of the model had not all been fully met.

Even so, the authors noted that the findings did indicate the heritability of liability to schizophrenia across samples to be substantial. However, they also qualified these findings by pointing out that heritability is not a property of the trait alone, but also of the population sampled and of environmental circumstances.

Another polygenic theory has been proposed by Cancro. His model assumes the existence of a continuum with regard to phenotypic expression in the schizophrenic syndrome, and does not assume that the phenotype is inherently abnormal. That is, different subgroups of schizophrenia are assumed to be involved in different combinations or patterns of phenotypes.

In addition, there are a number of different genes that can produce each phenotype depending on the environmental conditions present. Therefore, two individuals who share the same genetically loaded trait may have obtained that characteristic by different means. Obviously, Cancro's model is quite complex.

Not only does he have the possibility of a particular phenotype developing from different pathways, he also allows for the heterogeneity of the schizophrenic disorder by assuming that multiple traits are involved. Although these polygenic models contain several advantages over the monogenic models, they are also more loosely formulated and consequently more difficult to test.

Studies measuring the fit of theoretical models to the ever increasing amount of data are continually taking place. For instance, Kidd and Cavalli Sforza reviewed the then current research data and tested the goodness of fit of these data to both a monogenic and polygenic model, using a threshold approach. On the basis of their analysis, neither a monogenic nor polygenic model could be discarded, as both seemed to fit the available data equally well.

This led the authors to speculate that any intermediate model between the two extremes investigated would also fit the data satisfactorily. In fact, they postulated that any model of genetic heterogeneity would fit, with improvement in fit increasing as more parameters are included in the calculations.

Interestingly, when Mathysse and Kidd expanded on this approach, they found that neither of the two extreme genetic hypotheses were adequate when trying to account for the observed data.

The single major locus or monogenic model provided estimates that were too low, whereas the multifactorial model provided estimates that were too high relative to the observed incidences of schizophrenia among monozygotic twins, and offspring of parents who were both affected.

Only a model with two interacting genes could possibly account for the incidences observed in the input data. Even so, both the monogenic and multifactorial models did predict genetic heterogeneity among schizophrenics and provided some interesting perspectives on the disorder. When the input data were fitted according to the monogenic model, the genetic composition varied along with the frequency of the schizophrenia producing allele.

Every individual has two alleles or genes, one from each parent, at each locus. At low frequencies of the allele, the development of the disorder seems to be largely due to environmental factors. At relatively high frequencies of the allele, the proportion of heterozygotes increases whereas the number of phenocopies decreases.

The proportion of homozygotes remains low no matter what the allelic frequency. Consequently, the single major locus model predicts that genes play an extremely important role on those occasions when they are present in double dose, but those cases would be rare even among persons manifesting the disorder.

Under the polygenic model, however, Mathysse and Kidd found that 9.1 per cent of the schizophrenic population has such a high genetic risk that environmental factors play virtually no role in the development of the disorder. That is, this model proposes that in one of every 11 schizophrenics, the disorder is almost totally genetically determined. Obviously the results are still unclear and the conclusions remain tentative.

Arguments over genetic hypotheses will probably continue to lack meaning as long as virtually all models can account for the data. Until the research data becomes homogeneous, future investigations will be unable to rely on the use of statistical techniques alone to clarify which genetic model is most applicable to schizophrenia.

Although monogenic and polygenic models represent the genetic extremes of inheritance, there are also models of genetic heterogeneity that rest on the principle that more than one gene at one locus is needed to produce schizophrenia. Although appearing similar to polygenic theory, the two theories are different because polygenic models require that more than one locus is necessary for the development of the disorder.

The effects of different schizophrenia producing genes are not cumulative in genetic heterogeneity theory as they are in polygenic models, and the heterogeneity model does not propose a continuum. Instead, each gene is believed to be separate and distinct in its effects.

Consequently, a particular combination of genes is not required to produce the disorder, but rather the appropriate specific genes. A heterogeneity model had been proposed by Karlsson. He postulated a two gene theory in which a mechanism of modified dominance was involved. The theory allowed for the prediction of distinctive personality types from specific genotypes. However, there does not appear to have been any independent validation of this model.

Taking a different approach to trying to uncover the mode of inheritance in schizophrenia, Mitsuda examined different types of schizophrenics and then classified them according to outcome. Cases for the apparent mode of transmission were classified into three groups: dominant, recessive, and intermediate.

His findings suggested that the dominant gene group contained more prognostically favorable cases, whereas the recessive group had more that were unfavorable, and the intermediate group fell somewhere in between. As a result, he decided on the existence of a heterogeneity model for which three modes of genetic transmission were possible.

At present, theories of heterogeneity are very popular. However, if they are correct, they imply that the likelihood of discovering the mode of genetic transmission in schizophrenia is very low.

On the other hand, monogenic theories appear more inviting because they allow for the possibility of identifying exactly what is transmitted in schizophrenia, thereby offering hope that therapeutic and preventive techniques will be developed sometime in the future. In light of these differences, it may be more worthwhile to direct research toward trying to disprove the single mode theory, if only to provide some direction as to which model is "true."

It is apparent that no single model for the mode of genetic transmission in schizophrenia has been universally accepted. There are even disagreements among investigators espousing the same type of model. For example, among the monogenic theorists there is no clear consensus regarding the mode of action of a single gene.

However, there does seem to be agreement among the research results that a relatively low frequency of homozygotes exists in the population, and that the manifestation rate is very high for these individuals. At present, it is still not possible to decide on exactly what is inherited in schizophrenia and how it is transmitted.

However, the studies conducted to date, along with the improvements for research strategies that they foster, provide the necessary impetus for more sophisticated model building and testing. These in turn, provide reason for optimism that this area of investigation is worthwhile, and that it will lead to the eventual discovery of the mode of transmission in schizophrenia.

METHODOLOGICAL DIFFICULTIES

There are many methods that have been employed in the study of genetics and schizophrenia. Twin studies have been especially valuable for assessing the importance of biological factors in an illness. Of course, their value is still dependent on the validity of the diagnoses of zygosity and disorder. Despite their usefulness, they still present several problems for data collection and analysis. In particular, the usage of different definitions of concordance rates can be particularly problematic.

Monozygotic twins often differ on biological traits. Other factors such as obstetrical complications and different birth weights may interact with susceptibility to psychiatric disorders such as schizophrenia.

Separation studies, including studies of adopted children, cross fostering studies, studies of half siblings, and studies of monozygotic twins reared apart, have proven invaluable by demonstrating unequivocably that genetic factors are involved in the transmission of schizophrenia.

Even so, these studies in themselves cannot elucidate what genetic and biological factors are operating and how they contribute to the development of a schizophrenic disorder. Instead, the best contemporary use of separation studies may be to shed more light on the relative importance and specificity of certain environmental factors in relation to schizophrenia.

This could possibly be accomplished through the comparison of rearing environments with populations at high and low genetic risk. In this manner, a negative correlation between genetic risk and a specific environmental factor would suggest a causal role for that factor.

Nuclear families have constituted the most popular form of data for genetic studies. This type of data is easily related to the epidemiology of the disease, but often poses difficulties for genetic analyses and interpretations. Part of the difficulty stems from the researchers' failure to distinguish the parent—offspring and sibling—sibling relationships. Instead, risks are calculated only for first degree relatives, while the finer distinctions are ignored.

Failure to note the sex of both proband and relative is also common and may affect morbid risk calculations. Until these factors are taken into account and sufficient specifications by relationship are made, these types of studies will remain limited in utility and ability to discriminate between proposed genetic models and types of heterogeneity.

Another valuable but often neglected resource in the investigation of genetic factors in schizophrenia is the use of multigenerational pedigrees.

One major advantage of this approach is that it allows for the potential identification of homogeneous phenotypic subgroups on the basis that any rare disorder found in several members of one family is likely to have the same etiology in all. Pedigrees are amenable to the study of genetic linkage and the testing of genetic models.

Sampling procedures also pose major problems in this area of research. For example, Lewine, Watt, and Grubb point out that the current use of parent—child concordance rates may result in the use of a sample that consists of schizophrenics with an unusually strong genetic and environmental predisposition to the disorder. These authors have also demonstrated how the overrepresen tation of females among index parents may introduce systematic sex differences in schizophrenic characteristics as well as biased prevalence rates.

This in turn may have negative implications for the interpretations based on this type of research. For example, current risk studies rely on theoretical concepts and assessment procedures based for the most part on adult male populations. At present, it is still unknown how generalizable these conceptualizations are to female schizophrenics. It is difficult to ascertain if findings regarding the offspring of schizophrenic women are comparable to the findings for offspring of schizophrenic men.

Although we know that the risk for schizophrenia is the same for both men and women, the evidence indicating that schizophrenia may take different forms in men and women suggests that different etiological factors may be of importance for the sexes. The preponderance of a reactive, atypical, or schizo affective type of schizophrenia in women suggests that careful diagnoses of index parents and offspring according to schizophrenic subtype are necessary to clarify the nature of schizophrenia, including its possible heterogeneity. In addition, the fact that varying sampling procedures may result in differing concordance rates does not permit easy comparison of results among studies.

The importance of studying families over unrelated individuals seems to be well established. As such, it appears to be the optimal strategy for genetic research on schizophrenia at this time. In addition, it is apparent that many unknowns remain in this area of investigation.

Therefore, until such a time as our methods of analyses and diagnostic capabilities have improved sufficiently, one needs to remain aware of the tentativeness of our present conclusions as to the exact nature and transmission of a genetic factor in schizophrenia.

Anytime one delves into the schizophrenia literature, one is struck by the overwhelming amount of information available in this area. Given the wealth of knowledge that has accumulated in this field, it seems almost ironic that schizophrenia remains one of the most complex and least understood of all the psychological disorders.

It is obvious that none of the theoretical models proposed to date is sufficiently capable of explaining schizophrenia. Each one of them is lacking in certain respects. For example, most research studies dealing with genetic or family theories cannot explain the occurrence of schizophrenia as either a strictly environmental or genetic outcome. Instead, the evidence tends to suggest that both genetics and family process play a role in the development of the disorder.

Similarly, although biochemical and neurophysiological theories hold much promise with regard to deriving useful diagnostic categories, their focus appears too narrow to explain the full scope of the schizophrenic syndrome.

On the other hand, psychological and psychoanalytic theories appear comprehensive in their provision of meaningful ways for conceptualizing and attending to schizophrenic process.

Because no one can perceive the schizophrenic's subjective experience, these theories appear to try to understand its nature and etiology through an intuitive or empathic approach. Unfortunately, these theories are difficult to investigate and therefore it is difficult to obtain enough empirical data to support these positions.

In trying to compensate for these deficits, much of the emphasis in research designs has been placed on analysis and description leading to a wealth of confusing and often contradictory data. This condition appears to be typical of the early stages of development of any complex field of knowledge; accordingly, many would argue that a need for synthesis that would integrate apparently opposing theories now exists.

However, a synthesis of incomplete information would not necessarily lead to a better understanding of schizophrenia. By ignoring some of the dichotomies and polarities in the data, by narrowing the focus to trying to find the interconnections among the information presently available, some valuable information might be lost. At the same time, one wants to avoid becoming engulfed in spending too much time and effort in the analysis of small details, as is often the case with research designed to support a particular theory.

It all boils down to the dilemma of not being able to formulate a proper integrative principle or theoretical model until one has at least enough information to delineate the concept one is involved with, and one cannot obtain that information unless a great deal of varied research encompassing many different areas of investigation is conducted. In order to reach the goal of a more comprehensive theoretical framework for schizophrenia, it seems important to do several things.

First, it appears vital that theory, research, and intervention remain interconnected so that any discovery in one area will influence the other. At present, dangerous rifts among these areas appear to be developing, wherein practitioners apply interventions with little or no theoretical basis and theoreticians form postulates without any specific delineation for integration into treatment or research strategies.

Similarly, we cannot afford to dismiss any of the material presently available, but instead should try to appraise and evaluate it objectively without any preconceived ideas or bias stemming from our own theoretical models.

Chapter 44

Limitation of Genetic Engineering

It takes only our imagination to come up with other possible applications of genetic engineering in agriculture. But nothing can be accomplished until scientists have identified the relevant genes, figured out what they do, and figured out how the proteins they make work in the organism. This point is vitally important.

For example, scientists at one company tried to make a blue rose. They transferred the gene for delphinin, the blue pigment in delphiniums, into a rose and the rose made delphinin. But its flowers weren't blue.

The scientists didn't understand how a delphinium uses its pigment, or how it would be used in a rose. There are virtually identical genes in flies and humans, but they have profoundly different results.

Until scientists understand each step of some life process in one organism, they will never successfully transfer the trait to another organism.

The only successes achieved so far have involved either a single gene or a group of closely related genes whose function has been worked out in detail.

Some characteristics, such as humans' height, are affected by at least dozens of genes, some of which surely affect other attributes. We are very far from being able to genetically engineer complex traits. Unlike traditional plant and animal breeding, genetic engineering is not hit-or-miss. Genetic engineers do not have perfect control over the transferred genes, but they have much more control than the traditional breeders have.

LEGITIMATE CONCERNS ABOUT TRANSGENIC AGRICULTURE

Monkeying with Mother Nature

Some people think that this is an enterprise that should be left to God or to Mother Nature, that man was never intended to monkey around with other species' genes. But it can't be the basis of an argument. Whoever claims to know what God intends usually can't prove it, and can't be talked out of it.

Some people have religious or ethical concerns. They might point to Leviticus 19:19, which prohibits crossbreeding. Vegetarians may reasonably decide that their food should not contain genes derived from animals. Jews and Muslims may reasonably decide that their food must not contain any genes derived from a pig. Some religious scholars believe that a gene loses its identity when it is copied and the copy is inserted into a target species.

That point of view would remove some, but not all, of the religious objections to genetically modified plants. Other advances in biotechnology have drawn most of the attention of clerics and ethicists. These include cloning, organ transplantation, research using foetal tissue, etc.

Food Safety

Crops modified in any way might not be safe to eat, so any major change in the food supply should be tested. This applies to changes made by genetic engineering but it ought logically to apply even more to changes made by other techniques. To a great extent, genetic engineers know what they are doing.

There can be unanticipated consequences, but by comparison, all other methods of improving crops involve an element of luck. The conservative approach is to test all crops whose genetics has been modified in any significant way. An example of a possible safety issue was brought out clearly several years ago. Although soybeans are a good source of protein, soy protein is low quality.

It doesn't have enough of the essential amino acid methionine. So scientists in Nebraska planned to transfer a gene from a Brazil nut to a soybean to get better quality protein from soybeans for use as an animal feed. Unfortunately, some people are allergic to Brazil nuts and it turned out that the better quality protein was one of the Brazil nut allergens. Since this fact was quickly revealed by testing, the genetic modification project was abandoned. This example shows that testing for safety is necessary.

It also shows that such testing is being done and is working. But can new foods ever be tested enough for complete assurance of safety? Another way to develop crops with new traits is to cause random gene mutations and select for them. This is analogous to spraying on the pages of a recipe book, then following the distorted recipes to see which work well, and recopying those recipes. Breeders have induced mutations using radiation, chemicals and high temperatures.

Since the effect of mutation is random, it makes sense that crops developed by mutation ought to be even more thoroughly tested than crops developed by genetic engineers since the genetic engineers are not relying on luck to get their improved traits. Yet there is essentially no regulatory process for plant breeding in the United States, although Canada requires that all new types of cultivars be tested. Even conventional breeding techniques can accidentally create harmful foods. In a famous example, an improved variety of celery caused farm workers who picked the celery to become hypersensitive to sunlight.

In another example a potato variety, Lenape, was withdrawn from the U.S. market in the 1960s when it was found to contain dangerously high levels of potato toxins (solanidine glycosides). Even without mutations, there is a large pool of genetic variability in every variety or species. This means that unfavorable combinations are possible. In every instance of sexual reproduction the child gets some genes from each parent, in a random assortment.

If John and Jane have a few hundred different genes (and about 30,000 that are identical), their children will each inherit a different subset of John's genes and a different subset of Jane's genes. Nobody can predict the characteristics each child will inherit from its parents.

Sometimes, two apparently healthy parents have a child with a genetic disease. Similarly, sometimes two plants which bear nutritious food can have offspring which are more toxic. This is not an argument against having children or against breeding crops, so it ought not to be an argument against transferring genes by biotechnology.

Conditions of growth can also affect food properties. Certain inconspicuous fungi can turn a wholesome food into a poisonous food. Every year there are deaths from *ergot,* a fungus that infects wheat and rye, and from aflatoxin, caused by a mold that infects peanuts and corn. In summary, genetic engineered crops need to be tested for safety. In the US, transgenic crops are tested much more strictly than crops developed by traditional breeding.

So far the testing that has been carried out has been sufficient to protect the public. During the ten years that we have been eating transgenic foods, nobody has ever been exposed to unsafe genetic engineered food. Meanwhile there have been many thousands of deaths because of unsafe conventional food. So it seems to me that the issues of food safety are being better managed for genetic engineered foods than for conventional foods.

Environmental Concerns

The third thread of concern is for the wild environment. Suppose a gene from an unrelated species is transferred to a crop species and then the modified crop produces pollen which fertilizes a wild plant. Or suppose some of the crop's seeds are carried by birds or by wind into the wild. The wild plant could reproduce and the gene could become fixed in the wild population.

If it conferred an advantage, a wild plant that had been barely making it in the struggle for existence could turn into a dominant species. There are many examples of plants taking over an environment. Usually they are natural plants introduced from a distant continent. In the American south, kudzu is a decorative plant that escaped and is spreading out of control. In the northeast we see the same thing happening in marshes, being taken over by purple loosestrife. Pasture land in the American west is being invaded by cheat grass.

These plants have no natural enemies and can overrun an ecology and devastate it.

So suppose a genetic engineered crop has been given a gene which makes it hardier. Suppose it gives the plant a tolerance to salty soil, or to cold, or to dryness. It is reasonable to fear that if the crop's pollen fertilizes a wild relative, that relative could produce a race of super weeds. The solution to this concern is, again, testing.

Scientists must study the plants growing wild in the area, determine which are closely related to the modified crop, experiment to see if hybridization is possible, and require that the crop be grown only in conditions for which hybridization is very unlikely. Or else, determine that the trait, in the wild relative, will not matter much.

Sometimes this is fairly easy. We can be pretty sure that soybeans will not hybridize with wild relatives because they self-pollinate and because the wild relatives live only in Asia. Corn can only hybridize with its wild relative, teosinte, found only in Guatemala and southern Mexico.

Sugar beets are harvested before they produce flowers (unless they are being grown for seed) so they cannot pollinate other varieties. But for some other crops the testing should be much more extensive and in some cases it will not be allowable to grow the genetically modified crop in localities where wild close relatives are found.

Earlier we mentioned a variety of corn which could grow well in soils with high levels of aluminum.

That variety was developed in Mexico, but cannot be field-tested there, perhaps because of the fear that it could pollinate its relative, teosinte, and give teosinte an advantage over other local wild plants.

If careful analysis confirms this danger, there might still be a way around the problem. It is now possible to produce plants which are male-sterile. They produce no pollen, or only defective pollen. A farmer could then plant conventional corn as a pollen source and aluminum tolerant corn for his main crop. Since the plant with the extra gene is infertile, it would not be able to spread its gene by pollination. This still leaves the possibility that a seed could escape, carrying the special gene, but corn cannot live at all as a wild plant — it cannot reseed itself — so this avenue of gene transmission is much less likely. Transgenic salmon have been engineered to grow faster than their wild relatives.

There is a concern that they might escape from confinement pens and reproduce, or even cross with their wild relatives. Nobody can confidently predict the ecological effect of this. The transgenics could monopolize the wild salmon's food supply or be preferred as mates.

If they were effective at attracting mates but less prolific breeders, salmon populations could crash. To prevent all these possibilities, Aqua Bounty Inc., the company developing transgenic salmon, plans to use only sterile females in commercial production. Not every species that escapes into the wild will be a problem.

Fig. Apis Mellifera

Most crops will simply die out because they can't compete with hardier wild plants.

In one experiment, rapeseed plants, both transgenic and conventional, were grown in a field but never harvested.

Scientists then followed the subsequent history of the field for ten years. All the crops declined in numbers from year to year. After the fifth year, none of the genetically modified crops could be found at all, and after ten years there were only a few crop plants of any type remaining in the field. In a more colorful example, during the nineteenth century, a wealthy and eccentric man brought to the United States populations of each type of bird mentioned in the works of Shakespeare.

Only one species was able to establish itself. That species, however, was the starling, now found in large numbers in every part of the United States. Some species out of their natural place can enrich the environment. The European honey bee (Apis mellifera) was introduced to America in colonial times.

Besides its value as a honey maker, it is the principal agent of pollination for many staple crops, which are also European imports. Organic farmers have a different concern. They consider a genetic engineered crop to be automatically *non-organic* even if it is grown without pesticides or chemical fertilizers.

They have expressed the concern that their crops might be cross fertilized by pollen from a gene modified crop. Organic farmers have every right to be protected from this problem. It is no different, in principle, from the problem faced by the seed companies who grow seed for sale.

It is solved partly by keeping the different crops far apart, and partly by more active techniques, like barriers. There is another concern.

Suppose a crop is developed which is resistant to a certain insect or fungus. Evolution is like an arms race. Insects or fungi can evolve to overcome whatever defence has been built into the crop.

For example, cotton with the Bacillus thuringiensis toxin will eventually lead to insects evolving a resistance to Bt toxin. But Bt is used by organic growers to control certain insects.

If boll budworms evolve Bt resistance, organic cotton farmers will not have alternative controls. Non-organic farmers might turn to some other insecticide, but even they would like some strategy to delay the evolution of resistant insects. The solution to this problem is the so-called refuge strategy.

Fig. Bt Cotton

Instead of growing only Bt cotton in a large field, the farmer must grow a mixture of Bt cotton and conventional cotton. This is an EPA rule. The theory is that then the insect with a lucky mutation who can tolerate the Bt toxin will have no advantage over the other insects without the mutation so Darwinian selection will not tend to increase the numbers of such insects in a population.

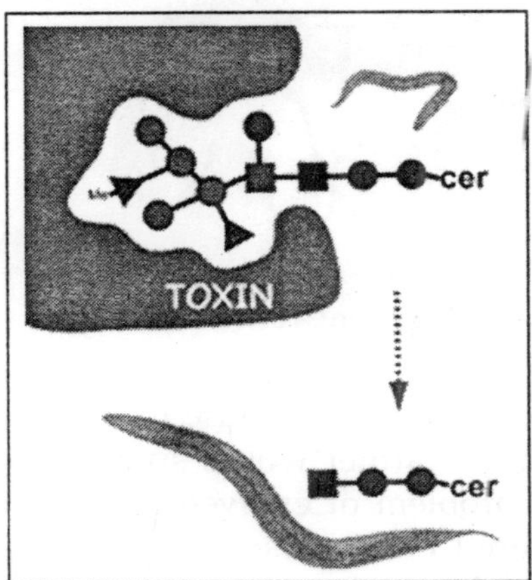

Fig. Bt Toxin

This strategy is not simple. What percentage of the cotton

in a field must be conventional, and how must the two types of plants be spaced? What about how the field is used in the following season? As more and more acres are sown with Bt crops, year after year, it seems as if the insects must eventually evolve the resistance. Eventually genetic engineers will develop better ways to delay the evolution of insect resistance. Many plants have natural insect defenses which they use only when they are being attacked.

Today's Bt crops express their toxin all the time, which gives their insect adversaries a constant environment in which to evolve. It would be much harder for insects to evolve a resistance to a varying environment.

So it would be better to control the gene for Bt toxin selectively. For example, if scientists could identify a control gene that turns on when the plant is attacked, they could use an an identical copy of that control gene to turn on the Bt toxin gene only when it was needed.

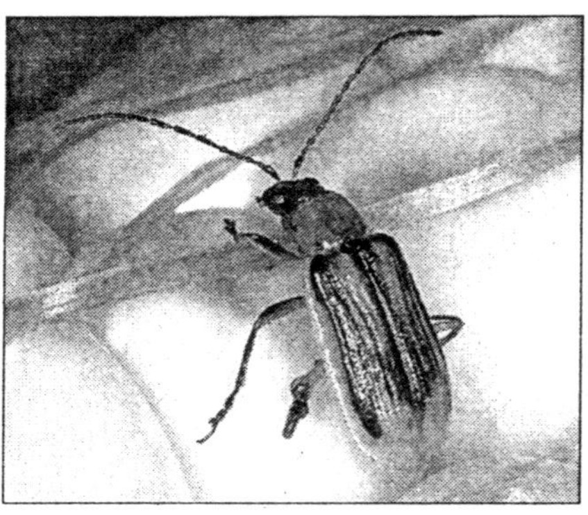

Fig. Rootworms (Beetles)

Even easier, a control gene could be used which would turn on the toxin gene in response to a cheap and harmless chemical which the farmer would spray only when deemed necessary. The problem of evolved resistance is not new to genetic engineered crops. Insects also evolve resistance to chemical pesticides and to other control methods. Some farmers used to protect corn from rootworms (beetles) by growing corn in a field used every other year for soybeans.

The corn rootworms soon evolved the habit of laying their eggs in soybean fields. Farmers and agricultural scientists are unlikely to ever find a perpetual solution to suppressing insect pests. One widely voiced concern about transgenic crops is quite ridiculous, the fear that they could transfer unwanted genes to the bacteria that live in our intestines. Many genetic engineered crops contain a gene for resistance to an antibiotic, such as kanamycin, which was transferred along with the useful genes.

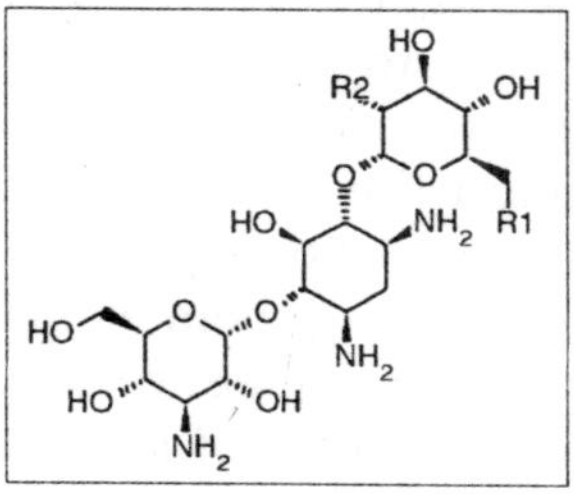

Fig. Kanamycin

The reason is that most genetic transfer attempts are unsuccessful in some way so the genetic engineers try to transfer the useful genes into hundreds of plant cells at once, hoping that a few cells will be successfully transformed.

But after a gene has been transferred into a plant cell, it takes much hard work to create a whole mature plant from that single cell and the genetic engineers want to avoid that work except for the cells that have been transformed.

But it is hard to tell when a gene transfer has worked if the gene only functions in the mature plant. The antibiotic resistance gene acts as a marker. When the transformed cells are treated with the antibiotic they survive, but cells that were not transformed die.

Only the marked cells are turned into mature plants for further experimentation. But the concern has been voiced that E. coli bacteria which live in our intestines could obtain these genes and become antibiotic resistant.

We certainly don't want that to happen, at least not by accident. We can't transfer genes by eating them, but, unfortunately, bacteria *can* take up DNA from their environment and incorporate it in their genome, although this phenomenon is extremely rare. It is billions of times more likely for bacteria to acquire a gene for antibiotic resistance by natural mutation. Each of us has a few such bacteria in our intestines now.

That's where the marker genes originally came from. If we are not taking the antibiotic medicine, the percentage of these bacteria will be quite negligible.

The only defence against both the natural evolution of antibiotic resistance and the circuitous route through transgenic crops is to minimize the use of antibiotics. Nevertheless, although the best scientific evidence is that antibiotic resistance genes would not be a problem, genetic engineers have stopped using them as marker genes.

The alternative marker gene now favored confers resistance to a herbicide instead of resistance to an antibiotic. Another kind of marker gene comes from a bioluminescent jellyfish. It promotes synthesis of a fluorescent protein. When this protein is exposed to ultraviolet light, it produces visible green light, clearly indicating that the desired gene has been transferred to the target organism. Also, after a genetic engineer has created a useful new variety of plant with both useful genes and marker genes it is possible to eliminate the marker genes from future generations of the plant using conventional cross-breeding.

ECONOMIC AND SOCIAL CONCERNS

There is one last concern often expressed. Although the gene transfer technology is available worldwide, some people worry that a few large companies would get control of world agriculture, and further, that small farmers in the poorer countries would be at an ever increasing disadvantage as their competition becomes ever more productive. The counter-argument for the first concern is that we already have antitrust laws in place.

The counter argument for the second concern is that the poor third-world farmers could also adopt more efficient farming practices. The experience of the last thirty years, the so-called green revolution for which Norman Borlaug received his Nobel Peace prize (1970), is that third world farmers can and do adopt new technologies. However, genetic engineering is yet one more technology which is making agriculture more dependent on large companies.

A Debt to Critics

The two most serious concerns about transgenic agriculture are food safety and environmental impact. So far the record of the technology has been enviable.

There have been no documented cases of any illness or any environmental damage. For this the scientists developing the technology must and do owe a debt of gratitude to the people who have raised doubts.

Responsible critics have suggested problems and the scientists have been able to take appropriate precautions, or have cancelled dangerous experiments. The anticipated mishaps didn't happen because they were anticipated. For example, if no critic had raised the possibility of allergies, would transgenic foods be tested for known or likely allergens? If no critic had raised the possibility of insects evolving resistance to Bacillus thuringiensis, would the Bt cotton be grown with non-transgenic cotton close by?

Would transgenic salmon farming be limited to sterile females if no critic had raised the possibility of escape and crossbreeding with wild stocks? It takes no credit away from the scientists to acknowledge that the enviable safety record of genetic engineering in agriculture derives as much from its critics as from its inventors.

Part II

New technology, proven to deliver advantages to farmer, consumer, and the environment but that there are reasons to be concerned because, like any new technology, it could be misused. Since the US has been the leader in adopting genetic engineering for agriculture, our government agencies have developed some standards for assuring food safety and environmental safety. Any genetic engineered product must meet these standards before it can be grown commercially.

FRUSTRATE TRANSGENIC AGRICULTURE

Although there are legitimate reasons to oppose genetic engineered agriculture, or at least to demand the most careful controls, there are a community of opponents who have taken their opposition beyond what is ethical.

A reasonable opposition expressing the concerns summarized earlier, but rather about an opposition for which the ends justify the means, including lies, vandalism, etc. These opponents must have some motive, and it seems that an alliance has emerged between at least four groups, each with its own agenda.

First there are people who sincerely believe that genetic engineering is ethically wrong, and that anything they do to stop it from happening is therefore right. Second there are foreign governments and their constituencies who are worried about American domination of agriculture. Closely related to this are advocates of organic agriculture who seem to be engendering public fear to make their own products more salable.

Third, there are environmental groups who have been misled by a radical fringe and have become willing to do anything to stop genetic engineering agriculture. The most conspicuous among these groups is Greenpeace. Finally, there are people opposed to capitalism or to large businesses dominating agriculture.

Opposition from environmental groups is particularly frustrating to me. Most measures which benefit the environment require people to give up something, and they don't like to do that.

Recycling is nearly painless and saves money, but many people won't make the effort. A few degrees adjustment of a thermostat could save vast amounts of energy, but most people would rather be comfortable.

We end up settling for half measures. But here is a technology that benefits the environment without asking people to give up anything, and its biggest opposition comes from environmental groups.

Misinformation about Food Safety

There is a deliberate campaign to frighten people about the safety of the food supply. This campaign has worked successfully in England and in much of Europe. Dr. Arpad Pusztai, who worked at the Rowett

Institute in Aberdeen, Scotland, performed an experiment. It began when a gene was transferred from a poisonous plant, the snowdrop, into a potato. The transferred gene specifies the production of a poisonous compound called lectin. Dr. Pusztai proceeded to experiment with rats. Some rats were fed with the raw potatoes which were genetically engineered to contain the poison.

The rats in the control group were fed ordinary raw potatoes and were also given the amount of lectin poison which the first group of rats would have gotten from eating the transgenic potatoes. Both groups of rats developed malformed organs, and there was no statistically significant difference between the rats who consumed the poisonous potatoes and those who consumed the poison. However, Dr. Pusztai claimed that his data showed that the rats who ate the genetically modified potatoes had more deformed organs.

No scientific journal would publish Dr. Pusztai's interpretation, and his institution would not support him. He hired an independent statistician to review his data, who also considered the data to show no difference between the two groups. But Dr. Puzstai held to his opinion. Eventually the disagreement became serious enough that his connection with the Rowett Institute was ended. The opponents of genetic engineering, mostly in England, have blown this result into a cause celebre.

Dr. Pusztai is portrayed as muzzled by the scientific establishment, although the British medical journal Lancet eventually published Pusztai's paper over the recommendations of its reviewers because of the widespread public interest. The British tabloid press covers this story continuously, with lurid photographs of deformed rat organs. The potatoes genetically engineered to be poisonous became synonymous with all transgenic food, called Frankenstein food in the tabloids.

There are varieties of potatoes, bred to be eaten, engineered to resist insects, viruses and fungi. All these varieties have been fed to rats and have never harmed them. The opponents of transgenic food have no explanation for that - they are content to use one probably misinterpreted experiment with potatoes nobody will ever eat, to stir up doubts about food safety.

During safety testing of transgenic tomatoes, Dr. Belinda Martineau discovered that when rats are fed huge amounts of tomato paste they can develop stomach lesions. It doesn't matter whether the tomatoes are transgenic or not. But before there were transgenic tomatoes, nobody had ever fed rats such large doses. Martineau's results were reported to the FDA but opponents routinely call this a "cover-up" of a health hazard with transgenic tomatoes.

The gist of the article is that dangerous untested foods are being foisted upon an unsuspecting American public, by mad scientists. Every genetically engineered crop has been tested for safety. The testing has been much more extensive than that for any other foods, including foods developed by radiation induced mutation. Dr. Billings knows this. He knows about the government testing rules that establish the safety of each individual crop. He wants genetically modified food to be tested as strictly as drugs are tested.

Billings' op-ed article contains only one 'fact' that most people would not have known before - the rest is either his opinion or just wrong. He says that transgenic soybeans have been shown to be deficient in a certain unidentified nutrient. It is not easy to track down the source of this 'fact', but I did it. The nutrient in question is a phyto-estrogen (also known as a phytosterol). Although phyto-estrogens are not essential to human health, there is some indication that they help prevent cancer. The study that indicates that transgenic soybeans are deficient in phyto-estrogens comes from Dr. Marc Lappe, who wrote the book "Against The Grain", a polemic against genetic engineering and especially against the Monsanto Co., the leading company in the field, which developed the soybeans in question.

Here is how Dr. Lappe established that genetic engineered soybeans are defective. Understand that there are dozens of varieties of soybeans with the herbicide tolerance trait and over a hundred varieties of conventional soybeans. Dr. Lappe compared one conventional variety with one transgenic variety. He found a 12 per cent difference. But individual soybean varieties vary by more than 100 per cent in their phyto-estrogen content. Also phyto-estrogen content is not stable.

It declines with storage, by much more than the 12 per cent difference. But Lappe at least reports his data along with his biased interpretation of it. Dr. Billings reports only the interpretation without any indication that it comes from a biased scientist whose own data show insignificant variations in a nutrient whose role in human health is not even firmly established.

Dr. Billings seeks only to mislead people. His purpose is to plant a little seed of doubt about food safety, hoping that it will fester in our minds, mingle with similar misinformation, and eventually become accepted fact. Not everything misleading is false. Propagandists are very skilled at making true statements that seem to imply something quite different. For example, they frequently say that the FDA's rules about testing genetically modified food are voluntary, implying that some testing doesn't get done.

Under current law, FDA has no authority to require safety tests for any food, transgenic or conventional, although it can prevent the sale of foods it considers unsafe. But, in fact, each developer of a transgenic crop has consulted with FDA and performed every test FDA suggested. They also like to say that safety test results are trade secrets. True, it would be legal to keep test results secret. But no developer has done it.

People allergic to Brazil nuts should not expect to have to avoid soybeans, but the allergen was identified by testing and therefore the modified soybeans were never created. Also, the scientists who demonstrated this published their results in a scientific journal, and this led the FDA to not allow any gene to be transferred into a food from a species known to cause allergies. That should be seen as evidence that the genetic engineers are responsible people, and that testing is working well. But the unethical opponents of biotechnology routinely present this episode, carefully worded, as if there was a near disaster, revealing gross problems with the current regulatory system.

Tryptophan Deaths

One of the most active anti-transgenic groups is *Mothers For Natural Law* which spreads the following half-truth - that, in 1989, 37 people died and thousands were paralyzed by consuming tryptophan made by genetic engineered bacteria. Half-truth because there were deaths and illnesses (eosinophilia myalgia syndrome) caused by tryptophan, sold by the "health food" industry.

Tryptophan is one of the twenty amino acids which are needed by every living thing. All bacteria already contain genes to make tryptophan and the tryptophan sold as a dietary supplement is made industrially using bacteria and is then purified. Showa Demko Ltd., the company whose tryptophan caused illnesses, genetically engineered bacteria to make *more* tryptophan than was needed for the bacteria's own life cycle.

The illnesses had nothing to do with the genetic engineering. Some cases of eosinophilia myalgia syndrome were traced back to Showa Demko's tryptophan manufactured as far back as 1983, years before the company used genetic engineered bacteria. It is now known that eosinophilia myalgia syndrome is caused by consuming excessively large doses of tryptophan, from whatever source.

Back in 1989 it was thought that the cases of eosinophilia myalgia syndrome had been caused by some kind of contamination.

All tryptophan molecules made by living things are chemically identical and there is no way that tryptophan made by genetically engineered bacteria could be different from tryptophan made by any other bacteria. The tragic epidemic of eosinophilia myalgia syndrome makes a very good argument for more scrutiny of the health food industry. Presenting it as an indictment of transgenic food is a huge distortion.

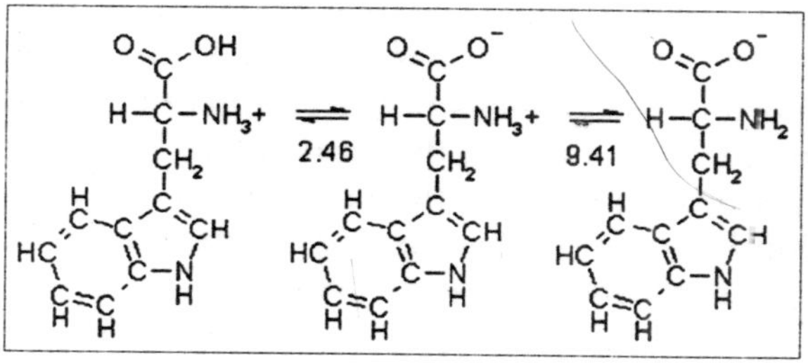

Fig. Tryptophan Molecules

Advocates of genetic modification of crops often say that it is not significantly different from ordinary breeding techniques. They say that virtually every crop is genetically modified and that people have been genetically modifying plants and animals for several thousand years. This is, of course, true, but are the genetic transformations now possible through biotechnology different from classical breeding in some fundamental way?

The opponents say that the new gene transfer techniques are completely different from anything that nature has ever allowed. Since this is only a matter of how the two sides define *fundamental,* it really isn't a case that illustrates an unethical behaviour by either side. There is one minor exception.

The opponents like to illustrate their case by pointing to a tomato with a gene from a fish. This example seems to be selected from all the myriad possibilities because it strikes a chord of negative emotion. We just don't think that anything from a fish belongs in a tomato. This poster child for the opponents is played up endlessly. Their flyers and posters show a tomato with fins, or sometimes a whole fish with a stem and a few leaves. Sometimes the tomato is a strawberry. One is supposed to think that these are typical examples of genetic engineering. It is possible to transfer a gene from a fish to a tomato plant. It was tried by DNA Plant Technology of Oakland, California.

The fish, an arctic flounder, can tolerate very cold water because its blood contains a natural antifreeze. The hope was that a tomato plant would also be cold tolerant. When the resulting plant was tested, it was a failure. The company abandoned the project and has no plan to try again. All the posters portray is a just-so-story, a product that doesn't exist. In fact, no plant product on the market today contains a gene from any kind of animal, with one exception - there is a gene from a luminescent jellyfish used as a marker, an indication that the gene transfer has been successful.

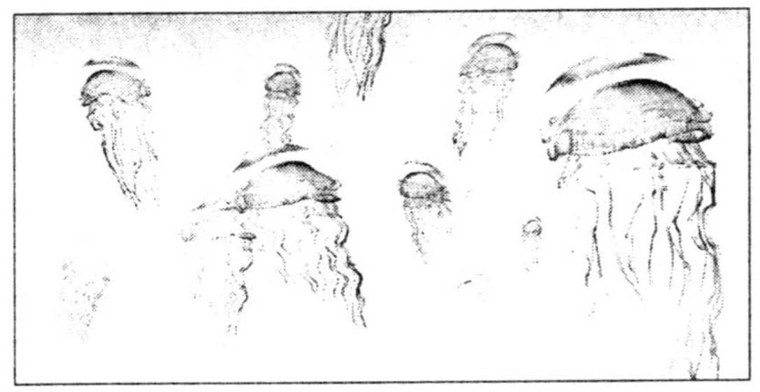

Fig. A Luminescent Jellyfish

But the story still bothers people even when they know it doesn't have much to do with anything we already eat. There is a feeling that it somehow goes against nature to make such huge changes in an organism's genes. It might be useful to examine a few cases from nature, which can be more complex than most of us imagine. At the very least, it will be interesting. There are natural examples of genetic engineering, and they are actually quite close to our lives.

Wheat is sometimes called the staff of life. Yet wheat has a complex genetic story. It is the result of three separate instances of natural genetic engineering. To introduce these changes, we need to explain that wild grasses similar to wheat have their genes dispersed among seven pairs of chromosomes.

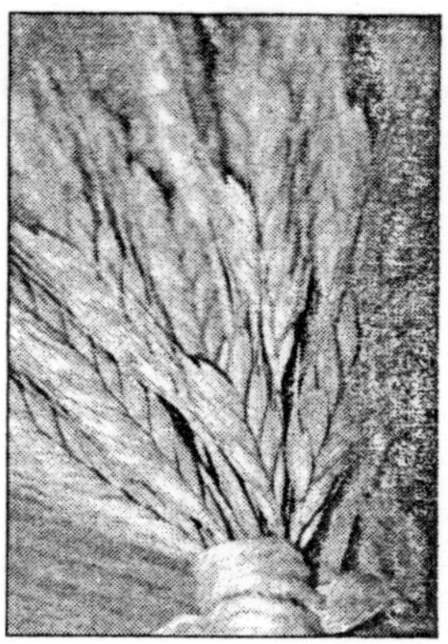

Fig. Einkorn Wheat (Triticum Monococcum)

One of the earliest known domestic wheat varieties is einkorn wheat (Triticum monococcum), which has seven chromosome pairs, like a wild grass. But another variety of wheat, emmer wheat, has 14 chromosome pairs.

Fig. Emmer Wheat

It resulted from an "impossible" cross species mating with another wild grass (Aegilops speltoids). This cross preceded modern biotechnology by several thousand years.

It either happened by itself or with the help of a Sumerian farmer. This new plant had new characteristics that breeders, with ordinary breeding and selection, exploited to produce many modern varieties, such as durum wheat, which has grains that are easy to separate from the hulls. But natural genetic engineering was not finished with wheat.

Fig. Aegilops Speltoids

Around the time of the Roman empire there was another "impossible" cross species mating with a third wild grass (Triticum tauscii). The resulting new variety of wheat, bread wheat (Triticum aestivum), has twenty one chromosome pairs, the complete genomes of three separate species of grasses.

Fig. Triticum Aestivum

This last mating brought in the genetic recipe for gluten, which makes dough springy and lets it hold together when yeast makes it rise. The record of these crosses is written in the genomes of wheat varieties and in analyses of grain from archaeological sites.

But the latest step in the series is a grain plant with twenty eight chromosome pairs. It is the result of a wheat-rye cross that happened with human help only very recently, but which made no use of the new gene transfer technology.

Wheat has been involved in three "impossible" cross species matings during its history as a human food, none relying on the modern DNA technology. But the DNA manipulation techniques themselves rely on methods developed by nature.

To cut a DNA molecule at a specific place, the genetic engineers rely on a collection of natural enzymes called restriction enzymes, each of which recognizes a specific site to make its cut.

To join two pieces of DNA, they rely on a natural enzyme called DNA ligase. To copy DNA they rely on the natural enzyme DNA polymerase. These enzymes are used by living cells to manipulate their DNA. In a few instances, they are even used to manipulate another creature's DNA. There is a species of bacteria, Agrobacter tumafaciens, whose way of life is to invade a plant and cause it to create a gall, a home for the bacterium.

It works its will on the plant by invading its cells and stitching a few of its own genes into the plant's DNA. In effect, Agrobacter tumafaciens is a natural genetic engineer who changes the genome of the infected plant so that it produces food and protection for the bacterium. How different is that from what human genetic engineers do? In fact, one of our ways to transfer a gene into a plant cell is to use A. tumafaciens as a "vector".

Fig. Fat Green Caterpillar

There is a fat green caterpillar, the tomato hornworm, that eats tomato plants, and there is a parasitic wasp that lays its eggs in the body of the caterpillar.

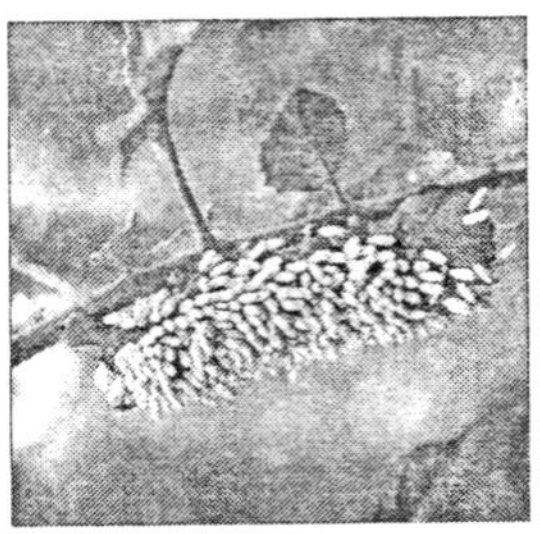

Fig Parasitic Wasp

The wasp larvae use the live caterpillar for food. But why doesn't the caterpillar's immune system attack the wasp larvae? Because the wasp has evolved a partnership with a virus. The virus is carried by the wasp into the caterpillar, where it goes to work changing the caterpillar's DNA, modifying the caterpillar's immune system to the benefit of the wasp larvae and the virus. We need to look no further than our own bodies for a very ancient example of a cross species mating. Within each of our cells there are tiny bodies called mitochondria, which produce the cells' energy. Each mitochondrion is the descendent of what must once have been a free living bacterium.

The mitochondria have their own DNA and they make their own enzymes. In fact, they would have all the machinery needed to run a cell, except that they, eons ago, transferred most of their genes into our nuclear genome.

These and other examples of natural DNA mixing across species and even between plants, animals, bacteria and viruses, show that nature invented genetic engineering before mankind did. Nature even goes to the exact opposite extreme. There is a species of fish that cannot reproduce except by a cross species mating.

Fig Poecilia Formosa

The Amazon molly (Poecilia formosa), a tiny fish just a

few inches long, is a species with no males. Every Amazon molly is a female. It bears its young alive, like its better known relative, the sailfin molly (Poecilia latipinna), which is commonly kept in home aquaria.

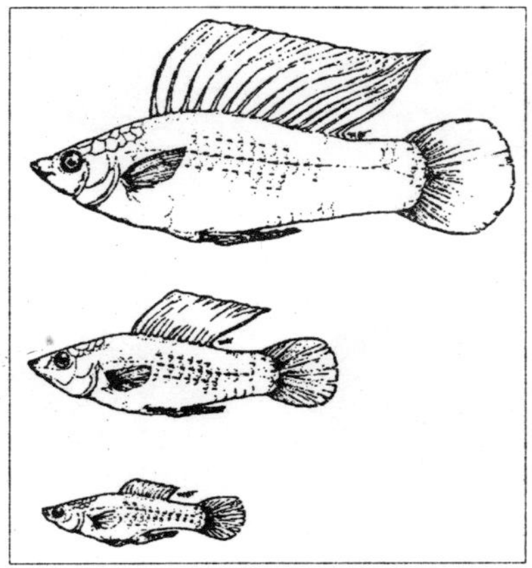

Fig. Poecilia Latipinna

How can a fish reproduce with no males? The Amazon molly borrows the services of a male sailfin molly.

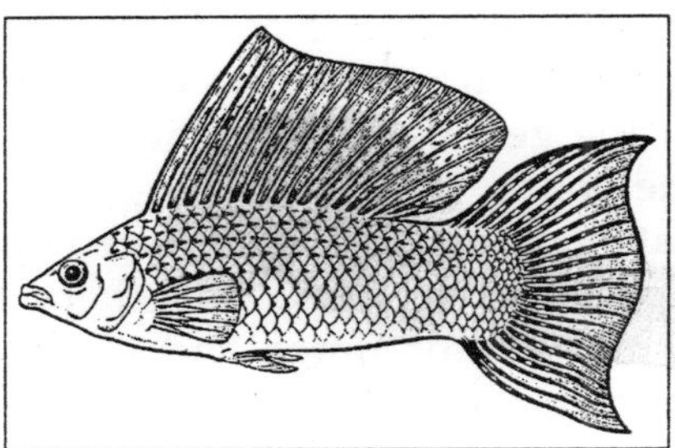

Fig. Sailfin Molly

She mates and the sailfin's sperm enter her eggs, causing them to begin development. But the male sailfin molly makes no genetic contribution to the developing embryo. The DNA in his sperm is wasted, which is why we can consider the Amazon molly a totally different species.

There are numerous other species all across the animal kingdom which have dispensed entirely with males, and reproduce by parthenogenesis, but it is certainly a surprise to find a species which relies on males of another species to fertilize its eggs.

Fig. Cupressus Dupreziana

But not so much of a surprise as to learn of a species of cypress in North Africa (Cupressus dupreziana) which plays the trick in reverse. The pollen of the cypress requires the female parts of a different tree to produce its seed cones. The structural and nutritional parts of the seed cones are built by the female, but the genetic component of the seeds comes entirely from the pollen.

It is true that modern methods can speed up the processes which transfer genes between species, genera, families, even kingdoms, by millions of times and channel them into directions of our own choosing. Ultimately what we consider to be natural is a personal decision. But that decision should not be affected by street theater. Nature can give us examples of almost anything we can imagine.

Index

A

B

C

D

S

T

U

V

W

Z